Grundlehren der mathematischen Wissenschaften 336

A Series of Comprehensive Studies in Mathematics

Peter M. Gruber

Convex and Discrete Geometry

Springer

Peter M. Gruber
Institute of Discrete Mathematics and Geometry
Vienna University of Technology
Wiedner Hauptstrasse 8-10
1040 Vienna, Austria
e-mail: *peter.gruber@tuwien.ac.at*

Library of Congress Control Number: 2007922936

Mathematics Subject Classification (2000): *Primary:* 52-XX, 11HXX
Secondary: 05Bxx, 05Cxx, 11Jxx, 28Axx, 46Bxx, 49Jxx, 51Mxx, 53Axx, 65Dxx, 90Cxx, 91Bxx

ISSN 0072-7830
ISBN 978-3-540-71132-2 Springer Berlin Heidelberg New York

Springer is a part of Springer Science+Business Media
springer.com

Typesetting by the author and SPi using a Springer LaTeX macro package
Cover design: *design & production* GmbH, Heidelberg

Printed on acid-free paper SPIN: 11966739 41/SPi 5 4 3 2 1 0

Preface

In this book we give an overview of major results, methods and ideas of convex and discrete geometry and their applications. Besides being a graduate-level introduction to the field, the book is a practical source of information and orientation for convex geometers. It should also be of use to people working in other areas of mathematics and in the applied fields.

We hope to convince the reader that convexity is one of those happy notions of mathematics which, like group or measure, satisfy a genuine demand, are sufficiently general to apply to numerous situations and, at the same time, sufficiently special to admit interesting, non-trivial results. It is our aim to present convexity as a branch of mathematics with a multitude of relations to other areas.

Convex geometry dates back to antiquity. Results and hints to problems which are of interest even today can already be found in the works of Archimedes, Euclid and Zenodorus. We mention the Platonic solids, the isoperimetric problem, rigidity of polytopal convex surfaces and the problem of the volume of pyramids as examples. Contributions to convexity in modern times started with the geometric and analytic work of Galileo, the Bernoullis, Cauchy and Steiner on the problems from antiquity. These problems were solved only in the nineteenth and early twentieth century by Cauchy, Schwarz and Dehn. Results without antecedents in antiquity include Euler's polytope formula and Brunn's inequality. Much of modern convexity came into being with Minkowski. Important later contributors are Blaschke, Hadwiger, Alexandrov, Pogorelov, and Klee, Groemer, Schneider, McMullen together with many further living mathematicians. Modern aspects of the subject include surface and curvature measures, the local theory of normed spaces, best and random approximation, affine-geometric features, valuations, combinatorial and algebraic polytope theory, algorithmic and complexity problems.

Kepler was the first to consider problems of discrete geometry, in particular packing of balls and tiling. His work was continued by Thue, but the systematic research began with Fejes Tóth in the late 1940s. The Hungarian school deals mainly with packing and covering problems. Amongst numerous other contributors we mention Rogers, Penrose and Sloane. Tiling problems are a classical and also a modern topic. The ball packing problem with its connections to number theory, coding and the theory of finite groups always was and still is of great importance. More recent is the research on arrangements, matroids and the relations to graph theory.

The geometry of numbers, the elegant older sister of discrete geometry, was created by Lagrange, Gauss, Korkin, Zolotarev, Fedorov, the leading figures Minkowski and Voronoĭ, by Blichfeldt, by Delone, Ryshkov and the Russian school, by Siegel, Hlawka, Schmidt, Davenport, Mahler, Rogers and others. Central problems of modern research are the theory of positive definite quadratic forms, including reduction, algorithmic questions and the lattice ball packing problem.

From tiny branches of geometry and number theory a hundred years ago, convexity, discrete geometry and geometry of numbers developed into well-established areas of mathematics. Now their doors are wide open to other parts of mathematics and a number of applied fields. These include algebraic geometry, number theory, in particular Diophantine approximation and algebraic number theory, theta series, error correcting codes, groups, functional analysis, in particular the local theory of normed spaces, the calculus of variations, eigenvalue theory in the context of partial differential equations, further areas of analysis such as geometric measure theory, potential theory, and also computational geometry, optimization and econometrics, crystallography, tomography and mathematical physics.

We start with convexity in the context of real functions. Then convex bodies in Euclidean space are investigated, making use of analytic tools and, in some cases, of discrete and combinatorial ideas. Next, various aspects of convex polytopes are studied. Finally, we consider geometry of numbers and discrete geometry, both from a rather geometric point of view. For more detailed descriptions of the contents of this book see the introductions of the individual chapters. Applications deal with measure theory, the calculus of variations, complex function theory, potential theory, numerical integration, Diophantine approximation, matrices, polynomials and systems of polynomials, isoperimetric problems of mathematical physics, crystallography, data transmission, optimization and other areas.

When writing the book, I became aware of the following phenomena in convex and discrete geometry. (a) Convex functions and bodies which have certain properties, have these properties often in a particularly strong form. (b) In complicated situations, the average object has often almost extremal properties. (c) In simple situations, extremal configurations are often regular or close to regular. The reader will find numerous examples confirming these statements.

In general typical, rather than refined results and proofs are presented, even if more sophisticated versions are available in the literature. For some results more than one proof is given. This was done when each proof sheds different light on the problem. Tools from other areas are used freely. The reader will note that the proofs vary a lot. While in the geometry of numbers and in some more analytic branches of convex geometry most proofs are crystal clear and complete, in other cases details are left out in order to make the ideas of the proofs better visible. Sometimes we used more intuitive arguments which, of course, can be made precise by inserting additional detail or by referring to known results. The reader should keep in mind that all this is typical of the various branches of convex and discrete geometry. Some proofs are longer than in the original literature. While most results are proved, there are some important theorems, the proofs of which are regrettably omitted. Some of the proofs given are complicated and we suggest skipping these at a first reading. There are plenty of comments, some stating the author's personal opinions. The emphasis is more on the systematic aspect of convexity theory. This means that many interesting results are not even mentioned. In several cases we study a notion in the context of convexity in one section (e.g. Jordan measure) and apply additional properties of it in another section (e.g. Fubini's theorem). We have tried to make the attributions to the best of our knowledge, but the history of convexity would form a complicated book. In spite of this, historical remarks and quotations are dispersed throughout the book. The selection of the material shows the author's view of convexity. Several sub-branches of convex and discrete geometry are not touched at all, for example axiomatic and abstract convexity, arrangements and matroids, and finite packings, others are barely mentioned.

I started to work in the geometry of numbers as a student and became fascinated by convex and discrete geometry slightly later. My interest was greatly increased by the study of the seminal little books of Blaschke, Fejes Tóth, Hadwiger and Rogers, and I have been working in these fields ever since.

For her great help in the preparation of the manuscript I am obliged to Edith Rosta and also to Franziska Berger who produced most of the figures. Kenneth Stephenson provided the figures in the context of Thurston's algorithm. The whole manuscript or part of it was read by Iskander Aliev, Keith Ball, Károly Böröczky Jr., Gábor Fejes Tóth, August Florian, Richard Gardner, Helmut Groemer, Christoph Haberl, Rajinder Hans-Gill, Martin Henk and his students Eva Linke and Matthias Henze, Jiří Matoušek, Peter McMullen, Matthias Reitzner, Rolf Schneider, Franz Schuster, Tony Thompson, Jörg Wills, Günter Ziegler and Chuanming Zong and his students. Useful suggestions are due to Peter Engel, Hendrik Lenstra, Peter Mani and Alexander Schrijver. I thank all these colleagues, students and friends for their efforts in correcting mathematical and linguistic errors, for pointing out relevant results and references which I had missed, and for their expert advice. Thanks in particular are due to Paul Goodey for his help.

I am indebted to numerous mathematicians for discussions, side remarks, questions, correct and false conjectures, references to the literature, orthodox and, sometimes, unorthodox views, and interesting lectures over many years. The book reflects much of this. Most of these friends, colleagues and students are mentioned in the book via their work. I finally remember with gratitude my teacher, colleague and friend Edmund Hlawka and my late senior friends László Fejes Tóth and Hans Zassenhaus.

Special thanks go to Springer. For their expert advice and help I thank, in particular, Catriona Byrne, Stefanie Zoeller and Joachim Heinze.

Vienna, January 2007 *Peter M. Gruber*

Contents

Convex Functions

Convex functions came to the face rather late. Early contributions are due to Stolz [972], Brunn [174], Hadamard [459] and Jensen [544] around the beginning of the twentieth century. The systematic study of convex functions dealing with continuity and differentiability properties, with variants of the fundamental notions and with inequalities began only in the twentieth century. The unpublished seminal lecture notes of Fenchel [334] in the early 1950s led to convex analysis, a careful study of analytic properties of convex functions related to optimization. Major contributions were made by Moreau, Rockafellar and Phelps.

Convex functions appear and are useful in many areas of mathematics, including the calculus of variations, control theory, inequalities and functional equations, optimization and econometrics. On the other hand, investigation of convex functions per se has led to a rich and voluminous theory.

This chapter contains basic geometric and analytic properties of convex functions of one and of several variables, related to convex geometry. A highlight is Alexandrov's theorem on second-order differentiability almost everywhere of a convex function. A further result is of Stone–Weierstrass type. Applications treat inequalities, the characterization of the gamma function by Bohr and Mollerup, and a sufficient condition in the calculus of variations due to Courant and Hilbert.

For more detailed expositions and references on convex functions with an emphasis on convexity, consult the articles and books of Beckenbach [87], Roberts and Varberg [841], Roberts [840] and also Giles [379], Van Tiel [1006] and Czerwik [234].

We will not treat convex functions in the sense of convex analysis and general analysis. For material in this direction, see Moreau [755], Rockafellar [843] and Phelps [800], Giles [379], Van Tiel [1006], Borwein and Lewis [158], Hiriart-Urruty and Lemaréchal [505], Magaril-Il'yaev and Tikhomirov [678] and Niculescu and Persson [770].

Throughout the book, vectors are columns, unless stated otherwise, but for convenience, we write vectors as rows as usual. If there is need to stress, that a row actually means a column, the superscript ^{T} is added.

1 Convex Functions of One Variable

Convex functions of one variable are important for analytic inequalities of various sorts, for functional equations and for special functions. The convexity of a function of one variable has far reaching analytic and geometric consequences. For example, each convex function is almost everywhere twice differentiable.

In this section we first consider continuity, affine support, and differentiability properties. Then classical inequalities are given. Finally, we present the characterization of the gamma function due to Bohr and Mollerup together with Artin's elegant proof.

For references to the literature, see the books and surveys cited above.

1.1 Preliminaries

This section contains the basic definitions of convex sets and convex functions and the relation between these two notions via epigraphs. The setting is Euclidean d-space $\mathbb{E}^d$.

Convex Sets and Convex Functions

Let C be a set in $\mathbb{E}^d$. The set C is *convex* if it contains with any two points x, y also the line segment $[x, y]$ with endpoints x, y. In other words, C is convex, if

$$(1-\lambda)x + \lambda y \in C \text{ for } x, y \in C,\ 0 \le \lambda \le 1.$$

C is *strictly convex*, if it is closed and

$$(1-\lambda)x + \lambda y \in \operatorname{int} C \text{ for } x, y \in C, x \neq y,\ 0 < \lambda < 1,$$

where int stands for interior. It is obvious that each strictly convex set is convex (Figs. 1.1 and 1.2).

Examples of convex sets are solid regular polytopes, Euclidean balls and ellipsoids.

Next, let $f : C \to \mathbb{R}$ be a real function on C. The function f is *convex* if C is convex and

$$f\big((1-\lambda)x + \lambda y\big) \le (1-\lambda)f(x) + \lambda f(y) \text{ for } x, y \in C,\ 0 \le \lambda \le 1.$$

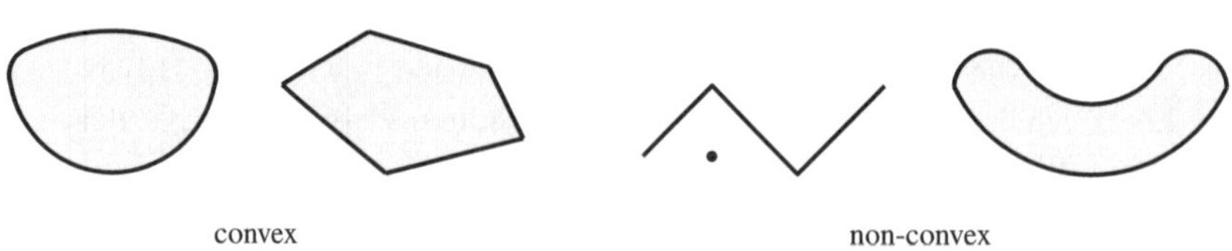

Fig. 1.1. Convex and non-convex sets

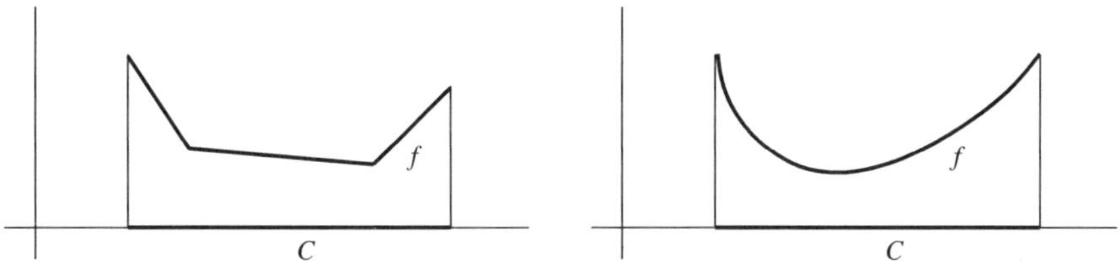

Fig. 1.2. Convex and strictly convex functions

Thus f is convex if any line segment in $\mathbb{E}^d \times \mathbb{R}$ with endpoints on the graph of f is on or above the graph of f. Call f *strictly convex* if C is convex (not necessarily strictly) and

$$f\big((1-\lambda)x + \lambda y\big) < (1-\lambda)f(x) + \lambda f(y) \text{ for } x, y \in C,\ x \neq y,\ 0 < \lambda < 1.$$

Equivalently, f is strictly convex if C is convex and if the relative interior of each proper line segment in $\mathbb{E}^d \times \mathbb{R}$ with endpoints on the graph of f is above the graph. Clearly, a strictly convex function is convex.

Examples of convex functions are norms, semi-norms, positive definite and semi-definite quadratic forms on $\mathbb{E}^d$. A positive definite quadratic form is strictly convex, the other examples are not strictly convex.

A function $f : C \to \mathbb{R}$ is called *concave*, respectively, *strictly concave* if $-f$ is convex, respectively, strictly convex.

Remark. We point out that in the above definitions $\mathbb{E}^d$ may be replaced by a linear or a linear topological space. Similar remarks hold for some of the notions and results below.

Epigraphs of Convex Functions

Let C be a set in $\mathbb{E}^d$ and $f : C \to \mathbb{R}$ a real function. The *epigraph* of f is the set

$$\text{epi } f = \big\{(x, t) \in \mathbb{E}^d \times \mathbb{R} : x \in C, t \geq f(x)\big\} \subseteq \mathbb{E}^d \times \mathbb{R} = \mathbb{E}^{d+1}.$$

The following, almost trivial result makes it possible to transfer information on convex sets to convex functions and vice versa (Fig. 1.3).

Proposition 1.1. *Let $C \subseteq \mathbb{E}^d$ and $f : C \to \mathbb{R}$. Then the following statements are equivalent:*

(i) *f is a convex function.*
(ii) epi f *is a convex set in* $\mathbb{E}^d \times \mathbb{R} = \mathbb{E}^{d+1}$.

Proof. Left to the reader. □

After having introduced convex sets and convex functions in the context of $\mathbb{E}^d$, we consider the case $d = 1$ in more detail.

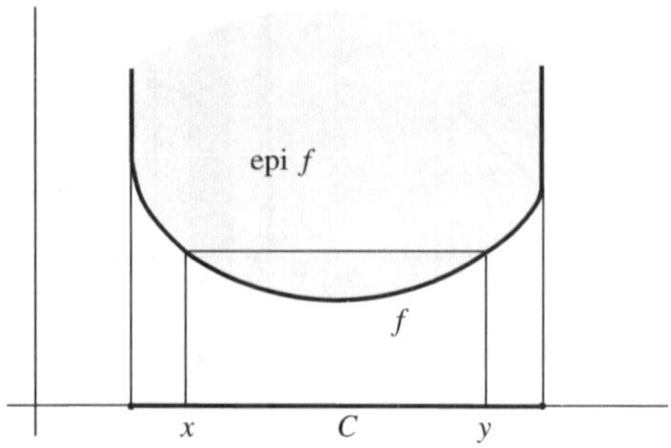

Fig. 1.3. Epigraph of a strictly convex function

1.2 Continuity, Support and Differentiability

Using tools from real analysis, continuity, affine support and differentiability properties of convex functions of one variable were investigated thoroughly in the early twentieth century. It is difficult to make precise attributions.

In the following we present basic pertinent results. It turns out that a convex function which is continuous or differentiable, is so in a particularly strong sense.

Let I and J denote (bounded or unbounded) intervals in $\mathbb{R}$. By int, cl and bd we mean interior, closure and boundary.

A Simple Preparatory Result

We start with the following simple, yet useful result (Fig. 1.4).

Lemma 1.1. *Let $f : I \to \mathbb{R}$ be convex. Then*

$$\frac{f(y) - f(x)}{y - x} \le \frac{f(z) - f(x)}{z - x} \le \frac{f(z) - f(y)}{z - y} \quad \textit{for } x, y, z \in I,\ x < y < z.$$

Proof. Let $x, y, z \in I$, $x < y < z$, and choose $0 < \lambda < 1$ such that $y = (1 - \lambda)x + \lambda z$. The convexity of f then implies the desired inequalities as follows:

$$\begin{aligned}\frac{f(y) - f(x)}{y - x} &= \frac{f\big((1-\lambda)x + \lambda z\big) - f(x)}{(1-\lambda)x + \lambda z - x}\\ &\le \frac{(1-\lambda) f(x) + \lambda f(z) - f(x)}{(1-\lambda)x + \lambda z - x} = \frac{f(z) - f(x)}{z - x},\\ \frac{f(z) - f(y)}{z - y} &= \frac{f(z) - f\big((1-\lambda)x + \lambda z\big)}{z - (1-\lambda)x - \lambda z}\\ &\ge \frac{f(z) - (1-\lambda) f(x) - \lambda f(z)}{z - (1-\lambda)x - \lambda z} = \frac{f(z) - f(x)}{z - x}. \qquad \square\end{aligned}$$

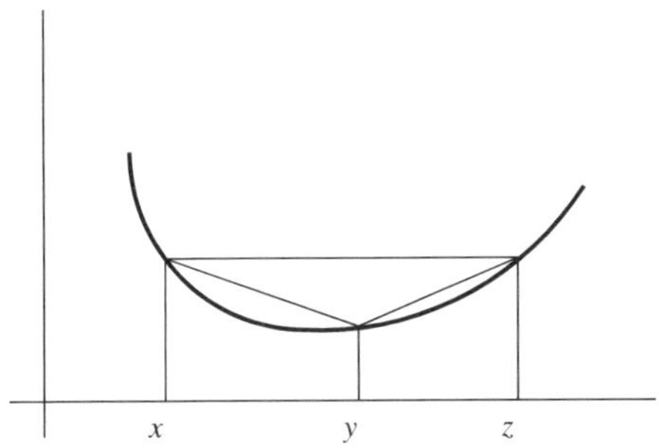

Fig. 1.4. Convex function

Continuity Properties

First, some needed terminology is introduced. Let $f : I \to \mathbb{R}$. The function f is *Lipschitz* on $J \subseteq I$ if there is a constant $L > 0$, a *Lipschitz constant* of f on J, such that

$$|f(x) - f(y)| \leq L|x - y| \text{ for } x, y \in J.$$

f is *absolutely continuous* on J, if for every $\varepsilon > 0$, there is a $\delta > 0$ such that if $[a_i, b_i],\ i = 1, \dots, n$, is any finite system of non-overlapping intervals in J of total length less than δ, then

$$\sum_i |f(b_i) - f(a_i)| < \varepsilon.$$

Clearly, any Lipschitz function is absolutely continuous and any absolutely continuous function is continuous. An example of a continuous function, which is not absolutely continuous, is $f : [0, 1] \to \mathbb{R}$, defined by $f(0) = 0, f(x) = x \sin(\frac{1}{x})$ for $0 < x \leq 1$.

The basic continuity properties of a convex function of one variable are the following.

Theorem 1.1. *Let $f : I \to \mathbb{R}$ be convex. Then f is Lipschitz on each compact interval in* int I. *Thus, in particular, f is absolutely continuous on each compact interval in* int I *and continuous on* int I.

Proof. Let J be a compact interval in int I. For the proof it is sufficient to show that f is Lipschitz on J. Choose $u, v, w, z \in I, u < v, w < z$, where u, v are to the left of J and w, z to the right. Now, let $x, y \in J, x < y$. Applying Lemma 1.1 to u, v, x and to v, x, y yields the inequality

$$\frac{f(v) - f(u)}{v - u} \leq \frac{f(y) - f(x)}{y - x}.$$

Similarly, applying it to x, y, w and to y, w, z shows that

$$\frac{f(y) - f(x)}{y - x} \leq \frac{f(z) - f(w)}{z - w}.$$

Together these inequalities imply that

$$\frac{|f(x)-f(y)|}{|x-y|} \leq \max\left\{\left|\frac{f(v)-f(u)}{v-u}\right|, \left|\frac{f(z)-f(w)}{z-w}\right|\right\} = L, \text{ say.} \quad \square$$

If a is an endpoint of I, a convex function $f : I \to \mathbb{R}$ may not be continuous at a. A simple example is provided by the function $f : [0, 1] \to \mathbb{R}$ defined by $f(0) = 1,\ f(x) = 0$ for $x \in (0, 1]$.

Support Properties

A function $f : I \to \mathbb{R}$ *has affine support* at a point $x \in I$ if there is an affine function $a : \mathbb{R} \to \mathbb{R}$ of the form $a(y) = f(x) + u(y - x)$ for $y \in \mathbb{R}$ where u is a suitable constant, such that

$$f(y) \geq a(y) = f(x) + u(y - x) \text{ for } y \in I.$$

The affine function a is called an *affine support* of f at x. The geometric notion of affine support is intimately connected with the notion of convexity.

Our first result shows that a convex function has affine support at each point in the interior of its interval of definition. This result may be considered as a 1-dimensional Hahn–Banach theorem. The corresponding d-dimensional result is Theorem 2.3 and Theorem 4.1 is the corresponding result for convex bodies.

Theorem 1.2. *Let $f : I \to \mathbb{R}$ be convex and $x \in \operatorname{int} I$. Then f has affine support at x.*

Proof. We may suppose that $x = 0$ and $f(0) = 0$. Let $w \in I, w \neq 0$. Then the convexity of f implies that

$$\begin{aligned} 0 &= (\lambda + \mu) f\Big(\frac{\lambda}{\lambda+\mu}(-\mu w) + \frac{\mu}{\lambda+\mu}(\lambda w)\Big) \\ &\leq \lambda f(-\mu w) + \mu f(\lambda w) \text{ for } \lambda, \mu > 0, \text{ where } \lambda w, -\mu w \in I, \end{aligned}$$

or

$$\frac{-f(-\mu w)}{\mu} \leq \frac{f(\lambda w)}{\lambda} \text{ for } \lambda, \mu > 0, \text{ where } \lambda w, -\mu w \in I.$$

The supremum (over μ) of the left-hand side is therefore less than or equal to the infimum (over λ) of the right-hand side. Hence we may choose $\alpha \in \mathbb{R}$ such that

$$\frac{-f(-\mu w)}{\mu} \leq \alpha \leq \frac{f(\lambda w)}{\lambda} \text{ for } \lambda, \mu > 0, \text{ where } \lambda w, -\mu w \in I.$$

Equivalently,

$$f(\lambda w) \geq \alpha\lambda \text{ for } \lambda \in \mathbb{R}, \text{ where } \lambda w \in I.$$

Thus $a(\lambda w) = \alpha\lambda$ for $\lambda \in \mathbb{R}$ is an affine support of f at $x = 0$. $\square$

A Characterization of Convex Functions

The concept of affine support can be used to define the convexity of a function, as the next result shows. Theorem 2.4 is the corresponding d-dimensional result and the corresponding result for convex bodies is Theorem 4.2.

Theorem 1.3. *Let I be open and $f : I \to \mathbb{R}$. Then the following statements are equivalent:*

(i) *f is convex.*
(ii) *f has affine support at each $x \in I$.*

Proof. (i) $\Rightarrow$ (ii) This follows from Theorem 1.2.

(ii) $\Rightarrow$ (i) If f has affine support at each $x \in I$, say $a_x(\cdot)$, then clearly,

$$f(y) = \sup\{a_x(y) : x \in I\} \text{ for } y \in I.$$

As the supremum of a family of affine and thus convex functions, f is also convex:

$$\begin{aligned} f\big((1-\lambda)y+\lambda z\big) &= \sup\big\{a_x\big((1-\lambda)y+\lambda z\big) : x \in I\big\} \\ &= \sup\big\{(1-\lambda)a_x(y)+\lambda a_x(z) : x \in I\big\} \\ &\le (1-\lambda)\sup\big\{a_x(y) : x \in I\big\} + \lambda \sup\{a_x(z) : x \in I\big\} \\ &= (1-\lambda)f(y)+\lambda f(z) \text{ for } y, z \in I, 0 \le \lambda \le 1. \quad \square \end{aligned}$$

It is sufficient in Theorem 1.3 to assume that f is affinely supported locally at each point of I.

First-Order Differentiability

In the following several well-known results from analysis will be used. A reference for these is [499].

A theorem of Lebesgue says that an absolutely continuous real function on an interval is almost everywhere differentiable. This combined with Theorem 1.1 shows that a convex function $f : I \to \mathbb{R}$ is almost everywhere differentiable. Yet, as we shall see below, the convexity of f yields an even stronger result: f is differentiable at each point of I with, at most, a countable set of exceptions.

In order to state this result in a precise form we need the notions of *left* and *right derivative* $f'_-(x)$ and $f'_+(x)$ of a function $f : I \to \mathbb{R}$ at a point $x \in I$:

$$f'_-(x) = \lim_{y \to x-0} \frac{f(y)-f(x)}{y-x} \quad (x \text{ not the left endpoint of } I),$$
$$f'_+(x) = \lim_{y \to x+0} \frac{f(y)-f(x)}{y-x} \quad (x \text{ not the right endpoint of } I).$$

The left and right derivatives may or may not exist.

Theorem 1.4. *Let I be open and $f : I \to \mathbb{R}$ convex. Then f'_- and f'_+ exist, are non-decreasing (not necessarily strictly), and $f'_- \le f'_+$ on I. f is differentiable precisely at those points $x \in I$ where f'_- is continuous. Hence $f'(x)$ exists for all $x \in I$ with a set of exceptions which is at most countable and f' is non-decreasing on its domain of definition.*

Proof. Before embarking on the proof we state a simple proposition, the proof of which is left to the interested reader.

(1) Consider a non-decreasing sequence of real continuous non-decreasing functions on I with limit g, say. Then g is continuous on the left.

The proof of the theorem is split into several steps. First,

(2) f'_-, f'_+ exist on I and

$$f'_-(x) \le f'_+(x) \le f'_-(y) \le f'_+(y) \text{ for } x, y \in I, x < y.$$

Let $x, y \in I, x < y$. By Lemma 1.1,

$$\frac{f(w) - f(x)}{w - x} \le \frac{f(z) - f(x)}{z - x} \le \frac{f(y) - f(x)}{y - x} \text{ for } w, z \in I, w < x < z < y,$$

and the first two expressions are non-decreasing in w, respectively, z. Thus their limits as $w \to x - 0$ and $z \to x + 0$ exist and we deduce that

(3)

$$f'_-(x), f'_+(x) \text{ exist and } f'_-(x) \le f'_+(x) \le \frac{f(y) - f(x)}{y - x}.$$

Similar arguments show that

(4)

$$f'_-(y), f'_+(y) \text{ exist and } \frac{f(y) - f(x)}{y - x} \le f'_-(y) \le f'_+(y).$$

Propositions (3) and (4) yield (2). This settles the assertion about f'_- and f'_+.

Second,

(5) f'_- is continuous on the left and f'_+ on the right.

The functions $g_n, n = 1, 2, \ldots$, defined by

$$g_n(x) = \frac{f(x - \frac{1}{n}) - f(x)}{-\frac{1}{n}} \text{ for } x \in I \text{ such that } x - \frac{1}{n} \in I,$$

are continuous and non-decreasing by Lemma 1.1 and, again by Lemma 1.1, form a non-decreasing sequence with limit f'_-. It thus follows from (1) that f'_- is continuous on the left. This establishes (5) for f'_-. The statement about f'_+ is shown similarly.

Third,

(6) $f'(x)$ exists precisely for those $x \in I$ for which f'_- is continuous.

Let $x \in I$ and assume first that f'_- is continuous at x. By (2), $f'_-(x) \leq f'_+(x) \leq f'_-(y)$ for all $y \in I$ with $x < y$. Letting $y \to x+0$, the continuity of f'_- at x yields $f'_-(x) = f'_+(x)$ which, in turn, shows that $f'(x)$ exists. Assume, second, that f'_- is not continuous at x. Then (2) and (5) imply

$$f'_-(x) < \lim_{y \to x+0} f'_-(y) \leq \lim_{y \to x+0} f'_+(y) = f'_+(x).$$

Hence $f'(x)$ does not exist. The proof of (6) is now complete.

The theorem finally follows from (2) and (6) on noting that a non-decreasing function on I has at most countably many points of discontinuity. □

A less precise extension of Theorem 1.4 to convex functions in d variables is due to Reidemeister 2.6. A precise extension to convex bodies in $\mathbb{E}^d$ is the Anderson–Klee theorem 5.1.

An important consequence of Theorem 1.4 is the following result.

Theorem 1.5. *Let I be open and $f : I \to \mathbb{R}$ convex. If f is differentiable on I, then f' is continuous, i.e. f is of class $\mathcal{C}^1$.*

This result can be extended to d dimensions, compare Theorem 2.8.

First-Order Differentiability and Affine Support

The following result shows that, for convex functions of one variable, the relation between differentiability and affine support is particularly simple.

Proposition 1.2. *Let $f : I \to \mathbb{R}$ be convex and $x \in \operatorname{int} I$. Then an affine function $a : \mathbb{R} \to \mathbb{R}$ of the form $a(y) = f(x) + u(y - x)$ for $y \in \mathbb{R}$ is an affine support of f at x if and only if*

$$f'_-(x) \leq u \leq f'_+(x).$$

Proof. Denote the coordinates in $\mathbb{E}^2$ by y and z. By Theorem 1.4, the left and right derivatives $f'_-(x)$ and $f'_+(x)$ exist. Their definitions show that the half-lines

(7) $$z = f(x) + f'_-(x)(y - x) \text{ for } y \leq x,$$

$$z = f(x) + f'_+(x)(y - x) \text{ for } y \geq x,$$

are the left and right half-tangents of the curve $z = f(y)$ at $y = x$ (Fig. 1.5).

As a consequence of Propositions (3) and (4) in the proof of Theorem 1.4 we have

(8) $$f(y) \geq f(x) + f'_-(x)(y - x) \text{ for } y \in I,\ y \leq x,$$

$$f(y) \geq f(x) + f'_+(x)(y - x) \text{ for } y \in I,\ y \geq x,$$

$$f'_-(x) \leq f'_+(x).$$

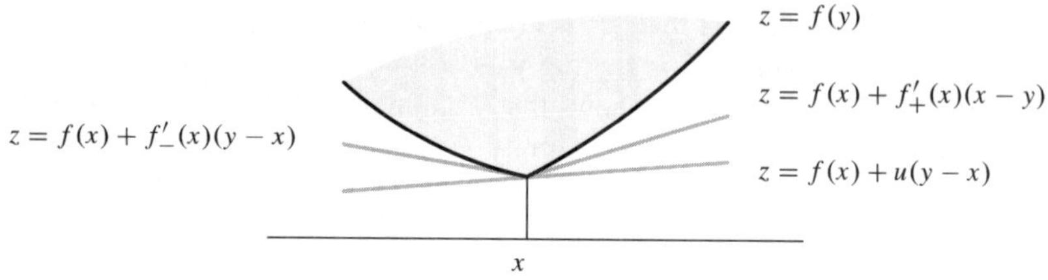

Fig. 1.5. Support and *left* and *right* side differentiability

It follows from (7) and (8) that a function of the form $z = f(x) + u(y - x)$ for $y \in \mathbb{R}$ is an affine support of f at x if and only if $f'_-(x) \le u \le f'_+(x)$. □

As an immediate consequence of this proposition we obtain the following result. The corresponding d-dimensional result is Theorem 2.7.

Theorem 1.6. *Let $f : I \to \mathbb{R}$ be convex and $x \in \operatorname{int} I$. Then the following statements are equivalent:*

(i) *f is differentiable at x.*
(ii) *f has unique affine support at x, say $a : \mathbb{R} \to \mathbb{R}$, where $a(y) = f(x) + u(y - x)$ for $y \in \mathbb{R}$ and $u = f'(x)$.*

Second-Order Differentiability

We need the following weak notion of twice differentiability: a function $f : I \to \mathbb{R}$ is *twice* (or is *second-order) differentiable almost everywhere* on I if there are sets $M, N \subseteq I$ of (Lebesgue) measure 0 such that

$$f'(x) = \lim_{y \to x} \frac{f(y) - f(x)}{y - x} \text{ exists for } x \in I \backslash M,$$

$$\text{and } \lim_{\substack{y \to x \\ y \in I \backslash M}} \frac{f'(y) - f'(x)}{y - x} \text{ exists for } x \in I \backslash (M \cup N).$$

The latter limit is denoted by $f''(x)$. (In Sect. 2.2, we will consider a slightly different notion of twice differentiability almost everywhere for functions of several variables.)

For the convenience of the reader, we define the *Bachmann–Landau symbols* $o(\cdot)$ and $O(\cdot)$: let $g, h : I \to \mathbb{R}$ and $x \in I$. Then we say that

$$g(y) = o\big(h(y)\big) \text{ as } y \to x,\, y \in I, \text{ if } \frac{|g(y)|}{|h(y)|} \to 0$$
$$\text{as } y \to x,\, y \in I,\, y \ne x,$$
$$g(y) = O\big(h(y)\big) \text{ as } y \to x,\, y \in I \text{ if } |g(y)| \le \text{const}\,|h(y)|$$
$$\text{for } y \in I, \text{ close to } x,$$

where const is a suitable positive constant.

The above definition of second-order differentiability is slightly weaker than the common definition. In this way it still embodies the main idea of the common second-order differentiability and has the advantage that it applies to general convex functions as the following result shows:

Theorem 1.7. *Let I be open and $f : I \to \mathbb{R}$ convex. Then f is twice differentiable almost everywhere on I. Moreover, for almost every $x \in I$,*

$$f(y) = f(x) + f'(x)(y-x) + \frac{1}{2} f''(x)(y-x)^2 + o(|y-x|^2) \text{ as } y \to x,\ y \in I.$$

Proof. We first state two well-known theorems of Lebesgue, see [499].

(9) A non-decreasing real function on I is almost everywhere differentiable.

(10) The derivative of an absolutely continuous function $g : J \to \mathbb{R}$ exists almost everywhere on J, is Lebesgue integrable and, for $x \in J$,

$$g(y) = g(x) + \int_x^y g'(t)\,dt \text{ for } y \in J.$$

By Theorem 1.4 above and (9), there are a countable set M and a set N of measure 0, both in I, such that the following statements hold: $f'(x)$ exists and equals $f'_-(x)$ for each $x \in I \backslash M$ and f'_- exists on I and is non-decreasing, and $f''_-(x)$ exists for each $x \in I \backslash N$. Thus,

$$\text{(11)} \qquad f''_-(x) = \lim_{y \to x} \frac{f'_-(y) - f'_-(x)}{y-x} = \lim_{\substack{y \to x \\ y \in I \backslash M}} \frac{f'_-(y) - f'_-(x)}{y-x}$$
$$= \lim_{\substack{y \to x \\ y \in I \backslash M}} \frac{f'(y) - f'(x)}{y-x} = f''(x) \text{ for } x \in I \backslash (M \cup N),$$

concluding the proof of the first assertion of the theorem.

To show the second assertion, note that (11) yields the following:

$$\text{(12)} \qquad f'(y) = f'(x) + f''(x)(y-x) + o(|y-x|)$$
$$\text{as } y \to x,\ y \in I \backslash M \text{ for } x \in I \backslash (M \cup N).$$

Since by Theorem 1.1 f is absolutely continuous, Proposition (10) shows that one may integrate (12) from x to y to obtain

$$f(y) = f(x) + f'(x)(y-x) + \frac{1}{2} f''(x)(y-x)^2 + o(|y-x|^2)$$
$$\text{as } y \to x,\ y \in I \backslash M \text{ for } x \in I \backslash (M \cup N).$$

Now note that the continuity of both sides of this equality permits us to cancel the restriction that $y \notin M$. □

Theorem 1.5 says that for any convex function f the differentiability of f implies that f' is continuous, i.e. f is of class $\mathcal{C}^1$. Examples show that there is no analogous result for f'': If f'' exists it need not be continuous.

1.3 Convexity Criteria

If a given function is known to be convex, the convexity may yield useful information. For examples, see the following section. Thus the question arises to find out whether a function is convex or not. While, in principle, Theorem 1.3 is a convexity criterion, it is of little practical value.

The property that a convex function has non-decreasing derivative on the set where the latter exists, and the property that a function which has affine support everywhere is convex, yield a simple, yet useful convexity criterion which will be stated below.

Convexity Criteria

Our aim is to prove the following result:

Theorem 1.8. *Let I be open and $f : I \to \mathbb{R}$ differentiable. Then the following statements are equivalent:*

(i) *f is convex.*

(ii) *f' is non-decreasing.*

Proof. (i)⇒(ii) This follows from Theorem 1.4.

(ii)⇒(i) If (ii) holds, then the first mean value theorem from calculus implies that for any $x \in I$,

$$f(y) = f(x) + f'\big(x + \vartheta(y - x)\big)(y - x) \geq f(x) + f'(x)(y - x)$$

for any $y \in I$ and suitable ϑ depending on x and y, where $0 < \vartheta < 1$.

Hence f has affine support at each $x \in I$ and thus is convex by Theorem 1.3. □

Corollary 1.1. *Let I be open and $f : I \to \mathbb{R}$ twice differentiable. Then the following are equivalent:*

(i) *f is convex.*

(ii) $f'' \geq 0$.

Remark. Simple arguments show that, for differentiable $f : I \to \mathbb{R}$, the assumption that f' is strictly non-decreasing implies that f is strictly convex. Similarly, if $f : I \to \mathbb{R}$ is twice differentiable, the assumption that $f'' > 0$ yields the strict convexity of f. Examples show that the converse implications do not hold generally.

1.4 Jensen's and Other Inequalities

Notions and results of convex geometry are valuable tools for inequalities and the related field of functional equations and inequalities. This agrees with the following quotation of Mitrinović which we have taken from Roberts and Varberg [842], p.188:

> ...it should be emphasized that the theory of convexity..., taken together with a few elementary devices, can be used to derive a large number of the most familiar and important inequalities of analysis.

Perhaps the most important elementary inequality dealing with convex functions is the inequality of Jensen [544], proved earlier by Hölder [519] for differentiable convex functions.

In this section it is proved by means of a simple convexity argument. As corollaries we obtain a series of classical inequalities. Let I be an interval in $\mathbb{R}$.

For a thorough treatment of Jensen's inequality and its consequences, see Kuczma [620] and Castillo and Ruiz Cobo [196]. See also Roberts and Varberg [841] and Roberts [840] for an overview of inequalities in the context of convex functions. A beautiful book on inequalities is Steele [953].

Jensen's Inequality

Jensen's inequality is as follows:

Theorem 1.9. *Let $f : I \to \mathbb{R}$ be convex, $x_1, \dots, x_n \in I$, and $\lambda_1, \dots, \lambda_n \geq 0$ such that $\lambda_1 + \dots + \lambda_n = 1$. Then $\lambda_1 x_1 + \dots + \lambda_n x_n \in I$ and*

$$f(\lambda_1 x_1 + \dots + \lambda_n x_n) \leq \lambda_1 f(x_1) + \dots + \lambda_n f(x_n).$$

We give two proofs.

Proof (by Induction). For $n = 1$ the assertion is trivial. Assume now that $n > 1$ and that the assertion holds for $n - 1$. We have to prove it for n. If $\lambda_n = 0$ the assertion reduces to the case $n - 1$ and thus holds by the induction assumption. If $\lambda_n = 1$, then $\lambda_1 = \dots = \lambda_{n-1} = 0$ and the assertion is true trivially. It remains to consider the case $0 < \lambda_n < 1$ and thus $0 < \lambda_1 + \dots + \lambda_{n-1} = 1 - \lambda_n < 1$. Then

$$\begin{aligned}
&\lambda_1 x_1 + \dots + \lambda_n x_n \\
&\quad = (1 - \lambda_n)\Big(\frac{\lambda_1}{1 - \lambda_n} x_1 + \dots + \frac{\lambda_{n-1}}{1 - \lambda_n} x_{n-1}\Big) + \lambda_n x_n \in I,
\end{aligned}$$

$$\begin{aligned}
&f(\lambda_1 x_1 + \dots + \lambda_n x_n) \\
&\quad = f\Big((1 - \lambda_n)\Big(\frac{\lambda_1}{1 - \lambda_n} x_1 + \dots + \frac{\lambda_{n-1}}{1 - \lambda_n} x_{n-1}\Big) + \lambda_n x_n\Big) \\
&\quad \leq (1 - \lambda_n) f\Big(\frac{\lambda_1}{1 - \lambda_n} x_1 + \dots + \frac{\lambda_{n-1}}{1 - \lambda_n} x_{n-1}\Big) + \lambda_n f(x_n) \\
&\quad \leq \lambda_1 f(x_1) + \dots + \lambda_{n-1} f(x_{n-1}) + \lambda_n f(x_n)
\end{aligned}$$

by the induction assumption, the convexity of I and the convexity of f. □

Proof (by affine support properties). By induction, $x_0 = \lambda_1 x_1 + \cdots + \lambda_n x_n \in I$. Let $a(x) = f(x_0) + u(x - x_0)$ be an affine support of f at x_0. Then $f(x_i) \geq a(x_i)$ for $i = 1, \ldots, n$, and thus

$$\begin{aligned}\lambda_1 f(x_1) + \cdots + \lambda_n f(x_n) &\geq \lambda_1 a(x_1) + \cdots + \lambda_n a(x_n)\\ &= (\lambda_1 + \cdots + \lambda_n) f(x_0) + u\big(\lambda_1 x_1 + \cdots + \lambda_n x_n - (\lambda_1 + \cdots + \lambda_n) x_0\big)\\ &= f(x_0) = f(\lambda_1 x_1 + \cdots + \lambda_n x_n). \quad \square\end{aligned}$$

Remark. If f is strictly convex and $\lambda_1, \ldots, \lambda_n > 0$, then equality holds in Jensen's inequality precisely in case when $x_1 = \cdots = x_n$.

Mechanical Interpretation on Jensen's Inequality

The center of gravity of the masses $\lambda_1, \ldots, \lambda_n$ at the points $\big(x_1, f(x_1)\big), \ldots, \big(x_n, f(x_n)\big)$ on the graph of f is the point

$$(x_c, y_c) = \big(\lambda_1 x_1 + \cdots + \lambda_n x_n, \lambda_1 f(x_1) + \cdots + \lambda_n f(x_n)\big).$$

It is contained in the convex polygon with vertices $\big(x_1, f(x_1)\big), \ldots, \big(x_n, f(x_n)\big)$ which, in turn, is contained in the epigraph of f (as can be shown). Thus (x_c, y_c) is also contained in the epigraph of f which is equivalent to Jensen's inequality.

Inequality Between the Arithmetic and the Geometric Mean

As a direct consequence of Jensen's inequality we have the following inequality.

Corollary 1.2. *Let* $x_1, \ldots, x_n \geq 0$ *and* $\lambda_1, \ldots, \lambda_n \geq 0$ *be such that* $\lambda_1 + \cdots + \lambda_n = 1$. *Then*

$$x_1^{\lambda_1} \cdots x_n^{\lambda_n} \leq \lambda_1 x_1 + \cdots + \lambda_n x_n.$$

In particular,

$$(x_1 \cdots x_n)^{\frac{1}{n}} \leq \frac{x_1 + \cdots + x_n}{n}.$$

Proof. We may suppose that $x_1, \ldots, x_n > 0$. Since $\exp : \mathbb{R} \to \mathbb{R}^+$ is convex by Theorem 1.8, an application of Jensen's inequality to $y_1 = \log x_1, \ldots, y_n = \log x_n$ then gives the desired inequality:

$$\begin{aligned}x_1^{\lambda_1} \cdots x_n^{\lambda_n} &= \exp\big(\log(x_1^{\lambda_1} \cdots x_n^{\lambda_n})\big) = \exp(\lambda_1 \log x_1 + \cdots + \lambda_n \log x_n)\\ &\leq \lambda_1 \exp(\log x_1) + \cdots + \lambda_n \exp(\log x_n) = \lambda_1 x_1 + \cdots + \lambda_n x_n. \quad \square\end{aligned}$$

Actually, exp is strictly convex.

Young's Inequality

The inequality of the arithmetic and the geometric mean, in turn, yields our next result.

Corollary 1.3. *Let $x, y \geq 0$ and $p, q > 1$ such that $\frac{1}{p} + \frac{1}{q} = 1$. Then*

$$xy \leq \frac{x^p}{p} + \frac{y^q}{q}.$$

Proof. For $x = 0$ or $y = 0$ this inequality is trivial. For $x, y > 0$ it is the special case $n = 2, x_1 = x^p, x_2 = y^q, \lambda_1 = \frac{1}{p}, \lambda_2 = \frac{1}{q}$ of the arithmetic–geometric mean inequality. □

Hölder's Inequality for Sums

The following result generalizes the Cauchy–Schwarz inequality for sums.

Corollary 1.4. *Let $x_1, y_1, \dots, x_n, y_n \geq 0$ and $p, q > 1$ such that $\frac{1}{p} + \frac{1}{q} = 1$. Then*

$$x_1 y_1 + \dots + x_n y_n \leq \left(x_1^p + \dots + x_n^p\right)^{\frac{1}{p}} \left(y_1^q + \dots + y_n^q\right)^{\frac{1}{q}}.$$

Proof. If all x_i or all y_i are 0, Hölder's inequality is trivial. Otherwise apply Young's inequality with

$$x = \frac{x_i}{\left(x_1^p + \dots + x_n^p\right)^{\frac{1}{p}}}, \quad y = \frac{y_i}{\left(y_1^q + \dots + y_n^q\right)^{\frac{1}{q}}}$$

for $i = 1, \dots, n$, sum from 1 to n and note that $\frac{1}{p} + \frac{1}{q} = 1$. □

Hölder's Inequality for Integrals

A generalization of the Cauchy–Schwarz inequality for integrals is the following inequality.

Corollary 1.5. *Let $f, g : I \to \mathbb{R}$ be non-negative, integrable, with non-vanishing integrals and let $p, q > 1$ such that $\frac{1}{p} + \frac{1}{q} = 1$. Then*

$$\int_I fg \, dx \leq \Big(\int_I f^p \, dx\Big)^{\frac{1}{p}} \Big(\int_I g^q \, dx\Big)^{\frac{1}{q}}.$$

Proof. By Young's inequality, we have

$$\frac{f}{(\int_I f^p \, dx)^{\frac{1}{p}}} \frac{g}{(\int_I g^q \, dx)^{\frac{1}{q}}} \leq \frac{f^p}{p \int_I f^p \, dx} + \frac{g^q}{q \int_I g^q \, dx}.$$

Integrate this inequality over I and note that $\frac{1}{p} + \frac{1}{q} = 1$. □

Minkowski's Inequality for Sums

Minkowski's inequality for sums or the *triangle inequality for the p-norm* is as follows.

Corollary 1.6. *Let* $x_1, y_1, \dots, x_n, y_n \geq 0$ *and* $p \geq 1$. *Then*

$$\left((x_1 + y_1)^p + \cdots + (x_n + y_n)^p\right)^{\frac{1}{p}} \leq \left(x_1^p + \cdots + x_n^p\right)^{\frac{1}{p}} + \left(y_1^p + \cdots + y_n^p\right)^{\frac{1}{p}}.$$

Proof. For $p = 1$ this inequality holds trivially. Assume now that $p > 1$. Let $q > 1$ such that $\frac{1}{p} + \frac{1}{q} = 1$. Then $(p-1)q = p$ and thus

$$\begin{aligned}
&(x_1 + y_1)^p + \cdots + (x_n + y_n)^p \\
&\quad = x_1(x_1 + y_1)^{p-1} + \cdots + x_n(x_n + y_n)^{p-1} \\
&\qquad + y_1(x_1 + y_1)^{p-1} + \cdots + y_n(x_n + y_n)^{p-1} \\
&\quad \leq \left(x_1^p + \cdots + x_n^p\right)^{\frac{1}{p}} \left((x_1 + y_1)^p + \cdots + (x_n + y_n)^p\right)^{\frac{1}{q}} \\
&\qquad + \left(y_1^p + \cdots + y_n^p\right)^{\frac{1}{p}} \left((x_1 + y_1)^p + \cdots + (x_n + y_n)^p\right)^{\frac{1}{q}} \\
&\quad = \left((x_1 + y_1)^p + \cdots + (x_n + y_n)^p\right)^{\frac{1}{q}} \left((x_1^p + \cdots + x_n^p)^{\frac{1}{p}} + (y_1^p + \cdots + y_n^p)^{\frac{1}{p}}\right)
\end{aligned}$$

by Hölder's inequality. □

Minkowski's Inequality for Integrals

A similar argument gives our last result.

Corollary 1.7. *Let* $f, g : I \to \mathbb{R}$ *be non-negative and integrable and let* $p \geq 1$. *Then*

$$\left(\int_I (f+g)^p dx\right)^{\frac{1}{p}} \leq \left(\int_I f^p dx\right)^{\frac{1}{p}} + \left(\int_I g^p dx\right)^{\frac{1}{p}}.$$

1.5 Bohr and Mollerup's Characterization of Γ

A given functional equation may have many solutions. To single out interesting special solutions, additional conditions must be imposed, for example continuity, measurability, boundedness or convexity conditions. In the case of *Cauchy's functional equation* of 1821,

$$f(x + y) = f(x) + f(y) \text{ for } x, y \in \mathbb{R},$$

many such results are known.

The characterization of the gamma function by Bohr and Mollerup [136] is also a result of this type. It will be presented below, together with Artin's [39] elegant proof.

For more information on functional equations and convex functions the reader may consult the books of Kuczma [620], Castillo and Ruiz Cobo [196] and Czerwik [234], to which we add the nice little treatise of Smítal [944].

The Gamma Function

Γ is defined by

$$\Gamma(x) = \int_0^{+\infty} t^{x-1} e^{-t} dt \text{ for } x > 0$$

and then extended, by analytic continuation, to the whole complex plane $\mathbb{C}$, except the points $0, -1, -2, \ldots,$ where it has poles of first-order. There are other ways to define Γ. See, e.g. formula (6) at the end of this section (Fig. 1.6).

We first collect some properties of Γ.

Theorem 1.10. Γ *has the following properties:*

(i) $\Gamma(1) = 1$.
(ii) $\Gamma(x+1) = x\Gamma(x)$ *for* $x > 0$, *i.e.* Γ *satisfies* Euler's functional equation.
(iii) Γ *is* logarithmic convex, *i.e.* $\log \Gamma$ *is convex for* $x > 0$.

Proof. (i) $\Gamma(1) = \int_0^{+\infty} e^{-t} dt = \lim_{s \to +\infty} \int_0^{s} e^{-t} dt = \lim_{s \to +\infty} \left(- e^{-t}\big|_0^s \right) = 1.$

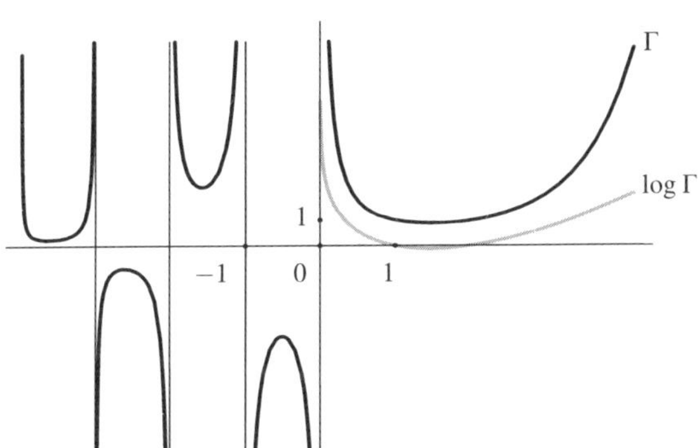

Fig. 1.6. Gamma and log-gamma function

(ii) Let $x > 0$. Then

$$\Gamma(x+1) = \int_0^{+\infty} t^x e^{-t} dt = \lim_{s\to+\infty} \int_0^s t^x e^{-t} dt$$
$$= \lim_{s\to+\infty} \left(-t^x e^{-t}\Big|_0^s + x \int_0^s t^{x-1} e^{-t} dt \right)$$
$$= x \lim_{s\to+\infty} \int_0^s t^{x-1} e^{-t} dt = x\Gamma(x).$$

(iii) Let $x, y > 0$ and $0 < \lambda < 1$. Then, putting $1 - \lambda = \frac{1}{p}$ and $\lambda = \frac{1}{q}$,

$$\log \Gamma\big((1-\lambda)x + \lambda y\big)$$
$$= \log \int_0^{+\infty} t^{(1-\lambda)x+\lambda y-1} e^{-t} dt = \log \int_0^{+\infty} \left(t^{x-1}e^{-t}\right)^{1-\lambda} \left(t^{y-1}e^{-t}\right)^{\lambda} dt$$
$$\leq \log \left(\Big(\int_0^{+\infty} t^{x-1} e^{-t} dt \Big)^{1-\lambda} \Big(\int_0^{+\infty} t^{y-1} e^{-t} dt \Big)^{\lambda} \right)$$
$$= (1-\lambda) \log \Gamma(x) + \lambda \log \Gamma(y)$$

by Hölder's inequality for integrals, see Corollary 1.5. □

Characterization of the Gamma Function

The functions $g : \mathbb{R}^+ \to \mathbb{R}$ which satisfy the conditions

$$g(1) = 1 \text{ and } g(n+1) = n\, g(n) \ (= n!) \text{ for } n = 1, 2, \ldots,$$

are precisely the functions which can be represented in the form $g(x) = \Gamma(x) + h(x)$ for $x > 0$, where $h : \mathbb{R}^+ \to \mathbb{R}$ is an arbitrary function with zeros at $1, 2, \ldots$. Similarly, the functions $g : \mathbb{R}^+ \to \mathbb{R}$, which satisfy the stronger property that

$$g(1) = 1 \text{ and } g(x+1) = x\, g(x) \text{ for } x > 0,$$

are precisely the functions of the form $g(x) = \Gamma(x)h(x)$ for $x > 0$, where $h : \mathbb{R}^+ \to \mathbb{R}$ is any function with period 1 and $h(1) = 1$. Among this large family of functions, Γ is singled out by the property of logarithmic convexity, as shown by Bohr and Mollerup [136]:

Theorem 1.11. *Let $g : \mathbb{R}^+ \to \mathbb{R}^+$ be a function having the Properties* (i)–(iii) *of Theorem* 1.10. *Then $g = \Gamma$ on $\mathbb{R}^+$.*

Proof. Properties (i) and (ii) imply that

(1) $g(n+1) = n!$ for $n = 0, 1, \ldots$

The main step of the proof is to show that Properties (i)–(iii) yield the formula

$$(2)\quad g(x) = \lim_{n\to\infty} \frac{n^x n!}{x(x+1)\cdots(x+n)} \quad \text{for } x > 0.$$

Assume first that $0 < x \le 1$. The logarithmic convexity (iii), together with the functional equation (ii) and (1), shows that

$$(3)\quad g(n+1+x) = g\big((1-x)(n+1) + x(n+2)\big) \\ \le g(n+1)^{1-x} g(n+2)^x = (n+1)^x n!,$$

$$(4)\quad n! = g(n+1) = g\big(x(n+x) + (1-x)(n+1+x)\big) \\ \le g(n+x)^x g(n+1+x)^{1-x} = (n+x)^{-x} g(n+1+x)^x g(n+1+x)^{1-x} \\ = (n+x)^{-x} g(n+1+x).$$

An immediate consequence of the functional equation (ii) is the identity

$$(5)\quad g(n+1+x) = (n+x)(n-1+x)\cdots x g(x).$$

Combining (3)–(5), we obtain the following inequalities:

$$\left(1+\frac{x}{n}\right)^x \le \frac{(n+x)(n-1+x)\cdots x g(x)}{n^x n!} \le \left(1+\frac{1}{n}\right)^x \text{ for } n = 1, 2, \ldots,$$

which, in turn, yields (2) in case $0 < x \le 1$.

Assume, second, that $x > 1$. Using the functional equation (ii), this can be reduced to the case already settled: choose an integer m such that $0 < x - m \le 1$. Then

$$\begin{aligned} g(x) &= (x-1)\cdots(x-m)g(x-m) \\ &= (x-1)\cdots(x-m) \lim_{n\to\infty} \frac{n^{x-m} n!}{(x-m)\cdots(x-m+n)} \\ &= \lim_{n\to\infty} \left(\frac{n^x n!}{x(x+1)\cdots(x+n)} \cdot \frac{(x+n)\cdots(x+n-m+1)}{n^m}\right) \\ &= \lim_{n\to\infty} \frac{n^x n!}{x(x+1)\cdots(x+n)} \end{aligned}$$

by (ii) and the already settled case of (2). Thus (2) also holds for $x > 1$, which concludes the proof of (2).

Since Γ has Properties (i)–(iii) and (2) was proved using only these properties, we see that

$$(6)\quad \Gamma(x) = g(x) = \lim_{n\to\infty} \frac{n^x n!}{x(x+1)\cdots(x+n)} \quad \text{for } x > 0. \qquad \square$$

Formula (6) was used by Euler in 1729 to introduce the gamma function. It is sometimes named after Gauss. Theorem 1.11 can be used to derive other properties of the gamma function. See, e.g. Webster [1016].

2 Convex Functions of Several Variables

Convex functions in d variables appear in several areas of mathematics, for example in optimization. Many of the general results for convex functions of one variable extend to convex functions in d variables. While in some cases the extensions are straightforward, for numerous results the proofs are essentially more difficult and require new ideas.

In the following we consider continuity, affine support, and differentiability properties, including Alexandrov's celebrated theorem on second-order differentiability almost everywhere. Then a Stone–Weierstrass type result is given, showing a relation between convex and continuous functions. As an application, we present a sufficient condition in the calculus of variations due to Hilbert and Courant.

Let C be a convex set in $\mathbb{E}^d$ with non-empty interior and $I \subseteq \mathbb{R}$ an interval.

For more information the reader may consult the books and surveys cited in the introduction of this chapter.

2.1 Continuity, Support and First-Order Differentiability, and a Heuristic Principle

The results in this section are direct extensions of the basic results on continuity, affine support and first-order differentiability of convex functions of one variable, including Jensen's inequality, which were presented in Sects. 1.2 and 1.3. In most cases the proofs are more involved. Finally there are some heuristic remarks concerning a sort of reinforcement principle.

Jensen's Inequality

As a useful tool we state the following generalization of Theorem 1.9. Its proof is verbatim the same as that of its 1-dimensional relative and thus is omitted.

Theorem 2.1. *Let $f : C \to \mathbb{R}$ be convex, $x_1, \dots, x_n \in C$, and $\lambda_1, \dots, \lambda_n \geq 0$ such that $\lambda_1 + \dots + \lambda_n = 1$. Then $\lambda_1 x_1 + \dots + \lambda_n x_n \in C$ and*

$$f(\lambda_1 x_1 + \dots + \lambda_n x_n) \leq \lambda_1 f(x_1) + \dots + \lambda_n f(x_n).$$

Continuity Properties

Let $f : C \to \mathbb{R}$. f is *Lipschitz* on a subset $D \subseteq C$ if there is a constant $L > 0$, a *Lipschitz constant* of f on D, such that

$$|f(x) - f(y)| \leq L\|x - y\| \text{ for } x, y \in D,$$

where $\|\cdot\|$ denotes the Euclidean norm on $\mathbb{E}^d$. f is *locally Lipschitz* at a point $x \in C$ if there is a neighborhood N of x such that f is Lipschitz on $C \cap N$. The corresponding Lipschitz constant may depend on x and N.

Theorem 2.2. *Let* $f : C \to \mathbb{R}$ *be convex. Then* f *is Lipschitz on each compact subset of* $\operatorname{int} C$*. Thus, in particular,* f *is continuous on* $\operatorname{int} C$.

Proof. It is sufficient to show the following.

(1) Let $x \in \operatorname{int} C$. Then f is locally Lipschitz at x.

The proof of (1) is divided into several steps. For $\varepsilon > 0$ let $N(\varepsilon)$ denote the ε-neighborhood of x.

First,

(2) There are $\alpha, \varepsilon > 0$ such that $N(2\varepsilon) \subseteq \operatorname{int} C$ and f is bounded above by α on $N(2\varepsilon)$.

For the proof of (2) it is sufficient to show that f is bounded above on some simplex in C which contains x in its interior. Let S be a simplex in C with vertices $x_1, \dots, x_{d+1}$, say, and $x \in \operatorname{int} S$. If $y \in S$, then $y = \lambda_1 x_1 + \cdots + \lambda_{d+1} x_{d+1}$, where $\lambda_1, \dots, \lambda_{d+1} \geq 0$, $\lambda_1 + \cdots + \lambda_{d+1} = 1$. Jensen's inequality then yields the desired upper bound:

$$\begin{aligned} f(y) &= f(\lambda_1 x_1 + \cdots + \lambda_{d+1} x_{d+1}) \leq \lambda_1 f(x_1) + \cdots + \lambda_{d+1} f(x_{d+1}) \\ &\leq |f(x_1)| + \cdots + |f(x_{d+1})| = \alpha, \text{ say.} \end{aligned}$$

Second,

(3) There is a $\gamma > 0$ such that $|f|$ is bounded above by γ on $N(2\varepsilon)$.

Let $y \in N(2\varepsilon)$. Then $2x - y = x - (y - x) \in N(2\varepsilon) \subseteq C$ and thus

$$f(x) = f\Big(\frac{1}{2} y + \frac{1}{2}(2x - y)\Big) \leq \frac{1}{2} f(y) + \frac{1}{2} f(2x - y)$$

by the convexity of f. This, together with (2), then shows that

$$\alpha \geq f(y) \geq 2f(x) - f(2x - y) \geq 2f(x) - \alpha.$$

Hence

$$|f(y)| \leq \max\{\alpha, |2f(x) - \alpha|\} = \gamma, \text{ say.}$$

Third,

(4) f is Lipschitz with Lipschitz constant $L = \dfrac{2\gamma}{\varepsilon}$ on $N(\varepsilon)$.

Let $y, z \in N(\varepsilon)$, $y \neq z$. Choose $w \in N(2\varepsilon)$ such that $z \in [y, w]$ and $\|w - z\| = \varepsilon$. Since the restriction of f to the line-segment $[y, w]$ is convex, Lemma 1.1, together with (3), yields the following:

$$\frac{f(z) - f(y)}{\|z - y\|} \leq \frac{f(w) - f(z)}{\|w - z\|} \leq \frac{2\gamma}{\varepsilon} \text{ or } f(z) - f(y) \leq \frac{2\gamma}{\varepsilon}\|z - y\|.$$

Clearly, the same conclusion holds with y and z exchanged. Thus

$$|f(z) - f(y)| \leq L\|z - y\| \text{ where } L = \frac{2\gamma}{\varepsilon}.$$

This concludes the proof of (4) and thus of (1). □

Remark. Note that continuity may not extend to $\operatorname{bd} C$ as the following example shows: let B^d be the solid Euclidean unit ball in $\mathbb{E}^d$ and $f : B^d \to \mathbb{R}$. Suppose that for a suitable real $\alpha > 0$ the restriction of f to $\operatorname{int} B^d$ is convex and bounded above by α and the restriction of f to $\operatorname{bd} B^d$ is bounded below by 2α, but otherwise arbitrary. Then f is convex and discontinuous at each point of $\operatorname{bd} B^d$.

Support and a Hahn–Banach Type Result

Support and the related separation properties of convex functions and convex sets are of great importance for convex analysis. Moreover, they are bridges between convex geometry and optimization and – to some extent – functional analysis. See, e.g. Hiriart-Urruty and Lemaréchal [505], Stoer and Witzgall [970] and Rudin [861].

A function $f : C \to \mathbb{R}$ *has affine support* at a point $x \in C$ if there is an affine function $a : \mathbb{E}^d \to \mathbb{R}$ of the form $a(y) = f(x) + u \cdot (y - x)$ for $y \in \mathbb{E}^d$, where u is a suitable vector in $\mathbb{E}^d$, such that

$$f(y) \geq a(y) = f(x) + u \cdot (y - x) \text{ for } y \in C.$$

a is called an *affine support* of f at x. The dot $\cdot$ denotes the usual inner product in $\mathbb{E}^d$.

The following result extends Theorem 1.2. It is a finite-dimensional version of the Hahn–Banach theorem. For the general Hahn–Banach theorem see, e.g. [861]. Our proof is by induction. It generalizes to the infinite dimensional case where, of course, induction has to be replaced by transfinite induction. A related result on the existence of support hyperplanes of a convex body is Theorem 4.1. Our proof of the latter result is essentially finite dimensional and thus basically different from the proof of the following result.

Theorem 2.3. *Let $f : C \to \mathbb{R}$ be convex and P an affine subspace in $\mathbb{E}^d$ through a point $x \in \operatorname{int} C$. Suppose that the restriction $f|P$ has an affine support a_P at x. Then f has an affine support a at x which extends a_P, i.e. $a|P = a_P$.*

Proof. We may assume that $x = o$ and $f(o) = 0$. Let $L = P$. Let dim stand for dimension. If $\dim L = d$, we are done. Otherwise it is sufficient to prove the following proposition:

(5) Let $k = \dim L, \dim L + 1, \dots, d$. Then there are a k-dimensional linear subspace $L_k \supseteq L$ and a linear function $l_k : L_k \to \mathbb{R}$ which affinely supports $f|L_k$ and extends $l_L = a_P$.

The proof of (5) is by induction. If $k = \dim L$, (5) holds by assumption. Suppose now that (5) holds for a $k < d$. Choose $w \in C \backslash L_k$. Then

$$\begin{aligned}\lambda l_k(y) + \mu l_k(z) &= (\lambda + \mu) l_k\Big(\frac{\lambda}{\lambda+\mu} y + \frac{\mu}{\lambda+\mu} z\Big)\\ &\le (\lambda+\mu) f\Big(\frac{\lambda}{\lambda+\mu} y + \frac{\mu}{\lambda+\mu} z\Big)\\ &= (\lambda+\mu) f\Big(\frac{\lambda}{\lambda+\mu}(y - \mu w) + \frac{\mu}{\lambda+\mu}(z + \lambda w)\Big)\\ &\le \lambda f(y-\mu w) + \mu f(z + \lambda w)\end{aligned}$$

for $y, z \in C \cap L_k$ and $\lambda, \mu > 0$ such that $y - \mu w, z + \lambda w \in C$

by the induction hypothesis and the convexity of f. Hence

$$\frac{l_k(y) - f(y - \mu w)}{\mu} \le \frac{f(z + \lambda w) - l_k(z)}{\lambda}$$

for $y, z \in C \cap L_k$ and $\lambda, \mu > 0$ such that $y - \mu w, z + \lambda w \in C$.

The supremum of the left-hand side of this inequality is therefore less than or equal to the infimum of the right-hand side. Thus there is an $\alpha \in \mathbb{R}$ such that the following inequalities hold:

$$\frac{l_k(y) - f(y - \mu w)}{\mu} \le \alpha \text{ for } y \in C \cap L_k \text{ and } \mu > 0 \text{ such that } y - \mu w \in C,$$

$$\alpha \le \frac{f(z + \lambda w) - l_k(z)}{\lambda} \text{ for } z \in C \cap L_k \text{ and } \lambda > 0 \text{ such that } z + \lambda w \in C.$$

This can also be expressed as follows:

(6) $f(z + \lambda w) \ge l_k(z) + \alpha\lambda$
for all $z \in C \cap L_k$ and $\lambda \in \mathbb{R}$ such that $z + \lambda w \in C$.

Let L_{k+1} be the $(k+1)$-dimensional subspace of $\mathbb{E}^d$ spanned by L_k and w, and let the linear function $l_{k+1} : L_{k+1} \to \mathbb{R}$ be defined by

$$l_{k+1}(z + \lambda w) = l_k(z) + \alpha\lambda \text{ for } z + \lambda w \in L_{k+1}.$$

(6) then shows that l_{k+1} affinely supports $f|L_{k+1}$ at o. The induction is complete, concluding the proof of (5) and thus of the theorem. □

A Characterization of Convex Functions

As a consequence of Theorem 2.3 and the proof of Theorem 1.3, we have the following equivalence. Clearly, this equivalence can be used to give an alternative definition of the notion of convex function.

Theorem 2.4. *Let C be open and $f : C \to \mathbb{R}$. Then the following are equivalent:*

(i) *f is convex.*

(ii) *f has affine support at each $x \in C$.*

First-Order Differentiability and Partial Derivatives

A function $f : C \to \mathbb{R}$ is *differentiable* at a point $x \in C$ *in the sense of Stolz* or *Fréchet* if there is a (necessarily unique) vector $u \in \mathbb{E}^d$, such that

$$f(y) = f(x) + u \cdot (y - x) + o(\|y - x\|) \text{ as } y \to x,\ y \in C.$$

A weaker notion of differentiability is differentiability in the sense of Gâteaux. Since both notions of differentiability coincide for convex functions in finite dimensions, only the former will be considered.

Theorem 2.5. *Let $f : C \to \mathbb{R}$ be convex and $x \in \operatorname{int} C$. Then the following statements are equivalent:*

(i) *f is differentiable at x.*
(ii) *The partial derivatives $f_{x_i}(x),\ i = 1, \dots, d$, exist.*

Proof. (i)⇒(ii) This is well known from calculus and easy to prove.

(ii)⇒(i) Let $u_i = f_{x_i}(x),\ i = 1, \dots, d$, let $u = (u_1, \dots, u_d)$ and let $\{b_1, \dots, b_d\}$ be the standard basis of $\mathbb{E}^d$. Then

$$f(x + tb_i) = f(x) + u_i t + o(|t|) \text{ as } t \to 0 \text{ for } i = 1, \dots, d.$$

Combined with Jensen's inequality, this shows that

$$\begin{aligned}
(7)\quad f(y) &= f\big(x + (y - x)\big)\\
&= f\Big(\frac{1}{d}\big(x + d(y_1 - x_1)b_1\big) + \dots + \frac{1}{d}\big(x + d(y_d - x_d)b_d\big)\Big)\\
&\le \frac{1}{d} f\big(x + d(y_1 - x_1)b_1\big) + \dots + \frac{1}{d} f\big(x + d(y_d - x_d)b_d\big)\\
&= f(x) + u_1(y_1 - x_1) + \dots + u_d(y_d - x_d)\\
&\quad + o(|y_1 - x_y|) + \dots + o(|y_d - x_d|)\\
&= f(x) + u \cdot (y - x) + o(\|y - x\|) \text{ as } y \to x,\ y \in C.
\end{aligned}$$

A similar argument yields the following:

$$(8)\quad f(2x - y) \le f(x) + u \cdot (x - y) + o(\|y - x\|) \text{ as } y \to x,\ 2x - y \in C.$$

The final inequality we are seeking follows from the convexity of f:

$$(9)\quad f(x) = f\Big(\frac{1}{2}y + \frac{1}{2}(2x - y)\Big) \le \frac{1}{2}f(y) + \frac{1}{2}f(2x - y) \text{ for } y,\ 2x - y \in C.$$

The inequalities (9) and (8) now imply that

$$\begin{aligned}
f(x) + u \cdot (y - x) &\le \frac{1}{2}f(y) + \frac{1}{2}f(2x - y) + u \cdot (y - x)\\
&\le \frac{1}{2}f(y) + \frac{1}{2}f(x) + \frac{1}{2}u \cdot (y - x) + o(\|y - x\|) \text{ as } y \to x,\ y \in C,
\end{aligned}$$

and thus

$$f(y) \geq f(x) + u \cdot (y - x) + o(\|y - x\|) \text{ as } y \to x,\ y \in C.$$

Together with the inequality (7) this finally implies that

$$f(y) = f(x) + u \cdot (y - x) + o(\|y - x\|) \text{ as } y \to x,\ y \in C. \quad \square$$

Simple examples show that in this result one may not replace convex by continuous.

The next result is due to Reidemeister [827].

Theorem 2.6. *Let C be open and $f : C \to \mathbb{R}$ convex. Then f is differentiable almost everywhere on C.*

Proof. We first show that

(10) The left-hand side and right-hand side partial derivatives $f_{x_i}^-, f_{x_i}^+, i = 1, \dots, d$, exist on C and are measurable.

Let $\{b_1, \dots, b_d\}$ be the standard basis of $\mathbb{E}^d$. For given i, consider the functions $g_n, h_n, n = 1, 2, \dots$, which are defined by

$$g_n(x) = \frac{f(x - \frac{1}{n} b_i) - f(x)}{-\frac{1}{n}} \text{ for } x \in C \text{ such that } x - \frac{1}{n} b_i \in C,$$

$$h_n(x) = \frac{f(x + \frac{1}{n} b_i) - f(x)}{\frac{1}{n}} \text{ for } x \in C \text{ such that } x + \frac{1}{n} b_i \in C.$$

By Theorem 1.4,

$$g_n \to f_{x_i}^-,\ h_n \to f_{x_i}^+ \text{ as } n \to \infty \text{ on } C.$$

As the pointwise limits of continuous functions, $f_{x_i}^-, f_{x_i}^+$ are measurable on C, concluding the proof of (10).

Second,

(11) $f_{x_i}, i = 1, \dots, d$, exist almost everywhere on C.

For given i, the set

$$\{x \in C : f_{x_i}^-(x) \neq f_{x_i}^+(x)\}$$

is measurable on C by (10). Fubini's theorem and Theorem 1.4 for convex functions of one variable together imply that this set has measure 0, concluding the proof of (11).

Having proved (11), Reidemeister's theorem is an immediate consequence of Theorem 2.5. $\square$

A different proof of this result can be obtained from Theorems 2.2 and 2.5 and Rademacher's theorem on the differentiability almost everywhere of Lipschitz continuous functions.

Remark. More precise measure-theoretic information on the size of the set of points at which a convex function is differentiable was given by Anderson and Klee [28] (in the guise of a result on convex bodies). Mazur [701] and Preiss and Zajíček [815, 816] employ the topological tool of Baire categories, respectively, the metric tool of porous sets to estimate the size of this differentiability set. From all three points of view, it is a large set. Compare Theorems 5.1 and 5.2 dealing with convex bodies and the discussion on first-order differentiability after these results. For a discussion of differentiability properties of convex functions on infinite dimensional spaces, see the book of Benyamini and Lindenstrauss [97].

First-Order Differentiability and Affine Support

Our first result is as follows:

Theorem 2.7. *Let $f : C \to \mathbb{R}$ be convex and $x \in \operatorname{int} C$. Then the following are equivalent:*

(i) *f is differentiable at x.*
(ii) *f has unique affine support at x, say $a : \mathbb{E}^d \to \mathbb{R}$, where $a(y) = f(x)+u\cdot(y-x)$ for $y \in \mathbb{E}^d$ and $u = \operatorname{grad} f(x) = \big(f_{x_1}(x), \dots, f_{x_d}(x)\big)$.*

Proof. By Theorem 1.4,

$$f_{x_i}^-(x) \text{ and } f_{x_i}^+(x),\ i = 1, \dots, d, \text{ exist.}$$

(i)⇒(ii) By Theorem 2.4 f has an affine support $a : \mathbb{E}^d \to \mathbb{R}$ at x, where $a(y) = f(x) + u \cdot (y - x)$ for $y \in \mathbb{E}^d$. Let $i = 1, \dots, d$. The restriction of f to the intersection of C with the line through x parallel to the ith coordinate axis is a convex function of one variable. This function has derivative

$$f_{x_i}(x) = f_{x_i}^-(x) = f_{x_i}^+(x)$$

at x by Theorem 2.5 and $f(x)+u_i(y_i - x_i)$ for $y_i \in \mathbb{R}$ is an affine support at x. Thus $u_i = f_{x_i}(x)$ by Proposition 1.2. Since this holds for $i = 1, \dots, d$, the affine support a is unique and has the desired form.

(ii)⇒(i) If f is not differentiable at x, then there is an index i such that $f_{x_i}(x)$ does not exist, see Theorem 2.5. Then

$$f_{x_i}^-(x) < f_{x_i}^+(x)$$

by Theorem 1.4. By Proposition 1.2, each affine function of the form $f(x)+u_i(y_i - x_i)$ for $y_i \in \mathbb{R}$ where

$$f_{x_i}^-(x) \le u_i \le f_{x_i}^+(x)$$

is an affine support of the restriction of f to the intersection of C and the line through x parallel to the ith coordinate axis. Each of these affine supports can be extended to an affine support of f at x by the Hahn–Banach type theorem 2.3. Hence f does not have a unique affine support at x. □

Theorem 2.8. *Let C be open and $f : C \to \mathbb{R}$ convex and differentiable on C. Then all partial derivatives of f are continuous, i.e. f is of class $\mathcal{C}^1$.*

Proof. It is sufficient to show the following:

(12) Let $x, x_1, x_2, \dots \in C$ be such that $x_1, x_2, \dots \to x$. Then $u_n = \operatorname{grad} f(x_n) \to \operatorname{grad} f(x)$.

Theorem 2.7 implies that

(13) $f(y) \geq f(x_n) + u_n \cdot (y - x_n)$ for $y \in C$ and $n = 1, 2, \dots$.

By Theorem 2.2, f is Lipschitz in a suitable neighborhood $N \subseteq C$ of x with Lipschitz constant L, say. By omitting finitely many x_n and changing notation, if necessary, we may assume that $x_n \in N$ for $n = 1, 2, \dots$. For each n, choose $y_n \in N$ such that $y_n - x_n \neq o$ and $y_n - x_n$ has the same direction as u_n. These remarks, together with (13), imply that

$$\|u_n\| \, \|y_n - x_n\| = u_n \cdot (y_n - x_n) \leq f(y_n) - f(x_n) \leq L\|y_n - x_n\|$$

and thus $\|u_n\| \leq L$, i.e. the sequence (u_n) is bounded. For the proof of (12) it is then sufficient to show the following:

(14) Let (u_{n_k}) be a convergent subsequence of (u_n) with limit v, say. Then $v = \operatorname{grad} f(x)$.

Since $u_{n_k} \to v$, $x_{n_k} \to x$, and f is continuous, (13) implies that $f(y) \geq f(x) + v \cdot (y - x)$ for each $y \in C$. Thus f is supported at x by the affine function a defined by $a(y) = f(x) + v \cdot (y - x)$ for $y \in \mathbb{E}^d$. Since f is differentiable at x, Theorem 2.7 implies that a is unique and $v = \operatorname{grad} f(x)$. □

Heuristic Observation

In Sect. 1.2 and in the present section we have encountered the phenomenon that a convex function which has a particular property such as differentiability, has it in a particularly pure form. This phenomenon occurs also in the context of the Venkov–Alexandrov–McMullen theorem 32.2, 32.3 which shows that a convex polytope which tiles by translation, is even a lattice tile. We think of these phenomena as special cases of the following heuristic proposition.

Heuristic Principle. *Consider a basic property which a convex function, a convex body or a convex polytope can have. Then, in many cases, a convex function, body or polytope which has this property, has an even stronger such property.*

2.2 Alexandrov's Theorem on Second-Order Differentiability

In the last section it was shown that a convex function is differentiable almost everywhere, and we remarked that the same holds with respect to Baire category and metric. A deep result of Alexandrov says that, in the sense of measure theory, even more is true: a convex function is twice differentiable, except on a set of measure zero. In contrast, results of Zamfirescu [1037, 1038] imply that, from the Baire category viewpoint, a typical convex function is twice differentiable only on a small set.

This section contains a proof of the theorem of Alexandrov [14] on second order differentiability of convex functions. It generalizes, in a non-trivial way, a result of Busemann and Feller [183] for $d = 2$ and, of course, Theorem 1.7 for $d = 1$.

Besides the original proof of Alexandrov, we mention a proof of Zajíček [1036] and one in the book of Evans and Gariepy [314].

For more information on differentiability properties of convex functions, respectively, smooth boundary points of convex bodies, see Schneider [904, 907, 908] and Gruber [431].

Second-Order Differentiability

A function $f : C \to \mathbb{R}$ is *twice* (or *second-order) differentiable* at a point $x \in \operatorname{int} C$, if there are a vector $u \in \mathbb{E}^d$, the *gradient* of f at x, and a real $d \times d$ matrix H, the *Hessian matrix* of f at x, such that

$$f(y) = f(x) + u \cdot (y - x) + \frac{1}{2}(y - x)^T H(y - x) + o(\|y - x\|^2) \text{ as } y \to x.$$

Alexandrov's Theorem

Following Zajíček [1036], we prove *Alexandrov's differentiability theorem* [14]:

Theorem 2.9. *Let C be open and $f : C \to \mathbb{R}$ convex. Then f is twice differentiable almost everywhere on C.*

Because the required tools are cited explicitly and the necessary definitions and explanations are incorporated into the proof, the latter looks longer than it actually is. A source for the tools is Mattila [696].

Proof. First, several tools are collected. The extension theorem of McShane shows that

(1) A Lipschitz mapping which maps a set $D \subseteq \mathbb{E}^d$ into $\mathbb{E}^d$ can be extended to a Lipschitz mapping of $\mathbb{E}^d$ into $\mathbb{E}^d$ with the same Lipschitz constant.

By Rademacher's differentiability theorem,

(2) A Lipschitz mapping which maps C into $\mathbb{E}^d$ is almost everywhere differentiable on C.

Call a mapping $K : C \to \mathbb{E}^d$ *differentiable (in the sense of Stolz or Fréchet)* at a point $x \in C$ if there is a real $d \times d$ matrix A, the *derivative* of K at x, such that

$$K(y) = K(x) + A(y - x) + o(\|y - x\|) \text{ as } y \to x,\ y \in C.$$

The next result is a version of Sard's theorem.

(3) Let $K : \mathbb{E}^d \to \mathbb{E}^d$ be a Lipschitz mapping and let $N \subseteq \mathbb{E}^d$ be the set of all points of $\mathbb{E}^d$ where K is differentiable but with *singular derivative*, i.e. $\det A = 0$. Then the image $K(N)$ of N has measure 0.

The final tool is well known.

(4) Let $K : \mathbb{E}^d \to \mathbb{E}^d$ be Lipschitz and let $N \subseteq \mathbb{E}^d$ have measure 0. Then $K(N)$ also has measure 0.

Second, the notion of subgradient and some of its properties will be considered. Let $x \in C$. Then the set $G(x)$ of all vectors $u \in \mathbb{E}^d$ for which the affine function $a : \mathbb{E}^d \to \mathbb{R}$ defined by $a(y) = f(x) + u \cdot (y - x)$ for $y \in \mathbb{E}^d$ is an affine support of f at x is the *subgradient* of f at x. The mapping $G : x \to G(x)$ for $x \in C$ is set-valued. If $G(x)$ is a singleton, say $G(x) = \{u\}$, then u is the *gradient* of f at x, see Theorem 2.7. Now we show that

(5) G is *monotone*, i.e.

$$(x - y) \cdot (u - v) \geq 0 \text{ for } x, y \in C, u \in G(x), v \in G(y).$$

The definitions of $G(x)$ and $G(y)$ imply that $f(y) \geq f(x) + u \cdot (y - x)$ and $f(x) \geq f(y) + v \cdot (x - y)$, respectively. Adding these inequalities yields (5). Let I denote the identity mapping, respectively, the $d \times d$ unit matrix.

(6) Let $D = \bigcup\{x + G(x) : x \in C\}$. Then the set-valued mapping $K = (I + G)^{-1} : D \to \mathbb{E}^d$ is Lipschitz and thus single-valued.

(Here $x + G(x) = \{x + u : u \in G(x)\}$, for $x + u \in D$ the set $K(x + u)$ is the set of all $t \in C$ with $x + u \in t + G(t)$, and when we say that K is *Lipschitz* this means that, if $w \in K(x + u)$ and $z \in K(y + v)$ then the following inequality holds: $\|z - w\| \leq \|y + v - x - u\|$.) To prove (6), let $x + u, y + v \in D$ and choose $w \in K(x + u), z \in K(y + v)$. Then $x + u \in w + G(w), y + v \in z + G(z)$. The monotonicity of G then shows that

$$(z - w) \cdot (y + v - z - x - u + w) \geq 0, \text{ or}$$

$$\|z - w\|^2 \leq (y + v - x - u) \cdot (z - w) \leq \|y + v - x - u\| \, \|z - w\|$$

by the Cauchy–Schwarz inequality, and thus $\|z - w\| \leq \|y + v - x - u\|$.

Third, G is said to be *continuous* at $x \in C$ if G is single-valued at x, say $G(x) = u$, and for any neighborhood of u the set $G(y)$ is contained in this neighborhood if $y \in C$ and $\|y - x\|$ is sufficiently small. By Reidemeister's theorem 2.6,

(7) f is differentiable on $C \backslash L$, where the set $L \subseteq C$ has measure 0.

Then

(8) G is continuous on $C \backslash L$.

To see this, the following must be shown: let $x \in C \backslash L$, $x_1, x_2, \dots \in C$ such that $x_n \to x$, and choose $u_1 \in G(x_1), u_2 \in G(x_2), \dots,$ arbitrarily. Then $u_n \to u =$

$G(x)$. The proof of this statement is verbatim the same as for the corresponding statement (12) in the proof of Theorem 2.8.

Fourth, call G *differentiable* at $x \in C$ if $G(x)$ is single-valued at x, say $G(x) = u$, and if there is a real $d \times d$ matrix H, the *derivative* of G at x, such that the set $G(y) - u - H(y - x)$ is contained in a neighborhood of o of radius $o(\|y - x\|)$ as $y \to x, y \in C$. The latter property will also be expressed in the form $v = u + H(y-x) + o(\|y-x\|)$ as $y \to x, y \in C$ uniformly for $v \in G(y)$, or in the form $G(y) = u + H(y-x) + o(\|y-x\|)$ as $y \to x, y \in C$. Clearly, if G is differentiable at x, then it is continuous at x.

(9) Let G be continuous at $x \in C$ and let the single-valued mapping $K = (I+G)^{-1} : D \to \mathbb{E}^d$ be differentiable at $w = x + u$ where $u = G(x)$, and such that its derivative A is non-singular. Then G is differentiable at x with derivative $H = A^{-1} - I$.

The differentiability of K at w implies that

(10) $K(z) - K(w) = A(z - w) + r$ for $z \in D$, where $\|r\|/\|z - w\|$ is arbitrarily small if $\|z - w\|$ is sufficiently small.

For a real $d \times d$ matrix $B = (b_{ik})$ define $\|B\| = (\sum b_{ik}^2)^{\frac{1}{2}}$. A result from linear algebra based on the Cauchy–Schwarz inequality then shows that $\|Bp\| \le \|B\| \, \|p\|$ for $p \in \mathbb{E}^d$. Thus $\|p\| = \|B^{-1}Bp\| \le \|B^{-1}\| \, \|Bp\|$, or

(11) $\|Bp\| \ge \dfrac{\|p\|}{\|B^{-1}\|}$ for $p \in \mathbb{E}^d$ and each non-singular $d \times d$ matrix B.

The following statement is a consequence of the continuity of G at x.

(12) Let $w = x + u \in D$, $u = G(x)$, $z = y + v \in D$, $y \in C$, $v \in G(y)$ and thus $x = K(w)$, $y = K(z)$. Then $\|z - w\|$ ($\le \|v - u\| + \|y - x\|$) is arbitrarily small, uniformly for $v \in G(y)$, if $\|y - x\|$ is sufficiently small.

For x, y, w, z as in (12), Propositions (10)–(12) show that

$$\begin{aligned}\|y - x\| &= \|K(z) - K(w)\| \ge \|A(z - w)\| - \|r\| \ge \frac{\|z - w\|}{\|A^{-1}\|} - \frac{\|z - w\|}{2\|A^{-1}\|} \\ &= \frac{\|z - w\|}{2\|A^{-1}\|} \text{ if } \|y - x\| \text{ is sufficiently small.}\end{aligned}$$

Combining this with (10) yields the following proposition and thus completes the proof of statement (9):

Let $w = x + u \in D$, $u = G(x)$, $z = y + v \in D$, $y \in C$, $v \in G(y)$ and thus $x = K(w)$, $y = K(z)$. Then $z - w = A^{-1}(y - x) + A^{-1}r$, or

$$\begin{aligned}v &= u - (y - x) + A^{-1}(y - x) + A^{-1}r \\ &= u + (A^{-1} - I)(y - x) + A^{-1}r,\end{aligned}$$

where $\|A^{-1}r\|/\|y - x\|$ is arbitrarily small, uniformly for $v \in G(y)$, if $\|y - x\|$ is sufficiently small.

Fifth, we show that

(13) G is differentiable almost everywhere on C.

By (6), $K = (I + G)^{-1} : D \to \mathbb{E}^d$ is Lipschitz and thus can be extended to a Lipschitz function of $\mathbb{E}^d$ into $\mathbb{E}^d$ by (1). Denote this extension also by K. Let M be the set of points in $\mathbb{E}^d$ where this K is not differentiable and by N the set where it is differentiable but with singular derivative. Propositions (2)–(4) then show that $K(M \cup N)$ has measure 0. The set L, from (7) and (8), is also of measure 0. For the proof of (13), it is thus sufficient to show that G is differentiable for any $x \in C \backslash \big(K(M \cup N) \cup L\big)$. For each such x Proposition (8) says that G is continuous at x and the above definitions of M and N imply that K is differentiable at $x + u$ with non-singular derivative (note that $x + u = (I + G)(x) = K^{-1}(x) \notin M \cup N$). Hence G is differentiable at x by (9). The proof of (13) is complete.

In the sixth, and final, step it will be shown that

(14) $f(y) = f(x) + u \cdot (y - x) + \frac{1}{2}(y - x)^T H(y - x) + o(\|y - x\|^2)$ for almost all $x \in C$ and each $y \in C$. Here $u = G(x)$ and H is the derivative of G at x.

The idea of the proof of (14) is to restrict f to a line segment on which it is differentiable almost everywhere and then represent it as the integral of its derivative.

f and G are differentiable at almost every point $x \in C$, see (7) and (13). Let x be such a point. Then, in particular,

(15) $G(y) = u + H(y - x) + o(\|y - x\|)$ for $y \to x$, $y \in C$, where $u = G(x)$ and H is the derivative of G at x.

In addition, (7) shows that, for almost every unit vector h and almost every $t \geq 0$ for which $x + th \in C$, the function f is differentiable at $x + th$. If h and t are such a unit vector and such a number, respectively, then

$$f(y) = f(x + th) + v \cdot \big(y - (x + th)\big) + o\big(\|y - (x + th)\|\big) \text{ as } y \to x + th,\ y \in C,$$

where $v = G(y + th)$. Thus, in particular, for $y = x + sh$,

$$f(x + sh) = f(x + th) + v \cdot h\,(s - t) + o(|s - t|) \text{ as } s \to t,\ x + sh \in C.$$

Hence, for $t \geq 0$ such that $x + th \in C$, the convex function $f(x + th)$ of one variable is differentiable at almost every t. Its derivative is $v \cdot h$, where $v = G(x + th)$. Since a convex function of one variable is absolutely continuous in the interior of its interval of definition by Theorem 1.1, a theorem of Lebesgue shows that integration of its derivative yields the original function, see the proof of Theorem 1.7. Hence

$$f(x + th) - f(x) = \int_0^t G(x + sh) \cdot h\, ds$$

$$= \int_0^t \big(u \cdot h + h^T H h\, s + o(s)\big)\, ds = u \cdot h\, t + \frac{1}{2} h^T H h\, t^2 + o(t^2) \text{ as } t \to 0$$

by (15), uniformly for all h. Hence

(16) $f(y) = f(x) + u \cdot (y - x) + \frac{1}{2}(y - x)^T H(y - x) + o(\|y - x\|^2)$ as $y \to x,\ y \in C,$

uniformly for y of the form $x + th$ and thus for almost all $y \in C$. Since $f(y)$ and $f(x) + u \cdot (y - x) + \frac{1}{2}(y - x)^T H(y - x)$ depend continuously on y, (16) holds for all $y \in C$. This concludes the proof of (14) and thus of Alexandrov's theorem. □

In contrast to the fact established in Theorem 2.8 that the existence of all first partial derivatives of a convex function implies that the function is of class $\mathcal{C}^1$, the existence of all second partial derivatives or the second order differentiability of a convex function does *not* guarantee that it is of class $\mathcal{C}^2$.

2.3 A Convexity Criterion

As for functions of one variable, it is of interest to ascertain when a function $f : C \to \mathbb{R}$ of several variables is convex.

Below we give a simple, yet useful convexity criterion due to Brunn [174] and Hadamard [460].

Hessians

Let $f : C \to \mathbb{R}$ have partial derivatives of second order, $f_{x_i,x_k}(x),\ i, k = 1, \dots, d$, at $x \in C$. As in Section 2.2, the $d \times d$ matrix

$$H = H(x) = \begin{pmatrix} f_{x_1,x_1}(x) & \dots & f_{x_1,x_d}(x) \\ f_{x_2,x_1}(x) & \dots & f_{x_2,x_d}(x) \\ \dots & \dots & \dots \\ f_{x_d,x_1}(x) & \dots & f_{x_d,x_d}(x) \end{pmatrix}$$

is called the *Hessian matrix* of f at x. Corresponding to it is the *Hessian (quadratic) form* of f at x, defined by

$$y \to \frac{1}{2} y^T H y = \frac{1}{2} \sum_{i,k=1}^{d} f_{x_i,x_k}(x)\, y_i y_k \text{ for } y \in \mathbb{E}^d.$$

The Convexity Criterion of Brunn and Hadamard

The following result for convex functions of d variables will be proved by retracing it back to the one-dimensional case and using the fact that a convex function in one variable of class $\mathcal{C}^2$ has non-negative second derivative.

Theorem 2.10. *Let C be open and $f : C \to \mathbb{R}$ of class $\mathcal{C}^2$. Then the following statements are equivalent:*

(i) *f is convex.*

(ii) *For any $x \in C$ the Hessian form $H(x)$ of f at x is positive semi-definite.*

Proof. (i)⇒(ii) It is sufficient to show the following:

(1) Let $x, y \in C, x \neq y$. Then $(y-x)^T H(x)(y-x) \geq 0$.

Let g be the convex function defined by $g(t) = f\big((1-t)x + ty\big)$ for $t \in \mathbb{R}$ such that $(1-t)x + ty = x + t(y-x) \in C$. Since f and thus g is of class $\mathcal{C}^2$, the chain rule for functions of d variables yields the following:

$$g'(t) \text{ exists and equals } \sum_{i=1}^{d} f_{x_i}\big(x + t(y-x)\big)(y_i - x_i),$$

$$g''(t) \text{ exists and equals } \sum_{i,k=1}^{d} f_{x_i,x_k}\big(x + t(y-x)\big)(y_i - x_i)(y_k - x_k).$$

Since g is convex and of class $\mathcal{C}^2$, Corollary 1.1 implies that, in particular, $g''(0) \geq 0$ and thus

$$\frac{1}{2}(y-x)^T H(x)(y-x) = \frac{1}{2}\sum_{i,k=1}^{d} f_{x_i,x_k}(x)(y_i - x_i)(y_k - x_k) = \frac{1}{2}g''(0) \geq 0,$$

concluding the proof of (1).

(ii)⇒(i) By Theorem 2.4, it is sufficient to show the following:

(2) Let $x \in C$. Then f has an affine support at x.

Since f is of class $\mathcal{C}^2$, Taylor's theorem, for functions of d variables, implies that

$$\begin{aligned} f(y) &= f(x) + u \cdot (y-x) + \frac{1}{2}\sum_{i,k=1}^{d} f_{x_i,x_k}\big(x + \vartheta(y-x)\big)(y_i - x_i)(y_k - x_k) \\ &= f(x) + u \cdot (y-x) + \frac{1}{2}(y-x)^T H\big(x + \vartheta(y-x)\big)(y-x) \\ &\geq f(x) + u \cdot (y-x) \text{ for } y \in C, \end{aligned}$$

where $u = \operatorname{grad} f(x)$, $0 < \vartheta < 1$ is chosen suitably, depending on y, and we have used the assumption that the Hessian form of f at $x + \vartheta(y-x)$ is positive semi-definite. This concludes the proof of (2). □

Remark. Essentially the same proof yields the following: If the Hessian form of f is positive definite for each $x \in C$, then f is strictly convex. The following simple example shows that the converse does not hold. Let $f(x) = x_1^4 + \cdots + x_d^4$ for $\|x\| \leq 1$, then f is strictly convex, but the Hessian form at o is the zero form and therefore not positive definite.

2.4 A Stone–Weierstrass Type Theorem

The problem, whether interesting special families of functions are dense in the class of real or complex continuous functions on a given space, has drawn attention ever since Weierstrass gave his well known approximation theorem.

In this section, we present a Stone-Weierstrass type result which shows that differences of convex functions are dense in the space of continuous functions.

A Stone–Weierstrass Type Theorem for Convex Functions

For the following result, see Alfsen [21].

Theorem 2.11. *Let C be compact and convex. Then the set $\mathcal{D}$ of all differences of continuous convex functions on C is dense in the space of all real continuous functions on C, endowed with the maximum norm.*

We present two proofs. The first one was proposed by Schneider [910] and makes use of the Weierstrass approximation theorem for several variables and the convexity criterion of Brunn and Hadamard. The second proof is based on a theorem of Stone of Stone–Weierstrass type.

Proof (using the Weierstrass approximation theorem). We need the following version of the approximation theorem:

> The family of all real polynomials in d variables on the cube $K = \{x : |x_i| \le 1\}$ is dense in the space of all real continuous functions on K.

For the proof of the theorem we may assume that $C \subseteq K$. Since each continuous real function on the compact set C can be continuously extended to K, it follows that

> The family of all (restrictions of) real polynomials on C is dense in the space of all real continuous functions on C.

This yields the theorem if we can show that

> Each polynomial p on C can be represented as the difference of two convex polynomials on C.

To see this, note that for sufficiently large $\lambda > 0$ the polynomials $p(x)+\lambda(x_1^2+\cdots+x_d^2)$ and $\lambda(x_1^2\cdots+x_d^2)$ both are convex on C by the convexity criterion of Brunn and Hadamard. □

Proof (based on a theorem of Stone). The required result of Stone is as follows:

(1) Let S be a compact space and $\mathcal{F}$ a family of real continuous functions on S having the following properties:

(i) $\mathcal{F}$ is closed under multiplication by real numbers, addition, and multiplication.

(ii) $\mathcal{F}$ *separates points*, i.e. for any $x, y \in S$, $x \neq y$, there is a function $f \in \mathcal{F}$ with $f(x) \neq f(y)$.

(iii) $\mathcal{F}$ *vanishes nowhere* on S, i.e. for any $x \in S$ there is a function $f \in \mathcal{F}$ with $f(x) \neq 0$.

Then $\mathcal{F}$ is dense in the space of all real continuous functions on S, endowed with the maximum norm.

Next, some simple properties of convex functions will be given:

(2) Let $f : C \to \mathbb{R}$ be convex and $f \geq 0$. Then f^2 is also convex.

The convexity and the non-negativity of f together with the fact that the function $t \to t^2$ for $t \geq 0$ is non-decreasing and convex, imply (2):

$$f\big((1-\lambda)x + \lambda y\big)^2 \leq \big((1-\lambda)f(x) + \lambda f(y)\big)^2 \leq (1-\lambda)f(x)^2 + \lambda f(y)^2$$
for $x, y \in C, 0 \leq \lambda \leq 1$.

The following proposition is an immediate consequence of (2), the fact that sums of convex functions are again convex, and of the simple identity $fg = \frac{1}{2}\big((f+g)^2 - (f^2+g^2)\big)$:

(3) Let $f, g : C \to \mathbb{R}$ be convex and $f, g \geq 0$. Then $fg \in \mathcal{D}$.

In the last part of the proof it will be shown that $\mathcal{D}$ has Properties (i)–(iii) in Stone's theorem (1).

(4) $\mathcal{D}$ has Property (i).

Only multiplication has to be justified. Let $f - g, h - k \in \mathcal{D}$. The convex functions $f, g, h, k : C \to \mathbb{R}$ are continuous on the compact set C by assumption and thus are bounded. After adding the same suitable constant to each of these functions and changing notation if this constant is $\neq 0$, we may assume that $f, g, h, k \geq 0$. Then $(f-g)(h-k) = fh + gk - fk - gh \in \mathcal{D}$ by (3), concluding the proof of (4).

(5) $\mathcal{D}$ has Properties (ii) and (iii).

This follows by considering, for example, affine functions on C, which are clearly convex.

Having proved (4) and (5), the theorem is an immediate consequence of Stone's theorem (1). □

Remark. The above proof shows that Theorem 2.11 actually holds in spaces which are more general than $\mathbb{E}^d$.

2.5 A Sufficient Condition of Courant and Hilbert in the Calculus of Variations

For many extremum problems necessary conditions for solutions are well known and easy to obtain. In general, it is more difficult to give sufficient conditions. Consider the following simple example: Let $f : I \to \mathbb{R}$ be differentiable. Then, if f attains a

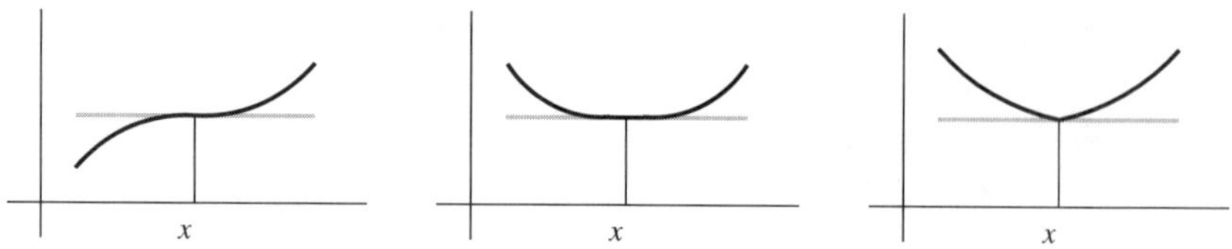

Fig. 2.1. Stationary, extreme and unique extreme points

local minimum at a point $x \in \operatorname{int} I$, necessarily $f'(x) = 0$. Conversely, the condition that $f'(x) = 0$ does not guarantee that f attains a local minimum at x, but together with the condition that f is convex, it does. If, in addition, we know that f is strictly convex, this local minimum is even the unique global minimum of f (Fig. 2.1).

A similar situation arises in the calculus of variations: Consider a variational problem and the corresponding Euler–Lagrange equation(s). A smooth solution of the variational problem necessarily satisfies the Euler–Lagrange equation(s). Conversely, if a function satisfies the Euler–Lagrange equation(s), this may not be sufficient for the function to be a solution of the variational problem. But it is sufficient if, in addition, certain convexity conditions are satisfied. The first result of this type seems to be due to Courant and Hilbert [227], p.186, which, in essence, is reproduced below.

For a wealth of more recent pertinent results and references, see the survey of Brechtken–Manderscheid and Heil [165].

A Sufficient Condition

A result of Courant and Hilbert is as follows, where $[a, b]$ is an interval in $\mathbb{R}$.

Theorem 2.12. *Let $f : [a, b] \times \mathbb{E}^2 \to \mathbb{R}$ be of class C^2 and assume that for each fixed $x \in [a, b]$ the function $(y, z) \to f(x, y, z)$ for $(y, z) \in \mathbb{E}^2$ is convex, respectively, strictly convex. Let $\alpha, \beta \in \mathbb{R}$ and assume that $y : [a, b] \to \mathbb{R}$ is a function of class C^1 such that $y(a) = \alpha$, $y(b) = \beta$. Then the following statements are equivalent:*

(i) *y is a minimizer, respectively, unique minimizer of the integral*

$$I(w) = \int_a^b f\big(x, w(x), w'(x)\big)\, dx$$

among all functions $w : [a, b] \to \mathbb{R}$ of class C^1 with $w(a) = \alpha$, $w(b) = \beta$.

(ii) *y satisfies the* Euler–Lagrange equation

$$f_y(x, y, y') - \frac{d}{dx} f_{y'}(x, y, y') = 0.$$

Proof. (i)⇒(ii) This is a standard result in the calculus of variations.

(ii)⇒(i) Assume first, that f satisfies the convexity condition. Then, for each $x \in [a, b]$, the expression $f\big(x, y(x) + s, y'(x) + t\big)$ is convex in (s, t) on $\mathbb{E}^2$ and Theorem 2.7 implies that, in particular,

(1) $f(x, y(x) + s, y'(x) + t)$
$\geq f(x, y(x), y'(x)) + f_y(x, y(x), y'(x))s + f_{y'}(x, y(x), y'(x))t$
for $(s, t) \in \mathbb{E}^2$.

Any function $w : [a, b] \to \mathbb{R}$ of class $\mathcal{C}^1$ with $w(a) = \alpha$, $w(b) = \beta$ can be represented in the form $w(x) = y(x) + s(x)$ for $x \in [a, b]$, where $s : [a, b] \to \mathbb{R}$ is of class $\mathcal{C}^1$ and satisfies $s(a) = s(b) = 0$. Thus (1) and (ii) yield the following:

$$
\begin{aligned}
(2) \qquad I(w) = I(y+s) &= \int_a^b f(x, y(x) + s(x), y'(x) + s'(x))\, dx \\
&\geq \int_a^b f(x, y(x), y'(x))\, dx \\
&\quad + \int_a^b f_y(x, y(x), y'(x))s(x)\, dx + \int_a^b f_{y'}(x, y(x), y'(x))s'(x)\, dx \\
&= I(y) + \int_a^b \Big\{\frac{d}{dx} f_{y'}(x, y(x), y'(x))\Big\}s(x)\, dx \\
&\quad + \int_a^b f_{y'}(x, y(x), y'(x))s'(x)\, dx \\
&= I(y) + \int_a^b \frac{d}{dx}\{f_{y'}(x, y(x), y'(x))s(x)\}dx \\
&= I(y) + f_{y'}(x, y(x), y'(x))s(x)\Big|_a^b = I(y).
\end{aligned}
$$

Since w was arbitrary, this shows that y is a minimizer.

Assume, second, that f satisfies the condition of strict convexity. Then, in (1), we have strict inequality unless $(s, t) = (0, 0)$. (Otherwise the graph of the function $(s, t) \to f(x, y(x)+s, y'(x)+t)$ and its affine support at $(s, t) = (0, 0)$ have a line-segment in common which contradicts the strict convexity.) Then strict inequality holds in (2) unless $s(x) = s'(x) = 0$ for all $x \in [a, b]$, i.e. $w = y$. This shows that y is the unique minimizer. □

The Background

In order to understand better what the result of Courant and Hilbert really means, note the following: consider the (infinite dimensional) space of functions

$$\mathcal{F} = \{w : [a, b] \to \mathbb{R},\ \text{where } w \in \mathcal{C}^2, w(a) = \alpha, w(b) = \beta\}.$$

$\mathcal{F}$ is convex. Using the convexity property of f, it is easy to show that the mapping

$$I : w \to I(w) = \int_a^b f\big(x, w(x), w'(x)\big)\, dx \in \mathbb{R}, \quad \text{for } w \in \mathcal{F}$$

is a convex function on $\mathcal{F}$. A solution y of the Euler–Lagrange equation is a stationary value of this mapping and thus a minimizer by convexity. If the mapping is strictly convex, then this minimizer is even the unique minimizer.

Convex Bodies

Sporadic results on convex bodies have appeared in the mathematical literature since antiquity, with an increasing rate in the nineteenth century. Systematic investigations started only in the nineteenth and the early twentieth century with the work of Cauchy, Steiner, Brunn and, in particular, Minkowski. Important contributors in the twentieth century were Blaschke, Hadwiger, Alexandrov and many contemporary mathematicians. The following quotation of Klee [593] shows roughly where, in mathematics, this area is located and what are some of its characteristics:

> The study of convex sets is a branch of geometry, analysis, and linear algebra that has numerous connections with other areas of mathematics and serves to unify many apparently diverse mathematical phenomena. It is also relevant to several areas of science and technology.

During the twentieth century the relationship of convex geometry with analytic flavor to other branches of mathematics and to applied areas increased greatly. We mention differential and Riemannian geometry, functional analysis, calculus of variations and control theory, optimization, geometric measure theory, inequalities, Fourier series and spherical harmonics, probability, and mathematical physics. Besides these relationships of a systematic character, there are minor connections to numerous other areas, including complex function theory of one and several variables, aspects of ordinary and partial differential equations, dynamical systems and potential theory.

In this chapter, we try to justify the following observation of Ball [53]:

> Although convexity is a simple property to formulate, convex bodies possess a surprisingly rich structure.

We present the major analytic aspects of convex geometry together with many applications. We begin with general properties of convex bodies, including some results of combinatorial geometry. Then the boundary structure of convex bodies is investigated. This comprises smooth, singular and extreme points. The natural topology on the space of convex bodies is introduced next and Blaschke's selection theorem proved. Mixed volumes and quermassintegrals are treated in the following section. The discussion of valuations is an important topic. Our exposition includes extension results, characterizations of volume and Hadwiger's functional theorem. The latter is applied to prove the principal kinematic formula of integral geometry. A central theme is the Brunn–Minkowski inequality, which leads to geometric and physical isoperimetric inequalities and to the concentration of measure phenomenon. The following section deals with Steiner symmetrization and Schwarz rearrangement which are valuable tools, for example for isoperimetric inequalities of mathematical physics. Area measures and the intrinsic metric of convex surfaces are then studied, including the existence and uniqueness problems of Minkowski and Weyl. We present solutions of these problems due to Alexandrov, Fenchel and Jessen, and Pogorelov. Then we give some hints to dynamical aspects of convex geometry dealing with the evolution of convex surfaces and billiards. Next come John's ellipsoid theorem and the reverse isoperimetric inequality. Then asymptotic best approximation of convex bodies is treated and applied to the isoperimetric problem for polytopes. Special convex bodies have always attracted interest. Here, simplices, balls and ellipsoids are considered. Finally, the space of convex bodies is studied from topological, measure, metric, group and lattice viewpoints.

Applications deal with complex function theory of several variables, Lyapunov's convexity theorem for vector-valued measures, Pontryagin's minimum principle, Birkhoff's convexity theorem on doubly stochastic matrices, a series of results from mathematical physics, in particular the theorem of Wulff on the form of crystals, and Choquet's characterization of vector lattices.

In this chapter we will often use convex polytopes, related notions and their simple properties, in particular approximation properties. The reader who is not familiar with convex polytopes may consult the introductory sections of the next chapter.

The reader who wants to get more detailed information is referred to the books and surveys of Blaschke [124], Bonnesen and Fenchel [149], Alexandrov [18], Eggleston [290], Hadwiger [466, 468], Santaló [881], Leichtweiss [640], Burago and Zalgaller [178], Schneider [907], Groemer [405], Thompson [994], Gardner [359, 360], Klain and Rota [587], Ball [53] and Magaril-Il'yaev and Tikhomirov [678]. In addition, we refer to parts I and IV of the Handbook of Convex Geometry [475], to Convexity and Its Applications [219] and to the collected or selected works of Minkowski [745], Blaschke [129] and Alexandrov [18, 19].

There are omissions. The theory of curvature and area measures will only be mentioned briefly. More, but by no means sufficient material deals with the local theory of normed spaces. For thorough representations of these areas, see Schneider [907], respectively, Pisier [802], Tomczak-Jaegermann [1001], Ball [53], and the pertinent surveys and chapters in the Handbook of Convex Geometry [475], the Handbook of the Geometry of Banach Spaces [477] and the monograph of Benyamini and Lindenstrauss [97]. Integral geometry and geometric probability are only touched. For information see Santaló [881] and Schneider and Weil [911].

3 Convex Sets, Convex Bodies and Convex Hulls

In this section the notions of convex sets, convex bodies and convex hulls are introduced. Several simple properties are presented, including Carathéodory's theorem on convex hulls. Next, a short excursion into combinatorial geometry will include the theorems of Helly, Radon and, again, Carathéodory, together with some applications. As an example, where the notion of convexity is used in an analytic context to clarify or to describe a situation, we present Hartogs' theorem on power series in d complex variables.

3.1 Basic Concepts and Simple Properties

We begin with the definitions of convex sets and convex bodies and investigate convex hulls, including Carathéodory's theorem. Then we consider convex cones and prove a simple decomposition result.

Convex Sets and Convex Bodies

The earliest explicit mention of the notion of convexity seems to be in the first four axioms in the book *On the Sphere and Cylinder* of Archimedes [35]. The third and the fourth axiom are as follows:

> 3. Similarly also there are certain finite surfaces, not in a plane themselves but having their extremities in a plane, and such that they will either lie wholly on the same side of the plane containing their extremities or will have no part on the other side.
> 4. I call *convex in the same direction* surfaces such that, if any two points on them are taken, either the straight lines between the points all fall upon the same side of the surface, or some fall on one and the same side while others fall along the surface itself, but none falls on the other side.

Archimedes thus actually gives two definitions of a convex surface. The first is by means of support properties, the second is the common one by means of line segments which are contained on the same side of the surface. In the sequel, we state the second definition for sets instead of surfaces in the usual form and show the equivalence of the two definitions later in Theorem 4.2.

A set C in $\mathbb{E}^d$ is *convex* if it has the following property:

$$(1-\lambda)x + \lambda y \in C \text{ for } x, y \in C,\ 0 \le \lambda \le 1.$$

It is *strictly convex* if it is closed and

$$(1-\lambda)x + \lambda y \in \operatorname{int} C \text{ for } x, y \in C,\ x \ne y,\ 0 < \lambda < 1,$$

where int stands for interior. A compact convex set is a *convex body*. It is a *proper* convex body, if its interior is non-empty, otherwise *improper*. A proper convex body in $\mathbb{E}^2$ is also called a *convex disc*. There exist many characterizations of convex sets, see the survey of Mani-Levitska [684] and the references there. Let $\mathcal{C} = \mathcal{C}(\mathbb{E}^d)$ be the *space of convex bodies* in $\mathbb{E}^d$ and $\mathcal{C}_p = \mathcal{C}_p(\mathbb{E}^d)$ the *space of proper convex bodies*.

Convex bodies and, more generally, convex sets are *the* object of research in convex geometry. They play a prominent role not only in convexity but also in many other areas of mathematics and its applications as will become clear in the following.

Convex Hulls

Given a set A in $\mathbb{E}^d$, its *convex hull*, conv A, is the intersection of all convex sets in $\mathbb{E}^d$ which contain A. Since the intersection of convex sets is always convex, conv A is convex and it is the smallest convex set in $\mathbb{E}^d$ with respect to set inclusion, which contains A. For the study of convex hulls we need the following concept: Let $x_1, \dots, x_n \in \mathbb{E}^d$. Then any point x of the form $x = \lambda_1 x_1 + \dots + \lambda_n x_n$, where $\lambda_1, \dots, \lambda_n \geq 0$ and $\lambda_1 + \dots + \lambda_n = 1$, is a *convex combination* of $x_1, \dots, x_n$.

Lemma 3.1. *Let $A \subseteq \mathbb{E}^d$. Then* conv A *is the set of all convex combinations of points of A.*

Proof. First, we show that

(1) The set of all convex combinations of points of A is convex.

Let $x = \lambda_1 x_1 + \dots + \lambda_m x_m$ and $y = \lambda_{m+1} x_{m+1} + \dots + \lambda_n x_n$ be two convex combinations of points of A and $0 \leq \lambda \leq 1$. Then

$$(1-\lambda)x + \lambda y = (1-\lambda)\lambda_1 x_1 + \dots + (1-\lambda)\lambda_m x_m + \lambda\lambda_{m+1} x_{m+1} + \dots + \lambda\lambda_n x_n.$$

Since the coefficients of $x_1, \dots, x_n$ all are non-negative and their sum is 1, the point $(1-\lambda)x + \lambda y$ is also a convex combination of points of A, concluding the proof of (1).

Second, the following will be shown, compare the proof of Jensen's inequality, see Theorems 1.9 and 2.1.

(2) Let $C \subseteq \mathbb{E}^d$ be convex. Then C contains all convex combinations of its points.

It is sufficient to prove, by induction, that C contains all convex combinations of any n of its points, $n = 1, 2, \dots$ This is trivial for $n = 1$. Assume now that $n > 1$ and that the statement holds for $n - 1$. We have to prove it for n. Let $x = \lambda_1 x_1 + \dots + \lambda_n x_n$ be a convex combination of $x_1, \dots, x_n \in C$. If $\lambda_n = 0$, then $x \in C$ by the induction hypothesis. If $\lambda_n = 1$, then trivially, $x = x_n \in C$. Assume finally that $0 < \lambda_n < 1$. Then $0 < \lambda_1 + \dots + \lambda_{n-1} = 1 - \lambda_n < 1$ and thus

$$\lambda_1 x_1 + \dots + \lambda_n x_n = (1-\lambda_n)\left(\frac{\lambda_1}{1-\lambda_n} x_1 + \dots + \frac{\lambda_{n-1}}{1-\lambda_n} x_{n-1}\right) + \lambda_n x_n \in C$$

by the induction hypothesis and the convexity of C. The proof of (2) is complete.

Since conv A is the smallest convex set containing A, Proposition (1) implies that conv A is contained in the set of all convex combinations of points of A. Conversely, the convex set conv A contains by (2) all convex combinations of its points and thus, a fortiori, all convex combinations of the points of A. □

Mechanical Interpretation of the Convex Hull

This lemma says that the convex hull of a finite set $\{x_1, \dots, x_n\}$ in $\mathbb{E}^d$ consists of all centres of gravity of (non-negative) masses $\lambda_1, \dots, \lambda_n$ at the points $x_1, \dots, x_n$.

In a hand-written appendix to his third proof of the fundamental theorem of algebra, which was printed in Werke 3, p.112, Gauss [363] gave an alternative description: the convex hull of $x_1, \dots, x_n$ consists of all points x which have the following property: assume that x supports positive mass, then there are masses at the points $x_1, \dots, x_n$, such that x is in equilibrium with respect to the gravitational pull exerted by the masses at $x_1, \dots, x_n$.

Carathéodory's theorem [190]

refines the above lemma:

Theorem 3.1. *Let $A \subseteq \mathbb{E}^d$. Then* conv *A is the set of all convex combinations of affinely independent points of A, i.e. the union of all* simplices *with vertices in A.*

We reproduce a proof due to Radon [821], see also Alexandroff and Hopf [9], p. 607.

Proof. Let $x \in \operatorname{conv} A$. By Lemma 3.1 we may represent x in the form $x = \lambda_1 x_1 + \dots + \lambda_n x_n$, where $x_1, \dots, x_n \in A$, $\lambda_1, \dots, \lambda_n > 0$, $\lambda_1 + \dots + \lambda_n = 1$ and n is minimal. We have to show that the points $x_1, \dots, x_n$ are affinely independent.

Assume not. Then there are numbers $\mu_1, \dots, \mu_n$, not all 0, such that

(3) $\mu_1 + \dots + \mu_n = 0$,

(4) $\mu_1 x_1 + \dots + \mu_n x_n = o$.

By (3) at least one μ_k is positive. Choose k such that λ_k/μ_k is minimal among all such k. Then

$$\lambda_i - \frac{\lambda_k}{\mu_k}\mu_i \geq 0 \text{ for } i = 1, \dots, n,\ \lambda_k - \frac{\lambda_k}{\mu_k}\mu_k = 0,$$
$$\left(\lambda_1 - \frac{\lambda_k}{\mu_k}\mu_1\right) + \dots + \left(\lambda_n - \frac{\lambda_k}{\mu_k}\mu_n\right) = 1$$

by (3). Hence (4) implies that

$$\begin{aligned} x = \lambda_1 x_1 + \dots + \lambda_n x_n &= \lambda_1 x_1 + \dots + \lambda_n x_n - \frac{\lambda_k}{\mu_k}(\mu_1 x_1 + \dots + \mu_n x_n) \\ = \left(\lambda_1 - \frac{\lambda_k}{\mu_k}\mu_1\right)x_1 + \dots &+ \left(\lambda_{k-1} - \frac{\lambda_k}{\mu_k}\mu_{k-1}\right)x_{k-1} + \left(\lambda_{k+1} - \frac{\lambda_k}{\mu_k}\mu_{k+1}\right)x_{k+1} \\ &+ \dots + \left(\lambda_n - \frac{\lambda_k}{\mu_k}\mu_n\right)x_n \end{aligned}$$

is a representation of x as a convex combination of at most $n - 1$ points of A. This contradicts our choice of n and thus concludes the proof. □

Remark. Carathéodory's theorem is a cornerstone of combinatorial convex geometry. For more information on Carathéodory's theorem and, more generally, on combinatorial geometry, see Sect. 3.2 and the surveys and books cited there.

The next result is a simple consequence of Carathéodory's theorem.

Corollary 3.1. *Let* $A \subseteq \mathbb{E}^d$ *be compact. Then* conv A *is compact.*

Proof. The set

$$\big\{(\lambda_1, \dots, \lambda_{d+1}, x_1, \dots, x_{d+1}) : \lambda_i \geq 0, \lambda_1 + \dots + \lambda_{d+1} = 1, x_j \in A\big\}$$

is a compact subset of $\mathbb{E}^{d+1+(d+1)d} = \mathbb{E}^{(d+1)^2}$. Hence its image under the continuous mapping

$$(\lambda_1, \dots, \lambda_{d+1}, x_1, \dots, x_{d+1}) \to \lambda_1 x_1 + \dots + \lambda_{d+1} x_{d+1}$$

of $\mathbb{E}^{(d+1)^2}$ into $\mathbb{E}^d$ is also compact. By Carathéodory's theorem this image is conv A. □

Closure and Interior

In the following some useful minor results on closure and interior of convex sets and on convex hulls of subsets of $\mathbb{E}^d$ are presented. Let $C \subseteq \mathbb{E}^d$ be convex. By relint C we mean the interior of C relative to the *affine hull* aff C of C, i.e. the smallest flat in $\mathbb{E}^d$ containing C.

Proposition 3.1. *Let* $C \subseteq \mathbb{E}^d$ *be convex. Then the following statements hold:*

(i) cl C *is convex.*

(ii) relint C *is convex.*

(iii) $C \subseteq$ cl relint C.

Let B^d denote the solid Euclidean unit ball in $\mathbb{E}^d$.

Proof. (i) To show that cl C is convex, let $x, y \in \operatorname{cl} C$ and $0 \leq \lambda \leq 1$. Choose sequences (x_n), (y_n) in C such that $x_n \to x$, $y_n \to y$ as $n \to \infty$. The convexity of C implies that $(1-\lambda)x_n + \lambda y_n \in C$. Since $(1-\lambda)x_n + \lambda y_n \to (1-\lambda)x + \lambda y$, it follows that $(1-\lambda)x + \lambda y \in \operatorname{cl} C$.

(ii) It is sufficient to consider the case where int $C \neq \emptyset$ and to show that int C is convex. Let $x, y \in \operatorname{int} C$ and $0 \leq \lambda \leq 1$. Choose $\delta > 0$ such that $x + \delta B^d, y + \delta B^d \subseteq \operatorname{int} C \subseteq C$. Then

$$(1-\lambda)x + \lambda y + \delta B^d = (1-\lambda)(x + \delta B^d) + \lambda(y + \delta B^d) \subseteq C$$

by the convexity of B^d and C. Thus $(1-\lambda)x + \lambda y \in \operatorname{int} C$.

(iii) It is sufficient to consider the case where int $C \neq \emptyset$ and to prove that $C \subseteq \operatorname{cl} \operatorname{int} C$. Let $x \in C$ and choose $y \in \operatorname{int} C$. Then there is a $\delta > 0$ such that $y + \delta B^d \subseteq \operatorname{int} C \subseteq C$. The convexity of C now implies that

$$(1-\lambda)x + \lambda y + \lambda\delta B^d = (1-\lambda)x + \lambda(y + \delta B^d) \subseteq C.$$

Thus $(1-\lambda)x + \lambda y \in \operatorname{int} C$ for $0 < \lambda \leq 1$. Since $(1-\lambda)x + \lambda y \to x$ as $\lambda \to 0$, it follows that $x \in \operatorname{cl} \operatorname{int} C$. □

Proposition 3.2. *Let $A \subseteq \mathbb{E}^d$ be bounded. Then* $\operatorname{cl}\operatorname{conv} A = \operatorname{conv}\operatorname{cl} A$.

Proof. Since $\operatorname{conv} A$ is convex, Proposition 3.1 shows that $\operatorname{cl}\operatorname{conv} A$ is convex too. $\operatorname{cl}\operatorname{conv} A$ is a closed set which contains A. Thus it also contains $\operatorname{cl} A$. Since $\operatorname{cl}\operatorname{conv} A$ is a convex set which contains $\operatorname{cl} A$, it also contains $\operatorname{conv}\operatorname{cl} A$.

The set $\operatorname{conv}\operatorname{cl} A$ is convex and contains A, thus it contains $\operatorname{conv} A$. Since by Corollary 3.1 $\operatorname{conv}\operatorname{cl} A$ is compact and thus closed, it contains $\operatorname{cl}\operatorname{conv} A$. □

Convex Cones

A notion which is important in several contexts, for example in ordered topological vector spaces and in linear optimization, is that of convex cones. A closed set C in $\mathbb{E}^d$ is a *closed convex cone* with *apex o* if it satisfies the following property:

$$\lambda x + \mu y \in C \text{ for all } x, y \in C,\ \lambda, \mu \geq 0.$$

Then, in particular, C is convex and contains, with each point x, also the ray $\{\lambda x : \lambda \geq 0\}$. An example is the positive (non-negative) orthant $\{x : x_i \geq 0\}$. A *closed convex cone* with *apex $a \in \mathbb{E}^d$* is simply the translate of a closed convex cone with apex o by the vector a. The *lineality space* L of a closed convex cone C with apex o is the linear subspace

$$L = C \cap (-C)$$

of $\mathbb{E}^d$. It is the largest linear subspace of $\mathbb{E}^d$ which is contained in C. The convex cone C is *pointed* if $L = \{o\}$. If H is a hyperplane containing only the point o of a pointed closed convex cone C and $p \in C, \neq o$, then $C \cap (H + p)$ is a convex body, sometimes called a *basis* of C. It *generates* C in the sense that

$$C = \bigcup \bigl\{\lambda\bigl(C \cap (H + p)\bigr) : \lambda \geq 0\bigr\}.$$

For later reference we prove the following simple result, where $L^\perp$ denotes the *orthogonal complement* of L, i.e. $L^\perp = \{y : x \cdot y = 0 \text{ for all } x \in L\}$. Clearly, $L^\perp$ is a subspace of $\mathbb{E}^d$ and $\mathbb{E}^d = L \oplus L^\perp$.

Proposition 3.3. *Let C be a closed convex cone in $\mathbb{E}^d$ with apex o and lineality space L. Then*

$$C = (C \cap L^\perp) \oplus L,$$

where $C \cap L^\perp$ is a pointed closed convex cone with apex o.

Proof. First, the equality will be shown. Let $x \in C$. Noting that $\mathbb{E}^d = L^\perp \oplus L$, we have $x = y + z$ for some $y \in L^\perp, z \in L$. Since C is a convex cone with apex o and $x \in C, -z \in -L = L \subseteq C$, it follows that $y = x - z \in C$. Hence $x = y + z$, where $y \in C \cap L^\perp$ and $z \in L$. Thus $x \in (C \cap L^\perp) \oplus L$. This shows that $C \subseteq (C \cap L^\perp) \oplus L$. If, conversely, $x = y + z \in (C \cap L^\perp) \oplus L$, where $y \in C \cap L^\perp, z \in L \subseteq C$, then, noting that C is a convex cone with apex o, it follows that $x = y + z \in C$. Thus $C \supseteq (C \cap L^\perp) \oplus L$, concluding the proof of the equality.

$C \cap L^\perp$ clearly is a closed convex cone with apex o. We show that it is pointed:

$$(C \cap L^\perp) \cap \big(-(C \cap L^\perp)\big) = (C \cap L^\perp) \cap (-C) \cap (-L^\perp)$$

$$= \big(C \cap (-C)\big) \cap (L^\perp \cap L^\perp) = L \cap L^\perp = \{o\}. \quad \square$$

Convex Cone Generated by a Set, Positive Hull

If $\{y_1, \dots, y_n\}$ is a finite set in $\mathbb{E}^d$, then by the *cone generated* by it, $\operatorname{cone}\{y_1, \dots, y_n\}$, or by its *positive hull* $\operatorname{pos}\{y_1, \dots, y_n\}$, we mean the set

$$\big\{\lambda_1 y_1 + \dots + \lambda_n y_n : \lambda_i \geq 0\big\}.$$

It is easy to see that this is a closed convex cone with apex o, in fact, the smallest such cone containing the set $\{y_1, \dots, y_n\}$. An open subset C of $\mathbb{E}^d$ is an *open convex cone* with *apex o* if

$$\lambda x + \mu y \in C \text{ for all } x, y \in C,\ \lambda, \mu > 0.$$

Then C is convex. Note that $o \notin C$, unless $C = \mathbb{E}^d$.

3.2 An Excursion into Combinatorial Geometry: The Theorems of Carathéodory, Helly and Radon

The following result of Kirchberger [585] of 1903 seems to be the first result of what is now called combinatorial geometry. It still conveys well the spirit of the latter. Consider a flock of black and white sheep. If among any four sheep the black ones can be separated from the white ones by a straight fence, then the black and the white sheep of the whole flock can thus be separated. The later theorems of Carathéodory [190], Radon [821] and, in particular, of Helly [490] attracted much more attention and led to a multitude of pertinent results, especially in the 1960s and 1970s. Some of these results were proved for spaces more general than $\mathbb{E}^d$.

We prove below the theorems of Radon, Carathéodory, and Helly. While we have proved Carathéodory's theorem in a similar way in Sect. 3.1, the present proof makes the relation between the theorems of Radon and Carathéodory more clear. These results, as well as certain other properties of convex sets, led to various attempts to extend convexity or, rather, combinatorial geometry to a more general context. The background of Helly's theorem is singular homology theory, see the interesting article of Debrunner [249].

For more information we refer to the surveys of Danzer, Grünbaum and Klee [241] and Eckhoff [283, 284] and to the monograph of Boltyanskiĭ, Martini and Soltan [145]. From the voluminous literature we cite the article of Tverberg [1003] on Radon's theorem. Other aspects of combinatorial geometry were treated by Goodman, Pollack and Wenger [386] and Matoušek [695]. Axiomatic and generalized convexity is dealt with by Van de Vel [1005] and Coppel [222].

The Theorems of Radon, Carathéodory and Helly

Above the theorem of Carathéodory was proved, following Radon. If the argument used by Radon is formulated properly, it is called Radon's theorem. The theorems of Radon, Carathéodory and Helly, are equivalent in the sense that each can be proved using any of the other ones. The equivalence is stated in several references, for example in [284], p. 430, but we were not able to locate in the literature a complete proof in the context of $\mathbb{E}^d$. Also here, we do not prove full equivalence.

Theorem 3.2. *The following statements hold:*

(i) Radon's Theorem. *Let $A \neq \emptyset$ be a set of at least $d+2$ points in $\mathbb{E}^d$. Then there are subsets B, C of A such that*

$$B \cap C = \emptyset, \ \text{conv}\, B \cap \text{conv}\, C \neq \emptyset.$$

(ii) Carathéodory's Theorem. *Let $A \neq \emptyset$ be a set in $\mathbb{E}^d$. Then* conv A *is the set of all convex combinations of affinely independent points in A.*

(iii) Helly's Theorem. *Let $\mathcal{F}$ be a family of convex bodies in $\mathbb{E}^d$. If any $d+1$ convex bodies in $\mathcal{F}$ have non-empty intersection, then the intersection of all convex bodies in $\mathcal{F}$ is non-empty.*

Our proof of the implication (i)⇒(ii), in essence, is that of Carathéodory's theorem in Sect. 3.1. The proof of the implication (ii)⇒(iii) is due to Rademacher and Schoenberg [819].

Proof. (i) It is sufficient to prove Radon's theorem for $A = \{x_1, \dots, x_{d+2}\} \subseteq \mathbb{E}^d$. There are $\mu_1, \dots, \mu_{d+2} \in \mathbb{R}$, not all 0, such that

$$\mu_1 + \dots + \mu_{d+2} = 0,$$
$$\mu_1 x_1 + \dots + \mu_{d+2} x_{d+2} = 0.$$

We clearly may assume that $\mu_1, \dots, \mu_k \geq 0$ and $-\lambda_{k+1} = \mu_{k+1}, \dots, -\lambda_{d+2} = \mu_{d+2} \leq 0$. Then

$$\mu_1 + \dots + \mu_k = \lambda_{k+1} + \dots + \lambda_{d+2} > 0,$$

$$\mu_1 x_1 + \dots + \mu_k x_k = \lambda_{k+1} x_{k+1} + \dots + \lambda_{d+2} x_{d+2}$$

and thus

$$\frac{\mu_1 x_1 + \dots + \mu_k x_k}{\mu_1 + \dots + \mu_k} = \frac{\lambda_{k+1} x_{k+1} + \dots + \lambda_{d+2} x_{d+2}}{\lambda_{k+1} + \dots + \lambda_{d+2}}.$$

Now put $B = \{x_1, \dots, x_k\}$, $C = \{x_{k+1}, \dots, x_{d+2}\}$.

(i)⇒(ii) Let $x \in \text{conv}\, A$. By Lemma 3.1, $x = \lambda_1 x_1 + \dots + \lambda_n x_n$, where $\lambda_1, \dots, \lambda_n > 0$, $\lambda_1 + \dots + \lambda_n = 1$, $x_1, \dots, x_n \in A$. We suppose that this representation is chosen such that n is minimal. We have to show that $x_1, \dots, x_n$ are affinely independent.

Suppose not. Then, by Radon's theorem in aff $\{x_1, \dots, x_n\}$, there are disjoint subsets of $\{x_1, \dots, x_n\}$, the convex hulls of which have non-empty intersection. By re-numbering, if necessary, we thus may assume that

$$\begin{aligned} &\nu_1 x_1 + \dots + \nu_m x_m = \nu_{m+1} x_{m+1} + \dots + \nu_n x_n, \text{ or } \nu_1 x_1 + \dots - \nu_n x_n = o, \\ &\quad \text{where } \nu_1, \dots, \nu_n \geq 0, \nu_1 + \dots + \nu_m = \nu_{m+1} + \dots + \nu_n = 1, \\ &\quad \text{and thus } \nu_1 + \dots - \nu_n = 0. \end{aligned}$$

Thus, up to notation, we have the same situation as in the proof of the Theorem 3.1 of Carathéodory which easily leads to a contradiction.

(ii)$\Rightarrow$(iii) Since the intersection of a family of compact sets is non-empty, if each finite subfamily has non-empty intersection, it is sufficient to prove (iii) for finite families $\mathcal{F} = \{C_1, \dots, C_n\}$, say, where $n \geq d + 2$.

Assume that (iii) does not hold for $\mathcal{F}$. Consider the function $\delta : \mathbb{E}^d \to \mathbb{R}$ defined by

$$\begin{aligned} \delta(x) &= \max\{\delta(x, C_i) : i = 1, \dots, n\} \text{ for } x \in \mathbb{E}^d, \\ &\text{where } \delta(x, C_i) = \min\{\|x - y\| : y \in C_i\}. \end{aligned}$$

Let δ assume its minimum at $p \in \mathbb{E}^d$, say. Since (iii) does not hold for $\mathcal{F}$, $\delta(p) > 0$. By re-numbering, if necessary, we may suppose that

$$\delta(p) = \delta(p, C_i) \text{ precisely for } i = 1, \dots, m \ (\leq n).$$

Choose $q_i \in C_i$ such that

$$\delta(p) = \delta(p, C_i) = \|p - q_i\| \text{ for } i = 1, \dots, m.$$

Then

$$p \in \operatorname{conv}\{q_1, \dots, q_m\},$$

since otherwise we could decrease $\delta(p, C_i)$ and thus $\delta(p)$ by moving p closer to $\operatorname{conv}\{q_1, \dots, q_m\}$. By Carathéodory's theorem there is a subset of $\{q_1, \dots, q_m\}$ of $k \leq d + 1$ points, such that p is in the convex hull of this subset. By re-numbering, if necessary, we may assume that this subset is the set $\{q_1, \dots, q_k\}$. Then

$$p = \lambda_1 q_1 + \dots + \lambda_k q_k, \text{ where } \lambda_1, \dots, \lambda_k \geq 0, \ \lambda_1 + \dots + \lambda_k = 1.$$

Clearly,

$$C_i \subseteq H_i^- = \{x : (x - p) \cdot (q_i - p) \geq \|q_i - p\|^2 \ (> 0)\} \text{ for } i = 1, \dots, k.$$

Since $k \leq d + 1$, we may choose a point $y \in C_1 \cap \dots \cap C_k \subseteq H_1^- \cap \dots \cap H_k^-$. Then

$$\begin{aligned} 0 &= (y - p) \cdot (p - p) = (y - p) \cdot (\lambda_1 q_1 + \dots + \lambda_k q_k - p) \\ &= (y - p) \cdot \big(\lambda_1 (q_1 - p) + \dots + \lambda_k (q_k - p)\big) \\ &= \lambda_1 (y - p) \cdot (q_1 - p) + \dots + \lambda_k (y - p) \cdot (q_k - p) > 0, \end{aligned}$$

which is the desired contradiction. □

Diameter and Circumradius: Jung's Theorem.

The following estimate of Jung [555] relates circumradius and diameter. We reproduce a proof which makes use of Helly's theorem.

The *circumradius* of a set A in $\mathbb{E}^d$ is the minimum radius of a Euclidean ball which contains the set. It is easy to see that the ball which contains A and whose radius is the circumradius is unique. It is called the *circumball* of A. The *diameter* $\operatorname{diam} A$ of A is the supremum of the distances between two points of the set.

Theorem 3.3. *Let $A \subseteq \mathbb{E}^d$ be bounded. Then A is contained in a (solid Euclidean) ball of radius*

$$\varrho = \Big(\frac{d}{2d+2}\Big)^{\frac{1}{2}} \operatorname{diam} A.$$

Proof. In the first step we show that

(1) The theorem holds for sets A consisting of $d+1$ or fewer points.

Let A be such a set and c the centre of a ball containing A and of minimum radius, say σ. We may assume that $c = o$. Let

$$\{x_1, \dots, x_n\} = \{x \in A : \|x\| = \sigma\}, \text{ where } n \le d+1.$$

Then $c \in \operatorname{conv}\{x_1, \dots, x_n\}$, since otherwise we could decrease σ by moving $c\ (= o)$ closer to $\operatorname{conv}\{x_1, \dots, x_n\}$. Hence

$$c = o = \lambda_1 x_1 + \dots + \lambda_n x_n, \text{ where } \lambda_1, \dots, \lambda_n \ge 0,\ \lambda_1 + \dots + \lambda_n = 1.$$

Then

$$1 - \lambda_k = \sum_{\substack{i=1\\ i\ne k}}^{n} \lambda_i \ge \sum_{\substack{i=1\\ i\ne k}}^{n} \lambda_i \frac{\|x_i - x_k\|^2}{(\operatorname{diam} A)^2} = \sum_{i=1}^{n} \lambda_i \frac{x_i^2 + x_k^2 - 2\, x_i \cdot x_k}{(\operatorname{diam} A)^2}$$

$$= \frac{1}{(\operatorname{diam} A)^2}\Big(2\sigma^2 - 2\Big(\sum_{i=1}^{n} \lambda_i\, x_i\Big) \cdot x_k\Big) = \frac{2\sigma^2}{(\operatorname{diam} A)^2} \text{ for } k = 1, \dots, n.$$

Now, summing over k yields

$$n - 1 \ge \frac{2n\sigma^2}{(\operatorname{diam} A)^2}, \text{ or } \sigma \le \Big(\frac{n-1}{2n}\Big)^{\frac{1}{2}} \operatorname{diam} A \le \Big(\frac{d}{2d+2}\Big)^{\frac{1}{2}} \operatorname{diam} A.$$

This concludes the proof of (1).

For general A consider, for each point of A, the ball with centre at this point and radius ϱ. By (1), any $d+1$ of these balls have non-empty intersection. Thus all these balls have non-empty intersection by Helly's theorem. Any ball with radius ϱ and centre at a point of this intersection then contains each point of A. □

The Centrepoint of a Finite Set

We give a second application of Helly's theorem. Let $A \subseteq \mathbb{E}^d$ be a finite set consisting of, say, n points. A point $x \in \mathbb{E}^d$ is a *centrepoint* of A if each closed halfspace which contains x contains at least $\frac{n}{d+1}$ points of A. Rado [820] proved the following result.

Theorem 3.4. *Each finite set in $\mathbb{E}^d$ has a centrepoint.*

Proof. Let $A \subseteq \mathbb{E}^d$ be a finite set with n points. Clearly, a point in $\mathbb{E}^d$ is a centrepoint of A if and only if it lies in each open halfspace H^o which contains more than $\frac{d}{d+1}n$ points of A. For the proof of the theorem it is thus sufficient to show that the open halfspaces H^o have non-empty intersection. This is certainly the case if the convex polytopes $\operatorname{conv}(A \cap H^o)$ have non-empty intersection. Given $d+1$ of these polytopes, each contains more than $\frac{d}{d+1}n$ points of A. If their intersection were empty, then each point of A is in the complement of one of these polytopes. Hence A is the union of these complements, but the union consists of less than n points. This contradiction shows that any $d+1$ of the polytopes have non-empty intersection. By Helly's theorem all these polytopes then have non-empty intersection. Each point of this intersection is a centrepoint. □

Remark. For extensions and the related ham-sandwich theorem see Matoušek [695] and the references cited there.

3.3 Hartogs' Theorem on Power Series

Given a power series in d complex variables,

$$(1) \quad \sum_{n_1,\dots,n_d=0}^{\infty} a_{n_1\cdots n_d} z_1^{n_1} \cdots z_d^{n_d} \text{ for } z = (z_1, \dots, z_d) \in \mathbb{C}^d,$$

the problem arises to determine the largest connected open set in $\mathbb{C}^d$ on which it converges, its *domain of convergence*. Considering the case $d = 1$, it is plausible to conjecture, that a domain of convergence is a polycylinder, i.e. a Cartesian product of d open circular discs. Surprisingly, this is wrong. A theorem of Hartogs gives a complete description of domains of convergence.

This section contains the definition of Reinhardt domains and Hartogs' theorem which says that the domains of convergence of power series in d complex variables are precisely the Reinhardt domains. Only part of the proof is given.

For a complete proof and for more information on analytic functions in several complex variables, see, e.g. L. Kaup and B. Kaup [568].

Reinhardt Domains and Domains of Convergence

In order to state Hartogs' theorem, we need the notion of a (*complete, logarithmically convex*) *Reinhardt domain*. This is an open, connected set G in $\mathbb{C}^d$ with the following properties:

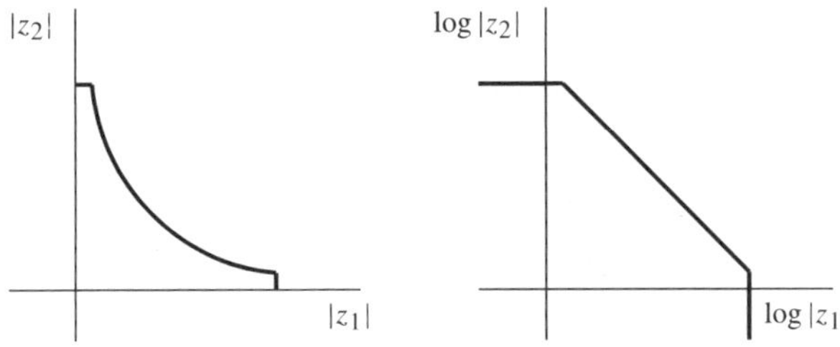

Fig. 3.1. Logarithmically convex sets in $\mathbb{C}^2$

(2) $z = (z_1, \dots, z_d) \in \mathbb{C}^d$, $(w_1, \dots, w_d) \in G$,
$|z_1| < |w_1|, \dots, |z_d| < |w_d| \Rightarrow z \in G$,

(3) $\{(\log|z_1|, \dots, \log|z_d|) : (z_1, \dots, z_d) \in G, z_i \neq 0\} \subseteq \mathbb{E}^d$ is convex.

For $d = 1$, Reinhardt domains are simply the open circular discs with centre 0 in $\mathbb{C}$ (Fig. 3.1).

Hartogs' Theorem

Hartogs [480] found the following characterization of domains of convergence of power series in d complex variables.

Theorem 3.5. *Let G be an open connected set in $\mathbb{C}^d$. Then the following statements are equivalent:*

(i) *G is the domain of convergence of a power series in d complex variables of the form* (1).
(ii) *G is a complete, logarithmically convex Reinhardt domain.*

We prove only the implication (i)⇒(ii). By const, a positive constant is meant. If const appears several times in the same context, this does not mean that it is always the same constant.

Proof. (i)⇒(ii) The main tool for the proof is the following *lemma of Abel*, see, e.g. [568]:

(4) Let $z, w \in \mathbb{C}^d$ be such that $|a_{n_1 \cdots n_d} w_1^{n_1} \cdots w_d^{n_d}| \leq \text{const}$ for $n_1, \dots, n_d = 0, 1, \dots$, and let $|z_i| < |w_i|$ for $i = 1, \dots, d$. Then $z \in G$.

Assume now that G satisfies (i). G is open and connected and, by (4), has property (2). In order to show that it also has property (3), it is sufficient to show the following.

(5) Let $x, y \in G$, $x_i, y_i \neq 0$ for $i = 1, \dots, d$, let $0 \leq \lambda \leq 1$, and let $z \in \mathbb{C}^d$ such that $\log|z_i| = (1 - \lambda)\log|x_i| + \lambda \log|y_i|$ for $i = 1, \dots, d$. Then $z \in G$.

Since G is open, we may choose $u, v \in G$ such that

(6) $|x_i| < |u_i|, |y_i| < |v_i|$ for $i = 1, \dots, d$.

From $u, v \in G$ it follows that

(7) $|a_{n_1 \cdots n_d} u_1^{n_1} \cdots u_d^{n_d}|, |a_{n_1 \cdots n_d} v_1^{n_1} \cdots v_d^{n_d}| \le$ const for $n_1, \dots, n_d = 0, 1, \dots$

Taking into account (5), Propositions (6) and (7) imply that

$|z_i| = |x_i|^{1-\lambda} |y_i|^{\lambda} < |u_i|^{1-\lambda} |v_i|^{\lambda} = w_i$, say, for $i = 1, \dots, d$, where
$|a_{n_1 \cdots n_d} w_1^{n_1} \cdots w_d^{n_d}| = |a_{n_1 \cdots n_d} u_1^{n_1} \cdots u_d^{n_d}|^{1-\lambda} |a_{n_1 \cdots n_d} v_1^{n_1} \cdots v_d^{n_d}|^{\lambda}$
$\le$ const for $n_1, \dots, n_d = 0, 1, \dots$

Hence $z \in G$ by Abel's lemma (4), concluding the proof of (5). Thus G has property (3) and the proof that G satisfies (ii) is complete. □

4 Support and Separation

Support and separation of convex sets play an important role in convex geometry, convex analysis, optimization, optimal control and functional analysis.

In this section, we first introduce the notions of support hyperplane and support function and show basic pertinent results. Then separation of convex sets and bodies will be considered. We mention oracles as a tool to specify convex bodies. To illustrate the usefulness of the former notions and results, Lyapunov's theorem on vector-valued measures and Pontryagin's minimum principle from optimal control theory are presented.

The reader who wants to get more information may consult the books cited in the introduction of this chapter. In addition, we refer to books on convex analysis and optimization, in particular to those of Rockafellar [843], Stoer and Witzgall [970], Hiriart-Urruty and Lemaréchal [505] and Borwein and Lewis [158].

4.1 Support Hyperplanes and Support Functions

Support hyperplanes and support functions are basic tools of convex geometry and are important in other areas.

In the following we first consider metric projection of $\mathbb{E}^d$ onto convex bodies. Next we show that, for any convex body, there is a support hyperplane through each of its boundary points. This leads to a characterization of convex bodies by means of support properties. Then support functions are introduced, a classical means to describe convex bodies. Finally, we characterize support functions as convex functions which are positive homogeneous of degree one.

For more precise information compare the references cited above.

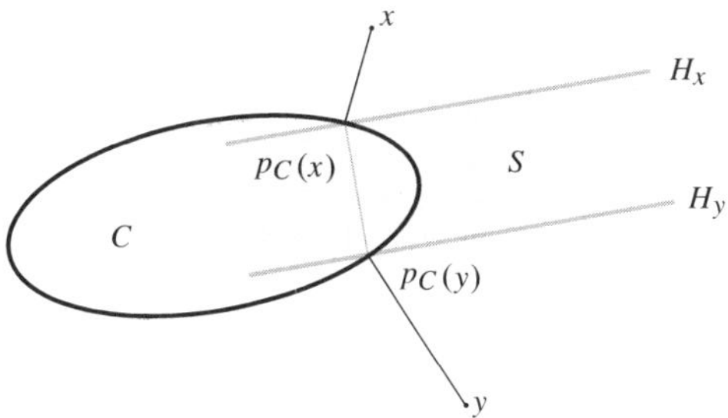

Fig. 4.1. Metric projection is non-expansive

Metric Projection

Let C be a closed convex set in $\mathbb{E}^d$. For each $x \in \mathbb{E}^d$, there is a unique point $p_C(x) \in C$ closest to it. Since C is closed, the existence is obvious. To see the uniqueness, assume that there are points $y, z \in C$, $y \neq z$, both having minimum distance from x. Then $\|y - x\| = \|z - x\|$ and therefore $\|\frac{1}{2}(y + z) - x\| < \|y - x\|, \|z - x\|$, noting that $y \neq z$. Since $\frac{1}{2}(y + z) \in C$ by the convexity of C, this contradicts our choice of y, z. The mapping $p_C : x \to p_C(x)$ of $\mathbb{E}^d$ onto C, thus obtained, is the *metric projection* of $\mathbb{E}^d$ onto C with respect to the Euclidean norm.

The following useful result is due to Busemann and Feller [183].

Lemma 4.1. *Let $C \subseteq \mathbb{E}^d$ be a closed convex set. Then the metric projection $p_C : \mathbb{E}^d \to C$ is* non-expansive, *i.e.*

$$\|p_C(x) - p_C(y)\| \leq \|x - y\| \text{ for } x, y \in \mathbb{E}^d.$$

Given a hyperplane H in $\mathbb{E}^d$, let H^+ and H^- denote the closed halfspaces determined by H. We say, H *separates* two sets, if one set is contained in H^+ and the other one in H^-.

Proof. Let $x, y \in \mathbb{E}^d$. We consider only the case where $x, y \notin C$ (Fig. 4.1). The other cases are treated similarly. If $p_C(x) = p_C(y)$, we are done. Assume then that $p_C(x) \neq p_C(y)$. Let S be the slab orthogonal to the line segment $[p_C(x), p_C(y)] \subseteq C$ and such that its boundary hyperplanes H_x and H_y contain $p_C(x)$ and $p_C(y)$, respectively. We claim that x and $p_C(y)$ are separated by H_x. Otherwise there is a point on $[p_C(x), p_C(y)]$ and thus in C which is closer to x than $p_C(x)$, which is impossible. Similarly, y and $p_C(x)$ are separated by H_y. Taken together, this means that x and y are on different sides of the slab S. Hence $\|x - y\|$ is at least equal to the width of S, that is $\|x - y\| \geq \|p_C(x) - p_C(y)\|$. □

Support Hyperplanes, Normal Vectors and Support Sets

Let $C \subseteq \mathbb{E}^d$ be closed and convex. A hyperplane $H = H_C(x)$ is a *support hyperplane* of C at a point x of the boundary bd C of C, if $x \in H_C(x)$ and C is contained in one of the two closed halfspaces determined by H. In this case, we denote the halfspace containing C by H^-, the other one by H^+. H^- is called a *support halfspace* of C at x. In general, we represent H in the form $H = \{z : u \cdot z = u \cdot x\}$, where u is a normal (unit) vector of H pointing into H^+. Then $H^- = \{z : u \cdot z \leq u \cdot x\}$ and $H^+ = \{z : u \cdot z \geq u \cdot x\}$. u is an *exterior normal (unit) vector* of H, of C or of $\operatorname{bd} C$ at x. Note that H may not be unique. The intersection $C \cap H$ is the *support set* of C determined by H or, with exterior normal (unit) vector u.

In Theorems 1.2 and 2.3, it was shown that a convex function has affine support at each point in the interior of its domain of definition. The following theorem is the corresponding result for convex sets.

Theorem 4.1. *Let $C \subseteq \mathbb{E}^d$ be a closed convex set. For each $x \in \operatorname{bd} C$, there is a support hyperplane $H_C(x)$ of C at x, not necessarily unique. If C is compact, then for each vector $u \in \mathbb{E}^d \setminus \{o\}$, there is a unique support hyperplane $H_C(u)$ of C with exterior normal vector u.*

Let S^{d-1} denote the Euclidean unit sphere in $\mathbb{E}^d$.

Proof. First, the following will be shown.

(1) Let $y \in \mathbb{E}^d \backslash C$. Then the hyperplane H through $p_C(y) \in \operatorname{bd} C$, orthogonal to $y - p_C(y)$, supports C at $p_C(y)$.

It is sufficient to show that H separates y and C. If this does not hold, there is a point $z \in C$ which is not separated from y by H. Then the line segment $[p_C(y), z] \subseteq C$ contains a point of C which is closer to y than $p_C(y)$. This contradicts the definition of $p_C(y)$ and thus concludes the proof of (1).

Next we claim the following.

(2) Let $H_n = \{z : u_n \cdot z = x_n \cdot u_n\}$ be support hyperplanes of C at the points $x_n \in \operatorname{bd} C$, $n = 1, 2, \dots$ Assume that $u_n \to u \in \mathbb{E}^d \setminus \{o\}$ and $x_n \to x$ ($\in \operatorname{bd} C$) as $n \to \infty$. Then $H = \{z : u \cdot z = u \cdot x\}$ is a support hyperplane of C at x.

Clearly, $x \in H$. It is sufficient to show that $C \subseteq H^-$. Let $z \in C$. Then $u_n \cdot z \leq u_n \cdot x_n$ for $n = 1, 2, \dots$ Letting $n \to \infty$, we see that $u \cdot z \leq u \cdot x$, or $z \in H^-$, concluding the proof of (2).

For the proof of the first assertion in the theorem, choose points $y_n \in \mathbb{E}^d \backslash C$, $n = 1, 2, \dots$, such that $y_n \to x$. By Lemma 4.1, $x_n = p_C(y_n)$ ($\in \operatorname{bd} C$) $\to x = p_C(x)$. Proposition (1) shows that, for $n = 1, 2, \dots$, there is a support hyperplane of C at x_n, say $H_n = \{z : u_n \cdot z = u_n \cdot x_n\}$, where $u_n \in S^{d-1}$. By considering a subsequence and re-numbering, if necessary, we may suppose that $u_n \to u \in S^{d-1}$, say. An application of (2) then implies that $H = \{z : u \cdot z = u \cdot x\}$ is a support hyperplane of C at x.

To see the second assertion, note that the compactness of C implies that $u \cdot x = \sup\{u \cdot z : z \in C\}$ for a suitable $x \in C$. Clearly, $x \in \operatorname{bd} C$ and $H = \{z : u \cdot z = u \cdot x\}$ is a support hyperplane of C (at x) with exterior normal vector u. □

Remark. The proofs of Theorem 2.3 on affine support of convex functions and of the corresponding Theorem 4.1 on support hyperplanes of closed convex sets are essentially different. The former proof is based on the linearity structure of $\mathbb{E}^d$ and can be extended easily to infinite dimensions by transfinite induction. The latter proof rests on the Euclidean structure of $\mathbb{E}^d$ and on compactness and its extension to infinite dimensions is not of much interest.

Results on convex functions *correspond*, in many cases, to results on convex sets or convex bodies and vice versa. Quite often it is possible to *transform* a result on convex functions into a result on convex sets or convex bodies. Examples deal with support, separation, and differentiability properties.

Characterization of Convex Sets by Support Properties

The definition of a support hyperplane of C at a point $x \in \operatorname{bd} C$ still makes sense if the assumption that C is convex is omitted. Thus we may speak of a *support hyperplane* of a closed set in $\mathbb{E}^d$ at a boundary point. The characterizations of convex functions by support properties in Theorems 1.3 and 2.4 correspond to the following result.

Theorem 4.2. *Let $C \subseteq \mathbb{E}^d$ be closed and let $\operatorname{int} C \neq \emptyset$. Then the following are equivalent:*

(i) *C is convex.*

(ii) *C has a support hyperplane $H_C(x)$ at each point $x \in \operatorname{bd} C$.*

Proof. (i)⇒(ii) This follows from Theorem 4.1.

(ii)⇒(i) The intersection of any family of convex sets is also convex. Thus it suffices to prove that

$$(3)\quad C = \bigcap_{x \in \operatorname{bd} C} H_C^-(x).$$

Since C is contained in the set on the right-hand side, it remains to show the following: Let $z \in \mathbb{E}^d \backslash C$. Then $z \notin H_C^-(x)$ for a suitable $x \in \operatorname{bd} C$. To see this, let $y \in \operatorname{int} C$ and choose $x \in [y, z] \cap \operatorname{bd} C$. By (ii) there is a support halfspace $H_C^-(x)$ of C at x. Then $y \in \operatorname{int} C \subseteq C \subseteq H_C^-(x)$ and therefore $z \notin H_C^-(x)$. □

Corollary 4.1. *Let $C \subseteq \mathbb{E}^d$ be closed and convex. Then*

$$C = \bigcap_{x \in \operatorname{bd} C} H_C^-(x).$$

How to Specify a Convex Body

While the definitions of convex sets and convex bodies are extremely simple, it is a highly non-trivial task to specify an arbitrary convex body so that one may obtain, from this specification, important information about the body, for example analytic or geometric information about its boundary, or information on its volume, surface area, diameter, width, etc. Classical tools which sometimes serve this purpose are support functions, distance and radial functions of a convex body. See, e.g. Schneider [907]. In the context of algorithmic convex geometry, convex bodies are specified by various membership oracles, see Sect. 4.2 and, for more information, Grötschel, Lovász and Schrijver [409]. Here, we consider support functions.

Support Functions and Norms

Let C be a convex body in $\mathbb{E}^d$. The position of a support hyperplane $H_C(u)$ of C with given exterior normal vector $u \neq o$ is determined by its *support function* $h_C : \mathbb{E}^d \to \mathbb{R}$ defined by

$$h_C(u) = \sup\{u \cdot y : y \in C\} \text{ for } u \in \mathbb{E}^d.$$

Clearly,

$$H_C(u) = \{x : u \cdot x = h_C(u)\},\ H_C^-(u) = \{x : u \cdot x \leq h_C(u)\}.$$

If $u \in S^{d-1}$, i.e. u is a unit vector, then $h_C(u)$ is the *signed distance* of the origin o to the support hyperplane $H_C(u)$. More precisely, the distance of o to $H_C(u)$ is equal to $h_C(u)$ if $o \in H_C(u)^-$ and equal to $-h_C(u)$ if $o \in H_C(u)^+$. Since C is the intersection of all its support halfspaces by Corollary 4.1, we have

$$\begin{aligned}(4)\quad C &= \{x : u \cdot x \leq h_C(u) \text{ for all } u \in \mathbb{E}^d\}\\ &= \{x : u \cdot x \leq h_C(u) \text{ for all } u \in S^{d-1}\}.\end{aligned}$$

Assume now that C is a proper convex body in $\mathbb{E}^d$ with centre at the origin o. We can assign to C a norm $\|\cdot\|_C$ on $\mathbb{E}^d$ as follows:

$$\|x\|_C = \inf\{\lambda > 0 : x \in \lambda C\} \text{ for } x \in \mathbb{E}^d.$$

The convex body C is the solid unit ball of this new norm on $\mathbb{E}^d$.

Without proof, we mention the following: for a proper convex body C with centre o, its *polar body*

$$C^* = \{y : x \cdot y \leq 1 \text{ for } x \in C\}$$

is also a proper, o-symmetric convex body. A simple proof, which is left to the reader, shows the following relations between the support functions and the norms corresponding to C and C^*:

$$h_C(u) = \|u\|_{C^*} \text{ and } h_{C^*}(u) = \|u\|_C \text{ for } u \in \mathbb{E}^d.$$

For more information on polar bodies, see Sect. 9.1.

Characterization of Support Functions

The next result reveals the simple nature of support functions.

Theorem 4.3. *Let* $h : \mathbb{E}^d \to \mathbb{R}$. *Then the following statements are equivalent:*

(i) *h is the support function of a (unique) convex body C, i.e. $h = h_C$.*

(ii) *h has the following properties:*

$h(\lambda u) = \lambda h(u)$ *for* $u \in \mathbb{E}^d, \lambda \geq 0$

$h(u+v) \leq h(u) + h(v)$ *for* $u, v \in \mathbb{E}^d$

Statement (ii) means that h is *positively homogeneous of degree* 1 and *subadditive* and thus, in particular, convex.

Proof. (i)⇒(ii) Let $h = h_C$, where C is a suitable convex body. Then

$$\begin{aligned} h_C(\lambda u) &= \sup\{\lambda u \cdot x : x \in C\} = \lambda \sup\{u \cdot x : x \in C\} = \lambda h_C(u), \\ h_C(u+v) &= \sup\{(u+v) \cdot x : x \in C\} \\ &\leq \sup\{u \cdot x : x \in C\} + \sup\{v \cdot x : x \in C\} = h_C(u) + h_C(v) \\ &\text{for } u, v \in \mathbb{E}^d, \lambda \geq 0. \end{aligned}$$

(ii)⇒(i) Define

$$(5) \quad C = \left\{x : v \cdot x \leq h(v) \text{ for all } v \in \mathbb{E}^d\right\} = \bigcap_{v \in \mathbb{E}^d} \{x : v \cdot x \leq h(v)\}.$$

Being an intersection of closed halfspaces, C is closed and convex. Taking $v = \pm b_1, \dots, \pm b_d$, where $\{b_1, \dots, b_d\}$ is the standard basis of $\mathbb{E}^d$, shows that C is bounded. If $C \neq \emptyset$, the definition (5), of C, implies that $h_C \leq h$ (note (4)). Thus, to finish the proof, it is sufficient to show that

$$(6) \quad C \neq \emptyset \text{ and } h \leq h_C.$$

Let $u \in \mathbb{E}^d \setminus \{o\}$. By (ii), the epigraph epi h, of h, is a closed convex cone in $\mathbb{E}^{d+1} = \mathbb{E}^d \times \mathbb{R}$ with apex at the origin $(o, 0)$, directed upwards and with non-empty interior. By Theorem 4.1, there is a support hyperplane H of epi h at the point $\big(u, h(u)\big) \in$ bd epi h, where $\big(u, h(u)\big) \neq (o, 0)$. Since epi h is a convex cone with non-empty interior and apex $(o, 0)$, it follows that H supports epi h also at $(o, 0)$. (Fig. 4.2)

The exterior normal vectors of H at $(o, 0)$ point below $\mathbb{E}^d$. Thus we may choose such a vector of the form $(x, -1)$. Hence $H = \{(v, s) : v \cdot x - s = 0\}$. Then $H^- = \{(v, s) : v \cdot x \leq s\} \supseteq$ epi h and thus, in particular, $v \cdot x \leq h(v)$ for each point $(v, h(v)) \in$ bd epi h and therefore $v \cdot x \leq h(v)$ for all $v \in \mathbb{E}^d$. Hence $x \in C$, by the definition of C in (5), and thus $C \neq \emptyset$. Since H is a support hyperplane of epi h at $(u, h(u))$, we have $u \cdot x = h(u)$. For $x \in C$, the definition of the support function h_C implies that $h_C(u) \geq u \cdot x$. Thus $h_C(u) \geq h(u)$. The proof of (6) is complete, concluding the proof of the implication (ii)⇒(i). □

Warning. We warn the reader: the above proof cannot be trivialized, since, a priori, it is not clear that the boundary hyperplanes of the halfspaces appearing in (5) all touch C.

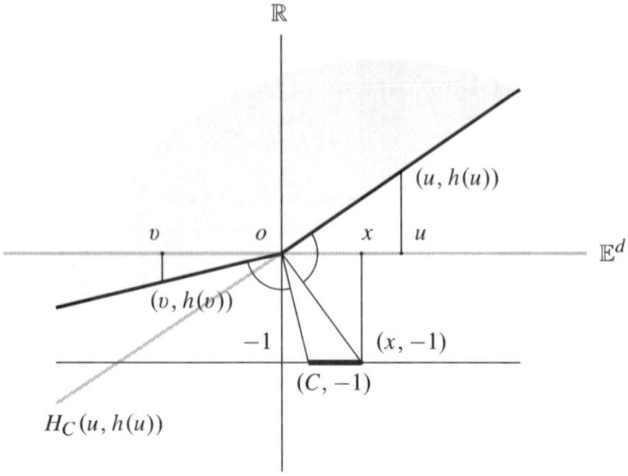

Fig. 4.2. A homogeneous convex function is a support function

4.2 Separation and Oracles

Separation of convex sets, in particular of convex polytopes and convex polyhedra, plays an important role in optimization and convex analysis.

In this section, we show a standard separation theorem which for most applications is sufficient, and describe oracles to specify convex bodies.

For more information on separation and oracles, see Stoer and Witzgall [970], Hiriart-Urruty and Lemaréchal [505], and Grötschel, Lovász and Schrijver [409].

Separation and Strong Separation

Convex sets C and D in $\mathbb{E}^d$ are *separated* if there is a hyperplane H such that $C \subseteq H^-$ and $D \subseteq H^+$ or vice versa, where H^+ and H^- are the closed halfspaces determined by H. Then H is called a *separating hyperplane* of C and D. The sets C and D are *strongly separated* if there is a closed slab S with $\operatorname{int} S \neq \emptyset$ such that $C \subseteq S^-$ and $D \subseteq S^+$ or vice versa. Here S^- and S^+ are the closed halfspaces, determined by the two boundary hyperplanes of S, not containing S. We call S a *separating slab* of C and D.

In the following, two separation results will be presented. The first result is a simple yet useful tool.

Proposition 4.1. *Let C, D be convex sets in $\mathbb{E}^d$. Then the following statements are equivalent:*

(i) *C and D are separated, respectively, strongly separated.*

(ii) *The convex set $C - D = \{x - y : x \in C, y \in D\}$ and $\{o\}$ are separated, respectively, strongly separated.*

Proof. We consider only the case of strong separation, the case of separation is similar, but simpler.

(i)⇒(ii) For the proof of the convexity of $C - D$, let $x, z \in C$, $y, w \in D$ and $0 \leq \lambda \leq 1$. Then

$$(1-\lambda)(x-y) + \lambda(z-w) = \big((1-\lambda)x + \lambda z\big) - \big((1-\lambda)y + \lambda w\big) \in C - D$$

by the convexity of C and D. Thus $C - D$ is convex. Next, let $S = \{x : \alpha \leq u \cdot x \leq \beta\}$, $\alpha < \beta$, be a slab which strongly separates C and D, say $C \subseteq \{x : u \cdot x \leq \alpha\}$ and $D \subseteq \{y : u \cdot y \geq \beta\}$. Then $C - D \subseteq \{x - y : u \cdot (x - y) \leq \alpha - \beta\}$. Thus $C - D$ and $\{o\}$ are separated by the slab $\{z : \alpha - \beta \leq u \cdot z \leq 0\}$.

(ii)⇒(i) Let the slab $S = \{z : -\gamma \leq u \cdot z \leq 0\}, \gamma > 0$, separate $\{o\}$ and $C - D$. Then $u \cdot (x - y) \leq -\gamma$, i.e. $u \cdot x + \gamma \leq u \cdot y$ for all $x \in C$ and $y \in D$. Let $\alpha = \sup\{u \cdot x : x \in C\}$. Then the slab $\{z : \alpha \leq u \cdot z \leq \alpha + \gamma\}$ separates C and D. □

Theorem 4.4. *Let $C, D \subseteq \mathbb{E}^d$ be convex. Then the following hold:*

(i) *Let C be compact, D closed and $C \cap D = \emptyset$. Then C and D are strongly separated.*
(ii) *Let* relint $C \cap$ relint $D = \emptyset$. *Then C and D are separated.*

Proof. (i) By the assumptions in (i), we may choose $p \in C, q \in D$ having minimum distance. Let $u = q - p \ (\neq o)$. Then the slab $\{x : u \cdot p \leq u \cdot x \leq u \cdot q\}$ separates C and D.

(ii) By Proposition 3.1, the sets relint C and relint D are convex. Since, by assumption, these sets are disjoint, Proposition 4.1 shows that $o \notin E =$ relint $C -$ relint D. We shall prove that $o \notin$ int cl E. Otherwise, there are points $x_1, \dots, x_{d+1} \in E$ such that o is an interior point of the simplex with vertices $x_1, \dots, x_{d+1}$. Since E is convex, by Proposition 4.1 we have $o \in E$, which is the desired contradiction. Thus $o \in$ bd cl E or $o \notin$ cl E. Since E is convex, cl E is convex by Proposition 3.1. It thus follows from Theorem 4.2, respectively, from (i), that $\{o\}$ and cl E can be separated by a hyperplane. Hence, a fortiori, $\{o\}$ and $E =$ relint $C -$ relint D can be separated by a hyperplane. Thus relint C and relint D can be separated by a hyperplane by Proposition 4.1. This, in turn, implies that cl relint C and cl relint D can be separated by a hyperplane. Now apply Proposition 3.1 to see that C and D can be separated by a hyperplane. □

Simple examples show that, in general, disjoint closed convex sets C and D cannot be strongly separated, but if C and D are convex polyhedra, this is possible. This fact is of importance in optimization (Fig. 4.3).

Strong and Weak Oracles to Specify Convex Bodies

Before considering oracles, some definitions are in order: Let C be a proper convex body in $\mathbb{E}^d$ and $\varepsilon > 0$. Define the *ε-neighbourhood* of C or the *parallel body* of C at distance ε and the *inner parallel body* of C at distance ε by

$$C_\varepsilon = C + \varepsilon B^d = \big\{x + \varepsilon y : x \in C, y \in B^d\big\},\ C_{-\varepsilon} = \big\{x : x + \varepsilon B^d \subseteq C\big\}.$$

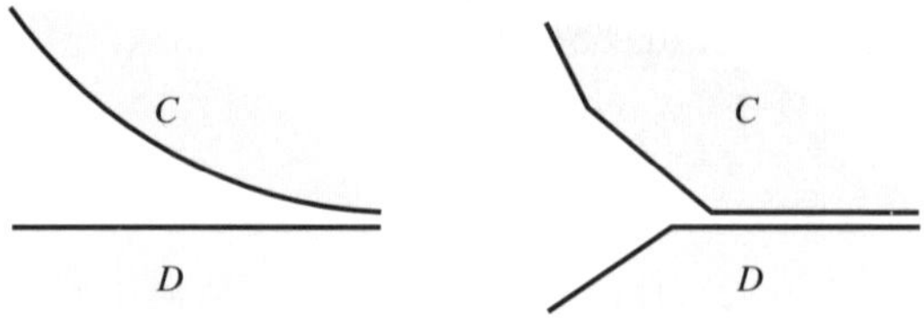

Fig. 4.3. Separated and strongly separated unbounded convex sets

Let $\mathbb{Q}$ be the field of rationals. A way to specify a convex body C in the context of algorithmic convex geometry is by means of oracles. The *strong* and the *weak separation oracles* are as follows:

Oracle 4.1. *Given a point $y \in \mathbb{E}^d$, the oracle says that $y \in C$ or specifies a vector $u \in \mathbb{E}^d$ such that $u \cdot x < u \cdot y$ for all $x \in C$.*

Oracle 4.2. *Given a point $y \in \mathbb{Q}^d$ and a rational $\varepsilon > 0$, the oracle says that $y \in C_\varepsilon$ or specifies a vector $u \in \mathbb{Q}^d$ with maximum norm $||u||_\infty = 1$ such that $u \cdot x < u \cdot y$ for all $x \in C_{-\varepsilon}$.*

The *strong* and the *weak membership oracles* are the following:

Oracle 4.3. *Given a point $y \in \mathbb{E}^d$, the oracle says that $y \in C$ or $y \notin C$.*

Oracle 4.4. *Given a point $y \in \mathbb{Q}^d$ and a rational $\varepsilon > 0$, the oracle says that $y \in C_\varepsilon$ or $y \notin C_{-\varepsilon}$.*

The weak oracles are shaped more to the need of real life algorithms used by computers.

4.3 Lyapunov's Convexity Theorem

Convex geometry has many applications in other areas of mathematics and in related fields. The applications are of many different types. In some cases the notion of convexity or other notions of convex geometry serve to describe or clarify a situation. An example is the Lyapunov convexity theorem 4.5. Sometimes it is a convexity condition which yields an interesting result, such as the Bohr–Mollerup characterization 1.11 of the gamma function or the sufficient condition of Courant and Hilbert in the calculus of variations, see Theorem 2.12. Finally, in some cases, methods or results of convex geometry are useful, sometimes indispensable, tools for proofs. For examples, see the proofs of Pontryagin's minimum principle 4.6, Birkhoff's theorem 5.7 and the isoperimetric inequalities of mathematical physics in Sects. 8.4 and 9.4.

In the following we present a proof of Lyapunov's convexity theorem on vector-valued measures.

For vector measures, a good reference is Diestel and Uhl [267]. More recent surveys on Lyapunov's convexity theorem are Olech [779], Hill [503] and E. Saab and P. Saab [870].

Lyapunov's Convexity Theorem

Let M be a set with a σ-algebra $\mathcal{M}$ of subsets. A finite signed measure ν on the measure space $\langle M, \mathcal{M}\rangle$ is *non-atomic* if, for every set $A \in \mathcal{M}$ with $\nu(A) \neq 0$, there is a set $B \in \mathcal{M}$ with $B \subseteq A$ and $\nu(B) \neq 0, \nu(A)$. Lyapunov [671] proved the following basic result.

Theorem 4.5. *Let $\mu_1, \ldots, \mu_d$ be d finite, signed, non-atomic measures on $\langle M, \mathcal{M}\rangle$. Then the range of the (finite, signed, non-atomic) vector-valued measure $\mu = (\mu_1, \ldots, \mu_d)$ on $\langle M, \mathcal{M}\rangle$, that is the set*

$$R(M) = \big\{\mu(B) = \big(\mu_1(B), \ldots, \mu_d(B)\big) : B \in \mathcal{M}\big\} \subseteq \mathbb{E}^d,$$

is compact and convex.

There exist several proofs of this result. We mention the short ingenious proof of Lindenstrauss [659]. Below we follow the more transparent proof of Artstein [40]. Let relbd stand for boundary of a set relative to its affine hull.

Proof. Before beginning with the proof, some useful notions and tools will be presented:

The *variation* $|\mu|$ of the vector-valued measure μ is a set function on $\langle M, \mathcal{M}\rangle$ defined by

$$|\mu|(A) = \sup\Big\{\sum_{i=1}^{n} \|\mu(A_i)\| : A_1, \ldots, A_n \subseteq A, \text{ disjoint}, \\ A_1, \ldots, A_n \in \mathcal{M},\ n = 1, 2, \ldots\Big\}.$$

$|\mu|$ is a finite, non-atomic measure on $\langle M, \mathcal{M}\rangle$. Thus, any set $A \in \mathcal{M}$ with $|\mu|(A) > 0$ contains sets $B \in \mathcal{M}$ for which $|\mu|(B)$ is positive and arbitrarily small. A finite signed measure ν on $\langle M, \mathcal{M}\rangle$ is *absolutely continuous* with respect to the measure $|\mu|$ if $\nu(B) = 0$ for each $B \in \mathcal{M}$ with $|\mu|(B) = 0$. This is equivalent to each of the following statements: first, for each $\varepsilon > 0$, there is a $\delta > 0$ such that $|\nu(B)| \leq \varepsilon$ for any $B \in \mathcal{M}$ with $|\mu|(B) \leq \delta$. Second, there exists a function $f : M \to \mathbb{R}$ which is integrable with respect to the measure $|\mu|$ such that

$$\nu(B) = \int_B f\, d|\mu| \text{ for } B \in \mathcal{M}.$$

f is the Radon–Nikodym derivative of ν with respect to $|\mu|$. Given $A \in \mathcal{M}$, let A^-, A^0, A^+ denote the following sets:

$$A^- = \{s \in A : f(s) < 0\},\ A^0 = \{s \in A : f(s) = 0\},\ A^+ = \{s \in A : f(s) > 0\},$$

all in $\mathcal{M}$. Clearly, $|\mu|(B) = 0$ for each set $B \in \mathcal{M}$, $B \subseteq A^+$, with $\nu(B) = 0$. That is, the restriction of $|\mu|$ to A^+ is absolutely continuous with respect to the restriction of ν to A^+ and similarly for A^-.

For the proof of the theorem it is sufficient to show the following:

(1) Let $x \in \operatorname{cl}\operatorname{conv} R(M)$. Then $x \in R(M)$.

For $x = o$ this is trivial. Thus we may suppose that $x \neq o$. Then the proof of (1) is split into several steps.

First, let

$$\mathcal{N} = \{A \in \mathcal{M} : x \in \operatorname{cl}\operatorname{conv} R(A)\}.$$

Order $\mathcal{N}$ by set inclusion (up to sets of $|\mu|$-measure 0). Then

(2) $\mathcal{N}$ has a minimal element.

In order to prove (2), the following will be shown first:

(3) Let $\{A_i,\ i = 1, 2, \dots\}$ be a countable decreasing chain in $\mathcal{N}$. Then $A_0 = \bigcap_i A_i$ is a lower bound of this chain in $\mathcal{N}$.

Clearly, A_0 is a lower bound of the chain. We have to show that $A_0 \in \mathcal{N}$, i.e. $x \in \operatorname{cl}\operatorname{conv} R(A_0)$. For this it is sufficient to prove that $\operatorname{cl}\operatorname{conv} R(A_0) \supseteq \bigcap_i \operatorname{cl}\operatorname{conv} R(A_i)$. For each measurable subset $A_i \cap B$ of A_i, where $B \in \mathcal{M}$, the definition of $|\mu|$ implies that

$$\|\mu(A_0 \cap B) - \mu(A_i \cap B)\| = \|\mu\big((A_i \backslash A_0) \cap B\big)\| \leq |\mu|(A_i \backslash A_0) = \varepsilon_i,$$

say. Thus $R(A_0) + \varepsilon_i B^d \supseteq R(A_i)$. Hence $\operatorname{cl}\operatorname{conv} R(A_0) + \varepsilon_i B^d \supseteq \operatorname{cl}\operatorname{conv} R(A_i)$. Since $\varepsilon_i \to 0$ (note that $A_0 = \bigcap_i A_i$ and $|\mu|$ is a finite measure), we see that $\operatorname{cl}\operatorname{conv} R(A_0) \supseteq \bigcap_i \operatorname{cl}\operatorname{conv} R(A_i)$. Since $x \in \operatorname{cl}\operatorname{conv} R(A_i)$ for each i, it finally follows that $x \in \operatorname{cl}\operatorname{conv} R(A_0)$, or $A_0 \in \mathcal{N}$. The proof of (3) is complete. Using (3), we next show the following refinement of (3):

(4) Let $\{A_\iota, \iota \in I\}$ be a chain in $\mathcal{N}$. Then it has a lower bound in $\mathcal{N}$.

If this chain has a smallest element, we are finished. Otherwise choose a decreasing countable sub-chain $\{A_{\iota_i}, i = 1, 2, \dots\}$ such that $\inf\{|\mu|(A_{\iota_i}) : i = 1, 2, \dots\} = \inf\{|\mu|(A_\iota) : \iota \in I\}$. Then $A_0 = \bigcap_i A_{\iota_i} \in \mathcal{N}$ by (3) and $|\mu|(A_0) = \inf\{|\mu|(A_\iota) : \iota \in I\}$. We have to show that $A_0 \subseteq A_\iota$ for each $\iota \in I$ (up to a set of $|\mu|$-measure 0). Let $\iota \in I$. If there is an i such that $\iota_i < \iota$, then clearly $A_0 \subseteq A_\iota$. Otherwise $A_\iota \subseteq A_{\iota_i}$ for each $i = 1, 2, \dots$ and thus $A_\iota \subseteq A_0$. Since $|\mu|(A_\iota) \geq |\mu|(A_0)$, this then implies that $A_\iota = A_0$ (up to a set of $|\mu|$-measure 0) and thus, a fortiori, $A_0 \subseteq A_\iota$. Having proved that $A_0 \in \mathcal{N}$ and $A_0 \subseteq A_\iota$ for $\iota \in I$, the proof of (4) is complete. (4) implies (2) by Zorn's lemma.

Second, the following statement will be shown:

(5) Let $A \in \mathcal{M}$, $u \in S^{d-1}$, and let ν be the finite, signed, non-atomic measure $\nu = u \cdot \mu = u_1\mu_1 + \cdots + u_d\mu_d$ on $\langle M, \mathcal{M}\rangle$. Then ν is absolutely continuous with respect to $|\mu|$ and

$$H(u) \cap \operatorname{cl}\operatorname{conv} R(A) = \mu(A^+) + \operatorname{cl}\operatorname{conv} R(A^0).$$

Here $H(u)$ is the support hyperplane of $\operatorname{cl}\operatorname{conv} R(A)$ with exterior normal vector u. The absolute continuity of ν with respect to $|\mu|$ is obvious. Clearly, $H(u)$ is the hyperplane

$$\begin{aligned} u \cdot y &= \sup\big\{u \cdot z : z \in \operatorname{cl}\operatorname{conv} R(A)\big\} = \sup\big\{u \cdot z : z \in R(A)\big\} \\ &= \sup\big\{\nu(C) : C \in \mathcal{M}, C \subseteq A\big\} \\ &= \sup\big\{\nu(A^+ \cap C) + \nu(A^0 \cap C) + \nu(A^- \cap C) : C \in \mathcal{M}, C \subseteq A\big\} = \nu(A^+). \end{aligned}$$

For the proof of the inclusion $H(u) \cap \operatorname{cl}\operatorname{conv} R(A) \supseteq \mu(A^+) + \operatorname{cl}\operatorname{conv} R(A^0)$, it is sufficient to show the following: Let $B \in \mathcal{M}$, $B \subseteq A^0$. Then $\mu(A^+) + \mu(B) \in H(u) \cap \operatorname{cl}\operatorname{conv} R(A)$. Clearly, $A^+ \cup B \in \mathcal{M}$ and $A^+ \cup B \subseteq A$. Hence

$$\begin{aligned} &\mu(A^+) + \mu(B) = \mu(A^+ \cup B) \in R(A) \subseteq \operatorname{cl}\operatorname{conv} R(A), \\ &u \cdot \big(\mu(A^+) + \mu(B)\big) = \nu(A^+) + \nu(B) = \nu(A^+). \end{aligned}$$

Thus $\mu(A^+) + \mu(B) \in H(u) \cap \operatorname{cl}\operatorname{conv} R(A)$, concluding the proof of the first inclusion. Next, the reverse inclusion $H(u) \cap \operatorname{cl}\operatorname{conv} R(A) \subseteq \mu(A^+) + \operatorname{cl}\operatorname{conv} R(A^0)$ will be shown. Since μ is a finite vector-valued measure, $R(A)$ is bounded. Thus $\operatorname{cl}\operatorname{conv} R(A) = \operatorname{conv}\operatorname{cl} R(A)$ by Proposition 3.2. For the proof of the reverse inclusion, it is thus sufficient to show the following: let $y \in H(u) \cap \operatorname{cl} R(A)$. Then $y \in \mu(A^+) + \operatorname{cl}\operatorname{conv} R(A^0)$. Clearly, $y_i \to y\big(\in H(u)\big)$ as $i \to \infty$ for suitable $y_i = \mu(A_i) \in R(A)$. Thus $u \cdot y_i = u \cdot \mu(A_i) = \nu(A_i) \to u \cdot y$ and thus

$$\begin{aligned} \nu(A_i) &= \nu(A^+ \cap A_i) + \nu(A^0 \cap A_i) + \nu(A^- \cap A_i) \\ &= \nu(A^+ \cap A_i) + \nu(A^- \cap A_i) \to u \cdot y = \nu(A^+). \end{aligned}$$

Hence $\nu(A^+ \cap A_i) \to \nu(A^+)$ and $\nu(A^- \cap A_i) \to 0$ or $\nu(A^+ \backslash A_i) \to 0$ and $\nu(A^- \cap A_i) \to 0$. Since $|\mu|$ is absolutely continuous with respect to $|\nu|$ on A^+ and also on A^-, it thus follows that $|\mu|(A^+ \backslash A_i) \to 0$ and $|\mu|(A^- \cap A_i) \to 0$. Hence $\mu(A^+ \backslash A_i), \mu(A^- \cap A_i) \to o$ and therefore

$$\begin{aligned} y &= \lim \mu(A_i) = \lim \big(\mu(A^+ \cap A_i) + \mu(A^0 \cap A_i) + \mu(A^- \cap A_i)\big) \\ &= \lim \big(\mu(A^+) - \mu(A^+ \backslash A_i) + \mu(A^0 \cap A_i) + \mu(A^- \cap A_i)\big) \\ &= \mu(A^+) + \lim \mu(A^0 \cap A_i) \in \mu(A^+) + \operatorname{cl} R(A^0) \\ &\subseteq \mu(A^+) + \operatorname{cl}\operatorname{conv} R(A^0). \end{aligned}$$

This concludes the proof of the reverse inclusion. The proof of (5) is complete.

Third, we shall prove the following:

(6) Let $A \in \mathcal{M}$ be minimal such that $x \in \operatorname{cl}\operatorname{conv} R(A)$. Then $x \in R(A)$.

Note that $x \neq o$ and $o, x \in \operatorname{cl}\operatorname{conv} R(A)$. We distinguish two cases. First case: $x \in \operatorname{relint}\operatorname{cl}\operatorname{conv} R(A)$. By Lemma 3.1 the point x is in the relative interior of a convex polytope with vertices $\mu(A_1), \dots, \mu(A_k) \in R(A)$, say. If $B \in \mathcal{M}$, $B \subseteq A$, is such that $|\mu|(B)$ is sufficiently small (such B exist since the measure $|\mu|$ is non-atomic), then $\|\mu(A_1 \backslash B) - \mu(A_1)\|, \dots, \|\mu(A_k \backslash B) - \mu(A_k)\| \leq |\mu|(B)$ are so small that x is still in the relative interior of the convex polytope with vertices $\mu(A_1 \backslash B), \dots, \mu(A_k \backslash B) \in R(A \backslash B) \subseteq R(A)$. Hence $x \in \operatorname{cl}\operatorname{conv} R(A \backslash B)$, in contradiction to the minimality of A. Hence the first case cannot hold. Second case: $x \in \operatorname{relbd} \operatorname{cl} \operatorname{conv} R(A)$. Then we may choose $u \in S^{d-1}$ such that $x \in H(u) \cap \operatorname{cl} \operatorname{conv} R(A)$ but $\operatorname{cl} \operatorname{conv} R(A) \not\subseteq H(u)$. By (5), $x - \mu(A^+) \in \operatorname{cl} \operatorname{conv} R(A^0)$. The set $A^0 \in \mathcal{M}$ is minimal such that $x - \mu(A^+) \in \operatorname{cl} \operatorname{conv} R(A^0)$. Otherwise $A \in \mathcal{M}$ would not be minimal such that $x \in \operatorname{cl} \operatorname{conv} R(A)$. Since $\operatorname{cl} \operatorname{conv} R(A) \not\subseteq H(u)$, Proposition (5) shows that the dimension of $\operatorname{cl} \operatorname{conv} R(A^0)$ is smaller than the dimension of $\operatorname{cl} \operatorname{conv} R(A)$. If $R(A^0) = \{o\}$ we have $x = \mu(A^+)$ and we are done. If not, we may repeat the above argument with $x - \mu(A^+)$ and A^0 instead of x and A to show that $x - \mu(A^+) - \mu(A^{0+}) \in \operatorname{cl} \operatorname{conv} R(A^{00})$, where the dimension of $\operatorname{cl} \operatorname{conv} R(A^{00})$ is smaller than the dimension of $\operatorname{cl} \operatorname{conv} R(A^0)$, etc. In any case, after finitely many repetitions we arrive at (6).

Propositions (2) and (6) finally yield (1), concluding the proof of the theorem. □

Remark. The range of a finite non-atomic vector-valued signed measure is actually a zonoid and vice versa, as shown by Rickert [835]. See also Bolker [141]. Zonoids are particular convex bodies. See Sect. 7.3 for a definition.

4.4 Pontryagin's Minimum Principle

The minimum principle, sometimes also called maximum principle, of Pontryagin and his collaborators Boltyanskiĭ, Gamkrelidze and Mishchenko [813] is a central result of control theory. It yields information on optimal controls.

In the following we present a version which makes essential use of support hyperplanes and which rests on convexity arguments, both finite and infinite dimensional.

For more information on control theory see, e.g. Pontryagin, Boltyanskiĭ, Gamkrelidze, Mishchenko [813] and Agrachev and Sachkov [2].

The Time Optimal Control Problem

Consider a system of linear differential equations with given initial value:

$$(1)\quad \dot{x} = Ax + Bu,\ x(0) = a,$$

where A and B are real $d \times d$, respectively, $d \times c$ matrices and $a \in \mathbb{E}^d$. A *control* $u : [0, +\infty) \to C$ is the parametrization of a curve which is contained in a given convex body C in $\mathbb{E}^c$. We suppose that each of its c components is measurable (with respect to Lebesgue measure on $\mathbb{R}$). Given a control u, by a *solution* of (1), we mean a continuous parametrization $x : [0, +\infty) \to \mathbb{E}^d$ of a curve, such that the components of x are almost everywhere differentiable and (1) holds almost everywhere on $[0, +\infty)$. A control u is said to *transfer* the *initial state* a to a state $b \in \mathbb{E}^d \setminus \{a\}$ in time t if $x(t) = b$ holds for the corresponding solution x of (1).

The *time optimal control problem* is to find a control u which transfers a to b in minimum time.

The Minimum Principle

The following result of Pontryagin et al. [813] sheds light on the time optimal control problem.

Theorem 4.6. *Assume that there is a control for* (1) *which transfers a to b. Then the following propositions hold:*

(i) *There is a time optimal control which transfers a to b.*

(ii) *Let u be a time optimal control which transfers a to b in minimum time T, say. Then there is a point $w \in \mathbb{E}^d$ such that for the solution of the initial value problem*

(2) $\dot{y} = -A^T y,\ y(T) = w,$

the following statement holds, where $w(t) = B^T y(t)$:

$$u(t) \in C \cap H_C\big(w(t)\big) \text{ for almost every } t \in [0, T].$$

For the geometric meaning of Proposition (ii), see the remarks after the proof of the theorem.

Proof. Before beginning the proof we collect some tools.

(3) The solutions of (1) and (2) can be presented as follows:

$$x(t) = e^{At}a + \int_0^t e^{A(t-s)}Bu(s)\,ds \text{ for } t \geq 0,$$

$$y(t) = e^{-A^T(t-T)}w\ (\neq o) \qquad \text{for } t \geq 0.$$

Here e^{At} is the non-singular $d \times d$ matrix $I + \frac{1}{1!}At + \frac{1}{2!}A^2t^2 + \cdots$. The integral is to be understood component-wise. For $t \geq 0$, consider the Hilbert space $\mathcal{L}^2 = \mathcal{L}^2([0,t], \mathbb{E}^c)$ of all vector-valued functions $u : [0,t] \to \mathbb{E}^c$ with measurable component functions such that u^2 is integrable on $[0,t]$. Besides the ordinary topology on $\mathcal{L}^2$, there is also the so-called weak topology. With respect to this topology the following statements hold.

(4) $C(t) = \{u \in \mathcal{L}^2 : u(s) \in C \text{ for } s \in [0,t]\} \subseteq \mathcal{L}^2$
is compact and, trivially, convex for $t \geq 0$.

(5) The mapping $u \to e^{At}a + \int_0^t e^{A(t-s)}Bu(s)\,ds$

is a continuous and, trivially, affine mapping of $\mathcal{L}^2$ into $\mathbb{E}^d$ for $t \geq 0$.

Next, we define and investigate the *set $R(t)$ reachable in time t*,

$$R(t) = \left\{e^{At}a + \int_0^t e^{A(t-s)}Bu(s)\,ds : u \in C(t)\right\} \subseteq \mathbb{E}^d,\ t \geq 0.$$

Note that a continuous affine image of a compact convex set is again compact and convex. Hence Propositions (4) and (5) yield the following property of $R(t)$:

(6) $R(t) \subseteq \mathbb{E}^d$ is compact and convex for $t \geq 0$.

The second property of $R(t)$, that will be used below, is a sort of weak right-hand side continuity:

(7) Let $t_1 \geq t_2 \geq \cdots \to T$ and $b \in R(t_n)$ for $n = 1, 2, \ldots$ Then $b \in R(T)$.

By definition,

$$b = e^{At_n} a + \int_0^{t_n} e^{A(t_n - s)} B u_n(s)\, ds \text{ for suitable } u_n \in C(t_n)$$

$$= \Big[e^{AT} a + \int_0^T e^{A(T-s)} B u_n(s)\, ds \Big]$$

$$+ \Big[\big(e^{At_n} - e^{AT}\big) a + \int_T^{t_n} e^{A(t_n - s)} B u_n(s)\, ds + \int_0^T \big(e^{A(t_n - s)} - e^{A(T-s)}\big) B u_n(s)\, ds \Big].$$

The quantity in the first bracket is contained in $R(T)$ and $R(T)$ is compact by (6). The quantity in the second bracket tends to o as $n \to \infty$. Hence $b \in R(T)$. The third required property of $R(t)$ is as follows:

(8) If $b \in \operatorname{int} R(T)$, then $b \in R(t)$ for all $t \leq T$ sufficiently close to T.

To see this, choose a simplex in $\operatorname{int} R(T)$ with vertices $x_1, \ldots, x_{d+1}$, say, which contains b in its interior. By the definition of $R(T)$, there are solutions $x_1(\cdot), \ldots, x_{d+1}(\cdot)$ of (1), corresponding to suitable controls in $C(T)$ which transfer a to $x_1, \ldots, x_{d+1}$ in time T. That is, $x_1(T) = x_1, \ldots, x_{d+1}(T) = x_{d+1}$. These solutions are continuous. Hence, for all $t < T$ sufficiently close to T, the points $x_1(t), \ldots, x_{d+1}(t)$ are the vertices of a simplex in the convex set $R(T)$ and this simplex still contains b. Hence $b \in R(t)$ by (6).

After these preparations, the proof of the theorem is rather easy. Let $T (\geq 0)$ be the infimum of all $t > 0$ such that $b \in R(t)$. By (7), $b \in R(T)$, concluding the proof of statement (i). Since $b \neq a$, a consequence of (i) is that $T > 0$. To show (ii), note first that $b \notin \operatorname{int} R(T)$. Otherwise $b \in R(t)$ for suitable $t < T$ by (8), in contradiction to the definition of T. Hence b is a boundary point of the convex body $R(T)$ (see (6)). Theorem 4.1 then shows that

(9) $(z - b) \cdot w \leq 0$ for $z \in R(T)$,

with a suitable $w \in \mathbb{E}^d \setminus \{o\}$. If $u \in C(T)$ is a control which transfers a to b in time T, Propositions (3) and (9) yield the following:

$$\int_0^T e^{A(T-s)} B\big(v(s) - u(s)\big)\, ds \cdot w \leq 0 \text{ for } v \in C(T),$$

i.e.

$$\int_0^T B^T e^{A^T(T-s)} w \cdot \big(u(s) - v(s)\big)\, ds \geq 0 \text{ for } v \in C(T),$$

i.e.

$$\int_0^T B^T y(s) \cdot \big(u(s) - v(s)\big)\, ds \geq 0 \text{ for } v \in C(T).$$

Since the latter inequality holds for all $v \in C(T)$, we see that for $w(s) = B^T y(s)$ $(\neq o)$,

(10) $w(s) \cdot \big(u(s) - v(s)\big) \geq 0$ for each $v \in C(T)$ and almost every $s \in [0, T]$.

Otherwise, there is a control $v \in C(T)$ such that the inner product in (10) is negative on a subset of the interval $[0, T]$ of positive measure. Clearly, for the control in $C(T)$ which coincides with this v on the subset and with u outside of it, the integral over the expression in (10) is negative. This is the required contradiction. (10) implies that

$$w(s) \cdot u(s) \geq w(s) \cdot v(s) \text{ for each } v \in C(T) \text{ and almost every } s \in [0, T],$$

i.e.

$$w(s) \cdot u(s) = \max\{w(s) \cdot v : v \in C\} \text{ for almost every } s \in [0, T],$$

i.e.

$$u(s) \in C \cap H_C\big(w(s)\big) \text{ for almost every } s \in [0, T]. \quad \square$$

A Geometric Interpretation of the Minimum Principle and Bang-Bang Controls

The above version of the minimum principle says that, for the time optimal control u, the following statement holds: for almost every t the point $u(t)$ is contained in the support set $C \cap H_C\big(w(t)\big)$ as the support hyperplane $H_C\big(w(t)\big)$ rolls continuously over C. If C is strictly convex, each support set consists of a single point. The time optimal control u then may be chosen as a continuous parametrization of a curve on $\operatorname{bd} C$. Suppose now that C is the unit cube in $\mathbb{E}^d$ and such that the support set $C \cap H_C\big(w(s)\big)$ is a vertex of C for all times $t \in [0, T]$, with a finite set of exceptions. Then there are times $0 = t_0 < t_1 < \cdots < t_n = T$, vertices $v_0, \dots, v_n$ of C and a time optimal control u such that

$$u(t) = v_k \text{ for } t \in (t_{k-1}, t_k),\ k = 1, \dots, n.$$

Then u is said to be a *bang-bang control*. Essentially this says, if each of the admissible strategies of a control problem varies independently in a given interval, time optimal controls consist of a finite sequence of pure strategies.

5 The Boundary of a Convex Body

The boundary of a convex body may be investigated from many different viewpoints.

One object of study concerns the smooth and the singular points. The size of the sets of the different types of singular points has been investigated from the measure-theoretic, the topological (Baire) and the metric (porosity) perspective. A related question concerns the first-order differentiability of the boundary. A deep result of Alexandrov deals with second-order differentiability.

The notion of extreme and the more restrictive notion of exposed point play an important role both in finite dimensions, for example, in convex analysis, and in infinite dimensions and have been studied intensively.

A further topic of research is Alexandrov's theory based on the notion of intrinsic or geodesic metric on the boundary of a convex body. This is a sort of differential geometry without differentiability assumptions. We also mention Schäffer's [882] investigations of the geometry of the unit sphere in finite-dimensional normed spaces.

In the following we investigate regular, singular and extreme points and first and second-order differentiability properties, including Alexandrov's theorem. An application deals with Birkhoff's theorem on doubly stochastic matrices.

For a more detailed discussion of smooth, singular and extreme points, see Schneider [907]. Some hints on the study of the geodesic metric will be made in Sect. 10.2. For a deeper study of smoothness and strict convexity in infinite-dimensional spaces in the context of type and cotype theory, see the books of Pisier [802] and Tomczak-Jaegermann [1001].

5.1 Smooth and Singular Boundary Points, Differentiability

In this section, we investigate the size of the set of singular boundary points of a given convex body, using Hausdorff measure, Hausdorff dimension, and Baire categories, and then give the convex body version of Alexandrov's differentiability theorem.

Smooth and Singular Boundary Points

Let C be a proper convex body and $x \in \operatorname{bd} C$. By Theorem 4.1 there is a support hyperplane of C at x. If it is unique, x is called a *smooth*, *regular* or *differentiable*, otherwise a *singular* boundary point of C. If all boundary points of C are smooth, C is said to be *smooth, differentiable*, or *regular*. The set

$$N_C(x) = \{u : u \cdot y \leq u \cdot x \text{ for all } y \in C\}$$

consists of the origin o and all exterior normal vectors of support hyperplanes of C at x. It is a closed convex cone with apex o, as will be shown in the proof of Theorem 5.1. $N_C(x)$ is called the *normal cone* of C at x. The point x is smooth if and only if the normal cone of C at x is simply a ray or, equivalently, has dimension 1.

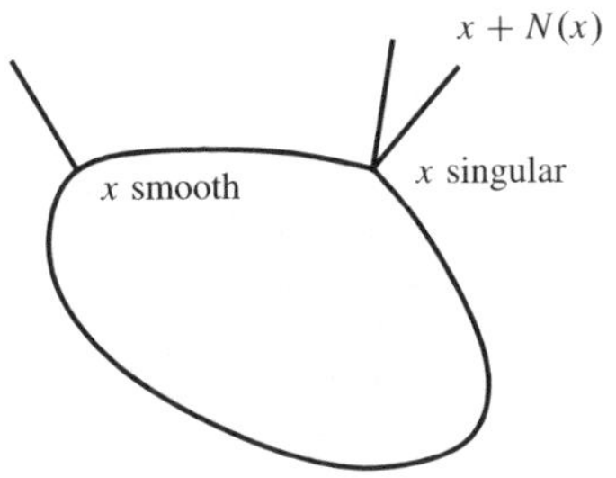

Fig. 5.1. Smooth and singular points, normal cones

Using the dimension of the normal cone, one may classify the singular boundary points (Fig. 5.1).

The natural question arises, to determine the size of the set of singular boundary points of C.

The Hausdorff Measure of the Set of Singular Points

A continuous curve of finite length λ in $\mathbb{E}^d$ can be covered by a sequence of Euclidean balls of arbitrarily small diameters such that the sum of the diameters is arbitrarily close to λ. This property was used by Carathéodory [190] to define the linear measure of more general sets. His idea was extended by Hausdorff [482] as follows: For $0 \leq s \leq d$, the *s-dimensional Hausdorff measure* $\mu_s(A)$ of a set $A \subseteq \mathbb{E}^d$ is defined by

$$\mu_s(A) = \lim_{\varepsilon \to +0} \left(\inf \left\{ \sum_{n=1}^{\infty} (\operatorname{diam} U_n)^s : U_n \subseteq \mathbb{E}^d,\ \operatorname{diam} U_n \leq \varepsilon,\ A \subseteq \bigcup_{n=1}^{\infty} U_n \right\} \right),$$

where for $U \subseteq \mathbb{E}^d$, $\operatorname{diam} U = \sup\{\|x - y\| : x, y \in U\}$ is the *diameter* of U. The above defined μ_s, actually, is not a measure but an outer measure. If A is Lebesgue measurable then, up to a constant depending on d, its Hausdorff measure $\mu_d(A)$ equals the Lebesgue measure $\mu(A)$ or $V(A)$ of A. If A is a Borel set in the boundary of a given proper convex body C, then $\mu_{d-1}(A)$ equals, up to a constant depending on d, the ordinary Lebesgue or Borel area measure $\sigma(A)$ or $S(A)$ of A. If K is a Jordan curve in $\mathbb{E}^d$, then $\mu_1(K)$ is its length. For more detailed information on measure theory, respectively, geometric measure theory see, e.g. Falconer [317] or Mattila [696].

Anderson and Klee [28] gave the following result, where a set has *σ-finite measure* if it is a countable union of sets of finite measure.

Theorem 5.1. *Let $C \in \mathcal{C}_p$. Then the set of singular points of* $\operatorname{bd} C$ *has σ-finite $(d-2)$-dimensional Hausdorff measure.*

Proof. In a first step we show the following.

(1) Let $x \in \operatorname{bd} C$. Then $N_C(x)$ is a closed convex cone with apex o.

To see that $N_C(x)$ is a convex cone with apex o, let $u, v \in N_C(x)$ and $\lambda, \mu \geq 0$. Then $u \cdot y \leq u \cdot x$ and $v \cdot y \leq v \cdot x$ for all $y \in C$. Thus $(\lambda u + \mu v) \cdot y \leq (\lambda u + \mu v) \cdot x$

for all $y \in C$ and therefore $\lambda u + \mu v \in N_C(x)$. To see that $N_C(x)$ is closed, let $u_1, u_2, \dots \in N_C(x)$, where $u_n \to u \in \mathbb{E}^d$, say. Then $u_n \cdot y \leq u_n \cdot x$ for all $y \in C$. Since $u_n \to u$, we obtain $u \cdot y \leq u \cdot x$ for all $y \in C$. Hence $u \in N_C(x)$, concluding the proof of (1).

Note that the following statement holds:

(2) Let $x \in \operatorname{bd} C$. Then x is singular if and only if $\dim N_C(x) \geq 2$.

Next, let $\mathcal{S}$ be the countable set of all simplices in $\mathbb{E}^d \backslash C$ of dimension $d-2$ with rational vertices and let $p_C : \mathbb{E}^d \to C$ denote the metric projection of $\mathbb{E}^d$ onto C, see Sect. 4.1. For the proof that

(3) $\{x \in \operatorname{bd} C : x \text{ singular}\} \subseteq \bigcup \{p_C(S) : S \in \mathcal{S}\}$,

let $x \in \operatorname{bd} C$ be singular. Then $\dim N_C(x) \geq 2$ by (2). Thus we may choose a simplex $S \in \mathcal{S}$ with $N_C(x) \cap S \neq \emptyset$. Then $x \in p_C(S)$, concluding the proof of (3).

Since by Lemma 4.1 the metric projection p_C is non-expansive, the simple property of the Hausdorff measure that non-expansive mappings do not increase the Hausdorff measure, implies that

(4) $\mu_{d-2}\big(p_C(S)\big) \leq \mu_{d-2}(S) < +\infty$ for each $S \in \mathcal{S}$.

Having proved (3) and (4), the theorem follows. □

Remark. Anderson and Klee actually proved a more precise result. Refinements of the latter are due to Zajíček [1036], Colesanti and Pucci [213] and Hug [527].

The Hausdorff Dimension of the Set of Singular Points

Given a set $A \subseteq \mathbb{E}^d$, its *Hausdorff dimension* $\dim_H A$ is defined by

$$\dim_H A = \sup\{s \geq 0 : \mu_s(A) = \infty\} = \inf\{s \geq 0 : \mu_s(A) = 0\},$$

see [317] or [696]. The theorem of Anderson and Klee then yields the following result.

Corollary 5.1. *Let $C \in \mathcal{C}_p$. Then the set of singular points of* bd C *has Hausdorff dimension at most $d-2$.*

The Baire Category of the Set of Singular Points

A *Baire space* is a topological space in which any meagre set has dense complement, where a set is *meagre* or *of first Baire category* if it is a countable union of subsets which are nowhere dense in the space. A set is *nowhere dense* if its closure has empty interior. Since a meagre subset can never exhaust a Baire space, meagre sets in a Baire space are considered to be small while the non-meagre sets are large. Particular non-meagre sets are the complements of meagre sets. When speaking of *most* or of *typical* elements of a Baire space, we mean all elements, with a meagre set of exceptions. To apply this topological instrument to distinguish between small and large subsets of a space, we need to know which spaces are Baire. Simple criteria are provided by versions of the *Baire category theorem*. We cite the following one:

each complete metric or locally compact space is Baire. The rational numbers with the usual topology are an example of a space which is *not* Baire.

Ever since their introduction by Baire [45] and Osgood [780], Baire categories have been applied successfully in real analysis. See, e.g. Oxtoby [782] and Holmes [520]. In recent years they found numerous applications in convex geometry, compare the surveys of Zamfirescu [1041] and Gruber [431], and Sect. 13.1.

Given a proper convex body, its boundary is compact and thus Baire by the category theorem. The finite-dimensional case of the *density theorem of Mazur* [701] is as follows.

Theorem 5.2. *Let $C \in \mathcal{C}_p$. Then the set of singular points of* $\operatorname{bd} C$ *is meagre in* $\operatorname{bd} C$.

Proof. For $n = 1, 2, \dots$, let

$$S_n = \Big\{x \in \operatorname{bd} C : \text{ there are support hyperplanes at } x \text{ with angle at least } \frac{1}{n}\Big\}.$$

A simple compactness argument shows that S_n is closed in $\operatorname{bd} C$. In order to show that

(5) S_n is nowhere dense in $\operatorname{bd} C$,

assume the contrary. Being closed, S_n then contains a relatively open subset G of $\operatorname{bd} C$. If B is a solid Euclidean ball of sufficiently small radius, there is a translate of B contained in C which touches $\operatorname{bd} C$ at a point $x \in G$. Then, clearly, C has a unique support hyperplane at x. Since this is in contradiction to $x \in G \subseteq S_n$, the proof of (5) is complete. Clearly,

(6) $\displaystyle\bigcup_{n=1}^{\infty} S_n = \big\{x \in \operatorname{bd} C : x \text{ singular}\big\}.$

The theorem now follows from (5) and (6). □

Differentiability

A $(d-1)$-dimensional manifold in $\mathbb{E}^d$ may be represented explicitly, implicitly or by means of parameters. The problem arises to move differentiability properties or results, from one representation to the others. In differential geometry this is done by means of refined versions of the theorems on implicit and inverse functions, using differentiability assumptions. Unfortunately, in general, such possibilities are not available in convex geometry. Thus one has to argue more carefully, even in simple cases. In the following we consider the problem of transferring first and second-order differentiability at given points from convex functions to convex bodies and vice versa.

First-Order Differentiability

Let D be an open convex set in $\mathbb{E}^d$ and $f : D \to \mathbb{R}$ a convex function. Clearly, notions such as support hyperplanes and regular and singular boundary points and the existence of support hyperplanes can be extended to the case of unbounded proper convex bodies such as the epigraph epi f of f.

The following simple result, essentially a version of Theorem 2.7, shows how to move first-order differentiability results from convex functions to convex sets and vice versa.

Lemma 5.1. *Let $D \subseteq \mathbb{E}^d$ be open and convex and $f : D \to \mathbb{R}$ convex. Then, given $x \in D$, the following are equivalent:*

(i) *f is differentiable at x.*
(ii) *$\big(x, f(x)\big)$ is a smooth boundary point of* epi f.

Proof. Left to the reader (use Theorem 2.7). □

Theorem 5.3. *Let $D \subseteq \mathbb{E}^d$ be open and convex and $f : D \to \mathbb{R}$ convex. Then f is differentiable at each point of D with a set of exceptions which is of σ-finite $(d-1)$-dimensional Hausdorff measure and meagre in D.*

Proof. Using the same proofs, Theorems 5.1 and 5.2 easily extend to the graph of f which is contained in the boundary of epi f in $\mathbb{E}^{d+1}$. Now apply Lemma 5.1 and note that subsets of bd epi f which are of σ-finite $(d-1)$-dimensional Hausdorff measure or are meagre project into such sets in D. □

This result is the analogue, for convex functions, of the Theorems 5.1 and 5.2 of Anderson and Klee and Mazur, respectively. For $d = 1$ the measure part says that f has at most countably many points of non-differentiability, which was also proved in Theorem 1.4. For general d, it is an essential refinement of Reidemeister's Theorem 2.6.

Second-Order Differentiability

When saying that the proper convex body C is *twice* or *second-order differentiable* at a point $x \in \operatorname{bd} C$, the following is meant: First, x is a smooth point of bd C. Next, let H be the unique support hyperplane of C at x and u the interior unit normal vector of C at x. Choose, in H, a Cartesian coordinate system with origin o at x. Together with u it yields a Cartesian coordinate system in $\mathbb{E}^d$. In this coordinate system, represent the lower side of bd C with respect to the last coordinate in the form

$$\big(y, g(y)\big) = y + g(y)u \text{ for } y \in \operatorname{relint} C'.$$

Here "$'$" is the orthogonal projection of $\mathbb{E}^d$ onto H and $g : \operatorname{relint} C' \to \mathbb{R}$ a convex function. We then require that, second, there is a positive semi-definite quadratic form r on H such that

$$g(y) = r(y) + o(\|y\|^2) \text{ as } y \to o.$$

(Note that x is the origin.) Clearly, this definition may be extended to unbounded proper convex bodies.

The following result makes it possible to transfer second-order differentiability results from convex functions to convex bodies and vice versa.

Lemma 5.2. *Let $D \subseteq \mathbb{E}^d$ be open and convex and $f : D \to \mathbb{R}$ convex. Then, given $x \in D$, the following statements are equivalent:*

(i) *f is twice differentiable at x in the sense of Theorem* 2.9,
(ii) epi *f is twice differentiable at $(x, f(x))$.*

Proof. (i)⇒(ii) We may suppose that $x = o$, $f(o) = 0$. By assumption, there are a vector $a \in \mathbb{E}^d$ and a positive semi-definite quadratic form $q(z) = z^T Az$ for $z \in \mathbb{E}^d$, such that

(7) $f(z) = a \cdot z + q(z) + o(\|z\|^2)$ as $z \to o\,(= x)$, $z \in \mathbb{E}^d$.

Let v be the unit normal vector of $\mathbb{E}^d$ pointing into int epi f. Let "$'$" denote the orthogonal projection of $\mathbb{E}^{d+1} = \mathbb{E}^d \times \mathbb{R}$ onto $\mathbb{E}^d$. Since, in particular, f is differentiable at o, Lemma 5.1 shows that o is a smooth point of bd epi f. Let H be the support hyperplane of epi f at o. Denote by u its unit normal vector pointing into int epi f. Let "$''$" denote the orthogonal projection of $\mathbb{E}^{d+1}$ onto H. Represent the lower side of bd epi f, with respect to the last coordinate of the Cartesian coordinate system determined by H, and u in the form

$$\big(y, g(y)\big) = y + g(y)u \text{ for } y \in (\operatorname{relint} \operatorname{epi} f)'' \subseteq H.$$

Since o is a smooth point of bd epi f, Lemma 5.1 shows that the convex function g is differentiable at o and thus (Fig. 5.2)

(8) $g(y) = o(\|y\|)$ as $y \to o$, $y \in H$.

Combining (7), $z = (y + g(y)u)' = y' + g(y)u'$ and (8), we see that

$$\begin{aligned} g(y) &= \big(f(z) - a \cdot z\big)(u \cdot v) = \big(q(z) + o(\|z\|^2)\big)(u \cdot v) \\ &= \big(q(y') + 2y'^T Au' g(y) + q(u')g(y)^2 + o(\|y\|^2)\big)(u \cdot v) \\ &= r(y) + o(\|y\|^2) \text{ as } y \to o, \end{aligned}$$

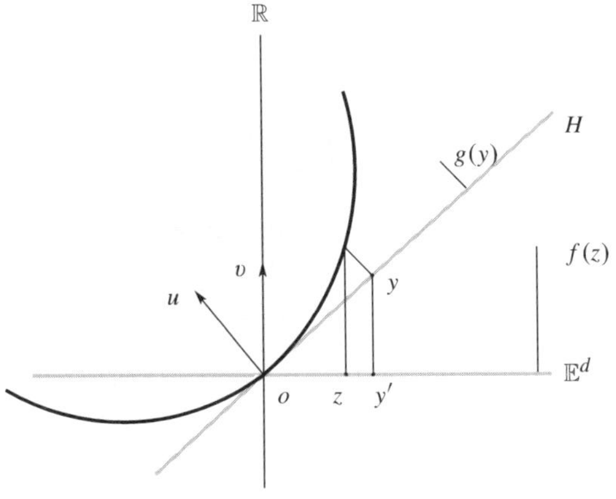

Fig. 5.2. Twice differentiability of functions and epigraphs

where $r(\cdot)$ is a suitable positive semi-definite quadratic form on H.

(ii)⇒(i) Let $x = o$, $f(o) = 0$, $u, v, H, r, {}''$, g be as before. By assumption,

(9) $g(y) = r(y) + o(\|y\|^2)$ as $y \to o$, $y \in H$,

where $r(y) = y^T B y$, for $y \in H$, is a positive semi-definite quadratic form on H. Lemma 5.1 shows that f is differentiable at o, that is

(10) $f(z) = a \cdot z + o(\|z\|)$ as $z \to o$, $z \in D$,

where a is a suitable vector in $\mathbb{E}^d$. Note that

(11) $y = \big(z + f(z)v\big)'' = z'' + f(z)v''$.

Propositions (9)–(11) yield the following:

$$\begin{aligned} f(z) &= \big(y + g(y)u\big) \cdot v = y \cdot v + g(y)\, u \cdot v \\ &= z'' \cdot v + f(z)\, v'' \cdot v + g\big(z'' + f(z)v''\big)\, u \cdot v, \end{aligned}$$

and then

$$\begin{aligned} f(z)(1 - v'' \cdot v) &= z'' \cdot v + r\big(z'' + f(z)v''\big)\, u \cdot v + o\big(\|z'' + f(z)v''\|^2\big)\, u \cdot v \\ &= z'' \cdot v + \big(z''^T B z'' + z''^T B v''\, 2 f(z) + v''^T B v'' f(z)^2\big) u \cdot v + o(\|z\|^2) \\ &= z'' \cdot v + \big(z''^T B z'' + z''^T B v''\, 2\, a \cdot z + v''^T B v'' (a \cdot z)^2\big) u \cdot v + o(\|z\|^2) \\ &= b \cdot z + q(z) + o(\|z\|^2) \text{ as } z \to o,\ z \in D. \end{aligned}$$

Here b and q are a suitable vector in $\mathbb{E}^d$ and a quadratic form on $\mathbb{E}^d$. Since f is convex, this can hold only if q is positive semi-definite. □

Using Lemma 5.2 and Alexandrov's Theorem 2.9 for convex functions, we obtain

Alexandrov's theorem on second-order differentiability of convex bodies:

Theorem 5.4. *Let $C \in \mathcal{C}_p$. Then* bd C *is almost everywhere twice differentiable.*

Proof. bd C can be covered by finitely many relative interiors of its lower sides with respect to suitable Cartesian coordinate systems. To the corresponding convex functions, apply Alexandrov's theorem 2.9 and Lemma 5.2, noting that a set on the relative interior of a lower side of bd C has $(d-1)$-dimensional Hausdorff measure 0 if its orthogonal projection into a hyperplane has $(d-1)$-dimensional Hausdorff measure 0. (A convex function on an open set is locally Lipschitz.) □

5.2 Extreme Points

Extreme points of convex bodies play an important role in convex analysis, convex geometry and functional analysis, for example in the context of the Krein–Milman theorem and Choquet theory. A refinement of the notion of extreme point is that of exposed point. Using this notion, Straszewicz [973] proved a result of Krein–Milman type.

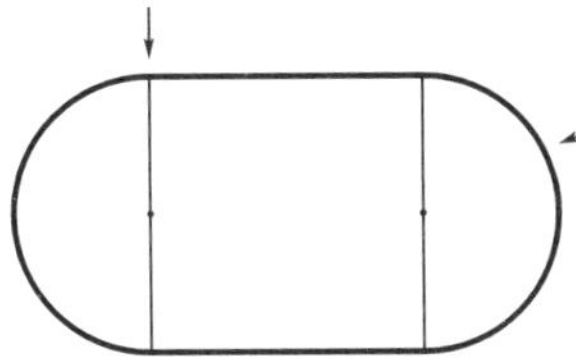

Fig. 5.3. Stadium with extreme and exposed points

In the sequel we prove a finite-dimensional version of the Krein–Milman theorem due to Minkowski and give a simple application dealing with maxima of convex functions.

For more results on extreme points and extreme sets, see Rockafellar [843], Stoer and Witzgall [970] and Hiriart-Urruty and Lemaréchal [505] in the finite-dimensional case and Holmes [520] in the context of functional analysis.

Extreme Points

A point of a convex body C is *extreme* if it is not a relative interior point of a line segment in C (see Fig. 5.3). Let $\operatorname{ext} C$ denote the set of all extreme points of C. Examples of extreme points are the vertices of a convex polytope and the boundary points of a solid Euclidean ball.

A Finite-Dimensional Krein–Milman Type Theorem

Minkowski [744], Sect. 12, proved the following finite-dimensional forerunner of the Krein–Milman theorem.

Theorem 5.5. *Let $C \in \mathcal{C}$. Then $\operatorname{ext} C$ is the smallest subset of C (with respect to inclusion) with convex hull C.*

The infinite-dimensional theorem of Krein–Milman is slightly weaker: $\operatorname{conv} \operatorname{ext} C$ may be a proper subset of C, but $\operatorname{cl} \operatorname{conv} \operatorname{ext} C$ always equals C. See, e.g. [861].

Proof. In a first step we will show, by induction on $n = \dim D$, that

(1) $D = \operatorname{conv} \operatorname{ext} D$ for each convex body D.

For $n = 0, 1$, this is trivial. Assume now that $n > 1$ and that (1) holds for convex bodies of dimensions $0, 1, \dots, n-1$. Let D be a convex body of dimension n. Since the definition of extreme points and the convex hull operation are independent of the dimension of the ambient space, we may assume that D is a proper convex body in $\mathbb{E}^n$. We have to show that, if $x \in D$, then $x \in \operatorname{conv} \operatorname{ext} D$. If $x \in \operatorname{bd} D$, then $x \in D \cap H$, where H is a support hyperplane of D at x, see Theorem 4.1. Clearly, $D \cap H$ is a convex body of dimension less than n. Thus $x \in \operatorname{conv} \operatorname{ext}(D \cap H)$ by the induction hypothesis. Since H is a support hyperplane of D, we have $\operatorname{ext}(D \cap H) \subseteq \operatorname{ext} D$.

Hence $x \in \operatorname{conv} \operatorname{ext} D$. If $x \in \operatorname{int} D$, choose $y, z \in \operatorname{bd} D$ such that $x \in [y, z]$. By the case just considered, $y, z \in \operatorname{conv} \operatorname{ext} D$ and thus $x \in \operatorname{conv} \operatorname{ext} D$. This concludes the proof of (1).

The second step is to show the following:

(2) Let $x \in \operatorname{ext} C$. Then $C \backslash \{x\}$ is convex.

If (2) did not hold, there are points $y, z \in C \backslash \{x\}$ such that $[y, z] \nsubseteq C \backslash \{x\}$. Since $[y, z] \subseteq C$ by the convexity of C, this can hold only if $x \in [y, z]$. Since $x \neq y, z$, the point x is not extreme, a contradiction. The proof of (2) is complete.

Finally, (1) and (2) yield the theorem. □

Remark. The following refinement of Theorem 5.5 is due to Straszewicz [973]: $C = \operatorname{cl} \operatorname{conv} \exp C$, where $\exp C$ is the set of all *exposed points* of C. These are the points $x \in C$ such that $\{x\} = C \cap H$ for a suitable support hyperplane of C at x. Clearly, each exposed point is extreme, but the converse does not hold generally. It is not difficult to see that the apex of a pointed closed convex cone C is an exposed and thus an extreme point of C.

Maxima of Convex Functions

A simple yet useful application of the above result is as follows:

Theorem 5.6. *Let $C \in \mathcal{C}$ and let $f : C \to \mathbb{R}$ be a continuous convex function. Then f attains its maximum m at an extreme point of C.*

Proof. Let $x \in C$ be such that $f(x) = m$. Theorem 5.5 and Lemma 3.1 show that $x = \lambda_1 x_1 + \cdots + \lambda_n x_n$ with suitable $x_1, \ldots, x_n \in \operatorname{ext} C$ and $\lambda_1, \ldots, \lambda_n > 0$, where $\lambda_1 + \cdots + \lambda_n = 1$. Jensen's inequality then yields the following:

$$m = f(x) \le \lambda_1 f(x_1) + \cdots + \lambda_n f(x_n) \le (\lambda_1 + \cdots + \lambda_n) m = m.$$

Thus equality holds throughout. Noting that $\lambda_1, \ldots, \lambda_n > 0$, it then follows that $f(x_1) = \cdots = f(x_n) = m$. □

5.3 Birkhoff's Theorem on Doubly Stochastic Matrices

Doubly stochastic $d \times d$ matrices have attracted a lot of interest, for example as the matrices of transition probabilities of discrete Markov chains or in the context of van der Waerden's conjecture on permanents of doubly stochastic matrices.

Using the notions of convex hull and extreme point, we give a precise description of the set of doubly stochastic $d \times d$ matrices due to Birkhoff.

For more information on van der Waerden's conjecture, see Egorychev [291]. For general information on doubly stochastic matrices compare Minc [729, 730] and Seneta [926].

Doubly Stochastic and Permutation Matrices

A real $d \times d$ matrix is *doubly stochastic* if all its entries are non-negative and the sum of the entries in each row and column equals 1. The set of all doubly stochastic $d \times d$ matrices is usually denoted Ω_d. It clearly may be considered a subset of $\mathbb{E}^{d^2}$. Particular cases include the $d \times d$ *permutation matrices*. These are $d \times d$ matrices such that, in each row and column, all entries are 0 with one exception which is 1. There are $d!$ $d \times d$ permutation matrices.

Birkhoff's Theorem

Because of the importance of doubly stochastic matrices, it is a natural problem to describe the set Ω_d of all doubly stochastic $d \times d$ matrices as a subset of $\mathbb{E}^{d^2}$. Birkhoff [119] gave the following satisfying result:

Theorem 5.7. *The set Ω_d of all doubly stochastic $d \times d$ matrices is the convex hull of the $d \times d$ permutation matrices and each permutation matrix is an extreme point of Ω_d.*

In other words, Ω_d is a convex polytope in $\mathbb{E}^{d^2}$, the vertices of which are precisely the $d \times d$ permutation matrices. There exist many proofs of Birkhoff's theorem. For a proof which makes use of tools from linear programming see, e.g. [841]. A simple geometric proof using properties of convex polytopes, will be given in Sect. 14.2. In the following we present a proof based on the *Frobenius–König theorem*, where by a *diagonal* of a $d \times d$ matrix we mean a sequence of d of its elements, one from each row and column. A closely related result is Hall's marriage theorem.

Lemma 5.3. *Let M be a real $d \times d$ matrix. Then the following are equivalent:*

(i) *Each diagonal of M contains* 0.
(ii) *M has a $p \times q$ zero submatrix with $p + q = d + 1$.*

Proof. For subsequences σ, τ of the sequence $(1, \dots, d)$ let $M_{\sigma\tau}$, $M_{\sigma\tau^c}$, $M_{\sigma^c\tau}$, and $M_{\sigma^c\tau^c}$ denote the matrices obtained from M by cancelling all rows and all columns with indices not in σ and not in τ, not in σ and in τ, in σ and not in τ, and in σ and in τ, respectively.

$\neg$(i)$\Rightarrow$ $\neg$(ii) Assume that there is a diagonal in M without 0. Let $M_{\sigma\tau}$ be a $p \times q$ zero submatrix of M. Then, in the columns with indices in τ, the entries of our diagonal must lie in $M_{\sigma^c\tau}$. Hence $q \le d - p$ or $p + q \le d$.

(i)$\Rightarrow$(ii) Here the proof is by induction on d. For $d = 1$ this implication clearly holds. Assume now that $d > 1$ and that the implication holds for $d - 1$. Let M be a real $d \times d$ matrix such that every diagonal contains 0. If M is the zero matrix, there is nothing to prove. Otherwise M has a non-zero entry, say m_{ij}. By assumption on M, every diagonal of $M_{i^c j^c}$ contains 0. The induction hypothesis then shows that $M_{i^c j^c}$ has an $r \times s$ zero submatrix with $r + s = d$. Thus M also has an $r \times (d - r)$ zero submatrix. To simplify, permute the rows and columns of M so that the resulting matrix N has the form

$$N = \begin{pmatrix} X, & O \\ Y, & Z \end{pmatrix}$$

where X is an $r \times r$ matrix and Z a $(d-r) \times (d-r)$ matrix. By assumption on M, every diagonal of N contains 0. Thus, if there is a diagonal of X which does not contain 0, then every diagonal of Z contains 0. This shows that all diagonals of X contain 0 or all diagonals of Z contain 0. We may suppose that the former is the case. (If the latter is the case, the proof is almost identical.) By the induction hypothesis, X has a $t \times u$ zero submatrix with $t + u = r + 1$. Then N, and therefore also M, has a $t \times (u + d - r)$ zero submatrix. Since $t + u + d - r = r + 1 + d - r = d + 1$, the induction is complete. □

Proof of the Theorem. Since a convex combination of doubly stochastic $d \times d$ matrices is also doubly stochastic, we see that

(1) Ω_d is convex.

The main step is to prove that

(2) Ω_d is the convex hull of the $d \times d$ permutation matrices.

Each doubly stochastic $d \times d$ matrix has n positive entries, where $d \le n \le d^2$. For the proof of (2) it is thus sufficient to show the following proposition by induction on n:

(3) Let $d \le n \le d^2$. Then each doubly stochastic $d \times d$ matrix with n positive entries is a convex combination of $d \times d$ permutation matrices.

Any doubly stochastic $d \times d$ matrix M with precisely d positive entries is a permutation matrix. Thus (3) holds for $n = d$. Assume now that $n > d$ and (3) holds for $n - 1$. Let M be a doubly stochastic $d \times d$ matrix with n positive entries. We show that M has a diagonal with positive entries. For assume not, that is, each diagonal of M contains 0. Lemma 5.3 then shows that M has a $p \times q$ zero submatrix with $p + q = d + 1$, say $M_{\sigma\tau}$. Since the non-zero entries of M in the rows with indices in σ are in $M_{\sigma\tau^c}$, the sum of the entries in each row of $M_{\sigma\tau^c}$ is 1. Therefore the sum of all entries of $M_{\sigma\tau^c}$ is p. Similarly, the sum of all entries of $M_{\sigma^c\tau}$ is q. Since $M_{\sigma\tau^c}$ and $M_{\sigma^c\tau}$ are disjoint submatrices of M, the sum of all entries of M is at least $p + q = d + 1$. This contradiction shows that M has a diagonal with positive entries. Since, by assumption, M has $n > d$ positive entries and the sum of all these n entries is d, among the d entries of the diagonal, there must be one entry less than 1. Let $0 < \mu < 1$ be the minimum of the entries of the diagonal and let P be the $d \times d$ permutation matrix with 1s precisely at the positions of the entries of the above diagonal. The $d \times d$ matrix $N = (1 - \mu)^{-1}(M - \mu P)$ is then doubly stochastic and has fewer than n positive entries. By the induction hypothesis, N is a convex combination of $d \times d$ permutation matrices. Since P is also a $d \times d$ permutation matrix, we finally also see that M is a convex combination of $d \times d$ permutation matrices. The induction, and thus the proof of (3), is complete. (3) implies (2).

Clearly,

(4) A $d \times d$ permutation matrix is not a convex combination of other $d \times d$ doubly stochastic matrices.

The theorem now is a consequence of Propositions (1), (2) and (4). □

6 Mixed Volumes and Quermassintegrals

Schneider [907] describes the Brunn–Minkowski theory in the preface of his monograph with the following words:

> Aiming at a brief characterization of the Brunn–Minkowski theory, one might say that it is the result of merging two elementary notions for point sets in Euclidean space: vector addition and volume.

The original problem of the Brunn–Minkowski theory is to obtain information on the volume

$$V(\lambda C + \mu D)$$

of the Minkowski linear combination $\lambda C + \mu D = \{\lambda x + \mu y : x \in C, y \in D\}$ of two convex bodies C, D for $\lambda, \mu \geq 0$, in terms of information on C and D. Prior to Minkowski, the only results of what is now the Brunn–Minkowski theory were Cauchy's [198] surface area formula, Steiner's [959] formula for the volume of parallel bodies of a convex body, the proof of the isoperimetric inequality by Schwarz [922] and the inequality of Brunn [173, 174] (–Minkowski). These results then were considered as interesting, but rather isolated contributions to geometry. Their fundamental importance became visible only after Minkowski [744] built, around them, a voluminous theory, now called after Brunn and himself. A central notion is that of mixed volumes. Important later contributors to this theory were Hadwiger [468] and Alexandrov [18]. Of contemporary mathematicians we mention Schneider [907]. The Brunn–Minkowski theory deals, amongst others, with mixed volumes, the corresponding measures, with developments around the Brunn–Minkowski inequality, geometric inequalities and other topics of an analytic flavour in convex geometry. Its numerous applications and relations to other areas range from isoperimetric problems of various sorts, including isoperimetric inequalities of mathematical physics, to crystallography, statistics and algebraic geometry.

In this section, we first give basic notions and preliminary results dealing with Minkowski addition and the Hausdorff metric. Then Blaschke's selection theorem is presented. Next, we consider Minkowski's theorem on mixed volumes and Steiner's formula for the volume of parallel bodies and study properties of mixed volumes and quermassintegrals, including Minkowski's inequalities, Cauchy's surface area formula and Kubota's formulae for quermassintegrals.

The notions and properties of volume and convex polytopes will be used several times in this section but will be treated in detail only in Sect. 7 and in the chapter "Convex Polytopes".

Further topics of the Brunn–Minkowski theory will be treated in Sects. 7–9. Major treatises and surveys on the Brunn–Minkowski theory were cited in the introduction of the present chapter, the most comprehensive reference is Schneider's monograph [907]. We also mention Sangwine-Yager [878] and Peri [790].

6.1 Minkowski Addition, Direct Sums, Hausdorff Metric, and Blaschke's Selection Theorem

Important ingredients of the Brunn–Minkowski theory are Minkowski addition and the Hausdorff metric. The natural topology on the space $\mathcal{C} = \mathcal{C}(\mathbb{E}^d)$ of all convex bodies in $\mathbb{E}^d$ is induced by the Hausdorff metric. The Blaschke selection theorem then says that this space is complete and locally compact.

In this section, these notions are introduced and Blaschke's selection theorem is proved. In addition, we present a result of the author which says that each proper convex body C can be represented as a direct sum of directly irreducible convex bodies $C_1, \dots, C_m$:

$$C = C_1 \oplus \cdots \oplus C_m,$$

where this representation is unique up to the order of summands.

For $\mathcal{C}$ and subspaces of it, various other metrics and notions of distance have been considered. See, e.g. [428] and the references cited there.

Minkowski Addition

There are several natural ways to define, on $\mathcal{C}$, geometrically interesting operations of addition and multiplication with (non-negative or general) real numbers. *Minkowski addition* and (ordinary) *multiplication with real numbers* are defined as follows:

$$C + D = \{x + y : x \in C, y \in D\} \text{ for } C, D \in \mathcal{C},$$
$$\lambda C = \{\lambda x : x \in C\} \text{ for } C \in \mathcal{C}, \lambda \in \mathbb{R}.$$

The following representation of $C + D$ gives a better idea of what $C + D$ really means. For $x \in \mathbb{E}^d$ the set $x + D\, (= \{x + y : y \in D\})$ is the *translate* of D by the vector x, where instead of $\{x\}$ we simply write x. Then

$$C + D = \bigcup_{x \in C} (x + D).$$

$C - D$ stands for $C + (-1)D$.

A Short Historical Excursion

Aristotle [37] asked the following question:

> Why does the sun, when it shines through a square, not produce rectangular forms but circles as is the case when it shines through wicker work?

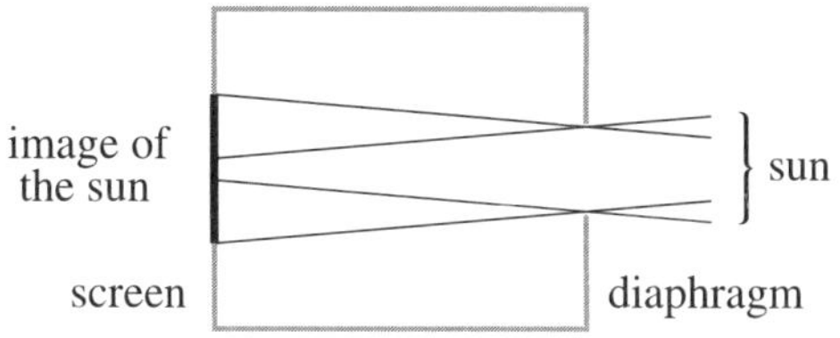

Fig. 6.1. Camera obscura

A similar question due to Tycho Brahe is about the form of the image of the sun on the screen of a camera obscura, depending on the shape of the diaphragm (Fig. 6.1). Implicitly using Minkowski addition, Kepler [575] solved Brahe's problem by showing – in our terminology – the following: the image of the sun is of the form $D + \lambda B^2$, where D is a convex disc, a translate of the diaphragm. See Scriba and Schreiber [923]. A similar phenomenon can be observed in sunshine at noon under a broad-leaved tree: the image of the sun on the ground consists of numerous rather round figures of the form $P + \lambda E$, where P is approximately of polygonal shape, not necessarily convex, and E an ellipse. See Schlichting and Ucke [890].

Properties of Minkowski Addition

The following simple properties of Minkowski addition will be used frequently.

Proposition 6.1. *Let $C, D \in \mathcal{C}$ and $\lambda \in \mathbb{R}$. Then $C + D, \lambda C \in \mathcal{C}$.*

Proof. We consider only $C + D$. To show the convexity of $C + D$, let $u + x, v + y \in C + D$ where $u, v \in C$, $x, y \in D$, and let $0 \le \lambda \le 1$. Then

$$(1 - \lambda)(u + x) + \lambda(v + y) = \big((1 - \lambda)u + \lambda v\big) + \big((1 - \lambda)x + \lambda y\big) \in C + D$$

by the convexity of C and D. Hence $C + D$ is convex. It remains to show that $C + D$ is compact. The Cartesian product $C \times D = \{(x, y) : x \in C, y \in D\} \subseteq \mathbb{E}^d \times \mathbb{E}^d = \mathbb{E}^{2d}$ is compact in $\mathbb{E}^{2d}$. Being the image of the compact set $C \times D$ under the continuous mapping $(x, y) \to x + y$ of $\mathbb{E}^{2d}$ onto $\mathbb{E}^d$, the set $C + D$ is also compact. □

The following result relates addition and multiplication with positive numbers of convex bodies to addition and multiplication with positive numbers of the corresponding support functions.

Proposition 6.2. *Let $C, D \in \mathcal{C}$ and $\lambda \ge 0$. Then*

$$h_{C+D} = h_C + h_D,\ h_{\lambda C} = \lambda h_C.$$

Proof. Again, we consider only $C + D$:

$$\begin{aligned} h_{C+D}(u) &= \sup\{u \cdot (x + y) : x \in C, y \in D\} \\ &= \sup\{u \cdot x : x \in C\} + \sup\{u \cdot y : y \in D\} \\ &= h_C(u) + h_D(u) \text{ for } u \in \mathbb{E}^d. \ \square \end{aligned}$$

We are now ready to state some simple algebraic properties of the space $\mathcal{C}$ of convex bodies in $\mathbb{E}^d$, endowed with the operations of Minkowski addition and multiplication with positive numbers.

Theorem 6.1. *The following claims hold:*

(i) *$\mathcal{C}$, endowed with Minkowski addition, is a commutative semigroup with cancellation law.*

(ii) *$\mathcal{C}$, endowed with Minkowski addition, and multiplication with non-negative numbers is a* convex cone, *i.e.* $C + D$, $\lambda C \in \mathcal{C}$ *for* $C, D \in \mathcal{C}, \lambda \geq 0$.

Proof. (i) Proposition 6.1 and the simple fact that Minkowski addition is associative and commutative, settle the first part of statement (i). For the proof that the cancellation law holds, let $B, C, D \in \mathcal{C}$ such that $B + D = C + D$. Then $h_B + h_D = h_C + h_D$ by Proposition 6.2. Therefore, $h_B = h_C$ which, in turn, implies that $B = C$ by Proposition (4) before Theorem 4.3.

(ii) This is simply a re-statement of Proposition 6.1. □

Direct Sum Decomposition of Convex Bodies

A convex body C is the *direct sum* of the convex bodies $C_1, \dots, C_m$,

$$C = C_1 \oplus \cdots \oplus C_m,$$

if

$$C = C_1 + \cdots + C_m \text{ and } \operatorname{lin} C_1 \oplus \cdots \oplus \operatorname{lin} C_m \text{ exists.}$$

By lin we mean the linear hull. The convex body C is *directly irreducible* if, in any direct decomposition of C, at most one summand is different from $\{o\}$.

Our aim is to prove the following result of the author [412], II. For alternative proofs and generalizations, see [412], Gale and Klee [352] and Kincses [584] and Schneider [907]. The following proof is essentially that of Kincses.

Theorem 6.2. *Let* $C \in \mathcal{C}_p$. *Then there are directly irreducible convex bodies* $C_1, \dots, C_m \in \mathcal{C}$, *such that*

$$C = C_1 \oplus \cdots \oplus C_m.$$

This decomposition is unique up to the order of the summands.

For the proof we need the following tool. For later reference we prove it in a slightly more general form than is needed below. $H_C(u)$ is the support hyperplane of C with exterior normal vector u.

Lemma 6.1. *Let* $C_1, \dots, C_m \in \mathcal{C}$, $\lambda_1, \dots, \lambda_m \geq 0$, *and* $u \in S^{d-1}$. *Then, for* $C = \lambda_1 C_1 + \cdots + \lambda_m C_m$, *the following holds:*

$$C \cap H_C(u) = \lambda_1 \big(C_1 \cap H_{C_1}(u)\big) + \cdots + \lambda_m \big(C_m \cap H_{C_m}(u)\big).$$

Proof. First, let $x \in C \cap H_C(u)$. Then $x = \lambda_1 x_1 + \cdots + \lambda_m x_m$ with suitable $x_i \in C_i$. Clearly, $u \cdot x_i \le h_{C_i}(u)$ for each i. In case $\lambda_i = 0$ we are free to choose $x_i \in C_i \cap H_{C_i}(u)$. It remains to show that, in case $\lambda_i > 0$, we also have $x_i \in C_i \cap H_{C_i}(u)$. If this did not hold, then there is a $\lambda_i > 0$ where $u \cdot x_i < h_{C_i}(u)$. Then,

$$\begin{aligned} h_C(u) &= u \cdot x = \lambda_1 u \cdot x_1 + \cdots + \lambda_m u \cdot x_m < \lambda_1 h_{C_1}(u) + \cdots + \lambda_m h_{C_m}(u) \\ &= h_{\lambda_1 C_1 + \cdots + \lambda_m C_m}(u) = h_C(u) \end{aligned}$$

by Proposition 6.2, which is a contradiction. Thus $x_i \in C_i \cap H_{C_i}(u)$ for $i = 1, \ldots, m$.

Second, let $x_i \in C_i \cap H_{C_i}(u)$ for $i = 1, \ldots, m$. Then

$$\begin{aligned} x &= \lambda_1 x_1 + \cdots + \lambda_m x_m \in \lambda_1 C_1 + \cdots + \lambda_m C_m = C, \\ u \cdot x &= \lambda_1 u \cdot x_1 + \cdots + \lambda_m u \cdot x_m = \lambda_1 h_{C_1}(u) + \cdots + \lambda_m h_{C_m}(u) \\ &= h_{\lambda_1 C_1 + \cdots + \lambda_m C_m}(u) = h_C(u), \text{ or } x \in H_C(u). \end{aligned}$$

Hence $x \in C \cap H_C(u)$. □

Proof of the Theorem. Existence. C is a proper convex body. If $C = C_1 \oplus \cdots \oplus C_m$, then each C_i is a proper convex body in its linear hull $\operatorname{lin} C_i$. Thus, if a body C_i can be directly decomposed, the decomposition takes place in $\operatorname{lin} C_i$. This then leads to a refinement of the given direct decomposition of C. Since a proper segment contained in a line through o is directly irreducible, the refinement process stops after a finite number of steps.

Uniqueness. Assume that

$$C = C_1 \oplus \cdots \oplus C_m = D_1 \oplus \cdots \oplus D_n,$$

where the C_i, D_j all have at least dimension 1 and are directly irreducible. Choose a unit vector u orthogonal to $\operatorname{lin} C_1$ such that the support set $(C_2 \oplus \cdots \oplus C_m) \cap H_{C_2 \oplus \cdots \oplus C_m}(u)$ consists of a single point s, say. Then $C \cap H_C(u) = C_1 + s$, and therefore,

$$C_1 + s = C \cap H_C(u) = \big(D_1 \cap H_{D_1}(u)\big) \oplus \cdots \oplus \big(D_n \cap H_{D_n}(u)\big)$$

by Lemma 6.1. If $s = s_1 + \cdots + s_n$ where $s_j \in \operatorname{lin} D_j$, then

$$C_1 = \big(D_1 \cap H_{D_1}(u) - s_1\big) + \cdots + \big(D_n \cap H_{D_n}(u) - s_n\big)$$

and since $\operatorname{lin}\big(D_j \cap H_{D_j}(u) - s_j\big) \subseteq \operatorname{lin} D_j$, the following direct sum of linear subspaces exists:

$$\operatorname{lin}\big(D_1 \cap H_{D_1}(u) - s_1\big) \oplus \cdots \oplus \operatorname{lin}\big(D_n \cap H_{D_n}(u) - s_n\big).$$

Thus

$$C_1 = \big(D_1 \cap H_{D_1}(u) - s_1\big) \oplus \cdots + \oplus \big(D_n \cap H_{D_n}(u) - s_n\big).$$

Since C_1 is directly irreducible, it coincides with one of these summands and thus is a subset of $D_j - s_j$, say. Changing the roles of the C_i and the D_j, we see that, similarly, D_j is a subset of a translate of C_i, say. This is possible only if $C_1 = C_i$. Hence $C_1 = D_j - s_j$. Continuing in this way, we obtain $m = n$ and $\{C_1, \dots, C_m\} = \{D_1 - s_1, \dots, D_m - s_m\}$. Hence

$$D_1 \oplus \cdots \oplus D_m - (s_1 + \cdots + s_m) = C_1 \oplus \cdots \oplus C_m = D_1 \oplus \cdots \oplus D_m.$$

This can hold only if $s_j = 0$ for all j. Hence $\{C_1, \dots, C_m\} = \{D_1, \dots, D_m\}$, concluding the proof of the uniqueness. □

Hausdorff Metric and the Natural Topology on $\mathcal{C}$

The space $\mathcal{C}$ of convex bodies in $\mathbb{E}^d$ is endowed with a natural topology. It can be introduced as the topology induced by the *Hausdorff metric* δ^H on $\mathcal{C}$, which is defined as follows:

$$\delta^H(C, D) = \max\left\{\max_{x\in C}\min_{y\in D}\|x-y\|,\ \max_{y\in D}\min_{x\in C}\|x-y\|\right\} \text{ for } C, D \in \mathcal{C}.$$

The metric δ^H was first defined by Hausdorff [481] in a more general context. A non-symmetric version of it was considered earlier by Pompeiu [812] and Blaschke [124] was the first to put it to use in convex geometry in his selection theorem, see below. The Hausdorff metric can be defined in different ways.

Proposition 6.3. *Let $C, D \in \mathcal{C}$. Then:*

(i) $\delta^H(C, D) = \inf\left\{\delta \ge 0 : C \subseteq D + \delta B^d, D \subseteq C + \delta B^d\right\}$

(ii) $\delta^H(C, D) = \max\left\{|h_C(u) - h_D(u)| : u \in S^{d-1}\right\}$

(iii) $\delta^H(C, D) =$ *maximum distance which a point of one of the bodies C, D can have from the other body*

Proof. Left to the reader. □

If we consider a topology on $\mathcal{C}$ or on a subspace of it, such as $\mathcal{C}_p$, it is always assumed that it is the topology induced by δ^H.

The Blaschke Selection Theorem

In many areas of mathematics there is need for results which guarantee that certain problems, in particular extremum problems, have solutions. Examples of such results are the Bolzano–Weierstrass theorem for sequences in $\mathbb{R}$, the Arzelà–Ascoli theorem for uniformly bounded equicontinuous families of functions and the selection Theorem 25.1 of Mahler for (point) lattices.

In convex geometry, the basic pertinent result is *Blaschke's selection theorem* [124] for convex bodies. It can be used to show that, for example, the isoperimetric problem for convex bodies has a solution. Here we give the following version of it, where a sequence of convex bodies is *bounded*, if it is contained in a suitable (solid Euclidean) ball.

It turns out that the selection theorems of Arzelà–Ascoli, Blaschke and Mahler are closely connected, see the remarks at the end of this section and Groemer's proof of Mahler's theorem outlined in Sect. 25.2.

Theorem 6.3. *Any bounded sequence of convex bodies in* $\mathbb{E}^d$ *contains a convergent subsequence.*

We present two proofs. The first proof, in essence, is due to Blaschke, the second one to Heil [487].

Proof (using ε-nets). Let $C_1, C_2, \dots \in \mathcal{C}$ be contained in a ball B. For the proof that the sequence $C_1, C_2, \dots,$ contains a convergent subsequence, the following will be shown first:

(1) The sequence $C_1, C_2, \dots,$ contains a subsequence $D_1, D_2, \dots,$ such that

$$\delta^H(D_m, D_n) \le \frac{1}{2^{\min\{m,n\}}} \text{ for } m, n = 1, 2, \dots$$

For the proof of (1) the main step is to prove that

(2) There are sequences $C_{11}, C_{12}, \dots; C_{21}, C_{22}, \dots; \dots,$
where $C_{11}, C_{12}, \dots$ is a subsequence of $C_1, C_2, \dots,$ and each subsequent sequence is a subsequence of the sequence preceding it, such that

$$\delta^H(C_{mi}, C_{mj}) \le \frac{1}{2^m} \text{ for } m = 1, 2, \dots, \text{ and } i, j = 1, 2, \dots$$

The first step of the induction is similar to the step from m to $m+1$, thus only the latter will be given. Let $m \ge 1$ and assume that the first m sequences have been constructed already and satisfy the inequality for $1, \dots, m$. Since the ball B is compact, it can be covered by a finite family of balls, each of radius $1/2^{m+2}$ with centre in B. To each convex body in B we associate all balls of this family which intersect it. Clearly, these balls cover the convex body. Since there are only finitely many subfamilies of this family of balls, there must be one which corresponds to each convex body from an infinite subsequence of $C_{m1}, C_{m2}, \dots,$ say $C_{m+1\,1}, C_{m+1\,2}, \dots$ Now, given $i, j = 1, 2, \dots,$ for any $x \in C_{m+1\,i}$ there is a ball in our subfamily which contains x. Hence $\|x - c\| \le 1/2^{m+2}$, where c is the centre of this ball. This ball also intersects $C_{m+1\,j}$. Thus we may choose $y \in C_{m+1\,j}$ with $\|y - c\| \le 1/2^{m+2}$. This shows that, for each $x \in C_{m+1\,i}$, there is $y \in C_{m+1\,j}$ with $\|x - y\| \le 1/2^{m+1}$. Similarly, for each $y \in C_{m+1\,j}$ there is $x \in C_{m+1\,i}$ with $\|x - y\| \le 1/2^{m+1}$. Thus

$$\delta^H\left(C_{m+1\,i}, C_{m+1\,j}\right) \le \frac{1}{2^{m+1}} \text{ for } i, j = 1, 2, \dots$$

by Proposition 6.3(iii). The induction is thus complete, concluding the proof of (2). By considering the diagonal sequence $D_1 = C_{11}, D_2 = C_{22}, \dots,$ we see that (1) is an immediate consequence of (2).

For the proof of the theorem it is sufficient to show that

$$\text{(3)}\quad D_1, D_2, \dots \to D, \text{ where } D = \bigcap_{n=1}^{\infty} \Big(D_n + \frac{1}{2^{n-1}} B^d\Big) \in \mathcal{C}.$$

(1) implies that

$$D_1 + \frac{1}{2} B^d \supseteq D_2,\ D_2 + \frac{1}{2^2} B^d \supseteq D_3, \dots$$

and thus,

$$\text{(4)}\quad D_1 + B^d \supseteq D_2 + \frac{1}{2} B^d \supseteq \dots$$

Being the intersection of a decreasing sequence of non-empty compact convex sets (see (3) and (4)), the set D is also non-empty, compact and convex, i.e. $D \in \mathcal{C}$. In order to prove that $D_1, D_2, \dots \to D$, let $\varepsilon > 0$. Then

$$\text{(5)}\quad D \subseteq D_n + \frac{1}{2^{n-1}} B^d \subseteq D_n + \varepsilon B^d \text{ for } n \geq 1 + \log_2 \frac{1}{\varepsilon}.$$

Let $G = \operatorname{int}(D + \varepsilon B^d)$. The intersection of the following decreasing sequence of compact sets

$$(D_1 + B^d)\backslash G \supseteq \Big(D_2 + \frac{1}{2} B^d\Big)\backslash G \supseteq \dots$$

is contained both in D (see (3)) and in $\mathbb{E}^d \backslash G$ and thus is empty. This implies that, from a certain index on, the sets in this sequence are empty. That is,

$$\text{(6)}\quad D_n \subseteq D_n + \frac{1}{2^{n-1}} B^d \subseteq G \subseteq D + \varepsilon B^d \text{ for all sufficiently large } n.$$

Since $\varepsilon > 0$ was arbitrary, (5) and (6) show that $D_1, D_2, \dots \to D$, concluding the proof of (3) and thus of the theorem. □

Proof (using the Arzelà–Ascoli theorem). A special case of the Arzelà–Ascoli theorem is as follows:

(7) Let B be a ball in $\mathbb{E}^d$ and $f_1, f_2, \dots : B \to \mathbb{R}$ a sequence of functions such that

$$|f_n(x)| \leq \text{const for } x \in B,\ n = 1, 2, \dots,$$
$$|f_n(x) - f_n(y)| \leq \|x - y\| \text{ for } x, y \in B,\ n = 1, 2, \dots$$

Then the sequence $f_1, f_2, \dots$ contains a uniformly convergent subsequence.

For the proof of the selection theorem we have to show the following:

(8) Let B be a ball and $C_1, C_2, \dots$ a sequence of convex bodies in B. Then this sequence contains a convergent subsequence.

Define $\delta_n : B \to \mathbb{R},\ n = 1, 2, \dots,$ by

$$\delta_n(x) = \mathrm{dist}(x, C_n) = \min\{\|x - u\| : u \in C_n\} \text{ for } x \in B.$$

In order to apply (7) to the functions δ_n, some properties have to be proved first. To see that

(9) δ_n is convex,

let $x, y \in B$ and $0 \leq \lambda \leq 1$. Choose $u, v \in C_n$ such that $\delta_n(x) = \|x - u\|$ and $\delta_n(y) = \|y - v\|$. Since $(1 - \lambda)u + \lambda v \in C_n$ by the convexity of C_n, it then follows that

$$\begin{aligned}\delta_n\big((1-\lambda)x + \lambda y\big) &\leq \|(1-\lambda)x + \lambda y - \big((1-\lambda)u + \lambda v\big)\| \\ &\leq (1-\lambda)\|x - u\| + \lambda\|y - v\| = (1-\lambda)\delta_n(x) + \lambda\delta_n(y),\end{aligned}$$

concluding the proof of (9). For the proof that

(10) $|\delta_n(x) - \delta_n(y)| \leq \|x - y\|$ for $x, y \in B$,

let $x, y \in B$ and choose $u, v \in C_n$ such that $\delta_n(x) = \|x - u\|$ and $\delta_n(y) = \|y - v\|$. Then

$$\delta_n(x) \leq \|x - v\| \leq \|x - y\| + \|y - v\| = \|x - y\| + \delta_n(y),$$

or $\delta_n(x) - \delta_n(y) \leq \|x - y\|$. Similarly, $\delta_n(y) - \delta_n(x) \leq \|x - y\|$ and the inequality (10) follows. (Alternatively, one may use Lemma 4.1.) Next,

(11) $|\delta_n(x)| \leq \mathrm{diam}\, B$ for $x \in B$.

Note that $\delta_n(x) = 0$ for $x \in C_n$ and take (10) into account.

Propositions (10), (11) and (7) imply that there is a subsequence $\delta_{n_1}, \delta_{n_2}, \dots,$ converging uniformly to a function $\delta_C : B \to \mathbb{R}$. As the uniform limit of a sequence of non-negative, convex and continuous functions on B, also δ_C is non-negative, convex and continuous on B. Hence

$$C = \big\{x \in B : \delta_C(x) = 0\big\} \in \mathcal{C} \text{ or } C = \emptyset.$$

If $C = \emptyset$, then $\delta_C(x) > 0$ for all $x \in B$. Hence $\delta_{n_j}(x) > 0$ for all sufficiently large j and all $x \in B$. This contradiction shows that

(12) $C \in \mathcal{C}$.

We finally show that

(13) $C_{n_1}, C_{n_2}, \dots \to C$.

Let $\varepsilon > 0$. Since δ_C is continuous on the compact set B and thus uniformly continuous and since δ_C is 0 precisely on C, there is $\delta > 0$ such that

$$\big\{x \in B : \delta_C(x) \leq \delta\big\} \subseteq C + \varepsilon B^d.$$

Since $\delta_C(x) \leq \delta_{n_k}(x) + \delta$ for all sufficiently large k,

(14) $C_{n_k} = \big\{x \in B : \delta_{n_k}(x) = 0\big\} \subseteq \big\{x \in B : \delta_C(x) \leq \delta\big\} \subseteq C + \varepsilon B^d$ for all sufficiently large k.

The definition of δ_{n_k} shows that

$$\{x \in B : \delta_{n_k}(x) \leq \varepsilon\} \subseteq C_{n_k} + \varepsilon B^d.$$

Since $\delta_{n_k}(x) \leq \delta_C(x) + \varepsilon$ for all sufficiently large k,

(15) $C = \{x \in B : \delta_C(x) = 0\} \subseteq \{x \in B : \delta_{n_k}(x) \leq \varepsilon\} \subseteq C_{n_k} + \varepsilon B^d$ for all sufficiently large k.

$\varepsilon > 0$ was chosen arbitrarily. Thus (14) and (15) together with Proposition 6.3 (i) imply (13).

Propositions (12) and (13) yield (8), concluding the proof of the theorem. □

Other Versions and Generalizations of the Selection Theorem

Sometimes the following versions of Blaschke's selection theorem are needed in the context of convex geometry. These are immediate consequences of the above form of the selection theorem. Here $\mathcal{C}$ is endowed with its natural topology and its subspace $\mathcal{C}_p$ of proper convex bodies with the corresponding induced topology.

Theorem 6.4. *The following claims hold:*

(i) *$\mathcal{C}$ is a locally compact space.*
(ii) *$\mathcal{C}_p$ is a locally compact space.*
(iii) *$\mathcal{C}$, endowed with the metric δ^H, is a boundedly compact complete metric space.*

Clearly, the Hausdorff metric can be defined for the space of compact sets in any metric space. The above proofs can easily be generalized to show that the Blaschke selection theorem holds for spaces of compact sets in boundedly compact metric spaces endowed with the Hausdorff metric, and even in more general situations. Such spaces of compact sets are sometimes called *hyperspaces*. An analogue of the selection theorem for hyperspaces was first proved by Vietoris [1010]. Related, more general results are due to Michael [722]. For hyperspaces, see Beer [88] and Illanes and Nadler [535].

Equivalence of Blaschke's Selection Theorem and the Arzelà–Ascoli Theorem

Using a general version of Blaschke's selection theorem, Bol [140] proved the classical Arzelà–Ascoli theorem. Conversely, Heil [487] proved a general version of Blaschke's selection theorem by means of the Arzelà–Ascoli theorem.

6.2 Minkowski's Theorem on Mixed Volumes and Steiner's Formula

A highlight of the early Brunn–Minkowski theory is Minkowski's [744] theorem on mixed volumes. It says that the volume of a linear combination of convex bodies is a polynomial in the coefficients of the linear combination. The coefficients of the polynomial are the mixed volumes. The following quotation from Hilbert's [502], p.XIX, obituary on Minkowski illustrates the meaning of mixed volumes.

> ...Thus the concept of mixed volume appears as the simplest generalization which comprises the notions of volume, surface area, total mean curvature as special cases. In this way the latter notions are related much more closely to each other. Thus we may expect now to obtain a deeper understanding than was possible before, of the mutual relations of these notions...

An important special case is Steiner's [959] formula for the volume of parallel bodies. The coefficients of the corresponding polynomial are, up to multiplicative constants, the so-called quermassintegrals. Mixed volumes and quermassintegrals played a dominant role in convex geometry throughout the twentieth century.

In this section we prove Minkowski's theorem on mixed volumes and Steiner's formula. In the following sections the notions of mixed volumes and quermassintegrals will be investigated in more detail.

For references, see the introduction of this chapter.

Minkowski's Theorem on Mixed Volumes

Theorem 6.5. *Let $C_1, \dots, C_m \in \mathcal{C}$. Then there are coefficients $V(C_{i_1}, \dots, C_{i_d})$, $1 \le i_1, \dots, i_d \le m$, called* mixed volumes, *which are symmetric in the indices and such that*

$$V(\lambda_1 C_1 + \cdots + \lambda_m C_m) = \sum_{i_1,\dots,i_d=1}^{m} V(C_{i_1}, \dots, C_{i_d}) \lambda_{i_1} \cdots \lambda_{i_d} \text{ for } \lambda_1, \dots, \lambda_m \ge 0.$$

The proof is by induction. It follows a clear line and, basically, is not difficult. To make it more transparent, it is split into several steps. The first tool is Lemma 6.1 above.

A *convex polytope* is the convex hull of a finite set in $\mathbb{E}^d$. Let $\mathcal{P}$ denote the space of all convex polytopes in $\mathbb{E}^d$. Given $P \in \mathcal{P}$, a *face* of P is the intersection of P with a support hyperplane. It is the convex hull of the intersection of the finite set which determines P with the support hyperplane and thus also a convex polytope. A face of dimension $d-1$ is called a *facet*. For more information, see Sect. 14.1

Lemma 6.2. *Let $P_1, \dots, P_m \in \mathcal{P}$. Then the following claims hold:*

(i) $P = \lambda_1 P_1 + \cdots + \lambda_m P_m \in \mathcal{P}$ *for* $\lambda_1, \dots, \lambda_m \ge 0$.

(ii) *There is a finite set $U \subseteq S^{d-1}$ such that for all $\lambda_1, \dots, \lambda_m \ge 0$, for which $P = \lambda_1 P_1 + \cdots + \lambda_m P_m$ is a proper convex polytope, the exterior unit normal vectors of the facets of P are contained in U.*

Proof. (i) We may assume that $\lambda_1, \dots, \lambda_m > 0$, otherwise consider fewer polytopes. Then the following will be shown.

(1) Let $e \in P$ be extreme and $e = \lambda_1 e_1 + \cdots + \lambda_m e_m$, where $e_i \in P_i$. Then each e_i is extreme in P_i.

If $e_i \in P_i$ is not extreme in P_i, then there is a line segment $S \subseteq P_i$ such that e_i is a relative interior point of S. Then e is a relatively interior point of the line segment $\lambda_1 e_1 + \cdots + \lambda_{i-1} e_{i-1} + \lambda_i S + \lambda_{i+1} e_{i+1} + \cdots + \lambda_m e_m \subseteq P$ and thus cannot be extreme in P. This contradiction completes the proof of (1). Since by Theorem 5.5 each P_i has only finitely many extreme points, its vertices, (1) implies that P has only finitely many extreme points. Since P is the convex hull of these by Theorem 5.5, it is a polytope.

(ii) We show the following:

(2) Let F_i be a face of P_i for $i = 1, \dots, m$, and $\lambda_1, \dots, \lambda_m \geq 0$, such that $\lambda_1 F_1 + \cdots + \lambda_m F_m$ has dimension $d - 1$ (and is contained in a hyperplane H). Then $F_1 + \cdots + F_m$ also has dimension $d - 1$ and is contained in a hyperplane parallel to H.

It is sufficient to show this under the assumption that $\lambda_i > 0$ and $o \in \operatorname{relint} F_i$ for each i. Then

$$\begin{aligned} H &= \operatorname{lin}(\lambda_1 F_1 + \cdots + \lambda_m F_m) = \operatorname{lin} \lambda_1 F_1 + \cdots + \operatorname{lin} \lambda_m F_m \\ &= \operatorname{lin} F_1 + \cdots + \operatorname{lin} F_m = \operatorname{lin}(F_1 + \cdots + F_m), \end{aligned}$$

concluding the proof of (2). Since each P_i has only finitely many faces, there is only a finite set U of unit normal vectors of hyperplanes H as in (2). Since the hyperplanes which contain facets of $P = \lambda_1 P_1 + \cdots + \lambda_m P_m$ are of this type by Lemma 6.1, the proof of (ii) is complete. □

Lemma 6.3. *Let p_n, $n = 1, 2, \dots$, be a sequence of real homogeneous polynomials in m variables of degree d and let p be a real function in m variables, defined for $\lambda_1, \dots, \lambda_m \geq 0$. Assume that*

(3) $p_n(\lambda_1, \dots, \lambda_m) \to p(\lambda_1, \dots, \lambda_m)$ *as* $n \to \infty$
for each m-tuple $\lambda_1, \dots, \lambda_m \geq 0$.

Then p is a homogeneous polynomial in m variables of degree d, restricted to $\lambda_1, \dots, \lambda_m \geq 0$.

Proof. First, the following will be shown:

(4) Let q_n, $n = 1, 2, \dots$, be a sequence of real polynomials in one variable of degree d and let q be a real function defined for $\lambda \geq 0$. Assume that

$$q_n(\lambda) \to q(\lambda) \text{ for each } \lambda \geq 0.$$

Then q is a polynomial in one variable of degree d, restricted to $\lambda \geq 0$.

Let $q_n(\lambda) = a_{0n} + a_{1n}\lambda + \cdots + a_{dn}\lambda^d$ for $\lambda \in \mathbb{R}$. Then

$$\begin{aligned} &a_{0n} &&= q_n(0) \to q(0), \\ &a_{0n} + a_{1n} \quad + \cdots + a_{dn} &&= q_n(1) \to q(1), \\ &a_{0n} + a_{1n} 2 + \cdots + a_{dn} 2^d &&= q_n(2) \to q(2), \\ &\dots\dots\dots\dots\dots\dots\dots && \\ &a_{0n} + a_{1n} d + \cdots + a_{dn} d^d &&= q_n(d) \to q(d). \end{aligned}$$

The coefficient matrices of these systems of linear equations in $a_{0n}, \dots, a_{dn}$, $n = 1, 2, \dots$, all are equal to a fixed Vandermonde matrix and thus non-singular. An application of Cramer's rule then shows that

$$a_{0n} \to a_0, \dots, a_{dn} \to a_d, \text{ say.}$$

Thus

$$q_n(\lambda) \to \begin{Bmatrix} a_0 + a_1\lambda + \cdots + a_d\lambda^d \\ q(\lambda) \end{Bmatrix} \text{ for each } \lambda \geq 0.$$

Hence $q(\lambda) = a_0 + a_1\lambda + \cdots + a_d\lambda^d$ for $\lambda \geq 0$, concluding the proof of (4).

We now prove the lemma by induction on m. In the following, all coefficients are supposed to be symmetric in the indices. For $m = 1$ let $p_n(\lambda) = a_n\lambda^d$. Then

$$a_n = p_n(1) \to p(1) = a, \text{ say,}$$

by assumption. Thus

$$p_n(\lambda) = a_n\lambda^d \to \begin{Bmatrix} a\ \lambda^d \\ p(\lambda) \end{Bmatrix} \text{ for each } \lambda \geq 0$$

by assumption. Hence $p(\lambda) = a\lambda^d$ for $\lambda \geq 0$. (The case $m = 1$ is also a consequence of Proposition (4).) Suppose now that $m > 1$ and that the lemma holds for $1, 2, \dots, m - 1$. Then

$$p_n(\lambda_1, \dots, \lambda_m) = p_{0n}(\lambda_2, \dots, \lambda_m) + p_{1n}(\lambda_2, \dots, \lambda_m)\lambda_1 + \cdots + p_{dn}(\lambda_2, \dots, \lambda_m)\lambda_1^d,$$

where

$$p_{in}(\lambda_2, \dots, \lambda_m)$$

is a homogeneous polynomial of degree $d - i$ in the variables $\lambda_2, \dots, \lambda_m$. Given $\lambda_2, \dots, \lambda_m \geq 0$, we have $p_n(\lambda_1, \lambda_2, \dots, \lambda_m) \to p(\lambda_1, \lambda_2, \dots, \lambda_m)$ for any $\lambda_1 \geq 0$ by assumption. An application of (4) then shows that, for the coefficients p_{in}, $i = 0, \dots, d$, (which are homogeneous polynomials in $\lambda_2, \dots, \lambda_m$ of degree $d - i$),

$$p_{in}(\lambda_2, \dots, \lambda_m) \text{ converges as } n \to \infty \text{ for } \lambda_2, \dots, \lambda_m \geq 0 \text{ and } i = 0, \dots, d.$$

Denote the limit by $q_i(\lambda_2, \dots, \lambda_m)$. The induction hypothesis, applied for $i = 0, \dots, d$, shows that

$$p_{in}(\lambda_2, \dots, \lambda_m) \to q_i(\lambda_2, \dots, \lambda_m) \text{ for } \lambda_2, \dots, \lambda_m \geq 0 \text{ and } i = 0, \dots, d,$$

where $q_i(\lambda_2, \dots, \lambda_m)$ is the restriction to $\lambda_2, \dots, \lambda_m \geq 0$ of a suitable homogeneous polynomial in $\lambda_2, \dots, \lambda_m$ of degree $d - i$. Then, clearly,

$$\begin{aligned} p_n(\lambda_1, \dots, \lambda_d) &= \sum_{i=0}^{d} p_{in}(\lambda_2, \dots, \lambda_m)\lambda_1^i \to \sum_{i=0}^{d} q_i(\lambda_2, \dots, \lambda_m)\lambda_1^i \\ &= q(\lambda_1, \dots, \lambda_m) \text{ for } \lambda_1 \geq 0 \text{ and } \lambda_2, \dots, \lambda_m \geq 0, \end{aligned}$$

say, where q is a homogeneous polynomial of degree d in $\lambda_1, \dots, \lambda_m$. Comparing this with (3), we see that $p(\lambda_1, \dots, \lambda_m) = q(\lambda_1, \dots, \lambda_m)$ for $\lambda_1, \dots, \lambda_m \geq 0$, concluding the induction and thus the proof of the lemma. □

Let $v(\cdot)$ denote $(d-1)$-dimensional volume.

Proof of the Theorem. First, the theorem is proved for convex polytopes by induction on d. The case $d = 1$ deals with line segments and is left to the reader. Assume now that $d > 1$ and that the theorem holds for $d - 1$. Let $P_1, \dots, P_m \in \mathcal{P}$ and choose U as in Lemma 6.2. Then the formula for the volume of convex polytopes, see Proposition (9) in Sect. 16.1, and Lemma 6.1 yield,

$$V(P = \lambda_1 P_1 + \dots + \lambda_m P_m) = \frac{1}{d} \sum_{u \in U} \{h_P(u)\, v(F) : F = P \cap H_P(u)\}$$
$$= \frac{1}{d} \sum_{u \in U} \{h_{\lambda_1 P_1 + \dots + \lambda_m P_m}(u)\, v(\lambda_1 F_1 + \dots + \lambda_m F_m) : F_i = P_i \cap H_{P_i}(u)\}$$
for $\lambda_1, \dots, \lambda_m \geq 0$.

Now note that $h_{\lambda_1 P_1 + \dots + \lambda_m P_m}(u) = \lambda_1 h_{P_1}(u) + \dots + \lambda_m h_{P_m}(u)$, by Proposition 6.2, and $v(\lambda_1 F_1 + \dots + \lambda_m F_m)$ is a homogeneous polynomial of degree $d - 1$ in $\lambda_1, \dots, \lambda_m$ by the induction hypothesis, where the coefficients are symmetric in their indices. ($F_1, \dots, F_m$ may be in different parallel hyperplanes, but this does not change the area of $\lambda_1 F_1 + \dots + \lambda_m F_m$.) Hence $V(P)$ is a homogeneous polynomial of degree d in $\lambda_1, \dots, \lambda_m$ for $\lambda_1, \dots, \lambda_m \geq 0$. By changing the coefficients suitably, if necessary, we may assume that the new coefficients are symmetric in their indices. This concludes the induction and thus proves the theorem for convex polytopes.

Second, we prove the theorem for convex bodies $C_1, \dots, C_m \in \mathcal{C}$. Choose sequences of convex polytopes $(P_{1n}), \dots, (P_{mn})$ such that $P_{in} \to C_i$ as $n \to \infty$ for $i = 1, \dots, m$, see the proof of Theorem 7.4. Since we have already proved the theorem for convex polytopes, there are homogeneous polynomials p_n of degree d in $\lambda_1, \dots, \lambda_m$, the coefficients of which are symmetric in their indices, such that

$$V(\lambda_1 P_{1n} + \dots + \lambda_m P_{mn}) = p_n(\lambda_1, \dots, \lambda_m)$$
$$\text{for } n = 1, 2, \dots \text{ and } \lambda_1, \dots, \lambda_m \geq 0.$$

Since $P_{in} \to C_i$ as $n \to \infty$, it follows for any m-tuple $\lambda_1, \dots, \lambda_m \geq 0$, that $\lambda_1 P_{1n} + \dots + \lambda_m P_{mn} \to \lambda_1 C_1 + \dots + \lambda_m C_m$. Thus, noting that by Theorem 7.5 the volume $V(\cdot)$ is continuous on $\mathcal{C}$, it follows that:

$$V(\lambda_1 P_{1n} + \dots + \lambda_m P_{mn}) \to V(\lambda_1 C_1 + \dots + \lambda_m C_m) \text{ as } n \to \infty \text{ for } \lambda_1, \dots, \lambda_m \geq 0.$$

Therefore the polynomials p_n satisfy

$$p_n(\lambda_1, \dots, \lambda_m) \to V(\lambda_1 C_1 + \dots + \lambda_m C_m) \text{ as } n \to \infty \text{ for } \lambda_1, \dots, \lambda_m \geq 0.$$

An application of Lemma 6.3 then shows that $V(\lambda_1 C_1 + \dots + \lambda_m C_m)$ itself is a homogeneous polynomial in $\lambda_1, \dots, \lambda_m$ of degree d. We clearly may write this polynomial with coefficients which are symmetric in their indices. Finally, denote the coefficient of $\lambda_{i_1} \cdots \lambda_{i_d}$ by $V(C_{i_1}, \dots, C_{i_d})$. □

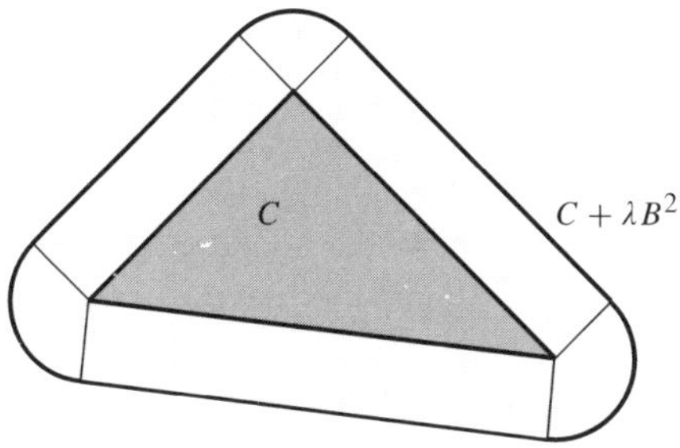

Fig. 6.2. Steiner's formula for parallel bodies

Steiner's Formula for Parallel Bodies

An important special case of Theorem 6.5 is the following result of Steiner [959] (see Fig. 6.2).

Theorem 6.6. *Let $C \in \mathcal{C}$. Then*

$$V(C + \lambda B^d) = W_0(C) + \binom{d}{1} W_1(C)\lambda + \cdots + \binom{d}{d} W_d(C)\lambda^d \quad \textit{for } \lambda \geq 0,$$

where

$$W_i(C) = V(\underbrace{C, \ldots, C}_{d-i}, \underbrace{B^d, \ldots, B^d}_{i}), \ i = 0, \ldots, d.$$

The polynomial on the right side is the *Steiner polynomial* and the quantities W_i are the *quermassintegrals* or the *mean projection measures* of the convex body C.

Remark. It is a natural, yet not intensively studied problem to extract geometric properties of convex bodies from their Steiner polynomials. An open conjecture of Teissier [993] relates the roots of the Steiner polynomial to the inradius and the circumradius of the corresponding convex body. For a series of interesting geometric results related to Steiner polynomials in the 2- and 3-dimensional case, see Hernández Cifre and Saorín [496] who also give some references to the literature.

Remark. For an explanation why quermassintegrals are called quermassintegrals, see a remark after Kubota's theorem 6.16.

6.3 Properties of Mixed Volumes

Mixed volumes are a seminal notion in convex geometry and, thus, have been investigated intensively from various viewpoints. Mixed area measures are localized versions of mixed volumes. Mixed volumes form a bridge between convex and algebraic geometry. For hints in this direction, see Sect. 19.5.

The present section contains a series of important properties of mixed volumes. We start with a representation of mixed volumes by means of volumes, then consider linearity, continuity, monotony, and valuation properties and, finally, prove Minkowski's inequalities.

For more information we refer to the books of Leichtweiss [640], Burago and Zalgaller [178] and Schneider [905] and the survey of Sangwine-Yager [878].

A Representation of Mixed Volumes

The following representation of mixed volumes gives insight into the meaning of mixed volumes. If, in a sum, a summand has a tilde, it is to be omitted. Summation is from 1 to d, taking into account the stated restrictions.

Theorem 6.7. *Let $C_1, \dots, C_d \in \mathcal{C}$. Then*

$$\begin{aligned}(1)\quad d\,!\,V(C_1,\dots,C_d) &= V(C_1+\dots+C_d)-\sum_{i_1} V(C_1+\dots+\widetilde{C_{i_1}}+\dots+C_d)\\ &\quad+\sum_{i_1<i_2} V(C_1+\dots+\widetilde{C_{i_1}}+\dots+\widetilde{C_{i_2}}+\dots+C_d)-\cdots\\ &\quad+(-1)^{d-1}\sum_i V(C_i).\end{aligned}$$

Proof. By Minkowski's theorem on mixed volumes, the sum appearing on the right side in (1) is equal to the following expression, where in each term we have omitted $V(C_{j_1}, \dots, C_{j_d})$.

$$(2)\quad \sum_{j_1,\dots,j_d} - \sum_{i_1}\ \sum_{j_1,\dots,j_d\neq i_1} + \sum_{i_1<i_2}\ \sum_{j_1,\dots,j_d\neq i_1,i_2} - \sum_{i_1<i_2<i_3}\ \sum_{j_1,\dots,j_d\neq i_1,i_2,i_3} + \cdots$$

Here the indices $j_1, \dots, j_d$ run from 1 to d. We consider the mixed volumes appearing in (2). First, take all mixed volumes whose indices coincide with $1, \dots, d$ up to their order. There are $d\,!$ such mixed volumes. They appear only in the first term and all have the same value. Their total value thus is $d\,!V(C_1, \dots, C_d)$. Secondly, take a mixed volume $V(C_{j_1}, \dots, C_{j_d})$, where the indices $j_1, \dots, j_d$ are not a permutation of $1, \dots, d$. Choose $i_1 < \dots < i_k$, k maximal, such that $j_1, \dots, j_d \neq i_1, \dots, i_k$. If this mixed volume and the mixed volumes corresponding to permutations of its indices together appear in the first sum l times, they appear in the second sum kl times, in the third sum $\binom{k}{2}l$ times, etc. Thus they appear in the expression (2), taken with the appropriate sign,

$$\left(1-\binom{k}{1}+\binom{k}{2}-\dots+(-1)^k\binom{k}{k}\right)l=(1-1)^k\,l=0$$

times. Since they all have the same value, their contribution to the expression in (2) thus is 0. This shows that all mixed volumes in the expression (2) cancel, except those with indices which are permutations of $1, \dots, d$, where the contribution of the latter is $d\,!\,V(C_1, \dots, C_d)$, as was shown first. □

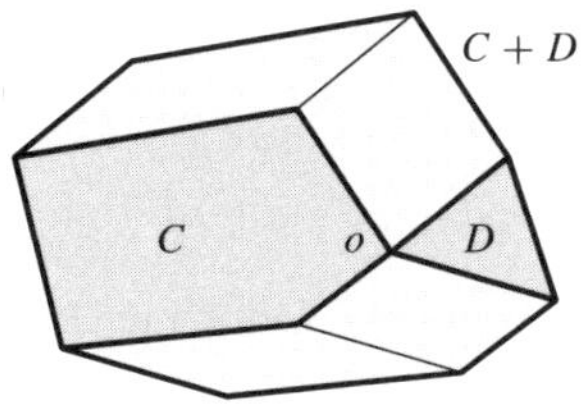

Fig. 6.3. Mixed volume $2V(C, D) = V(C + D) - V(C) - V(D)$

This result permits us to visualize mixed volumes, at least in dimension 2. Consider two convex bodies C and D. Then

$$2V(C, D) = V(C + D) - V(C) - V(D).$$

The shaded figure is the sum $C + D$ (see Fig. 6.3). The area of the lightly shaded part equals $2V(C, D)$.

Some of the following results are easy consequences of Theorem 6.7. In spite of this we present their common proofs.

The Notation for Mixed Volumes

The following result justifies the notation of mixed volumes.

Proposition 6.4. *Let $C_1, \dots, C_m \in \mathcal{C}$ and $j_1, \dots, j_d \in \{1, \dots, m\}$. Then $V(C_{j_1}, \dots, C_{j_d})$ depends only on $C_{j_1}, \dots, C_{j_d}$.*

Proof. Let $D_1, \dots, D_m \in \mathcal{C}$ such that $C_{j_1} = D_{j_1}, \dots, C_{j_d} = D_{j_d}$. Then, putting $\lambda_i = 0$ for all $i \neq j_1, \dots, j_d$,

$$\begin{aligned}
\sum_{i_1,\dots,i_d \in \{j_1,\dots,j_d\}} & V(C_{i_1}, \dots, C_{i_d}) \lambda_{i_1} \cdots \lambda_{i_d} \\
&= V(\lambda_{j_1} C_{j_1} + \cdots + \lambda_{j_d} C_{j_d}) \\
&= V(\lambda_{j_1} D_{j_1} + \cdots + \lambda_{j_d} D_{j_d}) \\
&= \sum_{i_1,\dots,i_d \in \{j_1,\dots,j_d\}} V(D_{i_1}, \dots, D_{i_d}) \lambda_{i_1} \cdots \lambda_{i_d} \\
&\text{for } \lambda_{j_1}, \dots, \lambda_{j_d} \geq 0.
\end{aligned}$$

Comparing coefficients, it follows in particular, that $V(C_{j_1}, \dots, C_{j_d}) = V(D_{j_1}, \dots, D_{j_d})$. □

Rigid Motions

Since volume is translation invariant and, more generally, rigid motion invariant, see Theorem 7.5, the following statements hold.

Proposition 6.5. *Let $C_1, \dots, C_d \in \mathcal{C}$. Then:*

(i) $V(C_1 + t_1, \dots, C_d + t_d) = V(C_1, \dots, C_d)$ *for all* $t_1, \dots, t_d \in \mathbb{E}^d$.
(ii) $V(mC_1, \dots, mC_d) = V(C_1, \dots, C_d)$ *for all rigid motions* $m : \mathbb{E}^d \to \mathbb{E}^d$.

Remark. More generally, we have the following: let $a : \mathbb{E}^d \to \mathbb{E}^d$ be an affinity with determinant $\det a \neq 0$. Then

$$V(aC_1, \dots, aC_d) = |\det a| V(C_1, \dots, C_d).$$

Linearity

Mixed volumes are linear in each variable with respect to non-negative linear combinations of convex bodies.

Proposition 6.6. *Let $C, D, D_2, \dots, D_d \in \mathcal{C}$. Then*

$$V(\lambda C + \mu D, D_2, \dots, D_d) = \lambda V(C, D_2, \dots, D_d) + \mu V(D, D_2, \dots, D_d)$$
for $\lambda, \mu \geq 0$.

Proof. Let $\lambda, \mu \geq 0$. The quantities

$$V\big(\lambda_1(\lambda C + \mu D) + \lambda_2 D_2 + \cdots + \lambda_d D_d\big),$$
$$V\big((\lambda_1\lambda)C + (\lambda_1\mu)D + \lambda_2 D_2 + \cdots + \lambda_d D_d\big)$$

have identical polynomial representations in $\lambda_1, \dots, \lambda_d$. The coefficient of $\lambda_1 \cdots \lambda_d$ in the first polynomial is

$$d\,!V(\lambda C + \mu D, D_2, \dots, D_d).$$

The coefficient of $\lambda_1 \cdots \lambda_d$ in the second polynomial can be obtained by representing the second quantity as a polynomial in $\lambda_1\lambda, \lambda_1\mu, \lambda_2, \dots, \lambda_d$ and then collecting $\lambda_1 \cdots \lambda_d$. Thus it is

$$d\,!\lambda V(C, D_2, \dots, D_d) + d\,!\mu V(D, D_2, \dots, D_d).$$

Since the coefficients coincide, the proof is complete. □

Continuity

Mixed volume are continuous in their entries.

Theorem 6.8. $V(\cdot, \dots, \cdot)$ *is continuous on* $\mathcal{C} \times \cdots \times \mathcal{C}$.

Proof. The following remark is clear:

> Let $p, p_n, n = 1, 2, \dots,$ be homogeneous polynomials of degree d in d variables such that

$$p_n(\lambda_1,\dots,\lambda_d) \to p(\lambda_1,\dots,\lambda_d) \text{ as } n \to \infty$$
$$\text{for each } d\text{-tuple } \lambda_1,\dots,\lambda_d \geq 0.$$

Assume that all coefficients $a_{i_1\dots i_d n}$ and $a_{i_1\dots i_d}$ of p_n and p, respectively, are symmetric in their indices $i_1,\dots,i_d$. Then

$$a_{i_1\dots i_d n} \to a_{i_1\dots i_d} \text{ for all } i_1,\dots,i_d \in \{1,\dots,d\}.$$

To show the theorem, consider sequences $(C_{1n}),\dots,(C_{dn})$ in $\mathcal{C}$ converging to $C_1,\dots,C_d \in \mathcal{C}$ say, respectively. Since for $\lambda_1,\dots,\lambda_d \geq 0$, we have $\lambda_1 C_{1n} + \dots + \lambda_d C_{dn} \to \lambda_1 C_1 + \dots + \lambda_d C_d$, the continuity of volume, which will be proved in Theorem 7.5, yields

$$V(\lambda_1 C_{1n} + \dots + \lambda_d C_{dn}) \to V(\lambda_1 C_1 + \dots + \lambda_d C_d)$$
$$\text{for } \lambda_1,\dots,\lambda_d \geq 0.$$

Now apply Minkowski's mixed volume Theorem 6.5 and the above remark for $a_{12\dots dn} = V(C_{1n},\dots,C_{dn})$ and $a_{12\dots d} = V(C_1,\dots,C_d)$. □

Monotony

Mixed volumes are non-decreasing in their entries.

Theorem 6.9. *Let $C_1,\dots,C_d, D_1,\dots, D_d \in \mathcal{C}$ such that $C_1 \subseteq D_1,\dots,C_d \subseteq D_d$. Then $V(C_1,\dots,C_d) \leq V(D_1,\dots,D_d)$.*

As examples show, equality does not mean that necessarily $C_i = D_i$ for all i. The proof of the theorem is based on several lemmas which are of interest themselves. Let $v(\cdot)$ denote $(d-1)$-dimensional volume. (For $(d-1)$-dimensional polytopes and thus for facets of d-dimensional polytopes it coincides with elementary area measure for polytopes, see Sect. 16.1.)

Lemma 6.4. *Let $C \in \mathcal{C}$ and $P \in \mathcal{P}$. For each facet F of P let u_F be the exterior unit normal vector of F. Then*

$$(3)\quad V(C,P,\dots,P) = \frac{1}{d} \sum_{F \text{ facet of } P} h_C(u_F)\, v(F).$$

Proof. First, the following will be shown:

(4) Let G be a face of P with $\dim G \leq d-2$. Then $V(G+\varepsilon B^d) = O(\varepsilon^2)$ as $\varepsilon \to +0$.

We may assume that $G \subseteq \mathbb{E}^{d-2}$, where $\mathbb{E}^d$ is represented in the form $\mathbb{E}^d = \mathbb{E}^{d-2} \times \mathbb{E}^2$. Let B^{d-2}, B^2 be the corresponding unit balls. Choose $\varrho > 0$ so large that $G + \varepsilon B^{d-2} \subseteq \varrho B^{d-2}$ for $0 < \varepsilon \leq 1$. Then $G + \varepsilon B^d \subseteq (G + \varepsilon B^{d-2}) \times \varepsilon B^2 \subseteq \varrho B^{d-2} \times \varepsilon B^2$ and thus $V(G+\varepsilon B^d) = O(\varepsilon^2)$ as $\varepsilon \to +0$.

Second, Minkowski's theorem 6.5 on mixed volumes shows that, in particular,

$$\begin{aligned} V(\varepsilon C + P) &= V(P, \dots, P) + d\,V(C, P, \dots, P)\varepsilon + \dots \\ &\quad + d\,V(C, \dots, C, P)\varepsilon^{d-1} + V(C, \dots, C)\varepsilon^d \\ &= V(P) + d\,V(C, P, \dots, P)\varepsilon + \dots + V(C)\varepsilon^d. \end{aligned}$$

Thus

$$(5)\quad V(C, P, \dots, P) = \frac{1}{d} \lim_{\varepsilon \to +0} \frac{V(\varepsilon C + P) - V(P)}{\varepsilon}.$$

Third, in the following we distinguish two cases. In the first case let $o \in C$. Choose $\sigma > 0$ such that $C \subseteq \sigma B^d$. The convex body $\varepsilon C + P$ then may be dissected into P, into cylinders, possibly slanting, with the facets as bases and into the remaining part of $\varepsilon C + P$. The cylinders can be described as follows: for a facet F of P with exterior unit normal vector u_F choose a point $p \in C \cap H_C(u_F)$. The corresponding cylinder then is $F + \varepsilon[o, p]$. The remaining part of $\varepsilon C + P$ is contained in the union of sets of the form $\varepsilon C + G \subseteq \varepsilon\sigma B^d + G$, where G ranges over the faces of P with $\dim G \leq d - 2$ and thus has volume $O(\varepsilon^2)$ by (4). Hence

$$(6)\quad V(\varepsilon C + P) = V(P) + \varepsilon \sum_{F \text{ facet of } P} h_C(u_F)\, v(F) + O(\varepsilon^2) \text{ as } \varepsilon \to 0.$$

Finally, (5) and (6) together yield (3) in case where $o \in C$.

In the second case let $o \notin C$. Choose $t \in \mathbb{E}^d$ such that $o \in C + t$. Then

$$\begin{aligned} V(C, P, \dots, P) &= V(C + t, P, \dots, P) \\ &= \frac{1}{d} \sum_{F \text{ facet of } P} h_{C+t}(u_F)\, v(F) \\ &= \frac{1}{d} \sum_{F \text{ facet of } P} h_C(u_F)\, v(F) + \frac{1}{d} \sum_{F \text{ facet of } P} t \cdot u_F\, v(F) \end{aligned}$$

by Proposition 6.5 and the first case. Now note that

$$\sum_{F \text{ facet of } P} t \cdot u_F\, v(F) = \|t\| \sum_{F \text{ facet of } P} \frac{t}{\|t\|} \cdot u_F\, v(F) = 0$$

since the latter sum equals the sum of the area of the projection of bd P onto the hyperplane through o orthogonal to t, taken once with a $+$ sign and once with a $-$ sign. This concludes the proof of (3) in case where $o \notin C$. □

For later reference we state the following consequence of (5), where for the Minkowski surface area, see Sect. 6.4.

Corollary 6.1. *Let $P \in \mathcal{P}$. Then $d\,V(B^d, P, \dots, P)$ equals the (Minkowski) surface area of P.*

The next result is a refinement of Lemma 6.4.

Lemma 6.5. *Let $C \in \mathcal{C}$, $P_2, \dots, P_d \in \mathcal{P}$, and let $U \subseteq S^{d-1}$ be a finite set containing all exterior unit normal vectors of the facets of all the polytopes $\lambda_2 P_2 + \dots + \lambda_d P_d$ with $\lambda_2, \dots, \lambda_d \geq 0$. U exists by Lemma 6.2. Let $v(\cdot, \dots, \cdot)$ denote the mixed volume in $d-1$ dimensions. Then*

$$(7)\quad V(C, P_2, \dots, P_d) = \frac{1}{d} \sum_{u \in U} h_C(u)\, v\big(P_2 \cap H_{P_2}(u), \dots, P_d \cap H_{P_d}(u)\big).$$

Proof. Let $P = \lambda_2 P_2 + \dots + \lambda_d P_d$. By Lemmas 6.4 and 6.1 and Theorem 6.5 for dimension $d-1$, we may express $V(C, P, \dots, P)$ in the following form:

$$\begin{aligned}(8)\quad V(C, P, \dots, P) &= \frac{1}{d} \sum_{u \in U} h_C(u)\, v\big(P \cap H_P(u)\big) \\ &= \frac{1}{d} \sum_{u \in U} h_C(u)\, v\big(\lambda_2\big(P_2 \cap H_{P_2}(u)\big) + \dots + \lambda_d\big(P_d \cap H_{P_d}(u)\big)\big) \\ &= \frac{1}{d} \sum_{u \in U} h_C(u) \sum_{i_2, \dots, i_d = 2}^{d} v\big(P_{i_2} \cap H_{P_{i_2}}(u), \dots, P_{i_d} \cap H_{P_{i_d}}(u)\big) \lambda_{i_2} \dots \lambda_{i_d}.\end{aligned}$$

Proposition 6.6 implies that

$$(9)\quad V(C, P, \dots, P) = \sum_{i_2, \dots, i_d = 2}^{d} V(C, P_{i_2}, \dots, P_{i_d}) \lambda_{i_2} \cdots \lambda_{i_d}.$$

Finally, equating the coefficients of $\lambda_2 \cdots \lambda_d$ in (8) and (9) yields (7). □

Proof (of the Theorem by induction on d). For $d = 1$ the result is trivial. Assume now that $d > 1$ and that the result holds for $d - 1$. It is sufficient to prove the following:

(10) Let $C, D, D_2, \dots, D_d \in \mathcal{C}$ such that $C \subseteq D$. Then
$V(C, D_2, \dots, D_d) \leq V(D, D_2, \dots, D_d)$.

Since by Theorem 6.8 mixed volumes are continuous, it suffices, for the proof of (10), to show the following special case of (10):

(11) Let $C, D \in \mathcal{C}$ such that $C \subseteq D$ and $P_2, \dots, P_d \in \mathcal{P}$. Then
$V(C, P_2, \dots, P_d) \leq V(D, P_2, \dots, P_d)$.

$C \subseteq D$ implies that $h_C \leq h_D$. Represent $V(C, P_2, \dots, P_d)$ and $V(D, P_2, \dots, P_d)$ as in Lemma 6.5. Since $h_C \leq h_D$ and, by induction,

$$v\big(P_2 \cap H_{P_2}(u), \dots, P_d \cap H_{P_d}(u)\big) \geq v(\{o\}, \dots, \{o\}) = 0,$$

the inequality $V(C, P_2, \dots, P_d) \leq V(D, P_2, \dots, P_d)$ follows. This proves (11) and thus (10) which, in turn, yields the theorem. □

Corollary 6.2. *Let $C_1, \dots, C_d \in \mathcal{C}$. Then $V(C_1, \dots, C_d) \geq 0$.*

Proof. Without loss of generality, $o \in C_1, \dots, C_d$. Then Theorem 6.9 shows that $V(C_1, \dots, C_d) \geq V\big(\{o\}, \dots, \{o\}\big) = 0$. □

The Valuation Property

Mixed volumes have the following weak additivity property.

Theorem 6.10. *Let $C, D, D_2, \dots, D_d \in \mathcal{C}$ such that $C \cup D \in \mathcal{C}$. Then*

$$\begin{aligned} &V(C \cup D, D_2, \dots, D_d) + V(C \cap D, D_2, \dots, D_d) \\ &\quad = V(C, D_2, \dots, D_d) + V(D, D_2, \dots, D_d). \end{aligned}$$

Note that here we only require that additivity holds for *certain* pairs C, D, whereas in other context, for example in measure theory, it is required that it holds for *all* pairs C, D. Given $D_2, \dots, D_d \in \mathcal{C}$, this weak additivity property of the mapping $C \to V(C, D_2, \dots, D_d)$ is expressed by saying that this mapping is a *valuation*. For more on valuations, see Sect. 7.

We present two proofs of this result. One is based on Minkowski's theorem on mixed volumes, the other one was suggested by Károly Böröczky, Jr. [157] and makes use of the continuity of mixed volumes and of Lemma 6.5.

Proof (using Minkowski's theorem on mixed volumes). The following simple proposition will be needed later:

(12) Let $C, D, E \in \mathcal{C}$ such that $C \cup D \in \mathcal{C}$. Then
$(C \cup D) + E = (C + E) \cup (D + E)$ and
$(C \cap D) + E = (C + E) \cap (D + E)$.

The first assertion is clear. Similarly, the inclusion $(C \cap D) + E \subseteq (C + E) \cap (D + E)$ is obvious. To show the reverse inclusion, let $x \in (C + E) \cap (D + E)$. Then $x = c + e = d + f$ with $c \in C, d \in D, e, f \in E$. Since $C \cup D$ is convex, there is a point $p = (1 - \lambda)c + \lambda d \in [c, d] \cap (C \cap D)$. Hence $x = (1 - \lambda)(c + e) + \lambda(d + f) = p + (1 - \lambda)e + \lambda f \in (C \cap D) + E$. Thus $(C + E) \cap (D + E) \subseteq (C \cap D) + E$, concluding the proof of (12).

To prove the theorem, note that

$$\begin{aligned} &V\big(\lambda(C \cup D) + \lambda_2 D_2 + \dots + \lambda_d D_d\big) \\ &\quad = \sum_{j=0}^{d} \binom{d}{j} \lambda^j \sum_{i_{j+1},\dots,i_d=2}^{d} V(\underbrace{C \cup D, \dots, C \cup D}_{j}, D_{i_{j+1}}, \dots, D_{i_d}) \lambda_{i_{j+1}} \cdots \lambda_{i_d}, \\ &V\big(\lambda(C \cap D) + \lambda_2 D_2 + \dots + \lambda_d D_d\big) \\ &\quad = \sum_{j=0}^{d} \binom{d}{j} \lambda^j \sum_{i_{j+1},\dots,i_d=2}^{d} V(\underbrace{C \cap D, \dots, C \cap D}_{j}, D_{i_{j+1}}, \dots, D_{i_d}) \lambda_{i_{j+1}} \cdots \lambda_{i_d}, \\ &V(\lambda C + \lambda_2 D_2 + \dots + \lambda_d D_d) \\ &\quad = \sum_{j=0}^{d} \binom{d}{j} \lambda^j \sum_{i_{j+1},\dots,i_d=2}^{d} V(\underbrace{C, \dots, C}_{j}, D_{i_{j+1}}, \dots, D_{i_d}) \lambda_{i_{j+1}} \cdots \lambda_{i_d}, \\ &V(\lambda D + \lambda_2 D_2 + \dots + \lambda_d D_d) \\ &\quad = \sum_{j=0}^{d} \binom{d}{j} \lambda^j \sum_{i_{j+1},\dots,i_d=2}^{d} V(\underbrace{D, \dots, D}_{j}, D_{i_{j+1}}, \dots, D_{i_d}) \lambda_{i_{j+1}} \cdots \lambda_{i_d}. \end{aligned}$$

By (12) the sum of the first two of these volumes equals the sum of the last two. Equating the coefficients of $\lambda\,\lambda_2\cdots\lambda_d$ on both sides of this equality then yields the theorem. □

Proof (using the continuity of mixed volumes and Lemma 6.5). Since the mixed volumes are continuous in their entries by Theorem 6.8, it is sufficient to prove the theorem in case where $D_2, \dots, D_d$ are polytopes.

We now show that, for the support functions of $C \cup D$ and $C \cap D$, we have the equalities,

$$h_{C\cap D} = \min\{h_C, h_D\} \text{ and } h_{C\cup D} = \max\{h_C, h_D\}.$$

Let $u \in \mathbb{E}^d \setminus \{o\}$. We may assume that $h_C(u) \le h_D(u)$. Choose $x \in C$, $y \in D$ such that $x \cdot u = h_C(u)$ and $y \cdot u = h_D(u)$. The line segment from x to y is contained in $C \cup D$ since $C \cup D$ is convex. Let z be the last point on this line segment contained in C. Then $z \in C \cap D$. Since

$$C \cap D \subseteq C \subseteq \{v : v \cdot u \le x \cdot u\} \subseteq \{v : v \cdot u \le z \cdot u\},$$

we have

$$h_{C\cap D}(u) \le h_C(u) \le z \cdot u \le h_{C\cap D}(u)$$

and thus

$$h_{C\cap D}(u) = h_C(u) = \min\{h_C(u), h_D(u)\}.$$

This proves the first equality. The proof of the second equality is even simpler and thus omitted.

These equalities yield the identity,

$$h_{C\cup D} + h_{C\cap D} = h_C + h_D.$$

Now use Lemma 6.5. □

Minkowski's Inequalities

Mixed volumes satisfy several inequalities, in particular the following *first and second inequality of Minkowski* [739].

Theorem 6.11. *Let $C, D \in \mathcal{C}$. Then:*

(i) $V(C, D, \dots, D)^d \ge V(C)V(D)^{d-1}$, *where for proper convex bodies C, D equality holds if and only if C and D are (positive) homothetic.*
(ii) $V(C, D, \dots, D)^2 \ge V(C, C, D, \dots, D)V(D)$.

The equality case in the second inequality is more complicated to formulate and to prove. It was settled by Bol [139], thereby confirming a conjecture of Minkowski.

Proof. The Brunn–Minkowski theorem 8.3 shows that

(13) The function $V\big((1-\lambda)C + \lambda D\big)^{\frac{1}{d}}, 0 \le \lambda \le 1$, is concave in λ. If C and D are proper, then this expression is linear if and only if C and D are homothetic.

Minkowski's theorem on mixed volumes implies that

$$(14)\quad V\big((1-\lambda)C+\lambda D\big)=\sum_{i=0}^{d}\binom{d}{i}(1-\lambda)^i\lambda^{d-i}V(\underbrace{C,\dots,C}_{i},\underbrace{D,\dots,D}_{d-i}).$$

(i) Let

$$f(\lambda)=V\big((1-\lambda)C+\lambda D\big)^{\frac{1}{d}} \text{ for } 0\le\lambda\le 1.$$

(14) shows that f is differentiable. By (13) f is concave and $f(0)=V(C)^{\frac{1}{d}}$, $f(1)=V(D)^{\frac{1}{d}}$. Thus $f'(1)\le V(D)^{\frac{1}{d}}-V(C)^{\frac{1}{d}}$, where equality holds if and only if f is linear, or, using (13), if and only if C and D are homothetic. Now calculate $f'(1)$ using (14). This yields (i), including the equality case.

(ii) By (13) f is concave. (14) shows that f is twice differentiable. Thus $f''(1)\le 0$. This yields (ii). □

The Alexandrov–Fenchel Inequality

There is a far reaching generalization of Minkowski's inequalities, the *inequality of Alexandrov* [11, 13] *and Fenchel* [333]:

Theorem 6.12. *Let $C, D, D_3, \dots, D_d \in \mathcal{C}$. Then,*

$$V(C,D,D_3,\dots,D_d)^2\ge V(C,C,D_3,\dots,D_d)\,V(D,D,D_3,\dots,D_d).$$

This inequality has attracted a great deal of interest in the last two or three decades. It is related to the Hodge index theorem in algebraic geometry, see Khovanskiĭ [582] and Teissier [991, 992]. Compare also the book of Ewald [315]. The equality case in the Alexandrov–Fenchel inequality has not yet been settled. The special case of mixed discriminants, where the equality case is known, was used by Egorychev [291] to prove van der Waerden's conjecture on permanents of doubly stochastic matrices. For more information, see Burago and Zalgaller [178], Sangwine-Yager [878] and Schneider [907].

6.4 Quermassintegrals and Intrinsic Volumes

Quermassintegrals or, with a different normalization, intrinsic volumes, are special mixed volumes. Originally they appeared in Steiner's [959] formula for the volume of parallel bodies of a convex body, see Theorem 6.6. These quantities are of fundamental importance in convex geometry and, in particular, in the context of valuations, see Sect. 7.3.

In the following we present a series of properties of quermassintegrals. Among these are Steiner formulae for quermassintegrals, the famous surface area formula of Cauchy and Kubota's formulae. The latter yield an explanation why quermassintegrals are called quermassintegrals. We also describe a problem of Blaschke to characterize the set of vectors of quermassintegrals. In Sect. 7.3 the functional theorems of

Hadwiger will be treated, in which the quermassintegrals are characterized as special valuations.

For more information we refer to Leichtweiss [640], Schneider [905, 908], Sangwine-Yager [878] and McMullen and Schneider [716].

Quermassintegrals and Intrinsic Volumes

Recall Steiner's theorem on parallel bodies, Theorem 6.6: Given a convex body $C \in \mathcal{C}$, we have,

$$V(C + \lambda B^d) = W_0(C) + \binom{d}{1} W_1(C)\lambda + \cdots + \binom{d}{d} W_d(C)\lambda^d \text{ for } \lambda \geq 0,$$

where the coefficients

$$W_i(C) = V(\underbrace{C, \dots, C}_{d-i}, \underbrace{B^d, \dots, B^d}_{i}),\ i = 0, \dots, d,$$

are the *quermassintegrals* of C. If C is contained in a subspace of $\mathbb{E}^d$, one can define the quermassintegrals of C both in $\mathbb{E}^d$ and in this subspace. Unfortunately the result is not the same as the following proposition shows, where $v(\cdot)$ is the volume and $w_i(\cdot), i = 0, \dots, d-1$, are the quermassintegrals in $d-1$ dimensions, $\kappa_0 = 1$, and $\kappa_i,\ i = 1, \dots, d$, is the i-dimensional volume of B^i.

Proposition 6.7. *Let $C \in \mathcal{C}(\mathbb{E}^{d-1})$ and embed $\mathbb{E}^{d-1}$ into $\mathbb{E}^d$ as usual (first $d-1$ coordinates). Then*

$$W_i(C) = \frac{i\,\kappa_i}{d\,\kappa_{i-1}} w_{i-1}(C) \text{ for } i = 1, \dots, d.$$

Proof. Let $u = (0, \dots, 0, 1)$. Then

$$\sum_{i=0}^{d} \binom{d}{i} W_i(C)\lambda^i = V(C + \lambda B^d) = \int_{-\lambda}^{\lambda} v\big((C + \lambda B^d) \cap (\mathbb{E}^{d-1} + tu)\big)\, dt$$

$$= \int_{-\lambda}^{\lambda} v\big(C + (\lambda^2 - t^2)^{\frac{1}{2}} B^{d-1}\big)\, dt = \int_{-\lambda}^{\lambda} \Big(\sum_{i=0}^{d-1} \binom{d-1}{i} w_i(C)(\lambda^2 - t^2)^{\frac{i}{2}}\Big)\, dt$$

$$= \sum_{i=0}^{d-1} \binom{d-1}{i} w_i(C) \int_{-\lambda}^{\lambda} (\lambda^2 - t^2)^{\frac{i}{2}}\, dt = \sum_{i=0}^{d-1} \binom{d-1}{i} w_i(C) \frac{\kappa_{i+1}}{\kappa_i} \lambda^{i+1}$$

for $\lambda \geq 0$

by Steiner's formula in $\mathbb{E}^d$, Steiner's formula in $\mathbb{E}^{d-1}$ and Fubini's theorem which is used to calculate the $(i+1)$-dimensional volume of the ball λB^{i+1}. Now equate coefficients. □

To remedy the situation that the quermassintegrals depend on the dimension of the embedding space, McMullen [708] introduced the *intrinsic volumes* $V_i(C)$, defined by

$$V_{d-i}(C) = \frac{\binom{d}{i}}{\kappa_i} W_i(C) \text{ for } C \in \mathcal{C} \text{ and } i = 0, \dots, d.$$

Proposition 6.7 shows that the intrinsic volumes depend only on the convex body and not on the dimension of the embedding space.

From Steiner's formula and the definition of the (Minkowski) surface area $S(\cdot)$ below we obtain

$$W_0(C) = V_d(C) = V(C),\ W_1(C) = \frac{2}{d} V_{d-1}(C) = \frac{1}{d} S(C),$$
$$W_d(C) = \kappa_d V_0(C) = \kappa_d.$$

In addition,

$$W_{d-1}(C) = \frac{\kappa_d}{2} w(C),\ \text{where } w(C) = \frac{2}{d\,\kappa_d} \int\limits_{S^{d-1}} h_C(u)\, d\sigma(u)$$

is the *mean width* of C. Here σ is the ordinary surface area measure. There is no particularly simple proof for this equality, so we prove it as a consequence of Hadwiger's functional theorem 7.9, see Corollary 7.1.

Minkowski's Surface Area

Minkowski [738, 739] introduced (see Fig. 6.4) the following notion of (*Minkowski*) *surface area* $S(C)$:

$$S(C) = \lim_{\varepsilon \to +0} \frac{V(C + \varepsilon B^d) - V(C)}{\varepsilon}.$$

Since, by Steiner's formula,

$$V(C + \varepsilon B^d) = V(C) + d\, W_1(C)\varepsilon + O(\varepsilon^2) \text{ as } \varepsilon \to +0,$$

this limit exists and equals $d\, W_1(C)$. Together with the Brunn–Minkowski theorem 8.1, this notion of surface area easily leads to the isoperimetric inequality,

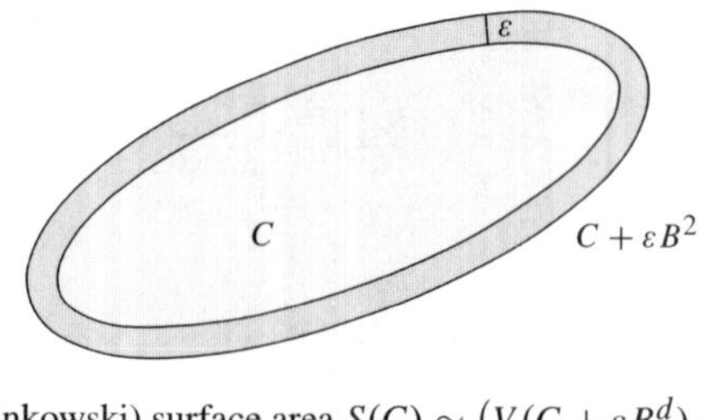

Fig. 6.4. (Minkowski) surface area $S(C) \sim \big(V(C + \varepsilon B^d) - V(C)\big)/\varepsilon$

including the equality case, see Theorem 8.7. Note that, if C has dimension $d-1$, then $S(C)$ equals twice the $(d-1)$-dimensional volume of C. If $d = 2$, we also write $P(C)$ for $S(C)$ and call $P(C)$ the *perimeter* of C.

We point out that this notion of surface area coincides with the elementary surface area of proper convex polytopes and with the common notion of area of differentiable surfaces in $\mathbb{E}^d$. For more complicated sets it is – if it exists – in general larger than common notions of surface area such as the Lebesgue or the Hausdorff surface area or the more recent notions of integral geometric surface areas, perimeter and currents. Such questions are not discussed in this book and we refer to Burago and Zalgaller [178] and Morgan [756].

Properties of Quermassintegrals

In Theorem 6.13 we collect results for quermassintegrals, proved in Sect. 6.3 for the more general case of mixed volumes.

Theorem 6.13. *The following statements hold for* $i = 0, \dots, d$:

(i) $W_i(\cdot)$ *is rigid motion invariant and thus, in particular, translation invariant, i.e.* $W_i(mC) = W_i(C)$ *for* $C \in \mathcal{C}$ *and each rigid motion* m *of* $\mathbb{E}^d$.
(ii) $W_i(\cdot)$ *is (positive) homogeneous of degree* $d-i$, *i.e.* $W_i(\lambda C) = \lambda^{d-i} W_i(C)\}$ *for* $C \in \mathcal{C}$ *and* $\lambda \geq 0$.
(iii) $W_i(\cdot)$ *is continuous on* $\mathcal{C}$.
(iv) $W_i(\cdot)$ *is non-decreasing on* $\mathcal{C}$ *with respect to set inclusion.*
(v) $W_i(\cdot)$ *is a valuation on* $\mathcal{C}$, *that is, if* $C, D \in \mathcal{C}$ *are such that* $C \cup D \in \mathcal{C}$, *then*

$$W_i(C \cup D) + W_i(C \cap D) = W_i(C) + W_i(D).$$

Proof. Properties (i)–(iv) are immediate consequences of Propositions 6.5, 6.6 and Theorems 6.9, 6.10. Property (v) follows from the first proof of Theorem 6.10 on putting $D_2 = \dots = D_d = B^d$, $\lambda_2 = \dots = \lambda_d = 1$ and comparing the coefficients of λ^i. □

Steiner Formulae for Quermassintegrals

Steiner's formula 6.6 for the volume of parallel bodies can be extended as follows.

Theorem 6.14. *Let* $C \in \mathcal{C}$. *Then*

$$W_i(C + \lambda B^d) = \sum_{k=0}^{d-i} \binom{d-i}{k} W_{i+k}(C)\lambda^k \text{ for } \lambda \geq 0 \text{ and } i = 0, \dots, d.$$

In particular,

$$S(C + \lambda B^d) = d\, W_1(C + \lambda B^d) = \sum_{k=0}^{d-1} d \binom{d-1}{k} W_{k+1}(C)\lambda^k \text{ for } \lambda \geq 0.$$

Proof. Applying Steiner's formula to $(C + \lambda B^d) + \mu B^d = C + (\lambda + \mu)B^d$ shows that

$$
\begin{aligned}
\sum_{i=0}^{d} \binom{d}{i} W_i(C + \lambda B^d)\mu^i &= \sum_{i=0}^{d} \binom{d}{i} W_i(C)(\lambda + \mu)^i \\
&= \sum_{i=0}^{d} \binom{d}{i} W_i(C)\Big(\lambda^i + \binom{i}{1}\lambda^{i-1}\mu + \dots + \binom{i}{i}\mu^i\Big) \\
&= \sum_{i=0}^{d} \Big(\binom{d}{i}\binom{i}{i} W_i(C) + \binom{d}{i+1}\binom{i+1}{i} W_{i+1}(C)\lambda + \cdots \\
&\quad + \binom{d}{d}\binom{d}{i} W_d(C)\lambda^{d-i}\Big)\mu^i \\
&= \sum_{i=0}^{d} \Big(\sum_{k=0}^{d-i} \binom{d}{i+k}\binom{i+k}{i} W_{i+k}(C)\lambda^k\Big)\mu^i \text{ for } \lambda, \mu \geq 0.
\end{aligned}
$$

Equating the coefficients of μ^i, we obtain that

$$
W_i(C + \lambda B^d) = \sum_{k=0}^{d-i} \frac{\binom{d}{i+k}\binom{i+k}{i}}{\binom{d}{i}} W_{i+k}(C)\lambda^k = \sum_{k=0}^{d-i} \binom{d-i}{k} W_{i+k}(C)\lambda^k
$$

for $\lambda \geq 0$. □

Cauchy's Surface Area Formula

Given $C \in \mathcal{C}$ and $u \in S^{d-1}$, let $C|u^\perp$ denote the orthogonal projection of C into the $(d-1)$-dimensional subspace $u^\perp = \{x : u \cdot x = 0\}$ orthogonal to u. Let σ be the ordinary surface area measure in $\mathbb{E}^d$. Then the *surface area formula of Cauchy* [198] is as follows:

Theorem 6.15. *Let $C \in \mathcal{C}$. Then*

(1) $S(C) = \dfrac{1}{\kappa_{d-1}} \displaystyle\int_{S^{d-1}} v(C|u^\perp)\, d\sigma(u).$

Proof. First, let $C = P$ be a proper convex polytope. If F is a facet of P, let u_F be its exterior unit normal vector. Since P is a proper convex polytope,

$$
S(P) = \sum_{F \text{ facet of } P} v(F),
$$

as pointed out above. Noting that $v(F)\,|u_F \cdot u| = v(F|u^\perp)$ for $u \in S^{d-1}$, integration over S^{d-1} shows that

$$\int_{S^{d-1}} v(F|u^{\perp})\,d\sigma(u) = v(F)\int_{S^{d-1}} |u_F\cdot u|\,d\sigma(u) = 2\kappa_{d-1}v(F).$$

Hence

$$S(P) = \sum_{F \text{ facet of } P} v(F) = \frac{1}{2\kappa_{d-1}}\int_{S^{d-1}} \Big(\sum_{F \text{ facet of } P} v(F|u^{\perp})\Big)\,d\sigma(u)$$

$$= \frac{1}{\kappa_{d-1}}\int_{S^{d-1}} v(P|u^{\perp})\,d\sigma(u).$$

Second, let C be a proper convex body. We may assume that $o \in \operatorname{int} C$. By the proof of Theorem 7.4, there is a sequence (P_n) of proper convex polytopes such that

$$P_n \subseteq C \subseteq \Big(1+\frac{1}{n}\Big)P_n \text{ and } P_n \to C \text{ as } n \to \infty.$$

$S(\cdot) = d\,W_1(\cdot)$ is continuous by Theorem 6.13 (iii). Thus

(2) $S(P_n) \to S(C)$ as $n \to \infty$.

The functions $v(P_n|u^{\perp}) : u \in S^{d-1}$ are continuous in u for each $n = 1, 2, \dots$ Since

$$v(P_n|u^{\perp}) \le v(C|u^{\perp}) \le \Big(1+\frac{1}{n}\Big)^{d-1} v(P_n|u^{\perp}) \text{ for } u \in S^{d-1},$$

and since $v(C|u^{\perp})$ is bounded on S^{d-1}, the function $v(C|u^{\perp}) : u \in S^{d-1}$ is the uniform limit of the continuous functions $v(P_n|u^{\perp})$. Thus it is continuous itself. Integration over S^{d-1} then shows that

$$\int_{S^{d-1}} v(P_n|u^{\perp})\,d\sigma(u) \le \int_{S^{d-1}} v(C|u^{\perp})\,d\sigma(u) \le \Big(1+\frac{1}{n}\Big)^{d-1}\int_{S^{d-1}} v(P_n|u^{\perp})\,d\sigma(u).$$

Since, by the first part of the proof,

$$S(P_n) = \frac{1}{\kappa_{d-1}}\int_{S^{d-1}} v(P_n|u^{\perp})\,d\sigma(u),$$

we conclude that

$$S(P_n) \to \frac{1}{\kappa_{d-1}}\int_{S^{d-1}} v(C|u^{\perp})\,d\sigma(u) \text{ as } n \to \infty.$$

Together with (2), this shows that

$$S(C) = \frac{1}{\kappa_{d-1}}\int_{S^{d-1}} v(C|u^{\perp})\,d\sigma(u).$$

Third, let C be a convex body of dimension $d-1$. We may assume that $C \subseteq v^\perp$ for suitable $v \in S^{d-1}$. Then

$$\frac{1}{\kappa_{d-1}} \int_{S^{d-1}} v(C|u^\perp)\, d\sigma(u) = \frac{1}{\kappa_{d-1}} \int_{S^{d-1}} |u \cdot v|\, d\sigma(u)\, v(C) = 2v(C) = S(C).$$

If, fourth, $\dim C < d-1$, then both sides in (1) are 0 and thus coincide. □

An Integral-Geometric Interpretation of Cauchy's Formula

Given a set $\mathcal{L}$ of lines in $\mathbb{E}^d$, a natural measure for $\mathcal{L}$ can be defined as follows: For any $(d-1)$-dimensional subspace $u^\perp$ of $\mathbb{E}^d$, where $u \in S^{d-1}$, consider the lines in $\mathcal{L}$ which are orthogonal to $u^\perp$. Let $v(u)$ denote the measure of the intersection of this set of lines with $u^\perp$ (if the intersection is measurable). The integral of $v(u)$ over S^{d-1} with respect to the ordinary surface area measure, then, is the *measure of the set $\mathcal{L}$ of lines* (if the integral exists). Clearly, this measure is rigid motion invariant. Cauchy's surface area formula now says that the surface area of a convex body C, that is the area of its boundary bd C, equals (up to a multiplicative constant) the integral of the function which assigns to each line the number of its intersection points with bd C. This interpretation extends to all sufficiently smooth surfaces in $\mathbb{E}^d$ and is the starting point for so-called *integral geometric surface area*. For more on integral geometry, see the standard monograph of Santaló [881] and Sect. 7.4. Geometric measure theory is treated by Falconer [317] and Mattila [696].

Kubota's Formulae for Quermassintegrals

Let $w_i(\cdot)$, $i = 0, \dots, d-1$, denote the quermassintegrals in $d-1$ dimensions. Then one may write Cauchy's formula in the following form:

$$W_1(C) = \frac{1}{d\,\kappa_{d-1}} \int_{S^{d-1}} w_0(C|u^\perp)\, d\sigma(u).$$

It was the idea of Kubota [619] to extend this formula to all quermassintegrals. We state the following special case.

Theorem 6.16. *Let $C \in \mathcal{C}$. Then*

$$(3)\quad W_i(C) = \frac{1}{d\,\kappa_{d-1}} \int_{S^{d-1}} w_{i-1}(C|u^\perp)\, d\sigma(u) \text{ for } i = 1, \dots, d.$$

Proof. First note that

$$(C + \lambda B^d)|u^\perp = C|u^\perp + \lambda B^d|u^\perp \text{ for } u \in S^{d-1}.$$

Steiner's formula in $d-1$ dimensions then shows that

$$v\big((C+\lambda B^d)|u^\perp\big) = v\big(C|u^\perp + \lambda B^d|u^\perp\big) = \sum_{i=0}^{d-1}\binom{d-1}{i} w_i(C|u^\perp)\lambda^i.$$

Since $C|u^\perp$ varies continuously with u and $w_i(\cdot)$ is continuous, we may integrate over S^{d-1} and Cauchy's surface area formula, applied to $C+\lambda B^d$, shows that

$$(4)\quad S(C+\lambda B^d) = \sum_{i=0}^{d-1}\binom{d-1}{i}\frac{1}{\kappa_{d-1}}\int\limits_{S^{d-1}} w_i(C|u^\perp)\,d\sigma(u)\,\lambda^i \text{ for } \lambda \ge 0.$$

Since $S(C+\lambda B^d) = d\,W_1(C+\lambda B^d)$, Steiner's formula for $W_1(\cdot)$ yields the following, see Theorem 6.14:

$$(5)\quad S(C+\lambda B^d) = d\sum_{i=0}^{d-1}\binom{d-1}{i} W_{i+1}(C)\lambda^i \text{ for } \lambda \ge 0.$$

Finally, equating coefficients in (4) and (5), implies (3). □

A Remark on the Proofs

In the proofs of Steiner's and Kubota's formulae for quermassintegrals, we have expressed the same quantity in two different ways as polynomials. Hence these polynomials must be identical, i.e. corresponding coefficients coincide. This finally yields the desired formulas. Expressing the same quantity in different ways and equating is a common method of proof in integral geometry, see Santaló [881].

Why are Quermassintegrals or Mean Projection Measures Called So?

Iterating (3) with respect to the dimension, yields the general formulae of Kubota. These express the quermassintegrals $W_i(C)$ of C as the mean of the $(d-i)$-dimensional volumes of the projections of C onto $(d-i)$-dimensional linear subspaces. These volumes are called Quermaße in German.

Blaschke's Problem for Quermassintegrals

We conclude this section with the following major problem which goes back to Blaschke [125].

Problem 6.1. *Determine the set* $\big\{\big(W_0(C),\dots,W_{d-1}(C)\big) : C \in \mathcal{C}\big\} \subseteq \mathbb{E}^d$.

In case $d=2$ the solution is given by the isoperimetric inequality: The set in question is $\{(P,A) : P^2 \ge 4\pi A,\ P, A \ge 0\}$, where P and A stand for perimeter and area. The problem is open for $d \ge 3$. For some references, see Hadwiger [468], Schneider [907], and Sangwine-Yager [877, 878].

A Problem of Santaló

Santaló proposed the following related question.

Problem 6.2. *Consider a finite set of geometric functionals $F_1, \dots, F_k$ for convex bodies in $\mathbb{E}^d$, for example the inradius r, the volume V and the diameter* diam. *Determine the set* $\big\{\big(F_1(C), \dots, F_k(C)\big) : C \in \mathcal{C}\big\} \subseteq \mathbb{E}^k$.

As is the case for the above problem of Blaschke, this problem is also difficult in this generality. For pertinent results see Hernández Cifre, Pastor, Salinas Martínez and Segura Gomis [497].

7 Valuations

Let $\mathcal{S}$ be a family of sets. A (*real*) *valuation* on $\mathcal{S}$ is a real function ϕ on $\mathcal{S}$ which is additive in the following sense:

$$\phi(C \cup D) + \phi(C \cap D) = \phi(C) + \phi(D) \text{ whenever } C, D, C \cup D, C \cap D \in \mathcal{S}, \\ \text{and } \phi(\emptyset) = 0 \text{ if } \emptyset \in \mathcal{S}.$$

In many cases of interest, $\mathcal{S}$ is *intersectional*, that is, $C \cap D \in \mathcal{S}$ for $C, D \in \mathcal{S}$. If $\mathcal{S}$ is the space $\mathcal{C}$ of convex bodies in $\mathbb{E}^d$ or a subspace of it such as the space $\mathcal{P}$ of convex polytopes, and in many other cases, the above additivity property is rather weak since we require additivity only for a small set of pairs C, D in $\mathcal{C}$. In case where $\mathcal{S}$ is the space of Jordan or of Lebesgue measurable sets in $\mathbb{E}^d$, the valuation property coincides with the common notion of additivity in measure theory. Examples of valuations on $\mathcal{C}$, $\mathcal{P}$ and on the space of lattice polytopes, are the volume, the quermassintegrals, affine surface area, the Dehn invariants in the context of Hilbert's third problem and the various lattice point enumerators. Mixed volumes also give rise to valuations on $\mathcal{C}$.

While special valuations have been investigated since antiquity, it seems that Blaschke [128], Sects. 41,43, was the first to consider valuations per se and he initiated their study. Then, his disciple Hadwiger started the systematic investigation of valuations, culminating in the functional theorems, see [468]. Important later contributions are due to Groemer, McMullen, Schneider, Betke and Kneser, Klain, Alesker, Ludwig and Reitzner, and others, see [108, 402, 587, 666, 713, 714, 716]. Amongst others, these contributions deal with Hilbert's third problem, McMullen's polytope algebra, lattice point enumerators and with characterizations and representations of certain important classes of valuations.

This section contains a small account of the rich theory of valuations. We start with extension results, introduce the elementary volume and Jordan measure, then give a characterization of the volume and, as a consequence, show Hadwiger's functional theorem. As an application of the functional theorem, the principal kinematic formula is proved.

For more information on valuations, see Hadwiger [468], McMullen and Schneider [716], McMullen [714], Klain and Rota [587] and Peri [790]. See also the very readable popular article of Rota [858]. Special valuations in the context of polytopes and lattice polytopes will be dealt with in Sects. 16.1 and 19.4.

7.1 Extension of Valuations

Given a valuation on a family of sets, the problem arises to extend this valuation to larger families of sets, for example to the (algebraic) lattice of sets generated by the given family.

This section contains the inclusion–exclusion formula and the extension results of Volland, rediscovered by Perles and Sallee in a more abstract form, and of Groemer.

The Inclusion–Exclusion Formula

Let $\mathcal{L}$, or more precisely, $\langle \mathcal{L}, \cap, \cup \rangle$, be a lattice of sets, where the lattice operations are the ordinary intersection and union. Given a valuation $\phi : \mathcal{L} \to \mathbb{R}$, iterating the relation

$$\phi(C \cup D) + \phi(C \cap D) = \phi(C) + \phi(D) \text{ for } C, D \in \mathcal{L},$$

easily leads to the equality,

$$(1)\quad \phi(C_1 \cup \dots \cup C_m) = \sum_i \phi(C_i) - \sum_{i<j} \phi(C_i \cap C_j) + \\ + \sum_{i<j<k} \phi(C_i \cap C_j \cap C_k) - \dots + (-1)^{m-1} \phi(C_1 \cap \dots \cap C_m) \\ \text{for } C_1, \dots, C_m \in \mathcal{L}.$$

The indices run from 1 to m. Sometimes it is helpful to rewrite (1) in a more concise form. We denote by I the ordered k-tuples $\emptyset \neq I = \{i_1, \dots, i_k\} \subseteq \{1, \dots, m\}$ where $k = 1, \dots, m,\ 1 \leq i_1 < i_2 < \dots < i_k \leq m$, and put

$$C_I = C_{i_1} \cap \dots \cap C_{i_k} \text{ and } |I| = k.$$

Then (1) may be written in the following form:

$$(2)\quad \phi(C_1 \cup \dots \cup C_m) = \sum_I (-1)^{|I|-1} \phi(C_I) \text{ for } C_1, \dots, C_m \in \mathcal{L}.$$

That ϕ has properties (1) or (2) is also expressed by saying that ϕ satisfies the *inclusion–exclusion formula* on $\mathcal{L}$. It will be seen in a moment, that this modest looking formula is crucial for the extension of valuations.

A General Extension Result

Let $\mathcal{S}$ be an intersectional family of sets and $\phi : \mathcal{S} \to \mathbb{R}$ a valuation. The problem arises, whether ϕ can be extended to a valuation on the lattice $\mathcal{L}(\mathcal{S})$ generated by $\mathcal{S}$. Since $\mathcal{S}$ is intersectional, $\mathcal{L}(\mathcal{S})$ consists of all possible unions of finitely many sets of $\mathcal{S}$. If ϕ can be extended to a valuation on $\mathcal{L}(\mathcal{S})$, then this extension satisfies the inclusion–exclusion formula on $\mathcal{L}$, as seen before. This shows that, in particular, the original valuation ϕ on $\mathcal{S}$ must satisfy the *inclusion–exclusion formula for an intersectional family* on $\mathcal{S}$:

$$\text{(3)}\quad \phi(C_1 \cup \cdots \cup C_m) = \sum_i \phi(C_i) - \sum_{i<j} \phi(C_i \cap C_j) + - \cdots$$
$$= \sum_I (-1)^{|I|-1} \phi(C_I), \text{ for } C_1, \dots, C_m, C_1 \cup \cdots \cup C_m \in \mathcal{S}.$$

Thus (3) is a necessary condition for a valuation ϕ on $\mathcal{S}$ to be extendible to a valuation on $\mathcal{L}(\mathcal{S})$. Surprisingly, this simple necessary condition for extension is also sufficient, as the following *first extension theorem of Volland* [1011] shows; see also Perles and Sallee [792].

Theorem 7.1. *Let $\mathcal{S}$ be an intersectional family of sets and $\phi : \mathcal{S} \to \mathbb{R}$ a valuation. Then the following claims are equivalent:*

(i) *ϕ satisfies the inclusion–exclusion formula on $\mathcal{S}$.*
(ii) *ϕ has a unique extension to a valuation on $\mathcal{L}(\mathcal{S})$.*

The proof follows Volland, but requires filling a gap.

Proof. Since the implication (ii)⇒(i) follows from what was said above, it is sufficient to prove that

(i)⇒(ii) In a first step it will be shown that

$$\text{(4)}\quad \sum_I (-1)^{|I|-1} \phi(C_I) = \sum_J (-1)^{|J|-1} \phi(D_J)$$
for $C_1, \dots, C_m, D_1, \dots, D_n \in \mathcal{S}$, where $C_1 \cup \cdots \cup C_m = D_1 \cup \cdots \cup D_n$.

Since $C_1 \cup \cdots \cup C_m = D_1 \cup \cdots \cup D_n$,

$$\sum_I (-1)^{|I|-1} \phi(C_I) = \sum_I (-1)^{|I|-1} \phi\big(C_I \cap (D_1 \cup \cdots \cup D_n)\big)$$
$$= \sum_I (-1)^{|I|-1} \phi\big((C_I \cap D_1) \cup \cdots \cup (C_I \cap D_n)\big)$$
$$= \sum_I (-1)^{|I|-1} \sum_J (-1)^{|J|-1} \phi(C_I \cap D_J)$$
$$= \sum_J (-1)^{|J|-1} \sum_I (-1)^{|I|-1} \phi(D_J \cap C_I) = \cdots = \sum_J (-1)^{|J|-1} \phi(D_J),$$

concluding the proof of (4).

Define a function $\phi : \mathcal{L}(\mathcal{S}) \to \mathbb{R}$ by

$$\text{(5)}\quad \phi(C_1 \cup \cdots \cup C_m) = \sum_I (-1)^{|I|-1} \phi(C_I) \text{ for } C_1, \dots, C_m \in \mathcal{S},$$

where on the right side ϕ means the given valuation on $\mathcal{S}$. By (4), this function ϕ is well defined and extends the given valuation. We show that

(6) ϕ is a valuation on $\mathcal{L}(\mathcal{S})$.

For the proof of (6) we will derive two identities. As a preparation for the proof of the first identity, we prove the following proposition.

(7) Let L be a non-empty finite set. Then $\sum\limits_{\substack{J,K\neq\emptyset\\ J\cup K=L}}(-1)^{|J|+|K|}=-(-1)^{|L|}$.

For simplicity, we omit, in all sums in the proof of (7), $(-1)^{|J|+|K|}$. The proof is by induction on $|L|$. For $|L|=1$, Proposition (7) is trivial. Assume now that $|L|=l>1$ and that (7) holds for $l-1$. We may suppose that $L=\{1,2,\dots,l\}$. Then

$$\sum_{\substack{J,K\neq\emptyset\\ J\cup K=L}}=\sum_{\substack{1\in J,K\\ J\cup K=L}}+\sum_{\substack{1\in J,\notin K\\ J\cup K=L}}+\sum_{\substack{1\notin J,\in K\\ J\cup K=L}}.$$

Next note that

$$\sum_{\substack{1\in J,K\\ J\cup K=L}}=\sum_{\substack{J=\{1\},1\in K\\ J\cup K=L}}+\sum_{\substack{1\in J,K=\{1\}\\ J\cup K=L}}+\sum_{\substack{\{1\}\subsetneq J,K\\ J\cup K=L}}\qquad =2(-1)^{|L|+1}-(-1)^{|L|-1+2},$$

$$\sum_{\substack{1\in J,\notin K\\ J\cup K=L}}=\sum_{\substack{J=\{1\},1\notin K\\ J\cup K=L}}+\sum_{\substack{\{1\}\subsetneq J,1\notin K\\ J\cup K=L}}\qquad =(-1)^{|L|}-(-1)^{|L|-1+1},$$

$$\sum_{\substack{1\notin J,1\in K\\ J\cup K=L}}=\qquad =(-1)^{|L|}-(-1)^{|L|-1+1},$$

by induction. Now we add to get (7). The induction is complete.

The first required identity is as follows:

(8) Let $C_1\cup\dots\cup C_m,\ D=D_1\cup\dots\cup D_n\in\mathcal{S}$. Then
$\phi\big((C_1\cup\dots\cup C_m)\cap(D_1\cup\dots\cup D_n)\big)=-\sum\limits_{I,J}(-1)^{|I|+|J|-1}\phi(C_I\cap D_J)$.

The proof is by induction on m. Assume first that $m=1$. Then

$$\begin{aligned}\phi\big(C_1\cap(D_1\cup\dots\cup D_n)\big)&=\phi\big((C_1\cap D_1)\cup\dots\cup(C_1\cap D_n)\big)\\ &=\sum_J(-1)^{|J|-1}\phi(C_1\cap D_J)=-\sum_{\substack{I=\{1\}\\ J\subseteq\{1,\dots,n\}}}(-1)^{|I|+|J|-1}\phi(C_I\cap D_J)\end{aligned}$$

by the definition of ϕ on $\mathcal{L}(\mathcal{S})$ and since $I=\{1\}$ is the only possibility for I. This settles (8) for $m=1$. Now let $m>1$ and assume that the identity (8) holds for $m-1$. Then

$$\begin{aligned}&\phi\big((C_1\cup\dots\cup C_m)\cap(D_1\cup\dots\cup D_n)\big)\\ &\quad=\phi\big((C_1\cap(D_1\cup\dots\cup D_n))\cup((C_2\cup\dots\cup C_m)\cap(D_1\cup\dots\cup D_n))\big)\\ &\quad=\phi\Big((C_1\cap D_1)\cup\dots\cup(C_1\cap D_n)\ \cup\bigcup_{\substack{i\in\{2,\dots,m\}\\ j\in\{1,\dots,n\}}}(C_i\cap D_j=E_{ij},\ \text{say})\Big)\end{aligned}$$

$$= \sum_{J\subseteq\{1,\dots,n\}} (-1)^{|J|-1}\phi(C_1\cap D_J)$$

$$+ \sum_{J\subseteq\{1,\dots,n\}} (-1)^{|J|} \sum_{K\subseteq\{2,\dots,m\}\times\{1,\dots,n\}} (-1)^{|K|-1}\phi(C_1\cap D_J\cap E_K)$$

$$+ \sum_{K\subseteq\{2,\dots,m\}\times\{1,\dots,n\}} (-1)^{|K|-1}\phi(E_K)$$

(by the definition of ϕ on $\mathcal{L}(\mathcal{S})$)

$$= - \sum_{\substack{I\subseteq\{1\}\\ J\subseteq\{1,\dots,n\}}} (-1)^{|I|+|J|-1}\phi(C_I\cap D_J)$$

$$+ \sum_{J\subseteq\{1,\dots,n\}} (-1)^{|J|}\phi\Big(\bigcup_{\substack{i\in\{2,\dots,m\}\\ j\in\{1,\dots,n\}}} (C_1\cap D_J\cap C_i\cap D_j)\Big)$$

$$+\phi\Big(\bigcup_{\substack{i\in\{2,\dots,m\}\\ j\in\{1,\dots,n\}}} (C_i\cap D_j)\Big)$$

(by the definition of ϕ on $\mathcal{L}(\mathcal{S})$)

$$= - \sum_{\substack{I\subseteq\{1\}\\ J\subseteq\{1,\dots,n\}}} (-1)^{|I|+|J|-1}\phi(C_I\cap D_J) - \sum_{\substack{I\subseteq\{2,\dots,m\}\\ J\subseteq\{1,\dots,n\}}} (-1)^{|I|+|J|-1}\phi(C_I\cap D_J)$$

$$- \sum_{J\subseteq\{1,\dots,n\}} (-1)^{|J|} \sum_{\substack{I\subseteq\{2,\dots,m\}\\ K\subseteq\{1,\dots,n\}}} (-1)^{|I|+|K|-1}\phi(C_1\cap C_I\cap D_J\cap D_K)$$

(by induction, applied twice)

$$= - \sum_{\substack{I\subseteq\{1\} \text{ or } I\subseteq\{2,\dots,m\}\\ J\subseteq\{1,\dots,n\}}} (-1)^{|I|+|J|-1}\phi(C_I\cap D_J)$$

$$- \sum_{\substack{I\cap\{1\}\neq\emptyset\\ I\cap\{2,\dots,m\}\neq\emptyset\\ J\subseteq\{1,\dots n\}\\ K\subseteq\{1,\dots,n\}}} (-1)^{|I|-1+|J|+|K|-1}\phi(C_I\cap D_{\underbrace{J\cup K}_{=L}})$$

$$= - \sum_{\substack{I\subseteq\{1\} \text{ or } I\subseteq\{2,\dots,m\}\\ J\subseteq\{1,\dots,n\}}} (-1)^{|I|+|J|-1}\phi(C_I\cap D_J)$$

$$+ \sum_{\substack{I\cap\{1\}\neq\emptyset\\ I\cap\{2,\dots,m\}\neq\emptyset}} (-1)^{|I|-1} \sum_{L\subseteq\{1,\dots,n\}} \underbrace{\Big(\sum_{J\cup K=L} (-1)^{|J|+|K|}\Big)}_{=-(-1)^{|L|}}\phi(C_I\cap D_L)$$

(by Proposition (7))

$$= - \sum_{\substack{I\subseteq\{1,\dots,m\}\\ L\subseteq\{1,\dots,n\}}} (-1)^{|I|+|L|-1}\phi(C_I\cap D_L).$$

The induction is complete, concluding the proof of the identity (8).

The second identity is as follows.

(9) Let $C_1, \ldots, C_m, D_1, \ldots, D_n \in \mathcal{S}$. Then

$$\begin{aligned}&\phi\big((C_1 \cup \cdots \cup C_m) \cup (D_1 \cup \cdots \cup D_n)\big)\\ &= \sum_I (-1)^{|I|-1}\phi(C_I) + \sum_{I,J}(-1)^{|I|+|J|-1}\phi(C_I \cap D_J) + \sum_J (-1)^{|J|-1}\phi(D_J)\\ &= \phi(C_1 \cup \cdots \cup C_m) + \sum_{I,J}(-1)^{|I|+|J|-1}\phi(C_I \cap D_J) + \phi(D_1 \cup \cdots \cup D_n).\end{aligned}$$

Now add the identities (8) and (9) to get $\phi(C \cup D) + \phi(C \cap D) = \phi(C) + \phi(D)$ for $C, D \in \mathcal{L}(\mathcal{S})$, concluding the proof of (6).

Thus we have obtained a valuation on $\mathcal{L}(\mathcal{S})$ which extends the given valuation on $\mathcal{S}$. Since each valuation on the lattice $\mathcal{L}(\mathcal{S})$, which extends the given valuation on $\mathcal{S}$, satisfies (5), the extension is unique. □

Remark. For far-reaching related results, see Groemer [402].

Extension of Valuations on Boxes and Convex Polytopes

A *box* in $\mathbb{E}^d$ is a set of the form $\{x : \alpha_i \le x_i \le \beta_i\}$. Its *edge-lengths* are $\beta_i - \alpha_i$, $i = 1, \ldots, d$. Let $\mathcal{B} = \mathcal{B}(\mathbb{E}^d)$ be the *space of boxes* in $\mathbb{E}^d$. For the definition of convex polytopes, see Sects. 6.2, 14.1 and let $\mathcal{P} = \mathcal{P}(\mathbb{E}^d)$ be the *space of convex polytopes* in $\mathbb{E}^d$. Both $\mathcal{B}$ and $\mathcal{P}$ are intersectional families of sets. The following *polytope extension theorem* is due to *Volland* [1011].

Theorem 7.2. *Let ϕ be a valuation on $\mathcal{B}$ or $\mathcal{P}$. Then ϕ satisfies the inclusion–exclusion formula on $\mathcal{B}$, respectively, $\mathcal{P}$ and thus can be extended uniquely to a valuation on $\mathcal{L}(\mathcal{B})$, the space of* polyboxes, *respectively, $\mathcal{L}(\mathcal{P})$, the space of* poly-convex polytopes *by Theorem 7.1.*

Proof. We consider only the case of convex polytopes, the case of boxes being analogous. By Volland's first extension theorem, it is sufficient to show that ϕ satisfies the inclusion–exclusion formula on $\mathcal{P}$:

(10) Let $P, P_1, \ldots, P_m \in \mathcal{P}(\mathbb{E}^d)$ such that $P = P_1 \cup \cdots \cup P_m$. Then

$$\phi(P) = \sum_I (-1)^{|I|-1}\phi(P_I).$$

The proof of (10) is by double induction on d and m. (10) holds trivially for $d = 0, 1, \ldots,$ and $m = 1$ and for $d = 0$ and $m = 1, 2, \ldots$ Assume now that $d > 0$ and $m > 1$ and that (10) holds for dimensions $0, 1, \ldots, d-1$ and in dimension d for $1, 2, \ldots, m-1$ polytopes. We have to establish it for d and m.

In the first step of the proof, we show that (10) holds in the following special cases:

(i) One of the polytopes $P_1, \ldots, P_m$ has dimension less than d, say P_m.

(ii) One of the polytopes $P_1, \ldots, P_m$ coincides with P, say P_1.

If P has dimension less than d, (10) holds by the induction hypothesis. Thus we may assume that P has dimension d. Then in each of the cases (i) and (ii) we have that $Q = P_1 \cup \cdots \cup P_{m-1} \in \mathcal{P}$ and $P = Q$ (note that P is convex) and thus, trivially, $P = Q \cup P_m$. Since ϕ is a valuation:

$$(11)\ \ \phi(P) = \phi(Q) + \phi(P_m) - \phi(Q \cap P_m).$$

Since $Q = P_1 \cup \cdots \cup P_{m-1}$ and $Q \cap P_m = (P_1 \cap P_m) \cup \cdots \cup (P_{m-1} \cap P_m)$, and (10) holds for d and $m - 1$ by induction, the equality (11) readily leads to (10).

In the second step, the general case is traced back to the cases (i) and (ii). If P_1 has dimension less than d or coincides with P, (10) holds since the cases (i) and (ii) are already settled. Thus we may suppose that P_1 has dimension d and is a proper subpolytope of P. Then we proceed as follows: Clearly, P_1 has facets which meet int P. Given such a facet, let H be the hyperplane containing it and let $H^-(\supseteq P_1)$ and H^+ be the closed halfspaces determined by H. Then $P^- = H^- \cap P$, $P^+ = H^+ \cap P$, $P^- \cup P^+ = P \in \mathcal{P}$ and since ϕ is a valuation, we have:

$$(12)\ \ \phi(P) = \phi(P^-) + \phi(P^+) - \phi(P^- \cap P^+).$$

Clearly, $P_i^+ = H^+ \cap P_i$, $P_i^- = H^- \cap P_i \in \mathcal{P}$ for $i = 1, \dots, m$. This leads to the representations

$$\begin{aligned}(13)\ \ & P^- = P_1^- \cup \cdots \cup P_m^-,\ P^+ = P_1^+ \cup \cdots \cup P_m^+,\\ & P^- \cap P^+ = (P_1^- \cap P_1^+) \cup \cdots \cup (P_m^- \cap P_m^+) \text{ and}\\ & \phi(P_I) = \phi(P_I^-) + \phi(P_I^+) - \phi(P_I^- \cap P_I^+) \text{ since } \phi \text{ is a valuation.}\end{aligned}$$

$\dim P_1^+ < d$ and $\dim(P^- \cap P^+) < d$. Thus case (i) and the induction hypothesis show that (7) holds for $\phi(P^+)$ and $\phi(P^- \cap P^+)$. Then (12) and (13) yield the following:

$$\begin{aligned}(14)\ \ \phi(P) - \sum_I (-1)^{|I|-1}\phi(P_I) &= \phi(P^-) - \sum_I (-1)^{|I|-1}\phi(P_I^-) + \phi(P^+)\\ &\quad - \sum_I (-1)^{|I|-1}\phi(P_I^+) - \phi(P^- \cap P^+) + \sum_I (-1)^{|I|-1}\phi(P_I^- \cap P_I^+)\\ &= \phi(P^-) - \sum_I (-1)^{|I|-1}\phi(P_I^-).\end{aligned}$$

By construction, $P_1 = P_1^- \subseteq P^-$. If $P_1^- = P^-$, then (13) and case (ii) show that the right-hand side in (14) is 0 and (10) holds. Otherwise the polytope $P_1 = P_1^-$, which has dimension d, is a proper subpolytope of P^-. Then we proceed again as before, etc. After finitely many repetitions of this procedure, we finally arrive at the following situation:

$$P^{-^k} = P_1^{-^k} \cup \cdots \cup P_m^{-^k},\ P_1 = P_1^{-^k} = P^{-^k},$$

$$(15)\ \ \phi(P) - \sum_I (-1)^{|I|-1}\phi(P_I) = \phi(P^{-^k}) - \sum_I (-1)^{|I|-1}\phi(P_I^{-^k}).$$

Case (ii) then shows that the right-hand side in (15) is 0. This concludes the proof of (10) and thus of the theorem. □

Extension of Valuations on Simplices

Ludwig and Reitzner [667] showed that each valuation on the space of all simplices in $\mathbb{E}^d$ has unique extensions to the spaces $\mathcal{P}$ and $\mathcal{L}(\mathcal{P})$ of all convex, respectively, all convex and polyconvex polytopes. This result yields an alternative way to define the elementary volume of polytopes, see Sect. 16.1.

Extension of Continuous Valuations on Convex Bodies

The following problem still seems to be open.

Problem 7.1. *Can every valuation on $\mathcal{C}$ be extended to a valuation on $\mathcal{L}(\mathcal{C})$, the space of* polyconvex bodies*?*

While the general extension problem is open, in important special cases, fortunately, the extension is possible. Considering the above proof of Volland's theorem, it is plausible that continuous valuations on $\mathcal{C}$ can be extended to valuations on $\mathcal{L}(\mathcal{C})$, where continuity is meant with respect to the topology induced by the Hausdorff metric. In fact, this is true, as the following *extension theorem of Groemer* [402] shows.

Theorem 7.3. *Let ϕ be a continuous valuation on $\mathcal{C}$. Then ϕ can be extended uniquely to a valuation on $\mathcal{L}(\mathcal{C})$.*

Proof. By Theorem 7.1 it is sufficient to show that ϕ satisfies the inclusion–exclusion formula (2). Let $C, C_1, \dots, C_m \in \mathcal{C}$ such that $C = C_1 \cup \dots \cup C_m$.

First, the following will be shown:

(16) Let $P_1, \dots, P_m \in \mathcal{P}$ such that $C_1 \subseteq \operatorname{int} P_1, \dots, C_m \subseteq \operatorname{int} P_m$. Then there are $Q_1, \dots, Q_m \in \mathcal{P}$ such that $C_1 \subseteq Q_1, \dots, C_m \subseteq Q_m$ and $Q_1 \cup \dots \cup Q_m \in \mathcal{P}$.

$P = P_1 \cup \dots \cup P_m$ is a not necessarily convex polytope containing the convex body C in its interior. Hence we may find a convex polytope Q with $C \subseteq Q \subseteq P$. Now let $Q_1 = P_1 \cap Q, \dots, Q_m = P_m \cap Q$.

By (16) we may choose m decreasing sequences of convex polytopes, say

(17) $(Q_{in}) \subseteq \mathcal{P}$ such that $Q_{i1} \supseteq Q_{i2} \supseteq \dots \to C_i$ as $n \to \infty$ for $i = 1, \dots, m$,

such that

(18) $Q_{1n} \cup \dots \cup Q_{mn} \in \mathcal{P}$.

Then

(19) $Q_{I1} \supseteq Q_{I2} \supseteq \dots \to C_I$ for all I.

Since, by Volland's polytope extension Theorem 7.2, ϕ satisfies, on $\mathcal{P}$, the inclusion–exclusion formula, it follows that

$$\phi(Q_{1n} \cup \cdots \cup Q_{mn}) = \sum_I (-1)^{|I|-1} \phi(Q_{In}).$$

Then, letting $n \to \infty$, the continuity of ϕ on $\mathcal{C}$ together with (17)–(19) implies that

$$\phi(C_1 \cup \cdots \cup C_m) = \sum_I (-1)^{|I|-1} \phi(C_I).$$

Thus the inclusion–exclusion formula is valid for ϕ on $\mathcal{C}$, concluding the proof of the theorem. □

The Euler Characteristic on $\mathcal{L}(\mathcal{C})$

The *Euler characteristic* on $\mathcal{L}(\mathcal{C})$ is a valuation χ which is defined as follows: First, let

$$\chi(C) = 1 \text{ for } C \in \mathcal{C} \text{ and } \chi(\emptyset) = 0.$$

This, clearly, is a continuous valuation on $\mathcal{C}$. By Groemer's extension theorem it extends uniquely to a valuation χ on $\mathcal{L}(\mathcal{C})$, the *Euler characteristic* on $\mathcal{L}(\mathcal{C})$. Since $\mathcal{L}(\mathcal{C})$ is a lattice, χ satisfies the inclusion–exclusion formula. Thus:

$$\begin{aligned}(20)\quad \chi(C_1 \cup \cdots \cup C_m) = m \; - \; &\#\{(i_1, i_2) : i_1 < i_2,\ C_{i_1} \cap C_{i_2} \neq \emptyset\}\\ &+ \#\{(i_1, i_2, i_3) : i_1 < i_2 < i_3,\ C_{i_1} \cap C_{i_2} \cap C_{i_3} \neq \emptyset\} - \cdots\\ &\text{for } C_1, \dots, C_m \in \mathcal{C},\end{aligned}$$

where # is the counting function.

7.2 Elementary Volume and Jordan Measure

Volume is one of the seminal concepts in convex geometry, where its theory is now part of the theory of valuations.

In this section, we define the notions of elementary volume of boxes and of volume or Jordan measure and establish some of their properties. In particular, it will be shown that both are valuations. Jordan measure plays an important role in several other sections. In some cases we use their properties not proved in the following, for example, Fubini's theorem or the substitution rule for multiple integrals. We also make some remarks about the problems of Busemann-Petty and Shephard and the slicing problem.

In Section 16.1, the elementary volume on the space of convex polytopes will be treated in a similar way.

Elementary Volume of Boxes

The *elementary volume* V on the space of boxes $\mathcal{B}$ is defined by

$$V(B) = \prod_i (\beta_i - \alpha_i) \text{ for } B = \{x : \alpha_i \leq x_i \leq \beta_i\} \in \mathcal{B}.$$

Easy arguments yield the following formulae to calculate the elementary volume of boxes, where $\mathbb{Z}^d$ is the *integer* (*point*) *lattice* in $\mathbb{E}^d$, that is the set of all points in $\mathbb{E}^d$ with integer coordinates:

(1) $V(B) = \lim_{n\to\infty} \frac{1}{n^d} \#\Big(B \cap \frac{1}{n}\mathbb{Z}^d\Big) = \lim_{n\to\infty} \frac{1}{n^d} \#\Big(\operatorname{int} B \cap \frac{1}{n}\mathbb{Z}^d\Big)$
for $B \in \mathcal{B}$.

The proof of the next result is left to the reader.

Proposition 7.1. *The elementary volume on $\mathcal{B}$ is a simple, (positive) homogeneous of degree d, translation invariant, non-decreasing, and continuous valuation.*

When we say that V is a *simple*, (*positive*) *homogeneous of degree k*, *translation invariant*, *rigid motion invariant*, *non-decreasing* or *monotone valuation* on $\mathcal{B}$, the following is meant:

$V(B) = 0$ for $B \in \mathcal{B}$, $\dim B < d$
$V(\lambda B) = \lambda^d V(B)$ for $B \in \mathcal{B}$, $\lambda \geq 0$
$V(B + t) = V(B)$ for $B \in \mathcal{B}$, $t \in \mathbb{E}^d$
$V(mB) = V(B)$ for $B \in \mathcal{B}$ and any rigid motion m of $\mathbb{E}^d$
$V(B) \leq V(C)$ for $B, C \in \mathcal{B}$, $B \subseteq C$
V or $-V$ is non-decreasing

Elementary Volume of Polyboxes

The elementary volume V on the space $\mathcal{B}$ of boxes is a valuation. Volland's polytope extension theorem 7.2 thus shows that it has a unique extension to a valuation V on the space $\mathcal{L}(\mathcal{B})$ of polyboxes, the *elementary volume* on $\mathcal{L}(\mathcal{B})$. Each polybox A may be represented in the form

$$A = B_1 \cup \cdots \cup B_m,$$

where the B_i are boxes with pairwise disjoint interiors. The inclusion-exclusion formula for V on $\mathcal{L}(\mathcal{B})$ and the fact that V is simple on $\mathcal{B}$, then yield

$$V(A) = V(B_1) + \cdots + V(B_m),$$

that is, V is *simply additive* on $\mathcal{L}(\mathcal{B})$. Using this representation of $V(A)$, Proposition (1) implies the following formulae for polyboxes:

$$(2)\quad V(A) = \lim_{n\to\infty} \frac{1}{n^d} \#\Big(A \cap \frac{1}{n}\,\mathbb{Z}^d\Big) = \lim_{n\to\infty} \frac{1}{n^d} \#\Big(\operatorname{int} A \cap \frac{1}{n}\,\mathbb{Z}^d\Big)$$
$$\text{for } A \in \mathcal{L}(\mathcal{B}).$$

As before, it is easy to prove the following result.

Proposition 7.2. *The elementary volume on $\mathcal{L}(\mathcal{B})$ is a simple, simply additive, homogeneous of degree d, translation invariant, non-decreasing valuation.*

Note that V is *not* continuous on $\mathcal{L}(\mathcal{B})$ if $\mathcal{L}(\mathcal{B})$ is endowed with the topology induced by the Hausdorff metric. To see this, let F_n, $n = 1, 2, \dots$, be finite sets such that $F_n \to [0, 1]^d$ with respect to the Hausdorff metric. Then $V(F_n) = 0$ for each n, but $V([0, 1]^d) = 1$.

Volume, Jordan Measure or Jordan Content

A set $J \subseteq \mathbb{E}^d$ is *Jordan*, *Riemann* or *Peano measurable*, if

$$\sup\big\{V(A) : A \in \mathcal{L}(\mathcal{B}),\, A \subseteq J\big\} = \inf\big\{V(B) : B \in \mathcal{L}(\mathcal{B}),\, J \subseteq B\big\}.$$

If J is Jordan measurable, then its *Jordan*, etc. *measure*, its *Jordan*, etc. *content*, or its *volume* $V(J)$ on $\mathbb{E}^d$ is this common value. In dimension 2 the volume is called *area* and we sometimes write $A(J)$ for $V(J)$. Let $\mathcal{J} = \mathcal{J}(\mathbb{E}^d)$ be the *family of Jordan measurable sets* in $\mathbb{E}^d$. Each Jordan measurable set is Lebesgue measurable, but not necessarily a Borel set. Its Jordan measure coincides with its Lebesgue measure. Note that Jordan measure, actually, is *not* a measure in the sense of measure theory since it lacks σ-additivity, and that $\mathcal{J}$ is a *proper* subfamily of the family of Lebesgue measurable sets in $\mathbb{E}^d$. As a consequence of (2) we obtain the formulae

$$(3)\quad V(J) = \lim_{n\to\infty} \frac{1}{n^d} \#\Big(J \cap \frac{1}{n}\,\mathbb{Z}^d\Big) = \lim_{n\to\infty} \frac{1}{n^d} \#\Big(\operatorname{int} J \cap \frac{1}{n}\,\mathbb{Z}^d\Big) \text{ for } J \in \mathcal{J}.$$

These formulae will be needed in proofs of Minkowski's fundamental theorem and the Minkowski–Hlawka theorem. Since no further use of Jordan measure for general sets will be made, we restrict our attention to convex and polyconvex bodies.

Volume of Convex Bodies

Convex bodies are Jordan measurable:

Theorem 7.4. $\mathcal{C} \subseteq \mathcal{J}$.

Proof. Let $C \in \mathcal{C}$. Assume, first, that $\operatorname{int} C \neq \emptyset$. Since Jordan measure and measurability of a set are invariant with respect to translations, we may suppose that $o \in \operatorname{int} C$. Choose a box $B \supseteq C$.

For $\varepsilon > 0$ the body $(1-\varepsilon)C$ is a compact subset of $\operatorname{int} C$. Hence there is a polybox A with $(1-\varepsilon)C \subseteq A \subseteq \operatorname{int} C \subseteq C$ and therefore,

$$A \subseteq C \subseteq \frac{1}{1-\varepsilon} A.$$

Since elementary volume on $\mathcal{L}(\mathcal{B})$ is non-decreasing and homogeneous of degree d, we have,

$$\begin{aligned} V(A) \leq V\Big(\frac{1}{1-\varepsilon} A\Big) &= \frac{1}{(1-\varepsilon)^d} V(A) \\ &= V(A) + O(\varepsilon)V(A) \leq V(A) + O(\varepsilon)V(B) \text{ as } \varepsilon \to 0. \end{aligned}$$

This implies that C is Jordan measurable.

Assume, second, that C is contained in a hyperplane. Since C is compact, it can be covered with polyboxes of arbitrarily small elementary volume. Thus C is Jordan measurable with $V(C) = 0$. □

Theorem 7.5. *The volume on $\mathcal{C}$ is a simple, homogeneous of degree d, non-decreasing, continuous, rigid motion invariant valuation.*

For the proof we need an auxiliary result about rotations: Let $\mathcal{SO}(d)$ denote the *special orthogonal group* of $\mathbb{E}^d$, i.e. the group of the orthogonal transformations of $\mathbb{E}^d$ with determinant 1. Let $B = \{b_1, \dots, b_d\}$ be the standard basis of $\mathbb{E}^d$ and denote by $\mathcal{SO}(d, B)$ the set of all transformations in $\mathcal{SO}(d)$ which fix at least $d-2$ vectors of the basis B. The required result is then as follows; for $d = 3$ it reduces to a version of a well-known result of Euler.

Lemma 7.1. *Let $A \in \mathcal{SO}(d)$. Then there are transformations $A_1, \dots, A_n \in \mathcal{SO}(d, B)$ such that $A = A_1 \cdots A_n$.*

Proof. The result holds trivially in case $d = 2$ since then $\mathcal{SO}(2) = \mathcal{SO}(2, B)$. Suppose now that $d > 2$ and that the result holds for dimension $d-1$.

Let $A \in \mathcal{SO}(d)$ but $A \notin \mathcal{SO}(d, B)$. Let $u = A b_d$ and assume without loss of generality that $u \neq \pm b_d$. Consider the orthogonal projection of u into $\mathbb{E}^{d-1}(= \operatorname{lin}\{b_1, \dots, b_{d-1}\})$ and let v be its unit normalization. There is $S \in \mathcal{SO}(d)$ with $Sv = b_{d-1}$ and $Sb_d = b_d$. Since $u \in \operatorname{lin}\{v, b_d\}$, we have $Su \in \operatorname{lin}\{b_{d-1}, b_d\}$. Let $T \in \mathcal{SO}(d, B)$ be such that $Tb_1 = b_1, \dots, Tb_{d-2} = b_{d-2}$ and $TSu = b_d$. Then $TSA\,b_d = TSu = b_d$.

Clearly, $S, TSA \in \mathcal{SO}(d)$ and both fix b_d. By the induction hypothesis, there are $S_1, \dots, S_l, T_1, \dots, T_m \in \mathcal{SO}(d, B)$ such that $S = S_1 \cdots S_l$ and $TSA = T_1 \cdots T_m$. Then

$$A = S^{-1} T^{-1} T_1 \cdots T_m = S_l^{-1} \cdots S_1^{-1} T^{-1} T_1 \cdots T_m.$$

This concludes the induction and the proof is complete. □

Proof of the Theorem. We first show that

(4) V is a valuation on $\mathcal{C}$.

Let $C, D \in \mathcal{C}$ such that $C \cup D \in \mathcal{C}$. In order to show that

(5) $V(C) + V(D) \leq V(C \cup D) + V(C \cap D)$,

choose sequences (P_n), (Q_n) in $\mathcal{L}(\mathcal{B})$ with

$$P_n \subseteq C,\ V(P_n) \to V(C) \text{ and } Q_n \subseteq D,\ V(Q_n) \to V(D) \text{ as } n \to \infty.$$

Then

$$P_n \cup Q_n \subseteq C \cup D,\ P_n \cap Q_n \subseteq C \cap D.$$

The fact that V is a valuation on $\mathcal{L}(\mathcal{B})$ and the definition of Jordan measure on $\mathcal{C}$ yield the following.

$$V(P_n) + V(Q_n) = V(P_n \cup Q_n) + V(P_n \cap Q_n) \leq V(C \cup D) + V(C \cap D).$$

Now, let $n \to \infty$ to get (5). The reverse inequality

(6) $V(C) + V(D) \geq V(C \cup D) + V(C \cap D)$

is obtained in a similar way by considering sequences (R_n), (S_n) in $\mathcal{L}(\mathcal{B})$ such that

$$R_n \supseteq C,\ V(R_n) \to V(C) \text{ and } S_n \supseteq D,\ V(S_n) \to V(D).$$

Having shown (5) and (6), the proof of (4) is complete.

The statement that

(7) V is simple

has been shown in the last part of the proof of Theorem 7.4 above, while the statement that

(8) V is homogeneous of degree d and non-decreasing,

follows from the corresponding properties of V on $\mathcal{L}(\mathcal{B})$ and the definition of Jordan measure on $\mathcal{C}$.

For the proof that

(9) V is continuous,

we have to show the following:

(10) Let $C, C_1, C_2, \dots \in \mathcal{C}$ such that $C_1, C_2, \dots \to C$. Then $V(C_1), V(C_2), \dots \to V(C)$.

Assume first, that $V(C) > 0$. Since volume and Hausdorff metric are translation invariant, we may assume that $o \in \operatorname{int} C$. Choose $\varrho > 0$ such that

$$\varrho B^d \subseteq C.$$

Since $C_1, C_2, \dots \to C$, we have the following: Let $\varepsilon > 0$. Then the inclusions

$$C_n \subseteq C + \varepsilon B^d,\ C \subseteq C_n + \varepsilon B^d$$

hold for all sufficiently large n. Hence

$$C_n \subseteq C + \frac{\varepsilon}{\varrho} C = \Big(1 + \frac{\varepsilon}{\varrho}\Big) C,$$
$$\Big(1 - \frac{\varepsilon}{\varrho}\Big) C + \frac{\varepsilon}{\varrho} C = C \subseteq C_n + \frac{\varepsilon}{\varrho} C.$$

Using support functions and arguing as in the proof of the cancellation law for convex bodies (Theorem 6.1), we obtain the inclusions

$$\Big(1 - \frac{\varepsilon}{\varrho}\Big) C \subseteq C_n \subseteq \Big(1 + \frac{\varepsilon}{\varrho}\Big) C$$

for all sufficiently large n. This, together with (8), yields the following:

$$\Big(1 - \frac{\varepsilon}{\varrho}\Big)^d V(C) \le V(C_n) \le \Big(1 + \frac{\varepsilon}{\varrho}\Big)^d V(C).$$
For given $\varepsilon > 0$ and all sufficiently large n,

which, in turn, implies (10) for $V(C) > 0$. Assume second, that $V(C) = 0$. Since $C_1, C_2, \dots \to C$, for any $\varepsilon > 0$

$$C_n \subseteq C + \varepsilon B^d$$

for all sufficiently large n. Thus, by Steiner's theorem on the volume of parallel bodies 6.6,

$$V(C_n) \le V(C + \varepsilon B^d) = V(C) + O(\varepsilon) = O(\varepsilon)$$
for given $\varepsilon > 0$ and all sufficiently large n.

This proves (10) for $V(C) = 0$. The proof of (10), and thus of (9), is complete.

It remains to show that

(11) V is rigid motion invariant.

Taking into account the definition of V, it is sufficient to show that any two congruent boxes have the same volume. To see this, note that, by Lemma 7.1, it is sufficient to prove the following, where, by a *dissection* of a proper convex polytope, we mean a representation of it as a union of finitely many proper convex polytopes with pairwise disjoint interiors:

(12) Let P, T be two congruent rectangles in $\mathbb{E}^2$. Then P can be dissected into finitely many convex polygons such that suitable translations of these polygons form a dissection of T.

In other words, P and T are *equidissectable with respect to (the group of) translations* (Fig. 7.1). Since equidissectability is transitive, as can be shown easily, the following argument leads to (12): first, P is equidissectable to a parallelogram Q, one pair of edges of which is parallel to a pair of edges of T. Q is equidissectable to a rectangle, the edges of which are parallel to those of T. R is equidissectable to a rectangle S, one side of which has length at least the length l of the corresponding parallel edges of T and at most twice this length. S and T are equidissectable. We omit the details. □

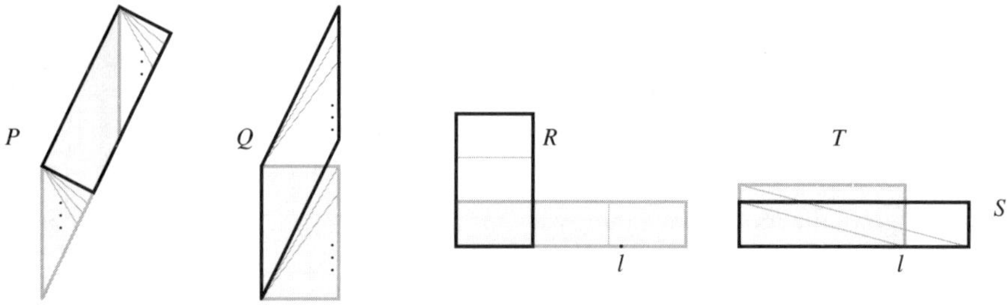

Fig. 7.1. Translative equidissectability of rectangles

Remark. The volume or Jordan measure on $\mathcal{C}$ can be extended in two ways to a valuation on $\mathcal{L}(\mathcal{C})$. First, by Groemer's extension theorem. Second, by the restriction of the Jordan measure on $\mathcal{J}$ which is a valuation on $\mathcal{J}$ to $\mathcal{L}(\mathcal{C}) \subseteq \mathcal{J}$. The uniqueness part of Groemer's theorem implies that both extensions coincide.

Comparison of Volumes by Sections and Projections; the Problems of Busemann-Petty and Shephard and the Slicing Problem

Given two convex bodies C, D, how do their volumes compare? More precisely, what conditions on sections or projections of C and D guarantee that $V(C) \leq V(D)$?

A particularly attractive problem in this context is the following *problem of Busemann-Petty* [184], where $v(\cdot)$ is the $(d-1)$-dimensional volume.

Problem 7.2. *Let $C, D \in \mathcal{C}_p$ be two o-symmetric convex bodies such that*

(13) $v(C \cap H) \leq v(D \cap H)$

for each $(d-1)$-dimensional linear subspace H of $\mathbb{E}^d$. Does it follow that $V(C) \leq V(D)$?

This problem has attracted a good deal of interest over the last two decades, so we give the main steps of its solution:

The answer is no for:

$d \geq 12$: Larman and Rogers [629]

$d \geq 10$: Ball [48]

$d \geq 7$: Giannopoulos [373], Bourgain [159]

$d \geq 5$: Papadimitrakis [785], Gardner [358], Zhang [1043]

The answer is yes for:

$d = 3$: Gardner [357]

$d = 4$: Zhang [1044]

A unified solution for all dimensions using Fourier analysis was given by Gardner, Koldobsky and Schlumprecht [361]. The answers are based on the notion of intersection body, introduced by Lutwak [668]. Use is made of the following interesting result of Lutwak: The solution of the Busemann-Petty problem in $\mathbb{E}^d$ is positive if and only if every symmetric proper convex body in $\mathbb{E}^d$ is an intersection body. For more precise information and additional references see the survey of Koldobsky and König [607] and Koldobsky's book [606]. It is a pity that the Busemann-Petty problem has a negative answer. Otherwise it would have made a deep and interesting result.

Equally intuitive is *Shephard's problem* [931]:

Problem 7.3. *Let $C, D \in \mathcal{C}_p$ be two o-symmetric convex bodies such that*

(14) $v(C|H) \leq v(D|H)$

for each $(d-1)$-dimensional linear subspace H of E^d, where $C|H$ denotes the orthogonal projection of C into H. Does it then follow that $V(C) \leq V(D)$?

The *projection theorem of Alexandrov* [11] says the following: Let C, D be centrally symmetric proper convex bodies such that for each hyperplane H the projections $C|H$ and $D|H$ have the same area. Then C and D coincide up to a translation. Considering this result, a positive answer to the above question seems plausible. Unfortunately, things go wrong as much as they possibly can: There are (even centrally symmetric) convex bodies C and D such that $v(C|H) < v(D|H)$ for all $(d-1)$-dimensional subspaces H of $\mathbb{E}^d$, yet $V(C) > V(D)$. Examples were provided by Petty [796] and Schneider [897]. Petty and Schneider also proved that the answer is positive, if the body D is a zonoid. For more information, consult Gardner's book [359].

In both cases, the question remains to determine precise additional conditions under which the problems of Busemann-Petty and Shephard have positive answers.

If C, D are proper, o-symmetric convex bodies, then both (13) and (14) imply that

$$V(C) \leq \sqrt{d}\, V(D).$$

This can be shown by the Corollary 11.2 of John's ellipsoid theorem 11.2, see Gardner [359], Theorems 4.2.13 and 8.2.13. While in the case of projections, no essential improvements are possible, $\sqrt{d}$ can be replaced by essentially smaller quantities in the case of sections. It is even possible that the *slicing problem* which has been investigated intensively in the local theory of normed spaces has a positive solution. We state the following version of it:

Problem 7.4. *Does there exist an absolute constant $c > 0$ such that the following inequality holds: Let $C, D \in \mathcal{C}_p$ be o-symmetric convex bodies such that*

$$v(C \cap H) \leq v(D \cap H)$$

for each $(d-1)$-dimensional subspace H. Then $V(C) \leq c\, V(D)$.

For more information see Gardner [359] and, in the local theory of normed spaces, Giannopoulos and Milman [374, 375].

7.3 Characterization of Volume and Hadwiger's Functional Theorem

Elementary volume on $\mathcal{B}$ and volume on $\mathcal{C}$ are valuations with particular properties, including simplicity, translation and rigid motion invariance, monotonicity and continuity. The valuation property and the specified properties are what one would expect from a notion of volume in an axiomatic theory. Thus it is a natural question to ask, whether the valuation property together with some of the specified properties characterize elementary volume on $\mathcal{B}$ and volume on $\mathcal{C}$.

In this section we give positive answers to the above questions. As an application, we prove Hadwiger's functional theorem, which by some mathematicians, for example by Rota, is considered to be one of the most beautiful and interesting theorems of all mathematics. Of its numerous applications, one in integral geometry is presented, the principal kinematic formula, see Theorem 7.10.

In Sect. 16.1 we will show that a valuation on the space $\mathcal{P}$ of convex polytopes, with certain additional properties, is a multiple of the elementary volume on $\mathcal{P}$. A result similar to the functional theorem, but for valuations on the space of lattice polytopes, is the Betke–Kneser theorem 19.6.

We follow Klain, see [586]. A different modern proof of Hadwiger's volume theorem is due to Chen [203].

Characterization of the Elementary Volume of Boxes

As a first characterization theorem we show the following result.

Theorem 7.6. *Let ϕ be a simple, translation invariant valuation on $\mathcal{B}$ which is monotone or continuous. Then $\phi = c\,V$, where c is a suitable constant.*

Proof. Let $c = \phi([0,1]^d)$. ϕ is a simple, translation invariant valuation on $\mathcal{B}$ and satisfies the inclusion–exclusion formula, by Volland's polytope extension theorem 7.2. Since, for each $n = 1, 2, \dots$, the unit cube $[0,1]^d$ can be dissected into n^d cubes, each a translate of the cube $[0, \frac{1}{n}]^d$, it follows that

$$\phi\big([0, \frac{1}{n}]^d\big) = \frac{1}{n^d}\phi([0,1]^d) = \frac{c}{n^d} = c\,V\big([0, \frac{1}{n}]^d\big) \text{ for } n = 1, 2, \dots$$

This, in turn, implies that

$$\phi(B) = c\,V(B) \text{ for each box } B \text{ with rational edge-lengths,}$$

on noting that each such box can be dissected into translates of the cube $[0, \frac{1}{n}]^d$ for suitable n. The monotonicity, respectively, the continuity then yields

$$\phi(B) = cV(B) \text{ for each } B \in \mathcal{B}. \quad \square$$

Example. The assumption that ϕ is monotone or continuous on $\mathcal{B}$ is essential, as the following valuation ψ on $\mathcal{B}(\mathbb{E}^1)$ shows: Let $f : \mathbb{R} \to \mathbb{R}$ be a non-continuous solution of Cauchy's functional equation

$$f(s+t) = f(s) + f(t) \text{ for } s, t \in \mathbb{R}.$$

Then ψ, defined by

$$\psi(B) = f(\beta - \alpha) \text{ for } B = [\alpha, \beta] \in \mathcal{B}(\mathbb{E}^1),$$

is a simple, translation invariant valuation on $\mathcal{B}(\mathbb{E}^1)$, but clearly not a multiple of the length.

Remark. A similar result holds for $\mathcal{L}(\mathcal{B})$, where, by the continuity of a valuation on $\mathcal{L}(\mathcal{B})$, we mean continuity of its restriction to $\mathcal{B}$.

First Characterization of the Volume of Convex Bodies

In the following we present a characterization of the volume on $\mathcal{C}$, see Hadwiger [462,464,468]. Versions of this result in the context of Jordan and Lebesgue measure, respectively, Riemann and Lebesgue integrals are well known.

Theorem 7.7. *Let ϕ be a simple, translation invariant, monotone valuation on $\mathcal{C}$. Then $\phi = c\,V$, where c is a suitable constant.*

Proof. By replacing ϕ by $-\phi$, if necessary, we may assume that

(1) ϕ is non-decreasing and thus non-negative on $\mathcal{C}$.

An application of the above characterization theorem to the restriction of ϕ to $\mathcal{B}$ shows that

(2) $\phi(B) = c\,V(B)$ for $B \in \mathcal{B}$,

where c is a suitable constant. By Theorem 7.2 ϕ has a unique extension to a valuation on $\mathcal{L}(\mathcal{B})$ which we also denote by ϕ. Similarly, $c\,V$ is a valuation on $\mathcal{L}(\mathcal{B})$ which extends the valuation $c\,V$ on $\mathcal{B}$, see Propositions 7.1, 7.2. Since ϕ and $c\,V$ coincide on $\mathcal{B}$ by (2), they must coincide on $\mathcal{L}(\mathcal{B})$ by Volland's polytope extension theorem 7.2:

(3) $\phi(A) = c\,V(A)$ for $A \in \mathcal{L}(\mathcal{B})$.

Volland's theorem, applied to the restriction of the valuation ϕ to the intersectional family $\mathcal{P}$, shows that ϕ satisfies the inclusion–exclusion formula on $\mathcal{P}$. Since ϕ is simple, we see that

(4) ϕ is simply additive on $\mathcal{P}$.

We now show that

(5) $\phi(A) \leq \phi(C)$ for $A \in \mathcal{L}(\mathcal{B}),\ C \in \mathcal{C},$ where $A \subseteq C$.

(Note that this is not trivial since in the theorem monotonicity is assumed only for $\mathcal{C}$.) Given A, C, we have $\operatorname{conv} A \in \mathcal{P}$. Then $A \subseteq \operatorname{conv} A \subseteq C$. Represent A in the form

$$A = B_1 \cup \dots \cup B_m, \text{ where } B_i \in \mathcal{B}$$

and the B_i have pairwise disjoint interiors. We may then represent conv A in the form

$$\text{conv}\, A = B_1 \cup \cdots \cup B_m \cup P_1 \cup \cdots \cup P_n \text{ where } P_j \in \mathcal{P}$$

and such that the boxes and convex polytopes $B_1, \ldots, P_n$ have pairwise disjoint interiors. Thus (4) and (1) imply the following:

$$\begin{aligned}\phi(A) &= \phi(B_1) + \cdots + \phi(B_m) \le \phi(B_1) + \cdots + \phi(B_m) + \phi(P_1) + \cdots + \phi(P_n)\\ &= \phi(\text{conv}\, A) \le \phi(C),\end{aligned}$$

concluding the proof of (5).

The next step is to prove the following counterpart of (5):

(6) $\phi(C) \le \phi(B)$ for $C \in \mathcal{C},\ B \in \mathcal{L}(\mathcal{B})$, where $C \subseteq \text{int}\, B$.

Given C, B, choose $P \in \mathcal{P}$, such that $C \subseteq P \subseteq B$. This is possible since $C \subseteq \text{int}\, B$. Represent B in the form

$$B = B_1 \cup \cdots \cup B_m, \text{ where } B_i \in \mathcal{B}$$

and the B_i have pairwise disjoint interiors. Then (1), (4) and (1) again yield the following:

$$\begin{aligned}\phi(C) &\le \phi(P) = \phi\big(P \cap (B_1 \cup \cdots \cup B_m)\big) = \phi\big((P \cap B_1) \cup \cdots \cup (P \cap B_m)\big)\\ &= \phi(P \cap B_1) + \cdots + \phi(P \cap B_m) \le \phi(B_1) + \cdots + \phi(B_m) = \phi(B),\end{aligned}$$

concluding the proof of (6).

In the last part of the proof we show that

(7) $\phi(C) = c\, V(C)$ for $C \in \mathcal{C}$.

Let $C \in \mathcal{C}$. Then (3), (5), (6) and (3) show that

$$\begin{aligned}\sup\{c\, V(A) : A \in \mathcal{L}(\mathcal{B}),\ A \subseteq C\} &\le \phi(C)\\ &\le \inf\{c\, V(B) : B \in \mathcal{L}(\mathcal{B}), C \subseteq \text{int}\, B\}.\end{aligned}$$

Since C is Jordan measurable, by Theorem 7.4, we have

$$\begin{aligned}\sup\{V(A) : A \in \mathcal{L}(\mathcal{B}), A \subseteq C\} &= V(C)\\ &= \inf\{V(B) : B \in \mathcal{L}(\mathcal{B}), C \subseteq \text{int}\, B\}.\end{aligned}$$

Hence $\phi(C) = c\, V(C)$, concluding the proof of (7) and thus of the theorem. □

Second Characterization of the Volume of Convex Bodies

Much more difficult than the proof of the first characterization of the volume is the proof of *Hadwiger's characterization of the volume*, see [462, 464, 468], where a rigid motion is *proper* if it has determinant 1.

Theorem 7.8. *Let ϕ be a simple, continuous valuation on $\mathcal{C}$ which is invariant with respect to proper rigid motions. Then $\phi = cV$, where c is a suitable constant.*

The following proof is due to Klain [586], see also [587]. It simplifies the elementary original proof of Hadwiger, but uses the non-elementary tool of spherical harmonics. A recent simplified elementary version of the original proof of Hadwiger is due to Chen [203].

First, some tools are put together. A convex polytope Z is a *zonotope* if it can be represented in the form

$$Z = S_1 + \cdots + S_n, \quad \text{where the } S_i \text{ are line segments.}$$

A convex body which is the limit of a convergent sequence of zonotopes is a *zonoid*. For more on zonotopes and zonoids, see Goodey and Weil [384].

Lemma 7.2. *Let C be an o-symmetric convex body with support function h_C such that the restriction of h_C to S^{d-1} is of class $\mathcal{C}^\infty$. Then there are zonoids Y and Z such that*

$$C + Y = Z.$$

Proof. If $f : S^{d-1} \to \mathbb{R}$ is a function of class $\mathcal{C}^\infty$, its *cosine transform* $\mathfrak{C} f : S^{d-1} \to \mathbb{R}$ is defined by

$$\mathfrak{C} f(u) = \int\limits_{S^{d-1}} |u \cdot v| f(v)\, d\sigma(v) \text{ for } u \in S^{d-1},$$

where σ is the ordinary surface area measure in $\mathbb{E}^d$. Using the Funk–Hecke theorem on spherical harmonics it can be shown that $\mathfrak{C}$ is a bijective linear operator on the space of all even functions $f : S^{d-1} \to \mathbb{R}$ of class $\mathcal{C}^\infty$ onto itself. See, e.g. [402].

Since $h_C|S^{d-1}$ is even and of class $\mathcal{C}^\infty$ by assumption, there is an even function $f : S^{d-1} \to \mathbb{R}$ of class $\mathcal{C}^\infty$ such that $h_C|S^{d-1} = \mathfrak{C} f$, that is,

$$h_C(u) = \int\limits_{S^{d-1}} |u \cdot v| f(v)\, d\sigma(v) \text{ for } u \in S^{d-1}.$$

Define $f^+, f^- : S^{d-1} \to \mathbb{R}$ by $f^+(v) = \max\{f(v), 0\}$, $f^-(v) = \max\{-f(v), 0\}$ for $v \in S^{d-1}$. Then $f = f^+ - f^-$ and thus

$$h_C(u) + \int\limits_{S^{d-1}} |u \cdot v| f^-(v)\, d\sigma(v) = \int\limits_{S^{d-1}} |u \cdot v| f^+(v)\, d\sigma(v) \text{ for } u \in S^{d-1}.$$

It is easy to check that the cosine transforms $\mathfrak{C} f^-$, $\mathfrak{C} f^+$ (homogeneously extended to $\mathbb{E}^d$ of degree 1) are positive homogeneous of degree 1 and convex. Thus the characterization theorem 4.3 for support functions shows that there are convex bodies Y, Z such that $h_Y = \mathfrak{C} f^-$, $h_Z = \mathfrak{C} f^+$ and therefore $h_C + h_Y = h_Z$, or

$$C + Y = Z.$$

Since the Riemann sums (actually functions of u) of the above parameter integrals for the support functions $h_Y = \mathcal{C}f^-$, $h_Z = \mathcal{C}f^+$ are linear combinations of support functions of line segments and thus of zonotopes, and since the Riemann sums converge uniformly to h_Y and h_Z, it follows that Y and Z are zonoids. □

The following tool is due to Sah [873].

Lemma 7.3. *Let S be a d-dimensional simplex. Then S can be dissected into finitely many convex polytopes, each symmetric with respect to a hyperplane.*

Proof. Let $F_1, \dots, F_{d+1}$ be the facets of S and let $c \in S$ be the centre of the unique inball of maximum radius of S. Let p_i be the point where the inball touches F_i. For $i < j$ let $P_{ij} = \text{conv}\{c, p_i, p_j, F_i \cap F_j\}$. Then each P_{ij} is a convex polytope, symmetric in the hyperplane through c and $F_i \cap F_j$, and $\{P_{12}, \dots, P_{d\,d+1}\}$ is a dissection of S. □

Proof of the Theorem. The main step of the proof is to show the following proposition:

(8) Let ψ be a simple, continuous valuation on $\mathcal{C}$ which is invariant with respect to proper rigid motions and such that $\psi([0,1]^d) = 0$. Then $\psi = 0$.

We prove (8) by induction on d. If $d = 1$, then ψ is simple and translation invariant. Thus $\psi([0,1]) = 0$ implies that $\psi([0, \frac{1}{n}]) = 0$ for $n = 1, 2, \dots$ This, in turn, shows that ψ vanishes on all compact line segments with rational endpoints. By continuity, ψ then vanishes on all compact line segments, that is, on $\mathcal{C}(\mathbb{E}^1)$, concluding the proof of (8) in case $d = 1$.

Assume now that $d > 1$ and that (8) holds in dimension $d - 1$. The proof of Proposition (8) for d is divided into a series of steps.

First, Volland's polytope extension theorem 7.2, applied to the valuation $\psi|\mathcal{P}$, shows that ψ satisfies the inclusion–exclusion formula on $\mathcal{P}$ and thus, being simple,

(9) ψ is simply additive on $\mathcal{P}$.

Second,

(10) $\psi(S) = \psi(-S)$ for each simplex S.

If d is even, S can be transformed into $-S$ by a proper rigid motion and (9) holds trivially. If d is odd, there is a dissection $\{P_1, \dots, P_n\}$ of S by Lemma 7.3 where each P_i is a convex polytope which is symmetric with respect to a hyperplane. Hence P_i can be transformed into $-P_i$ by a proper rigid motion. By the assumption in (8), ψ is invariant with respect to proper rigid motions. This together with (9) then yields (10).

Third,

(11) $\psi(W) = 0$ for each right cylinder $W \in \mathcal{C}$.

Embed $\mathbb{E}^{d-1}$ into $\mathbb{E}^d = \mathbb{E}^{d-1} \times \mathbb{R}$ as usual and define a function $\mu : \mathcal{C}(\mathbb{E}^{d-1}) \to \mathbb{R}$ by

$$\mu(C) = \psi(C \times [0,1]) \text{ for } C \in \mathcal{C}(\mathbb{E}^{d-1}) \subseteq \mathcal{C}(\mathbb{E}^d).$$

It is easy to see that μ is a simple, continuous valuation on $\mathcal{C}(\mathbb{E}^{d-1})$ which is invariant with respect to proper rigid motions in $\mathbb{E}^{d-1}$ and such that $\mu([0,1]^{d-1}) = 0$. Thus $\mu = 0$ on $\mathcal{C}(\mathbb{E}^{d-1})$ by the induction hypothesis. Hence $\psi(W) = 0$ for all right cylinders W of the form $W = C \times [0,1] \subseteq \mathbb{E}^{d-1} \times \mathbb{R} = \mathbb{E}^d$ with basis $C \in \mathcal{C}(\mathbb{E}^{d-1})$. Since ψ is invariant with respect to proper rigid motions in $\mathbb{E}^d$, this yields (11) for all right cylinders of height 1. Since ψ is simple and translation invariant, $\psi(W) = 0$ for each right cylinder of height $\frac{1}{n}$, $n = 1, 2, \dots$, and thus for all right cylinders of rational height. The continuity of ψ finally yields that (11) holds generally.

Fourth,

(12) $\psi(X) = 0$ for each slanting cylinder $X \in \mathcal{C}$ with polytope basis.

If X is long and thin, cut it with a hyperplane orthogonal to its cylindrical boundary into two pieces and glue the pieces together such as to obtain a right cylinder YW. Since ψ is translation invariant, (9) and (11) show that $\psi(X) = \psi(W) = 0$. If X is not long and thin, dissect it into finitely many long and thin slanting cylinders $X_1, \dots, X_n$, say. Since ψ satisfies (9), $\psi(X) = \psi(X_1) + \dots + \psi(X_n) = 0$ by what was just proved. This concludes the proof of (12).

Fifth,

(13) $\psi(P + L) = \psi(P)$ for each $P \in \mathcal{P}$ and any line segment L.

Let $L = [o, s]$ with $s \in \mathbb{E}^d$. If $\dim P \le d - 1$, then $P + L$ is a cylinder or of dimension $\le d - 1$. Using (12), respectively, the assumption that ψ is simple, we see that $\psi(P + L) = 0$. Clearly, $\psi(P) = 0$ too, by the simplicity of ψ. Hence $\psi(P + L) = \psi(P)$ in case $\dim P \le d - 1$. Assume now that $\dim P = d$. Let $F_1, \dots, F_n$ be the facets of P and denote the exterior unit normal vector of F_i by u_i. By renumbering, if necessary, we may suppose that $s \cdot u_i > 0$ precisely for $i = 1, \dots, m (< n)$. Then $P, F_1 + L, \dots, F_m + L$ form a dissection of $P + L$. (9) and (12) then show that

$$\psi(P + L) = \psi(P) + \psi(F_1 + L) + \dots + \psi(F_m + L) = \psi(P),$$

concluding the proof of (13).

Since a zonotope is a finite sum of line segments, a simple induction argument, starting with (13), shows that $\psi(P + Z) = \psi(P)$, $\psi(Z) = 0$ for each $P \in \mathcal{P}$ and each zonotope Z. Since, by assumption, ψ is continuous, this implies that

(14) $\psi(C + Z) = \psi(C)$, $\psi(Z) = 0$ for each $C \in \mathcal{C}$ and each zonoid Z.

Sixth,

(15) $\psi(C) = 0$ for each centrally symmetric convex body $C \in \mathcal{C}$ such that the restriction of h_C to S^{d-1} is of class $\mathcal{C}^\infty$.

By Lemma 7.2 there are zonoids Y and Z such that $C + Y = Z$. Hence $\psi(C) = \psi(C + Y) = \psi(Z) = 0$ by (14), which yields (15). Since the centrally symmetric convex bodies C with $h_C | S^{d-1}$ of class $\mathcal{C}^\infty$ are dense in the family of all centrally symmetric convex bodies, (15) and the continuity of ψ imply that

(16) $\psi(C) = 0$ for all centrally symmetric $C \in \mathcal{C}$.

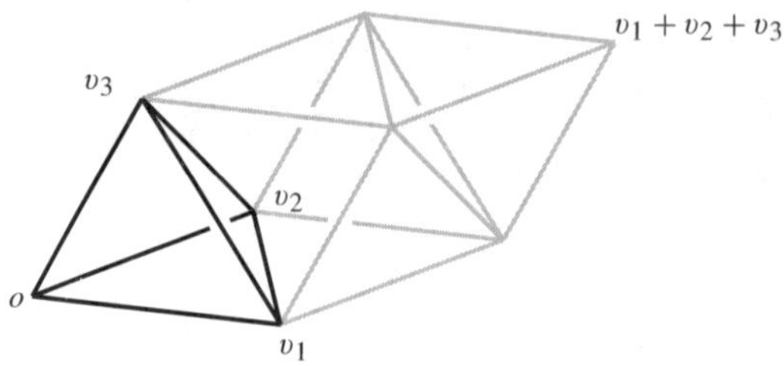

Fig. 7.2. Proof of Proposition (17)

Seventh,

(17) $\psi(S) = 0$ for each simplex S.

Since ψ is simple, (17) holds if $\dim S \le d-1$. Assume now that $\dim S = d$ and that o is a vertex of S. Let $v_1, \dots, v_d$ denote the other vertices of S and let $v = v_1 + \cdots + v_d$. Let $P = \{\lambda_1 v_1 + \cdots + \lambda_d v_d : 0 \le \lambda_i \le 1\}$. The hyperplanes through $v_1, \dots, v_d$ and $v - v_1, \dots, v - v_d$, respectively, dissect P into S, a centrally symmetric convex polytope Q, and $v - S$. (16), (9), (16) and (10) then imply that

$$0 = \psi(P) = \psi(S) + \psi(Q) + \psi(v - S) = 2\psi(S),$$

hence $\psi(S) = 0$, concluding the proof of (17) (see Fig. 7.2).

Finally, since each $P \in \mathcal{P}$ can be dissected into simplices, (9) and (17) show that $\psi(P) = 0$ for $P \in \mathcal{P}$. The continuity of ψ then implies that

$$\psi(C) = 0 \text{ for } C \in \mathcal{C}.$$

This concludes the induction. The proof of (8) is finished.

To complete the proof of the theorem, let $\psi : \mathcal{C} \to \mathbb{R}$ be defined by $\psi(C) = \phi(C) - cV(C)$ for $C \in \mathcal{C}$, where $c = \phi([0, 1]^d)$. ψ is a valuation which satisfies the assumptions in (8). Thus $\psi = 0$, i.e. $\phi = cV$, concluding the proof of the theorem. □

Remark. If, in this result, the group of proper rigid motions is replaced by the group of translations, then, besides the valuations of the form cV, many other valuations turn up. The situation on $\mathcal{C}$ thus is essentially more complicated than the situation on $\mathcal{B}$, see Theorem 7.6.

Hadwiger's Functional Theorem

The above characterizations of the volume will lead to an easy proof of *Hadwiger's* [462, 464, 468] celebrated *functional theorem*. A preliminary version of this result is due to Blaschke [128], Sect. 4, who also provided the basic idea of proof. Unfortunately, Blaschke's presentation was rather sloppy and it needed the genius of Hadwiger to recognize this treasure.

Theorem 7.9. *Let ϕ be a continuous valuation on $\mathcal{C}$ which is invariant with respect to proper rigid motions. Then*

$$\phi = c_0 W_0 + \cdots + c_d W_d \text{ with suitable constants } c_0, \ldots, c_d.$$

This result clearly shows the importance of the quermassintegrals $W_0, \ldots, W_d$ on $\mathcal{C}$. For information on quermassintegrals, see Sect. 6.4 and the references cited there.

Proof (by Induction on d). The functional theorem is trivial for $d = 0$. Assume now that $d > 0$ and that it holds for $d - 1$. We have to prove it for d. The restriction of ϕ to $\mathcal{C}(\mathbb{E}^{d-1})$ satisfies the assumption of the theorem. The induction hypothesis then shows that

$$(18)\quad \phi(C) = d_0 w_0(C) + \cdots + d_{d-1} w_{d-1}(C) \text{ for } C \in \mathcal{C}(\mathbb{E}^{d-1}),$$

where $d_0, \ldots, d_{d-1}$ are suitable constants and $w_0, \ldots, w_{d-1}$ are the quermassintegrals in $\mathbb{E}^{d-1}$. Note that

$$(19)\quad W_0(C) = 0,\ w_{i-1}(C) = \frac{d\,\kappa_{i-1}}{i\,\kappa_i} W_i(C)$$
$$\text{for } C \in \mathcal{C}(\mathbb{E}^{d-1}) \subseteq \mathcal{C}(\mathbb{E}^d = \mathbb{E}^{d-1} \times \mathbb{R}),$$

by Proposition 6.7. Consider the valuation

$$\psi = \phi - \sum_{i=1}^{d} d_{i-1} \frac{d\,\kappa_{i-1}}{i\,\kappa_i} W_i = \phi - \sum_{i=1}^{d} c_i W_i, \text{ say.}$$

$\psi(C) = 0$ for $C \in \mathcal{C}(\mathbb{E}^{d-1}) \subseteq \mathcal{C}(\mathbb{E}^d)$ by (18) and (19). Since ϕ is invariant with respect to proper rigid motions, by assumption, and the same is true of the quermassintegrals W_i, by Theorem 6.13, ψ is also invariant with respect to proper rigid motions. This shows that ψ is simple. Since ϕ is continuous, by assumption, and the quermassintegrals W_i are continuous by Theorem 6.13, the valuation ψ is continuous too. Thus ψ satisfies the assumptions of Hadwiger's characterization theorem for the volume which then implies that $\psi = c_0 V = c_0 W_0$ where c_0 is a suitable constant. Hence

$$\phi = \psi + \sum_{i=1}^{d} c_i W_i = \sum_{i=0}^{d} c_i W_i. \qquad \square$$

Remark. Hadwiger [462, 464] also proved a similar result, where, instead of continuity, monotony is assumed. An interesting recent result of Ludwig and Reitzner [666] shows that the valuations on $\mathcal{C}$ which are semi-continuous and invariant with respect to volume preserving affinities are precisely the linear combinations of the Euler characteristic, affine surface area and volume. Alesker [7, 8] determined the continuous rotation invariant and the continuous translation invariant valuations on $\mathcal{C}$. For further pertinent results and additional information, see McMullen and Schneider [716], McMullen [714] and Klain and Rota [587].

Mean Width and the (d-1)st Quermassintegral

As a simple consequence of Hadwiger's functional theorem we derive the following identity:

Corollary 7.1. *$W_{d-1}(C) = \frac{\kappa_d}{2} w(C)$ for $C \in \mathcal{C}$, where $w(C)$ is the mean width of C,*

$$w(C) = \frac{2}{d\,\kappa_d} \int\limits_{S^{d-1}} h_C(u)\, d\sigma(u).$$

Proof. The main step of the proof is to show that

(20) $w(\cdot)$ is a valuation.

To see this, let $C, D \in \mathcal{C}$ be such that $C \cup D \in \mathcal{C}$. In the second proof of Theorem 6.10 we showed the following equalities for the support functions of $C \cap D$ and $C \cup D$:

(21) $h_{C\cap D} = \min\{h_C, h_D\}$ and $h_{C\cup D} = \max\{h_C, h_D\}$.

Proposition (20) can now be obtained as follows:

$$\begin{aligned} &w(C) + w(D) \\ &= \frac{2}{d\,\kappa_d} \int\limits_{S^{d-1}} (h_C(u) + h_D(u))\, d\sigma(u) \\ &= \frac{2}{d\,\kappa_d} \int\limits_{S^{d-1}} \min\{h_C(u), h_D(u)\}\, d\sigma(u) + \frac{2}{d\,\kappa_d} \int\limits_{S^{d-1}} \max\{h_C(u), h_D(u)\}\, d\sigma(u) \\ &= w(C \cap D) + w(C \cup D) \end{aligned}$$

by (21), concluding the proof of (20).

w is a valuation by (20). It is easy to see that it is continuous, rigid motion invariant and positive homogeneous of degree 1. Hadwiger's functional theorem then shows that w is a multiple of W_{d-1}. Now let $C = B^d$ and note that $d\, W_{d-1}(B^d) = d\,\kappa_d$ by Steiner's formula and $w(B^d) = 2$ to determine the factor. □

7.4 The Principal Kinematic Formula of Integral Geometry

The following quote from Santaló [879], preface, gives an idea of what integral geometry is all about:

> To apply the idea of probability to random elements that are geometric objects (such as points, lines, geodesics, congruent sets, motions, or affinities), it is necessary, first, to define a measure for such sets of elements. Then, the evaluation of this measure for specific sets sometimes leads to remarkable consequences of a purely geometric character, in which the idea of probability turns out to be accidental. The definition of such a measure depends on the geometry with which we are dealing. According to Klein's famous Erlangen Program (1872), the criterion that distinguishes one

> geometry from another is the group of transformations under which the propositions remain valid. Thus, for the purposes of integral geometry, it seems to be natural to choose the measure in such a way that it remains invariant under the corresponding group of transformation.

The first reference of integral geometry and the closely related field of geometric probability, is the needle experiment of the naturalist Buffon [177]. Buffon found his formula in 1733, but it was published only in 1777. Principal contributors to this area are Cauchy, Barbier, Crofton and Czuber in the nineteenth century, Blaschke and his school, in particular his followers Santaló, Chern and Hadwiger, in the 1930s and later, and a number of contemporaries. See the collected works of Blaschke [129] and the books of Blaschke [128], Hadwiger [468], Stoka [971], Santaló [879], Ambartzumian [26], Schneider and Weil [911], Klain and Rota [587] and Beneš and Rataj [95].

A related, yet different, type of integral geometry deals with the Radon transform and its applications. See, e.g. Helgason [489], Gel'fand, Gindikin and Graev [368] and Palamodov [784].

In the following we state and prove the principal kinematic formula based on Hadwiger's functional theorem.

Measure on the Group of Rigid Motions

Let $\mathcal{M} = \mathcal{M}(\mathbb{E}^d)$ be the group of all (proper and improper) rigid motions in $\mathbb{E}^d$. Since every rigid motion is a (proper or improper) rotation, i.e. an element of the orthogonal group $\mathcal{O}(d)$, followed by a translation, we obtain an invariant measure μ on $\mathcal{M}$ as follows: Take the product of the invariant measure ν on the compact group $\mathcal{O}(d)$, normalized such that the whole group has measure 1, with the d-dimensional Lebesgue measure. See [760].

The Principal Kinematic Formula

permits us to calculate the integral

$$(1)\quad \int\limits_{\mathcal{M}} \chi(C \cap mD)\, d\mu(m),$$

where C, D are polyconvex bodies and χ is the Euler characteristic, see Sect. 7.1. It is due to Santaló [879] ($d = 2$), Blaschke [127, 128] ($d = 3$) and Chern and Yien [205] (general d). There exist many generalizations and extensions of it, for example the *general kinematic formula* which permits us to calculate the integral

$$\int\limits_{\mathcal{M}} W_i(C \cap mD)\, d\mu(m).$$

For convex bodies C, D the integral (1) is simply the measure of all rigid motions m such that the intersection $C \cap mD$ is not empty. If B is a convex body which contains C, the quotient

$$\frac{\int\limits_{\mathcal{M}} \chi(C \cap mD)\, d\mu(m)}{\int\limits_{\mathcal{M}} \chi(B \cap mD)\, d\mu(m)}$$

may be interpreted as the probability that a congruent copy of D which meets B also meets C. There are similar interpretations for χ replaced by the quermassintegrals W_i.

Our aim is to prove the *principal kinematic formula:*

Theorem 7.10. *Let $C, D \in \mathcal{L}(\mathcal{C})$. Then*

$$(2)\quad \int\limits_{\mathcal{M}} \chi(C \cap mD)\, d\mu(m) = \sum_{i=0}^{d} \frac{1}{\kappa_d} \binom{d}{i} W_i(C)\, W_{d-i}(D).$$

The quermassintegrals W_i, including χ, are continuous valuations on $\mathcal{C}$, see Theorem 6.13. Hence they can be extended in a unique way to valuations on $\mathcal{L}(\mathcal{C})$ by Groemer's extension theorem. These extensions are applied in (2).

Proof. We consider, first, the case $\mathcal{C}$. Define a function $\phi : \mathcal{C} \times \mathcal{C} \to \mathbb{R}$ by

$$(3)\qquad \begin{aligned} \phi(C, D) &= \mu\big(\{m \in \mathcal{M} : C \cap mD \neq \emptyset\}\big) = \int\limits_{\mathcal{M}} \chi(C \cap mD)\, d\mu(m) \\ &= \int\limits_{\mathcal{O}(d)} \Big(\int\limits_{\mathbb{E}^d} \chi\big(C \cap (rD + t)\big)\, dt \Big)\, d\nu(r) \\ &= \int\limits_{\mathcal{O}(d)} V(C - rD)\, d\nu(r) \text{ for } C, D \in \mathcal{C}, \end{aligned}$$

noting that $\{t : C \cap (rD + t) \neq \emptyset\} = C - rD$. Since $\{m \in \mathcal{M} : C \cap mD \neq \emptyset\}$ is compact in $\mathcal{M} = \mathcal{O}(d) \times \mathbb{E}^d$ and thus measurable, ϕ is well defined. In order to apply the functional theorem of Hadwiger to ϕ, several properties of ϕ have to be shown first.

The first property of ϕ is the following:

(4) $\phi(C, D) = \phi(D, C)$ for $C, D \in \mathcal{C}$.

χ is invariant with respect to rigid motions and the Haar measure μ is unimodular, that is, if in an integrand the variable m is replaced by m^{-1}, the integral does not change its value. See, e.g. [760], p.81. Thus (4) can be obtained as follows:

$$\begin{aligned} \phi(C, D) &= \int\limits_{\mathcal{M}} \chi(C \cap mD)\, d\mu(m) = \int\limits_{\mathcal{M}} \chi(m^{-1}C \cap D)\, d\mu(m) \\ &= \int\limits_{\mathcal{M}} \chi(D \cap m^{-1}C)\, d\mu(m) = \int\limits_{\mathcal{M}} \chi(D \cap mC)\, d\mu(m) = \phi(D, C). \end{aligned}$$

The second property of ϕ says that

(5) $\phi(\cdot, D)$ is a valuation on $\mathcal{C}$ for given $D \in \mathcal{C}$.

Let $B, C \in \mathcal{C}$ such that $B \cup C \in \mathcal{C}$. Since

$$(B \cup C) \cap mD = (B \cap mD) \cup (C \cap mD),$$
$$(B \cap C) \cap mD = (B \cap mD) \cap (C \cap mD),$$

and since χ is a valuation on $\mathcal{C}$, we obtain the equality

$$\chi\big((B \cup C) \cap mD\big) + \chi\big((B \cap D) \cap mD\big) = \chi(B \cap mD) + \chi(C \cap mD)$$

for $m \in \mathcal{M}$. Integrating this equality over $\mathcal{M}$ then shows that

$$\phi(B \cup C, D) + \phi(B \cap C, D) = \phi(B, D) + \phi(C, D),$$

concluding the proof of (5). Properties (4) and (5) yield the third property of ϕ, which says that

(6) $\phi(C, \cdot)$ is a valuation on $\mathcal{C}$ for given $C \in \mathcal{C}$.

The left invariance of the Haar measure μ yields the following fourth property of ϕ.

(7) $\phi(C, mD) = \phi(C, D)$ for $C, D \in \mathcal{C},\ m \in \mathcal{M}$.

More intricate is the proof of the fifth and last property,

(8) $\phi(\cdot, \cdot)$ is continuous on $\mathcal{C} \times \mathcal{C}$.

To see this, note that

$$\phi(C, D) = \int\limits_{\mathcal{O}(d)} V(C - rD)\, d\nu(r)$$

by (3). $V(C - rD)$ is continuous in (C, D, r), using the natural topology on $\mathcal{O}(d)$ (matrix norms). Since $\mathcal{O}(d)$ is compact, integration yields a function which is continuous in (C, D), concluding the proof of (8).

For given $C \in \mathcal{C}$ the function $\phi(C, \cdot)$ is a continuous valuation on $\mathcal{C}$ by (6) and (8). By (7) it is invariant with respect to proper rigid motions. An application of Hadwiger's functional theorem thus shows that

(9) $\phi(C, D) = c_0(C) W_0(D) + \cdots + c_d(C) W_d(D)$ for $C, D \in \mathcal{C}$ with suitable coefficients $c_0(C), \ldots, c_d(C)$.

We now investigate these coefficients and show first that

(10) $c_i(\cdot)$ is a valuation on $\mathcal{C}$ for $i = 0, \ldots, d$.

In the following, summation on i and j is from 0 to d. Let $B, C \in \mathcal{C}$ with $B \cup C \in \mathcal{C}$. Then

$$\phi(B \cup C, D) = \sum_i c_i(B \cup C)\, W_i(D),$$

$$\phi(B \cap C, D) = \sum_i c_i(B \cap C)\, W_i(D),$$

$$\phi(B, D) = \sum_i c_i(B)\, W_i(D),$$

$$\phi(C, D) = \sum_i c_i(C)\, W_i(D) \text{ for } D \in \mathcal{C}.$$

This, together with (5), shows that

$$0 = \sum_i \big(c_i(B \cup C) + c_i(B \cap C) - c_i(B) - c_i(C)\big)\, W_i(D) \text{ for } D \in \mathcal{C}.$$

Taking $D = B^0, B^1, \dots, B^d$, the vectors $\big(W_0(D), \dots, W_d(D)\big)$ form a basis of $\mathbb{E}^{d+1}$. Hence

$$c_i(B \cup C) + c_i(B \cap C) - c_i(B) - c_i(C) = 0,$$

concluding the proof of (10). Next:

(11) $c_i(\cdot)$ is continuous for $i = 0, \dots, d$.

Choose $C, C_1, C_2, \dots \in \mathcal{C}$ such that $C_n \to C$ as $n \to \infty$. Then

$$\sum_i c_i(C_n)\, W_i(D) = \phi(C_n, D) \to \phi(C, D) = \sum_i c_i(C)\, W_i(D) \text{ for any } D \in \mathcal{C}$$

by (9) and (8). Hence

$$\sum_i \big(c_i(C_n) - c_i(C)\big) W_i(D) \to 0 \text{ for any } D \in \mathcal{C}.$$

Taking $D = B^0, B^1, \dots, B^d$, the vectors $\big(W_0(D), \dots, W_d(D)\big)$ form a basis of $\mathbb{E}^{d+1}$. Hence

$$c_i(C_n) - c_i(C) \to 0.$$

The proof of (11) is complete. The last required property of the c_i says that

(12) $c_i(\cdot)$ is invariant with respect to proper rigid motions for $i = 0, \dots, d$.

Let $C \in \mathcal{C}$, $m \in M$. Then

$$\sum_i c_i(mC)\, W_i(D) = \phi(mC, D) = \phi(C, D) = \sum_i c_i(C)\, W_i(D)$$

by (9), (4) and (7), or

$$\sum_i \big(c_i(mC) - c_i(C)\big) W_i(D) = 0 \text{ for } D \in \mathcal{C}.$$

An argument that was used twice before, then shows that $c_i(mC) = c_i(C)$, concluding the proof of (12).

Having proved (10)–(12), Hadwiger's functional theorem shows that

(13) $c_i(\cdot) = c_{i0} W_0(\cdot) + \cdots + c_{id} W_d(\cdot)$ for $C \in \mathcal{C}$
with suitable coefficients c_{ij}.

In the last step of the proof, the coefficients c_{ij} in (13) will be determined. This is done by applying (3), (9) and (13) to the special convex bodies λB^d, μB^d, noting that ν is normalized and using Theorem 6.13 (ii):

$$\begin{aligned}\phi(\lambda B^d, \mu B^d) &= \int_{\mathcal{O}(d)} V\big(\lambda B^d - \mu\, r B^d\big)\, d\nu(r) = \int_{\mathcal{O}(d)} V\big((\lambda + \mu) B^d\big)\, d\nu(r)\\ &= V\big((\lambda+\mu)B^d\big) = (\lambda+\mu)^d V(B^d) = \kappa_d \sum_i \binom{d}{i} \lambda^i \mu^{d-i}\\ &= \sum_{i,j} c_{ij} W_i(\lambda B^d)\, W_j(\mu B^d) = \sum_{i,j} c_{ij} W_i(B^d)\, W_j(B^d)\, \lambda^{d-i} \mu^{d-j}\end{aligned}$$

Hence:

(14) $c_{ij} = 0$ for $i + j \neq d$ and $c_{i\,d-i} = \dfrac{\kappa_d \binom{d}{i}}{W_i(B^d)\, W_{d-i}(B^d)} = \dfrac{\binom{d}{i}}{\kappa_d}$

since $W_i(B^d) = \kappa_d$, by Steiner's theorem on the volume of parallel bodies 6.6.

Propositions (3), (9), (13) and (14) settle the theorem for $\mathcal{C}$.

To prove the theorem also for $\mathcal{L}(\mathcal{C})$, we proceed as follows: given a convex body $C \in \mathcal{C}$ and a rigid motion m, the valuations

$$\int_{\mathcal{M}} \chi(C \cap m\,\cdot)\, d\mu(m) \text{ and } W_{d-i}(\cdot)$$

on $\mathcal{C}$ are continuous and invariant with respect to proper rigid motions. By Groemer's extension theorem 7.3, these valuations have unique extensions to $\mathcal{L}(\mathcal{C})$. Since their values for $D_1 \cup \cdots \cup D_m \in \mathcal{L}(\mathcal{C})$, where $D_j \in \mathcal{C}$, are determined by the inclusion–exclusion formula, (2) continues to hold for the given C and all $D \in \mathcal{L}(\mathcal{C})$. Thus (2) holds for all $C \in \mathcal{C}$ and $D \in \mathcal{L}(\mathcal{C})$. Then, given $D \in \mathcal{L}(\mathcal{C})$, a similar argument shows that (2) holds for all $C \in \mathcal{L}(\mathcal{C})$ and the given $D \in \mathcal{L}(\mathcal{C})$. Thus (2) holds for all $C \in \mathcal{L}(\mathcal{C})$ and $D \in \mathcal{L}(\mathcal{D})$, concluding the proof of the theorem. □

Remark. For extensions of the principal kinematic formula to homogeneous spaces, see Howard [523] and Fu [345]. A comprehensive survey of more recent results on kinematic and Crofton type formulae in integral geometry is due to Hug and Schneider [528].

7.5 Hadwiger's Containment Problem

Consider the following

Problem 7.5. *Specify (simple necessary and sufficient) conditions such that two proper convex bodies satisfying these conditions have the property that one of the bodies is contained in a congruent copy of the other body.*

While, at present, a complete solution of this problem seems to be out of reach, there are interesting contributions to it.

Hadwiger's Sufficient Condition for Containment

As an application of the principal kinematic formula, we will prove the following result of Hadwiger [461], where A and P stand for area and perimeter.

Theorem 7.11. *Let $C, D \in \mathcal{C}_p(\mathbb{E}^2)$. If*

$$2\pi\left(A(C)+A(D)\right)-P(C)P(D)>0 \text{ and } A(C)\neq A(D),$$

then there is a rigid motion m such that either $C\subseteq \operatorname{int} mD$ or $D\subseteq \operatorname{int} mC$.

Proof. Assume first that C and D are proper convex polygons. If $C\cap mD=\emptyset$, then also $\operatorname{bd} C\cap m\operatorname{bd} D=\emptyset$. However, if $C\cap mD\neq\emptyset$, then there are two possibilities. The first possibility is that $\operatorname{bd} C\cap m\operatorname{bd} D\neq\emptyset$. If we disregard a set of rigid motions m of measure 0, then $\operatorname{bd} C\cap m\operatorname{bd} D$ consists of an even number of distinct points. The second possibility is that $\operatorname{bd} C\cap m\operatorname{bd} D=\emptyset$. Then $C\subseteq\operatorname{int} mD$ or $D\subseteq\operatorname{int} mC$.

Suppose now that no congruent copy of C is contained in $\operatorname{int} D$ and, similarly, with C and D exchanged. In terms of the Euler characteristic this implies the following:

$$\chi(C\cap mD)=0\Rightarrow\chi(\operatorname{bd} C\cap m\operatorname{bd} D)=0,$$
$$\chi(C\cap mD)=1\Rightarrow\chi(\operatorname{bd} C\cap m\operatorname{bd} D)\geq 2$$

for all $m\in\mathcal{M}$ with a set of exceptions of measure 0. Hence

$$\chi(\operatorname{bd} C\cap m\operatorname{bd} D)\geq 2\,\chi(C\cap mD)$$

for all $m\in\mathcal{M}$ with a set of exceptions of measure 0. Thus,

$$\int\limits_{\mathcal{M}}\chi(\operatorname{bd} C\cap m\operatorname{bd} D)\,d\mu(m)\geq 2\int\limits_{\mathcal{M}}\chi(C\cap mD)\,d\mu(m),$$

or, by the principal kinematic formula 7.10 for $d=2$ and the definition of W_i on $\mathcal{L}(\mathcal{C})$,

$$\chi(\operatorname{bd} C)A(\operatorname{bd} C) + \frac{2}{\pi}P(C)P(D) + A(\operatorname{bd} C)\chi(\operatorname{bd} D)$$
$$\geq 2\,\chi(C)A(D) + \frac{4}{\pi}P(C)P(D) + 2A(C)\chi(D).$$

Since $C, D \in \mathcal{C}$, we have $\chi(C) = \chi(D) = 1$ and $A(\operatorname{bd} C) = A(\operatorname{bd} D) = 0$. Hence

$$2\pi(A(C) + A(D)) - P(C)P(D) \leq 0.$$

Thus, if $2\pi(A(C) + A(D)) - P(C)P(D) > 0$, one of C, D contains a congruent copy of the other disc in its interior. This proves the theorem for convex polygons.

Assume, second, that C, D are proper convex discs such that

$$2\pi\bigl(A(C) + (A(D)\bigr) - P(C)P(D) > 0,\ A(C) \neq A(D),$$

say $A(C) > A(D)$. Choose convex polygons $Q \subseteq C$, $R \supseteq D$ such that

$$2\pi\bigl(A(Q) + A(R)\bigr) - P(Q)P(R) > 0,\ A(Q) > A(R).$$

By the first part of the proof a suitable congruent copy of R is then contained in int Q. Thus, a fortiori, a suitable congruent copy of D is contained in int C, concluding the proof of the theorem. □

Remark. It is an open question to extend Hadwiger's containment theorem to higher dimensions in a simple way. For ideas in this direction due to Zhou and Zhang, see the references in Klain and Rota [587].

8 The Brunn–Minkowski Inequality

In the Brunn–Minkowski inequality, the volume

$$V(C + D)$$

of the Minkowski sum $C+D = \{x+y : x \in C, y \in D\}$, of two convex bodies C, D, is estimated in terms of the volumes of C and D. Around the classical inequality, a rich theory with numerous applications developed over the course of the twentieth century.

In this section we first present different versions of the Brunn–Minkowski inequality, among them extensions to non-convex sets and integrals. Then, applications of the Brunn–Minkowski inequality to the classical isoperimetric inequality, sand piles, capillary surfaces, and Wulff's theorem from crystallography are given. Finally, we show that general Brunn–Minkowski or isoperimetric type inequalities lead to the concentration of measure phenomenon.

For references and related material, see Leichtweiss [640], Schneider [907], Ball [53] and, in particular, the comprehensive survey of Gardner [360] and the report of Barthe [77].

8.1 The Classical Brunn–Minkowski Inequality

The Brunn–Minkowski inequality was first proved by Brunn [173, 174] with a clever but rather vague proof. The equality case, in particular, was not settled satisfactorily, as was pointed out by Minkowski. Then, both Brunn [175] and Minkowski [735] provided correct proofs. At present several essentially different proofs are known, as well as a series of far-reaching extensions and applications.

Dinghas [270] commented on Brunn's proof as follows:

> The ingenious idea of Brunn to relate two convex sets by equal volume ratios ... saw substantial progress in the last twenty years. It did not yield only refinements of old, but produced also new results.

The proof that will be given below is a variant of the proof of Brunn [173, 174], respectively of its precise version due to Kneser and Süss [600]. It is by induction and makes use of the *idea of Brunn* which relates the d-dimensional case to the $(d-1)$-dimensional case by equal volume ratios: let $u \in S^{d-1}$ and $H(t) = \{x : u \cdot x = t\}$. For $0 \le s \le 1$ let $t_C(s)$ and $t_D(s)$ be such that the hyperplanes $H(t_C(s))$ and $H(t_D(s))$ divide the volume of C, respectively, D, in the ratio $s : 1-s$. Clearly,

$$C + D \supseteq \bigcup_{0 \le s \le 1} \big(C \cap H\big(t_C(s)\big) + D \cap H\big(t_D(s)\big)\big).$$

Now apply induction to $C \cap H\big(t_C(s)\big) + D \cap H\big(t_D(s)\big)$ and use Fubini's theorem. A different proof is due to Blaschke [124]. It makes use of a property of Steiner symmetrization, namely that

$$\operatorname{st}(C + D) \supseteq \operatorname{st} C + \operatorname{st} D.$$

This makes it possible to reduce the proof to the trivial case where C and D are balls. For Blaschke's proof, see Sect. 9.2.

The Classical Inequality

Our aim is to prove the *Brunn–Minkowski inequality*:

Theorem 8.1. *Let $C, D \in \mathcal{C}$. Then:*

(1) $V(C+D)^{\frac{1}{d}} \ge V(C)^{\frac{1}{d}} + V(D)^{\frac{1}{d}}$,

where equality holds if and only if one of C, D is a point, or C and D are improper and lie in parallel hyperplanes, or C and D are proper and positive homothetic.

Proof. The following inequality will be used below:

(2) $\big(v^{\frac{1}{d-1}} + w^{\frac{1}{d-1}}\big)^{d-1}\Big(\dfrac{V}{v} + \dfrac{W}{w}\Big) \ge (V^{\frac{1}{d}} + W^{\frac{1}{d}})^d$ for $v, w, V, W > 0$,

where equality holds if and only if

$$\frac{v}{V^{\frac{d-1}{d}}} = \frac{w}{W^{\frac{d-1}{d}}}.$$

To show (2), fix V, W and note that on each ray in the positive quadrant, $\{(v, w) : v, w > 0\}$, starting at o, the left side of the inequality (2) is constant. Thus, to determine the minimum of the left side, it is sufficient to determine its minimum on the open line segment $\{(v, 1-v) : 0 < v < 1\}$. Elementary calculus then yields (2).

The case when one or both convex bodies C, D are improper, is left to the reader. Thus, from now on, we assume that $V(C), V(D) > 0$. For $d = 1$ our result is trivial. Next, let $d > 1$ and assume that it holds for $d-1$. We have to prove it for d.

In the first part of the proof, we show the inequality (1). Let $u \in S^{d-1}$ and, for real t, let $H(t) = \{x : u \cdot x = t\}$ and $H^-(t) = \{x : u \cdot x \leq t\}$. Choose $\alpha_C < \beta_C$ such that $H(\alpha_C)$ and $H(\beta_C)$ are support hyperplanes of C and similarly for D. Then:

(3) $H(\alpha_C + \alpha_D)$ and $H(\beta_C + \beta_D)$ are support hyperplanes of $C + D$.

Let $v(\cdot)$ denote $(d-1)$-dimensional volume and put:

(4) $v_C(t) = v\big(C \cap H(t)\big)$, $V_C(t) = V\big(C \cap H^-(t)\big)$ for $\alpha_C \leq t \leq \beta_C$, and similarly for D.

The function $t \to V_C(t)/V(C)$, for $\alpha_C \leq t \leq \beta_C$, assumes the values $0, 1$ for $t = \alpha_C, \beta_C$, is continuous and strictly increasing for $\alpha_C \leq t \leq \beta_C$ and continuously differentiable with derivative $V_C'(t)/V(C) = v_C(t)/V(C) > 0$ for $\alpha_C < t < \beta_C$. Consider its inverse function $s \to t_C(s)$. Then:

(5) $t_C(\cdot)$ is defined for $0 \leq s \leq 1$,
$t_C(0) = \alpha_C,\ t_C(1) = \beta_C$,
$t_C(\cdot)$ is continuous for $0 \leq s \leq 1$,
$t_C(\cdot)$ is continuously differentiable with

$$t_C'(s) = \frac{V(C)}{v_C\big(t_C(s)\big)} > 0 \ \text{ for } 0 < s < 1.$$

Analogous statements hold for D.

Thus the function:

(6) $t_{C+D}(\cdot) = t_C(\cdot) + t_D(\cdot)$ is defined for $0 \leq s \leq 1$,
$t_{C+D}(0) = \alpha_C + \alpha_D,\ t_{C+D}(1) = \beta_C + \beta_D$,
$t_{C+D}(\cdot)$ is continuous for $0 \leq s \leq 1$,
$t_{C+D}(\cdot)$ is continuously differentiable with

$$t_{C+D}'(s) = \frac{V(C)}{v_C\big(t_C(s)\big)} + \frac{V(D)}{v_D\big(t_D(s)\big)} > 0 \text{ for } 0 < s < 1.$$

Since $H\big(t_{C+D}(s)\big) = H\big(t_C(s)\big) + H\big(t_D(s)\big)$ for $0 \leq s \leq 1$, we have:

(7) $(C + D) \cap H\big(t_{C+D}(s)\big) \supseteq C \cap H\big(t_C(s)\big) + D \cap H\big(t_D(s)\big)$ for $0 \leq s \leq 1$.

Now (3), Fubini's theorem, (6) and integration by substitution, (7), the induction hypothesis, (4), (6) and (2) together yield (1) as follows:

$$\text{(8)}\quad V(C + D) = \int\limits_{\alpha_C+\alpha_D}^{\beta_C+\beta_D} v\big((C + D) \cap H(t)\big)\, dt$$

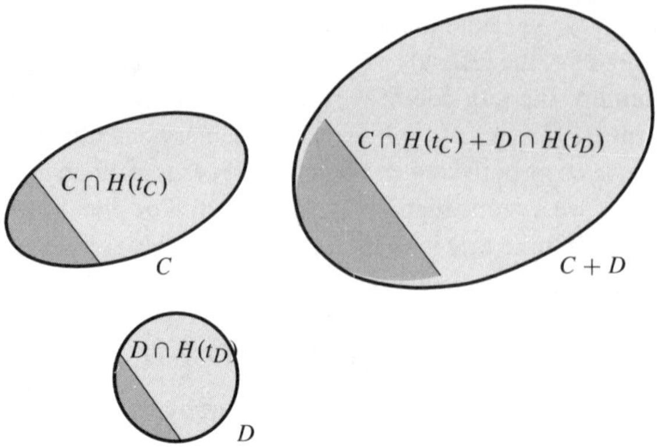

Fig. 8.1. Proof of the Brunn–Minkowski theorem; there is no misprint in the figure for $C+D$

$$
\begin{aligned}
&= \int_0^1 v\big((C+D)\cap H\big(t_{C+D}(s)\big)\big)t'_{C+D}(s)\,ds\\
&\geq \int_0^1 v\big(C\cap H\big(t_C(s)\big)+D\cap H\big(t_D(s)\big)\big)\,t'_{C+D}(s)\,ds\\
&\geq \int_0^1 \Big(v_C\big(t_C(s)\big)^{\frac{1}{d-1}}+v_D\big(t_D(s)\big)^{\frac{1}{d-1}}\Big)^{d-1}\left(\frac{V(C)}{v_C\big(t_C(s)\big)}+\frac{V(D)}{v_D\big(t_D(s)\big)}\right)ds\\
&\geq \int_0^1 \big(V(C)^{\frac{1}{d}}+V(D)^{\frac{1}{d}}\big)^d ds=\big(V(C)^{\frac{1}{d}}+V(D)^{\frac{1}{d}}\big)^d.
\end{aligned}
$$

In the second part of the proof, we assume that equality holds in (1). By translating C and D, if necessary, we may suppose that o is the centroid of both C and D. Let $u\in S^{d-1}$. Since, by assumption, there is equality in (1), we have equality throughout (8). Thus, in particular (see Fig. 8.1),

$$
\frac{v_C\big(t_C(s)\big)}{V(C)^{\frac{d-1}{d}}}=\frac{v_D\big(t_D(s)\big)}{V(D)^{\frac{d-1}{d}}}\quad\text{for } 0<s<1,
$$

by (2). An application of (5) then shows that

$$
\frac{t'_C(s)}{V(C)^{\frac{1}{d}}}=\frac{t'_D(s)}{V(D)^{\frac{1}{d}}}\quad\text{for } 0<s<1.
$$

This, in turn, implies that

$$(9)\quad \frac{t_C(s)}{V(C)^{\frac{1}{d}}} = \frac{t_D(s)}{V(D)^{\frac{1}{d}}} + \text{const} \ \text{ for } 0 \le s \le 1,$$

where we have used the continuity of t_C, t_D for $0 \le s \le 1$. Since o is the centroid of C, Fubini's theorem, (4) and (5) show that

$$\begin{aligned} 0 &= \int\limits_C u \cdot x\,dx = \int\limits_{\alpha_C}^{\beta_C} t\,v\big(C \cap H(t)\big)\,dt = \int\limits_{\alpha_C}^{\beta_C} t\,v_C(t)\,dt \\ &= \int\limits_0^1 t_C(s)\,v_C\big(t_C(s)\big)\,t_C'(s)\,ds = \int\limits_0^1 t_C(s)\,ds\,V(C), \end{aligned}$$

and similarly for D.

The constant in (9) is thus 0. Using support functions for C and D, Proposition (9) implies that

$$h_D(u) = \beta_D = t_D(1) = \Big(\frac{V(D)}{V(C)}\Big)^{\frac{1}{d}} t_C(1) = \Big(\frac{V(D)}{V(C)}\Big)^{\frac{1}{d}} \beta_C = \Big(\frac{V(D)}{V(C)}\Big)^{\frac{1}{d}} h_C(u).$$

Since $u \in S^{d-1}$ was arbitrary, it follows that

$$D = \Big(\frac{V(D)}{V(C)}\Big)^{\frac{1}{d}} C.$$

This settles the equality case. □

Other Common Versions of the Classical Brunn–Minkowski Inequality

The above version of the Brunn–Minkowski inequality readily yields the following results.

Theorem 8.2. *Let $C, D \in \mathcal{C}$. Then*

$$V\big((1-\lambda)C + \lambda D\big)^{\frac{1}{d}} \ge (1-\lambda)V(C)^{\frac{1}{d}} + \lambda V(D)^{\frac{1}{d}} \ \textit{for } 0 \le \lambda \le 1,$$

where there is equality for some λ with $0 < \lambda < 1$ if and only if C or D is a point, or C and D are improper and lie in parallel hyperplanes, or proper and homothetic.

Theorem 8.3. *Let $C, D \in \mathcal{C}$. Then the function*

$$\lambda \to V\big((1-\lambda)C + \lambda D\big)^{\frac{1}{d}} \ \textit{is strictly concave for } 0 \le \lambda \le 1,$$

unless C or D is a point, or C and D are either improper and lie in parallel hyperplanes, or proper and homothetic. In the latter cases this function is linear.

The following result, in essence, is the Brunn–Minkowski inequality in $d-1$ dimensions. It has applications in the geometry of numbers, see, e.g. Gruber [411].

Theorem 8.4. *Let $C \in \mathcal{C}_p$, let $u \in S^{d-1}$ and, for real λ, let $H(\lambda) = \{x : u \cdot x = \lambda\}$. Let $H(\alpha)$ and $H(\beta)$ be support hyperplanes of C where $\alpha < \beta$. Then the function*

$$\lambda \to v\big(C \cap H(\lambda)\big)^{\frac{1}{d-1}} \text{ is concave for } \alpha \le \lambda \le \beta.$$

If C is the convex hull of two homothetic convex bodies in parallel hyperplanes, this function is linear.

8.2 The Brunn–Minkowski Inequality for Non-Convex Sets

In view of its numerous applications, it makes sense to try to extend the Brunn–Minkowski inequality for convex bodies to more general classes of sets. Extensions to very general classes, including the class of all sets in $\mathbb{E}^d$, where inner and outer Lebesgue measure are used, have been given by Lyusternik [672], Schmidt [891], Henstock and Macbeath [494] and Ohmann [776]. For Brunn–Minkowski type inequalities on metric probability spaces and the concentration of measure phenomenon, see Sect. 8.6.

In the following we reproduce Ohmann's proof for the case of compact sets. It uses the idea of Brunn described above in a skilful way.

For more information the reader is referred to the above references and to the books and surveys of Hadwiger [468], Schneider [907] and Gardner [360].

The Brunn–Minkowski Inequality for Compact Sets

Our aim is to prove the following result.

Theorem 8.5. *Let S, T be non-empty compact sets in $\mathbb{E}^d$. Then:*

(1) $V(S+T)^{\frac{1}{d}} \ge V(S)^{\frac{1}{d}} + V(T)^{\frac{1}{d}}$.

Proof. In the following, summation or product over i is from 1 to d. The proof consists of two steps.

In the first step, (1) is proved for polyboxes by induction:

(2) Let $A = A_1 \cup \dots \cup A_m$, $B = B_1 \cup \dots \cup B_n$ be polyboxes where A_i and B_j are boxes. Then

$$V(A+B)^{\frac{1}{d}} \ge V(A)^{\frac{1}{d}} + V(B)^{\frac{1}{d}}.$$

For the proof of (2) it is sufficient to consider the case where the boxes $A_1, \dots, A_m$ are proper and have pairwise disjoint interiors.

If $m + n = 2$, let $\alpha_i, \beta_i > 0$ be the edge-lengths of the boxes A, B. Then

$$\begin{aligned}\frac{V(A)^{\frac{1}{d}} + V(B)^{\frac{1}{d}}}{V(A+B)^{\frac{1}{d}}} &= \frac{(\prod_i \alpha_i)^{\frac{1}{d}} + (\prod_i \beta_i)^{\frac{1}{d}}}{\big(\prod_i (\alpha_i + \beta_i)\big)^{\frac{1}{d}}} = \prod_i \Big(\frac{\alpha_i}{\alpha_i + \beta_i}\Big)^{\frac{1}{d}} + \prod_i \Big(\frac{\beta_i}{\alpha_i + \beta_i}\Big)^{\frac{1}{d}} \\ &\le \frac{1}{d}\Big(\sum_i \frac{\alpha_i}{\alpha_i + \beta_i} + \sum_i \frac{\beta_i}{\alpha_i + \beta_i}\Big) = 1\end{aligned}$$

by the inequality of the geometric and arithmetic mean, see Corollary 1.2. The proof of (2) for $m+n=2$ is complete. Assume now that $m+n>2$ and thus in particular $m \geq 2$, say, and that (2) holds for $2, \dots, m+n-1$ boxes. Since the boxes $A_1, \dots, A_m$ have pairwise disjoint interiors, there is a hyperplane H_A parallel to a coordinate hyperplane such that in each of the closed halfspaces H_A^+, H_A^- there is at least one of the boxes $A_1, \dots, A_m$. Then, among the boxes $A_1 \cap H_A^+, \dots, A_m \cap H_A^+$, there are $0 < m^+ < m$ proper boxes, say $A_1^+, \dots, A_{m^+}^+$ and among the boxes $A_1 \cap H_A^-, \dots, A_m \cap H_A^-$ there are $0 < m^- < m$ proper boxes, say $A_1^-, \dots, A_{m^-}^-$. Let H_B be a hyperplane parallel to H_A such that

$$(3) \quad \begin{aligned} V(A \cap H_A^+) &= \alpha V(A), \quad V(A \cap H_A^-) = (1-\alpha)V(A), \\ V(B \cap H_B^+) &= \alpha V(B), \quad V(B \cap H_B^-) = (1-\alpha)V(B), \end{aligned}$$

where $0 < \alpha < 1$. The polyboxes $B \cap H_B^+, B \cap H_B^-$ each consist of at most n boxes. Now, noting that the hyperplane $H_A + H_B$ separates the polyboxes $A \cap H_A^+ + B \cap H_B^+$ and $A \cap H_A^- + B \cap H_B^-$, the induction hypothesis and (3) yield the following:

$$\begin{aligned} V(A+B) &= V\big((A \cap H_A^+) \cup (A \cap H_A^-) + (B \cap H_B^+) \cup (B \cap H_B^-)\big) \\ &\geq V\big(A \cap H_A^+ + B \cap H_B^+\big) + V\big(A \cap H_A^- + B \cap H_B^-\big) \\ &\geq V\big(A_1^+ \cup \dots \cup A_{m^+}^+ + B \cap H_B^+\big) + V\big(A_1^- \cup \dots \cup A_{m^-}^- + B \cap H_B^-\big) \\ &\geq \big(V(A_1^+ \cup \dots \cup A_{m^+}^+)^{\frac{1}{d}} + V(B \cap H_B^+)^{\frac{1}{d}}\big)^d \\ &\quad + \big(V(A_1^- \cup \dots \cup A_{m^-}^-)^{\frac{1}{d}} + V(B \cap H_B^+)^{\frac{1}{d}}\big)^d \\ &= \big(V(A \cap H_A^+)^{\frac{1}{d}} + V(B \cap H_B^+)^{\frac{1}{d}}\big)^d + \big(V(A \cap H_A^-)^{\frac{1}{d}} + V(B \cap H_B^-)^{\frac{1}{d}}\big)^d \\ &= \big(V(A)^{\frac{1}{d}} + V(B)^{\frac{1}{d}}\big)^d. \end{aligned}$$

Thus the induction is complete, concluding the proof of (2).

To prove (1) for compact sets S, T, consider decreasing sequences of polyboxes with intersections S, respectively, T and use (2). □

Remark. Using monotone limits, the above version of the Brunn–Minkowski inequality can easily be extended to Lebesgue measurable sets, and even to non-measurable sets using inner measures and measurable kernels. Some caution is needed in the statement of these results for the following reasons. By an example of Sierpiński [939], the sum of two Lebesgue measurable sets in $\mathbb{E}^d$ need not be Lebesgue measurable. Similarly, the sum of two Borel sets in $\mathbb{E}^d$ need not be Borel, but it is Lebesgue measurable, see Erdös and Stone [309].

8.3 The Classical Isoperimetric and the Isodiametric Inequality and Generalized Surface Area

Besides geometric isoperimetric inequalities, a large body of isoperimetric inequalities of mathematical physics has been obtained by means of methods of convex geometry, in particular by means of the Brunn–Minkowski inequality. A different tool for proving both geometric and physical isoperimetric inequalities is symmetrization, see Sects. 8.4 and 9.4.

In the following we present the classical geometric isoperimetric and the isodiametric inequality. Then we consider an additional norm on $\mathbb{E}^d$. In the normed space thus obtained, we define the notion of generalized surface area and state the corresponding isoperimetric inequality.

For information on notions of surface area in geometric measure theory and the corresponding isoperimetric problems, including isoperimetric problems on manifolds, see Burago and Zalgaller [178], Talenti [986], Morgan [756], Chavel [201] and Ritoré and Ros [839]. These notions and problems are not touched here.

The basic reference for generalized surface area in finite-dimensional normed spaces and related matters is Thompson's monograph [994]. For pertinent material in the context of the local theory of normed spaces, see Sect. 8.6 and the references cited there, in particular the book of Ledoux [634]. Closer to convex geometry is Ball's [50] reverse isoperimetric inequality, see Theorem 11.3.

The Classical Isoperimetric Inequality

A common version of the isoperimetric inequality says that, amongst all convex bodies in $\mathbb{E}^d$ of given volume, it is precisely the Euclidean balls that have minimum surface area.

The planar case of this problem is mentioned in the Aeneid of Vergil [1009], book 1, verses 367, 368: Dido fled with a group of followers from Tyrus to escape the fate of her husband Sychaeus, who was killed by her brother Pygmalion. Close to where now is Tunis

> they bought as much land – and called it Byrsa – as could be encircled with a bull's hide.

Geometric results in antiquity concerning the isoperimetric problem are due to Archimedes (lost, but referred to by other authors) and Zenodorus. The contributions of Galilei in the renaissance are in the spirit of Zenodorus, while, in the eighteenth century, analytic attempts to prove the isoperimetric inequality drew upon tools from the calculus of variations. More geometric were the four hinge method and Steiner symmetrization, introduced to the isoperimetric problem in the late eighteenth and the first half of the nineteenth century. The arguments of the analysts and the geometric arguments of Steiner in the nineteenth century to prove the isoperimetric inequality all made use of the implicit unproven assumption that there was a solution and thus did not lead to a complete solution, as noted by Dirichlet.

Steiner, in particular, proved that a planar convex body which is not a circle is not a solution of the isoperimetric problem. He thought that this implies that the circular discs were the solutions. To show his error more clearly, we present Pólya's [810] pseudo-argument for the assertion that 1 is the largest positive integer: For any positive integer $n > 1$ there are integers larger than n, for example n^2. Thus n cannot be the largest positive integer. The only possible candidate thus is 1. Therefore 1 is the largest positive integer.

The importance of Steiner, in the context of the isoperimetric problem, lies in the methodological ideas. The first complete proofs of the geometric isoperimetric inequality were given by Edler [285] ($d = 2$) and Schwarz [922] ($d = 3$). The latter based his proof on the calculus of variations and used ideas of Weierstrass. For the history of the classical isoperimetric problem, see Gericke [370] and Danilova [236].

As will be seen below, Minkowski surface area and the Brunn–Minkowski theorem are ideal tools for an *easy* proof of the isoperimetric inequality for convex bodies. The same proof holds also for much more general sets, but the isoperimetric inequality thus obtained is not really of interest: Since the Minkowski surface area of complicated sets in general is rather large, such sets readily satisfy the isoperimetric inequality. For example, the Minkowski surface area of the set of rational points in the unit cube is infinite, whereas for any reasonable notion of surface area, a countable set should have area 0.

A good deal of the modern theory of geometric isoperimetric problems took place outside convexity in areas where other notions of area measures were available, for example, integral geometric concepts of area and the notions of perimeter and currents. Instead of $\mathbb{E}^d$ or S^{d-1} Riemannian and more general manifolds have been considered.

Other developments related to the geometric isoperimetric inequality deal with the so-called concentration of measure phenomenon in the context of the local theory of normed spaces.

The following word of Poincaré applies well to the geometric isoperimetric problem:

> There are no solved problems, there are only more-or-less solved problems.

Below we reproduce Hurwitz's [532] proof for $d = 2$ for Jordan domains, by means of Fourier series and Minkowski's [738] proof for general d for convex bodies. Ingredients of Minkowski's proof include Steiner's formula for parallel bodies, the Brunn–Minkowski inequality and Minkowski's notion of surface area. For Blaschke's proof involving Steiner symmetrization, see Sect. 9.2.

Hurwitz's Proof for Planar Jordan Curves by Means of Fourier Series

Hurwitz [532] derived the following version of the isoperimetric inequality. To simplify the argument a little bit our assumptions are stronger than actually needed. For more information on applications of Fourier series in convex geometry, see the monographs of Groemer [405] and Koldobsky [606] and the proceedings on Fourier analysis and convexity [343].

Theorem 8.6. *Let K be a closed Jordan curve in $\mathbb{E}^2$ of class C^2 with length L and let A denote the area of the Jordan domain bounded by K. Then*

$$L^2 \geq 4\pi A,$$

where equality holds if and only if K is a circle.

Proof. We may assume that K is positively oriented, has length 2π, with parametrization $(x, y) : [0, 2\pi] \to \mathbb{E}^2$, where the parameter is the arc-length s. Since the arc-length s is the parameter,

$$x'(s)^2 + y'(s)^2 = 1 \text{ for } 0 \le s \le 2\pi.$$

Integration implies that

$$\int_0^{2\pi} x'(t)^2 dt + \int_0^{2\pi} y'(t)^2 dt = 2\pi.$$

Consider the Fourier series for $x(\cdot)$ and $y(\cdot)$,

$$x(s) = \frac{a_0}{2} + \sum_{n=1}^{\infty}(a_n \cos ns + b_n \sin ns),\ y(s) = \frac{c_0}{2} + \sum_{n=1}^{\infty}(c_n \cos ns + d_n \sin ns).$$

Then

$$x'(s) = \sum_{n=1}^{\infty}(nb_n \cos ns - na_n \sin ns),\ y'(s) = \sum_{n=1}^{\infty}(nd_n \cos ns - nc_n \sin ns).$$

The formula of Leibniz to calculate the area of a planar set bounded by a closed Jordan curve of class C^1 and a version of Parseval's theorem then yield the following.

$$A = \int_0^{2\pi} x(t)y'(t)dt = \pi \sum_{n=1}^{\infty} n(a_n d_n - b_n c_n)$$

and

$$L^2 = 4\pi^2 = 2\pi \left(\int_0^{2\pi} x'(t)^2 dt + \int_0^{2\pi} y'(t)^2 dt \right) = 2\pi^2 \sum_{n=1}^{\infty} n^2 (a_n^2 + b_n^2 + c_n^2 + d_n^2).$$

Thus

$$\begin{aligned} L^2 - 4\pi A &= 2\pi^2 \sum_{n=1}^{\infty} \left(n^2(a_n^2 + b_n^2 + c_n^2 + d_n^2) - 2n(a_n d_n - b_n c_n)\right) \\ &= 2\pi^2 \sum_{n=1}^{\infty} \left((na_n - d_n)^2 + (nb_n + c_n)^2 + (n^2 - 1)(c_n^2 + d_n^2)\right) \ge 0, \end{aligned}$$

where equality holds if and only if

$$c_n = d_n = a_n = b_n = 0 \text{ for } n \ge 2,\ a_1 = d_1, b_1 = -c_1.$$

In this case

$$\begin{aligned} x(s) &= \frac{a_0}{2} + a_1 \cos s + b_1 \sin s, \\ y(s) &= \frac{c_0}{2} - b_1 \cos s + a_1 \sin s. \end{aligned}$$

This is the parametrization of a circle – note that

$$\left(x(s) - \frac{a_0}{2}\right)^2 + \left(y(s) - \frac{c_0}{2}\right)^2 = a_1^2 + b_1^2. \quad \square$$

The Isoperimetric Inequality for Convex Bodies

In the following we present three related proofs of the isoperimetric inequality for convex bodies. All three proofs make use of Minkowski's notion of surface area, see Sect. 6.4. The first proof is a simple application of the Brunn–Minkowski inequality. It yields the isoperimetric inequality, but does not settle the equality case. The second proof, which is due to Minkowski [738], is based on the Brunn–Minkowski inequality, including the equality case, and Steiner's formula for the volume of parallel bodies. It shows when there is equality. It is curious to note that Brunn [173], p.31, thought that

> This theorem [the Brunn–Minkowski theorem] cannot be used for a proof of the maximal property of the ball.

The third proof makes use of Minkowski's first inequality and also settles the equality case. Since the proof of Minkowski's first inequality is based on the Brunn–Minkowski theorem, it is not surprising, that the isoperimetric inequality for convex bodies can be derived from it.

The *isoperimetric inequality* for convex bodies is as follows.

Theorem 8.7. *Let $C \in \mathcal{C}_p(\mathbb{E}^d)$. Then*

$$\frac{S(C)^d}{V(C)^{d-1}} \geq \frac{S(B^d)^d}{V(B^d)^{d-1}},$$

where equality holds if and only if C is a solid Euclidean ball.

Proof (using the Brunn–Minkowski theorem). The Brunn–Minkowski theorem shows that

$$V(C+\lambda B^d)^{\frac{1}{d}} \geq V(C)^{\frac{1}{d}} + \lambda V(B^d)^{\frac{1}{d}} \text{ for } \lambda > 0.$$

Thus

$$\frac{V(C+\lambda B^d) - V(C)}{\lambda} \geq d\, V(C)^{\frac{d-1}{d}} V(B^d)^{\frac{1}{d}} + O(\lambda) \text{ as } \lambda \to 0.$$

For $\lambda \to 0$, we get

$$S(C) \geq V(C)^{\frac{d-1}{d}} \frac{d\, V(B^d)}{V(B^d)^{\frac{d-1}{d}}}, \text{ or } \frac{S(C)^d}{V(C)^{d-1}} \geq \frac{S(B^d)^d}{V(B^d)^{d-1}}. \quad \square$$

Proof (using the Brunn–Minkowski theorem and Steiner's theorem for the volume of parallel bodies). By Steiner's theorem:

(1) $V\big((1-\lambda)C + \lambda B^d\big)$ is a positive polynomial in λ for $0 \leq \lambda \leq 1$, and $V\big((1-\lambda)C + \lambda B^d\big) = (1-\lambda)^d V(C) + (1-\lambda)^{d-1}\lambda S(C) + O(\lambda^2)$ as $\lambda \to +0$.

As a consequence of the Brunn–Minkowski theorem, we see that

(2) The function $f(\lambda) = V\big((1-\lambda)C + \lambda B^d\big)^{\frac{1}{d}} - (1-\lambda)V(C)^{\frac{1}{d}} - \lambda V(B^d)^{\frac{1}{d}}$ for $0 \le \lambda \le 1$ with $f(0) = f(1) = 0$ is strictly concave, unless C is a ball, in which case it is the zero function.

Since, by (1), f is differentiable for $0 \le \lambda \le 1$, Proposition (2) shows that

(3) $f'(0) \ge 0$, where equality holds if and only if C is a ball.

By (1) and (2),

$$\begin{aligned} f'(\lambda) =& \frac{1}{d} V\big((1-\lambda)C + \lambda B^d\big)^{\frac{1}{d}-1} \big\{ -d(1-\lambda)^{d-1} V(C) \\ &+ \big(-(d-1)(1-\lambda)^{d-2}\lambda + (1-\lambda)^{d-1}\big) S(C) + o(\lambda) \big\} \\ &+ V(C)^{\frac{1}{d}} - V(B^d)^{\frac{1}{d}} \text{ as } \lambda \to 0+ . \end{aligned}$$

Hence (3) implies that

$$\frac{1}{d} V(C)^{\frac{1}{d}-1} \big\{ -dV(C) + S(C) \big\} + V(C)^{\frac{1}{d}} - V(B^d)^{\frac{1}{d}} \ge 0,$$

or, equivalently,

$$\frac{S(C)}{dV(C)^{\frac{d-1}{d}}} \ge V(B^d)^{\frac{1}{d}} = \frac{S(B^d)}{dV(B^d)^{\frac{d-1}{d}}},$$

where equality holds if and only if C is a ball. □

Proof (using Minkowski's first inequality). Note that, for $C \in \mathcal{C}_p$, we have $S(C) = d\,W_1(C) = d\,V(B^d, C \dots, C)$. By Minkowski's first inequality we have, $V(B^d, C, \dots, C)^d \ge V(B^d) V(C)^{d-1}$, where equality holds precisely in case where C is a Euclidean ball. Taking into account that $S(B^d) = d\,V(B^d)$ we thus obtain that

$$\frac{S(C)^d}{V(C)^{d-1}} \ge d^d V(B^d) = \frac{d^d V(B^d)^d}{V(B^d)^{d-1}} = \frac{S(B^d)^d}{V(B^d)^{d-1}},$$

where equality holds precisely in case where C is a ball. □

The Isodiametric Inequality

This inequality is due to Bieberbach [113]. We obtain it as a simple application of the Brunn–Minkowski theorem. See also Sect. 9.2

Theorem 8.8. *Let $C \in \mathcal{C}$. Then*

$$V(C) \le \Big(\frac{1}{2} \operatorname{diam} C\Big)^d V(B^d),$$

where equality holds if and only if C is a solid Euclidean ball.

Proof. We may suppose that $V(C) > 0$. We consider two cases. First, let C be centrally symmetric. Without loss of generality, we may assume that o is the centre of C. Then $C \subseteq (\frac{1}{2}\operatorname{diam} C)B^d$ and thus

$$V(C) \leq \Big(\frac{1}{2}\operatorname{diam} C\Big)^d V(B^d),$$

where equality holds if and only if $C = (\frac{1}{2}\operatorname{diam} C)B^d$.

Second, let C be not centrally symmetric. The Brunn–Minkowski theorem 8.2 then implies that

$$V(C)^{\frac{1}{d}} = \frac{1}{2}V(C)^{\frac{1}{d}} + \frac{1}{2}V(-C)^{\frac{1}{d}} < V\Big(\frac{1}{2}(C - C)\Big)^{\frac{1}{d}} = \frac{1}{2}V(C - C)^{\frac{1}{d}},$$

and thus:

(4) $V(C) < \dfrac{1}{2^d}V(C - C).$

Since $C - C$ is symmetric:

(5) $V(C - C) \leq \Big(\dfrac{1}{2}\operatorname{diam}(C - C)\Big)^d V(B^d)$

by the first case. Next note that

(6)

$$\begin{aligned}\operatorname{diam}(C - C) &= \max\{\|(u - v) - (x - y)\| : u, v, x, y \in C\}\\ &\leq \max\{\|u - x\| : u, x \in C\} + \max\{\|v - y\| : v, y \in C\}\\ &= 2\operatorname{diam} C.\end{aligned}$$

Now combine (4)–(6) to see that

$$V(C) < \Big(\frac{1}{2}\operatorname{diam} C\Big)^d V(B^d). \quad \square$$

The isodiametric inequality can be used to show that d-dimensional Hausdorff measure coincides, up to a multiplicative constant, with the Lebesgue measure in $\mathbb{E}^d$, see Morgan [756].

Urysohn's inequality

A refinement of the isodiametric inequality is the following *inequality of Urysohn* [1004] where, instead of the diameter, the mean width is used:

Theorem 8.9. *Let $C \in \mathcal{C}_p$. Then*

$$V(C) \leq \Big(\frac{1}{2}w(C)\Big)^d V(B^d),$$

where equality holds if and only if C is a solid Euclidean ball.

Proof.

$$\Big(\frac{1}{2}w(C)\Big)^d V(B^d)^d = W_{d-1}(C)^d = V(C, B^d, \dots, B^d)^d \geq V(C)V(B^d)^{d-1},$$

where we have used Corollary 7.1, the relation $W_{d-1}(C) = V(C, B^d, \dots, B^d)$ and Minkowski's first inequality, see Theorem 6.11. □

Stability of Geometric Inequalities

Given a geometric inequality for which equality holds for special convex bodies, for example for balls, ellipsoids, or simplices, the following stability problem arises.

Problem 8.1. *Let C be a convex body for which in a given geometric inequality there is equality up to $\varepsilon > 0$. How far does C deviate, in terms of ε, from convex bodies for which the equality sign holds?*

There is a body of interesting pertinent results, see the survey of Groemer [403].

Generalized Surface Area

Besides the common Euclidean norm, consider a second norm on $\mathbb{E}^d$. In this new normed space or *Minkowski space* the natural notion of volume is the ordinary volume V (Haar measure), possibly up to a multiplicative constant. Likewise, surface area is determined in each hyperplane up to a multiplicative constant, but now the constant may depend on the hyperplane. Several different natural proposals for this dependence have been made by Busemann [181], Holmes and Thompson [521] and Benson [96]. Busemann, for example, considers $(d-1)$-dimensional Hausdorff measure with respect to the new norm. These proposals amount to the introduction of an o-symmetric convex body I, the *isoperimetrix*, which may be obtained from the unit ball B in a variety of ways. For Busemann the isoperimetrix is the polar of the intersection body of B. For Holmes–Thompson it is the projection body of the polar of B. The *generalized surface area* $S_I(C)$ of a convex body C then is defined to be

$$S_I(C) = \lim_{\varepsilon \to +0} \frac{V(C + \varepsilon I) - V(C)}{\varepsilon}.$$

Compare the definition of Minkowski surface area in Sect. 6.4. Since, by Minkowski's theorem on mixed volumes,

$$V(C + \varepsilon I) = V(C) + d\,V(I, C, \dots, C)\varepsilon + O(\varepsilon^2) \text{ as } \varepsilon \to +0,$$

the generalized surface area exists and is equal to $d\,V(I, C \dots, C)$.

The Isoperimetric Inequality in a Minkowski Space

The analog of the classical isoperimetric inequality in a Minkowski space is as follows:

Theorem 8.10. *Let $C \in \mathcal{C}_p(\mathbb{E}^d)$ where $\mathbb{E}^d$ is endowed with a new norm and a corresponding isoperimetrix I. Then*

$$\frac{S_I(C)^d}{V(C)^{d-1}} \geq \frac{S_I(I)^d}{V(I)^{d-1}},$$

where equality holds if and only if C is homothetic to I.

If, in the three proofs of the classical isoperimetric inequality given above, the Euclidean unit ball B^d is replaced by I and instead of Steiner's formula for the volume of parallel bodies Minkowski's theorem on mixed volumes is used, we obtain this more general result.

8.4 Sand Piles, Capillary Surfaces and Wulff's Theorem in Crystallography

As remarked Sect. 8.3, the Brunn–Minkowski theorem has several applications to isoperimetric problems of mathematical physics.

In the following we present three pertinent results, a problem on the maximum volume of a sand pile with given area of the base, a problem on the minimum volume of a capillary surface and Wulff's theorem on the form of crystals.

For general references on isoperimetric inequalities of mathematical physics, see the books and surveys cited in the introduction to Sect. 9.4.

Sand Piles of Maximum Volume

What is the shape of a closed planar set of given area which supports a sand pile of maximum volume? The same mathematical problem arises also in Nádai's [761] theory of plasticity in the following form: For what cross-section of given area has a perfectly plastic rod maximum torsional rigidity? Compare Sect. 9.4 for the corresponding question for elastic rods. Leavitt and Ungar [633] proved the following result using the method of inner parallel sets. This method can also be used to prove the geometric isoperimetric inequality, see, e.g. Fejes Tóth [329].

Theorem 8.11. *Among all compact sets B in $\mathbb{E}^2$ of given area it is precisely the circular discs, up to sets of measure* 0, *that support sand piles of maximum volume.*

Proof. Let α be the maximum angle with the horizontal which a sand pile can sustain, the so-called *glide angle* of the sand. The height of a sand pile on B at a point $x \in B$ then is at most $\delta(x)\tan\alpha$, where $\delta(x) = \text{dist}(x, \text{bd}\, B) = \min\{\|x - y\| : y \in \text{bd}\, B\}$. Thus we have the following estimate for the volume V of a sand pile on B:

$$(1)\quad V \leq \tan\alpha \int_B \delta(x)\, dx.$$

Let $\varrho > 0$ be the maximum radius of a circular disc inscribed to B. Define the *inner parallel set of B at distance δ*, where $0 \leq \delta \leq \varrho$, by:

(2) $B_{-\delta} = \{x \in B : x + \delta B^2 \subseteq B\} = \{x \in B : \delta(x) \geq \delta\}.$

Clearly, $B_{-\delta}$ is compact, and since $B_{-\delta} + \delta B^2 \subseteq B$ for $0 \leq \delta \leq \varrho$, the Brunn–Minkowski theorem for compact sets shows that

(3) $A(B_{-\delta})^{\frac{1}{2}} + A(\delta B^2)^{\frac{1}{2}} \leq A\big(B_{-\delta} + \delta B^2\big)^{\frac{1}{2}} \leq A(B)^{\frac{1}{2}}$, and thus,
$A(B_{-\delta}) \leq \big(A(B)^{\frac{1}{2}} - \delta\pi^{\frac{1}{2}}\big)^2$ for $0 \leq \delta \leq \varrho$.

Combining (2), Fubini's theorem applied to the integral in (1), and (3), we obtain the following:

$$\begin{aligned}
V &\leq \tan\alpha \int\limits_B \delta(x)\,dx = \tan\alpha \int\limits_0^\varrho A(B_{-\delta})\,d\delta \\
&\leq \tan\alpha \int\limits_0^\varrho \big(A(B)^{\frac{1}{2}} - \delta\pi^{\frac{1}{2}}\big)^2 d\delta = -\frac{\tan\alpha}{3\pi^{\frac{1}{2}}}\big(A(B)^{\frac{1}{2}} - \delta\pi^{\frac{1}{2}}\big)^3\Big|_0^\varrho \\
&= -\frac{\tan\alpha}{3\pi^{\frac{1}{2}}}\big(A(B)^{\frac{1}{2}} - \varrho\pi^{\frac{1}{2}}\big)^3 + \frac{\tan\alpha\, A(B)^{\frac{3}{2}}}{3\pi^{\frac{1}{2}}} \leq \frac{\tan\alpha\, A(B)^{\frac{3}{2}}}{3\pi^{\frac{1}{2}}},
\end{aligned}$$

where in the last inequality there is equality if and only if $A(B) = \varrho^2\pi$, i.e. when B coincides with its maximum inscribed circular disc, up to a set of measure 0. Thus we have proved the following: The volume of a sand pile on B is bounded above by $\tan\alpha\, A(B)^{\frac{3}{2}}/3\pi^{\frac{1}{2}}$ and this bound can be attained, if at all, only when B is a circular disc up to a set of measure 0. A simple check shows that, in fact, there is equality if B is a circular disc – consider the circular cone with angle α with the horizontal. □

Equilibrium Capillary Surfaces

Let B be a compact body in $\mathbb{E}^2$ bounded by a closed Jordan curve K of class $\mathcal{C}^2$. Let $\mathbb{E}^2$ be embedded into $\mathbb{E}^3$ as usual (first two coordinates). Consider a vertical cylindrical container with cross-section B and filled with water up to the level of B. Due to forces in the surface of the water, the surface deviates from B close to the boundary of the container. If $u(x)$, $x \in B \cup K$, describes this deviation, then the following statements are well known (see Fig. 8.2):

(4) u is of class $\mathcal{C}^1$ on B, of class $\mathcal{C}^2$ on int B, and it is the unique such function satisfying

$$\operatorname{div}\frac{\operatorname{grad} u}{\big(1 + (\operatorname{grad} u)^2\big)^{\frac{1}{2}}} = \kappa\, u \text{ on int } B, \quad \frac{(\operatorname{grad} u)\cdot n}{\big(1 + (\operatorname{grad} u)^2\big)^{\frac{1}{2}}} = \cos\alpha \text{ on } K.$$

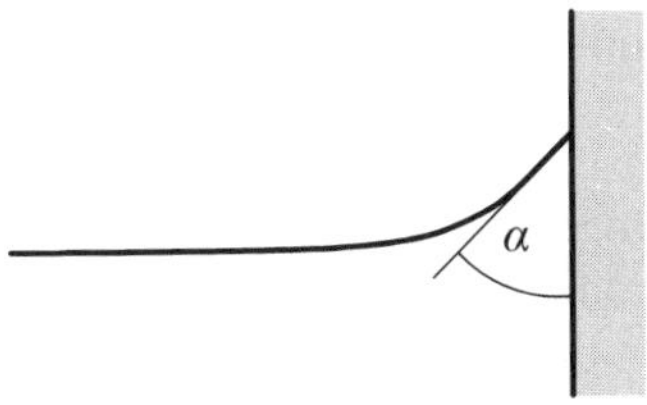

Fig. 8.2. Equilibrium capillary surface

Here $\kappa > 0$ is a physical constant, $n(x)$ is the exterior unit normal vector of K at $x \in K$, and $0 < \alpha < \frac{\pi}{2}$ is the angle between the water surface and the wall of the container.

The following result of Finn [336], p.236, gives information about the *capillary volume*

$$(5) \quad \int_B u(x)\,dx.$$

Theorem 8.12. *Among all compact bodies B in $\mathbb{E}^2$ of given area and bounded by closed Jordan curves K of class $\mathcal{C}^2$, precisely the circular discs yield the minimum capillary volume.*

Proof. Let $x = x(s)$, $s \in [0, L]$, be a parametrization of K where the parameter is the arc-length. Then x is of class $\mathcal{C}^1$.

We will apply the following version of Green's integral formula from calculus:

(6) Let $v : B \to \mathbb{E}^2$ be a continuous vector field such that $v|\operatorname{int} B$ is of class $\mathcal{C}^2$. Then

$$\int_B \operatorname{div} v(x)\,dx = \int_0^L v\big(x(s)\big) \cdot n\big(x(s)\big)\,ds.$$

Now, let u be the unique function satisfying (4). Then (4), (6) and the isoperimetric inequality of Theorem 8.6 imply that

$$\int_B u(x)\,dx = \frac{1}{\kappa}\int_B \operatorname{div}\frac{\operatorname{grad} u(x)}{\big(1+\big(\operatorname{grad} u(x)\big)^2\big)^{\frac{1}{2}}}\,dx$$

$$= \frac{1}{\kappa}\int_0^L \frac{\operatorname{grad} u\big(x(s)\big)\cdot n\big(x(s)\big)}{\big(1+\big(\operatorname{grad} u\big(x(s)\big)\big)^2\big)^{\frac{1}{2}}}\,ds$$

$$= \frac{\cos\alpha}{\kappa}\int_0^L ds = \frac{\cos\alpha}{\kappa}L \geq \frac{\sqrt{4\pi}\,\cos\alpha}{\kappa}A(B)^{\frac{1}{2}},$$

where equality holds if and only if B is a circular disc. □

Wulff's Theorem in Crystallography

Why do crystals have such particular forms? Since the late eighteenth century it was the belief of many crystallographers that underlying each crystal there is a point lattice, where any lattice parallelotope contains a certain set of atoms, ions or molecules, and any two such parallelotopes coincide up to translation. Compare also the discussion in Sect. 21. Any facet of the crystal is contained in a 2-dimensional lattice plane. The free surface energy per unit area of a facet (whatever this means) is small if the corresponding lattice plane is densely populated by lattice points and large otherwise. Thus, only for a small set of normal directions, the free surface energies per unit area in the corresponding lattice planes are small. Since real crystals minimize their total free surface energy according to Gibbs, Curie and Wulff, this explains why crystals have the form of particular polytopes. More precisely, we have the following *theorem of Wulff* [1031]. A first proof of it is due to Dinghas [268]. In our version of Wulff's theorem, no compatibility condition for the free surface energies per unit area is needed. For this reason the proof is slightly more difficult than otherwise.

There are various other versions of Wulff's theorem, for example those of Busemann [180], Fonseca [339] and Fonseca and Müller [340]. See also the surveys of Taylor [990], McCann [702] and Gardner [360] and the book of Dobrushin, Kotecký and Shlosman [275]. For applications to statistical mechanics and combinatorics, compare the report of Shlosman [932].

Theorem 8.13. *Let* $u_1, \dots, u_n \in S^{d-1}$ *(the exterior unit normal vectors of the facets of the crystal) be such that* $\mathbb{E}^d = \{\alpha_1 u_1 + \dots + \alpha_n u_n : \alpha_i \geq 0\}$. *Let* $\varepsilon_1, \dots, \varepsilon_n > 0$ *(the free surface energies per unit area of the facets). Let* K *be the convex polytope*

$$K = \{x : u_i \cdot x \leq \varepsilon_i \text{ for } i = 1, \dots n\}$$

(the crystal). Then, amongst all convex polytopes with volume equal to $V(K)$, *and such that the exterior normal vectors of their facets belong to* $\{u_1, \dots, u_n\}$, *those with minimum total free surface energy are precisely the translates of* K.

The proof of Dinghas is difficult to understand. Possibly, our proof is what Dinghas had in mind. It is related to Minkowski's proof of the isoperimetric inequality, see Theorem 8.7. Let $v(\cdot)$ denote the $(d-1)$-dimensional area measure.

Proof. We may assume that each of the halfspaces $\{x : u_i \cdot x \leq \varepsilon_i\}$, $i = 1, \dots, m$, contributes a ($(d-1)$-dimensional) facet to K, say

$$F_i = \{x : u_i \cdot x = \varepsilon_i\} \cap K,$$

while each of the halfspaces $\{x : u_i \cdot x \leq \varepsilon_i\}$, $i = m+1, \dots, n$, has the property that

$$F_i = \{x : u_i \cdot x = \delta_i \varepsilon_i\} \cap K$$

is a face of K with $\dim F_i < d-1$, where $0 \leq \delta_i \leq 1$ is chosen such that the hyperplane $\{x : x \cdot u_i = \delta_i \varepsilon_i\}$ supports K.

Let P be a convex polytope with $V(P) = V(K)$ such that all its facets are among the faces

$$G_i = \{x : u_i \cdot x = h_P(u_i)\} \cap P.$$

For the proof of the theorem we have to show the following:

(7) The total free surface energy of P is greater than that of K, unless P is a translate of K.

To see this, we first show that:

(8) $V\big((1-\lambda)P + \lambda K\big)$ is a positive polynomial in λ for $0 \le \lambda \le 1$, and

$$V\big((1-\lambda)P + \lambda K\big) = (1-\lambda)^d V(P) + (1-\lambda)^{d-1}\lambda\Big(\sum_{\substack{i=1\\ G_i \text{ facet of } P}}^{m} \varepsilon_i\, v(G_i) + \sum_{\substack{i=m+1\\ G_i \text{ facet of } P}}^{n} \delta_i\, \varepsilon_i\, v(G_i)\Big) + O(\lambda^2) \text{ as } \lambda \to 0.$$

The first statement follows from Minkowski's (see Fig. 8.3) theorem 6.5 on mixed volumes. To see the second, note that $(1-\lambda)P + \lambda K$ can be dissected into the polytope $(1-\lambda)P$, cylinders with basis $(1-\lambda)G_i$ and height $\lambda\varepsilon_i$ if $i \le m$, respectively, $\lambda\delta_i\varepsilon_i$ if $i > m$, where G_i is a facet of P, and polytopes of total volume $O(\lambda^2)$.

Second, since $V(P) = V(K)$, the theorem of Brunn–Minkowski implies that

(9) The function $f(\lambda) = V\big((1-\lambda)P + \lambda K\big)^{\frac{1}{d}} - (1-\lambda)V(P)^{\frac{1}{d}} - \lambda V(K)^{\frac{1}{d}}$ for $0 \le \lambda \le 1$ with $f(0) = f(1) = 0$ is strictly concave, unless P is a translate of K, in which case it is identically 0.

Note that $V(P) = V(K)$. Since, by (8), f is differentiable for $0 \le \lambda \le 1$, Proposition (9) implies that

(10) $f'(0) \ge 0$, where equality holds if and only if P is a translate of K.

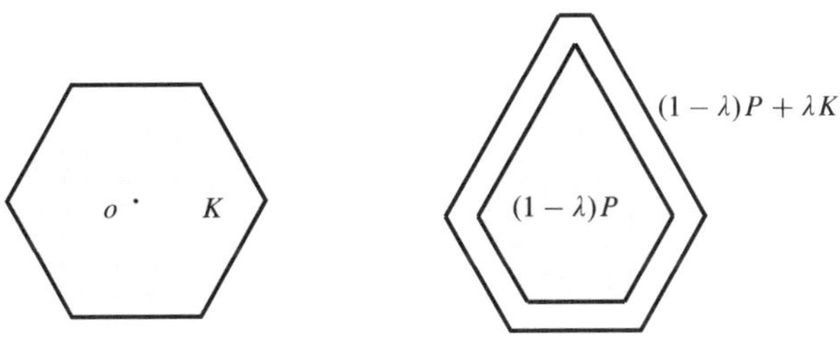

Fig. 8.3. Proof of Wulff's theorem

Third, by (8) and (9),

$$f'(\lambda) = \frac{1}{d} V\big((1-\lambda)P + \lambda K\big)^{\frac{1}{d}-1}\big\{ - d(1-\lambda)^{d-1} V(P) \\ + \big(- (d-1)(1-\lambda)^{d-2}\lambda + (1-\lambda)^{d-1}\big)\big(\sum + \sum\big) + o(\lambda)\big\} \\ + V(P)^{\frac{1}{d}} - V(K)^{\frac{1}{d}} \text{ as } \lambda \to 0.$$

Since $V(P) = V(K)$, we thus have,

$$\text{(11)} \quad f'(0) = \frac{1}{d} V(P)^{\frac{1}{d}-1}\big\{ - dV(P) + \sum + \sum \big\} \geq 0$$

by (10) or, equivalently,

$$\text{(12)} \quad \sum_{\substack{i=1 \\ G_i \text{ facet of } P}}^{m} \varepsilon_i v(G_i) + \sum_{\substack{i=m+1 \\ G_i \text{ facet of } P}}^{n} \delta_i \, \varepsilon_i \, v(G_i) \\ \geq d\, V(P) = d\, V(K) = \sum_{i=1}^{m} \varepsilon_i \, v(F_i).$$

Therefore

$$\text{(13)} \quad \sum_{\substack{i=1 \\ G_i \text{ facet of } P}}^{n} \varepsilon_i \, v(G_i) \geq \sum_{i=1}^{m} \varepsilon_i \, v(F_i).$$

If there is equality in (13), there must be equality in (12) and thus in (11). Then (10) shows that P is a translate of K. This concludes the proof of (7) and thus of the theorem. □

Remark. The evolution of curves in the plane or on other surfaces has attracted much interest in recent years. An example is the curvature flow where the velocity of a point of the curve is proportional to the curvature at this point and the direction is orthogonal to the curve. See Sect. 10.3. In recent years, Wulff's theorem gave rise to several pertinent results. See, e.g. the articles of Almgren and Taylor [24] and Yazaki [1032].

Packing of Balls and Wulff Polytopes

A problem due to Wills [1027] where Wulff's theorem plays a role is the following: Consider a lattice L which provides a packing of the unit ball B^d. Let P be a convex polytope, with vertices in L, and for $n \in \mathbb{N}$ consider the finite packings $\{B^d + l : l \in L \cap nP\}$. It turns out that the so-called parametric density $\delta_\rho(B^d, L \cap nP)$ of these packings approaches the density $\delta(B^d, L) = V(B^d)/d(L)$ of the lattice packing $\{B^d + l : l \in L\}$ rapidly as $n \to \infty$ if P is close to a certain Wulff polytope. Tools for the proof are Minkowski's theorem on mixed volumes and Ehrhart's polynomiality theorem for lattice polytopes. For more information and references, see Böröczky [155], in particular Sect. 13.

8.5 The Prékopa–Leindler Inequality and the Multiplicative Brunn–Minkowski Inequality

This section contains modern developments in the context of the Brunn–Minkowski inequality.

The Prékopa–Leindler Inequality

The following *inequality of Prékopa* [817] *and Leindler* [645] may be considered as an extension of the Brunn–Minkowski inequality to integrals. It leads to the so-called multiplicative Brunn–Minkowski inequality which is equivalent to the ordinary Brunn–Minkowski inequality in the more general case of compact sets, see below.

Theorem 8.14. *Let f, g, h be non-negative Borel functions on $\mathbb{E}^d$ and $0 < \lambda < 1$ such that*

(1) $h\big((1-\lambda)x + \lambda y\big) \geq f(x)^{1-\lambda} g(y)^{\lambda}$ *for all* $x, y \in \mathbb{E}^d$.

Then

(2) $$\int_{\mathbb{E}^d} h(x)\,dx \geq \Big(\int_{\mathbb{E}^d} f(x)\,dx\Big)^{1-\lambda}\Big(\int_{\mathbb{E}^d} g(x)\,dx\Big)^{\lambda}.$$

Before beginning with the proof, some remarks are in order. If f, g are given and if we set $k = f^{1-\lambda} g^{\lambda}$, then Hölder's inequality for integrals says that

$$\int_{\mathbb{E}^d} k(x)\,dx \leq \Big(\int_{\mathbb{E}^d} f(x)\,dx\Big)^{1-\lambda}\Big(\int_{\mathbb{E}^d} g(x)\,dx\Big)^{\lambda},$$

see Corollary 1.5 for the case $d = 1$. Thus the Prékopa–Leindler inequality seems to be Hölder's inequality, backwards. The difference is that h has to satisfy the inequality (1) for *all* pairs (x, y), and not just for the pairs (x, x). Thus h has to be larger than k.

The following proof is due to Brascamp and Lieb [163]. Let $|\cdot|$ denote Lebesgue measure on $\mathbb{R}$.

Proof (by induction on d). As a preparation for the proof in case $d = 1$, we show the following 1-dimensional version of the Brunn–Minkowski theorem. It is an immediate consequence of Theorem 8.5, but we prefer to give an independent proof:

(3) Let $R, S, T \subseteq \mathbb{R}$ be non-empty Borel sets and $0 < \lambda < 1$ such that $R \supseteq (1-\lambda)S + \lambda T$. Then

$$|R| \geq (1-\lambda)|S| + \lambda|T|.$$

Since the measure of a Borel set is the supremum of the measures of its compact subsets, it is sufficient to prove (3) for compact sets R, S, T. Then, by shifting S, T suitably, we may suppose that 0 is the right endpoint of S and the left endpoint of T. The set $(1-\lambda)S + \lambda T$ then includes the sets $(1-\lambda)S$ and λT which have only 0 in common. Hence $|R| \geq |(1-\lambda)S + \lambda T| \geq (1-\lambda)|S| + \lambda|T|$, concluding the proof of (3).

Now, let $d = 1$. Then f, g, h are non-negative Borel functions on $\mathbb{R}$ satisfying (1). Since the integral of a non-negative measurable function on $\mathbb{R}$ is the supremum of the integrals of its bounded, non-negative measurable minorants, we may assume that f, g, h are bounded. Excluding the case where f or g is a zero function, and taking into account that the assumption in (1) and the inequality (2) have the same homogeneity, we may assume that $\sup f = \sup g = 1$. By Fubini's theorem we then can write the integrals of f and g in the form

$$\int_{\mathbb{R}} f(x)\,dx = \int_0^1 |\{x : f(x) \geq t\}|\,dt, \quad \int_{\mathbb{R}} g(x)\,dx = \int_0^1 |\{y : g(y) \geq t\}|\,dt.$$

If $f(x) \geq t$ and $g(y) \geq t$, then $h((1-\lambda)x + \lambda y) \geq t$ by (1). Thus,

$$\{z : h(z) \geq t\} \supseteq (1-\lambda)\{x : f(x) \geq t\} + \lambda\{y : g(y) \geq t\}.$$

For $0 \leq t < 1$ the sets on the right-hand side are non-empty Borel sets in $\mathbb{R}$. Thus

$$|\{z : h(z) \geq t\}| \geq (1-\lambda)|\{x : f(x) \geq t\}| + \lambda|\{y : g(y) \geq t\}|$$

by (3). Integration from 0 to 1 and the inequality between the arithmetic and the geometric mean, see Corollary 1.2, then yield (2) in case $d = 1$:

$$\begin{aligned}
\int_{\mathbb{R}} h(z)\,dz &\geq \int_0^1 |\{z : h(z) \geq t\}|\,dt \\
&\geq (1-\lambda)\int_0^1 |\{x : f(x) \geq t\}|\,dt + \lambda \int_0^1 |\{y : g(y) \geq t\}|\,dt \\
&= (1-\lambda)\int_{\mathbb{R}} f(x)\,dx + \lambda \int_{\mathbb{R}} g(y)\,dy \\
&\geq \Big(\int_{\mathbb{R}} f(x)\,dx\Big)^{1-\lambda} \Big(\int_{\mathbb{R}} g(y)\,dy\Big)^{\lambda}.
\end{aligned}$$

Assume finally, that $d > 1$ and that the theorem holds for $d - 1$. Let f, g, h be non-negative Borel functions on $\mathbb{E}^d$ satisfying (1). Embed $\mathbb{E}^{d-1}$ into $\mathbb{E}^d$ as usual, i.e. $\mathbb{E}^d = \mathbb{E}^{d-1} \times \mathbb{R}$. Let $u = (0, \dots, 0, 1) \in \mathbb{E}^d$. Let $s, t \in \mathbb{R}$ and $r = (1-\lambda)s + \lambda t$. By (1), the non-negative measurable functions $f(x + su), g(y + tu), h(z + ru)$ on $\mathbb{E}^{d-1}$ satisfy the following condition:

$$h\big((1-\lambda)x+\lambda y+\big((1-\lambda)s+\lambda t\big)u\big)\ge f(x+su)^{1-\lambda}g(y+tu)^{\lambda}\text{ for } x,y\in\mathbb{E}^{d-1}.$$

The induction hypothesis thus implies that for the functions F, G, H defined by

$$F(s)=\int\limits_{\mathbb{E}^{d-1}} f(x+su)\,dx,\ G(t)=\int\limits_{\mathbb{E}^{d-1}} g(y+tu)\,dy,\ H(r)=\int\limits_{\mathbb{E}^{d-1}} h(z+ru)\,dz$$

for $r, s, t \in \mathbb{R}$, the following statement holds:

F, G, H are non-negative measurable functions on $\mathbb{R}$ such that $H\big((1-\lambda)s+\lambda t\big)\ge F(s)^{1-\lambda}G(t)^{\lambda}$ for $s,t\in\mathbb{R}$.

The case $d = 1$ of the Prékopa–Leindler inequality, applied to F, G, H and Fubini's theorem then yield inequality (1) for d. The induction and thus the proof of the theorem is complete. □

A Multiplicative Version of the Brunn–Minkowski Inequality

As a consequence of the Prékopa–Leindler inequality we have the following result.

Theorem 8.15. *Let S, T be non-empty compact sets in $\mathbb{E}^d$. Then*

$$V\big((1-\lambda)S+\lambda T\big)\ge V(S)^{1-\lambda}V(T)^{\lambda} \textit{ for } 0\le\lambda\le 1.$$

Proof. We may assume that $0<\lambda<1$. Now apply the Prékopa–Leindler inequality to the characteristic functions of S, T, and the compact set $R=(1-\lambda)S+\lambda T$. □

Using inner measures, this theorem can easily be extended to arbitrary sets by monotone limits.

Equivalence of the Ordinary and the Multiplicative Version

We will prove the following result:

Proposition 8.1. *The (ordinary) Brunn–Minkowski inequality for compact sets and its multiplicative version are equivalent in the sense that each easily implies the other.*

Proof. We show that for non-empty compact sets S, T in $\mathbb{E}^d$ and $0<\lambda<1$ the following statements are equivalent:

(i) The ordinary Brunn–Minkowski inequality holds.

(ii) The multiplicative Brunn–Minkowski inequality holds.

(i)⇒(ii) By the ordinary Brunn–Minkowski inequality,

$$V\big((1-\lambda)S+\lambda T\big)^{\frac{1}{d}}\ge(1-\lambda)V(S)^{\frac{1}{d}}+\lambda V(T)^{\frac{1}{d}}.$$

An application of the inequality of the arithmetic and the geometric mean to the right hand-side expression shows that the latter is bounded below by $V(S)^{\frac{1-\lambda}{d}}V(T)^{\frac{\lambda}{d}}$. Now raise both sides to the dth power to get

$$V\big((1-\lambda)S+\lambda T\big)\geq V(S)^{1-\lambda}V(T)^{\lambda}.$$

(ii)⇒(i) If $V(S)=0$ or $V(T)=0$, the ordinary Brunn–Minkowski inequality clearly holds. Assume now that $V(S), V(T)>0$. Define:

$$(4)\quad U=\frac{1}{V(S)^{\frac{1}{d}}}S,\ V=\frac{1}{V(T)^{\frac{1}{d}}}T,\ \mu=\frac{\lambda V(T)^{\frac{1}{d}}}{(1-\lambda)V(S)^{\frac{1}{d}}+\lambda V(T)^{\frac{1}{d}}}.$$

Then

$$(5)\quad (1-\mu)U+\mu V=\frac{(1-\lambda)S+\lambda T}{(1-\lambda)V(S)^{\frac{1}{d}}+\lambda V(T)^{\frac{1}{d}}}.$$

The multiplicative version of the Brunn–Minkowski inequality applied to U, V, μ shows that

$$V\big((1-\mu)U+\mu V\big)\geq V(U)^{1-\mu}V(V)^{\mu}\geq 1$$

by (4). Now use (5) to obtain

$$V\big((1-\lambda)S+\lambda T\big)^{\frac{1}{d}}\geq (1-\lambda)V(S)^{\frac{1}{d}}+\lambda V(T)^{\frac{1}{d}}.\quad\square$$

8.6 General Isoperimetric Inequalities and Concentration of Measure

There are natural extensions of the isoperimetric inequality to the sphere S^d and the hyperbolic space H^d. In many cases, these extensions are based on symmetrization arguments or on Brunn–Minkowski type inequalities, see Lévy [653] and, in particular, Schmidt [891] and Dinghas [269]. More recently, these isoperimetric inequalities were extended even further in the context of metric probability spaces, a study initiated by Milman. Some of these results are rather surprising and are well described by "concentration of measure".

This section gives a description of the general isoperimetric problem in metric probability spaces. Then the cases of the sphere and of the Gaussian measure on $\mathbb{E}^d$ are discussed. Our exposition follows Ball [53].

For more material on general isoperimetric inequalities and on concentration of measure we refer to Burago and Zalgaller [178], Ball [53], Ledoux [634], Schechtman [885] and the references cited below.

The Euclidean Case

Let C be a compact set in $\mathbb{E}^d$. The Brunn–Minkowski theorem for compact sets then gives the following: Let $\varrho\geq 0$ such that $V(\varrho B^d)=V(C)$. Then:

$$(1)\quad V(C+\varepsilon B^d)\geq\big(V(C)^{\frac{1}{d}}+V(\varepsilon B^d)^{\frac{1}{d}}\big)^d=\big(V(\varrho B^d)^{\frac{1}{d}}+V(\varepsilon B^d)^{\frac{1}{d}}\big)^d$$
$$=V(\varrho B^d+\varepsilon B^d)\text{ for }\varepsilon\geq 0.$$

Using the definition of Minkowski surface area, this inequality readily yields the isoperimetric inequality. Expressed otherwise, (1) says that the volume of the ε-*neighbourhood* $C_\varepsilon = C + \varepsilon B^d$ of the compact set C is at least as large as the volume of the ε-neighbourhood of a (solid Euclidean) ball of the same volume as C. This result may be considered as a sort of isoperimetric inequality, where the notion of surface area is not needed. This is the starting point for general isoperimetric inequalities in metric probability spaces.

The Problem on Metric Probability Spaces

Let $\langle M, \delta, \mu\rangle$ be a space M with a metric δ and thus a topology and a Borel probability measure μ. By the ε-*neighbourhood* C_ε of a compact set C in M we mean the set

$$C_\varepsilon = \{x \in M : \delta_C(x) \leq \varepsilon\}, \text{ where } \delta_C(x) = \operatorname{dist}(x, C) = \min\{\delta(x, y) : y \in C\}.$$

Then the following *general isoperimetric* or *Brunn–Minkowski problem* arises:

Problem 8.2. *Given $\alpha, \varepsilon > 0$, for what compact sets C in M with $\mu(C) = \alpha$ has the ε-neighbourhood C_ε of C minimum measure?*

The Spherical Case

Let S^{d-1} be endowed with its *chordal metric*, i.e. the metric inherited from $\mathbb{E}^d$, and let S be the normalized area measure on S^{d-1}. An *isoperimetric inequality for* S^{d-1} of Lévy [653], p.269, and Schmidt [891] is as follows: Let $\alpha > 0$. Then, for all compact sets $C \subseteq S^{d-1}$ with $S(C) = \alpha$, we have,

$$S(C_\varepsilon) \geq S(K_\varepsilon) \text{ for } \varepsilon \geq 0,$$

where K is a spherical cap with $S(K) = \alpha$. The equality case was studied by Dinghas [269].

This result has the following surprising consequence: Let $\alpha = \frac{1}{2}$, so that C has the measure of a hemisphere H. Then, for each $\varepsilon \geq 0$, we have that $S(C_\varepsilon) \geq S(H_\varepsilon)$. An easy proof shows that the complement of H_ε has area measure about $e^{-\frac{1}{2}d\varepsilon^2}$. Hence $S(C_\varepsilon)$ is about $1 - e^{-\frac{1}{2}d\varepsilon^2}$. Thus, for any given $\varepsilon > 0$ and large d almost all of S^{d-1} lies within distance ε of any given compact set C in S^{d-1} of measure $\frac{1}{2}$. In other words, for large d the area of S^{d-1} is concentrated close to any compact set of measure $\frac{1}{2}$. As a consequence, for large d the area of S^{d-1} is concentrated near any great circle.

The situation just described becomes even more striking if it is interpreted in terms of a Lipschitz function $f : S^{d-1} \to \mathbb{R}$ with Lipschitz constant 1. There is a number m, the *median* of f, such that there are compact sets $C, D \subseteq S^{d-1}$ with $S(C \cap D) = 0$, $S(C) = S(D) = \frac{1}{2}$ and $C \cup D = S^{d-1}$, where $f(u) \leq m$ for $u \in C$ and $f(v) \geq m$ for $v \in D$. Then $S(C_\varepsilon)$ and $S(D_\varepsilon)$ both have area about $1 - e^{-\frac{1}{2}d\varepsilon^2}$.

Since f has Lipschitz constant 1, $f(u) \le m + \varepsilon$ for $u \in C_\varepsilon$ and $f(v) \ge m - \varepsilon$ for $v \in D_\varepsilon$. This shows that for large d

$$S\big(\{u \in S^{d-1} : |f(u) - m| > \varepsilon\}\big) \text{ is at most about } 2e^{-\frac{1}{2}d\varepsilon^2}.$$

In other words, for large d a real function on S^{d-1} with Lipschitz constant 1 is nearly equal to its median on most of S^{d-1}.

Heuristic Observations

This surprising phenomenon appears in one form or another in many results of the local theory of normed spaces, as initiated by Milman and his collaborators. Milman [726] expressed this in his unique way as follows:

> This phenomenon led to a complete reversal of our intuition on high-dimensional results. Instead of a chaotic diversity with an increase in dimension, which previous intuition suggested, we observe well organized and simple patterns of behaviour.

For similar situations dealing with approximation of convex bodies, the Minkowski–Hlawka theorem and Siegel's mean value formula, see Sects. 11.2 and 24.2.

The Case of Gaussian Measure

Let $\mathbb{E}^d$ be endowed with the standard Gaussian probability measure μ. It has density:

$$(2)\quad \delta(x) = \frac{1}{(2\pi)^{\frac{d}{2}}} e^{-\frac{1}{2}\|x\|^2} \text{ for } x \in \mathbb{E}^d.$$

Borell [150] showed that, in this space, we have the following Brunn–Minkowski type result. Let $\alpha > 0$. Then for all closed sets $C \subseteq \mathbb{E}^d$ with $\mu(C) = \alpha$,

$$\mu(C_\varepsilon) \ge \mu(H_\varepsilon^+) \text{ for } \varepsilon \ge 0,$$

where H^+ is a closed halfspace in $\mathbb{E}^d$ with $\mu(H^+) = \alpha$.

In particular, if $\alpha = \frac{1}{2}$, then for H^+ we may take the halfspace $\{x : x_1 \le 0\}$. Then

$$\begin{aligned}
\mu(\mathbb{E}^d \backslash H_\varepsilon^+) &= \frac{1}{(2\pi)^{\frac{d}{2}}} \int\limits_{x_1 \ge \varepsilon} e^{-\frac{x_1^2}{2} - \cdots - \frac{x_d^2}{2}}\, dx = \frac{1}{\sqrt{2\pi}} \int\limits_{\varepsilon}^{+\infty} e^{-\frac{x_1^2}{2}}\, dx_1 \\
&= \frac{1}{\sqrt{2\pi}} \int\limits_0^{+\infty} e^{\frac{-(t+\varepsilon)^2}{2}}\, dt = e^{\frac{-\varepsilon^2}{2}} \frac{1}{\sqrt{2\pi}} \int\limits_0^{+\infty} e^{-\frac{t^2}{2} - t\varepsilon}\, dt \\
&\le e^{\frac{-\varepsilon^2}{2}} \frac{1}{\sqrt{2\pi}} \int\limits_0^{+\infty} e^{-\frac{t^2}{2}}\, dt = e^{-\frac{\varepsilon^2}{2}} \text{ for } \varepsilon \ge 0.
\end{aligned}$$

Hence

$$\mu(C_\varepsilon) \ge 1 - e^{-\frac{\varepsilon^2}{2}} \text{ for } \varepsilon \ge 0.$$

Using the Prékopa–Leindler inequality 8.14 with $\lambda = \frac{1}{2}$, Maurey [699] gave a simple proof of the following weaker estimate. A version of Maurey's argument for the Euclidean case appeared in an article of Arias de Reyna, Ball and Villa [36], see also Matoušek's book [695].

Theorem 8.16. *Let $\mathbb{E}^d$ be endowed with the standard Gaussian measure μ. Then, if $C \subseteq \mathbb{E}^d$ is closed with $\mu(C) > 0$:*

$$(3)\quad \int\limits_{\mathbb{E}^d} e^{\frac{1}{4}\delta_C(x)^2}\, d\mu(x) \le \frac{1}{\mu(C)},$$

where $\delta_C(x) = \operatorname{dist}(x, C) = \min\{\|x-y\| : y \in C\}$. *If, in particular,* $\mu(C) = \frac{1}{2}$, *then*

$$(4)\quad \mu(C_\varepsilon) \ge 1 - 2e^{-\frac{\varepsilon^2}{4}}.$$

Proof. Recall the Prékopa–Leindler theorem and let

$$f(x) = e^{\frac{1}{4}\delta_C(x)^2}\delta(x),\ g(x) = \mathbb{1}_C(x)\,\delta(x),\ h(x) = \delta(x) \text{ for } x \in \mathbb{E}^d.$$

Here, $\delta(\cdot)$ is the density of the Gaussian measure μ, see (2), and $\mathbb{1}_C$ the characteristic function of C. We have to prove that

$$(5)\quad \int\limits_{\mathbb{E}^d} e^{\frac{1}{4}\delta_C(x)^2}\, d\mu(x)\ \mu(C) \le 1,$$

or

$$(6)\quad \int\limits_{\mathbb{E}^d} f(x)\,dx \int\limits_{\mathbb{E}^d} g(x)\,dx \le \Big(\int\limits_{\mathbb{E}^d} h(x)\,dx\Big)^2.$$

It is thus enough to check that $f, g, h,\ \lambda = \frac{1}{2}$ satisfy the assumption of the Prékopa–Leindler inequality, i.e.

$$(7)\quad f(x)\,g(y) \le h\Big(\frac{x+y}{2}\Big)^2 \text{ for } x, y \in \mathbb{E}^d.$$

It is sufficient to show (7) for $y \in C$, since otherwise $g(y) = 0$. But in this case $\delta_C(x) \le \|x-y\|$. Hence

$$\begin{aligned}(2\pi)^d f(x)\,g(y) &= e^{\frac{1}{4}\delta_C(x)^2} e^{-\frac{1}{2}x^2} e^{-\frac{1}{2}y^2}\\ &\le e^{\frac{1}{4}\|x-y\|^2 - \frac{1}{2}\|x\|^2 - \frac{1}{2}\|y\|^2} = e^{-\frac{1}{4}\|x+y\|^2}\\ &= \big(e^{-\frac{1}{2}\|\frac{x+y}{2}\|^2}\big)^2 = (2\pi)^d h\Big(\frac{x+y}{2}\Big)^2.\end{aligned}$$

This settles (7). By the Prékopa–Leindler theorem, (7) yields (6) which, in turn, implies (5), concluding the proof of (3).

To obtain the inequality (4) from (3), note that, in case $\mu(C) = \frac{1}{2}$, the inequality (3) implies

$$\int\limits_{\mathbb{E}^d} e^{\frac{1}{4}\delta_C(x)^2}\, d\mu(x) \le 2.$$

Clearly, the integral here is at least

$$\mu(\{x : \delta_C(x) > \varepsilon\})e^{\frac{1}{4}\varepsilon^2}.$$

Hence

$$\mu(\{x : \delta_C(x) > \varepsilon\}) \le 2e^{-\frac{1}{4}\varepsilon^2},$$

which immediately yields (4). □

9 Symmetrization

A convex body is *symmetric with respect to a group of transformations* if it is invariant under each transformation of the group. The group may consist of orthogonal, affine or projective transformations, or, if the convex body is a polytope, of combinatorial transformations.

For the rich modern theory of symmetric convex and non-convex polytopes which can be traced back to antiquity (Platonic solids), we refer to Coxeter [230, 232], Robertson [842], Johnson [550] and McMullen and Schulte [717].

Here, our objective is the following: A more symmetric convex body has in many cases better geometric or analytic properties. It is thus of interest to consider symmetrization methods which transform convex bodies into more symmetric ones. The known symmetrization methods all have the useful property that they decrease, respectively, increase salient geometric quantities such as the quermassintegrals, in particular the surface area, diameter, width, inradius and circumradius. More important for applications is the fact that they increase, respectively, decrease electrostatic capacity, torsional rigidity and the first principal frequency of membranes.

In this section we first study Steiner symmetrization and use it to prove the isodiametric, the isoperimetric and the Brunn–Minkowski inequalities. Then Schwarz symmetrization and rearrangement of functions are investigated. Our applications concern the isoperimetric inequalities of mathematical physics. In particular, we consider torsional rigidity of rods and the first principal frequency of membranes. Finally, we investigate central symmetrization and prove the inequality of Rogers and Shephard. It has applications in the geometry of numbers and in discrete geometry, see Sects. 30.1 and 30.3.

For more information on the geometric aspects of symmetrization, see the books of Hadwiger [468], Leichtweiss [640], Gardner [359], and the surveys of Lindenstrauss and Milman [660], Sangwine-Yager [878] and Talenti [986]. For pertinent results in mathematical physics we refer to the books of Pólya and Szegö [811], Bandle [64] and the survey of Talenti [987].

9.1 Steiner Symmetrization

In the following we define Steiner symmetrization, prove several of its properties and the sphericity theorem of Gross. Blaschke (or Minkowski) symmetrization is mentioned.

For more information, see the references mentioned above.

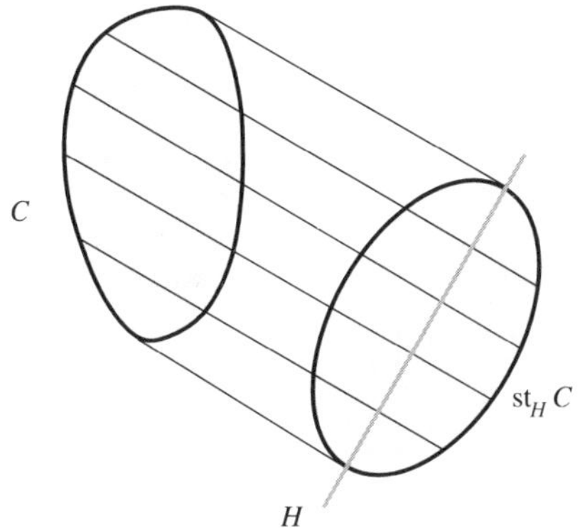

Fig. 9.1. Steiner symmetrization

Steiner Symmetrization

Let C be a convex body and H a hyperplane in $\mathbb{E}^d$. The *Steiner symmetral* $\mathrm{st}\,C = \mathrm{st}_H\,C$ of C (see Fig. 9.1) with respect to H is defined as follows: For each straight line L orthogonal to H and such that $C \cap L \neq \emptyset$, shift the line segment $C \cap L$ along L until its midpoint is in H. The union of all line segments thus obtained is $\mathrm{st}\,C$. Clearly, $\mathrm{st}\,C$ is symmetric with respect to (mirror) reflection in H.

According to Danilova [236], L'Huillier [654] and an anonymous author, possibly Gergonne [369], anticipated Steiner symmetrization in a vague form.

Basic Properties of Steiner Symmetrization

We first collect a series of simple results on Steiner symmetrization. Given a convex body C, the *inradius* $r(C)$ and the *circumradius* $R(C)$ are the maximum radius of a (solid Euclidean) ball contained in C and the minimum radius of a ball containing C, respectively.

Proposition 9.1. *Steiner symmetrization of convex bodies with respect to a given hyperplane H has the following properties:*

(i) $\mathrm{st}\,C \in \mathcal{C}$ *for* $C \in \mathcal{C}$
(ii) $\mathrm{st}\,\lambda C = \lambda\,\mathrm{st}\,C$ *(up to translations) for* $\lambda \geq 0,\ C \in \mathcal{C}$
(iii) $\mathrm{st}(C + D) \supseteq \mathrm{st}\,C + \mathrm{st}\,D$ *(up to translations) for* $C, D \in \mathcal{C}$
(iv) $\mathrm{st}\,C \subseteq \mathrm{st}\,D$ *for* $C, D \in \mathcal{C},\ C \subseteq D$,
(v) $\mathrm{st} : \mathcal{C}_p \to \mathcal{C}_p$ *is continuous*
(vi) $V(\mathrm{st}\,C) = V(C)$ *for* $C \in \mathcal{C}$
(vii) $S(\mathrm{st}\,C) \leq S(C)$ *for* $C \in \mathcal{C}$
(viii) $\mathrm{diam}\,\mathrm{st}\,C \leq \mathrm{diam}\,C$ *for* $C \in \mathcal{C}$
(ix) $r(\mathrm{st}\,C) \geq r(C),\ R(\mathrm{st}\,C) \leq R(C)$ *for* $C \in \mathcal{C}$

Note that $\mathrm{st}: \mathcal{C} \to \mathcal{C}$ is *not* continuous. Consider, for example, the sequence of line segments $\left[o, \left(\frac{1}{n}, 1\right)\right] \in \mathcal{C}(\mathbb{E}^2)$, $n = 1, 2, \dots$ Clearly,

$$\left[o, \left(\frac{1}{n}, 1\right)\right] \to [o, (0, 1)] \text{ as } n \to \infty,$$

but for Steiner symmetrization in the first coordinate axis we have that

$$\mathrm{st}\left[o, \left(\frac{1}{n}, 1\right)\right] = \left[o, \left(\frac{1}{n}, 0\right)\right] \to \{o\} \neq \left[\left(0, -\frac{1}{2}\right), \left(0, \frac{1}{2}\right)\right] = \mathrm{st}\,[o, (0, 1)]\,.$$

Proof. We may assume that $o \in H$. Let L be the line through o orthogonal to the hyperplane H.

(i) The proof that $\mathrm{st}\, C$ is compact is left to the reader. To show that $\mathrm{st}\, C$ is convex, let $x, y \in \mathrm{st}\, C$. Consider the convex trapezoid

$$T = \mathrm{conv}\,\big((C \cap (L + x)) \cup (C \cap (L + y))\big) \subseteq C.$$

Since $x, y \in \mathrm{st}\, T$ and since $\mathrm{st}\, T$ is also a convex (see Fig. 9.2) trapezoid and is contained in $\mathrm{st}\, C$, it follows that $[x, y] \subseteq \mathrm{st}\, T \subseteq \mathrm{st}\, C$.

(ii) Trivial.

(iii) Let $x \in \mathrm{st}\, C$, $y \in \mathrm{st}\, D$ or, equivalently, $x = h + l$, $y = k + m$, where $h, k \in H$ and $l, m \in L$ are such that $\|l\| \leq \frac{1}{2}$ length $\big(C \cap (L + x)\big)$ and $\|m\| \leq \frac{1}{2}$ length $\big(D \cap (L + y)\big)$. Then $x + y = (h + k) + (l + m)$ where $h + k \in H$ (note that $o \in H$) and $l + m \in L$ (note that $o \in L$), where

$$\begin{aligned}
\|l + m\| &\leq \|l\| + \|m\| \\
&\leq \frac{1}{2}\big(\text{length}\,\big(C \cap (L + x)\big) + \text{ length}\,\big(D \cap (L + y)\big)\big) \\
&= \frac{1}{2} \text{ length}\,\big(C \cap (L + x) + D \cap (L + y)\big) \\
&\leq \frac{1}{2} \text{ length}\,\big((C + D) \cap (L + x + y)\big).
\end{aligned}$$

Hence $x + y \in \mathrm{st}(C + D)$.

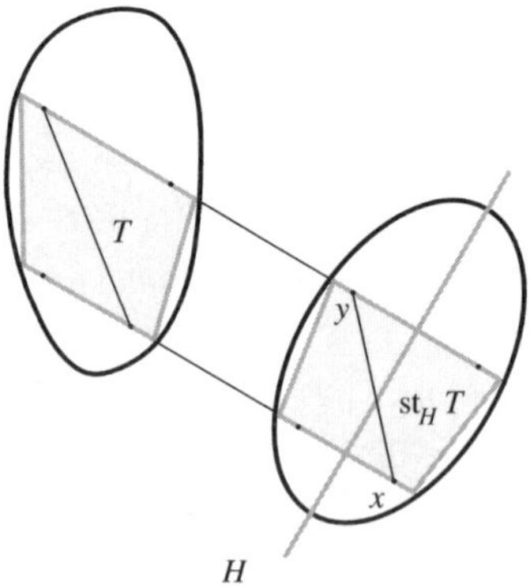

Fig. 9.2. Steiner symmetrization preserves convexity

(iv) Trivial.

(v) Let $C, C_1, C_2, \dots \in \mathcal{C}_p$ be such that $C_1, C_2, \dots \to C$. We have to show that $\operatorname{st} C_1, \operatorname{st} C_2, \dots \to \operatorname{st} C$. We clearly may suppose that $o \in \operatorname{int} C$. The assumption that $C_1, C_2, \dots \to C$ is then equivalent to the following: Let $\varepsilon > 0$, then $(1-\varepsilon)C \subseteq C_n \subseteq (1+\varepsilon)C$ for all sufficiently large n. By (ii) and (iv), this implies the following: Let $\varepsilon > 0$, then $(1-\varepsilon) \operatorname{st} C \subseteq \operatorname{st} C_n \subseteq (1+\varepsilon) \operatorname{st} C$ for all sufficiently large n. Since $o \in \operatorname{int} \operatorname{st} C$, this implies that $\operatorname{st} C_1, \operatorname{st} C_2, \dots \to \operatorname{st} C$.

(vi) This property is an immediate consequence of the definition of the Steiner symmetrization, (i) and Fubini's theorem.

(vii) By (iii) and (ii), $\operatorname{st}(C + \varepsilon B^d) \supseteq \operatorname{st} C + \varepsilon \operatorname{st} B^d = \operatorname{st} C + \varepsilon B^d$ for $\varepsilon > 0$. Hence

$$\begin{aligned}\frac{V(\operatorname{st} C + \varepsilon B^d) - V(\operatorname{st} C)}{\varepsilon} &\le \frac{V\big(\operatorname{st}(C + \varepsilon B^d)\big) - V(\operatorname{st} C)}{\varepsilon}\\ &= \frac{V(C + \varepsilon B^d) - V(C)}{\varepsilon} \quad \text{for } \varepsilon > 0,\end{aligned}$$

by (iv) and (vi). Now let $\varepsilon \to +0$ and note the definition of surface area in Sect. 6.4.

(viii) Let $x, y \in \operatorname{st} C$ and let T and $\operatorname{st} T$ be as in the proof of (i). Then, at least one of the diagonals of the trapezoid $T \subseteq C$ has length greater than or equal to $\|x - y\|$.

(ix) Let $c + \varrho B^d \subseteq C$. Then $\operatorname{st}(c + \varrho B^d) \subseteq \operatorname{st} C$ by (iv) and thus $b + \varrho B^d \subseteq \operatorname{st} C$ for suitable b. Hence $r(C) \le r(\operatorname{st} C)$. The inequality $R(\operatorname{st} C) \le R(C)$ is shown similarly, using balls containing C. □

Remark. Proposition 9.1(iii) can be refined as follows: Let $C, D \in \mathcal{C}_p$. Then

$$\operatorname{st}(C + D) \supseteq \operatorname{st} C + \operatorname{st} D,$$

where equality holds if and only if C and D are homothetic. Propositions (vi) and (vii) admit the following generalization and refinement: Let $C \in \mathcal{C}$, then

$$W_i(\operatorname{st} C) \le W_i(C) \text{ for } i = 0, 1, \dots, d,$$

where for $C \in \mathcal{C}_p$ and $i = 1, \dots, d-1$ equality holds if and only if C is symmetric in a hyperplane parallel to H. See the references at the beginning of Sect. 9.

Polar Bodies and Steiner Symmetrization

Let C be a convex body with $o \in \operatorname{int} C$. Its *polar body* C^* (with respect to o) is defined by

$$C^* = \{y : x \cdot y \le 1 \text{ for } x \in C\}.$$

It is easy to see that C^* is also a convex body with $o \in \operatorname{int} C^*$. Polar bodies play a role in geometric inequalities, the local theory of normed spaces, the geometry of numbers and other areas. The following property was noted by Ball [47] and Meyer and Pajor [720]. It can be used to prove the Blaschke–Santaló inequality, see Theorem 9.5. Let $v(\cdot)$ denote $(d-1)$-dimensional volume.

Proposition 9.2. *Let $C \in \mathcal{C}_p$ be o-symmetric. Then, Steiner symmetrization, with respect to a given hyperplane through o, satisfies the following inequality:*

$$V\big((\operatorname{st} C)^*\big) \geq V(C^*).$$

Proof. We may suppose that this hyperplane is $\mathbb{E}^{d-1} = \{x : x_d = 0\}$. Then

$$\begin{aligned}
\operatorname{st} C &= \Big\{\Big(x, \frac{1}{2}(s-t)\Big) : (x,r), (x,s) \in C\Big\},\\
C^* &= \big\{(y,t) : x \cdot y + st \leq 1 \text{ for } (x,s) \in C\big\},\\
(\operatorname{st} C)^* &= \Big\{(w,t) : x \cdot w + \frac{1}{2}(r-s)t \leq 1 \text{ for } (x,r), (x,s) \in C\Big\}.
\end{aligned}$$

For a set $A \subseteq \mathbb{E}^d$ and $t \in \mathbb{R}$, let $A(t) = \{v \in \mathbb{E}^{d-1} : (v,t) \in A\}$. Then

$$\begin{aligned}
&\frac{1}{2}\big(C^*(t) + C^*(-t)\big)\\
&\quad = \Big\{\frac{1}{2}(y+z) : x \cdot y + rt \leq 1,\ w \cdot z + s(-t) \leq 1 \text{ for } (x,r), (w,s) \in C\Big\}\\
&\quad \subseteq \Big\{\frac{1}{2}(y+z) : x \cdot y + rt \leq 1,\ x \cdot z + s(-t) \leq 1 \text{ for } (x,r), (x,s) \in C\Big\}\\
&\quad \subseteq \Big\{\frac{1}{2}(y+z) : x \cdot \frac{1}{2}(y+z) + \frac{1}{2}(r-s)t \leq 1 \text{ for } (x,r), (x,s) \in C\Big\}\\
&\quad = \Big\{w : x \cdot w + \frac{1}{2}(r-s)t \leq 1 \text{ for } (x,r), (x,s) \in C\Big\}\\
&\quad = (\operatorname{st} C)^*(t) \text{ for } t \in \mathbb{R}.
\end{aligned}$$

Since C is o-symmetric, C^* is also o-symmetric. Thus $C^*(t) = -C^*(-t)$ and therefore $v\big(C^*(t)\big) = v\big(C^*(-t)\big)$. The Brunn–Minkowski inequality in $d-1$ dimensions then shows that

$$\begin{aligned}
v\big((\operatorname{st} C)^*(t)\big) &\geq v\Big(\frac{1}{2}\big(C^*(t) + C^*(-t)\big)\Big)\\
&\geq \Big(\frac{1}{2} v\big(C^*(t)\big)^{\frac{1}{d-1}} + \frac{1}{2} v\big(C^*(-t)\big)^{\frac{1}{d-1}}\Big)^{d-1} = v\big(C^*(t)\big) \text{ for } t \in \mathbb{R}.
\end{aligned}$$

Now, integrating over t, Fubini's theorem yields $V\big((\operatorname{st} C)^*\big) \geq V(C^*)$. □

The Sphericity Theorem of Gross

The following highly intuitive *sphericity theorem of Gross* [407] and its corollary were used by Blaschke [124] for easy proofs of the isodiametric, the isoperimetric and the Brunn–Minkowski inequalities. Let $\kappa_d = V(B^d)$.

Theorem 9.1. *Let $C \in \mathcal{C}_p$ with $V(C) = V(B^d)$. Then, there is a sequence $C_1, C_2, \dots$ of convex bodies, each obtained from C by finitely many successive Steiner symmetrizations with respect to hyperplanes through o, such that*

$$C_1, C_2, \dots \to B^d.$$

Proof. For $D \in \mathcal{C}$, let $\varrho(D)$ denote the minimum radius of a ball with centre o which contains D. Let $\mathcal{S} = \mathcal{S}(C)$ be the family of all convex bodies which can be obtained from C by finitely many successive Steiner symmetrizations with respect to hyperplanes through o. Let

$$\sigma = \inf\{\varrho(D) : D \in \mathcal{S}\}.$$

There is a sequence (C_n) in $\mathcal{S}$ such that

(1) $\varrho(C_n) \to \sigma$.

Since $C \subseteq \varrho(C)B^d$, an application of Proposition 9.1(iv) shows that $C_n \subseteq \varrho(C)B^d$ for $n = 1, 2, \dots$ Now apply Blaschke's selection theorem 6.3. Then, by considering a suitable subsequence of (C_n) and renumbering, if necessary, we may assume that

(2) $C_n \to C_0 \in \mathcal{C}$, say.

Clearly, $\varrho(\cdot)$ is continuous on $\mathcal{C}$. It thus follows from (2) and (1) that $\varrho(C_n) \to \varrho(C_0) = \sigma$. We assert that

(3) $C_0 = \sigma B^d$.

For, if not, $C_0 \subsetneq \sigma B^d$ (note that $\varrho(C_0) = \sigma$). Thus there is a calotta of σB^d which is disjoint from C_0, where a *calotta* of B^d is the intersection of B^d with a closed half-space. Given a calotta of σB^d, we may cover $\operatorname{bd} \sigma B^d$ by finitely many mirror images of it in hyperplanes through the origin o. For suitable hyperplanes $H_1, \dots, H_k$ through o, the convex body:

(4) $D_0 = \mathrm{st}_{H_k} \mathrm{st}_{H_{k-1}} \cdots \mathrm{st}_{H_1} C_0 \in \mathcal{C}_p$

is then contained in $\operatorname{int} \sigma B^d$ and therefore:

(5) $\varrho(D_0) < \sigma$.

Clearly,

(6) $D_n = \mathrm{st}_{H_k} \mathrm{st}_{H_{k-1}} \cdots \mathrm{st}_{H_1} C_n \in \mathcal{S}$ for $n = 1, 2, \dots$.

Thus $\varrho(D_n) \geq \sigma$. Since $C_n \to C_0$ by (2), it follows from (4), (6) and Proposition 9.1(v) that $D_n \to D_0$. The continuity of $\varrho(\cdot)$, together with (5), then shows that $\varrho(D_n) < \sigma$ for all sufficiently large n. Since $D_n \in \mathcal{S}$ this is in contradiction to the definition of σ. The proof of (3) is complete.

Since $C_n \in \mathcal{S}$, Propositions (2) and (3) imply that $C_1, C_2, \dots \to \sigma B^d$. Since $V(C_n) = V(C) = V(B^d)$ for each n and volume is continuous, this is possible only if $\sigma = 1$. The proof (3), and thus of the theorem, is complete. □

Corollary 9.1. *Let* $C, D \in \mathcal{C}_p$. *Then, there are Steiner symmetrizations* st_{H_1}, $\mathrm{st}_{H_2}, \dots,$ *with respect to hyperplanes* $H_1, H_2, \dots$ *through o, such that*

$$C_n = \mathrm{st}_{H_n} \cdots \mathrm{st}_{H_1} C \to \left(\frac{V(C)}{\kappa_d}\right)^{\frac{1}{d}} B^d,$$

$$D_n = \mathrm{st}_{H_n} \cdots \mathrm{st}_{H_1} D \to \left(\frac{V(D)}{\kappa_d}\right)^{\frac{1}{d}} B^d.$$

Proof. Let $\varepsilon > 0$. By the sphericity theorem, there are hyperplanes $H_1, \dots, H_{n_1}$, through o, such that

$$\operatorname{st}_{H_{n_1}} \cdots \operatorname{st}_{H_1} C \subseteq (1+\varepsilon)\left(\frac{V(C)}{\kappa_d}\right)^{\frac{1}{d}} B^d.$$

Again, by the sphericity theorem, there are hyperplanes $H_{n_1+1}, \dots, H_{n_2}$, through o, such that

$$\operatorname{st}_{H_{n_2}} \cdots \operatorname{st}_{H_{n_1+1}} (\operatorname{st}_{H_{n_1}} \cdots \operatorname{st}_{H_1} D) \subseteq (1+\varepsilon)\left(\frac{V(D)}{\kappa_d}\right)^{\frac{1}{d}} B^d,$$

while maintaining

$$\operatorname{st}_{H_{n_2}} \cdots \operatorname{st}_{H_{n_1+1}} (\operatorname{st}_{H_{n_1}} \cdots \operatorname{st}_{H_1} C) \subseteq (1+\varepsilon)\left(\frac{V(C)}{\kappa_d}\right)^{\frac{1}{d}} B^d.$$

Applying the same argument with $\varepsilon/2$ to the convex bodies $\operatorname{st}_{H_{n_2}} \cdots \operatorname{st}_{H_1} C$, $\operatorname{st}_{H_{n_2}} \cdots \operatorname{st}_{H_1} D \in \mathcal{C}_p$, etc., finally yields the corollary. □

Remark. Considering the theorem and its corollary, the question arises, how fast do suitable iterated Steiner symmetrals of C approximate ϱB^d? The first result in this direction was proved by Hadwiger [463]. A more precise estimate is due to Bourgain, Lindenstrauss and Milman [160]. There are an absolute constant $\alpha > 0$ and a function $\alpha(\varepsilon) > 0$ such that the following statement holds: Let C be a convex body C of volume $V(B^d)$ and $\varepsilon > 0$. Then, by at most $\alpha\, d \log d + \alpha(\varepsilon)\, d$ Steiner symmetrizations of the body or the polar body, one obtains a convex body D with

$$(1-\varepsilon)B^d \subseteq D \subseteq (1+\varepsilon)B^d.$$

For a pertinent result involving random Steiner symmetrizations, see Mani-Levitska [683]. A sharp isomorphic result is due to Klartag and Milman [589]. A lower bound for the distance of iterated Steiner symmetrals of C from ϱB^d was given by Bianchi and Gronchi [112]. See [660, 878] for additional references.

Blaschke Symmetrization

Blaschke [124], p.103, introduced the following concept of symmetrization. Given a convex body C and a hyperplane H in $\mathbb{E}^d$, let C^H be the (mirror) reflection of C in H. Then the *Blaschke symmetrization* of C with respect to H is the convex body

$$\frac{1}{2}C + \frac{1}{2}C^H.$$

In recent years the Blaschke symmetrization has also been called *Minkowski symmetrization.* A surprisingly sharp sphericity result for this symmetrization is due to Klartag [588].

9.2 The Isodiametric, Isoperimetric, Brunn–Minkowski, Blaschke–Santaló and Mahler Inequalities

If C is a proper convex body but not a ball, then a suitable Steiner symmetral of C has smaller isoperimetric quotient than C. This observation led Steiner [958, 959], erroneously, to believe that Steiner symmetrization shows that balls have minimum isoperimetric quotient. The gap in this proof of the isoperimetric inequality was filled by Blaschke. He showed that Steiner symmetrization together with the Blaschke selection theorem, or rather the sphericity theorem of Gross, also provide easy proofs of the isodiametric, the isoperimetric and the Brunn–Minkowski inequalities.

Below we first present the proofs of Gross [407] and Blaschke [124] of these inequalities, but do not discuss the equality cases. Then the proof of Ball [47] and Meyer and Pajor [720] of the Blaschke–Santaló inequality will be given, again, excluding a discussion of the equality case. As a counterpart of the Blaschke–Santaló inequality, we present an inequality of Mahler. Let $\kappa_d = V(B^d)$.

The Isodiametric Inequality

A simple proof using d Steiner symmetrizations yields the following result of Bieberbach [113], a refinement of which was given by Urysohn [1004]. For a different proof, see Sect. 8.3.

Theorem 9.2. *Let $C \in \mathcal{C}$. Then $V(C) \le \frac{1}{2^d}(\operatorname{diam} C)^d V(B^d)$.*

Proof. By symmetrizing C with respect to each coordinate hyperplane we obtain a convex body D which is symmetric in the origin o and with $V(D) = V(C)$ and $\operatorname{diam} D \le \operatorname{diam} C$, see Proposition 9.1(vi,viii). Then $D \subseteq (\frac{1}{2}\operatorname{diam} D)B^d$. □

The Isoperimetric Inequality

An equally simple proof yields the next result.

Theorem 9.3. *Let $C \in \mathcal{C}_p$. Then $\dfrac{S(C)^d}{V(C)^{d-1}} \ge \dfrac{S(B^d)^d}{V(B^d)^{d-1}}$.*

Proof. By Corollary 9.1, there are hyperplanes $H_1, H_2, \dots$ through o, such that

$$\mathrm{st}_{H_n} \cdots \mathrm{st}_{H_1} C \to \left(\frac{V(C)}{\kappa_d}\right)^{\frac{1}{d}} B^d.$$

Now using Proposition 9.1(vii) and the continuity of the surface area $S(\cdot)$, see Theorem 6.13, we obtain the desired inequality:

$$S(C) \ge S(\mathrm{st}_{H_1} C) \ge \cdots \to S\left(\left(\frac{V(C)}{\kappa_d}\right)^{\frac{1}{d}} B^d\right) = \frac{V(C)^{\frac{d-1}{d}}}{\kappa_d^{\frac{d-1}{d}}} S(B^d). \quad \square$$

The Brunn–Minkowski Inequality

The following proof reduces the Brunn–Minkowski inequality to the trivial case where both bodies are balls.

Theorem 9.4. *Let* $C, D \in \mathcal{C}$. *Then* $V(C+D)^{\frac{1}{d}} \geq V(C)^{\frac{1}{d}} + V(D)^{\frac{1}{d}}$.

Proof. It is sufficient to consider the case where $C, D \in \mathcal{C}_p$. By Corollary 9.1, there are hyperplanes $H_1, H_2, \dots$ through o, such that

$$\operatorname{st}_{H_n} \cdots \operatorname{st}_{H_1} C \to \left(\frac{V(C)}{\kappa_d}\right)^{\frac{1}{d}} B^d,$$

$$\operatorname{st}_{H_n} \cdots \operatorname{st}_{H_1} D \to \left(\frac{V(D)}{\kappa_d}\right)^{\frac{1}{d}} B^d.$$

Thus Proposition 9.1(vi,iii) and the continuity of V, see Theorem 7.5, yield the following.

$$V(C+D)^{\frac{1}{d}} = V\left(\operatorname{st}_{H_1}(C+D)\right)^{\frac{1}{d}} \geq V(\operatorname{st}_{H_1} C + \operatorname{st}_{H_1} D)^{\frac{1}{d}} \geq \cdots$$

$$\to V\left(\left(\frac{V(C)}{\kappa_d}\right)^{\frac{1}{d}} B^d + \left(\frac{V(D)}{\kappa_d}\right)^{\frac{1}{d}} B^d\right)^{\frac{1}{d}} = V\left(\frac{1}{\kappa_d^{\frac{1}{d}}}\left(V(C)^{\frac{1}{d}} + V(D)^{\frac{1}{d}}\right) B^d\right)^{\frac{1}{d}}$$

$$= V(C)^{\frac{1}{d}} + V(D)^{\frac{1}{d}}. \quad \square$$

The Blaschke–Santaló Inequality

Blaschke [126] proved the following result for $d = 3$. It was extended to general d by Santaló [880].

Theorem 9.5. *Let* $C \in \mathcal{C}_p$ *be o-symmetric. Then*

$$V(C)V(C^*) \leq \kappa_d^2.$$

Proof. By the Corollary 9.1 to the sphericity theorem of Gross, there are hyperplanes $H_1, H_2, \dots$ through o, such that

$$C_n = \operatorname{st}_{H_n} \cdots \operatorname{st}_{H_1} C \to \left(\frac{V(C)}{\kappa_d}\right)^{\frac{1}{d}} B^d.$$

An easy argument for polar bodies then shows that

$$C_n^* \to \left(\frac{\kappa_d}{V(C)}\right)^{\frac{1}{d}} B^d.$$

Thus

$$\begin{aligned} V(C)V(C^*) &\leq V(\operatorname{st}_{H_1} C)V\left((\operatorname{st}_{H_1} C)^*\right) = V(C_1)V(C_1^*) \\ &\leq V(\operatorname{st}_{H_2} C_1)V\left((\operatorname{st}_{H_2} C_1)^*\right) = V(C_2)V(C_2^*) \leq \dots \\ &\to V(B^d)^2 = \kappa_d^2. \end{aligned}$$

by Proposition 9.2 and the continuity of volume on $\mathcal{C}$, see Theorem 7.5. $\square$

Mahler's Inequality

The following inequality of Mahler [679] is a useful tool for successive minima, see Theorem 23.2:

Theorem 9.6. *Let $C \in \mathcal{C}_p$ be o-symmetric. Then*

$$\frac{4^d}{d\,!} \le V(C)V(C^*).$$

Proof. Since a non-singular linear transformation does not change the product $V(C)V(C^*)$, we may assume the following. The o-symmetric cross-polytope $O = \{x : |x_1| + \cdots + |x_d| \le 1\}$ is inscribed in C and has maximum volume among all such cross-polytopes. Since O has maximum volume, C is contained in the cube $K = \{x : |x_i| \le 1\}$. Thus $O \subseteq C \subseteq K$. Polarity then yields $K^* \subseteq C^* \subseteq O^*$. Note that $K^* = O$. Hence

$$\frac{(2^d)^2}{(d\,!)^2} = V(O)^2 = V(O)V(K^*) \le V(C)V(C^*),$$

concluding the proof. □

A much stronger version of this simple result was conjectured also by Mahler:

Conjecture 9.1. *Let $C \in \mathcal{C}_p$ be o-symmetric. Then*

$$\frac{4^d}{d\,!} \le V(C)V(C^*).$$

Remark. It is plausible that the product $V(C)V(C^*)$ is minimum if C is the cube $\{x : |x_i| \le 1\}$. Then C^* is the cross-polytope $\{x : |x_1| + \cdots + |x_d| \le 1\}$ and we have, $V(C)V(C^*) = 4^d/d\,!$. This seems to have led Mahler to the above conjecture. Mahler proved it for $d = 2$. While the conjecture is open for $d \ge 3$, substantial progress has been achieved by Bourgain and Milman [161] who proved that there is an absolute constant α such that

$$\frac{\alpha^d}{d^d} \le V(C)V(C^*).$$

A simple proof of a slightly weaker result is due to Kuperberg [622]. A refined version of the estimate of Bourgain and Milman was given by Kuperberg [624]. For special convex polytopes a proof of the conjecture is due to Lopez and Reisner [663]. In [623] Kuperberg stated a conjecture related to Mahler's conjecture. For more information we refer to Lindenstrauss and Milman [660].

9.3 Schwarz Symmetrization and Rearrangement of Functions

A relative of Steiner's symmetrization is Schwarz's symmetrization. It was introduced by Schwarz [922] as a tool for the solution of the geometric isoperimetric problem. While of interest in convex geometry, its real importance is in the context of isoperimetric inequalities of mathematical physics, where a version of it, the spherical rearrangement of functions, is an indispensable tool.

In this section Schwarz symmetrization of convex bodies and spherically symmetric rearrangement of real functions are treated. Without proof we state a result on rearrangements. Applications are dealt with in the following sections.

For references to the literature on rearrangement of functions and its applications to isoperimetric inequalities of mathematical physics and partial differential equations, consult the books of Bandle [64] and Kawohl [569] and the survey of Talenti [987]. Unfortunately, a modern treatment of this important topic with detailed proofs, still seems to be missing.

Schwarz Symmetrization

Schwarz symmetrization or *Schwarz rounding* of a convex body C with respect to a given line L is defined as follows. For each hyperplane H orthogonal to L and which meets C, replace $C \cap H$ by the $(d-1)$-dimensional ball in H with centre at $H \cap L$ and $(d-1)$-dimensional volume equal to that of $C \cap H$. The union of all balls thus obtained is then the Schwarz symmetrization $\operatorname{sch} C = \operatorname{sch}_L C$ of C with respect to L.

The version 8.4 of the Brunn–Minkowski theorem implies that $\operatorname{sch} C \in \mathcal{C}$. Refining the argument that led to Corollary 9.1, shows that there are hyperplanes $H_1, H_2, \dots$, all containing L, such that

$$\operatorname{st}_{H_n} \cdots \operatorname{st}_{H_1} C \to \operatorname{sch}_L C.$$

As a consequence, many properties of Steiner symmetrization also hold for Schwarz symmetrization; in particular, the properties listed in Proposition 9.1.

Rearrangement of Functions

It is clear that Schwarz symmetrization extends to non-convex sets (see Fig. 9.3). Given a Borel function $f : \mathbb{E}^d \to \mathbb{R}$, we apply Schwarz symmetrization to the set $B = \{(x,t) : f(x) \geq t\} \subseteq \mathbb{E}^d \times \mathbb{R}$ with respect to the line $\{o\} \times \mathbb{R}$: Let

$$B(t) = \{(x,t) : f(x) \geq t\} \subseteq \mathbb{E}^d \times \{t\} \text{ for } t \in \mathbb{R}.$$

Since f, and thus B, is Borel, each set $B(t)$ is also Borel. If $B(t)$ has infinite measure, let $B^r(t) = \mathbb{E}^d \times \{t\}$; otherwise let $B^r(t)$ be the ball in $\mathbb{E}^d \times \{t\}$ with centre at (o,t) and measure equal to that of $B(t)$. The *rearrangement* $f^r : \mathbb{E}^d \to \mathbb{R}$ of f then is defined by

$$f^r(x) = \sup\{t : (x,t) \in B^r(t)\} \text{ for } x \in \mathbb{E}^d.$$

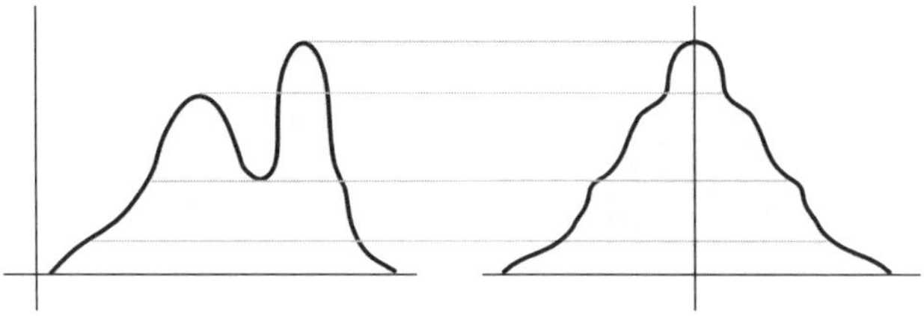

Fig. 9.3. Rearrangement of functions

The following *rearrangement theorem* is indispensable for certain isoperimetric inequalities of mathematical physics, see the next section. While this result was known and used at least since Faber [316] and Krahn [615], a complete proof requires tools of more recent geometric measure theory. See Talenti [987] for its history and references to proofs. Compare also Bandle [64]. Here no proof is presented.

Theorem 9.7. *Let* $f : \mathbb{E}^d \to [0, +\infty)$ *be locally Lipschitz and such that* $f(x) \to 0$ *as* $\|x\| \to \infty$. *Then* $f^r : \mathbb{E}^d \to [0, +\infty)$ *is also locally Lipschitz,* $f^r(x) \to 0$ *as* $\|x\| \to \infty$, *and*

$$\int\limits_{\mathbb{E}^d} \big(\operatorname{grad} f(x)\big)^2 dx \geq \int\limits_{\mathbb{E}^d} \big(\operatorname{grad} f^r(x)\big)^2 dx.$$

(Since f and f^r are locally Lipschitz, $(\operatorname{grad} f)^2$ and $(\operatorname{grad} f^r)^2$ are measurable and exist almost everywhere, see [696].)

9.4 Torsional Rigidity and Minimum Principal Frequency

We begin with the following quotation from Bandle [64], preface.

> The study of 'isoperimetric inequalities' in a broader sense began with the conjecture of St Venant in 1856. Investigating the torsion of elastic prisms, he observed that of all cross-sections of given area the circle has the maximal torsional rigidity. This conjecture was proved by Pólya in 1948. Lord Rayleigh conjectured that of all membranes with given area, the circle has the smallest principal frequency. This statement was proved independently by Faber and Krahn around 1923.

There is a large body of isoperimetric type inequalities in mathematical physics dealing with eigenvalues of partial differential equations. In physical terms these inequalities concern principal frequencies of membranes, torsional rigidity, bending of beams, electrostatic capacity, etc. Major contributors are Saint-Venant, Faber, Krahn, Pólya and Szegö, Osserman, Hersch, Payne, Talenti and others.

The above rearrangement theorem will be applied to give the best upper, respectively, lower, bound for the torsional rigidity of an elastic cylindrical rod and for the first principal frequency of a clamped membrane.

For more information, see the surveys and books of Pólya and Szegö [811], Bandle [64], Mossino [757], Hersch [498], Payne [786] and Talenti [987].

Torsional Rigidity

Let K be a smooth closed Jordan curve in $\mathbb{E}^2$. It is the boundary of a compact body B, where by a *body* we mean a compact set which equals the closure of its interior. Consider a cylindrical rod of homogeneous elastic material with cross-section B. The *torsional rigidity* $T(B)$ of the rod is the torque required for a unit angle of twist per unit length under the assumption that the shear modulus is 1. It can be expressed in the form

$$(1)\quad T(B)=\int_B \big(\operatorname{grad} u(x)\big)^2 dx,$$

where $u : B \to \mathbb{R}$ is the solution of the following boundary value problem, where

$$\Delta = \frac{\partial^2}{\partial x_1^2}+\frac{\partial^2}{\partial x_2^2}$$

is the Laplace operator:

$$\begin{aligned}(2)\quad & \Delta u + 2 = 0 \text{ in int } B\\ & u|K = 0\\ & u|B \text{ is continuous on } B\\ & u|\operatorname{int} B \text{ is of class } \mathcal{C}^2\end{aligned}$$

The following result was conjectured by Saint-Venant [875] and first proved by Pólya [809] a century later.

Theorem 9.8. *Let K be a closed Jordan curve of class $\mathcal{C}^1$ and B the compact body in $\mathbb{E}^2$ bounded by K and of area $A(B)$. Then*

$$T(B) \le \frac{A(B)^2}{2\pi}.$$

Equality is attained if B is a circular disc.

Proof. First, a different representation of $T(B)$ will be given:

$$(3)\quad T(B)=\sup\left\{\frac{\left(2\int_B f(x)\,dx\right)^2}{\int_B \big(\operatorname{grad} f(x)\big)^2 dx} : \right.$$

$$\left. f : B \to [0,+\infty) \text{ locally Lipschitz, } f \neq 0,\ f|K = 0\right\},$$

where the supremum is attained precisely for f of the form $f = u \neq 0$, u as in (2).

Let $\frac{\partial}{\partial n}$ denote the derivative in the direction of the exterior unit normal vector of K and let $x(\cdot) : [0, L] \to \mathbb{E}^2$ be a parametrization of K where the arc-length s is the parameter and L the length of K. Recall the following formula of Green for integrals.

$$\int_B \big(\operatorname{grad} f(x)\big) \cdot \big(\operatorname{grad} u(x)\big)\, dx + \int_B f(x)\Delta u(x)\, dx = \int_0^L f\big(x(s)\big)\frac{\partial u}{\partial n}\big(x(s)\big)\, dx,$$

where f and u are as above.

To prove (3), let f be chosen as in (3). Then (2), Green's formula, the equality $f|K = 0$, the Cauchy–Schwarz inequality for integrals and (1) yield the following:

$$\begin{aligned}
2\int_B f(x)\, dx &= -\int_B f(x)\Delta u(x)\, dx \\
&= \int_B \big(\operatorname{grad} f(x)\big) \cdot \big(\operatorname{grad} u(x)\big)\, dx - \int_0^L f\big(x(s)\big)\frac{\partial u}{\partial n}\big(x(s)\big)\, ds \\
&\le \int_B \|\operatorname{grad} f(x)\|\, \|\operatorname{grad} u(x)\|\, dx \\
&\le \left(\int_B \big(\operatorname{grad} f(x)\big)^2 dx \int_B \big(\operatorname{grad} u(x)\big)^2 dx\right)^{\frac{1}{2}} \\
&= \left(\int_B \big(\operatorname{grad} f(x)\big)^2 dx\, T(B)\right)^{\frac{1}{2}}.
\end{aligned}$$

Hence

$$T(B) \ge \frac{\big(2\int_B f(x)\, dx\big)^2}{\int_B \big(\operatorname{grad} f(x)\big)^2 dx}.$$

Since (2), Green's formula, $u|K = 0$ and (1) show that

$$\begin{aligned}
2\int_B u(x)\, dx &= -\int_B u(x)\Delta u(x)\, dx \\
&= \int_B \big(\operatorname{grad} u(x)\big)^2 dx - \int_0^L u\big(x(s)\big)\frac{\partial u}{\partial n}\big(x(s)\big)\, ds \\
&= \int_B \big(\operatorname{grad} u(x)\big)^2 dx = T(B),
\end{aligned}$$

we obtain

$$T(B) = \frac{\big(2\int_B u(x)\, dx\big)^2}{\int_B \big(\operatorname{grad} u(x)\big)^2 dx}.$$

The proof of (3) is thus complete.

Now, using (3) and the rearrangement theorem, it follows that

$$T(B) = \sup \left\{ \frac{\left(2 \int\limits_B f(x)\,dx\right)^2}{\int\limits_B \left(\operatorname{grad} f(x)\right)^2 dx} : \dots \right\}$$

$$\leq \sup \left\{ \frac{\left(2 \int\limits_D f^r(x)\,dx\right)^2}{\int\limits_D \left(\operatorname{grad} f^r(x)\right)^2 dx} : \dots \right\} \leq T(D),$$

where D is the circular disc with centre o and $A(D) = A(B)$. Its radius is $\varrho = \left(\frac{1}{\pi} A(B)\right)^{1/2}$. The solution of the boundary value problem,

$$\begin{aligned} &\Delta v + 2 = 0 \text{ in } \operatorname{int} D \\ &v | \operatorname{bd} D = 0 \\ &v \text{ is continuous on } D \\ &v | \operatorname{int} D \text{ is of class } C^2 \end{aligned}$$

is given by

$$v(x) = \frac{\varrho^2}{2} - \frac{\|x\|^2}{2} \text{ for } x \in D.$$

Thus

$$T(D) = 2 \int\limits_D \left(\operatorname{grad} v(x)\right)^2 dx = \frac{A(D)^2}{2\pi} = \frac{A(B)^2}{2\pi}. \quad \square$$

Remark. It is well known that, in the estimate for $T(B)$, equality holds precisely in case when B is a circular disc. For recent results on rods consisting of plastic material and for additional information, see Talenti [987].

First Principal Frequency of a Clamped Membrane

Let K be a smooth closed Jordan curve in $\mathbb{E}^2$ and let B be the compact body with K as boundary. Consider an elastic, homogeneous vibrating membrane on B clamped along K. Small vertical vibrations of this membrane are described by a function $v(x,t) : B \times \mathbb{R} \to \mathbb{R}$ which satisfies the *wave equation*,

$$\begin{aligned} (4) \quad &c\,\Delta v = v_{tt} \text{ in } \operatorname{int} B \times \mathbb{R} \\ &v | K \times \mathbb{R} = 0 \\ &v \text{ is continuous on } B \times \mathbb{R} \\ &v | \operatorname{int} B \times \mathbb{R} \text{ is of class } C^2 \end{aligned}$$

where $c > 0$ is a constant depending on the membrane. If v is a non-trivial solution of (4) of the form $v(x,t) = u(x)e^{i\omega t}$, then u is a solution of the following eigenvalue problem:

(5) $\Delta u + \lambda u = 0$ in int B, where $\lambda = \dfrac{\omega^2}{c}$
$u|K = 0$
u is continuous on B
$u|\,\mathrm{int}\,B$ is of class $\mathcal{C}^2$

Before stating and proving Rayleigh's theorem, we collect several well-known results on this eigenvalue problem. A non-trivial solution of (5) has the following properties:

(6) $u > 0$ in int B, $\dfrac{\partial u}{\partial n} > 0$ on K.

The eigenvalue problem (5) has non-trivial solutions precisely for a sequence of λ's,

$$0 < \lambda_1 \leq \lambda_2 \leq \cdots \to \infty .$$

Then

$$0 < \omega_1 = \sqrt{c\lambda_1} \leq \omega_2 = \sqrt{c\lambda_2} \leq \cdots \to \infty$$

are called the *principal frequencies* of the membrane.

Rayleigh [824] conjectured that, among all homogeneous membranes of given area, it is precisely the circular membranes which have the smallest first principal frequency. First proofs of this conjecture are due to Faber [316] and Krahn [615].

Theorem 9.9. *Among all homogeneous elastic membranes clamped on smooth closed Jordan curves and of given area, the circular membranes have the smallest first principal frequency.*

Proof. We use the same notation as in the proof of Theorem 9.8. Let K be a smooth closed Jordan curve which bounds a compact body B in $\mathbb{E}^2$ of given area. We first prove that λ_1 can be expressed as the infimum of *Rayleigh quotients* as follows:

(7) $$\lambda_1(B) = \inf \left\{ \frac{\int_B \left(\operatorname{grad} f(x)\right)^2 dx}{\int_B f(x)^2 dx} : f : B \to [0, +\infty) \text{ locally Lipschitz}, \ f \neq 0, \ f|K = 0 \right\},$$

where the infimum is attained precisely for f of the form $f = \mathrm{const}\, u$, const > 0, u as in (5).

Let f be as in (7). By (5) and (6) we have $f = gu$, where g is locally Lipschitz. Then (5), Green's integral formula and the property that $u|K = 0$ together show that

$$\begin{aligned}
&\int\limits_B \left(\left(\operatorname{grad} f(x)\right)^2 - \lambda f(x)^2\right) dx \\
&\quad = \int\limits_B \left((\operatorname{grad} gu)^2 + g^2 u \,\Delta u\right) dx \\
&\quad = \int\limits_B \{g^2(\operatorname{grad} u)^2 + 2gu \operatorname{grad} g \cdot \operatorname{grad} u + u^2(\operatorname{grad} g)^2 \\
&\qquad\quad - \operatorname{grad}(g^2u) \cdot \operatorname{grad} u\} dx + \int\limits_K g^2 u \frac{\partial u}{\partial n}\, dt \\
&\quad = \int\limits_B \{g^2(\operatorname{grad} u)^2 + 2gu \operatorname{grad} g \cdot \operatorname{grad} u + u^2(\operatorname{grad} g)^2 \\
&\qquad\quad - g^2(\operatorname{grad} u)^2 - 2gu \operatorname{grad} g \cdot \operatorname{grad} u\}\, dx \\
&\quad = \int\limits_B u^2(\operatorname{grad} g)^2\, dx \geq 0.
\end{aligned}$$

Hence,

$$\lambda \leq \frac{\int\limits_B \left(\operatorname{grad} f(x)\right)^2 dx}{\int\limits_B f(x)^2 dx},$$

where equality holds if and only if $\operatorname{grad} g(x) = o$, i.e. $g = \text{const}$ or $f = \text{const}\, u$. The proof of (7) is complete.

Proposition (7) and the rearrangement theorem finally imply the desired inequality:

$$\begin{aligned}
\lambda_1(B) &= \frac{\int\limits_B \left(\operatorname{grad} u(x)\right)^2 dx}{\int\limits_B u(x)^2 dx} \geq \frac{\int\limits_D \left(\operatorname{grad} u^r(x)\right)^2 dx}{\int\limits_D u^r(x)^2 dx} \\
&\geq \inf \left\{ \frac{\int\limits_D \left(\operatorname{grad} h(x)\right)^2 dx}{\int\limits_D h(x)^2 dx} : h \dots \right\} = \lambda_1(D). \quad \square
\end{aligned}$$

Remark. It can be shown that equality holds precisely for circular membranes. The first principal frequency for circular membranes can be expressed in terms of Bessel functions. For refinements, generalizations and related modern material we refer to Kac [558], Bandle [64], Payne [786], Protter [818] and Talenti [987]. Many of the pertinent results are related to the following question.

Problem 9.1. *What information about a compact body B in $\mathbb{E}^d$ can be obtained from the sequence of eigenvalues of the eigenvalue problem* (5)*?*

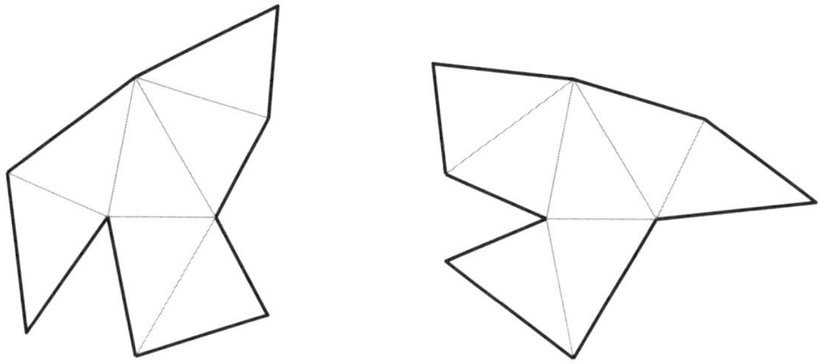

Fig. 9.4. Isospectral membranes

Weyl [1019] showed that the area of a plane membrane is determined by the sequence of its principal frequencies. This led to the speculation that, perhaps, the shape is also determined by this sequence. Quite unexpectedly, Gordon, Webb and Wolpert [388] constructed examples of essentially distinct non-convex *isospectral membranes* (see Fig. 9.4), i.e. membranes with the same sequence of principal frequencies. In other words, one *cannot* hear the shape of a drum. See also the article of Buser, Conway, Doyle and Semmler [185]. The figure shows a pair of isospectral membranes specified by McDonald and Meyers [703].

9.5 Central Symmetrization and the Rogers–Shephard Inequality

Given a convex body C, there are several possibilities to assign to C a convex body which is centrally symmetric with respect to some point. Here the following possibility will be considered: The *difference body* D of C, defined by

$$D = C - C = \{x - y : x, y \in C\},$$

is convex and symmetric with respect to o. The convex body $\frac{1}{2}D$ is called the *central symmetral* of C.

The difference body is important for, e.g. the isodiametric inequality, for packing and tiling. See Sects. 8.3, 30 and 32.2. In this section, we give tight lower and upper estimates for the volume of the difference body of a given convex body in terms of the volume of the original body. The lower estimate is an immediate consequence of the Brunn–Minkowski inequality. The upper estimate is the Rogers–Shephard inequality [852]. For applications of this inequality to density estimates for lattice and non-lattice packing, see Sects. 30.1 and 30.3. A simple proof of a more general result is due to Chakerian [199].

Estimates for $V(C - C)$

Our aim is to show the *inequality of Rogers and Shephard* without considering the equality cases. This is the right-hand inequality in the following result.

Theorem 9.10. *Let $C \in \mathcal{C}_p$. Then the volume of the difference body $D = C - C$ satisfies the inequalities,*

$$2^d V(C) \le V(D) \le \binom{2d}{d} V(C).$$

Proof. Lower estimate: The Brunn–Minkowski theorem easily yields

$$\begin{aligned} V(D) &= V(C - C) = V\big(C + (-C)\big) \ge \big(V(C)^{\frac{1}{d}} + V(-C)^{\frac{1}{d}}\big)^d \\ &= \big(2V(C)^{\frac{1}{d}}\big)^d = 2^d V(C). \end{aligned}$$

Upper estimate: Let $\mathbb{1}_C$ be the characteristic function of C. Unless indicated otherwise, integration is over $\mathbb{E}^d$. By changing the order of integration we have:

$$\begin{aligned} (1)\quad \int \Big(\int \mathbb{1}_C(y - x)\mathbb{1}_C(y)\, dy \Big)\, dx &= \int \mathbb{1}_C(y) \Big(\int \mathbb{1}_C(y - x)\, dx \Big)\, dy \\ &= \int \mathbb{1}_C(y) V(C)\, dy = V(C)^2. \end{aligned}$$

For each x, the integral

$$\int \mathbb{1}_C(y - x)\mathbb{1}_C(y)\, dy$$

is 0, unless there is a point y such that y and $y - x$ both belong to C. In this case $x = y - (y - x) \in C - C = D$. Hence (1) can be written in the form:

$$(2)\quad \int\limits_D \Big(\int \mathbb{1}_C(y - x)\mathbb{1}_C(y)\, dy \Big)\, dx = V(C)^2.$$

For each point $x \in D \backslash \{o\}$, let $\lambda = \lambda(x) \in (0, 1]$ be such that $z = \lambda^{-1} x \in \operatorname{bd} D \subseteq D$. Since $z \in D = C - C$, there are points $p, q \in C$ with $z = p - q$. By convexity,

$$(1 - \lambda)C + \lambda p \subseteq C,\ (1 - \lambda)C + \lambda q + x \subseteq C + x.$$

Since $\lambda p - (\lambda q + x) = \lambda(p - q) - x = \lambda z - x = o$, the sets $(1 - \lambda)C + \lambda p$ and $(1 - \lambda)C + \lambda q + x$ coincide and so both are contained in $C \cap (C + x)$. Thus:

$$\begin{aligned} (3)\quad \int \mathbb{1}_C(y - x)\mathbb{1}_C(y)\, dy &= V\big(C \cap (C + x)\big) \ge V\big((1 - \lambda)C\big) \\ &= (1 - \lambda)^d V(C) = \big(1 - \lambda(x)\big)^d V(C). \end{aligned}$$

Substituting this into (2) implies that

$$V(C)^2 \ge \int\limits_D \big(1 - \lambda(x)\big)^d V(C)\, dx.$$

Dividing by $V(C)$ and noticing that

$$\big(1 - \lambda(x)\big)^d = \int\limits_{\lambda(x)}^{1} d(1 - t)^{d-1} dt,$$

we see that

$$(4)\quad V(C) \geq \int_D (1-\lambda(x))^d dx = \int_D \left(\int_{\lambda(x)}^1 d\,(1-t)^{d-1} dt \right) dx$$
$$= \int_0^1 \left(\int_{\substack{x\in D\\ \lambda(x)\leq t}} d\,(1-t)^{d-1} dx \right) dt.$$

Since $z = \lambda(x)^{-1}x \in \mathrm{bd}\, D$, it follows that $x \in \lambda(x)D$. The condition that $\lambda(x) \leq t$ is thus equivalent to the condition that $x \in tD$. Consequently,

$$\int_0^1 \left(\int_{\substack{x\in D\\ \lambda(x)\leq t}} d\,(1-t)^{d-1} dx \right) dt = \int_0^1 d\,(1-t)^{d-1} V(tD)\, dt$$
$$= V(D) \int_0^1 d\,(1-t)^{d-1} t^d dt = \frac{(d\,!)^2}{(2d)\,!} V(D).$$

Substituting this back into (4) yields the desired upper bound for $V(D)$. □

Remark. $V(D) = 2^d V(C)$ holds precisely when C is centrally symmetric and $V(D) = \binom{2d}{d} V(C)$ holds precisely when C is a simplex, see [852]. In the proof of the latter result the following characterization of simplices is needed: C is a simplex if and only if $C \cap (C + x)$ is a non-negative homothetic image of C for each $x \in \mathbb{E}^d$ for which $C \cap (C + x) \neq \emptyset$. This characterization of Rogers and Shephard [852] of simplices slightly refines a classical characterization of simplices due to Choquet [208]. Compare Sect. 12.1.

For a stability version of the Rogers–Shephard inequality, see Böröczky Jr. [156].

A modern characterization of centrally symmetric convex bodies was given by Montegano [751].

10 Problems of Minkowski and Weyl and Some Dynamics

The original versions of *Minkowski's problem* [736, 739] and of *Weyl's problem* [1020] are as follows, where a *closed convex surface* is the boundary of a proper convex body.

Problem 10.1. *Given a positive function κ on S^{d-1}, is there a convex body C with Gauss curvature κ (as a function of the exterior unit normal vector)?*

Problem 10.2. *Given a Riemannian metric on S^{d-1}, is there a closed convex surface S in $\mathbb{E}^d$ such that S^{d-1} with the given Riemannian metric is isometric to S if the latter is endowed with its geodesic or intrinsic metric, i.e. distance in S is measured along geodesic segments.*

Both problems influenced the development of differential geometry and convex geometry throughout the twentieth century. An important additional aspect is that of the uniqueness of the convex body C and of the convex surface S, respectively. In the context of convex geometry, both problems have been solved satisfactorily at the right level of generality, including the uniqueness problems. Many of the pertinent results go back to Alexandrov and Pogorelov in the context of Alexandrov's differential geometry of not-necessarily differentiable convex surfaces.

In the following, the notion of area measure of a convex body will be introduced and, using this, the solution of Alexandrov [12] and Fenchel and Jessen [335] of Minkowski's problem, including uniqueness, will be presented. Then we consider the notion of intrinsic metric and state, without proofs, Alexandrov's [15] realization and Pogorelov's [803] uniqueness or rigidity theorem which together solve Weyl's problem for general convex surfaces in $\mathbb{E}^3$. The known proofs of these results are long and full of tedious technical details.

For more information on the above problems, the reader is referred to the books of Busemann [182], Pogorelov [806] and Schneider [907], to the surveys of Ivanova-Karatopraklieva and Sabitov [538, 539] and the references below.

In addition to investigations of fixed convex bodies or classes of convex bodies, there are results in convex geometry of a dynamical type, dealing with rigidity and deformation of closed convex surfaces, convex surfaces with boundary, non-convex polytopal spheres and frameworks. For rigidity we refer to Sects. 10.2, 17.1 and 17.2. Early results on the evolution of convex polytopal spheres are related to crystal growth, see the references in Sect. 8.4 dealing with Wulff's theorem. In recent years numerous results on deformation of closed convex surfaces by flows of different types have been studied in the context of differential geometry and partial differential equations. In some cases the surfaces shrink to a point and before collapsing their form becomes more and more spherical. These contributions may be considered convexity results, at least in some cases.

A different dynamical aspect in convex geometry is provided by billiards. Billiards have been studied in mathematics at least since Birkhoff and have attracted a lot of interest in the context of dynamical systems. There are results dealing with periodic and dense trajectories, with caustics and with the behaviour induced by boundary points of the billiard tables with particular curvature properties. Several of these investigations also belong to convex geometry.

In the following we try to convey some of the dynamical flavor of evolution and billiards by considering evolution of closed convex surfaces by curvature driven flows and convex caustics in billiard tables.

10.1 Area Measure and Minkowski's Problem

In this section, we define and investigate the notion of (surface) area measure of convex bodies. Then, given a Borel measure σ on S^{d-1} with certain properties, existence and uniqueness of a convex body with area measure σ is proved. This is then used to define Blaschke addition of proper convex bodies. We follow Alexandrov [10], Pogorelov [806] and Schneider [907].

The Reverse Normal Image

Let $C \in \mathcal{C}_p$ be given. Given a Borel set $B \subseteq S^{d-1}$, the *reverse normal* or *reverse spherical image* $n_C^{-1}(B)$ of B in bd C is defined by

$$n_C^{-1}(B) = \big\{x \in \text{bd}\, C : \text{ there is an exterior normal} \\ \text{unit vector of bd}\, C \text{ at } x \text{ contained in } B\big\}.$$

Let μ_{d-1} be the $(d-1)$-dimensional Hausdorff measure in $\mathbb{E}^d$. A subset of bd C is *measurable* if it is measurable with respect to $\mu_{d-1}|\,\text{bd}\, C$.

Proposition 10.1. *The reverse normal image of a Borel set in* S^{d-1} *is measurable in* bd C.

Proof. Let $\mathcal{M}$ be the family of all sets B in S^{d-1} for which $n_C^{-1}(B)$ is measurable. We have to show that $\mathcal{M}$ contains all Borel sets in S^{d-1}. This will be done in three steps.

First, the following will be shown:

(1) Let $B \subseteq S^{d-1}$ be compact. Then $n_C^{-1}(B)$ is compact in bd C and thus $B \in \mathcal{M}$.

To see this, let $x_1, x_2, \dots \in n_C^{-1}(B)$ such that $x_1, x_2, \dots \to x \in \text{bd}\, C$, say. We have to show that $x \in n_C^{-1}(B)$. Choose $u_1, u_2, \dots \in B \subseteq S^{d-1}$ for which $x_n \in n_C^{-1}(u_n)$ for $n = 1, 2, \dots$ By considering a suitable subsequence and renumbering, if necessary, we may suppose that $u_1, u_2, \dots \to u \in S^{d-1}$, say. Since B is compact, $u \in B$. The body C is contained in each of the support halfspaces $\{z : u_n \cdot z \le u_n \cdot x_n\}$. Since $u_n \to u$, $x_n \to x$, it follows that C is contained in the halfspace $\{z : u \cdot z \le u \cdot x\}$ and x is a common boundary point of this halfspace and of C. Hence this halfspace supports C and thus $x \in n_C^{-1}(u) \subseteq n_C^{-1}(B)$, concluding the proof of (1).

Second, we claim the following:

(2) Let $B \in \mathcal{M}$. Then $B^c = S^{d-1} \backslash B \in \mathcal{M}$.

At each point of $n_C^{-1}(B) \cap n_C^{-1}(B^c)$ there are two different exterior unit normal vectors of bd C, one in B, the other one in B^c. Hence each point of $n_C^{-1}(B) \cap n_C^{-1}(B^c)$ is singular. The theorem of Anderson and Klee 5.1 then shows that $\mu_{d-1}\big(n_C^{-1}(B) \cap n_C^{-1}(B^c)\big) = 0$. The set $n_C^{-1}(B^c)$ thus differs from the set $S^{d-1} \backslash n_C^{-1}(B)$ which is measurable by the assumption in (2), by a set of measure 0 and therefore is measurable itself. The proof of (2) is complete.

Third, the following statement holds:

(3) Let $B_1, B_2, \dots \in \mathcal{M}$. Then $B = B_1 \cup B_2 \cup \dots \in \mathcal{M}$.

This follows from the identity $n_C^{-1}(B) = n_C^{-1}(B_1) \cup n_C^{-1}(B_2) \cup \cdots$

Having proved (1) – (3), it follows that $\mathcal{M}$ is a σ-algebra of subsets of S^{d-1} containing all compact sets in S^{d-1}. It thus contains the smallest such σ-algebra, that is the family of all Borel sets in S^{d-1}. □

The Area Measure of Order $d-1$

Let $C \in \mathcal{C}_p$. The *area measure* σ_C *of order* $d-1$ is the Borel measure on S^{d-1} defined by

$$\sigma_C(B) = \mu_{d-1}\big(n_C^{-1}(B)\big) \text{ for each Borel set } B \subseteq S^{d-1}.$$

If $C = P$ is a convex polytope with facets $F_1, \dots, F_n$, and exterior unit normal vectors $u_1, \dots, u_n$, the area measure σ_P is concentrated at the points $u_1, \dots, u_n$ and

$$\sigma_P(\{u_i\}) = \mu_{d-1}(F_i) = v(F_i).$$

More generally,

$$\sigma_P(B) = \sum_{u_i \in B} v(F_i) \text{ for each Borel set } B \subseteq S^{d-1}.$$

If C is of class $\mathcal{C}^2$ with positive Gauss curvature κ_C (as a function of the exterior unit normal vector of bd C), then, as is well known,

$$\kappa_C(u) = \lim \frac{\mu_{d-1}(B)}{\sigma_C(B)} \text{ as the Borel set } B \subseteq S^{d-1} \text{ shrinks down to } \{u\}.$$

In other words, κ_C is the Radon–Nikodym derivative of μ_{d-1} with respect to σ_C. This can also be expressed in the form

$$\sigma_C(B) = \int_B \frac{d\mu_{d-1}(u)}{\kappa_C(u)} \text{ for each Borel set } B \subseteq S^{d-1}.$$

Let $\sigma, \sigma_1, \sigma_2, \dots$ be Borel measures on S^{d-1}. The measures $\sigma_1, \sigma_2, \dots$ are said to *converge weakly* to the measure σ if

$$\int_{S^{d-1}} f(u)\, d\sigma_n(u) \to \int_{S^{d-1}} f(u)\, d\sigma(u) \text{ for any continuous function } f : S^{d-1} \to \mathbb{R}.$$

The following result relates convergence of a sequence of convex bodies to the weak convergence of the corresponding sequence of area measures.

Proposition 10.2. *Let $C, C_1, C_2, \dots \in \mathcal{C}_p$ be such that $C_1, C_2, \dots \to C$. Then the area measures $\sigma_{C_1}, \sigma_{C_2}, \dots$ converge weakly to the area measure σ_C.*

Proof. We first present two tools. The Hausdorff metric and the Blaschke selection theorem may be extended easily to the space of all compact sets in $\mathbb{E}^d$, see the remarks after the proof of Blaschke's selection theorem. The first tool is as follows:

(4) Let $F_0, F_1, F_2, \dots \subseteq \text{bd}\, C$ be compact such that $F_1, F_2, \dots \to F_0$. Then $\limsup_{n\to\infty} \mu_{d-1}(F_n) \le \mu_{d-1}(F_0)$.

Since F_0 is compact, it is the intersection of the decreasing sequence (N_m), where

$$N_m = \Big(F_0 + \frac{1}{m}B^d\Big) \cap \operatorname{bd} C.$$

Hence

(5) $\mu_{d-1}(F_0) = \limsup_{m\to\infty} \mu_{d-1}(N_m)$.

Since each N_m is a neighbourhood of F_0 in $\operatorname{bd} C$, we see that for given m we have, $F_n \subseteq N_m$ for all sufficiently large n. Therefore

$$\limsup_{n\to\infty} \mu_{d-1}(F_n) \le \mu_{d-1}(N_m) \text{ for } m = 1, 2, \dots$$

This together with (5) yields (4).

The second tool is the following special case of the Portmanteau theorem, see Bauer [82].

(6) Let $\tau, \tau_1, \tau_2, \dots,$ be Borel probability measures on S^{d-1}. Then the following statements are equivalent:

(i) $\tau_1, \tau_2, \dots$ converge weakly to τ.

(ii) $\limsup_{n\to\infty} \tau_n(F) \le \tau(F)$ for each compact set $F \subseteq S^{d-1}$.

For the proof of Proposition 10.2 we may suppose that $o \in \operatorname{int} C$ and $o \in \operatorname{int} C_n$ for $n = 1, 2, \dots$ Let $\varrho : \mathbb{E}^d\backslash\{o\} \to \operatorname{bd} C$ be the radial mapping of $\mathbb{E}^d\backslash\{o\}$ onto $\operatorname{bd} C$ with centre o. Since $o \in \operatorname{int} C$ and $C_1, C_2, \dots \to C$, it is not difficult to show that

(7) The mappings $\varrho : \operatorname{bd} C_n \to \operatorname{bd} C$ and their inverses are Lipschitz with Lipschitz constants converging to 1.

We will apply (6). Let $F \subseteq S^{d-1}$ be compact. By (1) and since ϱ is continuous, the sets $\varrho\big(n_{C_n}^{-1}(F)\big) \subseteq \operatorname{bd} C$ are compact. By Blaschke's selection theorem for compact sets, we may assume by considering a suitable subsequence and renumbering, if necessary, that

(8) $\lim_{n\to\infty} \sigma_{C_n}(F)$ exists and equals the limit superior of the original sequence, and

(9) $\varrho\big(n_{C_n}^{-1}(F)\big) \to F_0$, say, where F_0 is compact in $\operatorname{bd} C$.

In order to show that

(10) $F_0 \subseteq n_C^{-1}(F) \subseteq \operatorname{bd} C$,

let $x \in F_0$. By (9) there are points $y_n \in \varrho\big(n_{C_n}^{-1}(F)\big) \subseteq \operatorname{bd} C$ with $y_n \to x$. Choose $x_n \in n_{C_n}^{-1}(F) \subseteq \operatorname{bd} C_n$ such that $\varrho(x_n) = y_n$. By (7) and since $y_n \to x$, and $\varrho(x) = x$, we have that $x_n \to x$. Let $u_n \in F$ such that $x_n \in n_{C_n}^{-1}(u_n)$. By considering a suitable subsequence and renumbering, if necessary, we may suppose that $u_n \to u \in S^{d-1}$, say. Since F is compact, $u \in F$. C_n is contained in the support halfspace $\{z : u_n \cdot z \le u_n \cdot x_n\}$. Now, noting that $C_n \to C, u_n \to u \in F \subseteq S^{d-1}$, and $x_n \to x \in \operatorname{bd} C$, the halfspace $\{z : u \cdot z \le u \cdot x\}$ is a support halfspace of C at x. Thus $x \in n_C^{-1}(u) \subseteq n_C^{-1}(F)$, concluding the proof of (10).

It follows from (4), (9) and (10) that

$$\limsup_{n\to\infty} \mu_{d-1}\big(\varrho\big(n_{C_n}^{-1}(F)\big)\big) \le \mu_{d-1}\big(n_C^{-1}(F)\big).$$

Since

$$\mu_{d-1}\big(n_C^{-1}(S^{d-1})\big) = \mu_{d-1}\big(\varrho\big(n_{C_n}^{-1}(S^{d-1})\big)\big) = \mu_{d-1}(\operatorname{bd} C),$$

the measures $\mu_{d-1}\big(\varrho(n_{C_n}^{-1}(\cdot)\big)$ are probability measures on S^{d-1} up to some constant. Thus (6) implies that the measures $\mu_{d-1}\big(\varrho\big(n_{C_n}^{-1}(\cdot)\big)\big)$ converge weakly to the measure $\mu_{d-1}\big(n_C^{-1}(\cdot)\big)$. Taking into account (7), this implies that the area measures $\sigma_{C_n}(\cdot) = \mu_{d-1}\big(n_{C_n}^{-1}(\cdot)\big)$ converge weakly to the area measure $\sigma_C(\cdot) = \mu_{d-1}\big(n_C^{-1}(\cdot)\big)$. □

Corollary 10.1. *Let $C, D \in \mathcal{C}_p$. Then*

$$V(C, D, \dots, D) = \frac{1}{d} \int\limits_{S^{d-1}} h_C(u)\, d\sigma_D(u).$$

Proof. Choose convex polytopes $P_n \in \mathcal{P}_p$, $n = 1, 2, \dots$, such that $P_n \to D$. By Proposition 10.2 the area measures σ_{P_n} converge weakly to the area measure σ_D. Hence, in particular:

$$(11)\quad \int\limits_{S^{d-1}} h_C(u)\, d\sigma_{P_n}(u) \to \int\limits_{S^{d-1}} h_C(u)\, d\sigma_D(u).$$

Lemma 6.5 and the definition of the area measure of polytopes show that

$$(12)\quad V(C, P_n, \dots, P_n) = \frac{1}{d} \sum_{F \text{ facet of } P_n} h_C(u_F)\, v(F) = \frac{1}{d} \int\limits_{S^{d-1}} h_C(u)\, d\sigma_{P_n}(u).$$

According to Theorem 6.8 mixed volumes are continuous in their entries. Since $P_n \to D$, we thus have:

$$(13)\quad V(C, P_n, \dots, P_n) \to V(C, D, \dots, D).$$

The corollary is now an immediate consequence of Propositions (11)–(13). □

Alexandrov's and Fenchel–Jessen's Generalization of Minkowski's Theorem

Minkowski [736, 739] proved the following result: Let $\kappa : S^{d-1} \to \mathbb{R}^+$ be a continuous function such that

$$(14)\quad \int\limits_{S^{d-1}} \frac{u}{\kappa(u)}\, d\mu_{d-1}(u) = o \text{ (componentwise)}.$$

Then, there is a proper convex body C with area measure σ_C, unique up to translation, such that

$$\sigma_C(B) = \int_B \frac{d\mu_{d-1}(u)}{\kappa(u)} \text{ for each Borel set } B \subseteq S^{d-1}.$$

This equality shows that κ is the following Radon–Nikodym derivative.

$$\kappa = \frac{d\mu_{d-1}}{d\sigma_C}.$$

κ is called the *generalized Gauss curvature* of C. For the question whether or, more precisely, when κ is the ordinary Gauss curvature of C, compare the references cited in the remarks at the end of this section.

The extension of Alexandrov [12] and Fenchel and Jessen [335] of Minkowski's theorem is as follows.

Theorem 10.1. *Let σ be a Borel measure on S^{d-1}. Then the following are equivalent:*

(i) *σ is not concentrated on a great circle of S^{d-1} and*

$$\int_{S^{d-1}} u\, d\sigma(u) = o \text{ (componentwise).}$$

(ii) *There is a proper convex body C, unique up to translation, with area measure σ.*

Busemann [182], p.60, praised this result with the words:

> ... we have here a first example of a deeper theorem of differential geometry in the large proved for a geometrically natural class of surfaces, i.e. without smoothness requirements necessitated by the methods rather than the problem.

Proof. The proof rests on the corresponding result of Minkowski for convex polytopes, see Theorem 18.2.

(i)⇒(ii) We first prove the existence of C. By a *spherically convex set* on S^{d-1} we mean the intersection of S^{d-1} with a convex cone with apex o. For $m = 1, 2, \dots$, decompose S^{d-1} into finitely many pairwise disjoint spherically convex sets, each of diameter at most $1/m$. Let $S_1, \dots, S_n$ denote those among these sets which have positive σ-measure. Here, and in the following, when i appears as an index it would be better to write mi instead, but we do not do it in order to avoid clumsy notation. Let

$$(15)\quad \varrho_i u_i = \frac{1}{\sigma(S_i)} \int_{S_i} u\, d\sigma(u),$$

where $0 < \varrho_i \leq 1$ and $u_i \in S^{d-1}$, $i = 1, \dots, n\ (= n(m))$.

$\varrho_i u_i$ is the centroid of S_i with respect to the measure σ. Since $\varrho_i u_i \in \operatorname{conv} S_i$ and $\operatorname{diam} S_i \leq \frac{1}{m}$, an elementary argument shows that

$$(16)\quad 1 - \frac{1}{2m^2} \leq \varrho_i \leq 1.$$

Let σ_m be the discrete Borel measure on S^{d-1} which is defined as follows:

$$\sigma_m(B) = \sum_{u_i \in B} \sigma(S_i)\varrho_i \text{ for each Borel set } B \subseteq S^{d-1}.$$

We show that

(17) σ_m converges weakly to σ as $m \to \infty$.

Let $g : S^{d-1} \to \mathbb{R}$ be continuous. Then

$$\int_{S^{d-1}} g(u)\, d\sigma_m(u) - \int_{S^{d-1}} g(u)\, d\sigma(u)$$
$$= \sum_{i=1}^{n} \Big(\sigma(S_i)\varrho_i g(u_i) - \int_{S_i} g(u)\, d\sigma(u)\Big) = \sum_{i=1}^{n} \int_{S_i} \big(\varrho_i g(u_i) - g(u)\big)\, d\sigma(u).$$

By (16),

$$|\varrho_i g(u_i) - g(u)| \leq |g(u_i) - g(u)| + \max_{u \in S^{d-1}} \{|g(u)|\} \frac{1}{2m^2}.$$

If $u \in S_i$, then $\|u_i - u\| \leq \operatorname{diam} S_i \leq \frac{1}{m}$. Since g is uniformly continuous on S^{d-1}, we see that

$$\int_{S^{d-1}} g(u)\, d\sigma_m(u) - \int_{S^{d-1}} g(u)\, d\sigma(u) \to 0 \text{ as } m \to \infty,$$

concluding the proof of (17).

By (i) and (15),

$$o = \int_{S^{d-1}} u\, d\sigma(u) = \sum_{i=1}^{n} \int_{S_i} u\, d\sigma(u) = \sum_{i=1}^{n} \varrho_i \sigma(S_i) u_i$$
$$= \sum_{i=1}^{n} \alpha_i u_i, \text{ where } \alpha_i = \varrho_i \sigma(S_i) > 0.$$

It follows from (i) that there is $0 < \tau < \frac{\pi}{2}$ such that each calotta of radius τ on S^{d-1} has positive measure. Thus, if m is sufficiently large, each open halfsphere contains one of the sets $S_1, \dots, S_n$ and, hence, a point among $u_1, \dots, u_n$. Thus we may apply Minkowski's theorem for polytopes, Theorem 18.2, to obtain, for all sufficiently large m, a polytope $P_m \in \mathcal{P}_p$ with area measure $\sigma_{P_m} = \sigma_m$ and facet areas $\alpha_i = \varrho_i \sigma(S_i)$. For the surface area of P_m, it follows from (17) that

(18) $$S(P_m) = \sum_{i=1}^{n} \alpha_i = \sum_{i=1}^{n} \varrho_i \sigma(S_i) = \sigma_m\left(S^{d-1}\right) \to \sigma\left(S^{d-1}\right) > 0.$$

The isoperimetric inequality then shows that

(19) The sequence $\big(V(P_m)\big)$ is bounded above.

By translating P_m, if necessary, we may suppose that

(20) $o \in P_m$ for all m.

We next show that

(21) The sequence of polytopes (P_m) is bounded.

For all $\alpha \in \mathbb{R}$ let $\alpha^+ = \max\{\alpha, 0\}$. Let $x = \|x\| v \in P_m$. Since $h_{P_m}(u) \geq h_{[o,x]}(u) = \|x\|(u \cdot v)^+$ for $u \in S^{d-1}$ by (20), it follows from (17) that

$$
\begin{aligned}
(22)\quad V(P_m) &= \tfrac{1}{d} \sum_{i=1}^{n} h_{P_m}(u_i)\alpha_i \geq \tfrac{\|x\|}{d} \sum_{i=1}^{n} (u_i \cdot v)^+ \alpha_i \\
&= \frac{\|x\|}{d} \int\limits_{S^{d-1}} (u \cdot v)^+ d\sigma_m(u) \ \Big(\to \frac{\|x\|}{d} \int\limits_{S^{d-1}} (u \cdot v)^+ d\sigma(u) \Big) \\
&> \beta \|x\| \text{ for all sufficiently large } m \text{ and } x \in P_m.
\end{aligned}
$$

Here, $\beta > 0$ is independent of v and thus of x. (19) and (22) together yield Proposition (21).

For the proof that

(23) The sequence $\big(V(P_m)\big)$ is bounded below by a positive constant,

it is sufficient to take, in (22), $x \in P_m$ such that $\|x\| \geq \gamma > 0$ for all sufficiently large m, where γ is a suitable positive constant. This is possible by (18).

Blaschke's selection theorem, the continuity of the volume on $\mathcal{C}$, see Theorem 7.5, and Propositions (21) and (23) yield the following. By taking a suitable subsequence of (P_m) and renumbering, if necessary,

$$
P_m \to C, \ \text{say, where } C \in \mathcal{C}_p.
$$

By Proposition 10.2,

$$
\sigma_m = \sigma_{P_m} \text{ then converges weakly to } \sigma_C.
$$

Comparing this with (17) implies that $\sigma = \sigma_C$. This settles the existence of the convex body C.

To show that C is unique up to translation, assume that $\sigma = \sigma_D$ for a convex body $D \in \mathcal{C}_p$. Corollary 10.1 then shows that

$$
\begin{aligned}
V(C, D, \dots, D) &= \frac{1}{d} \int\limits_{S^{d-1}} h_C(u)\, d\sigma_D(u) = \frac{1}{d} \int\limits_{S^{d-1}} h_C(u)\, d\sigma_C(u) \\
&= V(C, C, \dots, C) = V(C).
\end{aligned}
$$

Thus Minkowski's first inequality, see Theorem 6.11, shows that

$$V(C)^d = V(C, D, \dots D)^d \geq V(C)V(D)^{d-1}, \text{ or } V(C) \geq V(D).$$

Similarly, $V(D) \geq V(C)$. Hence

$$V(C) = V(D) = V(C, D, \dots, D).$$

Thus, in Minkowski's first inequality, we have equality. This, in turn, implies that C and D are homothetic, see Theorem 6.11. Since $V(C) = V(D)$, we see that D is a translate of C.

(ii)⇒(i) Choose convex polytopes $P_m \in \mathcal{P}_p,\ m = 1, 2, \dots,$ such that $P_m \to C$. By Proposition 10.2, the area measures σ_{P_m} converge weakly to the area measure $\sigma = \sigma_C$. Hence, in particular,

$$\int\limits_{S^{d-1}} u \, d\sigma_{P_m}(u) \to \int\limits_{S^{d-1}} u \, d\sigma_C(u) \quad \text{as } m \to \infty \text{ (componentwise)}.$$

By Minkowski's theorem for polytopes,

$$\int\limits_{S^{d-1}} u \, d\sigma_{P_m}(u) = o.$$

Hence

$$\int\limits_{S^{d-1}} u \, d\sigma_C(u) = o.$$

If σ_C were concentrated on a great circle, say the equator of S^{d-1}, let B be the open northern hemisphere. Then $\sigma_C(B) = 0$. On the other hand, $n_C^{-1}(B)$ consists of all points of $\operatorname{bd} C$ with exterior unit normal vectors in B. Thus $\sigma_C(B) = \mu_{d-1}\big(n_C^{-1}(B)\big) > 0$, a contradiction. □

Related Open Problems

The given solution of Minkowski's problem is rather satisfying, but it does not settle the following question. If $\kappa : S^{d-1} \to \mathbb{R}^+$ is of class $\mathcal{C}^\alpha$, to what class does the corresponding convex body C belong and if it is of class $\mathcal{C}^\beta$ with $\beta \geq 2$, is κ then its ordinary Gauss curvature? There is a large body of pertinent results, see the surveys in Pogorelov [806], Gluck [381], Su [976] and Schneider [908]. Selected references are Pogorelov [806], Caffarelli [186, 187] and Jerison [545].

Blaschke Multiplication and Blaschke Addition

Theorem 10.1 led Blaschke [124], p.112, to introduce, besides multiplication with reals and Minkowski addition, a second type of multiplication with reals and addition for convex bodies which may be described as follows. Given $\lambda \geq 0$ and two convex bodies $C, D \in \mathcal{C}_p$, the *Blaschke product* $\lambda \cdot_B C$ of λ and C and the *Blaschke sum* $C +_B D$ of C and D are defined by the equalities

$$\lambda \cdot_B C = \lambda^{\frac{1}{d-1}} C, \quad \sigma_{C +_B D} = \sigma_C + \sigma_D.$$

Here a convex body is considered only up to translation. For more information and references to applications, see Grünbaum [453], Schneider [908] and A. Thompson [994]. Two recent articles dealing with Blaschke sums are Campi, Colesanti and Gronchi [188] and Goodey, Kiderlen and Weil [383].

Christoffel's Problem

If, instead of Gauss curvature, respectively, the corresponding area measure, the mean curvature or other elementary symmetric functions of the principal curvatures, respectively, the corresponding measures are considered, the analogous problem is called *Christoffel's problem*, see Su [976] and Schneider [908].

10.2 Intrinsic Metric, Weyl's Problem and Rigidity of Convex Surfaces

In a metric space where any two points can be connected with a continuous curve of finite length, besides the given metric a second metric can be defined, the intrinsic or geodesic metric. The systematic study of the intrinsic metric of the boundary of a convex body in $\mathbb{E}^3$ was initiated by Alexandrov. The Weyl problem in this setting is to specify necessary and sufficient conditions such that a given metric space, endowed with its intrinsic metric, is isometric to the boundary of a suitable convex body in $\mathbb{E}^3$, if the latter is also endowed with its intrinsic metric. A solution of this problem is due to Alexandrov, and Pogorelov proved that the convex body is unique up to rigid motions.

In this section we first define the notion of intrinsic metric of a metric space, describe Weyl's problem and state Alexandrov's solution of it. Then rigidity of convex surfaces is considered and Pogorelov's uniqueness or rigidity result stated. No proofs are given.

For more information, see Alexandrov [15, 19], Busemann [182], Alexandrov and Zalgaller [20] and Pogorelov [805].

Intrinsic Metric of a Metric Space

We follow Alexandrov [15]. Let $\langle M, \delta\rangle$ be a metric space. Given a curve K in M by means of a parametrization $x : [a, b] \to M$, its *length* is defined by

$$\sup\left\{\sum_{i=1}^{n} \delta\big(x(t_{i-1}), x(t_i)\big) : n = 1, 2, \ldots, a = t_0 \leq t_1 \leq \cdots \leq t_n = b\right\}.$$

Suppose now that any two points of M can be connected by a continuous curve of finite length. Then the *intrinsic* or *geodesic metric* ϱ of the metric space $\langle M, \delta\rangle$ is defined as follows:

> $\varrho(x, y)$ is the infimum of the lengths of the continuous curves in M connecting x, y for any $x, y \in M$.

In general the metrics δ and ϱ will be different, but they induce the same topology on M. A continuous curve in M connecting two points x, y of M and of length $\varrho(x, y)$ is called a *geodesic segment* with endpoints x, y. It is non-trivial to show that, for sufficiently differentiable closed convex surfaces, or, more generally, for sufficiently smooth manifolds, this notion of geodesic segment coincides with the corresponding differential geometric notion. For a hint to a proof, see Gruber [423].

Weyl's Problem and Alexandrov's Realization Theorem

The following version of *Weyl's* [1020] *problem* was studied in convex geometry, where a *metric* $(d-1)$-*sphere* is a metric space homeomorphic to S^{d-1}.

Problem 10.3. *Let M be a metric $(d-1)$-sphere. Is there a closed convex surface S in $\mathbb{E}^d$ such that M and S are isometric, if both are endowed with their intrinsic metric? In other words, can M be realized by a closed convex surface?*

There exist many contributions to Weyl's problem, mainly in differential geometry. In the following we state without proof Alexandrov's [15] solution in the context of convex geometry.

Let M be a metric 2-sphere endowed with its intrinsic metric. If two geodesic segments have a common endpoint it is possible to define a notion of *lower angle* between them. The intrinsic metric on M is then said to have *positive curvature* if each point of M has a neighbourhood with the following property: for each triangle in this neighbourhood with geodesic segments as edges, the sum of the lower angles between its edges is at least π. *Alexandrov's* fundamental *realization theorem* now is as follows.

Theorem 10.2. *Let M be a metric 2-sphere with positive curvature. Then M is isometric to a closed convex surface S in $\mathbb{E}^3$ where both M and S are endowed with their intrinsic metrics.*

Rigidity of Convex Surfaces and Pogorelov's Rigidity Theorem

The question arises, whether the convex surface S in Alexandrov's realization theorem is unique up to rigid motions? In other words, if S and T are closed convex surfaces in $\mathbb{E}^3$ which are isometric if both are endowed with their intrinsic metric, do they coincide up to rigid motions? If this is the case, S is called *rigid*.

This problem for polytopal surfaces goes back to antiquity, see the discussion in Sect. 17.1. In the context of differential geometry the first pertinent result seems to be Liebmann's [657] theorem on the rigidity of spheres within the class of sufficiently smooth closed convex surfaces. In his proof the following result, already conjectured by Minding, is shown first. A sufficiently smooth closed convex surface with constant Gauss curvature is a sphere. Then he makes use of the fact that the Gauss curvature is determined by the intrinsic metric. For some references, see the book by Su [976]. We point out the remarkable and totally unexpected result of Nash [766] and Kuiper [621] that a sufficiently smooth closed convex surface is $\mathcal{C}^1$-diffeomorphically isometric to a topological sphere which is folded such as to form a set of arbitrarily small (Euclidean) diameter. It is this Nash who won the Nobel prize in economics in 1995, see Milnor [727].

The final rigidity result for closed convex surfaces in convex geometry is the following *rigidity theorem of Pogorelov* [803]:

Theorem 10.3. *Let S and T be two closed convex surfaces in $\mathbb{E}^3$ which are isometric with respect to their intrinsic metrics. Then S and T are congruent.*

So far no simple proof of this result is available. Surveys of the relevant literature are due to Pogorelov [805], Su [976] and Ivanova-Karatopraklieva and Sabitov [538, 539].

10.3 Evolution of Convex Surfaces and Convex Billiards

While, in a majority of results in convex geometry, fixed objects, in particular fixed convex bodies are studied, there are several groups of results in the last decades which deal with moving objects, for example with deformation of surfaces. Of these investigations with a dynamical aspect we mention the following:

flexible polytopal spheres and frameworks
evolution of convex surfaces by curvature driven flows
billiards

In the following we first cite some results on evolution of closed convex surfaces by flows which are driven by the mean and the Gauss curvature. Then caustics of convex billiard tables are considered. No proofs are given.

For more information on rigidity and flexibility of closed convex surfaces, polytopal spheres and frameworks, see Sects. 10.2, 17.1 and 17.2 and the references given there.

Evolution of Convex Curves and Surfaces by Curvature Driven Flows

All convex curves and surfaces considered in the following are assumed to be sufficiently differentiable.

If S_0 is a closed convex curve in $\mathbb{E}^2$, the problem is to find a family $\{S_t : t \geq 0\}$ of closed convex curves given by a sufficiently differentiable function $x(s,t)$ where, for fixed t, the expression $x(s,t)$ is an arc-length parametrization of S_t such that, for fixed s, the point $x(s,t)$ moves in time t in the direction of the inner unit normal vector of S_t at $x(s,t)$ with speed equal to the curvature of S_t at $x(s,t)$. In other words, $x(s,t)$ is the solution of the parabolic initial value problem

$$x_t(s,t) = x_{ss}(s,t)$$
$$x(s,0) = x(s),$$

where $x(s)$ is a given arc-length parametrization of S_0.

This problem, clearly, can be extended to dimensions $d \geq 2$ with the flow driven by the mean curvature, the Gauss curvature, a function of these, or by some other function of the principal curvatures. For simplicity, we consider only the mean and the Gauss curvature.

The following result was proved for $d = 2$ by Gage [349, 350] and Gage and Hamilton [351] and for $d \geq 3$ by Huisken [529]. Andrews [31] gave a generalization providing a simpler proof of Huisken's result. For the question of singularities which might evolve, see Huisken and Sinestrari [530] and White [1024]. For a selection of generalizations, related material and references, we refer to Andrews [33], the book of Chou and Zhu [210] and the reports of White [1022, 1023].

Theorem 10.4. *A given (sufficiently differentiable) closed convex surface S_0 in $\mathbb{E}^d$ is deformed by the mean curvature driven flow in the interior normal direction into a family $\{S_t : t \geq 0\}$ of closed convex surfaces shrinking to a point. If rescaled by suitable homotheties, these convex surfaces tend to S^{d-1}.*

An analogous result holds for the flow driven by the Gauss curvature. For $d = 2$ it is the same result as before. For general d it was proved by Chou [209] and Andrews [32]. See also Andrews [33] for a multitude of references. For affine evolutions compare Leichtweiss [643].

Convex Billiards and Caustics

A *(convex) billiard table* B in $\mathbb{E}^d$ is a proper convex body (see Fig. 10.1). A *billiard ball* in B is a point which moves with constant speed along a (straight) line in int B until it hits bd B. If the point where it hits bd B is a regular boundary point of B, the billiard ball is reflected in the usual way and moves again with the same speed along a line in int B, etc. If the billiard ball hits bd B at a singular point, it stops there. The curve described by a billiard ball is called a *billiard trajectory*.

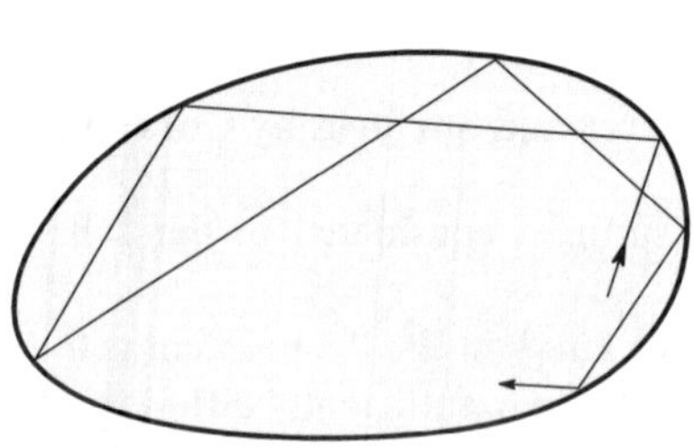

Fig. 10.1. Billiard

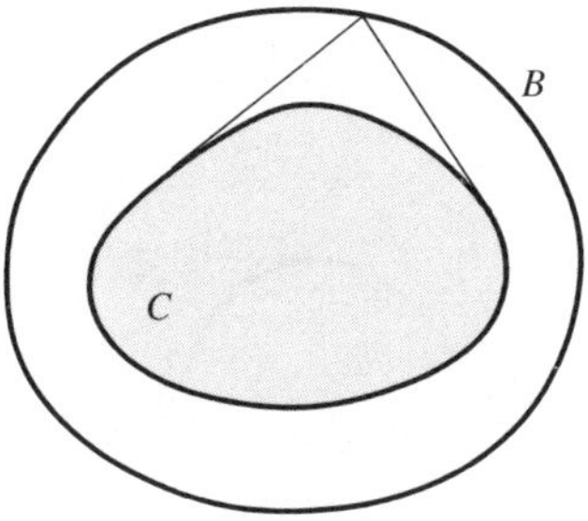

Fig. 10.2. Gardener construction of a billiard table, given a caustic

Billiards have been considered from various viewpoints in the context of ergodic theory, dynamical systems, partial differential equations, mechanics, physics, number theory and geometry. For some information, see the books of Birkhoff [120], Arnold [38], Lazutkin [632], Cornfeld, Fomin and Sinai [224], Katok, Strelcyn, Ledrappier and Przytycki [567], Gal'perin and Zemlyakov [354], Kozlov and Treshchëv [614], Petkov and Stoyanov [795], and Tabachnikov [985].

We consider *convex caustics* from a geometric viewpoint. These are convex bodies C in $\operatorname{int} B$ such that any trajectory which touches C once, touches it again after each reflection. Minasian [728] showed the following. Let C be a caustic of a planar billiard table B. Then there is a closed inelastic string such that $\operatorname{bd} B$ is obtained by wrapping the string around C, pulling it tight at a point and moving this point around C while keeping the string tight. Conversely, given a planar convex disc C, each convex disc B obtained in this way is a billiard table with caustic C. This nice result shows that the planar billiard tables with a given caustic may be obtained from the caustic by a generalization of the common gardener (see Fig. 10.2) construction of ellipses.

Lazutkin [631] related the problem of eigenvalues of the Laplace operator on a planar billiard table B to the existence of convex caustics and showed that, for billiard tables of class $\mathcal{C}^{553}$ and with positive curvature, there exists a large family of caustics. 553 was reduced to 6 by Douady [277]. The problem about the non-existence of caustics was studied by Mather [694] and Hubacher [524]. In [422] the author showed that in the sense of Baire categories, there is only a meagre set of billiard tables in $\mathbb{E}^2$ which have caustics.

Refining a result of Berger [98], Gruber [433] proved the following result. Its proof relies on Alexandrov's differentiability theorem.

Theorem 10.5. *Among all convex billiard tables in $\mathbb{E}^d$, $d \geq 3$, it is only the solid ellipsoids that have convex caustics. The caustics are precisely the confocal solid ellipsoids contained in their interiors and, moreover, the intersection of all confocal ellipsoids.*

For further results on caustics of planar billiard tables we refer to the articles of Gutkin and Katok [458], Knill [603] and Gutkin [457].

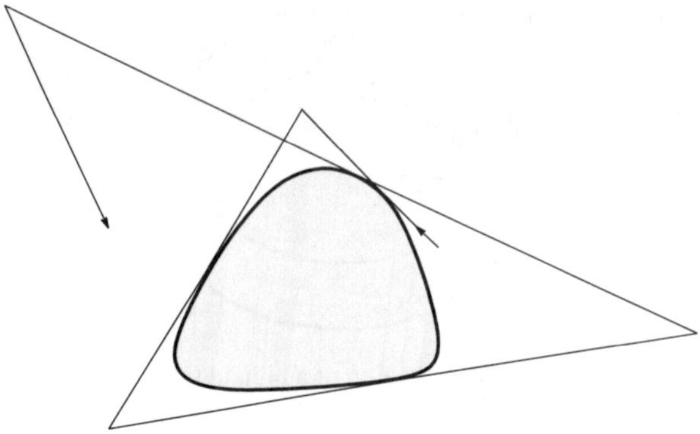

Fig. 10.3. Outer billiard

Outer Billiards

Let B be a planar, strictly convex body (see Fig. 10.3). Given $x_0 \in \mathbb{E}^2 \setminus B$, consider the support line of B through x_0 such that B is on the left side of this line if viewed from x_0. Let x_1 be the mirror image of x_0 with respect to the touching point. The dynamical system $x_0 \rightarrow x_1$ is called the *outer billiard* determined by B. In our context outer billiards are important for the approximation of planar convex bodies, see Sect. 11.2 and the article of Tabachnikov [984] cited there. The latter proved a result on outer billiards which yields an asymptotic series development for best area approximation of a planar convex body by circumscribed polygons as the number of edges tends to infinity.

11 Approximation of Convex Bodies and Its Applications

Most approximation results in convex geometry belong to one of the following types:

Approximation by special convex bodies, such as ellipsoids, simplices, boxes, or by special classes of convex bodies, for example centrally symmetric convex bodies, analytic convex bodies, or zonotopes

Asymptotic best approximation by convex polytopes with n vertices or facets as $n \rightarrow \infty$

Approximation of convex bodies by random polytopes, i.e. the convex hull of n random points

Asymptotic approximation by random polytopes as $n \rightarrow \infty$

There are many sporadic results of the first type scattered throughout the convexity literature. Pertinent results of a more systematic nature can be found in the context of the maximum and minimum ellipsoids in the local theory of normed spaces.

The first asymptotic formulae for best approximation of a convex body with respect to a metric were given by L. Fejes Tóth [329] for $d = 2$ in the early 1950s. Asymptotic formulae for general d were first proved by Schneider [905] for the

Hausdorff metric δ^H and then by Gruber [427] for the symmetric difference metric δ^V. These results gave rise to a long series of further investigations.

Early contributions to random approximation are due to Crofton, Czuber and Blaschke, the first asymptotic results were proved by Rényi and Sulanke [831] in case $d = 2$. Modern results for general d are due, amongst others, to Bárány and Buchta [69] and Reitzner [829].

In this section we first consider John's characterization of the ellipsoid of maximum volume contained in a convex body and use it to prove Ball's reverse isoperimetric inequality. Then the author's asymptotic formula for the best volume approximation of a convex body by circumscribed convex polytopes is given as the number of facets tends to infinity. This is then applied to the isoperimetric problem for convex polytopes.

For references concerning John's theorem and asymptotic best approximation, see the following sections. Recent surveys on random approximation are due to the author [435] and Schneider [909].

11.1 John's Ellipsoid Theorem and Ball's Reverse Isoperimetric Inequality

To a given convex body one may assign several ellipsoids in a canonical way. One example is the ellipsoid of inertia, for other examples, see the articles of Milman [725] and Lutwak, Yang and Zhang [670] and the book of Pisier [802]. Among the ellipsoids which are inscribed, respectively, circumscribed to a proper convex body in $\mathbb{E}^d$, there is precisely one of maximum, respectively, minimum volume. Simple proofs for this result are due to Löwner (unpublished), Behrend [90] ($d = 2$) and Danzer, Laugwitz and Lenz [242] (general d). John [549] characterized inscribed ellipsoids of maximum volume, a complement being due to Ball [51]. Both results are of interest in convex geometry and in the geometry of normed spaces. John's theorem implies, in particular, that for any origin symmetric convex body there is an ellipsoid which approximates it up to a factor $\sqrt{d}$.

Below these results are proved for convex bodies which are symmetric in o. The proof of the characterization result is taken from Gruber and Schuster [452]. It is based on the idea of Voronoĭ to identify symmetric, positive definite $d \times d$ matrices, respectively, positive definite quadratic forms on $\mathbb{E}^d$ with points in $\mathbb{E}^{\frac{1}{2}d(d+1)}$, compare Sect. 29.4. This idea was applied earlier in the same context by the author [421]. We give two classical applications, one to the group of affinities which map a convex body onto itself and one to the Banach–Mazur distance between norms on $\mathbb{E}^d$. A third application is the reverse isoperimetric inequality. For convex bodies C, the isoperimetric quotient

$$\frac{S(C)^d}{V(C)^{d-1}}$$

is bounded below by the isoperimetric quotient of the unit ball B^d, but it is clearly not bounded above. Behrend [89], for $d = 2$, and Ball [50], for general d, asked whether for any given convex body C there is an affine image, the isoperimetric quotient of which is bounded above in terms of d, and what is the worst case.

For more information on John's theorem and its aftermath, see the article of Ball [53] and the surveys of Johnson and Lindenstrauss [552], Ball [54], and Giannopoulos and Milman [374] in the Handbook of the Geometry of Banach Spaces. For information on other John type and so-called minimum position results see Gruber [445].

Uniqueness and John's Characterization of Inscribed Ellipsoids of Maximum Volume

Our aim here is to show the results of Löwner, Behrend [90], Danzer, Laugwitz and Lenz [242], and of John [549], Pełczyński [788], and Ball [51], respectively, for o-symmetric convex bodies.

Given $d \times d$ matrices $A = (a_{ij})$, $B = (b_{ij})$, define their *inner product* by $A \cdot B = \sum a_{ij} b_{ij}$, where summation on i and j is from 1 to d. The dot $\cdot$ denotes also the ordinary inner product in $\mathbb{E}^d$. For $u, v \in \mathbb{E}^d$, let $u \otimes v$ be the $d \times d$ matrix $u\, v^T$. Then $(u \otimes v)\, x = (v \cdot x)\, u$ for $u, v, x \in \mathbb{E}^d$. I is the $d \times d$ unit matrix.

Theorem 11.1. *Let $C \in \mathcal{C}_p$ be symmetric in o. Then there is a unique ellipsoid of maximum volume (necessarily symmetric in o) amongst all ellipsoids contained in C.*

Theorem 11.2. *Let $C \in \mathcal{C}_p$ be symmetric in o and $B^d \subseteq C$. Then the following statements are equivalent:*

(i) *B^d is the unique ellipsoid of maximum volume amongst all ellipsoids in C.*
(ii) *There are $u_i \in B^d \cap \operatorname{bd} C$ and $\lambda_i > 0$, $i = 1, \dots, m$, where $d \le m \le \frac{1}{2}d(d+1)$ such that*

$$I = \sum_i \lambda_i\, u_i \otimes u_i, \ \sum_k \lambda_k = d.$$

Here, and in the following, summation on i and k is from 1 to m. It is not difficult to extend both theorems to convex bodies C which are not necessarily symmetric in o.

Without loss of generality, we assume, in the following, that all ellipsoids are symmetric in o. Before beginning the proof, we state two tools:

(1) Each non-singular $d \times d$ matrix M can be represented in the form $M = AR$, where A is a symmetric, positive definite and R an orthogonal $d \times d$ matrix.

(Take $A = (MM^T)^{\frac{1}{2}}$ and $R = A^{-1}M$, see [367], p.112.) Identify a symmetric $d \times d$ matrix $A = (a_{ij})$ with the point $(a_{11}, \dots, a_{1d}, a_{22}, \dots, a_{2d}, \dots, a_{dd}) \in \mathbb{E}^{\frac{1}{2}d(d+1)}$. Then, the set of all symmetric positive definite $d \times d$ matrices is (represented by) an open convex cone $\mathcal{P}$ in $\mathbb{E}^{\frac{1}{2}d(d+1)}$ with apex at the origin. The set

(2) $\mathcal{D} = \{A \in \mathcal{P} : \det A \ge 1\}$ is a closed, smooth and strictly convex set in $\mathcal{P}$ with non-empty interior.

This follows from the implicit function theorem and *Minkowski's determinant inequality*,

$\left(\det(A+B)\right)^{\frac{1}{d}} \geq \left(\det A\right)^{\frac{1}{d}} + \left(\det B\right)^{\frac{1}{d}}$ for $A, B \in \mathcal{P}$, where equality holds precisely in case where A, B are proportional.

A proof of this inequality can be obtained by diagonalizing A and B simultaneously and then using the following inequality which is a consequence of the arithmetic–geometric mean inequality,

$\left((x_1 + y_1) \cdots (x_d + y_d)\right)^{\frac{1}{d}} \geq (x_1 \cdots x_d)^{\frac{1}{d}} + (y_1 \cdots y_d)^{\frac{1}{d}}$ for $x_i, y_i \geq 0$, where equality holds if and only if $x = (x_1, \dots, x_d)$, $y = (y_1, \dots, y_d)$ are linearly dependent.

See also Sect. 29.4.

Proof of Theorem 11.1. A simple compactness argument yields the existence of at least one ellipsoid in C of maximum volume.

To see uniqueness, assume that, on the contrary, there are two distinct maximum volume ellipsoids in C. Without loss of generality we may assume that their volumes are equal to that of B^d. By (1) these ellipsoids can be represented in the form AB^d, BB^d with suitable $d \times d$-matrices $A, B \in \mathcal{P}$, where $A \neq B$ and $\det A = \det B = 1$. Then $\frac{1}{2}(A + B)B^d \subseteq C$ by the convexity of C. Since $\det\left(\frac{1}{2}(A + B)\right) > 1$ by (2), the ellipsoid $\frac{1}{2}(A + B)B^d$ has greater volume than the maximum volume ellipsoids AB^d, BB^d, a contradiction. □

Proof of Theorem 11.2. (i)⇒(ii) By (1), the family of all ellipsoids in C is represented by the following set of symmetric positive definite matrices in $\mathcal{P}$.

$$\mathcal{E} = \left\{A \in \mathcal{P} : Au \cdot v = A \cdot u \otimes v \leq h_C(v) \text{ for all } u, v \in \operatorname{bd} B^d\right\}.$$

Clearly, $\mathcal{E}$ is the intersection of the family of the closed halfspaces

(3) $\left\{A \in \mathbb{E}^{\frac{1}{2}d(d+1)} : A \cdot u \otimes v \leq h_C(v)\right\} : u, v \in \operatorname{bd} B^d,$

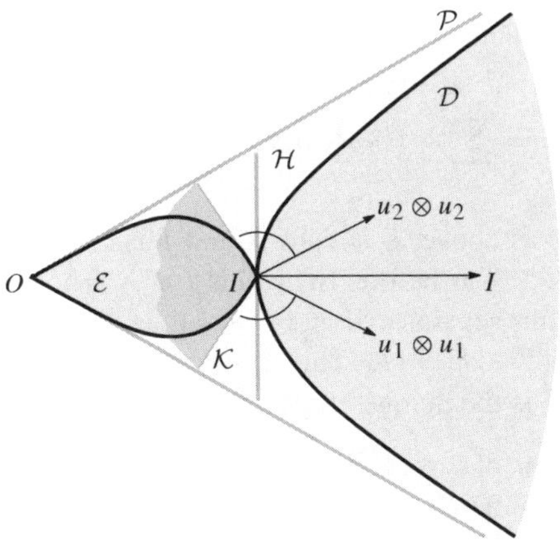

Fig. 11.1. Proof of John's theorem

with $\mathcal{P}$. Thus, it is the intersection of a closed convex set with $\mathcal{P}$ and therefore convex. By (i), $\det A < 1$ for all $A \in \mathcal{E} \setminus \{I\}$. This, together with (2), yields the following:

(4) $\mathcal{E}$ is convex, $\mathcal{D} \cap \mathcal{E} = \{I\}$ and $\mathcal{E}$ is separated from $\mathcal{D}$ by the unique support hyperplane $\mathcal{H}$ of $\mathcal{D}$ at I.

The *support cone* $\mathcal{K}$ of $\mathcal{E}$ at I is the closed convex cone with apex I generated by $\mathcal{E}$. Since $\mathcal{E}$ is the intersection of the closed halfspaces in (3) and $\mathcal{P}$, and since these halfspaces vary continuously as u, v range over $\operatorname{bd} B^d$, the support cone $\mathcal{K}$ is the intersection of those halfspaces in (3) which contain the apex I of $\mathcal{K}$ on their boundaries, i.e. $I \cdot u \otimes v = u \cdot v = h_C(v)$. Since $u \cdot v \leq 1, h_C(v) \geq 1$ for $u, v \in \operatorname{bd} B^d$, the equality $I \cdot u \otimes v = h_C(v)$ holds precisely in case where $u = v$ and $v \in \operatorname{bd} B^d \cap \operatorname{bd} C$. Thus $\mathcal{K}$ is the intersection of the halfspaces

$$\big\{A \in \mathbb{E}^{\frac{1}{2}d(d+1)} : A \cdot u \otimes u \leq 1\big\} : u \in \operatorname{bd} B^d \cap \operatorname{bd} C = B^d \cap \operatorname{bd} C.$$

The normal cone $\mathcal{N}$, of $\mathcal{E}$ at I, is the polar cone of $\mathcal{K} - I$ and thus is generated by the exterior normals $u \otimes u,\ u \in B^d \cap \operatorname{bd} C$, of these halfspaces,

(5) $\mathcal{N} = \operatorname{pos}\big\{u \otimes u : u \in B^d \cap \operatorname{bd} C\big\}$.

The cone $\mathcal{K}$ has apex I and, by (4), is separated form $\mathcal{D}$ by the hyperplane $\mathcal{H}$. The normal I of $\mathcal{H}$ points away from $\mathcal{K}$ and thus is contained in the normal cone $\mathcal{N}$. Noting (5), Carathéodory's theorem then implies that there are $u_i \in B^d \cap \operatorname{bd} C$, $\lambda_i > 0, i = 1, \dots, m$, where $m \leq \frac{1}{2}d(d+1)$, such that

$$I = \sum_i \lambda_i\, u_i \otimes u_i.$$

This, in turn, shows that

$$d = \operatorname{trace} I = \sum_i \lambda_i \operatorname{trace} u_i \otimes u_i = \sum_i \lambda_i.$$

For the proof that $m \geq d$, it is sufficient to show that $\operatorname{lin}\{u_1, \dots, u_m\} = \mathbb{E}^d$. If this were not the case, we could choose a unit vector u orthogonal to $u_1, \dots, u_m$ to obtain the contradiction

$$1 = u^2 = Iu \cdot u = \sum_i \lambda_i \big((u_i \otimes u_i)\,u\big) \cdot u = \sum_i \lambda_i \big((u_i \cdot u)\,u_i\big) \cdot u = 0.$$

(ii)⇒(i) Let $\mathcal{E}$ be as above. $\mathcal{E}$ is convex and I is a boundary point of it by (ii). Thus we may define $\mathcal{K}, \mathcal{N}$ as before. (ii) yields $I \in \mathcal{N}$. The hyperplane $\mathcal{H}$ through I and orthogonal to I thus separates $\mathcal{K}$ and $\mathcal{D}$ and thus, a fortiori, $\mathcal{D}$ and $\mathcal{E}$. Since $\mathcal{D}$ is strictly convex by (2), $\mathcal{D} \cap \mathcal{E} = \{I\}$. This shows that $\det A < 1$ for each $A \in \mathcal{E} \setminus \{I\}$, or, in other words, B^d is the unique ellipsoid in C with maximum volume. □

Remark. For the proof of a more general implication (i)⇒(ii), see Giannopoulos, Perissinaki and Tsolomitis [376]. Proofs of John's theorem (see Fig. 11.1) in the non-symmetric case and of the generalized version of Giannopoulos, Perissinaki and Tsolomitis in the spirit of the above proof are outlined in Gruber and Schuster [452].

The number of common points of bd C and the inscribed ellipsoid of maximum volume is precisely $\frac{1}{2}d(d+1)$ for most convex bodies C which are symmetric in o. See Gruber [421] and, for a different proof, Rudelson [860].

The Group of Affinities of a Convex Body

Danzer, Laugwitz and Lenz [242], proved the following result.

Corollary 11.1. *Let $C \in \mathcal{C}_p$ be symmetric in o. Then the group $\mathfrak{L}$ of all linear transformations which map C onto itself is a subgroup of the orthogonal group with respect to a suitable inner product.*

Proof. By replacing the given inner product on $\mathbb{E}^d$ by a suitable new inner product, if necessary, we may assume that the unique ellipsoid of maximum volume in C is the unit ball with respect to the new inner product. Denote the latter by B^d. Let $T \in \mathfrak{L}$. Since $TC = C$, the transformation T is volume-preserving. Thus TB^d is also an ellipsoid of maximum volume in $TC = C$. The uniqueness part of Theorem 11.1 then shows that $TB^d = B^d$, i.e. T is an orthogonal transformation with respect to the inner product corresponding to B^d. □

Remark. A similar result holds for general convex bodies and affinities.

The Banach–Mazur Compactum

On the space $\mathcal{N} = \mathcal{N}(\mathbb{E}^d)$ of all norms on $\mathbb{E}^d$ define the *Banach–Mazur distance* δ^{BM} as follows, where for a norm $|\cdot|$ the corresponding unit ball is denoted by $B_{|\cdot|}$.

$$\delta^{BM}(|\cdot|, [\![\cdot]\!]) = \inf\{\lambda > 1 : \exists T : \mathbb{E}^d \to \mathbb{E}^d \text{ linear}, B_{|\cdot|} \subseteq TB_{[\![\cdot]\!]} \subseteq \lambda B_{|\cdot|}\}$$
$$\text{for } |\cdot|, [\![\cdot]\!] \in \mathcal{N}.$$

δ^{BM} does not distinguish between isometric norms, is symmetric and $\log \delta^{BM}$ satisfies the triangle inequality. There are other ways to define δ^{BM}. $\mathcal{N}$, endowed with the distance δ^{BM}, is a compact space, called the *Banach–Mazur compactum*. It has attracted a lot of interest in the local theory of normed spaces, see the reports of Gluskin [382] and Szarek [981] and the book of Tomczak-Jaegermann [1001]. One of the difficult open questions in this area is to determine the diameter of $\mathcal{N}$. Here, the following simple estimate due to John [549] is given, where $\|\cdot\|$ denotes the Euclidean norm on $\mathbb{E}^d$.

Corollary 11.2. $\delta^{BM}(\|\cdot\|, |\cdot|) \le \sqrt{d}$ *for any* $|\cdot| \in \mathcal{N}(\mathbb{E}^d)$.

Proof. Assume that T is chosen such that $B^d \subseteq TB_{|\cdot|}$ is the ellipsoid of maximum volume in $C = TB_{|\cdot|}$. Choose u_i, λ_i according to John's ellipsoid theorem. Then

$$(6) \quad I = \sum_i \lambda_i u_i \otimes u_i, \sum_i \lambda_i = d.$$

Note that $u_i \in B^d \cap \operatorname{bd} C$. Hence C is contained in the support halfspace of B^d at u_i, and thus $u_i \cdot x \leq 1$ for $x \in C = T B_{|\cdot|}$. Represent x by (6) in the form

$$x = Ix = \sum_i \lambda_i \, (u_i \otimes u_i) \, x = \sum_i \lambda_i \, (u_i \cdot x) \, u_i.$$

Then

$$x^2 = x \cdot x = \sum_i \lambda_i \, (u_i \cdot x)^2 \leq \sum_i \lambda_i = d \text{ for } x \in T B_{|\cdot|},$$

again by (6). Thus $T B_{|\cdot|} \subseteq \sqrt{d}\, B^d$. □

Ball's Reverse Isoperimetric Inequality

Ball [50] proved the following result. The simpler 2-dimensional result was given previously by Behrend [89].

Theorem 11.3. *Let $C \in \mathcal{C}_p$ be symmetric in o. Then there is a non-singular linear transform T such that*

$$\frac{S(TC)^d}{V(TC)^{d-1}} \leq (2d)^d.$$

This cannot be improved if C is a cube.

Ball's proof shows that the isoperimetric quotient of TC is small, if B^d is the ellipsoid of maximum volume in TC. As shown by Barthe [76], the equality sign is needed precisely in the case where C is a parallelotope.

Proof. We need the following version of the inequality of Brascamp and Lieb [162] due to Ball [49], see also Barthe [76].

(7) Let $u_i \in S^{d-1}$, $\lambda_i > 0$, $i = 1, \dots, m$, such that

$$I = \sum_i \lambda_i \, u_i \otimes u_i, \ \sum_i \lambda_i = d$$

and let f_i, $i = 1, \dots, m$, be non-negative measurable functions on $\mathbb{R}$. Then

$$\int\limits_{\mathbb{E}^d} \prod_i f_i (u_i \cdot x)^{\lambda_i} dx \leq \prod_i \Big(\int\limits_{\mathbb{R}} f_i(t)\, dt \Big)^{\lambda_i}.$$

Choose a linear transformation T such that B^d is the ellipsoid of maximum volume in TC. We will show that

(8) $V(TC) \leq 2^d$.

Take u_i, λ_i as in John's theorem and consider the convex body $D = \{x : |u_i \cdot x| \leq 1, i = 1, \dots, m\}$. For each i let f_i be the characteristic function of the interval $[-1, 1]$. Then the function

$$x \to \prod_i f_i(u_i \cdot x)^{\lambda_i}$$

is the characteristic function of D. Now integrate and use (7) to see that $V(D) \leq 2^d$. Since $TC \subseteq D$, this implies (8).

The definition of surface area in Sect. 6.4, the inclusion $B^d \subseteq TC$ and (8) together yield the following:

$$\begin{aligned} S(TC) &= \lim_{\varepsilon \to +0} \frac{V(TC + \varepsilon B^d) - V(TC)}{\varepsilon} \\ &\leq \lim_{\varepsilon \to +0} \frac{V(TC + \varepsilon TC) - V(TC)}{\varepsilon} = dV(TC) \leq 2dV(TC)^{\frac{d-1}{d}}. \end{aligned}$$

This readily yields the desired upper estimate for the isoperimetric quotient of TC.

The simple proof that equality holds if C is a cube is omitted. □

11.2 Asymptotic Best Approximation, the Isoperimetric Problem for Polytopes, and a Heuristic Principle

Given a metric $\delta(\cdot, \cdot)$ or some other measure of distance on $\mathcal{C}$, a convex body C and a class of convex polytopes $\mathcal{Q}_n$ such as the class $\mathcal{P}^c_{(n)} = \mathcal{P}^c_{(n)}(C)$ of all convex polytopes which are circumscribed to C and have at most n facets, or the class $\mathcal{P}^i_n = \mathcal{P}^i_n(C)$ of all convex polytopes with at most n vertices which are inscribed into C, the following problems arise. First, to determine or estimate the quantity

$$\delta(C, \mathcal{Q}_n) = \min\{\delta(C, P) : P \in \mathcal{Q}_n\}.$$

Second, to describe the polytopes $P \in \mathcal{Q}_n$ for which the infimum is attained, the *best approximating polytopes* of C in $\mathcal{Q}_n$ with respect to the given metric $\delta(\cdot, \cdot)$. Using Blaschke's selection theorem it is easy to show that, for the common metrics and polytope classes $\mathcal{Q}_n$, best approximating polytopes exist. While precise answers to these problems are out of reach, it is possible to give satisfying results as $n \to \infty$.

In this section the author's asymptotic formula for $\delta^V(C, \mathcal{P}^c_{(n)})$ will be derived, using Zador's Theorem 33.2 for $\alpha = 2$. As an application, the isoperimetric problem for convex polytopes is considered. A comparison of asymptotic best and random approximation will show that, in high dimensions, the difference is negligible. This will lead to a vague heuristic principle.

For more information we refer to the book of Fejes Tóth [329] and the surveys [417,429,434].

The Asymptotic Formula for $\delta^V(C, \mathcal{P}^c_{(n)})$

Our aim is to show the following result of Gruber [427], where $\delta^V(\cdot, \cdot)$ is the *symmetric difference metric* on $\mathcal{C}_p$,

$$\delta^V(C, D) = V(C \Delta D) = V\big((C \setminus D) \cup (D \setminus C)\big) \text{ for } C, D \in \mathcal{C}_p.$$

Theorem 11.4. *Let $C \in \mathcal{C}_p$ be of class $\mathcal{C}^2$ with Gauss curvature $\kappa_C > 0$. Then there is a constant $\delta = \delta_{2,d-1} > 0$, depending only on d, such that*

$$(1)\quad \delta^V(C, \mathcal{P}^c_{(n)}) \sim \frac{\delta}{2}\Bigg(\int\limits_{\operatorname{bd} C} \kappa_C(x)^{\frac{1}{d+1}}\, d\sigma(x)\Bigg)^{\frac{d+1}{d-1}} \frac{1}{n^{\frac{2}{d-1}}} \text{ as } n \to \infty.$$

Here, σ is the ordinary surface area measure on $\operatorname{bd} C$. The integral in the asymptotic formula (1) is the *affine surface area* of C, a notion from affine differential geometry. It is invariant with respect to volume preserving affinities of the convex body C. The constant $\delta = \delta_{2,d-1}$ was introduced in [427], where we wrote div (for Dirichlet and Voronoĭ) instead of δ. It is related to Dirichlet–Voronoĭ tilings, see Sect. 32.1. We have,

$$\delta_{2,2} = \frac{5}{18\sqrt{3}},\quad \delta_{2,d-1} \sim \frac{d}{2\pi e} \text{ as } d \to \infty.$$

For the value of $\delta_{2,2}$, see Fejes Tóth [329] and the author [425, 427]. For a proof of the asymptotic formula for $\delta_{2,d-1}$, see Proposition 33.1.

Proof. Let $\lambda > 1$. We start with some preparations. For $p \in \operatorname{bd} C$, let H be the unique support hyperplane of C at p. Choose a Cartesian coordinate system in H with origin at p. Together with the interior unit normal vector of $\operatorname{bd} C$ at p, it forms a Cartesian coordinate system in $\mathbb{E}^d$. The lower part of $\operatorname{bd} C$ with respect to the last coordinate then can be represented in the form:

$$\big(s, f(s)\big) : s \in C',$$

where "$'$" denotes the orthogonal projection of $\mathbb{E}^d$ onto H and f is a convex function on C' such that $f|\operatorname{relint} C'$ is of class $\mathcal{C}^2$. For $u \in \operatorname{relint} C'$, define the quadratic form $q_u = q_{pu}$ by

$$q_u(s) = \sum_{i,j} f_{x_i,x_j}(u) s_i s_j \text{ for } s = (s_1, \dots, s_{d-1}) \in H.$$

Let $q_p = q_{pp}$. The Gauss curvature $\kappa_C(u)$ at the point $x = \big(u, f(u)\big) \in \operatorname{bd} C$, $u \in \operatorname{relint} C'$, is then given by

$$\kappa_C(u) = \frac{\det q_u}{\big(1 + \big(\operatorname{grad} f(u)\big)^2\big)^{\frac{d+1}{2}}}.$$

We also write $\kappa_C(x)$ for $\kappa_C(u)$. Since $\kappa_C > 0$, the quadratic forms q_u are all positive definite. Since f is of class $\mathcal{C}^2$, their coefficients are continuous. Hence we may choose an open convex neighbourhood $U' \subseteq C'$ of p in H such that

$$\frac{1}{\lambda} q_p(s) \le q_u(s) \le \lambda q_p(s) \text{ for } s \in H,\ u \in U'$$

$$\frac{1}{\lambda} \det q_p \le \det q_u \le \lambda \det q_p \text{ for } u \in U'$$

$$\frac{1}{\lambda} \kappa_C(u) \le \det q_p \le \lambda \kappa_C(u) \text{ for } u \in U'$$

Let U be the set on the lower part of bd C which projects onto U'. Clearly, U is a neighbourhood of p in bd C.

The following inequality is well known.

$$(2)\quad \left(\frac{1}{m}\left(\sigma_1^{\frac{d+1}{d-1}}+\cdots+\sigma_m^{\frac{d+1}{d-1}}\right)\right)^{\frac{d-1}{d+1}} \geq \frac{1}{m}(\sigma_1+\cdots+\sigma_m) \text{ for } \sigma_1,\ldots,\sigma_m>0.$$

By means of a suitable linear transformation, Zador's Theorem 33.2 for $\alpha = 2$ yields the following asymptotic formula:

(3) Let $J \subseteq H$ be Jordan measurable with $v(J) > 0$ and q a positive definite quadratic form on H. Then

$$\inf_{\substack{S\subseteq H\\ \#S=m}} \int_J \min_{t\in S}\{q(s-t)\}\,ds \sim \delta\, v(J)^{\frac{d+1}{d-1}} (\det q)^{\frac{1}{d-1}} \frac{1}{m^{\frac{2}{d-1}}} \text{ as } m\to\infty,$$

where $\delta > 0$ is a constant depending only on d.

After these preparations, the first step is to show that

$$(4)\quad \delta^V(C,\mathcal{P}^c_{(n)}) \geq \frac{\delta}{2\lambda^{\frac{4d}{d-1}}}\left(\int_{\mathrm{bd}\,C} \kappa_C(x)^{\frac{1}{d+1}}\,d\sigma(x)\right)^{\frac{d+1}{d-1}} \frac{1}{n^{\frac{2}{d-1}}}$$

for all sufficiently large n.

The open neighbourhoods U considered above cover the compact set bd C. Thus there is a finite subcover. By Lebesgue's covering lemma (see, e.g. [572], p.154), each set of sufficiently small diameter in bd C is then contained in one of the neighbourhoods of the subcover. Thus we may choose finitely many small pieces J_i, $i = 1,\ldots,l$, in bd C, points p_i and neighbourhoods U_i of p_i in bd C, support hyperplanes H_i of C at p_i, convex functions f_i, and quadratic forms $q_u = q_{p_i u}, q_i = q_{p_i\,p_i}$ as in the preparations, such that the following statements hold:

(5) The sets J_i are compact and pairwise disjoint, and their projections $J_i' \subseteq U_i \subseteq \mathrm{relint}\, C'$ are Jordan measurable

$$(6)\quad \frac{1}{\lambda}\, q_i(s) \leq q_u(s) \leq \lambda\, q_i(s) \text{ for } s\in H_i,\ u\in U_i'$$

$$(7)\quad \frac{1}{\lambda}\det q_i \leq \det q_u \leq \lambda \det q_i \text{ for } u\in U_i'$$

$$(8)\quad \frac{1}{\lambda}\kappa_C(u) \leq \det q_i \leq \lambda\,\kappa_C(u) \text{ for } u\in U_i'$$

$$(9)\quad \sum_i \int_{J_i'} \kappa_C(u)^{\frac{1}{d+1}}\,du \geq \frac{1}{\lambda}\int_{\mathrm{bd}\,C} \kappa_C(x)^{\frac{1}{d+1}}\,d\sigma(x)$$

Let $P_n \in \mathcal{P}^c_{(n)}$, $n = d+1,\ldots,$ be a sequence of best approximating convex polytopes of C. Since $\delta^V(C,P_n)\to 0$ and C is strictly convex (note that $\kappa_C > 0$), the maximum of the diameters of the facets of P_n tends to 0 as $n \to \infty$. This, together with (5), implies the following:

(10) $\delta^V(C, P_n) \geq \sum_i \{\text{volume of the subset of } P_n \text{ below } J_i\}$
for all sufficiently large n.

Let F_{nik}, $k = 1, \dots, m_{ni}$, be the facets of P_n below C such that $F'_{nik} \cap J'_i \neq \emptyset$. Clearly,

(11) $m_{ni} \to \infty$ as $n \to \infty$,

(12) $m_{n1} + \dots + m_{nl} \leq n$ for all sufficiently large n.

Let $s_{nik} \in \text{relint}\, C'$ be the projection into H_i of the point where F_{nik} touches C. Then the

(13) volume of the subset of P_n below J_i

$$= \sum_k \int\limits_{F'_{nik} \cap J'_i} \{f_i(s) - f_i(s_{nik}) - \text{grad}\, f_i(s_{nik}) \cdot (s - s_{nik})\}\, ds.$$

Since f_i is of class $\mathcal{C}^2$, the remainder term in Taylor's formula shows that the integrand here is

$$\frac{1}{2} q_{s_{nik}+\xi(s-s_{nik})}(s - s_{nik}) \text{ for } s \in F'_{nik} \cap J'_i$$

with suitable $\xi \in [0, 1]$ depending on s. Since the maximum of the diameters of the facets of P_n tends to 0 as $n \to \infty$, (5) shows that, for sufficiently large n, the sets F'_{nik} which meet J'_i are all contained in U'_i. For such n, it follows from (6) that the integrand is bounded below by

$$\frac{1}{2\lambda} q_i(s - s_{nik}) \text{ for } s \in F'_{nik} \cap J'_i.$$

Thus (10), (13), the lower bound for the integrand in (13), (11), (3), (2), (8), (12) and (9) yield (4), where summation on i is from 1 to l and on k from 1 to m_i:

$$\delta^V(C, \mathcal{P}^c_{(n)}) = \delta^V(C, P_n) \geq \frac{1}{2\lambda} \sum_i \sum_k \int\limits_{F'_{nik} \cap J'_i} q_i(s - s_{nik})\, ds$$

$$\geq \frac{1}{2\lambda} \sum_i \int\limits_{J'_i} \min_{k=1,\dots,m_{ni}} \{q_i(s - s_{nik})\}\, ds$$

$$\geq \frac{1}{2\lambda} \sum_i \inf_{\substack{S \subseteq H \\ \#S = m_{ni}}} \int\limits_{J'_i} \min_{t \in S}\{q_i(s - t)\}\, ds$$

$$\geq \frac{\delta}{2\lambda^2} \sum_i v(J'_i)^{\frac{d+1}{d-1}} (\det q_i)^{\frac{1}{d-1}} \frac{1}{m_{ni}^{\frac{2}{d-1}}}$$

$$= \frac{\delta}{2\lambda^2}\left\{\frac{1}{m_{n1}+\cdots+m_{nl}}\sum_i \left(v(J_i')(\det q_i)^{\frac{1}{d+1}}\frac{1}{m_{ni}}\right)^{\frac{d+1}{d-1}} m_{ni}\right\}$$
$$\times (m_{n1}+\cdots+m_{nl})$$
$$\geq \frac{\delta}{2\lambda^2}\left\{\frac{1}{m_{n1}+\cdots+m_{nl}}\sum_i v(J_i')(\det q_i)^{\frac{1}{d+1}}\frac{1}{m_{ni}} m_{ni}\right\}^{\frac{d+1}{d-1}}$$
$$\times (m_{n1}+\cdots+m_{nl})$$
$$= \frac{\delta}{2\lambda^2}\left\{\sum_i v(J_i')(\det q_i)^{\frac{1}{d+1}}\right\}^{\frac{d+1}{d-1}} \frac{1}{(m_{n1}+\cdots+m_{nl})^{\frac{2}{d-1}}}$$
$$\geq \frac{\delta}{2\lambda^{2+\frac{d+1}{d-1}}}\left(\sum_i \int\limits_{J_i'} \kappa_C(u)^{\frac{1}{d+1}}\,du\right)^{\frac{d+1}{d-1}} \frac{1}{n^{\frac{2}{d-1}}}$$
$$\geq \frac{\delta}{2\lambda^{2+2\frac{d+1}{d-1}}}\left(\int\limits_{\operatorname{bd} C} \kappa_C(x)^{\frac{1}{d+1}}\,d\sigma(x)\right)^{\frac{d+1}{d-1}} \frac{1}{n^{\frac{2}{d-1}}} \text{ for all sufficiently large } n.$$

In the second step it will be shown that

(14) $\delta^V(C, \mathcal{P}^c_{(n)}) \leq \dfrac{\lambda^{\frac{3d+2}{d-1}}\delta}{2}\left(\displaystyle\int\limits_{\operatorname{bd} C} \kappa_C(x)^{\frac{1}{d+1}}\,d\sigma(x)\right)^{\frac{d+1}{d-1}} \dfrac{1}{n^{\frac{2}{d-1}}}$

for all sufficiently large n.

By the same argument as at the beginning of the proof of (4), we may dissect bd C into finitely many pieces $K_i, i = 1, \dots, l$, and choose slightly larger sets L_i where $K_i \subseteq L_i$ in bd C, points p_i with neighbourhoods U_i in bd C, support hyperplanes H_i of C at p_i, functions f_i, quadratic forms $q_u = q_{p_i u}, q_i = q_{p_i\, p_i}$ such that the following statements hold:

(15) The sets K_i are compact sets which dissect bd C, $K_i' \subseteq L_i' \subseteq \operatorname{relint} U_i'$ are Jordan measurable, L_i' is open and U_i' convex

(16) $v(L_i') \leq \lambda v(K_i')$

(17) Propositions (6) and (8) hold

Next, convex polytopes Q_n will be constructed, for all sufficiently large n, which have at most n facets and are circumscribed to C. Let

(18) $\tau_i = \displaystyle\int\limits_{K_i} \kappa_C(x)^{\frac{1}{d+1}}\,d\sigma(x)\left(\int\limits_{\operatorname{bd} C} \kappa_C(x)^{\frac{1}{d+1}}\,d\sigma(x)\right)^{-1}, \; m_{ni} = \lfloor \tau_i n \rfloor.$

Then

(19) $m_{ni} \to +\infty$ as $n \to \infty$

(20) $m_{ni} \geq \frac{1}{\lambda}\tau_i n$ for all sufficiently large n

(21) $m_{n1} + \cdots + m_{nl} \leq n$

Choose points $s_{nik} \in L_i'$, $k = 1, \dots, m_{ni}$, such that

$$(22) \int\limits_{L_i'} \min_{k=1,\dots,m_{ni}} \{q_i(s - s_{nik})\}\, ds = \inf_{\substack{S \subseteq L_i' \\ \#S = m_{ni}}} \int\limits_{L_i'} \min_{t \in S}\{q_i(s - t)\}\, ds.$$

Let t_{nik} be the point on the lower part of $\operatorname{bd} C$ which projects onto s_{nik}. For sufficiently large n, the points t_{nik} are distributed rather densely over $\operatorname{bd} C$. Thus the intersection of the support halfspaces of C at these points is a convex polytope, say Q_n, where Q_n is circumscribed to C and – by (21) – has at most n facets. Clearly, $Q_n \to C$ as $n \to \infty$. Then (15) implies that

$$\begin{aligned}(23)\quad \delta^V(C, Q_n) &\leq \sum_i \{\text{volume of the subset of } Q_n \text{ below } L_i\} \\ &\leq \sum_i \int\limits_{L_i'} \min_{k=1,\dots,m_{ni}} \{f_i(s) - f_i(s_{nik}) \\ &\qquad - \operatorname{grad} f_i(s_{nik}) \cdot (s - s_{nik})\}\, ds\end{aligned}$$

for all sufficiently large n.

The expression in $\{\cdot\}$ equals

$$\frac{1}{2} q_{s_{nik} + \xi(s - s_{nik})}(s - s_{nik})$$

with suitable $\xi \in [0, 1]$, which by (17) and (6) is at most

$$\frac{\lambda}{2} q_i(s - s_{nik}).$$

Hence (23), (19), (3), (16), (20), (17) and (8), (18) and (15) show (14):

$$\begin{aligned}\delta^V(C, \mathcal{P}^c_{(n)}) \leq \delta^V(C, Q_n) &\leq \frac{\lambda}{2} \sum_i \int\limits_{L_i'} \min_{k=1,\dots,m_{ni}} \{q_i(s - s_{nik})\} ds \\ &\leq \frac{\lambda^2 \delta}{2} \sum_i v(L_i')^{\frac{d+1}{d-1}} (\det q_i)^{\frac{1}{d-1}} \frac{1}{m_{ni}^{\frac{2}{d-1}}} \\ &\leq \frac{\lambda^{2 + \frac{d+1}{d-1} + \frac{2}{d-1}} \delta}{2} \sum_i v(K_i')^{\frac{d+1}{d-1}} (\det q_i)^{\frac{1}{d-1}} \frac{1}{\tau_i^{\frac{2}{d-1}}} \frac{1}{n^{\frac{2}{d-1}}}\end{aligned}$$

$$= \frac{\lambda^{\frac{3d+1}{d-1}}\delta}{2}\sum_i \Big(v(K_i')(\det q_i)^{\frac{1}{d+1}}\Big)^{\frac{d+1}{d-1}}\frac{1}{\tau_i^{\frac{2}{d-1}}}\frac{1}{n^{\frac{2}{d-1}}}$$

$$\le \frac{\lambda^{\frac{3d+1}{d-1}+\frac{1}{d-1}}\delta}{2}\sum_i \Bigg(\int_{K_i'}\kappa_C(u)^{\frac{1}{d+1}}\,du\Bigg)^{\frac{d+1}{d-1}}\frac{1}{\tau_i^{\frac{2}{d-1}}}\frac{1}{n^{\frac{2}{d-1}}}$$

$$\le \frac{\lambda^{\frac{3d+2}{d-1}}\delta}{2}\sum_i \Bigg(\int_{K_i}\kappa_C(x)^{\frac{1}{d+1}}\,d\sigma(x)\Bigg)\Bigg(\int_{\mathrm{bd}\,C}\kappa_C(x)^{\frac{1}{d+1}}\,d\sigma(x)\Bigg)^{\frac{2}{d-1}}\frac{1}{n^{\frac{2}{d-1}}}$$

$$= \frac{\lambda^{\frac{3d+2}{d-1}}\delta}{2}\Bigg(\int_{\mathrm{bd}\,C}\kappa_C(x)^{\frac{1}{d+1}}\,d\sigma(x)\Bigg)^{\frac{d+1}{d-1}}\frac{1}{n^{\frac{2}{d-1}}}$$

for all sufficiently large n.

Having proved (4) and (14) for any $\lambda > 1$, the asymptotic formula (1) follows. □

Related Open Problems

In the context of this result, the following problems arise:

Eliminate the assumption that $\kappa_C > 0$. This was done by Böröczky [153]. A more coherent proof would be desirable.

Prove an asymptotic formula for a wider class of convex bodies. Originally the affine surface area was defined only for sufficiently smooth convex bodies. Approximation and other problems in convexity led to extensions to all convex bodies by Petty [797], Lutwak [669], Leichtweiss [641] and Schütt and Werner [921]. These generalizations all coincide as was proved by Schütt [919] and Leichtweiss [642]. See also Leichtweiss [644]. It seems feasible to extend Theorem 11.4, in the above form, to all convex bodies, using the generalized affine surface area. The case $d = 2$ was settled by Ludwig [665]. For general d, compare the corresponding result of Schütt [920] for approximation with random polytopes. Unfortunately, the generalized affine surface area is 0 for most convex bodies. Moreover, the irregularity Theorem 13.2 and its corollary imply that there is no other non-trivial asymptotic formula which holds for most convex bodies.

Prove more precise asymptotic formulae or even asymptotic series developments for $\delta^V(C, \mathcal{P}^c_{(n)})$ under suitable smoothness assumptions for C. For $d = 2$ a first step in this direction is due to Ludwig [664] and Tabachnikov [984] proved an asymptotic series development for $\delta^V(C, \mathcal{P}^c_{(n)})$. For general d Böröczky [154] and Gruber [440,441] gave estimates for the error term in the asymptotic formula (1) and in similar formulae.

Prove analogous results, given the number of edges, 2-faces, etc. of the approximating polytopes instead of the number of vertices or facets.

Prove results of this type for other measures of distance, for example for the deviation with respect to the surface area or other quermassintegrals.

Gruber [439] showed that for $d = 3$ best approximating polytopes have asymptotically affine regular hexagonal facets. For some information on the form of the best approximating polytopes for general d, see [443]. More precise results in higher dimensions would be desirable.

An Isoperimetric Problem for Convex Polytopes

Let $\mathcal{P}_{(n)}$, $n = d+1, \ldots$, denote the set of all proper convex polytopes in $\mathbb{E}^d$ with at most n facets. Then the problems arise to determine

$$\inf\left\{\frac{S(P)^d}{V(P)^{d-1}} : P \in \mathcal{P}_{(n)}\right\}$$

and to describe the polytopes $P_n \in \mathcal{P}_{(n)}$ with minimum isoperimetric quotient. As a consequence of the above approximation theorem, we have the following result.

Theorem 11.5. *Let $P_n \in \mathcal{P}_{(n)}$, $n = d+1, \ldots,$ be polytopes with minimum isoperimetric quotient amongst all polytopes in $\mathcal{P}_{(n)}$. Then there is a constant $\delta = \delta_{2,d-1} > 0$, depending only on d, such that*

$$\frac{S(P_n)^d}{V(P_n)^{d-1}} \sim d^d V(B^d) + \frac{d^d\,\delta}{2} S(B^d)^{\frac{d+1}{d-1}} \frac{1}{n^{\frac{2}{d-1}}} \text{ as } n \to \infty.$$

Proof. By the corollary of Lindelöf's theorem 18.4, each of the polytopes P_n is circumscribed to a ball. Since homotheties do not change the isoperimetric quotient, we may assume that each P_n is circumscribed to B^d. For such a polytope the volume equals $1/d$ times its surface area. Hence

$$(24) \quad \frac{S(P_n)^d}{V(P_n)^{d-1}} = d^d V(P_n) = d^d V(B^d) + d^d\big(V(P_n) - V(B^d)\big)$$
$$= d^d V(B^d) + d^d \delta^V(C, P_n).$$

P_n minimizes the isoperimetric quotient among all polytopes in $\mathcal{P}_{(n)}$, and is circumscribed to B^d. Thus, in particular, it minimizes the isoperimetric quotient among all polytopes in $\mathcal{P}_{(n)}$ circumscribed to B^d. Taking into account (24), we see that P_n minimizes the symmetric difference $\delta^V(C, P_n)$ among all polytopes in $\mathcal{P}_{(n)}$ which are circumscribed to B^d. Thus P_n is best approximating of B^d among all polytopes in $\mathcal{P}^c_{(n)}(B^d)$. Now apply Theorem 11.4 to the equality (24). □

Remark. For refinements and information on the form of P_n, see the author's articles [439] ($d = 3$) and [443] (general d). The results in these papers deal with normed spaces and make use of the generalized surface area as treated in Sect. 8.3. One of the tools used there is a result of Diskant [274] which extends Lindelöf's theorem to normed spaces.

Heuristic Observations

The asymptotic formula for best approximation of C by inscribed convex polytopes is as follows, see Gruber [427].

$$\delta^V(C, \mathcal{P}_n^i) \sim \frac{\gamma}{2} A(C)^{\frac{d+1}{d-1}} \frac{1}{n^{\frac{2}{d-1}}} \text{ as } n \to \infty,$$

where $\gamma = \gamma_{2,d-1} > 0$ is a suitable constant depending on d, and

$$A(C) = \int\limits_{\text{bd}\, C} \kappa_C(x)^{\frac{1}{d+1}} d\sigma(x)$$

the affine surface area of C. A result of Bárány [67] and Schütt [920] on random polytopes states the following, where $E\big(\delta^V(C, Q_k)\big)$ stands for the expectation of the difference of the volume of C and the volume of the convex hull Q_k of k random points uniformly distributed in C,

$$E\big(\delta^V(C, Q_k)\big) \sim c_d A(C) \frac{1}{k^{\frac{2}{d+1}}} \text{ as } k \to \infty.$$

Here $c_d > 0$ is a suitable constant depending on d. These two asymptotic formulae seem to say that random approximation is less efficient than best approximation (put $k = n$), but, as observed by Bárány, this is the wrong comparison to make. Being the convex hull of k random points in C, the random polytope Q_k in general has less than k vertices. Actually, for the expectation $E\big(v(Q_k)\big)$ of the number of vertices of Q_k, we have

$$E\big(v(Q_k)\big) \sim c_d A(C)\, k^{\frac{d-1}{d+1}} \text{ as } k \to \infty.$$

Denote this expectation by n. Then,

$$E\big(\delta^V(C, Q_k)\big) \sim c_d^{\frac{d+1}{d-1}} A(C)^{\frac{d+1}{d-1}} \frac{1}{n^{\frac{2}{d-1}}} \text{ as } n \to \infty.$$

For $d = 2, 3$ Bárány [68] proved an even stronger result. Since Mankiewicz and Schütt [686] showed that

$$\frac{c_d^{\frac{d+1}{d-1}}}{\frac{1}{2}\gamma_{2,d-1}} \to 1 \text{ as } d \to \infty,$$

we see that for large d, random approximation is almost as good as best approximation.

This is an example of the following vague principle.

Heuristic Principle. *In many complicated situations, for example in high dimensions or depending on many parameters, average configurations are almost extremal or attain almost the mean value or the median.*

For a particularly striking result on medians of functions in the context of the concentration of measure phenomenon, see Sect. 8.6. Examples of extremal results of this type seem to be also provided by the Minkowski–Hlawka inequality and Siegel's mean value theorem in the geometry of numbers. See the discussions in Sects. 24.2 and 30.3. In some cases such results hold up to absolute constants.

For information on random approximation see the surveys of Buchta [176], Schneider [906] and the author [435].

12 Special Convex Bodies

Special objects of mathematics have attracted interest since antiquity, early examples are the primes, the conics and the Platonic and Archimedean solids. The interest in special objects seems deep rooted in human nature. Mathematical reasons for this include the following: In many cases special objects exhibit interesting properties in a particularly pure or strong form, for example, they may be regular, extremal, or symmetric in a certain sense. Frequently such objects are solutions of very general problems and the same object may appear in rather different mathematical theories.

In this section we consider simplices, balls and ellipsoids.

For relevant references and surveys, see below. In addition, we refer to the treatises of Coxeter [230, 232] and McMullen and Schulte [717] on regular polytopes. From the vast literature on symmetry in geometry we mention the books of Robertson [842], Ziegler [1047] and Johnson [550]. A book dedicated to cubes is due to Zong [1051]. See also the survey [1050].

12.1 Simplices and Choquet's Theorem on Vector Lattices

Simplices, i.e. convex hulls of affinely independent sets in $\mathbb{E}^d$, are important in several branches of mathematics. They are building blocks for cell complexes in algebraic topology and their infinite-dimensional version appears in Choquet theory. In convex geometry, simplices are basic for combinatorial polytope theory and in numerous geometric inequalities the extremal bodies are simplices. Many different characterizations are known.

In the following a characterization of simplices due to Choquet, and its refinement by Rogers and Shephard will be presented. The former makes it possible to single out the vector lattices among the ordered topological vector spaces (of finite dimensions). This result was Choquet's starting point for Choquet theory (in infinite dimensions). Without proof we mention that the Rogers–Shephard characterization of simplices is used to settle the equality case in the Rogers–Shephard inequality for the volume of difference bodies, see Theorem 9.10.

For more information consult the survey of Heil and Martini [488] which contains a multitude of references to other surveys and books, and the reports of Fonf, Lindenstrauss and Phelps [338] and Soltan [947].

Characterizations of Simplices of Choquet and Rogers–Shephard

Choquet [208] ((i)⇔(ii)) and Rogers and Shephard [852] ((i)⇔(iii)) proved the following seminal result.

Theorem 12.1. *Let $C \in \mathcal{C}$. Then the following statements are equivalent:*

(i) *C is a simplex.*

(ii) *For any $\lambda, \mu \geq 0$ and $x, y \in \mathbb{E}^d$ with $(\lambda C + x) \cap (\mu C + y) \neq \emptyset$, there are $\nu \geq 0$ and $z \in \mathbb{E}^d$ such that*

$$(\lambda C + x) \cap (\mu C + y) = \nu C + z.$$

(iii) *For any $x, y \in \mathbb{E}^d$ with $(C + x) \cap (C + y) \neq \emptyset$, there are $\nu \geq 0$ and $z \in \mathbb{E}^d$ such that*

$$(C + x) \cap (C + y) = \nu C + z.$$

An *exposed point* of a convex body C is a point $p \in C$, such that there is a support hyperplane H of C with $H \cap C = \{p\}$.

Proof. The implications (i)⇒(ii) and (ii)⇒(iii) are easy to prove (consider the simplex $\{x : 0 \leq x_i,\ x_1 + \cdots + x_d \leq 1\}$), or trivial. Thus it is sufficient to show that

(iii)⇒(i) We may suppose that $\dim C = d$. The proof is by induction on d. For $d = 0, 1$ each convex body and thus, in particular, C is a simplex. Assume now that $d > 1$ and that the implication (iii)⇒(i) holds for dimensions $0, 1, \dots, d-1$.

C contains an exposed boundary point, take for example a point where a circumscribed sphere touches C. After a translation, if necessary, we may suppose that this point is the origin o. Choose a support hyperplane H of C with $C \cap H = \{o\}$. Since C is proper, the smooth points are dense in $\operatorname{bd} C$, see Theorem 5.1. Thus we may choose a smooth point $p \in \operatorname{bd} C$ such that the open line segment with endpoints o, p is contained in $\operatorname{int} C$. In order to see that

(1) $C \cap (C - \lambda p) = (1 - \lambda) C$ for $0 < \lambda < 1$,

note first that, for $0 < \lambda < 1$, the intersection $C \cap (C - \lambda p)$ is non-empty and thus homothetic to C by (iii). Considering the line segment $[o, p] \subseteq C$ then shows that the homothety has centre o and factor $1 - \lambda$, concluding the proof of (1).

Let K be the unique support hyperplane of C at p and K^- the corresponding support halfspace. Then we have the following.

(2) $C = \operatorname{conv}(\{o\} \cup F)$, where $F = C \cap K$.

For the proof of (2), it is sufficient to show that $C = (\operatorname{cone} C) \cap K^-$, where cone $C = \bigcup \{\mu C : \mu \geq 0\}$. Clearly, $C \subseteq (\operatorname{cone} C) \cap K^-$. To see the reverse inclusion, let $x \in (\operatorname{cone} C) \cap K^-$, $x \notin [o, p]$. There are points $y, z \in C$ such that $y = \alpha x$ with suitable $\alpha > 0$ and $z - p = \beta(x - p)$ with suitable $\beta > 0$. For the latter we have used the fact that p is a smooth boundary point of C. If $0 < \lambda < 1$ is sufficiently close to 1, then the line segments $[o, y]$ and $[p, z]$-λp meet at a point w, say. Since

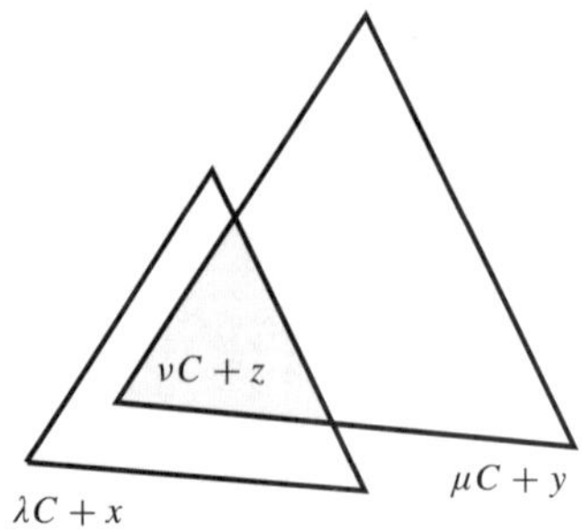

Fig. 12.1. Choquet simplex

$w \in C \cap (C - \lambda p) = (1 - \lambda)C$, (see (1)) and $w = (1 - \lambda)x$, we have that $x \in C$. Thus $(\text{cone } C) \cap K^- \subseteq C$. The proof of (2) is complete.

For each vector x parallel to the hyperplane K and such that $F \cap (F + x) \neq \emptyset$, the set $F \cap (F + x)$ is a face of $C \cap (C + x)$, and $C \cap (C + x)$ is homothetic to C by (iii). Hence $F \cap (F + x)$ is homothetic to F. The induction hypothesis then shows that F is a simplex. Since $o \notin K$, Proposition (2) finally implies that C is also a simplex, concluding the proof of (i). □

Remark. A compact convex set C in an infinite-dimensional topological vector space which satisfies property (iii) is called a *Choquet simplex* (see Fig. 12.1). Choquet simplices are basic in Choquet theory and thus in measure theory in infinite dimensions.

Topological Vector Lattices in Finite Dimensions

Let V be a real vector space and $\preceq$ a (*partial*) *ordering* of V in the following sense:

$$\begin{aligned} &x \preceq x && \text{for } x \in V \\ &x \preceq y,\, y \preceq x \Rightarrow x = y && \text{for } x, y \in V \\ &x \preceq y,\, y \preceq z \Rightarrow x \preceq z && \text{for } x, y, z \in V \end{aligned}$$

$\langle V, \preceq\rangle$ is an *ordered vector space* if, in addition, the ordering is compatible with the operations in V, i.e.

$$\begin{aligned} &x \preceq y \Rightarrow x + z \preceq y + z && \text{for } x, y, z \in V \\ &x \preceq y \Rightarrow \lambda x \preceq \lambda y && \text{for } x, y \in V,\ \lambda \geq 0 \end{aligned}$$

Then $K = \{x \in V : o \preceq x\}$ is a convex cone with apex o such that $K \cap (-K) = \{o\}$, that is, K is pointed. K is called the *positive cone* (see Fig. 12.2) of $\langle V, \preceq\rangle$. Conversely, if K is a pointed convex cone in V with apex o, then the definition

(3) $x \preceq y$ if $y - x \in K$ or, equivalently, $y \in K + x$ for $x, y \in V$

makes V into an ordered vector space with positive cone K. The ordered vector space $\langle V, \preceq\rangle$ is a *vector lattice*, if for any $x, y \in V$ there is a greatest lower bound $x \wedge y$ and a smallest upper bound $x \vee y$ of the set $\{x, y\}$ in V, that is,

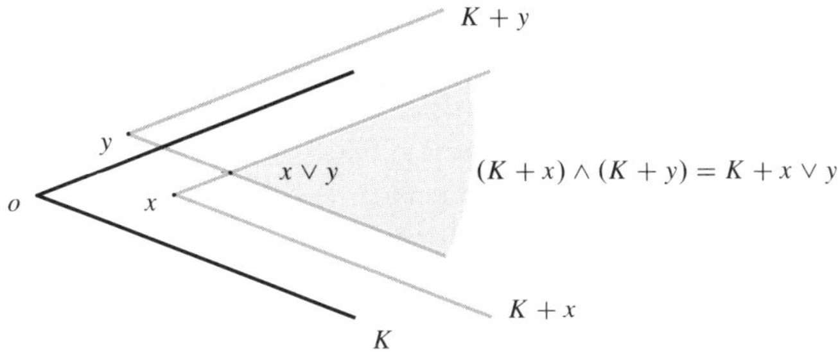

Fig. 12.2. Positive cone, vector lattice

$$x \wedge y \preceq x, y \text{ and if } z \preceq x, y, \text{ then } z \preceq x \wedge y,$$

and similarly for $x \vee y$.

V is a *topological vector space* if V is Hausdorff and the mappings $(x, y) \to x + y$ and $(\lambda, x) \to \lambda x$ are continuous. Up to isomorphisms, $\mathbb{E}^d$ is the only d-dimensional (real) topological vector space (forget the norm, but retain the topology).

If the vector space V is ordered and topological, it is called an *ordered topological vector space* if the ordering is compatible with the topology in the sense that

$$o \preceq A \Rightarrow o \preceq \operatorname{cl} A \text{ for } A \subseteq V.$$

This is equivalent to the requirement that the positive cone is closed.

In the following we consider the finite-dimensional case. Let K be a pointed closed convex cone in $\mathbb{E}^d$ with apex o. Then there is a hyperplane H with $o \notin H$ and such that $C = H \cap K$ is a convex body. Clearly, $K = \operatorname{pos} C = \{\lambda x : \lambda \geq 0, x \in C\}$. K is a *simplicial cone* if C is a simplex. It is easy to see that this definition is independent of the choice of H.

Choquet's Characterization of Topological Vector Lattices

We conclude this section with a finite-dimensional case of *Choquet's theorem on vector lattices*, see [208].

Theorem 12.2. *Let K be a pointed closed convex cone in $\mathbb{E}^d$ with apex o which makes $\mathbb{E}^d$ into an ordered topological vector space. Then the following statements are equivalent:*

(i) *$\mathbb{E}^d$ is a vector lattice with positive cone K.*
(ii) *K is a simplicial cone with* $\dim K = d$.

Proof. (i)⇒(ii) In the following the above properties and definitions will be applied several times without explicit reference. The first step is to show the following:

(4) Let $x, y \in \mathbb{E}^d$. Then $(K + x) \cap (K + y) = K + x \vee y$.

To show that $(K+x) \cap (K+y) \subseteq K + x \vee y$, let $w \in (K+x) \cap (K+y)$. Then $w \in K+x$ and $w \in K+y$ and thus $x, y \preceq w$ and therefore $x \vee y \preceq w$, or $w \in K + x \vee y$. To show the reverse inclusion, let $w \in K + x \vee y$. Then $x \vee y \preceq w$ and thus $x, y \preceq w$. This yields $w \in (K+x) \cap (K+y)$. The proof of (4) is complete.

Choose a hyperplane H as in the remarks before the theorem and let $C = H \cap K$. Then the following statement holds:

(5) Let $\lambda, \mu \geq 0$ and $x, y \in \mathbb{E}^d$ with $(\lambda C + x) \cap (\mu C + y) \neq \emptyset$. Then $(\lambda C + x) \cap (\mu C + y) = \nu C + x \vee y$ for suitable $\nu \geq 0$.

Let G be the hyperplane parallel to H which contains $\lambda C + x$ and $\mu C + y$. Then

$$\lambda C + x = (K+x) \cap G, \ \ \mu C + y = (K+y) \cap G$$

and thus,

$$(\lambda C + x) \cap (\mu C + y) = (K+x) \cap (K+y) \cap G$$
$$mm = (K + x \vee y) \cap G = \nu C + x \vee y \text{for suitable} \nu \geq 0$$

by (4). This concludes the proof of (5).

(5) Together with Theorem 12.2 of Choquet and Rogers and Shephard implies that C is a simplex. Hence K is a simplicial cone. To see that $\dim K = d$, consider a basis $\{b_1, \dots, b_d\}$ of $\mathbb{E}^d$ and note that for $z = o \wedge b_1 \wedge \dots \wedge b_d$ the cone $K + z$ contains $o, b_1, \dots, b_d$. Thus $\dim K = \dim(K+z) = d$.

(ii)⇒(i) This is easy to prove on noting that after a suitable linear transformation we may assume that $K = \{x : 0 \leq x_i\}$. □

Remark. Choquet actually proved his result in infinite dimensions using Choquet simplices instead of (finite dimensional) simplices. A generalization to vector spaces without any topology is due to Kendall [574]. See the surveys by Peressini [789] and Rosenthal [857].

12.2 A Characterization of Balls by Their Gravitational Fields

During the twentieth century a multitude of different characterizations of (Euclidean) balls and spheres have been given. Besides elementary characterizations, characterizations by extremal properties, in particular by properties of isoperimetric type, and other characterizations in the context of convexity, there is a voluminous body of differential geometric characterizations. Interesting sporadic characterizations of balls have their origin in other branches of mathematics, for example in potential theory and, even outside of mathematics.

In the sequel we consider a characterization of balls by their Newtonian gravitational fields. We consider the case $d = 3$, but the result can be extended easily to any $d \geq 2$, where, for $d = 2$, logarithmic potentials have to be used.

Surveys of characterizations of balls are due to Bonnesen and Fenchel [149], Giering [377], Burago and Zalgaller [178], Bigalke [114] and Heil and Martini [488]. We add two references, one related to cartography by Gruber [426] and one to electrostatics by Mendez and Reichel [719]. For a characterization of balls using topological tools see Montejano [750].

The Newtonian Gravitational Field and its Potential

Let C be a compact set in $\mathbb{E}^3$ consisting of homogeneous matter. Up to a multiplicative constant, its *(Newtonian) gravitational field* g is given by

$$g(y) = \int_C \frac{x-y}{\|x-y\|^3}\, dx \text{ for } y \in \mathbb{E}^3\backslash C,$$

where the integral is to be understood componentwise. The corresponding *(Newtonian) gravitational potential* P then is given by

$$P(y) = \int_C \frac{dx}{\|x-y\|} \text{ for } y \in \mathbb{E}^3\backslash C,$$

up to an additive constant. Note that $g = \operatorname{grad} P$.

The Problem and One Answer

A natural question to ask is whether C is determined uniquely by the gravitational field g or the gravitational potential P. Pertinent results are due to Novikov [774] and Shahgholian [927]. A special case of the latter's result is the following (earlier) result of Aharonov, Schiffer and Zalcman [3].

Theorem 12.3. *Let* $C = \operatorname{cl}\operatorname{int} C \subseteq \mathbb{E}^3$ *be a compact body consisting of homogeneous matter such that* $\mathbb{E}^3\backslash C$ *is connected. If the gravitational field of* C *coincides in* $\mathbb{E}^3\backslash C$ *with the gravitational field of a suitable point mass, then* C *is a ball.*

The following proof relies heavily on results from potential and measure theory for which we refer to Wermer [1018] and Bauer [82].

Proof. We clearly may suppose that the point mass is located at the origin o. Then, by assumption,

$$\int_C \frac{x-y}{\|x-y\|^3}\, dx = -\frac{\alpha\, y}{\|y\|^3} \text{ for } y \in \mathbb{E}^3\backslash C,$$

where $\alpha > 0$ is a suitable constant. Since the fields coincide on the open connected set $\mathbb{E}^3\backslash C$, the corresponding potentials coincide on $\mathbb{E}^3\backslash C$ up to an additive constant, i.e.

$$\int_C \frac{dx}{\|y-x\|} = \frac{\alpha}{\|y\|} + \beta \text{ for } y \in \mathbb{E}^3\backslash C,$$

where β is a suitable constant. Letting $\|y\| \to +\infty$, it follows that $\beta = 0$. Thus,

$$(1) \quad \int_C \frac{dx}{\|y-x\|} = \frac{\alpha}{\|y\|} \text{ for } y \in \mathbb{E}^3\backslash C.$$

The integral here may be considered as the convolution of the locally integrable function $1/\|x\|$ on $\mathbb{E}^3\backslash\{o\}$ with the characteristic function of C which is bounded, measurable and has compact support. It thus defines a bounded continuous function on *all* of $\mathbb{E}^3$. Then

(2) $o \in \operatorname{int} C$,

since otherwise the function $\alpha/\|y\|$ is unbounded on $\mathbb{E}^3\backslash C$ and thus the above integral is unbounded by (1), a contradiction.

A real function u on a domain in $\mathbb{E}^3$ of class $\mathcal{C}^2$ which satisfies the Laplace equation

$$\Delta u = \frac{\partial^2 u}{\partial x_1^2} + \frac{\partial^2 u}{\partial x_2^2} + \frac{\partial^2 u}{\partial x_3^2} = 0$$

is *harmonic*. We now show the following:

(3) Let u be a harmonic function on a neighbourhood of C. Then

$$\int_C u(x)\,dx = \alpha\, u(o),$$

where α is as in (1). Modify u outside C to be of class $\mathcal{C}^2$ with compact support S, preserving the harmonicity on and near C. Since u is of class $\mathcal{C}^2$ and has compact support, a potential theoretic result shows that

(4) $u(x) = -\dfrac{1}{4\pi}\displaystyle\int_S \frac{\Delta u(y)}{\|x-y\|}\,dy$ for $x \in \mathbb{E}^3$,

see Wermer [1018], p.13. Thus,

$$\int_C u(x)\,dx = -\frac{1}{4\pi}\iint_{C\,S} \frac{\Delta u(y)}{\|x-y\|}\,dy\,dx = -\frac{1}{4\pi}\int_S \Big(\Delta u(y)\int_C \frac{dx}{\|x-y\|}\Big)\,dy$$
$$= -\frac{1}{4\pi}\int_S \Delta u(y)\,\frac{\alpha}{\|y\|}\,dy = \alpha\, u(o)$$

by (4), Fubini's theorem, (1) and (4) again, concluding the proof of (3).

Now, noting (2), choose $p \in \operatorname{bd} C$ closest to o and take points $p_1, p_2, \dots \in \mathbb{E}^3\backslash C$ with $p_n \to p$. Each of the functions v_n defined by

$$v_n(x) = \frac{\|x\|^2 - \|p_n\|^2}{\|x-p_n\|^3} \text{ for } x \in \mathbb{E}^3\backslash\{p_n\}$$

is harmonic on C and $v_n(x) \to v(x)$ for $x \in C\backslash\{p\}$, where v is a harmonic function defined by

$$v(x) = \frac{\|x\|^2 - \|p\|^2}{\|x-p\|^3} \text{ for } x \in \mathbb{E}^3\backslash\{p\}.$$

The sequence $\big(v_n(\cdot)\big)$ converges pointwise and thus stochastically to $v(\cdot)$ on $C\backslash\{p\}$ and is uniformly integrable on C. Thus

$$(5)\quad \int\limits_C v(x)\,dx = \lim_{n\to\infty} \int\limits_C v_n(x)\,dx = \lim_{n\to\infty} \alpha\, v_n(o) = \alpha\, v(o)$$

by Proposition (3). For the notions of stochastic convergence and uniform integrability and for the first equality in (5) see, e.g. Bauer [82], Sects. 20, 21.

Finally, let a harmonic function w be defined by

$$w(x) = 1 + \|p\| v(x) \text{ for } x \in \mathbb{E}^3 \backslash \{p\}$$

and let B be the ball $\|p\| B^3$. Since $o \in \operatorname{int} C$ by (2), our choice of p shows that $B \subseteq C$. Thus

$$(6)\quad \alpha\, w(o) = \alpha\big(1 + \|p\| v(o)\big) = 0 = \int\limits_C \big(1 + \|p\| v(x)\big)\,dx = \int\limits_C w(x)\,dx$$

by the definitions of w and v, (3) applied to $u = 1$ and Proposition (5). The mean value property of harmonic functions implies that

$$(7)\quad \int\limits_B w(x)\,dx = \gamma\ w(o) = 0,$$

where $\gamma > 0$ is a suitable multiplicative constant, see Wermer [1018], Appendix. From (6) and (7) we conclude that

$$0 = \int\limits_C w(x)\,dx = \int\limits_B w(x)\,dx + \int\limits_{C\backslash B} w(x)\,dx = \int\limits_{C\backslash B} w(x)\,dx.$$

Since $w(x) > 0$ for $x \in C\backslash B$, it follows that $C\backslash B$ has measure 0. Noting that $C = \operatorname{cl}\operatorname{int} C$ by assumption, this means that $C\backslash B = \emptyset$, or $C \subseteq B$. The reverse inclusion being obvious, $C = B$ follows. □

12.3 Blaschke's Characterization of Ellipsoids and Its Applications

Ellipsoids and Euclidean spaces play an important role in many branches of mathematics. Among these are convex and differential geometry, the local theory of normed spaces, functional analysis, approximation, operator and potential theory, dynamical systems, combinatorial optimization and mechanics. The first characterizations of ellipsoids in convex geometry go back to Brunn and Blaschke, and at present there is still interest for these characterizations. One may distinguish between characterizations based on:

Affine and projective transformations
Sections
Projections and illuminations
Extremal properties
Other geometric and analytic properties

In the following we consider a characterization of ellipsoids due to Blaschke [123, 124] by the property that the shadow boundaries under parallel illumination all are planar. As a consequence, two characterizations of Euclidean spaces are given. A remark concerns geometric stability problems. Finally, we define Radon norms.

For more information we refer to the books and surveys of Bonnesen and Fenchel [149], Day [248], Laugwitz [630], Gruber and Höbinger [446], Petty [798], Amir [27], Istrătescu [537], Heil and Martini [488], Lindenstrauss and Milman [660], Li, Simon and Zhao [655], Thompson [994], Martini [692], and Deutsch [263].

A Version of Blaschke's Characterization of Ellipsoids

Among all characterizations of ellipsoids in convex geometry, Blaschke's [123, 124] characterization mentioned above has the largest number of applications. In some of these the following slightly refined version due to Marchaud [689] is used.

Theorem 12.4. *Let C be a proper convex body in $\mathbb{E}^d, d \geq 3$. Then the following statements are equivalent:*

(i) *C is an ellipsoid.*

(ii) *For each line L through o there is a hyperplane H such that*

(1) $C + L = C \cap H + L.$

In other words, the *shadow boundary* $\operatorname{bd}(C+L) \cap C$ of C under illumination parallel to L contains the "planar curve" $\operatorname{bd}(C + L) \cap H$. Before beginning with the proof we state two auxiliary results, the proofs of which are left to the reader but take some effort.

Lemma 12.1. *Let $C \in \mathcal{C}_p$ and $k \in \{2, \dots, d\}$. Then the following assertions are equivalent:*

(i) *C is an ellipsoid.*

(ii) *For each k-dimensional plane H which meets* $\operatorname{int} C$, *the intersection $C \cap H$ is an ellipsoid.*

Lemma 12.2. *Let $D \in \mathcal{C}_p(\mathbb{E}^2)$ and $w \in \operatorname{int} D$ such that for any pair of parallel support lines S, T of D there are points $s \in D \cap S, t \in D \cap T$ with $w \in [s, t]$. Then w is the centre of C.*

Proof of the Theorem. The implication (i)⇒(ii) is easy. We show only that

(ii)⇒(i) A simple argument implies that the intersection of C with any 3-dimensional plane which meets $\operatorname{int} C$ also has property (ii). If the implication (ii)⇒(i) holds in case $d = 3$, the first lemma implies that C is an ellipsoid. It is thus sufficient to consider the case

$$d = 3.$$

Our first proposition is as follows:

(2) Let U be a support plane of C. Then $C \cap U$ is not a line segment.

To see this, assume the contrary. For each illumination of C parallel to a line L parallel to U but not to the line segment $C \cap U$, the plane H must contain this line segment. If the direction of L differs only slightly from the direction of the line segment, then H is almost parallel to L and (1) cannot hold, a contradiction.

As a consequence of Corollary 11.1 we may assume that, after a suitable translation and the choice of a suitable inner product in $\mathbb{E}^3$,

(3) the group of affinities in $\mathbb{E}^3$ which leave C invariant (as a whole), is a subgroup of the orthogonal group.

Let M be a line and U a plane not parallel to M. An *affine reflection* of $\mathbb{E}^3$ in M parallel to U is an affinity which reflects each plane parallel to U in its intersection point with M. If M is orthogonal to U we speak of an *orthogonal reflection* of $\mathbb{E}^3$ in M. Now the following will be shown:

(4) The group of affinities in $\mathbb{E}^3$ which leave C invariant, contains the orthogonal reflections in all lines through o, except for a set of lines which is at most countable.

Let U, V be a pair of parallel support planes of C which are not parallel to any of the at most countably many 2-faces in $\operatorname{bd} C$. Then $C \cap U, C \cap V$ consist of one point each by (2), say u and v, respectively. Let W be a plane between U, V and let w be the intersection point of W and the line M through u, v. For any illumination of C parallel to U, V, W, the plane H must contain u, v and thus also w. This implies that the convex disc $D = C \cap W$ satisfies the assumptions of the second lemma and thus has centre w. Since W was an arbitrary plane between U, V, it follows that C is invariant with respect to the affine reflection in the line M, parallel to U. Noting (3) and our choice of U, V, we obtain (4).

A simple compactness argument then shows that

the group of affinities in $\mathbb{E}^d$ which leave C invariant contains the orthogonal reflections in all lines through o.

This readily implies that the intersection of C with any plane through o is a circular disc, which in turn shows that C is a ball with centre o with respect to the chosen inner product. □

Remark. It turns out that, for a characterization of ellipsoids, only illuminations in a rather small set of directions are needed, see [446].

Geometric Stability Problems

Considering the various stability problems in the mathematical literature, the following geometric problem is quite natural.

Problem 12.1. *Consider a geometric property which characterizes certain convex bodies. How well can a convex body which satisfies this property approximately, be approximated by convex bodies which satisfy this property exactly?*

The first such result in convex geometry seems to be due to Groemer [404]. A stability result related to Blaschke's characterization of ellipsoids was given by the author [434].

Kakutani's Characterization of Euclidean Norms

As a consequence of Blaschke's characterization of ellipsoids, we will prove the following characterization of Euclidean norms by Kakutani [559].

Theorem 12.5. *Let $|\cdot|$ be a norm on $\mathbb{E}^d, d \geq 3$, and $k \in \{2, \dots, d-1\}$. Then the following statements are equivalent:*

(i) *$|\cdot|$ is Euclidean.*

(ii) *For each k-dimensional linear subspace S of $\mathbb{E}^d$ there is a (linear) projection $p_S : \mathbb{E}^d \rightarrow S$ with norm $|p_S| = \sup\{|p_S(x)| : x \in \mathbb{E}^d, |x| \leq 1\}$ equal to* 1.

The following proof was proposed by Klee [591]. It makes use of polarity. Here polarity is defined slightly more generally than in Sect. 9.1.

$$C^* = \{y : x \cdot y \leq 1 \text{ for } y \in C\} \text{ for convex } C \subseteq \mathbb{E}^d,\ o \in C.$$

The properties of polarity needed in the proof of Kakutani's theorem and one further property are collected together in Lemma 12.3, the proof of which is left to the interested reader.

Lemma 12.3. *The following properties hold:*

(i) *S linear subspace of $\mathbb{E}^d \Rightarrow S^* = S^{\perp}$*

(ii) *$C \subseteq \mathbb{E}^d$ convex, $o \in C \Rightarrow C^{**} = \operatorname{cl} C$*

(iii) *$C, D \subseteq \mathbb{E}^d$ convex, $o \in C \cap D \Rightarrow (C \cap D)^* = \operatorname{cl}\operatorname{conv}(C^* \cup D^*)$*

(iv) *$E \subseteq \mathbb{E}^d$ ellipsoid with centre $o \Rightarrow E^*$ ellipsoid with centre o*

(v) *$C \subseteq \mathbb{E}^d$ convex, $o \in C$, S a linear subspace of $E^d \Rightarrow C + S = \operatorname{cl}\operatorname{conv}(C \cup S)$*

Proof of the Theorem. It is easy to show that (i)⇒(ii).

(ii)⇒(i) By Lemma 12.1 it is sufficient to consider the case

$$d = k + 1, \text{ or } k = d - 1.$$

Clearly, statement (ii) can be expressed as follows, where $B = \{x : |x| \leq 1\}$ is the solid unit ball of the norm $|\cdot|$.

(5) For each hyperplane S through o, there is a line T through o such that

$$B \cap S = (B + T) \cap S.$$

Note that

$$\begin{aligned} B^* + S^* &= \operatorname{cl}\operatorname{conv}(B^* \cup S^*) = (B \cap S)^* = \big((B + T) \cap S\big)^* \\ &= \operatorname{cl}\operatorname{conv}\big((B + T)^* \cup S^*\big) = (B + T)^* + S^* \\ &= \big(\operatorname{cl}\operatorname{conv}(B \cup T)\big)^* + S^* = \big(\operatorname{cl}\operatorname{conv}(B^{**} \cup T^{**})\big)^* + S^* \\ &= (B^* \cap T^*)^{**} + S^* = (B^* \cap T^*) + S^*, \end{aligned}$$

by properties (v), (iii) of the lemma, Proposition (5) and, again, properties (iii), (v), (ii), (iii) and (ii). Thus property (i) and Proposition (5) imply that

For each line $L(=S^*)$ through o there is a hyperplane $H(=T^*)$ through o such that

$$B^* + L = B^* \cap H + L.$$

Blaschke's characterization of ellipsoids then shows that B^* is an ellipsoid. Since o is in the centre of B, also B^* has centre o. An application of property (iv) then implies that B is an ellipsoid, concluding the proof of Statement (i) of the theorem. □

Orthogonality in Normed Spaces

A well-known notion of orthogonality in normed spaces which goes back to a question of Carathéodory on Finsler spaces is as follows. Given a normed space with norm $|\cdot|$, a vector x is *orthogonal* to a vector y, in symbols

$$x \perp y, \text{ if } |x| \leq |x + \lambda y| \text{ for all } \lambda \in \mathbb{R}.$$

Expressed geometrically, $x \perp y$ means that the line through x parallel to the vector y supports the ball $\{z : |z| \leq |x|\}$ at the boundary point x. This notion of orthogonality is sometimes named for Birkhoff [118] or James [541]. Orthogonality is important in the context of approximation and still attracts interest, see the pertinent literature in the books of Amir [27] and Istrătescu [537].

The following result is due to Blaschke [123, 124] and Birkhoff [118].

Theorem 12.6. *Let $|\cdot|$ be a norm on $\mathbb{E}^d$, $d \geq 3$, and let $\perp$ be the corresponding notion of orthogonality. Then the following are equivalent:*

(i) *$|\cdot|$ is Euclidean.*
(ii) *$\perp$ is symmetric, i.e. $x \perp y$ implies $y \perp x$ for $x, y \in \mathbb{E}^d$.*

Proof. (i)⇒(ii) This is trivial.

(ii)⇒(i) Let $B = \{x : |x| \leq 1\}$. We will prove that

(6) For each line L through o there is a hyperplane H through o such that

$$B + L = B \cap H + L.$$

Let bd $B \cap L = \{\pm p\}$, say, and let $H + p$ be a support hyperplane of B at p. For the proof that $B + L = B \cap H + L$, it is sufficient to show that any line of the form $L + q$, where $q \in B$, meets $B \cap H$. Given such a line, let $(L + q) \cap H = \{r\}$. Since $r \in H$ and $H + p$ supports B at p, we have $|p| \leq |p + \lambda r|$ for all $\lambda \in \mathbb{R}$ or $p \perp r$. Thus $r \perp p$ by (ii), or $|r| \leq |r + \mu p|$ for all $\mu \in \mathbb{R}$, or $|r| \leq |s|$ for all $s \in L + r = L + q$. Hence, in particular, $|r| \leq |q| \leq 1$, or $r \in B$. This concludes the proof of (6).

Having proved (6), Blaschke's ellipsoid theorem shows that B is an ellipsoid, concluding the proof of (i). □

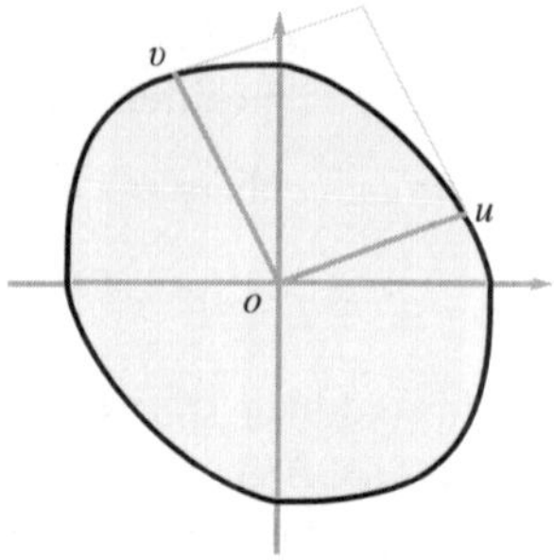

Fig. 12.3. Symmetry of orthogonality with respect to a Radon norm

Remark. For $d = 2$, the so-called Radon norms, which include the Euclidean norms, are characterized by property (ii). There are several other properties which also characterize Euclidean norms in dimension $d \geq 3$ and Radon norms in case $d = 2$, see the survey of Gruber [420]. Thus it makes sense, to consider Radon norms as the 2-dimensional equivalent of Euclidean norms in dimension $d \geq 3$. For a stability result with respect to orthogonality both for Euclidean and Radon norms, see Gruber [434].

Radon norms can be constructed as follows: Take a continuous curve in the unit square $[0, 1]^2$ which connects the points $(0, 1)$ and $(1, 0)$ and such that the curve together with the line segments $[o, (0, 1)]$ and $[o, (1, 0)]$ is the boundary of a convex disc. Consider its polar curve, which is also contained in $[0, 1]^2$ and connects the points $(0, 1)$ and $(1, 0)$. Rotate the polar curve by $\pi/2$ about the origin o in the positive direction. The given curve, the rotated polar curve and their reflections in o form the boundary of a convex disc with centre o. Now apply a non-singular linear transform to this convex disc. This, then, is the unit disc of a Radon norm and each Radon norm (see Fig. 12.3) can be obtained in this fashion.

13 The Space of Convex Bodies

The space $\mathcal{C}$ of convex bodies and subspaces of it such as the space $\mathcal{C}_p$ of proper convex bodies, have been investigated from the viewpoint of topological and metric spaces, lattices and groups. In spite of a multitude of results, we believe that the work is only at its beginning. In addition to Baire category results and metric estimates, many results deal with structure preserving mappings, which turn out to be few and surprisingly simple.

It seems that, with respect to their natural topologies, the spaces $\mathcal{C}$ and $\mathcal{C}_p$ are homogeneous, but we are not aware of a proof. In contrast, the results which will be given in the following indicate that, with respect to the group, the lattice and metric structures, both $\mathcal{C}$ and $\mathcal{C}_p$ are far from being homogeneous. If sometime in the future, there will be local versions of the results on structure preserving mappings, we think that these will show that, still, neighbourhoods of generic pairs of convex bodies in $\mathcal{C}$ or $\mathcal{C}_p$ are totally different from the group, lattice and metric viewpoint.

In the following we first deal with the topology, considering Baire category results. Then, a result is presented which shows difficulties with the introduction of measures. Then characterizations of the isometries of the metric spaces $\langle \mathcal{C}, \delta^H \rangle$ and $\langle \mathcal{C}_p, \delta^V \rangle$ are stated without proofs. Finally, we consider the algebraic structure of $\mathcal{C}$. A characterization of homomorphisms with additional properties of $\langle \mathcal{C}, + \rangle$ into $\langle \mathbb{E}^d, + \rangle$ is presented. Its proof makes use of spherical harmonics. Results on endomorphisms of the semigroup $\langle \mathcal{C}, + \rangle$ and the lattice $\langle \mathcal{C}, \wedge, \vee \rangle$ are stated last.

For more information and additional references, see [428] and the references and books cited below.

13.1 Baire Categories

A version of Blaschke's selection theorem says that the spaces $\mathcal{C}$ and $\mathcal{C}_p$, endowed with their natural topologies (which are induced by, e.g. the metric δ^H), are locally compact, see Theorem 6.4. Thus both are Baire according to a modern form of Baire's category theorem. This means that each meagre set has dense complement. By most or typical convex bodies we mean all convex bodies with a meagre set of exceptions. For these notions, see Sect. 5.1.

The first Baire category result dealing with spaces of convex bodies is due to Klee [590]. It says that most convex bodies are smooth and strictly convex. For unclear reasons it was soon forgotten. Its re-discovery by Gruber [414] some 20 years later led to a voluminous body of results, see the surveys of Zamfirescu [1039, 1041] and the author [431]. These results treat:

Differentiability properties
Geodesics
Billiards, normals and mirrors
Approximation
Contact points
Shadow boundaries
Metric projection
Fixed points and attractors
Cut loci and conjugate points
Packing and covering

It is interesting to note that, sometimes, Baire type convexity results are in contrast to results of differential geometry. For example, a result of the author [423] says that for most proper convex bodies C in $\mathbb{E}^3$ there is no closed geodesic on bd C while on each sufficiently smooth proper convex body C in $\mathbb{E}^3$ there are infinitely many closed geodesics on bd C according to a famous theorem of Bangert [65] and Hingston [504].

In the following we prove the result of Klee mentioned above. Then an irregularity criterion will be shown and applied to the approximation of convex bodies.

What Does the Boundary of a Typical Convex Body Look Like?

An easy proof will yield the following answer.

Theorem 13.1. *Most proper convex bodies are smooth and strictly convex.*

Proof. Smoothness: For $n = 1, 2, \dots,$ let

$$\begin{aligned}\mathcal{C}_n =& \big\{C \in \mathcal{C}_p : \exists\, p \in \operatorname{bd} C,\ u, v \in S^{d-1} \text{ such that} \\ & \|u - v\| \geq \frac{1}{n},\ C \subseteq \{x : x \cdot u \leq p \cdot u\}, \{x : x \cdot v \leq p \cdot v\}\big\}.\end{aligned}$$

Simple compactness arguments show that

(1) $\mathcal{C}_n$ is closed in $\mathcal{C}_p$.

(It is sufficient to show that $C_1, C_2, \dots \in \mathcal{C}_n,\ C \in \mathcal{C}_p,\ C_1, C_2, \dots \to C$ implies that $C \in \mathcal{C}_n$ too.) To see that

(2) $\operatorname{int} \mathcal{C}_n = \emptyset$,

assume the contrary. Since the smooth convex bodies are dense in $\mathcal{C}_p$, the set $\mathcal{C}_n$ then would contain a smooth convex body, but this is incompatible with the definition of $\mathcal{C}_n$. (1) and (2) imply that $\mathcal{C}_n$ is nowhere dense. Hence

$$\bigcup_{n=1}^{\infty} \mathcal{C}_n \text{ is meagre.}$$

To conclude the proof of the smoothness assertion, note that

$$\big\{C \in \mathcal{C}_p : C \text{ is not smooth }\big\} = \bigcup_{n=1}^{\infty} \mathcal{C}_n.$$

Strict convexity: Replacing $\mathcal{C}_n$ by

$$\mathcal{D}_n = \big\{C \in \mathcal{C}_p : \exists\, p, q \in \operatorname{bd} C : \|p - q\| \geq \frac{1}{n},\ [p, q] \subseteq \operatorname{bd} C\big\},$$

the proof is similar to the proof in the smoothness case. □

Remark. This result has been refined and generalized in the following directions.

Most convex bodies are not of class $\mathcal{C}^{1+\varepsilon}$, see Gruber [423] ($\varepsilon = 1$) and Klima and Netuka [599] ($\varepsilon > 0$).

Most convex bodies are of class $\mathcal{C}^1$, but have quite unexpected curvature properties. For a multitude of pertinent results, mainly due to Zamfirescu, see the surveys [431, 1041].

Zamfirescu [1040] proved that all convex bodies are of class $\mathcal{C}^1$ and strictly convex, with a countable union of porous sets of exceptions. A porous set is meagre but the converse does not hold generally. For a definition see [431].

An Irregularity Criterion

If an approximation or iteration procedure is very fast or very slow for a dense set of elements of a space, what can be said for typical elements? The following result of Gruber [418] gives a Baire type answer.

Theorem 13.2. *Let $\mathcal{B}$ be a Baire space. Then the following statements hold:*

(i) *Let $\alpha_1, \alpha_2, \dots > 0$ and let $f_1, f_2, \dots : \mathcal{B} \to [0, +\infty)$ be continuous functions such that the set $\{x \in \mathcal{B} : f_n(x) = o(\alpha_n) \text{ as } n \to \infty\}$ is dense in $\mathcal{B}$. Then for most $x \in \mathcal{B}$ the inequality $f_n(x) < \alpha_n$ holds for infinitely many n.*

(ii) *Let $\beta_1, \beta_2, \dots > 0$ and let $g_1, g_2, \dots : \mathcal{B} \to [0, +\infty)$ be continuous functions such that the set $\{x \in \mathcal{B} : \beta_n = o(g_n(x)) \text{ as } n \to \infty\}$ is dense in $\mathcal{B}$. Then for most $x \in \mathcal{B}$ the inequality $\beta_n < g_n(x)$ holds for infinitely many n.*

Proof. (i) Since the functions f_n are continuous, the sets $\{x \in \mathcal{B} : f_n(x) \geq \alpha_n\}$ are closed. Hence

$$\mathcal{B}_n = \{x \in \mathcal{B} : f_n(x) \geq \alpha_n,\ f_{n+1}(x) \geq \alpha_{n+1}, \dots\} \text{ is closed.}$$

The assumption in (i) implies that $\operatorname{int} \mathcal{B}_n = \emptyset$. Thus, $\mathcal{B}_n$ is nowhere dense and therefore,

$$\bigcup_{n=1}^{\infty} \mathcal{B}_n = \{x \in \mathcal{B} : f_n(x) \geq \alpha_n \text{ for all but finitely many } n\} \text{ is meagre.}$$

This implies (i) on noting that

$$\mathcal{B} \backslash \bigcup_{n=1}^{\infty} \mathcal{B}_n = \{x \in \mathcal{B} : f_n(x) < \alpha_n \text{ for infinitely many } n\}.$$

(ii) Replacing $\alpha_n, f_n, \geq, <$ by $\beta_n, g_n, \leq, >$, the proof is similar. □

Remark. Clearly, instead of assuming that f_n and g_n are continuous, it is sufficient to assume that the f_n are upper and the g_n are lower semi-continuous.

For Most Convex Bodies Asymptotic Best Approximation is Irregular

Among numerous applications of the irregularity criterion, we consider one in the context of asymptotic best approximation of proper convex bodies. For other applications, see [418, 431].

Let δ^V be the symmetric difference metric. Given $C \in \mathcal{C}_p$, let $\mathcal{P}^c_{(n)} = \mathcal{P}^c_{(n)}(C)$ be the family of all convex polytopes with at most n facets which are circumscribed to C. Let

$$\delta^V(C, \mathcal{P}^c_{(n)}) = \inf\{\delta^V(C, P) : P \in \mathcal{P}^c_{(n)}\}.$$

The infimum is attained and the convex polytopes for which it is attained are called best approximating. For more information, see Sect. 11.2.

Corollary 13.1. *Let $\varphi, \psi : \mathbb{N} \to \mathbb{R}^+$ be such that $0 < \varphi(n), \psi(n) = o(n^{-\frac{2}{d-1}})$ as $n \to \infty$. Then for most convex bodies $C \in \mathcal{C}_p$ we have*

$$\delta^V(C, \mathcal{P}^c_{(n)}) < \varphi(n) \text{ for infinitely many } n,$$
$$\delta^V(C, \mathcal{P}^c_{(n)}) > \psi(n) \text{ for infinitely many } n.$$

Proof. Simple arguments dealing with best approximating polytopes show that

$$\delta^V\big(C, \mathcal{P}^c_{(n)}(C)\big) \text{ is continuous in } C (\in \mathcal{C}_p).$$

Trivially,

$$\delta^V\big(P, \mathcal{P}^c_{(n)}(P)\big) = 0 \text{ for } n \text{ sufficiently large, for each } P \in \mathcal{P}_p,$$

where $\mathcal{P}_p = \mathcal{P}_p(\mathbb{E}^d)$ is the space of all proper convex polytopes in $\mathbb{E}^d$, and the Theorem 11.4 on asymptotic best volume approximation shows that

$$\delta^V\big(C, \mathcal{P}^c_{(n)}(C)\big) \sim \frac{\text{const}}{n^{\frac{2}{d-1}}} \text{ as } n \to \infty$$
$$\text{for each } C \in \mathcal{C}_p \text{ of class } \mathcal{C}^2 \text{ with positive Gauss curvature,}$$

where const > 0 is a constant depending on C. Now apply the above irregularity result. □

13.2 Measures on $\mathcal{C}$?

As seen above, the topological concept of Baire categories is an effective tool to distinguish between small (meagre) and large (non-meagre) sets in $\mathcal{C}$ and $\mathcal{C}_p$. Considering this, the following problem arises:

Problem 13.1. *Define a geometrically useful measure on $\mathcal{C}$ or on $\mathcal{C}_p$ which is easy to handle.*

The spaces $\mathcal{C}$ and $\mathcal{C}_p$ are locally compact with respect to their common topologies. Thus there should be many measures available on these spaces. Unfortunately this is not so, at least so far. A conjecture of the author that Hausdorff measures with respect to the metric δ^H might do, was readily disproved by Schneider [901]. More general is a negative result of Bandt and Baraki [66]. In view of these results which indicate that a solution of the above problem might be difficult, it seems to be worth while to study the following problem:

Problem 13.2. *Given an interesting subset $\mathcal{D}$ of $\mathcal{C}$ or $\mathcal{C}_p$, for example the set of all proper convex bodies of class $\mathcal{C}^k$, or the set of all proper convex bodies with singular points, find non-decreasing functions $h, k : [0, +\infty) \to [0, +\infty)$ such that for the corresponding Hausdorff measures μ_h, μ_k with respect to a given metric δ on $\mathcal{C}$ or $\mathcal{C}_p$ we have*

$$\mu_h(\mathcal{D}) = 0, \ \ \mu_k(\mathcal{D}) > 0.$$

Here,

$$\mu_h(\mathcal{A}) = \lim_{\varepsilon \to +0} \left(\inf \left\{ \sum_{n=1}^{\infty} h(\operatorname{diam} \mathcal{U}_n) : \mathcal{U}_n \subseteq \mathcal{D},\ \operatorname{diam} \mathcal{U}_n \leq \varepsilon,\ \mathcal{A} \subseteq \bigcup_{n=1}^{\infty} \mathcal{U}_n \right\} \right),$$

where diam is the diameter with respect to the given metric δ.

In this section we prove the result of Bandt and Baraki.

Non-Existence of Isometry-Invariant Measures on $\langle \mathcal{C}, \delta^H \rangle$

A Borel measure μ on $\mathcal{C}$ is *isometry-invariant* with respect to δ^H if

> $\mu(\mathcal{D}) = \mu\big(I(\mathcal{D})\big)$ for all Borel sets $\mathcal{D} \subseteq \mathcal{C}$ and each isometry $I : \mathcal{C} \to \mathcal{C}$ with respect to δ^H.

These isometries have been determined by Gruber and Lettl [449], see Theorem 13.4 below. This result shows that there are few isometries of $\langle \mathcal{C}, \delta^H \rangle$ into itself. Thus the condition that a measure μ on $\mathcal{C}$ is isometry-invariant is not too restrictive. In spite of this we have the following negative result of Bandt and Baraki [66].

Theorem 13.3. *Let $d > 1$. Then there is no positive σ-finite Borel measure on $\mathcal{C}$ which is invariant with respect to all isometries of $\langle \mathcal{C}, \delta^H \rangle$ into itself.*

Proof. Assume that there is a positive σ-finite Borel measure μ on $\mathcal{C}$ which is isometry-invariant with respect to δ^H. Let

$$\mathcal{C}_n = \mathcal{C}(nB^d) = \{C \in \mathcal{C} : C \subseteq nB^d\} \text{ for } n = 1, 2, \dots$$

Since $\mathcal{C}$ is the union of the compact sets $\mathcal{C}_n$ and μ is positive, there is an n with

(1) $\mu(\mathcal{C}_n) > 0$.

For this n we have the following:

(2) $\mathcal{C}$ contains uncountably many pairwise disjoint isometric copies of $\mathcal{C}_n$.

To see this, we first show that

(3) the sets $\mathcal{C}_n + [o, p] = \{C + [o, p] : C \in \mathcal{C}_n\},\ p \in 3nS^{d-1}$, are pairwise disjoint.

If (3) did not hold, there are $p, q \in 3nS^{d-1},\ p \neq q$, such that $C+[o,p] = D+[o,q]$ for suitable $C, D \in \mathcal{C}_n$. Choose a linear form l on $\mathbb{E}^d$ such that $l(p) = 1,\ l(q) = 0$. Let $r \in C$ be such that

$$l(r) = \min\{l(x) : x \in C\}.$$

Choose $s \in D$ and $0 \leq \lambda \leq 1$, such that $r = s + \lambda q$. Choose $t \in C$ and $0 \leq \nu \leq 1$, such that $s + q = t + \nu p$. Then $t + \nu p - q + \lambda q = r$ and thus

$$l(t) + \nu = l(t + \nu p - q + \lambda q) = l(r) \leq l(t).$$

Hence $\nu = 0$ and thus $s + q = t$ and it follows that

$$3n = \|q\| = \|t - s\| \le \|t\| + \|s\| \le 2n.$$

This contradiction concludes the proof of (3). Now, noticing that the sets $\mathcal{C}_n + [o, p]$, $p \in 3nS^{d-1}$, all are isometric to $\mathcal{C}_n$, the proof of (2) is complete.

A simple, well-known measure-theoretic result says that a measure cannot be σ-finite if there is an uncountable family of pairwise disjoint sets with positive measure. This together with the isometry-invariance of μ and statements (1) and (2) implies that μ is not σ-finite, which yields the desired contradiction. □

13.3 On the Metric Structure of $\mathcal{C}$

There is a long list of metrics and other notions of distance on $\mathcal{C}$ and certain subspaces of it, including $\mathcal{C}_p$, that have been studied. See, e.g. [428,429]. The most import ones are the Hausdorff metric δ^H on $\mathcal{C}$, the symmetric difference metric δ^V on $\mathcal{C}_p$, and the Banach–Mazur distance on the space of all (equivalence classes with respect to non-singular linear transformations of) proper, o-symmetric convex bodies. While a good deal of all articles in convex geometry make use of such metrics, the metric spaces $\langle \mathcal{C}, \delta^H \rangle$, etc. per se have rarely been investigated. Of what is known, we mention estimates for ε-nets of $\langle \mathcal{C}(B^d), \delta^H \rangle$, $\langle \mathcal{C}_p(B^d), \delta^V \rangle$ and characterizations of isometries.

In the following two characterizations of isometries will be stated. The proofs are rather long and technical and thus are omitted. For related results, see the author [428].

Description of the Isometries of $\langle \mathcal{C}, \delta^H \rangle$

Refining earlier work of Schneider [902], Gruber and Lettl [449] showed the following result.

Theorem 13.4. *Let $I : \mathcal{C} \to \mathcal{C}$ be a mapping. Then the following statements are equivalent:*

(i) *I is an isometry with respect to δ^H.*
(ii) *There are a rigid motion $m : \mathbb{E}^d \to \mathbb{E}^d$ and a convex body D, such that*

$$I(C) = m\,C + D \text{ for } C \in \mathcal{C}.$$

Remark. The Hausdorff metric δ^H clearly may be extended to the space $\mathcal{K} = \mathcal{K}(\mathbb{E}^d)$ of all compact subsets of $\mathbb{E}^d$. Generalizing the above result, Gruber and Lettl [448] characterized the isometries of $\langle \mathcal{K}, \delta^H \rangle$.

Description of the Isometries of $\langle \mathcal{C}_p, \delta^V \rangle$

A result of the author [415] is as follows.

Theorem 13.5. *Let $I : \mathcal{C}_p \to \mathcal{C}_p$ be a mapping. Then the following statements are equivalent:*

(i) *I is an isometry with respect to δ^V.*
(ii) *There is a volume-preserving affinity $a : \mathbb{E}^d \to \mathbb{E}^d$ such that*

$$I(C) = a\,C \text{ for } C \in \mathcal{C}_p.$$

Remark. A more general local version of this result has been obtained by Weisshaupt [1017].

13.4 On the Algebraic Structure of $\mathcal{C}$

The space $\mathcal{C}$ of convex bodies is an Abelian semigroup with respect to Minkowski addition $+$ on which the non-negative reals operate. Since $\langle\mathcal{C},+\rangle$ satisfies the cancellation law, it can be embedded into an Abelian group or, more precisely, into a vector space over $\mathbb{R}$. The embedding can be achieved by considering equivalence classes of pairs of convex bodies (as in the construction of the integers from the natural numbers) or via support functions. For some references to the large pertinent literature, see [428]. A different line of research deals with the characterization of homomorphisms with additional properties of $\langle\mathcal{C},+\rangle$ into itself, into $\langle\mathbb{E}^d,+\rangle$ and into $\langle\mathbb{R},+\rangle$. Major contributions of this type are due to Schneider.

The definitions

$$C\wedge D=C\cap D,\;\; C\vee D=\operatorname{conv}(C\cup D) \text{ for } C,D\in\mathcal{C}$$

make $\mathcal{C}$ into an atomic lattice $\langle\mathcal{C},\wedge,\vee\rangle$. It was investigated mainly by Belgian mathematicians. In addition, a characterization of the endomorphisms of $\langle\mathcal{C},\wedge,\vee\rangle$ has been given.

Linearity and lattice properties of $\mathcal{C}$ were used as axioms for so-called convexity spaces with the aim to raise convex geometry to a more general level of abstraction. Other attempts to define convexity spaces are based on combinatorial results such as the theorems of Carathéodory, Helly and Radon. See Sect. 3.2 and the references cited there.

In the following a result of Schneider [898] on homomorphisms of $\langle\mathcal{C},+\rangle$ into $\langle\mathbb{E}^d,+\rangle$ is presented first. It deals with the Steiner point or curvature centroid of convex bodies. Besides the centroid, the centres of the inscribed ellipsoid of maximum volume and the circumscribed ellipsoid of minimum volume and other points, the curvature centroid is one of the points which are assigned in a natural way to a convex body. Tools for Schneider's proof are spherical harmonics. We state the needed definitions and some properties of the latter. Finally, two results on endomorphisms of $\langle\mathcal{C},+\rangle$ and $\langle\mathcal{C},\wedge,\vee\rangle$ due to Schneider [900] and Gruber [424], respectively, are given without proof.

For more information and references to the original literature, see the articles and surveys of Schneider [899], McMullen and Schneider [716], Saint-Pierre [874] and the author [428] and the books of Schneider [907], Sect. 3.4 and Groemer [405], Sect. 5.8.

Spherical Harmonics

Spherical harmonics are important tools of analysis and, in particular, of pure and applied potential theory. Their first applications to problems of convex geometry date back to Hurwitz [532] and Minkowski [741]. For expositions in the spirit of geometry, see Seidel [924] and, in particular, the monograph of Groemer [405]. The Proceedings on Fourier analysis and convexity [343] and the book of Koldobsky [606] also contain many results in convexity and the geometry of numbers based on Fourier series, Fourier transforms, and spherical harmonics.

In the following we give the definitions and properties that will be used below.

A polynomial $h : \mathbb{E}^d \to \mathbb{R}$ is *harmonic* if it satisfies the Laplace equation

$$\Delta h = \frac{\partial^2 h}{\partial x_1^2} + \cdots + \frac{\partial^2 h}{\partial x_d^2} = 0.$$

The restriction of a homogeneous harmonic polynomial $h : \mathbb{E}^d \to \mathbb{R}$ to the unit sphere S^{d-1} is a *spherical harmonic*. For $n = 0, 1, \ldots,$ let $\mathcal{H}_n^d$ be the linear space of all spherical harmonics in d variables of degree n and let

$$\mathcal{H}^d = \mathcal{H}_0^d \oplus \mathcal{H}_1^d \oplus \mathcal{H}_2^d \oplus \cdots$$

be the linear space of all finite sums of spherical harmonics in d variables. Let σ denote the ordinary surface area measure in $\mathbb{E}^d$. By a *rotation* in $\mathbb{E}^d$ an orthogonal transformation with determinant 1 is meant and $\kappa_d = V(B^d)$.

Proposition 13.1. *We have the following statements:*

(i) $\dim \mathcal{H}_n^d = \dfrac{2n+d-2}{n+d-2}\dbinom{n+d-2}{d-2}$.

(ii) $\langle h, k\rangle = \displaystyle\int_{S^{d-1}} h(u)k(u)\,d\sigma(u) = 0$ *for* $h \in \mathcal{H}_m^d, k \in \mathcal{H}_n^d, m \neq n$.

(iii) *The spherical harmonics in* $\mathcal{H}_1^d$ *are the functions of the form*

$$u \to a \cdot u \text{ for } u \in S^{d-1}, \text{ where } a \in \mathbb{E}^d.$$

For the norm $\|\cdot\|_2$ *on* $\mathcal{H}_n^d$ *related to the inner product* $\langle\cdot,\cdot\rangle$, *we have*

$$\|u_i\|_2 = \kappa_d^{\frac{1}{2}} \text{ for } u = (u_1, \ldots, u_d) \in S^{d-1}.$$

(iv) *Let* $h \in \mathcal{H}_n^d$ *and* $r : \mathbb{E}^d \to \mathbb{E}^d$ *a rotation. Then* $rh \in \mathcal{H}_n^d$, *where* rh *is defined by* $rh(u) = h\big(r^{-1}(u)\big)$ *for* $u \in S^{d-1}$.

(v) *If* $\mathcal{H}$ *is a linear subspace of* $\mathcal{H}_n^d$ *that is invariant under rotations, then* $\mathcal{H} = \{0\}$ *or* $\mathcal{H}_n^d$.

(vi) *Let* $h \in \mathcal{H}^d$. *Then* $h + \alpha$ *is (the restriction to* S^{d-1} *of) the support function of a suitable convex body if* $\alpha > 0$ *is sufficiently large.*

(vii) *The family of all convex bodies the support functions of which are finite sums of spherical harmonics, i.e. are in* $\mathcal{H}^d$, *is dense in* $\mathcal{C}$.

The Steiner Point

The *Steiner point* or *curvature centroid* s_C of a convex body C is the point

$$s_C = \frac{1}{\kappa_d} \int\limits_{S^{d-1}} h_C(u) u \, d\sigma(u),$$

where the integral is to be understood componentwise. If C is a proper convex body of class $\mathcal{C}^2$, it can be shown that s_C is the centroid of mass distributed over bd C with density equal to the Gauss curvature of bd C. A nice application of the Steiner point to approximate matching of shapes was given by Aichholzer, Alt and Rote [5].

Homomorphisms of $\langle \mathcal{C}, + \rangle$ into $\langle \mathbb{E}^d, + \rangle$ and a Characterization of the Steiner Point

The following result was proved by Schneider [898].

Theorem 13.6. *Let* $s : \mathcal{C} \to \mathbb{E}^d$ *be a mapping. Then the following statements are equivalent:*

(i) $s(C) = s_C$ *for* $C \in \mathcal{C}$.

(ii) s *satisfies the properties:*

- (a) $s(C + D) = s(C) + s(D)$ *for* $C, D \in \mathcal{C}$
- (b) $s(mC) = ms(C)$ *for* $C \in \mathcal{C}$ *and all rigid motions* $m : \mathbb{E}^d \to \mathbb{E}^d$
- (c) $s(\cdot)$ *is continuous*

Proof. (i)$\Rightarrow$(ii) This is a simple exercise.

(ii)$\Rightarrow$(i) The first step is to show the following:

(1) Let $H : \mathcal{H}^d \to \mathbb{E}^d$ be a linear mapping such that $H(rh) = rH(h)$ for $h \in \mathcal{H}^d$ and each rotation $r : \mathbb{E}^d \to \mathbb{E}^d$. Then

$$H(h) = \beta \int\limits_{S^{d-1}} h(u) u \, d\sigma(u) \text{ for } h \in \mathcal{H}^d,$$

where β is a real constant depending only on d and H and the integral is considered componentwise.

To prove this, let $h = h_0 + h_1 + \cdots + h_m$ where $h_n \in \mathcal{H}_n^d$. From Proposition 13.1(ii) and the fact that each coordinate u_i of $u \in S^{d-1}$ is a spherical harmonic of degree 1 in $\mathcal{H}_1^d$ by Proposition 13.1(iii), it follows that

$$\int\limits_{S^{d-1}} h(u) u \, d\sigma(u) = \int\limits_{S^{d-1}} h_1(u) u \, d\sigma(u).$$

For the proof of (1) it is thus sufficient to show that

(2) $H(h_n) = \beta \int\limits_{S^{d-1}} h_1(u) u \, d\sigma(u)$ for $n = 1$ and $= o$ for $n \neq 1$.

We distinguish three cases:

$n = 0$: Since by the assumption of (1) we have, $H(h_0) = H(rh_0) = rH(h_0)$ for any rotation r, the point $H(h_0) \in \mathbb{E}^d$ is rotation invariant and thus is equal to o, concluding the proof of (2) for $n = 0$.

$n > 1$: Then

$$\dim \mathcal{H}_n^d = \frac{2n+d-2}{n+d-2}\binom{n+d-2}{d-2} \geq 2n+d-2 > d$$

by Proposition 13.1(i). Consider the linear subspace $\mathcal{H} = \{h \in \mathcal{H}_n^d : H(h) = o\}$ of $\mathcal{H}_n^d$. By the assumption of (1), $r\mathcal{H} = \mathcal{H}$ for each rotation r. Hence $\mathcal{H} = \{0\}$ or $\mathcal{H}_n^d$ by Proposition 13.1(v). Since $\mathcal{H}$ is the kernel of the linear mapping $H : \mathcal{H}_n^d \to \mathbb{E}^d$ and $\dim \mathcal{H}_n^d > d$, it follows that $\dim \mathcal{H} > 0$ and thus $\mathcal{H} = \mathcal{H}_n^d$ or, equivalently, $H(h) = o$ for each $h \in \mathcal{H}_n^d$. The proof of (2) for $n > 1$ is complete.

$n = 1$: By Proposition 13.1(iii), every $h \in \mathcal{H}_1^d$ is of the form $h(u) = a \cdot u$ for $u \in \mathbb{E}^d$ where $a \in \mathbb{E}^d$. Thus one can define a linear transformation $l : \mathbb{E}^d \to \mathbb{E}^d$ by

$$l(a) = H(a \cdot u) \in \mathbb{E}^d \text{ for } a \in \mathbb{E}^d.$$

Using the assumption in (1), it then follows that

$$\begin{aligned}(3)\quad (lr)(a) &= l\big(r(a)\big) = H\big((ra)\cdot u\big) = H\big(a\cdot(r^{-1}u)\big) = H\big(r(a\cdot u)\big)\\ &= rH(a\cdot u) = r\big(l(a)\big) = (rl)(a) \text{ for } a \in \mathbb{E}^d \text{ and all rotations } r,\end{aligned}$$

where in the expression $r(a \cdot u)$ it is assumed that r operates on u – note that $a \cdot u \in \mathcal{H}_1^d$. Hence the linear transformation $l : \mathbb{E}^d \to \mathbb{E}^d$ commutes with each rotation $r : \mathbb{E}^d \to \mathbb{E}^d$. This will be used to show that

(4) $l(a) = \gamma a$ for $a \in \mathbb{E}^d$, where γ is a suitable constant.

If every $a \in \mathbb{E}^d$ is an eigenvector of the linear transformation $l : \mathbb{E}^d \to \mathbb{E}^d$, then all eigenvalues are the same and Proposition (4) holds. Otherwise choose an $a \in \mathbb{E}^d$ such that $l(a)$ is not a multiple of a. Let r be a rotation such that $ra = a$ but $rl(a) \neq l(a)$. Then $rl(a) = lr(a) = l(a)$ by (3). This contradiction concludes the proof of (4). We now show that (2) holds for h_1 where $h_1(u) = a \cdot u$ according to Proposition 13.1(iii). Noting the definition of $l(\cdot)$, (4) and since for the Steiner point of the convex body $\{a\}$ we have $s_{\{a\}} = a$, it then follows that

$$\begin{aligned}H(h_1) &= H(a\cdot u) = l(a) = \gamma a = \gamma s_{\{a\}} = \frac{\gamma}{\kappa_d}\int\limits_{S^{d-1}} h_{\{a\}}(u)u\,d\sigma(u)\\ &= \frac{\gamma}{\kappa_d}\int\limits_{S^{d-1}} (a\cdot u)u\,d\sigma(u) = \frac{\gamma}{\kappa_k}\int\limits_{S^{d-1}} h_1(u)u\,d\sigma(u),\end{aligned}$$

where $h_{\{a\}}(u) = a \cdot u$ is the support function of the convex body $\{a\}$. This proves (2) for $n = 1$ where $\beta = \gamma/\kappa_d$. The proof of (2) and thus of (1) is complete.

In the second step a particular linear map $H : \mathcal{H}^d \to \mathbb{E}^d$ is constructed which satisfies the assumptions of (1). Let $s : \mathcal{C} \to \mathbb{E}^d$ satisfy (ii). Then

(5) s is positive homogeneous of degree 1.

To see this let $C \in \mathcal{C}$. By property (a), $s(C) = s(\frac{1}{l}C) + \cdots + s(\frac{1}{l}C)$ or $s(\frac{1}{l}C) = \frac{1}{l}s(C)$ for $l = 1, 2, \ldots$ This in turn implies that $s(\frac{k}{l}C) = \frac{k}{l}s(C)$ for $k, l = 1, 2, \ldots$ Hence $s(\lambda C) = \lambda s(C)$ for $\lambda \geq 0$ by property (c). The proof of (5) is complete. Now the construction of H is as follows: let $h \in \mathcal{H}^d$. Then $h = h_0 + \cdots + h_m$ where $h_n \in \mathcal{H}_n^d$. By Proposition 13.1(vi) there is a constant $\alpha > 0$ such that each of the functions $h_n + \alpha$ is (the restriction to S^{d-1} of) a support function of a convex body, say C_n. Now define

(6) $H(h) = s(C_0) + \cdots + s(C_m) - (m+1)s(\alpha B^d)$.

It is easy to see that $H(h)$ does not depend on the particular choice of α as long as all functions $h_n + \alpha$ are support functions of convex bodies. Thus $H(h)$ is well defined and maps $\mathcal{H}^d$ into $\mathbb{E}^d$. Next,

(7) H satisfies the assumptions of (1).

The definition of H in (6) together with (5) and property (a) imply that H is linear. By its definition H satisfies the equality $H(rh) = rH(h)$ for all $h \in \mathcal{H}^d$ and all rotations r. The proof of (7) is complete.

In the third step the aim is to show the following proposition.

(8) Let $C \in \mathcal{C}$ be such that $h_C = h_0 + \cdots + h_m$ with suitable $h_n \in \mathcal{H}_n^d$. Then $s(C) = s_C$.

Apply (6) with $h = h_C$ and note property (a). Then

$$H(h_C) + (m+1)s(\alpha B^d) = s(C_0) + \cdots + s(C_m) = s(C_0 + \cdots + C_m).$$

From $h_C + (m+1)\alpha = (h_0 + \alpha) + \cdots + (h_m + \alpha)$ it follows that $C + (m+1)\alpha B^d = C_0 + \cdots + C_m$. Thus property (a) shows that

$$\begin{aligned} s(C) + (n+1)s(\alpha B^d) &= s(C) + s\big((n+1)\alpha B^d\big) = s\big(C + (n+1)\alpha B^d\big) \\ &= s(C_0 + \cdots + C_n). \end{aligned}$$

It follows that $H(h_C) = s(C)$. Since by (7) H satisfies the assumptions in (1), Proposition (1) then shows that

(9) $$s(C) = H(h_C) = \beta \int\limits_{S^{d-1}} h_C(u)u\, d\sigma(u).$$

To determine β, let $e = (1, 0, \ldots, 0) \in \mathbb{E}^d$ and let $t : \mathbb{E}^d \to \mathbb{E}^d$ be the translation $x \to x + e$ for $x \in \mathbb{E}^d$. Then property (b) yields

$$s(\{e\}) + s(\{e\}) = s(\{e\} + \{e\}) = s\big(t(\{e\})\big) = ts(\{e\}) = s(\{e\}) + e.$$

Hence $s(\{e\}) = e$ and, applying (9) in the special case where $C = \{e\}$ and thus $h_{\{e\}}(u) = e \cdot u \in \mathcal{H}_1^d$, we find that

$$e = \beta \int\limits_{S^{d-1}} (e \cdot u)u\, d\sigma(u) = \beta e \int\limits_{S^{d-1}} u_1^2\, d\sigma(u) = \beta \kappa_d e$$

by Proposition 13.1(iii). Thus $\beta = 1/\kappa_d$ and (9) shows that

$$s(C) = \frac{1}{\kappa_d} \int\limits_{S^{d-1}} h_C(u) u \, d\sigma(u) = s_C,$$

concluding the proof of (8).

In the last step note that $s(\cdot)$ and s_C are continuous in C. Hence (8) implies that $s(C) = s_C$ for all $C \in \mathcal{C}$, i.e. Proposition (ii) holds. □

Remark. For a slight refinement of this result, see Posicel'skiĭ [814]. The problem remains whether one can relax the properties (b) and (c).

Description of the Continuous Endomorphisms of the Convex Cone $\langle \mathcal{C}, + \rangle$

The next result is due to Schneider [900].

Theorem 13.7. *Let $d \geq 3$ and let $E : \mathcal{C} \to \mathcal{C}$ be a mapping. Then the following statements are equivalent:*

(i) *$E(C) = C + \lambda(C - C)$ for $C \in \mathcal{C}$, where $\lambda \geq 0$ is a constant.*

(ii) *E has the following properties:*

  (a) *$E(C + D) = E(C) + E(D)$ for $C, D \in \mathcal{C}$*
  (b) *$E(aC) = aE(C)$ for $C \in \mathcal{C}$ and all surjective affinities $a : \mathbb{E}^d \to \mathbb{E}^d$*
  (c) *E is continuous*

Description of the Endomorphisms of the Lattice $\langle \mathcal{C}, \wedge, \vee \rangle$

By an *endomorphism* of the lattice $\langle \mathcal{C}, \wedge, \vee \rangle$ a mapping $E : \mathcal{C} \to \mathcal{C}$ is meant for which

$$E(C \wedge D) = E(C) \wedge E(D), \; E(C \vee D) = E(C) \vee E(D) \text{ for } C, D \in \mathcal{C}.$$

As a final result of this section we state a description by Gruber [424] of the endomorphisms of $\langle \mathcal{C}, \wedge, \vee \rangle$.

Theorem 13.8. *Let $d \geq 2$ and let $E : \mathcal{C} \to \mathcal{C}$ be a mapping. Then the following statements are equivalent:*

(i) *E is an endomorphism of $\langle \mathcal{C}, \wedge, \vee \rangle$.*

(ii) *For E one of the following hold:*

  (a) *$E(C) = D$ for $C \in \mathcal{C}$, where D is a fixed convex body*
  (b) *There is a surjective affinity $a : \mathbb{E}^d \to \mathbb{E}^d$ such that $E(C) = aC$ for $C \in \mathcal{C}$*

Remark. For a description of the endomorphisms in case $d = 1$, see [424].

Convex Polytopes

The early history of convex polytopes is lost. About 2000 BC convex polytopes appeared in a mathematical context in the Sumerian civilization, in Babylonia and in Egypt. Sources are the Moscow papyrus and the Rhind papyrus. Some of the regular polytopes were already known by then. A basic problem was to calculate the volumes of truncated pyramids. This was needed to determine the number of bricks for fortifications and buildings. Babylonians sometimes did the calculations correctly, sometimes not, while Egyptians used the right formula. For pertinent information the author is obliged to the assyriologist Hermann Hunger [531]. In the fifth century BC Democritos also discovered this formula and Eudoxos proved it, using the method of exhaustion. Theaitetos developed a theory of regular polytopes, later treated by Plato in the dialogue Timaios. Euclid, around 300 BC, considered metric properties of polytopes, the volume problem, including the exhaustion method, and the five regular polytopes, the Platonic solids. Zenodoros, who lived sometime between 200 BC and 90 AD, studied the isoperimetric problem for polygons and polytopes and Pappos, about 300 AD, dealt with the semi-regular polytopes of Archimedes. In the renaissance the study of convex polytopes was in the hands of artists such as Uccello, Pacioli, da Vinci, Dürer, and Jamnitzer. Then it went back to mathematics. Kepler investigated the regular and the semi-regular polytopes and planar tilings. Descartes considered convex polytopes from a metric point of view, almost arriving at Euler's polytope formula, discovered by Euler only hundred years later. Contributions to polytope theory in the late eighteenth and the nineteenth century are due to Legendre, Cauchy, Steiner, Schläfli and others. At the turn of the nineteenth and in the twentieth century important results were given by Minkowski, Dehn, Sommerville, Steinitz, Coxeter and numerous contemporaries. At present, emphasis is on the combinatorial, algorithmic, and algebraic aspects. Modern relations to other areas date back to Newton (polynomials), Fourier (linear optimization), Dirichlet, Minkowski and Voronoĭ (quadratic forms) and Fedorov (crystallography). In recent decades polytope theory was strongly stimulated and, in part, re-oriented by linear optimization, computer science and algebraic geometry. Polytope theory, in turn, had a certain impact on these areas. For the history, see Federico [318] and Malkevitch [681].

The material in this chapter is arranged as follows. After some preliminaries and the introduction of the face lattice, combinatorial properties of convex polytopes are considered, beginning with Euler's polytope formula. In Sect. 14 we treat the elementary volume as a valuation, and Hilbert's third problem. Next, Cauchy's rigidity theorem for polytopal convex surfaces and rigidity of frameworks are discussed. Then classical results of Minkowski, Alexandrov and Lindelöf are studied. Lindelöf's results deals with the isoperimetric problem for polytopes. Section 14 treats lattice polytopes, including results of Ehrhart, Reeve and Macdonald and the Betke–Kneser valuation theorem. Applications of lattice polytopes deal with irreducibility of polynomials and the Minding–Bernstein theorem on the number of zeros of systems of polynomial equations. Finally we present an account of linear optimization, including aspects of integer linear optimization.

For additional material the reader may wish to consult the books of Alexandrov [16], Grünbaum [453], McMullen and Shephard [718], Brøndsted [171], Ewald [315] and Ziegler [1045], the survey of Bayer and Lee [83] and other surveys in the Handbooks of Convex Geometry [475] and Discrete and Computational Geometry [476].

Regular polytopes and related topics will not be considered. For these we refer to Coxeter [230, 232], Robertson [842], McMullen and Schulte [717] and Johnson [551]. For McMullen's algebra of polytopes, see [713].

14 Preliminaries and the Face Lattice

The simple concept of a convex polytope embodies a wealth of mathematical structure and problems and, consequently, yields numerous results. The elementary theory of convex polytopes deals with faces and normal cones, duality, in particular polarity, separation and other simple notions. It was developed in the late eighteenth, the nineteenth and the early twentieth century. Some of the results are difficult to attribute. In part this is due to the large number of contributors.

In this section we first give basic definitions, and then show the equivalence of the notions of $\mathcal{V}$-polytopes and $\mathcal{H}$-polytopes and, similarly, of $\mathcal{V}$- and $\mathcal{H}$-polyhedra. We conclude with a short study of the face lattice of a convex polytope using polarity.

For more information, see the books cited earlier, to which we add Schneider [907] and Schrijver [915].

14.1 Basic Concepts and Simple Properties of Convex Polytopes

In the following we introduce the notion of convex polytopes and describe two alternative ways to specify convex polytopes: as convex hulls ($\mathcal{V}$-polytopes) and as intersections of halfspaces ($\mathcal{H}$-polytopes). An example deals with a result of Gauss on zeros of polynomials.

Convex Polytopes and Faces

A *convex polytope* P in $\mathbb{E}^d$ is the convex hull of a finite, possibly empty, set in $\mathbb{E}^d$. If $P = \operatorname{conv}\{x_1, \dots, x_n\}$, then the extreme points of P are among the points $x_1, \dots, x_n$ by Minkowski's theorem 5.5 on extreme points. Thus, a convex polytope has only finitely many extreme points. If, conversely, a convex body has only finitely many extreme points, then it is the convex hull of these, again by Minkowski's theorem. Hence, the convex polytopes are precisely the convex bodies with finitely many extreme points. The intersection of P with a support hyperplane H is a *face* of P. It is not difficult to prove that

(1) $$H \cap P = \operatorname{conv}(H \cap \{x_1, \dots, x_n\}).$$

This shows that a face of P is again a convex polytope. The faces of dimension 0 are the *vertices* of P. These are precisely the extreme, actually the exposed points of P. The faces of dimension 1 are the *edges* of P and the faces of dimension $\dim P - 1$ the *facets*. The empty set $\emptyset$ and P itself are called the *improper faces*, all other faces are *proper*. Since a face of P is the convex hull of those points among $x_1, \dots, x_n$ which are contained in it, P has only finitely many faces. Since each boundary point of P is contained in a support hyperplane by Theorem 4.1, $\operatorname{bd} P$ is the union of all proper faces of P. By $\mathcal{F}(P)$ we mean the family of all faces of P, including $\emptyset$ and P. The *space of all convex polytopes* in $\mathbb{E}^d$ is denoted by $\mathcal{P} = \mathcal{P}(\mathbb{E}^d)$ and $\mathcal{P}_p = \mathcal{P}_p(\mathbb{E}^d)$ is its sub-space consisting of all *proper* convex polytopes, that is those with non-empty interior.

Gauss's Theorem on the Zeros of the Derivative of a Polynomial

In an appendix to his third proof of the fundamental theorem of algebra, Gauss [363] proved the following result.

Theorem 14.1. *Let p be a polynomial in one complex variable. Then the zeros of its derivative p' are contained in the convex polygon determined by the zeros of p.*

Proof. Let $z_1, \dots, z_n \in \mathbb{C}$ be the zeros of p, each written according to its multiplicity. Then

$$p(z) = a(z - z_1) \cdots (z - z_n) \text{ for } z \in \mathbb{C}$$

with suitable $a \in \mathbb{C}$. Let $z \neq z_1, \dots, z_n$. Dividing the derivative $p'(z)$ by $p(z)$ implies that

(2) $$\frac{p'(z)}{p(z)} = \frac{1}{z - z_1} + \cdots + \frac{1}{z - z_n} = \frac{\bar{z} - \bar{z}_1}{|z - z_1|^2} + \cdots + \frac{\bar{z} - \bar{z}_n}{|z - z_n|^2}.$$

Assume now that z is a zero of p'. We have to show that $z \in \operatorname{conv}\{z_1, \dots, z_n\}$. If z is equal to one of $z_1, \dots, z_n$, this holds trivially. Otherwise (2) shows that

$$z = \frac{\frac{1}{|z-z_1|^2} z_1 + \cdots + \frac{1}{|z-z_n|^2} z_n}{\frac{1}{|z-z_1|^2} + \cdots + \frac{1}{|z-z_n|^2}}.$$

Thus z is a convex combination of $z_1, \dots, z_n$. □

$\mathcal{V}$-Polytopes and $\mathcal{H}$-Polytopes

Convex polytopes as defined earlier are also called *convex $\mathcal{V}$-polytopes*. Here $\mathcal{V}$ stands for vertices. Dually, a *convex $\mathcal{H}$-polyhedron* is the intersection of finitely many closed halfspaces. A bounded convex $\mathcal{H}$-polyhedron is called a *convex $\mathcal{H}$-polytope*.

A formal proof of the following folk theorem is due to Weyl [1021], see also Minkowski [744], Sect. 4.

Theorem 14.2. *Let $P \subseteq \mathbb{E}^d$. Then the following statements are equivalent:*

(i) *P is a convex $\mathcal{V}$-polytope.*

(ii) *P is a convex $\mathcal{H}$-polytope.*

Proof. (i)⇒(ii) We may suppose that $\dim P = d$. P has only finitely many faces. Since each boundary point of P is contained in a support hyperplane of P by Theorem 4.1, bd P is the union of its proper faces. Connecting an interior point of P with a line segment which misses all faces of dimension at most $d-2$ with an exterior point, each point where it intersects the boundary of P must be contained in a face of dimension $d-1$, i.e. in a facet. Thus P must have facets. Let H_i, $i = 1, \dots, m$, be the hyperplanes containing the facets of P and H_i^- the corresponding support halfspaces. We claim that

$$P = H_1^- \cap \dots \cap H_m^-. \tag{3}$$

The inclusion $P \subseteq H_1^- \cap \dots \cap H_m^-$ is trivial. To show the reverse inclusion, let $x \in \mathbb{E}^d \backslash P$. For each of the finitely many faces of P of dimension at most $d-2$, consider the affine hull of the face and x. Choose a point $y \in \operatorname{int} P$ which is contained in none of these affine hulls. The intersection of the line segment $[x, y]$ with bd P then is a point $z \in \operatorname{bd} P$ which is contained in none of these affine hulls and thus in none of the faces of P of dimension at most $d-2$. Since bd P is the union of all faces, z is contained in a suitable facet and thus in one of the hyperplanes, say H_i. Then $x \notin H_i^-$ and therefore $x \notin H_1^- \cap \dots \cap H_m^-$. Hence $P \supseteq H_1^- \cap \dots \cap H_m^-$, concluding the proof of (3).

(ii)⇒(i) Let $P = H_1^- \cap \dots \cap H_m^-$ be bounded, where each H_i^-, $i = 1, \cdots, m$, is a halfspace with boundary hyperplane H_i. Clearly, P is a convex body. By Minkowski's theorem 5.5, P is the convex hull of its extreme points. To conclude the proof that P is a convex polytope, it is thus sufficient to show that P has only finitely many extreme points. To see this, it is sufficient to prove the following proposition:

(4) Let e be an extreme point of P. Then e is the intersection of a sub-family of $\{H_1, \dots, H_m\}$.

By re-indexing, if necessary, we may assume that

$$e \in H_1, \dots, H_k, \operatorname{int} H_{k+1}^-, \dots, \operatorname{int} H_m^-.$$

It is sufficient to show that

(5) $$\{e\} = H_1 \cap \cdots \cap H_k.$$

If this did not hold, the flat $H = H_1 \cap \cdots \cap H_k$ has dimension at least 1 and we could choose $u, v \in H$, int $H_{k+1}^-, \dots,$ int H_m^-, $u, v \neq e$, and such that $e = \frac{1}{2}(u + v)$. Then $u, v \in P$ and we obtain a contradiction to the assumption that e is an extreme point of P. This proves (5) and thus concludes the proof of (4). □

Corollary 14.1. *Let $P, Q \in \mathcal{P}$. Then $P \cap Q \in \mathcal{P}$.*

14.2 Extension to Convex Polyhedra and Birkhoff's Theorem

Many combinatorial and geometric results on convex polytopes have natural extensions to convex polyhedra.

In the following we generalize some of the definitions and the main result of the last section to convex polyhedra. As a tool, which will be needed later, we give a simple representation of normal cones of polyhedra. An application of the latter is a short geometric proof of Birkhoff's theorem 5.7 on doubly stochastic matrices.

$\mathcal{V}$-Polyhedra and $\mathcal{H}$-Polyhedra

A set P in $\mathbb{E}^d$ is a *convex $\mathcal{V}$-polyhedron* if there are finite sets $\{x_1, \dots, x_m\}$ and $\{y_1, \dots, y_n\}$ in $\mathbb{E}^d$ such that

$$\begin{aligned} P &= \operatorname{conv}\{x_1, \dots, x_m\} + \operatorname{pos}\{y_1, \dots, y_n\} \\ &= \big\{\lambda_1 x_1 + \cdots + \lambda_m x_m : \lambda_i \geq 0,\ \lambda_1 + \cdots + \lambda_m = 1\big\} \\ &\quad + \big\{\mu_1 y_1 + \cdots + \mu_n y_n : \mu_j \geq 0\big\} \\ &= Q + \bigcup \{\mu R : \mu \geq 0\} = Q + C, \end{aligned}$$

where

$$\begin{aligned} Q &= \big\{\lambda_1 x_1 + \cdots + \lambda_m x_m : \lambda_i \geq 0,\ \lambda_1 + \cdots + \lambda_m = 1\big\}, \\ R &= \big\{\mu_1 y_1 + \cdots + \mu_n y_n : \mu_j \geq 0,\ \mu_1 + \cdots + \mu_n = 1\big\}, \\ C &= \big\{\mu_1 y_1 + \cdots + \mu_n y_n : \mu_j \geq 0\big\}. \end{aligned}$$

Thus P is the sum of a convex polytope Q and a closed convex cone C with apex o and therefore a closed convex set.

A set P in $\mathbb{E}^d$ is a *convex $\mathcal{H}$-polyhedron* if it is the intersection of finitely many closed halfspaces, that is

$$P = \{x \in \mathbb{E}^d : Ax \leq b\},$$

where A is a real $m \times d$ matrix, $b \in \mathbb{E}^m$, and the inequality is to be understood componentwise. This definition abundantly meets the requirements of linear optimization. A bounded convex $\mathcal{H}$-polyhedron is a convex $\mathcal{H}$-polytope.

Convex cones which are polyhedra are called *polyhedral convex cones*. These may be represented as *convex $\mathcal{V}$-cones* and *convex $\mathcal{H}$-cones*. Clearly, polyhedral convex cones are closed.

Support hyperplanes and the notion of (bounded or unbounded) *faces*, in particular of *vertices, edges* and *facets*, of convex polyhedra are introduced as in the case of convex polytopes.

We now show that the notions of convex $\mathcal{V}$- and $\mathcal{H}$-polyhedra coincide, as in the case of convex polytopes.

Theorem 14.3. *Let $P \subseteq \mathbb{E}^d$. Then the following statements are equivalent:*

(i) *P is a convex $\mathcal{V}$-polyhedron.*
(ii) *P is a convex $\mathcal{H}$-polyhedron.*

Proof. The theorem will be proved in several steps:

(1) Let $C \subseteq \mathbb{E}^d$. Then the following are equivalent:

(i) C is a pointed convex $\mathcal{V}$-cone with apex o.
(ii) C is a pointed convex $\mathcal{H}$-cone with apex o.

(i)$\Rightarrow$(ii) Let $C = \{\mu_1 y_1 + \cdots + \mu_n y_n : \mu_j \geq 0\} = \bigcup\{\mu T : \mu \geq 0\}$, where $T = \{\mu_1 y_1 + \cdots + \mu_n y_n : \mu_j \geq 0,\ \mu_1 + \cdots + \mu_n = 1\}$ is a convex polytope. If $o \in T$, it must be a vertex of T. To see this, note that C is pointed, hence o is an extreme point of C and thus of T and therefore a vertex of T. Hence o is among the points $y_1, \dots, y_n$. Removing o, the convex hull of the remaining points is a convex polytope U such that $C = \bigcup\{\mu U : \mu \geq 0\}$ and $o \notin U$. Let H be a hyperplane which strictly separates o and U and let V be the radial projection of U into H with centre o. V is a convex $\mathcal{V}$-polytope and thus a convex $\mathcal{H}$-polytope in H by Theorem 14.2. Since $C = \bigcup\{\mu V : \mu \geq 0\}$, we easily see that C is a convex $\mathcal{H}$-polyhedron.

(ii)$\Rightarrow$(i) Let $K = \{x : |x_i| \leq 1\}$. Since C is pointed, o is an extreme point of the convex polytope $C \cap K$ and thus a vertex. Hence there is a support hyperplane H of $C \cap K$ with $(C \cap K) \cap H = \{o\}$. Since C is a cone with apex o, it follows that $C \cap H = \{o\}$. Let $p \in C \setminus \{o\}$. Then the $\mathcal{H}$-polyhedron $C \cap (H + p)$ is bounded. Otherwise it contains a ray $p + S$ with $o \in S \subseteq H$, say. Since C is a cone with apex o, we have $\mu(p + S) = \mu p + S \subseteq C$ for all $\mu > 0$. Since C is closed, also $S = 0p + S \subseteq C$, in contradiction to $C \cap H = \{o\}$. Since $C \cap (H + p)$ is bounded, it is an $\mathcal{H}$- and thus a $\mathcal{V}$-polytope by Theorem 14.2, say $C \cap (H + p) = \{\mu_1 y_1 + \cdots + \mu_n y_n : \mu_j \geq 0,\ \mu_1 + \cdots + \mu_n = 1\}$ with suitable $y_1, \dots, y_n \in C$. Since C is a convex cone with apex o and $C \cap H = \{o\}$, each ray in C with endpoint o meets $H + p$. Together this shows that

$$C = \bigcup\big\{\mu\big(C \cap (H + p)\big) : \mu \geq 0\big\} = \{\mu_1 y_1 + \cdots + \mu_n y_n : \mu_j \geq 0\}.$$

The proof of (1) is complete.

(2) Let $C \subseteq \mathbb{E}^d$. Then the following are equivalent:

(i) C is a convex $\mathcal{V}$-cone with apex o.
(ii) C is a convex $\mathcal{H}$-cone with apex o.

If (i) or (ii) hold, then C is a closed convex cone with apex o. Hence Proposition 3.3 shows that

$$C = (C \cap L^{\perp}) \oplus L,$$

where L is the linearity space of C and $C \cap L^{\perp}$ is a pointed closed convex cone with apex o. We may assume that $L = \{x : x_1 = \cdots = x_c = 0\}$, $L^{\perp} = \{x : x_{c+1} = \cdots = x_d = 0\}$. Then it is easy to see that

$$C \text{ is a convex } \left\{ \begin{matrix} \mathcal{V} \\ \mathcal{H} \end{matrix} \right\}\text{-cone} \Longleftrightarrow C \cap L^{\perp} \text{ is a convex } \left\{ \begin{matrix} \mathcal{V} \\ \mathcal{H} \end{matrix} \right\}\text{-cone.}$$

An application of (1) then implies (2).

(3) Let $P \subseteq \mathbb{E}^d$. Then the following are equivalent:

(i) P is a convex $\mathcal{V}$-polyhedron.

(ii) P is a convex $\mathcal{H}$-polyhedron.

Embed $\mathbb{E}^d$ into $\mathbb{E}^{d+1}$ as usual and let $u = (o, 1) \in \mathbb{E}^{d+1}$. Consider $P + u \subseteq \mathbb{E}^{d+1}$ and let C be the smallest closed convex cone with apex o containing $P + u$.

(i)⇒(ii) Let $P = \text{conv}\{x_1, \dots, x_m\} + \text{pos}\{y_1, \dots, y_n\}$. Then

$$\begin{aligned} C = \big\{ &(\lambda_1 x_1 + \cdots + \lambda_m x_m, \lambda) + \mu_1 y_1 + \cdots + \mu_n y_n : \\ &\lambda_i \geq 0,\ \lambda_1 + \cdots + \lambda_m = \lambda,\ \mu_j \geq 0 \big\} \end{aligned}$$

is a convex $\mathcal{V}$- and thus a convex $\mathcal{H}$-cone by (2). Then

$$P + u = C \cap \big\{ (x, z) : z \geq 1, -z \geq -1 \big\}$$

is a convex $\mathcal{H}$-polyhedron. This, in turn, shows that P is a convex $\mathcal{H}$-polyhedron.

(ii)⇒(i) Let $P = \{x : Ax \leq b\}$. Then

$$C = \big\{ (x, z) : z \geq 0,\ Ax - bz \leq o \big\}$$

is an $\mathcal{H}$-cone and thus a $\mathcal{V}$-cone by (2), say

$$C = \text{pos}\big\{ (x_1, 1), \dots, (x_m, 1), y_1, \dots, y_n \big\}$$

with suitable $x_i, y_i \in \mathbb{E}^d$. Then

$$\begin{aligned} P + u &= C \cap (\mathbb{E}^d + u) \\ &= \big\{ (\lambda_1 x_1 + \cdots + \lambda_m x_m, 1) + \mu_1 y_1 + \cdots + \mu_n y_n : \\ &\qquad \lambda_i \geq 0,\ \lambda_1 + \cdots + \lambda_m = 1,\ \mu_j \geq 0 \big\} \end{aligned}$$

and thus

$$\begin{aligned} P = \big\{ &\lambda_1 x_1 + \cdots + \lambda_m x_m + \mu_1 y_1 + \cdots + \mu_n y_n : \\ &\lambda_i \geq 0,\ \lambda_1 + \cdots + \lambda_m = 1,\ \mu_j \geq 0 \big\}. \end{aligned}$$

The proof of (3) and thus of the theorem is complete. □

Generalized Convex Polyhedra

For the geometric theory of positive definite quadratic forms, see Sect. 29.4, for Dirichlet-Voronoĭ tilings, see Sect. 32.1, and in other contexts, a more general notion of convex polyhedron is needed: call a closed convex set in $\mathbb{E}^d$ a *generalized convex polyhedron* if its intersection with any convex polytope is a convex polytope. A generalized convex polyhedron looks locally like a convex polyhedron, but may have countably many different faces. An example of a generalized convex polygon is the set

$$\operatorname{conv}\left\{(u, v) \in \mathbb{Z}^2 : v \geq u^2\right\} \subseteq \mathbb{E}^2.$$

Normal Cones of Polytopes and Polyhedra

Let P be a convex polyhedron and p a boundary point of P. The *normal cone* $N_P(p)$ of P at p is the closed convex cone of all exterior normal vectors of support hyperplanes of P at p, that is,

$$N_P(p) = \left\{u : u \cdot x \leq u \cdot p \text{ for all } x \in P\right\}.$$

For later reference we need the following result, where the a_i are the row vectors of the $n \times d$ matrix A, the β_i the components of the vector $b \in \mathbb{E}^n$, and the inequality $Ax \leq b$ is to be understood componentwise. We consider the row vectors a_i as vectors in $\mathbb{E}^d$ and thus write $a_i \cdot u$ instead of $a_i^T \cdot u$ for the inner product. Using the matrix product and considering a_i as a row vector, we also write the inner product in the form $a_i u$.

Proposition 14.1. *Let $P = \{x : Ax \leq b\}$ be a convex polyhedron and $p \in \operatorname{bd} P$. Assume that $a_i p = \beta_i$ precisely for $i = 1, \dots, k$. Then*

$$N_P(p) = \operatorname{pos}\left\{a_1, \dots, a_k\right\}. \tag{4}$$

If p is a vertex of P, then

$$\dim N_P(p) = d. \tag{5}$$

Proof. Since $N_P(p)$ depends on P only locally and is translation invariant, we may assume that $p = o$. Then

$$N_P(p) = N_C(o), \text{ where } C \text{ is the closed convex cone } \{x : Bx \leq o\}$$

and B is the $k \times d$ matrix consisting of the rows $a_1, \dots, a_k$. Since each of the hyperplanes $\{x : a_i \cdot x = a_i x = 0\}$ supports C at o, it follows that $N_C(o) \supseteq \{a_1, \dots, a_k\}$ and thus

$$N_C(o) \supseteq \operatorname{pos}\left\{a_1, \dots, a_k\right\}.$$

To show that equality holds, assume that, on the contrary, there is $c \in N_C(o) \setminus \operatorname{pos}\{a_1, \dots, a_k\}$. Choose a hyperplane $\{x : u \cdot x = \alpha\}$ which strictly separates c and the closed convex cone $\operatorname{pos}\{a_1, \dots, a_k\}$, say $u \cdot c > \alpha > u \cdot x$ for each x of the form $x = \lambda_1 a_1 + \cdots + \lambda_k a_k$, $\lambda_i \geq 0$. Since this holds for all $\lambda_i \geq 0$, it follows that $u \cdot c > 0 \geq u \cdot a_i = a_i \cdot u$ for $i = 1, \ldots, k$ and thus $u \in C$. From $u \in C$ and $c \in N_C(o)$ we conclude that $u \cdot c \leq 0$, a contradiction. Hence, equality holds and the proof of (4) is complete.

Now suppose that p is a vertex of P. If (5) did not hold, then by (4) we may choose $x \neq o$ such that

$$a_i \cdot (p \pm x) = a_i \cdot p \pm a_i \cdot x = a_i \cdot p + 0 = \beta_i \text{ for } i = 1, \ldots, k,$$

while still

$$a_i \cdot (p \pm x) < \beta_i \text{ for } i = k+1, \ldots, m.$$

Then $p \pm x \in P$. Hence p is not extreme and thus cannot be a vertex of P. This contradiction concludes the proof of (5). □

Birkhoff's Theorem on Doubly Stochastic Matrices

Proposition 14.1 yields an easy geometric proof of Birkhoff's theorem on doubly stochastic matrices, compare Barvinok [80]. For definitions and a different proof see Sect. 5.3.

Theorem 14.4. *The set Ω_d of all doubly stochastic $d \times d$ matrices is a convex polytope in $\mathbb{E}^{d^2}$, the vertices of which are precisely the $d \times d$ permutation matrices.*

Proof. If $d = 1$, the result is obvious. Assume now that $d > 1$ and that it holds for $d - 1$.

Ω_d may be interpreted as the subset of $\mathbb{E}^{d^2}$ defined by the following equalities and inequalities

$$\sum_i x_{ij} = 1,\ \sum_j x_{ij} = 1,\ x_{ij} \geq 0 \text{ for } i, j = 1, \ldots, d.$$

Since this subset is bounded, it is a convex polytope. Consider the affine sub-space $\mathcal{S}$ of $\mathbb{E}^{d^2}$ defined by the $2d$ hyperplanes

$$\sum_i x_{ij} = 1,\ \sum_j x_{ij} = 1 \text{ for } i, j = 1, \ldots, d.$$

To see that $\dim \mathcal{S} = (d-1)^2$, note that among the coefficient matrices

$$\begin{pmatrix} 1, 0, \ldots, 0 \\ 1, 0, \ldots, 0 \\ \cdots\cdots\cdots \\ 1, 0, \ldots, 0 \end{pmatrix}, \ldots, \begin{pmatrix} 0, 0, \ldots, 1 \\ 0, 0, \ldots, 1 \\ \cdots\cdots\cdots \\ 0, 0, \ldots, 1 \end{pmatrix}, \begin{pmatrix} 1, 1, \ldots, 1 \\ 0, 0, \ldots, 0 \\ \cdots\cdots\cdots \\ 0, 0, \ldots, 0 \end{pmatrix}, \ldots, \begin{pmatrix} 0, 0, \ldots, 0 \\ 0, 0, \ldots, 0 \\ \cdots\cdots\cdots \\ 1, 1, \ldots, 1 \end{pmatrix}$$

of the $2d$ hyperplanes there are precisely $2d - 1$ linearly independent ones. Ω_d is a proper convex polytope in the $(d-1)^2$-dimensional affine sub-space $\mathcal{S}$, defined by the

d^2 inequalities $x_{ij} \geq 0$. Now let $V = (v_{ij})$ be a vertex of Ω_d. By Proposition 14.1 V is the intersection of dim $\Omega_d = (d-1)^2$ defining support planes of Ω_d. Thus $v_{ij} = 0$ for some $(d-1)^2$ entries of V. The doubly stochastic matrix V cannot have a row consisting only of zeros and if every row would contain at least two non-zero entries, V would contain at most $d(d-2) < (d-1)^2$ zero entries, a contradiction. Therefore V has a row with precisely one non-zero entry. This entry must then be 1. In the column which contains this entry, all other entries are 0. Cancelling this row and column we get a $(d-1) \times (d-1)$ doubly stochastic matrix W, say. If W were not a vertex of Ω_{d-1}, it could be represented as the midpoint of two distinct matrices in Ω_{d-1}. This also implies that V could be represented as the midpoint of two distinct matrices in Ω_d, a contradiction. Thus W is a vertex of Ω_{d-1} and therefore a permutation matrix in Ω_{d-1} by induction. This then implies that V is a permutation matrix in Ω_d.

Conversely, each permutation matrix in Ω_d is extreme, and since Ω_d is a polytope, it is a vertex. The induction is complete. □

14.3 The Face Lattice

The family of all faces of a convex polytope is an (algebraic) lattice with special properties.

In this section, we study this lattice and show that it is atomic, co-atomic and complemented. A tool to show this is polarity.

For more information we refer to Grünbaum [453], Schrijver [915] and Ziegler [1045]. A standard treatise on lattice theory is Grätzer [390].

The Face Lattice of a Convex Polytope

Given $P \in \mathcal{P}$, denote by $\mathcal{F}(P)$ the family of all faces of P, including the improper faces $\emptyset$ and P. In a first result we describe the simple relation between faces and faces of faces.

Theorem 14.5. *Let $P \in \mathcal{P}$. Then the following statements hold:*

(i) *Let $F \in \mathcal{F}(P)$ and $G \in \mathcal{F}(F)$. Then $G \in \mathcal{F}(P)$.*
(ii) *Let $F, G \in \mathcal{F}(P)$ and $G \subseteq F$. Then $G \in \mathcal{F}(F)$.*

Proof. We may assume that $o \in G \subsetneq F$.

(i) Choose

$$H_F = \{x : u \cdot x = 0\},\ H = \{x \in H_F : v \cdot x = 0\},$$

such that H_F is a support hyperplane of P in $\mathbb{E}^d$ and H a support hyperplane of F in H_F with

$$F = H_F \cap P,\ G = H \cap F.$$

Here u and v are exterior normal vectors. For $\delta > 0$ the hyperplane

$$H_G = \{x : (u + \delta v) \cdot x = 0\}$$

intersects F along G. If $\delta > 0$ is sufficiently small, then all vertices of P not in G are in int H_G^-. Hence H_G is a support hyperplane of P with $G = H_G \cap P$. This shows that G is a face of P.

(ii) If $G = \emptyset, F$, we are done. If not, choose support hyperplanes H_F, H_G of P such that $F = H_F \cap P$, $G = H_G \cap P$. This, together with $G \subsetneq F$, implies that $H_F \neq H_G$ and $H_F \cap H_G$ is a hyperplane in H_F. Since H_G supports P and $F \subseteq P$, H_F and $F \not\subseteq H_G$, the hyperplane $H_F \cap H_G$ in H_F supports F. Since $G = H_G \cap P$, $G \subseteq F \subseteq P$, and $F \subseteq H_F$, we have $G = H_G \cap P = H_G \cap F = H_G \cap H_F \cap F$. Hence G is a face of F. □

The next result shows that $\mathcal{F}(P)$ is a lattice.

Theorem 14.6. *Let $P \in \mathcal{P}$ and define binary operations $\wedge$ (intersection) and $\vee$ (join) on $\mathcal{F}(P)$ as follows:*

$$(1)\quad F \wedge G = F \cap G,$$
$$F \vee G = \bigcap \big\{H \in \mathcal{F}(P) : F, G \subseteq H\big\} \text{ for } F, G \in \mathcal{F}(P).$$

Then $\big\langle \mathcal{F}(P), \wedge, \vee \big\rangle$ is a lattice with zero $\emptyset$ and unit element P, the face lattice *of P.*

Proof. A finite family of sets, which is closed with respect to intersection and contains the empty set $\emptyset$ and the union of all sets, is a lattice with respect to the operations $\wedge, \vee$ as defined in (1). The zero element of this lattice is $\emptyset$ and the unit element is the union of all sets. For the proof of the theorem it is thus sufficient to show the following.

$$(2)\qquad \text{Let } F, G \in \mathcal{F}(P). \text{ Then } F \cap G \in \mathcal{F}(P).$$

If $F \cap G = \emptyset$ or $F = G$ or $F = P$ or $G = P$, we are done. Otherwise we may assume that $o \in F \cap G$. Choose support hyperplanes

$$H_F = \{x : u \cdot x = 0\},\ H_G = \{x : v \cdot x = 0\}$$

of P, such that

$$F = H_F \cap P,\ G = H_G \cap P.$$

Since $F \neq G$, we have $H_F \neq H_G$. Thus $H_F \cap H_G$ is a support plane (of dimension $d - 2$) of P with

$$F \cap G = (H_F \cap H_G) \cap P.$$

The hyperplane

$$H = \{x : (u + v) \cdot x = 0\}$$

is a support hyperplane and

$$H \cap P = (H_F \cap H_G) \cap P = F \cap G,$$

concluding the proof of (2) and thus of the theorem. □

The Face Lattices of P and P^* are Anti-Isomorphic

We first present a tool, where

$$P^* = \{y : x \cdot y \leq 1 \text{ for } x \in P\}$$

is the polar of P, see Sect. 9.1.

Proposition 14.2. *Let $P \in \mathcal{P}$ such that $o \in \operatorname{int} P$. Then the following statements hold:*

(i) $P^* \in \mathcal{P}$.
(ii) $P^{**} = P$.

Proof. (i) Represent P in the form $P = \operatorname{conv}\{x_1, \dots, x_n\}$. Then

$$\begin{aligned} P^* &= \{y : (\lambda_1 x_1 + \cdots + \lambda_n x_n) \cdot y \leq 1 \text{ for } \lambda_i \geq 0,\ \lambda_1 + \cdots + \lambda_n = 1\} \\ &= \{y : x_i \cdot y \leq 1 \text{ for } i = 1, \dots, n\} = \bigcap_{i=1}^{n} \{y : x_i \cdot y \leq 1\} \end{aligned}$$

is an intersection of finitely many closed halfspaces. Since $o \in \operatorname{int} P$, there is $\varrho > 0$ such that $\varrho B^d \subseteq P$. The definition of polarity then easily yields $P^* \subseteq (\varrho B^d)^* = (1/\varrho) B^d$. Thus P^* is bounded. Since P^* is the intersection of finitely many halfspaces and is bounded, Theorem 14.2 implies that $P^* \in \mathcal{P}$.

(ii) To show that $P \subseteq P^{**}$, let $x \in P$. Then $x \cdot y \leq 1$ for all $y \in P^*$ by the definition of P^*. This, in turn, implies that $x \in P^{**}$ by the definition of P^{**}. To show that $P \supseteq P^{**}$, let $x \in \mathbb{E}^d \backslash P$. The separation theorem 4.4 then provides a point $y \in \mathbb{E}^d$ such that $x \cdot y > 1$ while $z \cdot y \leq 1$ for all $z \in P$. Hence $y \in P^*$. From $x \cdot y > 1$ we then conclude that $x \notin P^{**}$ by the definition of P^{**}. The proof that $P = P^{**}$ is complete. □

The following result relates the face lattices of P and P^*, where $\dim \emptyset = -1$.

Theorem 14.7. *Let $P \in \mathcal{P}$ such that $o \in \operatorname{int} P$. Define a mapping $^\diamond = {}^{\diamond_P}$ by*

$$F^\diamond = \{y \in P^* : x \cdot y = 1 \textit{ for } x \in F\} \textit{ for } F \in \mathcal{F}(P),$$

where, in particular, $\emptyset^\diamond = P^*$ and $P^\diamond = \emptyset$. Then the following statements hold:

(i) $^\diamond$ *is a one-to-one mapping of $\mathcal{F}(P)$ onto $\mathcal{F}(P^*)$.*
(ii) $\dim F^\diamond + \dim F = d - 1$ *for each $F \in \mathcal{F}(P)$.*
(iii) $^\diamond$ *is an* anti-isomorphism *of the lattice $\langle \mathcal{F}(P), \wedge, \vee \rangle$ onto the lattice $\langle \mathcal{F}(P^*), \wedge, \vee \rangle$.*

By the latter we mean that $^\diamond$ is one-to-one and onto and

$$(F \wedge G)^\diamond = F^\diamond \vee G^\diamond \text{ and } (F \vee G)^\diamond = F^\diamond \wedge G^\diamond \text{ for } F, G \in \mathcal{F}(P).$$

Proof. The first step is to show the following:

(3) Let $F \in \mathcal{F}(P)$. Then $F^\lozenge \in \mathcal{F}(P^*)$.

Let $F = \operatorname{conv}\{x_1, \dots, x_k\}$. Then

$$F^\lozenge = \big\{y \in P^* : (\lambda_1 x_1 + \dots + \lambda_k x_k) \cdot y = 1 \text{ for } \lambda_i \geq 0,\ \lambda_1 + \dots + \lambda_k = 1\big\}$$
$$= \big\{y \in P^* : x_i \cdot y = 1 \text{ for } i = 1, \dots, k\big\} = \bigcap_{i=1}^{k} \{y \in P^* : x_i \cdot y = 1\}$$
$$= P^* \cap \bigcap_{i=1}^{k} \{y : x_i \cdot y = 1\}.$$

(If $F = \emptyset$, this shows that $F^\lozenge = P^*$ if, as usual, the intersection of an empty family of subsets of $\mathbb{E}^d$ is defined to be $\mathbb{E}^d$.) Each of the hyperplanes $\{y : x_i \cdot y = 1\}$ supports P^* (or does not meet P^*) by the definition of P^*. Thus $F^\lozenge$ is the intersection of k faces of P^* and hence is a face of P^* by Theorem 14.6, applied to P^*.

In the second step we prove that

(4) $\quad {}^\lozenge = {}^{\lozenge_P}$ is one-to-one and onto.

For this, it is sufficient to show that

(5) $\quad {}^{\lozenge_{P^*}}\,{}^{\lozenge_P} =$ identity and, dually, ${}^{\lozenge_P}\,{}^{\lozenge_{P^*}} =$ identity.

Let $F \in \mathcal{F}(P)$. We have to prove the equality $F = F^{\lozenge\lozenge}$. To show the inclusion $F \subseteq F^{\lozenge\lozenge}$, let $x \in F$. The definition of $F^\lozenge$ then implies that $x \cdot y = 1$ for all $y \in F^\lozenge$. Hence $x \in F^{\lozenge\lozenge}$ by the definition of $F^{\lozenge\lozenge}$. To show the reverse inclusion $F \supseteq F^{\lozenge\lozenge}$, let $x \in P = P^{**}$, $x \notin F$. Consider a support hyperplane $H = \{z : z \cdot y = 1\}$ of P such that $F = H \cap P$. Then $z \cdot y \leq 1$ for all $z \in P$. Hence $y \in P^*$. Further, $z \cdot y = 1$ precisely for those $z \in P$ for which $z \in F$ and thus, in particular, $x \cdot y < 1$. By the definition of $F^\lozenge$ we then have $y \in F^\lozenge$. This, together with $x \in P^{**}$ and $x \cdot y < 1$, finally yields that $x \notin F^{\lozenge\lozenge}$ by the definition of $F^{\lozenge\lozenge}$. The proof of the equality $F = F^{\lozenge\lozenge}$ and thus of (5) is complete. (5) implies (4).

The definition of ${}^{\lozenge_P}$ and proposition (4) readily imply the next statement.

(6) Let $F, G \in \mathcal{F}(P)$, $F \subsetneqq G$. Then $F^\lozenge \supsetneqq G^\lozenge$.

In the third step of the proof the following will be shown.

(7) Let $F \in \mathcal{F}(P)$. Then $\dim F^\lozenge = d - 1 - \dim F$.

If $F = \emptyset$ or P, then $F^\lozenge = P^*$, resp. $\emptyset$ and (7) holds trivially. Assume now that $F \neq \emptyset, P$. An argument similar to the one in the proof of Theorem 14.6 shows that F is a (proper or non-proper) face of a facet of P. Using this, a simple proof by induction implies that there is a sequence of faces of P, say $\emptyset = F_{-1}, F_0, \dots, F = F_k, \dots, F_{d-1}, F_d = P \in \mathcal{F}(P)$, such that

(8) $\emptyset = F_{-1} \subsetneqq F_0 \subsetneqq \dots \subsetneqq F = F_k \subsetneqq \dots \subsetneqq F_{d-1} \subsetneqq F_d = P$,
where $\dim F_i = i$.

Then (6) implies that

(9) $\quad P = F_{-1}^{\Diamond} \supsetneq F_0^{\Diamond} \supsetneq \cdots \supsetneq F^{\Diamond} = F_k^{\Diamond} \supsetneq \cdots \supsetneq F_{d-1}^{\Diamond} \supsetneq F_d^{\Diamond} = \emptyset.$

Propositions (4) and (6) and the fact that for faces $G, H \in \mathcal{F}(P)$ with $G \subsetneq H$ holds $\dim G < \dim H$ together imply that the sequences of inclusions (8) and (9) are compatible only if $\dim F_i^{\Diamond} = d - 1 - \dim F_i$ for $i = -1, 0, \dots, d$. Proposition (7) is the special case where $i = k$.

Finally, (i) and (ii) hold by (3), (4) and (7). Since by (3), (4) and (7) the mappings $^{\Diamond_P}$ and $^{\Diamond_{P^*}} = {}^{\Diamond_P^{-1}}$ are onto, one-to-one and inclusion reversing, both are anti-isomorphisms. This proves (iii). □

The Face Lattice of P is Atomic, Co-Atomic and Complemented

First, the necessary lattice-theoretic terminology is introduced. Given a lattice $\mathcal{L} = \langle \mathcal{L}, \wedge, \vee \rangle$ with 0 and 1, an *atom* of $\mathcal{L}$ is an element $a \neq 0$, such that there is no element of $\mathcal{L}$ strictly between 0 and a. $\mathcal{L}$ is *atomic* if each element of $\mathcal{L}$ is the join of finitely many atoms. The dual notions of *co-atom* and *co-atomic* are defined similarly with $0, \vee$ exchanged by $1, \wedge$. The lattice $\mathcal{L}$ is *complemented* if for each $l \in \mathcal{L}$ there is an element $m \in \mathcal{L}$ such that $l \wedge m = 0$ and $l \vee m = 1$.

Our aim is to show the following properties of the face lattice.

Theorem 14.8. *Let $P \in \mathcal{P}$. Then the following statements hold:*

(i) $\langle \mathcal{F}(P), \wedge, \vee \rangle$ *is atomic and co-atomic.*
(ii) $\langle \mathcal{F}(P), \wedge, \vee \rangle$ *is complemented.*

Proof. We may assume that $o \in \operatorname{int} P$. Let $P = \operatorname{conv}\{v_1, \dots, v_n\}$, where the v_i are the vertices of P. Let $F_1, \dots, F_m$ be the facets of P.

(i) Clearly, $v_1, \dots, v_n$ are the atoms and $F_1, \dots, F_m$ the co-atoms of $\mathcal{F}(P)$. Given $F \in \mathcal{F}(P)$, F is the convex hull of the vertices contained in it, say $v_1, \dots, v_k$. Then F is the smallest set in $\mathcal{F}(P)$ with respect to inclusion which contains $v_1, \dots, v_k$. Hence $F = v_1 \vee \cdots \vee v_k$. That is, $\mathcal{F}(P)$ is atomic. To see that it is co-atomic, let $F \in \mathcal{F}(P)$ and consider the face $F^{\Diamond} \in \mathcal{F}(P^*)$. We have just proved that the face lattice of any convex polytope is atomic. Thus, in particular, $\mathcal{F}(P^*)$ is atomic. This shows that $F^{\Diamond} = w_1 \vee \cdots \vee w_l$ for suitable vertices $w_1, \dots, w_l$ of P^*. Now apply $^{\Diamond_{P^*}}$ and take into account Theorem 14.7 and that $^{\Diamond_P \Diamond_{P^*}}$ is the identity to see that

$$F = F^{\Diamond_P \Diamond_{P^*}} = (w_1 \vee \cdots \vee w_k)^{\Diamond_{P^*}} = w_1^{\Diamond_{P^*}} \wedge \cdots \wedge w_l^{\Diamond_{P^*}}.$$

Since w_i is a vertex of P^*, the set $w_i^{\Diamond_{P^*}}$ is a face of P of dimension $d-1$ by Theorem 14.7 (i), i.e. a facet of P. Thus $\mathcal{F}(P)$ is co-atomic.

(ii) Let $F \in \mathcal{F}(P)$. If $F = \emptyset$ or P, then $G = P$, respectively, $\emptyset$ is a complement of F. Assume now that $F \neq \emptyset, P$. Let $G \in \mathcal{F}(P)$ be a maximal face disjoint from F. Clearly, $F \wedge G = \emptyset$. For the proof that $F \vee G = P$, it is sufficient to show that the set of vertices in F and G is not contained in a hyperplane. Assume that, on the contrary, there is a hyperplane H_{FG} which contains all vertices of F and G and thus F and G. Let H_G be a support hyperplane of P with $G = H_G \cap P$. Clearly $H_G \neq H_{FG}$. Rotate H_G keeping the $(d-2)$-dimensional plane $H_G \cap H_{FG}$ fixed to the first position, say H, where it contains a further vertex v of P, not in H_{FG}. Then H is a support hyperplane of P and $H \cap P$ is a face E of P with $E \supsetneq G$ and such that $E \cap H_{FG} = G$. Hence $E \cap F = \emptyset$. This contradicts the maximality of G. □

14.4 Convex Polytopes and Simplicial Complexes

A (finite) *simplicial complex* $\mathcal{C}$ in $\mathbb{E}^d$ is a family of finitely many simplices in $\mathbb{E}^d$ such that for any $S \in \mathcal{C}$ each face of S is also in $\mathcal{C}$ and for any $S, T \in \mathcal{C}$, the intersection $S \cap T$ is a face of both S and T. In topology a (convex or non-convex) polytope is defined to be the union of all simplices of a simplicial complex in $\mathbb{E}^d$, see Alexandroff and Hopf [9] or Maunder [698]. The problem arises, whether a convex polytope P in the sense of convex geometry can be obtained in this way. In some cases even more is demanded, the vertices of the simplices all should be vertices of P. If the latter holds, we speak of a *simplification* of P. That simplifications always exist seems to be well known. Anyhow, this result is used by several authors, including Macdonald [675], Ehrhart [292, 293] and Betke and Kneser [108] without further comment. Thus it is important in our context, see Sects. 19.1, 19.2 and 19.4. For convex polytopes a proof was communicated to the author by Peter Mani-Levitska [685] and only then we found the proof of Edmonds [286] in the literature. Edmonds's idea is also described by Lee [635, 636]. Since these proofs are essentially different, and each contains an interesting idea, both are presented.

Convex Polytopes have Simplifications

Our aim is to prove the following result.

Theorem 14.9. *Let $P \in \mathcal{P}$. Then P has a simplification.*

The *k-skeleton* k-skel P of a convex polytope P in $\mathbb{E}^d$ is the union of its faces of dimension $\leq k$. By a *simplification* of k-skel P we mean simplifications of the faces of P of dimension $\leq k$ which fit together at common sub-faces. Let $f_0(P)$ denote the number of vertices of P.

Proof (by Edmonds). We may assume that P is proper. By induction, the following will be shown.

(6) Let $k \in \{0, 1, \dots, d\}$. Then each face of P of dimension $\leq k$ has a simplification. These simplifications together form a simplification of k-skel P.

Consider a linear ordering of the set of vertices of P. Clearly, (6) holds for $k = 0$. Assume now, that $k > 0$ and that (6) holds for $k - 1$. We construct a simplification of k-skel P as follows. Let F be a face of P with $\dim F = k$. By the induction assumption, all proper faces of F have a simplification. Let p be the first vertex of

P in F. For any proper face G of F, which is disjoint from $\{p\}$, consider the convex hulls of p and the simplices of the simplification of G. Since p is not in the affine hull of G, these convex hulls are all simplices. As G ranges over the proper faces of F which do not contain p, the simplices thus obtained form a simplification of F.

If H is a proper face of F which contains p, then the simplification of F just constructed, if restricted to H, forms a simplification of H. Being the first vertex of P in F, the vertex p is also the first vertex of P in H. Hence the earlier simplification of H coincides with the simplification of H constructed in an earlier step of the induction.

The simplifications of the faces of P of dimension k thus fit together if such faces have a face of dimension less than k in common. Hence the simplifications of the faces of P of dimension k together form a simplification of k-skel P.

The induction and thus the proof of (6) is complete. The theorem is the case $k = d$ of (6). □

Proof (by Mani-Levitska). It is sufficient to prove the Theorem for proper convex polytopes. The proof is by (a strange) induction where the dimension d is variable.

(7) Let $k \in \mathbb{N}$ and let P be a proper convex polytope in (some) $\mathbb{E}^d$ such that $k = f_0(P) - d$. Then P has a simplification.

For $k = 1$, $f_0(P) - d = 1$, i.e. $f_0(P) = d + 1$. Since P is proper, it is a simplex and (7) holds trivially. Assume now that $k > 1$ and that (7) holds for $k-1$. We have to establish it for k. Let P be a proper convex polytope in $\mathbb{E}^d$ such that $f_0(P) - d = k$. Then $f_0(P) = d + k > d + 1$. Embed $\mathbb{E}^d$ into $\mathbb{E}^{d+1}$ as usual (first d coordinates) and let " $'$ " denote the orthogonal projection of $\mathbb{E}^{d+1}$ onto $\mathbb{E}^d$. Choose a proper convex polytope Q in $\mathbb{E}^{d+1}$ such that $Q' = P$, $f_0(Q) = f_0(P)$ and such that Q has precisely one vertex above each vertex of P and no other vertices. Let L be the lower side of bd Q with respect to the last coordinate. L is the union of certain faces of Q. Since $f_0(Q) - \dim Q = f_0(P) - (d + 1) < k$, the proper convex polytope Q by the induction assumption has a simplification, say $\mathcal{C}$. Then $\{S' : S \in \mathcal{C}, S \subseteq L\}$ is a simplification of P.

This concludes the induction and thus proves the theorem. □

15 Combinatorial Theory of Convex Polytopes

Euler's polytope formula of 1752, praised by Klee [592] as the first landmark in the combinatorial theory of convex polytopes, led to a voluminous literature both in topology and convex geometry. Major contributors in convex geometry are Schläfli, Eberhard, Brückner, Schoute, Dehn, Sommerville, Steinitz, Hadwiger, Alexandrov and numerous living mathematicians. Investigations deal with f-vectors, graphs and boundary complexes, algorithms and matroids. Several relations to linear optimization are known.

In this section we consider the Euler polytope formula and its converse due to Steinitz for $d = 3$, shellings and the Euler polytope formula for general d and its modern aftermath. Then the problem whether a graph can be realized as the edge graph of a convex polytope is studied. Balinski's theorem and the Perles–Blind–Mani theorem for simple graphs are given as well as the Steinitz representation theorem. Next we touch the problem whether a polytopal complex can be realized as the boundary complex of a convex polytope. Finally, combinatorial types of convex polytopes are discussed.

For more information we refer to the books of Grünbaum [453], McMullen and Shephard [718], Brøndsted [171], Ziegler [1045], Richter-Gebert [833] and Matoušek [695], the historical treatise of Federico [318], the proceedings [346], [562] and the surveys of Klee and Kleinschmidt [595] and Bayer and Lee [83].

15.1 Euler's Polytope Formula and Its Converse by Steinitz for $d = 3$

A result which readily implies Euler's [312, 313] polytope formula was given by Descartes around 1630. The original manuscript was lost, but a handwritten copy by Leibniz of 1676 survived. It was found in 1860 in the Royal Library of Hanover, immediately published and inserted into the collected works of Descartes [261]. See the comment of Federico [318]. In their fundamental treatise on topology Alexandroff and Hopf [9], p.1, write,

> ... - the discovery of the Euler (more correctly: Descartes-Euler) polytope theorem may be considered the first important event in topology (1752).

Below we define f-vectors and state the basic problem on f-vectors. Then two versions of the Euler polytope formula are presented, one for convex polytopes in $\mathbb{E}^3$, the other one for planar graphs. Finally, the converse of Euler's formula by Steinitz is presented. Since graphs will be used extensively in Sects. 15.4 and 34.1, the necessary graph-theoretic terminology is described in some detail.

The f-Vector of a Convex Polytope

Let $P \in \mathcal{P}$ be a convex polytope. Its *f-vector* $f = f(P)$ is the d-tuple

$$f = (f_0, f_1, \dots, f_{d-1}),$$

where $f_i = f_i(P)$ is the number of i-dimensional faces of P. In particular, f_0 is the number of vertices of P. A basic task of combinatorial polytope theory is the following.

Problem 15.1. *Characterize among all d-tuples of positive integers the f-vectors of proper convex polytopes in $\mathbb{E}^d$.*

For $d = 1, 2$, this problem is trivial. For $d = 3$ its solution follows from Euler's polytope formula and its converse by Steinitz, see later. For $d > 3$ it is open, but a lot of far-reaching contributions have been given, see the discussion in Sect. 15.2.

Euler's Polytope Formula for $d = 3$

Theorem 15.1. *Let P be a proper convex polytope in $\mathbb{E}^3$ with v vertices, e edges and f facets, i.e. $f(P) = (v, e, f)$. Then*

$$v - e + f = 2,\ v \le 2f - 4,\ f \le 2v - 4.$$

Euler's original proof lacked an argument of the type of Jordan's curve theorem for polygons which then was not available. Since the curve theorem for polygons is easy to show (see, e.g. Benson [96]), this criticism of Euler's proof should not be taken too seriously. The first rigorous proof seems to have been that of Legendre [639]. Before presenting it, we show the following lemma, actually a special case of the Gauss–Bonnet theorem. By a *convex spherical polygon* on the 2-dimensional unit sphere S^2, we mean the intersection of S^2 with a closed convex polyhedral cone with apex o.

Lemma 15.1. *The area of a convex spherical polygon on S^2, the boundary of which consists of n great circular arcs and with interior angles $\alpha_1, \dots, \alpha_n$ at the vertices, is $\alpha_1 + \dots + \alpha_n - (n-2)\pi$.*

Proof. Since the polygon may be dissected into convex spherical triangles, it is sufficient to prove the lemma for convex spherical triangles. Let T be a spherical triangle on S^2 with interior angles α, β, γ. Consider the three great circles containing the edges of T. Each pair of these great circles determines two spherical 2-gons, one containing T, the other one containing $-T$. The interior angles of these six 2-gons are $\alpha, \alpha, \beta, \beta, \gamma, \gamma$. Hence these 2-gons have areas (Fig. 15.1)

$$\frac{\alpha}{2\pi}4\pi = 2\alpha, 2\alpha, 2\beta, 2\beta, 2\gamma, 2\gamma.$$

The 2-gons cover each of T and $-T$ three times and the remaining parts of S^2 once. Thus we have for the area $A(T)$ of T,

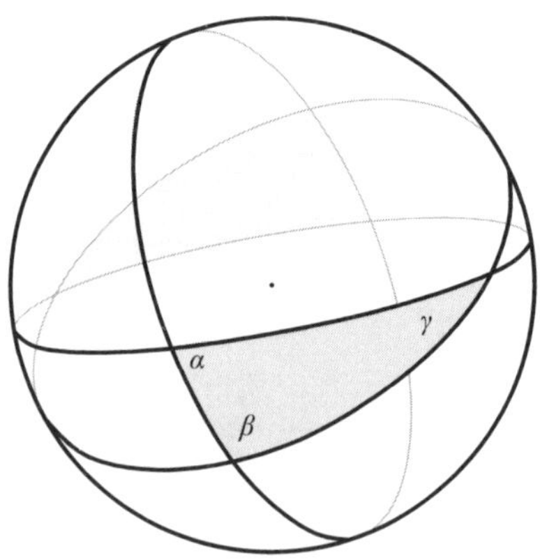

Fig. 15.1. Area of a spherical triangle

$$4A(T) + 4\pi = 2\alpha + 2\alpha + 2\beta + 2\beta + 2\gamma + 2\gamma, \text{ or } A(T) = \alpha + \beta + \gamma - \pi. \quad \square$$

Proof of the polytope formula. We may assume that $o \in \operatorname{int} P$. Projecting bd P from o radially onto S^2 yields a one-fold covering of S^2 by f convex spherical polygons. Apply the above lemma to each of these polygons. Note that the sum of the interior angles of such polygons at a common vertex is 2π and that each edge is an edge of precisely two of these polygons, and add. Then

$$(1) \qquad A(S^2) = 4\pi = 2\pi v - 2\pi e + 2\pi f, \text{ or } v - e + f = 2.$$

Since each edge is incident with two vertices and each vertex with at least three edges, it follows that

$$3v \le 2e, \text{ similarly, } 3f \le 2e.$$

By (1),

$$-2v = -2e + 2f - 4, \ -2f = -2e + 2v - 4.$$

Adding then yields that

$$v \le 2f - 4, \ f \le 2v - 4. \quad \square$$

Abstract Graphs, Graphs and Realizations

Before proceeding to the graph version of Euler's formula, some notation will be introduced.

A (finite) *abstract graph* $\mathcal{G}$ consists of two sets, the *set of vertices* $\mathcal{V} = \mathcal{V}(\mathcal{G}) = \{1, \dots, v\}$ and the *set of edges* $\mathcal{E} = \mathcal{E}(\mathcal{G})$ which consists of two-element subsets of $\mathcal{V}$. If $\{u, w\}$ is an edge, we write also uw for it and call w a *neighbour* of u and vice versa. If a vertex is contained in an edge or an edge contains a vertex then the vertex and the edge are *incident*. If two edges have a vertex in common, they are *incident* at this vertex. A *path* is a sequence of edges such that each edge is incident with the preceding edge at one vertex and with the next edge at the other vertex. We may write a path also as a sequence $w_1 w_2 \cdots w_k$ of vertices. We say that it *connects* the first and the last vertex of this sequence. A path is a *cycle* if the first and the last vertex coincide. A cycle which consists of k edges is a *k-cycle*. $\mathcal{G}$ is *connected* if any two distinct of its vertices are connected by a path. If $\mathcal{G}$ has at least $k + 1$ vertices, it is *k-connected* if any two distinct vertices can be connected by k paths which, pairwise, have only these vertices in common. With some effort it can be shown that this is equivalent to the following. The graph which is obtained from $\mathcal{G}$ by deleting any set of $k - 1$ vertices and the edges incident with these, is still connected.

More geometrically, by definition, a (*finite*) *graph* consists of a finite point set in some space, the *set of vertices* and, for each of a set of pairs of distinct vertices, of a continuous curve which connects these vertices and contains no vertex in its relative interior. These curves form the *set of edges*. The notions and notations which were defined earlier for abstract graphs are defined for graphs in the obvious way. Two graphs $\mathcal{G}, \mathcal{H}$, one or both of which may be abstract, are *isomorphic* if there is a bijection between $\mathcal{V}(\mathcal{G})$ and $\mathcal{V}(\mathcal{H})$ which maps (endpoints of) edges onto (endpoints of) edges in both directions.

A graph in $\mathbb{E}^2$ (or $\mathbb{C}$ or $\mathbb{C} \cup \{\infty\}$) is *planar* if no two edges of it cross. More generally, an abstract graph or a graph is called *planar* if it is isomorphic to a planar graph in $\mathbb{E}^2$. The isomorphic image in $\mathbb{E}^2$ then is a *planar realization* of the given, possibly abstract graph. There are simple criteria for planarity. Let $\mathcal{G}$ be a planar connected graph in $\mathbb{E}^2$. Omitting from $\mathbb{E}^2$ the vertices and edges of $\mathcal{G}$ leaves a finite system of connected open sets, the *countries* of $\mathcal{G}$, often called *domains* of $\mathcal{G}$. Since $\mathcal{G}$ is connected, the bounded countries are simply connected. If $\mathcal{G}$ is considered as embedded into $\mathbb{C} \cup \{\infty\}$, also the unbounded country is simply connected. It is easy to see that each planar graph in $\mathbb{E}^2$ is isomorphic to a planar graph in $\mathbb{E}^2$ the edges of which are polygonal curves.

The *edge graph* $\mathcal{G}(P)$ of a proper convex polytope P is the graph consisting of the vertices and the edges of P. If $\dim P = 3$, then $\mathcal{G}(P)$ is planar.

For more information on geometric graph theory, see Nishizeki and Chiba [772], Bollobás [142], Mohar and Thomassen [748] and Felsner [332].

Euler's Formula for Graphs in $\mathbb{E}^2$

We will show the following result.

Theorem 15.2. *Let $\mathcal{G}$ be a connected planar graph in $\mathbb{E}^2$ with v vertices, e edges and f countries. Then*

$$v - e + f = 2.$$

Proof. Define the *Euler characteristic* $\chi(\mathcal{H})$ of a connected planar graph $\mathcal{H}$ in $\mathbb{E}^2$ as the number of vertices minus the number of edges plus the number of countries of $\mathcal{H}$, including the unbounded country. For simplicity, we assume that all graphs considered have polygonal curves as edges.

The following proposition is well-known.

(2) Let a graph $\mathcal{K}$ in $\mathbb{E}^2$ arise from a connected planar graph $\mathcal{H}$ in $\mathbb{E}^2$ by one of the following two operations:

(i) Consider a polygonal curve connecting a vertex of $\mathcal{H}$ with a point in a country of $\mathcal{H}$ and which contains no vertex or point on an edge of $\mathcal{H}$ in its relative interior. Add this curve to the edges of $\mathcal{H}$ and its endpoint to the vertices.

(ii) Consider a polygonal curve connecting two vertices of $\mathcal{H}$ and which contains no vertex or point on an edge of $\mathcal{H}$ in its relative interior. Add this curve to the edges of $\mathcal{H}$.

Then $\chi(\mathcal{K}) = \chi(\mathcal{H})$.

Since $\mathcal{G}$ is connected, it can be constructed by finitely many operations as described in (2), beginning with a graph consisting of a single vertex which has Euler characteristic 2. Proposition (2) then shows that $\chi(\mathcal{G}) = 2$. □

Corollaries of Euler's Formula for Graphs in $\mathbb{E}^2$

The following consequences of Euler's formula will be needed later.

Corollary 15.1. *Let $\mathcal{H}$ be a planar graph in $\mathbb{E}^2$ with v vertices, e edges and f countries. If $\mathcal{H}$ contains no 3-cycle, then*

$$e \leq 2v - 4. \tag{3}$$

Here equality holds if and only if $\mathcal{H}$ is 2-connected and each country of $\mathcal{H}$ has a boundary 4-cycle.

Proof. Assume, first, that $\mathcal{H}$ is 2-connected. Then each edge is on the boundary of two distinct countries of $\mathcal{H}$. Since the boundary cycles of the countries are at least 4-cycles, we see that

$$4f \leq 2e,$$

where equality holds if and only if each country of $\mathcal{H}$ has a boundary 4-cycle. Since

$$8 = 4v - 4e + 4f,$$

by Euler's formula, addition yields (3), where equality holds if and only if each country of $\mathcal{H}$ has a boundary 4-cycle.

If $\mathcal{H}$ is not 2-connected, there is inequality in (3) as can be shown by induction on v by dissecting $\mathcal{H}$ into two or more sub-graphs which have, pairwise, at most one vertex in common. □

Corollary 15.2. *Let $\mathcal{H}$ be a planar 3-connected graph in $\mathbb{E}^2$. Then there is a vertex incident with three edges or a country with a boundary 3-cycle.*

Proof. Let v, e, f denote the numbers of vertices, edges and countries of $\mathcal{H}$, respectively. Since $\mathcal{H}$ is 3-connected and each country has a boundary cycle consisting of at least 3 vertices, we may represent v and f in the form

$$v = v_3 + v_4 + \cdots, \quad f = f_3 + f_4 + \cdots,$$

where v_k is the number of vertices which are incident with precisely k edges and f_k is the number of countries the boundary cycle of which consists of precisely k edges. Each edge is incident with precisely two vertices and, noting that $\mathcal{H}$ is 3-connected, each edge is on the boundary of precisely two countries. Hence

$$2e = 3v_3 + 4v_4 + \cdots = 3f_3 + 4f_4 + \cdots$$

Euler's formula then yields,

$$\begin{aligned} 8 = 4v - 4e + 4f &= 4v_3 + 4v_4 + 4v_3 + \cdots - 3v_3 - 4v_4 - 5v_5 - \cdots \\ &\quad - 3f_3 - 4f_4 - 5f_5 - \cdots + 4f_3 + 4f_4 + 4f_5 + \cdots \\ &= v_3 - v_5 - 2v_6 - \cdots + f_3 - f_5 - 2f_6 - \cdots \leq v_3 + f_3. \quad \square \end{aligned}$$

The Converse of Euler's Polytope Formula for $d = 3$ by Steinitz

Steinitz [962, 965] proved the following converse of Euler's formula, perhaps conjectured already by Euler. Together with Euler's polytope formula, it yields a characterization of the f-vectors of convex polytopes in $\mathbb{E}^3$.

Theorem 15.3. *Let v, e, f be positive integers such that*

$$v - e + f = 2, \ v \le 2f - 4, \ f \le 2v - 4.$$

Then there is a proper convex polytope in $\mathbb{E}^3$ with v vertices, e edges and f facets.

Proof. Clearly, we have the following proposition.

(4) Let P_0 be a proper convex polytope in $\mathbb{E}^3$ with f-vector (v_0, e_0, f_0). Assume that P has at least one vertex incident with three edges and at least one triangular facet. By cutting off this vertex, resp. by pasting a suitable triangular pyramid to P at this facet we obtain proper polytopes with f-vectors $(v_0 + 2, e_0 + 3, f_0 + 1)$, and $(v_0 + 1, e_0 + 3, f_0 + 2)$, respectively, satisfying the same assumption.

A consequence of this proposition is the following.

(5) Let P_0 be as in (4) with f-vector (v_0, e_0, f_0). By repeating the cutting process i times and the pasting process j times, we obtain a proper convex polytope with f-vector $(v_0 + 2i + j, e_0 + 3i + 3j, f_0 + i + 2j)$.

To prove the theorem note that, by assumption,

$$2v - f - 4, \ 2f - v - 4$$

are non-negative integers. Their difference is divisible by 3. Thus they give the same remainder on division by 3, say $r \in \{0, 1, 2\}$. Hence there are non-negative integers i, j, such that

$$2v - f - 4 = 3i + r, \ 2f - v - 4 = 3j + r$$

and thus

$$v = (4 + r) + 2i + j, \ f = (4 + r) + i + 2j. \tag{6}$$

Take for P_0 a pyramid with basis a convex $(3 + r)$-gon. The f-vector of P_0 is the vector

$$(4 + r, 6 + 2r, 4 + r).$$

Applying (5) to this P_0 with the present values of i and j, we obtain a convex polytope P with v vertices and f facets, see (6). By Euler's polytope formula, P has $e = v + f - 2$ edges. □

15.2 Shelling and Euler's Formula for General *d*

The first attempt to extend Euler's polytope formula to general d goes back to Schläfli [888] in 1850 in a book which was published only in 1901. His proof that, for the f-vector of a proper convex polytope P in $\mathbb{E}^d$, we have,

(1) $$f_0 - f_1 + - \cdots + (-1)^{d-1} f_{d-1} = 1 - (-1)^d$$

made use of an argument of the following type. The boundary of P can be built up from a given facet F_1 by successively adding the other facets in a suitable order, say $F_2, \dots, F_m$, such that $(F_1 \cup \cdots \cup F_{i-1}) \cap F_i$ is homeomorphic to a $(d-2)$-dimensional convex polytope for $i = 2, \dots, m-1$. Schläfli, and later other mathematicians, seem to have thought that the existence of such a shelling was obvious; see the references in Grünbaum [453]. After many unsuccessful attacks, in particular in the 1960s, this was proved in an ingenious way by Bruggesser and Mani [172] in 1970. Before that, the only elementary proof of the formula (1) was that of Hadwiger [467].

A topological proof of (1), by Poincaré [807, 808], also had serious gaps. Rigorous proofs in the context of topology were given only in the 1930s when the necessary algebraic-topological machinery was available. For this information I am indebted to Matthias Kreck [617].

The shelling result of Bruggesser and Mani, which yields a simple proof of (1), is also a tool to prove the equations of Dehn [251] and Sommerville [948] for f-vectors and is used in the striking proof of the upper bound conjecture by McMullen [705].

The problem of characterizing the f-vectors of convex polytopes is settled for $d = 2$ where it is trivial, and for $d = 3$, see the preceding section. For $d > 3$ it is far from a solution, but there are important contributions towards it. For simplicial convex polytopes a characterization was proposed by McMullen [707] in the form of his celebrated g-conjecture. The g-conjecture was proved by Billera and Lee [116] and Stanley [951], see also McMullen [715].

In this section we present the shelling result of Bruggesser and Mani and show how it leads to Euler's polytope formula (1) for general d. Then other relations for f-vectors are discussed.

Shellings are treated by Björner and Björner et al. [121, 122] and from the voluminous literature on f-vectors we cite the books of Grünbaum [453], McMullen and Shephard [718], Brøndsted [171] and Ziegler [1045] and the surveys of Bayer and Lee [83] and Billera and Björner [115].

Polytopal Complexes and Shellings

A (finite) *abstract complex* $\mathcal{C}$ is a family of subsets of $\{1, \dots, v\}$, the *faces* of $\mathcal{C}$, such that $\emptyset, \{1\}, \dots, \{v\} \in \mathcal{C}$ and for any two faces $F, G \in \mathcal{C}$ also $F \cap G \in \mathcal{C}$. The faces $\{1\}, \dots, \{v\}$ are called the *vertices* of $\mathcal{C}$.

Closer to convexity, we define a (finite) *polytopal complex* $\mathcal{C}$ to be a finite family of convex polytopes in $\mathbb{E}^d$, called the *faces* of $\mathcal{C}$, such that any (geometric) face of a polytope in $\mathcal{C}$ belongs also to $\mathcal{C}$ and the intersection of any two polytopes in $\mathcal{C}$ is a (geometric) face of both of them, and thus is contained in $\mathcal{C}$. (What we call a polytopal complex is often called a polyhedral complex. The reason for our terminology is that we want to stress that our complexes consist of convex polytopes and not of, possibly, unbounded convex polyhedra.) The *underlying set* of $\mathcal{C}$ is the union of all polytopes in $\mathcal{C}$. The *dimension* of $\mathcal{C}$, $\dim \mathcal{C}$, is the maximum dimension of a polytope in $\mathcal{C}$. The complex $\mathcal{C}$ is *pure* if any polytope in $\mathcal{C}$ is a (geometric) face of a polytope in $\mathcal{C}$ of dimension $\dim \mathcal{C}$. Two (abstract or geometric) complexes are *isomorphic* if there is a bijection between their (abstract or geometric) vertices which preserves their (abstract or geometric) faces in both directions.

Given a convex polytope P in $\mathbb{E}^d$, the family of all its (geometric) faces, including $\emptyset$ and P, is a polytopal complex, the *complex* $\mathcal{C} = \mathcal{C}(P)$ *of* P. Also the family of all its proper (geometric) faces, including $\emptyset$, is a polytopal complex, the *boundary complex* $\mathcal{C}(\text{relbd}\, P)$ *of* P. See Theorems 14.5 and 14.6.

By a *shelling* of a convex polytope P or, more precisely, of its boundary complex $\mathcal{C}(\text{relbd}\, P)$, we mean a linear ordering $F_1, \dots, F_m$ of the facets of P such that either $\dim P = 0, 1$, or which otherwise satisfies the following conditions:

(i) (The boundary complex of) F_1 has a shelling.
(ii) For $i = 2, \dots, m$ we have $(F_1 \cup \dots \cup F_{i-1}) \cap F_i = G_1 \cup \dots \cup G_j$, where $j \leq k$ and $G_1 = G_{i1}, \dots, G_k = G_{ik}$ is a shelling of F_i.

The Bruggesser-Mani Shelling Theorem

Before stating this result, we present a simple property of shellings.

Proposition 15.1. *Let $G_1, \dots, G_k$ be a shelling of a convex polytope F. Then $G_k, \dots, G_1$ is also a shelling of F.*

Proof (by induction on $d = \dim F$). The assertion clearly holds for $d = 0, 1$. Assume now that $d > 1$ and that it holds in dimension $d - 1$. We prove it for F. First, each G_i and thus, in particular, G_k is shellable by (i) and (ii). Second, let $i < k$. By (ii),

$$(G_1 \cup \dots \cup G_{i-1}) \cap G_i = H_1 \cup \dots \cup H_j, \tag{2}$$

where $j \leq n$ and $H_1, \dots, H_n$ is a shelling of G_i. For each facet H_l of G_i there is a unique facet G_m of F such that $H_l = G_i \cap G_m$. Thus (2) implies that

$$G_i \cap (G_{i+1} \cup \dots \cup G_k) = H_{j+1} \cup \dots \cup H_n.$$

Now note that $H_n, \dots, H_1$ is also a shelling of G_i by induction. The induction is complete. □

The *Bruggesser-Mani shelling theorem* is as follows:

Theorem 15.4. *Convex polytopes are shellable.*

Proof. Actually, we prove slightly more.

(3) Let $P \in \mathcal{P}_p$ and let $p \in \mathbb{E}^d \backslash P$ be a point not contained in the affine hull of any facet of P. Then there is a shelling of P where the facets of P which are visible from p come first.

Clearly, (3) holds in case $d = 1$. Assume now that $d > 1$ and (3) holds for $d - 1$. Choose a line L through p which meets int P, is not parallel to any facet of P, and intersects the affine hulls of different facets of P in distinct points. Orient L such that p is in the positive direction relative to P. Starting at the point where L leaves P, move along L in the positive direction to infinity and, from there, return again along L but from the opposite direction (Fig. 15.2).

Order the facets of P as they become visible on our flight before we reach infinity and as they disappear on the horizon on the flight back. This gives an ordering $F_1, \dots, F_m$ of the facets of P such that

(4) In the ordering $F_1, \dots, F_m$ the facets visible from p come first.

We will show that

(5) $F_1, \dots, F_m$ is a shelling of P.

F_1 is shellable by induction. If F_i, $i > 1$, appears in the ordering before we reach infinity, then

$$(F_1 \cup \dots \cup F_{i-1}) \cap F_i$$

consists precisely of those facets of F_i which are visible from p_i, where $\{p_i\} = L \cap \operatorname{aff} F_i$. By induction, these facets appear first in a shelling of F_i. If F_i appears after we have reached infinity, then

$$(F_1 \cup \dots \cup F_{i-1}) \cap F_i$$

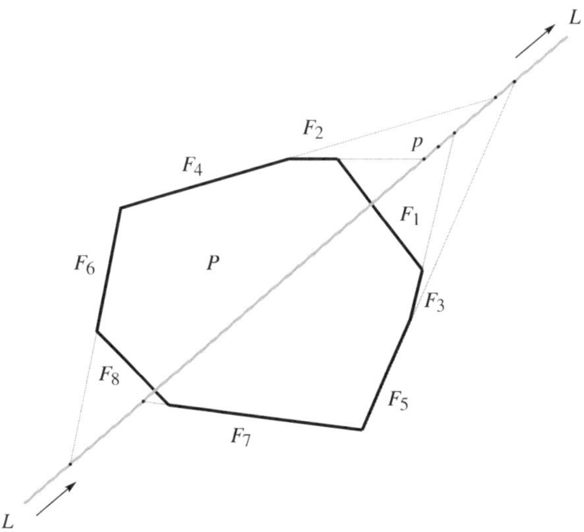

Fig. 15.2. Shelling

consists precisely of those facets of F_i which are not visible from p_i. By induction, these facets come last in a shelling of F_i. The earlier proposition then says that these facets come first in a different shelling of F_i. Hence $F_1, \dots, F_m$, in fact, is a shelling of P, concluding the proof of (5). Having proved (4) and (5), the induction and thus the proof of (3) is complete. □

The Euler Characteristic of Polytopal Complexes

In Sect. 7.1 we have defined the Euler characteristic χ on the lattice of polytopes, i.e. for finite unions of convex polytopes. It is the unique valuation which is 1 for convex polytopes.

Here, the *Euler characteristic* $\chi(\mathcal{C})$ *of a polytopal complex* $\mathcal{C}$ is defined by

$$\chi(\mathcal{C}) = f_0 - f_1 + - \cdots ,$$

where $f_i = f_i(\mathcal{C}), i = 0, 1, \dots, d$, is the number of convex polytopes in $\mathcal{C}$ of dimension i. f_0 is the number of vertices, f_1 the number of edges, etc. If $\mathcal{C}$ and $\mathcal{D}$ are complexes such that $\mathcal{C} \cup \mathcal{D}$ is also a complex, that is, $P \cap Q \in \mathcal{C} \cap \mathcal{D}$ for $P \in \mathcal{C}$ and $Q \in \mathcal{D}$, then the following *additivity property* holds, as can be shown easily,

$$\chi(\mathcal{C} \cup \mathcal{D}) + \chi(\mathcal{C} \cap \mathcal{D}) = \chi(\mathcal{C}) + \chi(\mathcal{D}).$$

In other words, χ is a valuation on the family of polytopal complexes.

It can be shown that the Euler characteristic (in the present sense) of a polytopal complex and the Euler characteristic (in the sense of Sect. 7.1) of the underlying polytope coincide.

The Euler polytope formula for general *d*

shows that the Euler characteristic of the boundary complex of a convex polytope can be expressed in a very simple form.

Theorem 15.5. *Let $P \in \mathcal{P}_p$. Then*

$$\chi\big(\mathcal{C}(\operatorname{bd} P)\big) = f_0 - f_1 + - \cdots + (-1)^{d-1} f_{d-1} = 1 - (-1)^d, \chi\big(\mathcal{C}(P)\big) = 1.$$

Proof. It is sufficient to show the following, where $\mathcal{C}(F_1 \cup \cdots \cup F_i)$ is the polytopal complex consisting of the facets $F_1, \dots, F_i$ and all their faces:

(6) Let $P \in \mathcal{P}_p$ and let $F_1, \dots, F_m,\ m = f_{d-1}$, be a shelling of P. Then

(i) $\chi\big(\mathcal{C}(F_i)\big) = 1$ for $i = 1, \dots, m$.

(ii) $$\chi\big(\mathcal{C}(F_1 \cup \cdots \cup F_i)\big) = \begin{cases} 1 & \text{for } i = 1, \dots, m-1. \\ 1 - (-1)^d & \text{for } i = m. \end{cases}$$

(iii) $\chi\big(\mathcal{C}(P)\big) = 1$.

The proof of (6) is by induction. (6) is clear for $d = 1$. Assume now that $d > 1$ and that (6) holds for $d - 1$. By induction, (iii) implies that

(7) $\chi\bigl(\mathcal{C}(F_i)\bigr) = 1$ for $i = 1, \dots, m$,

settling (i). The assertion for $i = 1$ in (ii) is a consequence of (i). Assume next that $1 < i < m$. Then

$$\begin{aligned}
(8)\quad & \chi\bigl(\mathcal{C}(F_1 \cup \dots \cup F_i)\bigr) \\
&= \chi\bigl(\mathcal{C}(F_1 \cup \dots \cup F_{i-1})\bigr) + \chi\bigl(\mathcal{C}(F_i)\bigr) - \chi\bigl(\mathcal{C}((F_1 \cup \dots \cup F_{i-1}) \cap F_i)\bigr) \\
&= \chi\bigl(\mathcal{C}(F_1 \cup \dots \cup F_{i-1})\bigr) + 1 - \chi\bigl(\mathcal{C}(G_1 \cup \dots \cup G_j)\bigr) \\
&= \chi\bigl(\mathcal{C}(F_1 \cup \dots \cup F_{i-1})\bigr) + 1 - 1 \\
&= \chi\bigl(\mathcal{C}(F_1 \cup \dots \cup F_{i-1})\bigr) = \dots = \chi\bigl(\mathcal{C}(F_1)\bigr) = 1
\end{aligned}$$

by the additivity of χ, where $j < k$, $G_1, \dots, G_k$ is a shelling of F_i and we have used induction on (ii) and (7). Assume next that $i = m$. Then

$$\begin{aligned}
& \chi\bigl(\mathcal{C}(F_1 \cup \dots \cup F_m)\bigr) \\
&= \chi\bigl(\mathcal{C}(F_1 \cup \dots \cup F_{m-1})\bigr) + \chi\bigl(\mathcal{C}(F_m)\bigr) - \chi\bigl(\mathcal{C}(F_1 \cup \dots \cup F_{m-1}) \cap F_m\bigr) \\
&= 1 + 1 - \bigl(1 - (-1)^{d-1}\bigr) = 1 + (-1)^d
\end{aligned}$$

by the additivity of χ, (8), (7), noting that $\mathcal{C}((F_1 \cup \dots \cup F_{m-1}) \cap F_m)$ is the boundary complex of F_m, and induction on (ii). The proof of (ii) is complete. To show (iii), note that $\mathcal{C}(P)$ is obtained from $\mathcal{C}(F_1 \cup \dots \cup F_m)$ by adding P. Hence the definition of χ shows that

$$\chi\bigl(\mathcal{C}(P)\bigr) = \chi\bigl(\mathcal{C}(F_1 \cup \dots \cup F_m)\bigr) + (-1)^d = 1 - (-1)^d + (-1)^d = 1.$$

This settles (iii). The induction and thus the proof of (6) is complete. □

Remark. It can be shown that the Euler polytope formula is the only linear relation which is satisfied by the f-vectors of all d-dimensional convex polytopes.

The Dehn–Sommerville Equations and McMullen's g-Theorem

At present a characterization of all f-vectors of proper convex polytopes, i.e. a solution of Problem 15.1 for $d > 3$, seems to be out of reach. But for *simplicial* and *simple convex polytopes* this can be done. These are convex polytopes all facets of which are simplices, resp. convex polytopes, each vertex of which is incident with precisely d edges. We consider only the simplicial case.

Dehn [251] ($d = 5$) and Sommerville [948] (general d) specified $\lfloor \frac{d+1}{2} \rfloor$ independent linear equations, including the Euler polytope formula (1), which are satisfied by the f-vectors of all proper simplicial convex polytopes. It can be shown that there is no linear relation independent of the Dehn–Sommerville equations which is satisfied by the f-vectors of all proper simplicial convex polytopes, see Grünbaum [453], Sect. 9.2.

McMullen [707] stated in 1970 a characterization of the set of all f-vectors of proper simplicial convex polytopes in $\mathbb{E}^d$, the *g-conjecture*. This conjecture was confirmed by the efforts of Stanley [951], who used heavy algebraic machinery (necessity of McMullen's conditions) and Billera and Lee [116] (sufficiency). See also McMullen [715].

The Lower and the Upper Bound Theorem

It is a natural question to determine the best lower and upper bounds for $f_i(P)$, $i = 1, \dots, d-1$, for all convex polytopes P with given $n = f_0(P)$ (or $f_k(P)$ for given k). The best lower bounds are attained by so-called stacked polytopes, as shown by Barnette [73, 74]. In particular,

$$f_{d-1}(P) \geq (d-1)n - (d+1)(d-2).$$

The best upper bounds are attained by so-called neighbourly polytopes. This was shown by McMullen [705], thereby confirming Motzkin's [758] *upper bound conjecture*. In particular,

$$f_{d-1}(P) \leq 2 \sum_{i=0}^{\frac{d}{2}}{}^{*} \binom{n-d-1+i}{i},$$

where

$$\sum_{i=0}^{\frac{d}{2}}{}^{*} a_i = \begin{cases} a_0 + \cdots + a_{\frac{d-1}{2}} & \text{for } d \text{ odd}, \\ a_0 + \cdots + a_{\frac{d-2}{2}} + \frac{1}{2} a_{\frac{d}{2}} & \text{for } d \text{ even}. \end{cases}$$

15.3 Steinitz' Polytope Representation Theorem for $d = 3$

The edge graph of a proper convex polytope P in $\mathbb{E}^3$ is planar and 3-connected. To see the former property, project the 1-skeleton of P from an exterior point of P which is sufficiently close to a given relative interior point of a facet F of P onto a plane parallel to F on the far side of P. For the latter property, see Theorem 15.8 later. The proof that, conversely, a planar, 3-connected graph can be realized as the edge graph of a convex polytope in $\mathbb{E}^3$ is due to Steinitz [963, 965]. Grünbaum [453], Sect. 13.1, stated in 1967 that this is

> The most important and deepest known result on 3-polytopes ...

So far, it has resisted extension to higher dimensions. Since the determination of the face lattices of 4-dimensional convex polytopes is NP-hard, almost surely there is no such extension. In spite of this, the Steinitz theorem led to a collection of non-trivial results of modern polytope theory centred around the following questions.

Problem 15.2. *Characterize, among all graphs or in a given family of graphs, those which can be realized as edge graphs of convex polytopes.*

Problem 15.3. *Given the edge graph of a convex polytope, what can be said about the polytope?*

In this section we give a short proof of the Steinitz theorem using the Koebe–Brightwell–Scheinerman representation theorem 34.1 for planar graphs. Section 15.4 contains further contributions to these problems.

For additional information the reader may consult the books of Grünbaum [453], Ziegler [1045], Richter-Gebert [833] and Matoušek [695] and the surveys of Bayer and Lee [83] and Kalai [561].

Steinitz' Representation Theorem for Convex Polytopes in $\mathbb{E}^3$

We shall prove the following basic result.

Theorem 15.6. *Let $\mathcal{G}$ be a planar 3-connected graph. Then $\mathcal{G}$ is isomorphic to the edge graph of a convex polytope in $\mathbb{E}^3$.*

Proof. For terminology, see Sect. 34.1. An application of the Koebe–Brightwell–Scheinerman theorem 34.1 together with a suitable Möbius transformation and stereographic projection shows that there is a primal-dual circle representation of $\mathcal{G}$ on the unit sphere S^2 with the following properties.

The country circle of the outer country is the equator.
All other country circles are in the southern hemisphere.
Three of these touch the equator at points which are $\frac{2\pi}{3}$ apart.

From now on we use this primal-dual circle representation of $\mathcal{G}$.

For each country circle choose a closed halfspace with the country circle in its boundary plane such that the halfspace of the equator circle contains the south pole of S^2 and the other halfspaces the origin o. The intersection of these halfspaces is then a convex polytope P. The countries, resp. the country circles of $\mathcal{G}$ correspond to the facets of P. Let $v \in S^2$ be a vertex (of the representation) of $\mathcal{G}$ on S^2. The country circles of the countries with vertex v form a ring around v like a string of beads, possibly of different sizes, on the vertex circle of v. The latter intersects these country circles orthogonally. All other country circles are outside this ring. This shows that the boundary planes of the halfspaces corresponding to the circles of the ring meet in a point v_P, say, radially above v and v_P is an interior point of all other halfspaces. Thus v_P is a vertex of P. If C is a country circle of the ring, the facet F of P determined by C has two edges which contain v_P and are tangent to C and thus to S^2. If $vw \cdots z$ is a cycle of $\mathcal{G}$ around C, then $v_P w_P \cdots z_P$ is a cycle of edges of F and thus is *the* cycle of edges of F. Each edge of F is tangent to S^2 at the point where it touches the incircle C of F. Since, by construction, the edge graph of P is obtained by radial projection of the representation of $\mathcal{G}$ (in S^2), the proof is complete. □

Different Ways to Represent a Convex Polytope

The above proof of the Steinitz representation theorem shows that any convex polytope in $\mathbb{E}^3$ is (combinatorially) isomorphic to a convex polytope all edges of which touch the unit ball B^3. A far-reaching generalization of this result is due to Schramm [914] in which B^3 is replaced by any smooth and strictly convex body in $\mathbb{E}^3$.

A problem of Steinitz [964] asks whether each convex polytope in $\mathbb{E}^d$ is *circumscribable* that is, it is isomorphic to a convex polytope circumscribed to B^d and such that each of the facets of this polytope touches B^d. A similar problem of Steiner [961] deals with *inscribable* convex polytopes, i.e. there is an isomorphic convex polytope contained in B^d such that all its vertices are on the boundary of B^d. Steinitz [964] proved the existence of a non-circumscribable convex polytope in $\mathbb{E}^3$. Schulte [917] proved the existence of non-inscribable convex polytopes in $\mathbb{E}^d$ for $d \geq 4$. For more information, see Grünbaum [453], Grünbaum and Shephard [455] and Florian [337].

Klee asked whether each convex polytope in $\mathbb{E}^d$ is *rationally representable*, that is, it is isomorphic to a convex polytope in $\mathbb{E}^d$, all vertices of which have rational coordinates. The affirmative answer for $d = 3$ follows from a proof of the Steinitz representation theorem where all steps may be carried out in the rational space $\mathbb{Q}^3$, see Grünbaum [453]. For $d \geq 8$ Perles proved that there are convex polytopes which cannot be represented rationally, see Grünbaum [453], p. 94. Richter-Gebert [833] could show that this holds already for $d \geq 4$. For more information compare Bayer and Lee [83] and Richter-Gebert [833].

15.4 Graphs, Complexes, and Convex Polytopes for General *d*

There are many connections between graphs and complexes on the one hand and convex polytopes on the other hand. A first such result is Steinitz' representation theorem for $d = 3$, see the preceding section where we also mentioned two basic problems. Important later contributions are due to Grünbaum, Perles, Blind, Mani, Kalai and others.

One of the great problems in this context is the following:

Problem 15.4. *Characterize, among all polytopal complexes, those which are isomorphic to boundary complexes of convex polytopes of dimension d.*

This problem remains open, except for $d = 3$, in which case the Steinitz representation theorem provides an answer. More accessible is the following

Problem 15.5. *Given suitable sub-complexes of the boundary complex of a convex polytope, what can be said about the polytope?*

In this section we first show that there is an algorithm to decide whether an abstract graph can be realized as the edge graph of a convex polytope. Then Balinski's theorem on the connectivity of edge graphs is presented. Next, we give the theorem of Perles, Blind and Mani together with Kalai's proof, which says that, for simple convex polytopes, the edge graph determines the combinatorial structure of the polytope. For general convex polytopes, no such result can hold. Finally, we give a short report on related problems for complexes instead of graphs.

For more information, see the books of Grünbaum [453] and Ziegler [1045].

Is a Given Graph Isomorphic to an Edge Graph?

A first answer to this question is the following result.

Theorem 15.7. *There is an algorithm to decide whether a given abstract graph can be realized as the edge graph of a proper convex polytope in* $\mathbb{E}^d$.

The following proof is shaped along Grünbaum's [453], p. 92, proof of a similar result on complexes.

Proof. A *statement in elementary algebra* is any expression constructed according to the usual rules and involving only the symbols

$$+, -, \cdot, =, <, (,), [,], 0, 1, \vee, \wedge, \neg, \forall, \exists$$

and real variables. The quantifiers $\forall$ and $\exists$ act only on real variables. A theorem of Tarski [989] is then as follows.

(1) Every statement in elementary algebra which contains no free variables (i.e. such that each variable is bound by $\forall$ or $\exists$) is effectively decidable. That is, there is an algorithm to decide whether any such statement is true or false.

Now, given an abstract graph $\mathcal{G} = \langle \mathcal{V}, \mathcal{E} \rangle$, the question as to whether it is realizable as the edge graph of a proper convex polytope in $\mathbb{E}^d$ may be put in the following form, in which (1) is applicable.

(2) There are reals $x_{ij},\ i \in \mathcal{V},\ j = 1, \dots, d$, such that there are reals u_{ij}, and $\alpha_i,\ i \in \mathcal{V},\ j = 1, \dots, d$, where

$$\sum_{j=1}^{d} u_{ij} x_{ij} = \alpha_i \text{ for } i \in \mathcal{V},$$

$$\sum_{j=1}^{d} u_{ij} x_{kj} < \alpha_i \text{ for } i, k \in \mathcal{V} \text{ and } k \neq i,$$

and such that for every subset $\{l, m\} \subseteq \mathcal{V}$ the following statements are equivalent:

(i) $\{l, m\} \in \mathcal{E}$,

(ii) There are reals $v_j,\ j = 1, \dots, d$, and β, where

$$\sum_{j=1}^{d} v_j x_{ij} \begin{cases} = \beta & \text{for } i \in \{l, m\} \\ < \beta & \text{for } i \in \mathcal{V}, i \notin \{l, m\}, \end{cases}$$

and such that for all reals $y_1, \dots, y_d$ there are reals $\lambda_i,\ i \in \mathcal{V}$, for which $\lambda_1 + \dots + \lambda_n = 0$ and

$$y_j = \sum \lambda_i x_{ij} \text{ for } j = 1, \cdots, d.$$

The first condition makes sure that the points $(x_{i1}, \dots, x_{id})$ are vertices of a convex polytope, in the second condition it is tested whether $\mathcal{G}$ is the edge graph of this polytope and the third condition makes sure that it is proper.

The theorem is now a consequence of (1) and (2). □

Remark. Two convex polytopes are *combinatorially equivalent* if there is a bijection between the sets of vertices of these polytopes which maps (the sets of vertices of) faces onto (the sets of vertices of) faces in both directions. A convex polytope is 2-*neighbourly*, if any two of its vertices are connected by an edge. A special case of 2-neighbourly polytopes are cyclic polytopes in $\mathbb{E}^d$ where $d \geq 4$. *Cyclic polytopes* are the convex hulls of $n \geq d + 1$ points on the *moment curve*

$$\{(t, t^2, \dots, t^d) : t \in \mathbb{R}\} \ (d \geq 3).$$

(Sometimes, convex polytopes are called cyclic, if they are combinatorially equivalent to cyclic polytopes as defined here.) There are 2-neighbourly polytopes which are *not* combinatorially equivalent to a cyclic polytope. See, e.g. Grünbaum [453], p. 124. A *complete graph*, that is a graph in which any two vertices are connected by an edge, thus can be realized as the edge graph of combinatorially different proper convex polytopes. It clearly can always be realized as the edge graph of a simplex.

In spite of the above theorem, the problem to characterize, in a simple way, the graphs which can be realized as edge graphs of convex polytopes remains open.

Edge Graphs of Convex Polytopes are d-Connected

Recall that a graph is *d-connected* if the deletion of any $d - 1$ of its vertices and of the edges incident with these vertices leaves it connected. We show the following theorem of Balinski [**?**].

Theorem 15.8. *Let P be a proper convex polytope in $\mathbb{E}^d$. Then the edge graph of P is d-connected.*

The following proof is taken from Grünbaum [453]. For references to other proofs, see Ziegler [1045].

Proof (by induction on d). The theorem is trivial for $d = 1, 2$. Assume then that $d > 2$ and that it holds for $1, 2, \dots, d - 1$.

Let $v_1, \dots, v_{d-1} \in V = \mathcal{V}(P)$. We have to show that the graph arising from the edge graph of P by deleting $v_1, \dots, v_{d-1}$ and the edges incident with these vertices, is still connected.

Since P is the disjoint union of all relative interiors of its faces (including the improper face P) , the point $\frac{1}{d-1}(v_1 + \dots + v_{d-1})$ is in the relative interior of a face of P, say F. We distinguish two cases.

First, $F \neq P$. Choose $u \in S^{d-1}$ and $\alpha \in \mathbb{R}$ such that $F = \{x \in P : u \cdot x = \alpha\}$ and $P \subseteq \{x : u \cdot x \leq \alpha\}$. Since $\frac{1}{d-1}(v_1 + \dots + \dots + v_{d-1}) \in F$ and $v_1, \dots, v_{d-1} \in P$, we see that $v_1, \dots, v_{d-1} \in F$. Choose $\beta < \alpha$ such that $P \subseteq \{x : u \cdot x \geq \beta\}$ and

$G = \{x \in P : u \cdot x = \beta\}$ is a face $\neq \emptyset$ of P. A vertex $v \in V \backslash \{v_1, \ldots, v_{d-1}\}$ is contained in G or is connected by a not deleted edge to a vertex $w \in V$ with $u \cdot w < u \cdot v$. Continuing, we see that each vertex of $V \backslash \{v_1, \ldots, v_{d-1}\}$ either is contained in G or can be connected by a path consisting of not deleted edges, to a vertex in G. By induction, the graph of G is connected. The graph obtained from the edge graph of P by deleting $v_1, \ldots, v_{d-1}$ and the edges incident with these vertices, is thus connected.

Second, $F = P$. Then $\frac{1}{d-1}(v_1 + \cdots + v_{d-1}) \in \operatorname{int} P$. Choose $u \in S^{d-1}$ and $\alpha \in \mathbb{R}$ such that the hyperplane $\{x : u \cdot x = \alpha\}$ contains $v_1, \ldots, v_{d-1}$ and another vertex v of P. Choose $\beta < \alpha < \gamma$ such that the hyperplanes $\{x : u \cdot x = \beta\}$ and $\{x : u \cdot x = \gamma\}$ support P at facets G and H, say. By the same argument as earlier, we see that each vertex in $V \backslash \{v_1, \ldots, v_{d-1}\}$ is connected to a vertex in G or H by a path consisting of not deleted edges. In particular, v is connected by such paths to a vertex in G as well as to a vertex in H. By induction, the graphs of G and H are connected. The graph obtained from the edge graph of P by deleting $v_1, \ldots, v_{d-1}$ and the edges incident with these vertices, is thus connected.

The induction is complete, concluding the proof. □

Edge Graphs of Simple Polytopes Determine the Combinatorial Structure; the Perles–Blind–Mani Theorem

Let $\mathcal{G}$ be a graph, possibly abstract. $\mathcal{G}$ is *i-regular* if any vertex is incident with precisely i edges of $\mathcal{G}$. If $\mathcal{G}$ is d-regular then, using Theorem 15.7, it is possible, at least in principle, to find out whether it can be realized as the edge graph of a proper *simple* convex polytope P in $\mathbb{E}^d$. If this is the case, the question arises whether P is determined uniquely (up to combinatorial isomorphisms). Surprisingly, the answer is yes, as shown by Blind and Mani-Levitska [133]. This confirms a conjecture of Perles [791]. Below we reproduce Kalai's [560] elegant proof of this result. See also Ziegler [1045].

Theorem 15.9. *Let a graph $\mathcal{G}$ be isomorphic to the edge graph $\mathcal{G}(P)$ of a proper simple convex polytope P in $\mathbb{E}^d$. Then the combinatorial structure of P is determined by $\mathcal{G}$.*

In other words, given a set of vertices, we can read off from $\mathcal{G}$ whether their convex hull is a face of P or not.

Proof. The notions which will be introduced in the following for $\mathcal{G}(P)$ also apply to $\mathcal{G}$.

Let $\mathcal{O}$ be an *acyclic orientation* of $\mathcal{G}(P)$, that is an edge orientation with no oriented cycle. Since $\mathcal{O}$ contains no oriented cycle, it induces a partial order on $\mathcal{V}(\mathcal{G}(P)) = \mathcal{V}(P)$. For $v, w \in \mathcal{V}(\mathcal{G}(P))$, let $v \preceq_{\mathcal{O}} w$ if there is a path from v to w, oriented from v to w. A vertex of a sub-graph of $\mathcal{G}(P)$ is a *sink* of the sub-graph with respect to $\mathcal{O}$, if there is no edge of this sub-graph incident with v and oriented away from v. $\mathcal{O}$ is *good* if every sub-graph of $\mathcal{G}(P)$ of the form $\mathcal{G}(F)$ where F is a face of P, has precisely one sink.

The existence of good acyclic orientations of $\mathcal{G}(P)$ is easy to see. Choose $u \in S^{d-1}$ such that the linear form $x \to u \cdot x$ assumes different values at different vertices of P. Then orient an edge vw of P from v to w if $u \cdot v < u \cdot w$ and from w to v if $u \cdot v > u \cdot w$.

We now characterize, among all acyclic orientations of $\mathcal{G}(P)$, the good ones. Let $\mathcal{O}$ be an acyclic orientation. The *in-degree* of a vertex v of $\mathcal{G}(P)$ is the number of edges incident with v and oriented towards it. Let $h_i^{\mathcal{O}}$ be the number of vertices of $\mathcal{G}(P)$ with in-degree i, where $i = 0, 1, \dots, d$, and let

$$f^{\mathcal{O}} = h_0^{\mathcal{O}} + 2h_1^{\mathcal{O}} + 2^2 h_2^{\mathcal{O}} + \cdots + 2^d h_d^{\mathcal{O}}.$$

If a vertex v of $\mathcal{G}(P)$ has in-degree i, then v is a sink of 2^i faces of P. (Since P is simple, any set of $j \leq i$ edges of P incident with v determines a face of dimension j of P which contains these edges but no further edge incident with v.) Let f be the number of non-empty faces of P. Since each face of P has at least one sink,

$$f^{\mathcal{O}} \geq f, \text{ and } \mathcal{O} \text{ is good if and only if } f^{\mathcal{O}} = f.$$

Thus,

(3) among all acyclic orientations $\mathcal{O}$ of $\mathcal{G}(P)$, the good orientations are precisely those with minimum $f^{\mathcal{O}}$.

It is easy to see that each face F of the simple polytope P is simple in aff F, i.e. F is i-regular, where $i = \dim F$. Clearly, $\mathcal{G}(F)$ is a sub-graph of $\mathcal{G}(P)$. We next characterize among all sub-graphs of $\mathcal{G}(P)$ those which are the edge graphs of faces of P:

(4) Let $\mathcal{H}$ be an *induced* sub-graph of $\mathcal{G}(P)$ that is, it contains all edges of $\mathcal{G}(P)$ which are incident only with vertices of $\mathcal{H}$. Then the following statements are equivalent:

(i) $\mathcal{H} = \mathcal{G}(F)$ where F is a face of P.

(ii) $\mathcal{H}$ is connected, i-regular for some i and initial with respect to some good acyclic orientation $\mathcal{O}$ of $\mathcal{G}(P)$.

Here, when saying that the induced sub-graph $\mathcal{H}$ is *initial*, we mean that each edge of $\mathcal{G}(P)$ which is incident with precisely one vertex of $\mathcal{H}$ (and thus is not an edge of $\mathcal{H}$) is oriented away from this vertex.

(i)$\Rightarrow$(ii) Clearly, $\mathcal{H}$ is connected and i-regular where $i = \dim F$. Let $u \in S^{d-1}$ such that the linear function $x \to u \cdot x$ assumes different values at different vertices of P and such that the values at the vertices of F are smaller than the values at the other vertices of P. The orientation $\mathcal{O}$ which is determined by this linear form is an acyclic good orientation and $\mathcal{H}$ is initial with respect to it.

(ii)$\Rightarrow$(i) Let v be a sink of $\mathcal{H}$ with respect to $\mathcal{O}$. There are i edges in $\mathcal{H}$ incident with v and oriented towards it. Thus v is a sink of the face F of P of dimension i containing these i edges. Since $\mathcal{O}$ is good, v is the unique sink of F and thus all vertices of F are $\preceq_{\mathcal{O}} v$. Being initial, $\mathcal{H}$ contains all vertices $\preceq_{\mathcal{O}} v$. Hence $\mathcal{V}(F) \subseteq$

$\mathcal{V}(\mathcal{H})$. Since both $\mathcal{G}(F)$ and $\mathcal{H}$ are i-regular, induced and connected, this can happen only if $\mathcal{V}(F) = \mathcal{V}(H)$ and thus $\mathcal{G}(F) = \mathcal{H}$. The proof of (4) is complete.

By (3), all good acyclic orientations of $\mathcal{G}$ can be determined just by considering $\mathcal{G}$. Then (4) permits us to determine those subsets of $\mathcal{V}(\mathcal{G}) = \mathcal{V}(P)$, which correspond to faces of P just by inspection of $\mathcal{G}$ and its good acyclic orientations. □

Remark. There are non-isomorphic convex polytopes with isomorphic edge graphs, see the remarks after Theorem 15.7. Thus the Perles–Blind–Mani theorem cannot be extended to all convex polytopes.

The actual construction of the combinatorial structure by the above proof obviously is prohibitive (one has to compute all orderings of $\mathcal{V}(P)$). A more effective, but still exponential algorithm is due to Achatz and Kleinschmidt [1], but see Joswig, Kaibel and Körner [554].

An extension of the Perles–Blind–Mani theorem to other families of convex polytopes is due to Joswig [553].

The Boundary Complex

There is an algorithm to decide whether a polytopal complex is isomorphic to the boundary complex of a convex polytope in $\mathbb{E}^d$, see Grünbaum [453], p. 92, but there is no feasible characterization of such polytopal complexes. In other words, Problem 15.4 remains unsolved. Considering the universality results of Mnëv and Richter-Gebert which will be described in Sect. 15.5, and algorithmic hardness results, the chances that Problem 15.4 has a positive solution are low.

The *k-skeleton*, $k = 0, 1, \ldots, d$, of a convex polytope P is the complex consisting of all faces of P of dimension at most k. Sometimes the union of these faces is called the k-skeleton, see the proof of Theorem 14.9. Considering the k-skeleton of a convex polytope, what information on the boundary complex does it provide? An extension of an old result of Whitney [1025] says that the $(d-2)$-skeleton of a proper convex polytope P in $\mathbb{E}^d$ determines the boundary complex of P, see Kalai [561].

15.5 Combinatorial Types of Convex Polytopes

Two natural questions on polytopes are the following.

Problem 15.6. *Given a convex polytope P in $\mathbb{E}^d$, describe the space of all convex polytopes which have the same face structure as P. This space is the combinatorial type of P.*

Problem 15.7. *Enumerate the essentially different convex polytopes in $\mathbb{E}^d$. That is, enumerate the combinatorial types of convex polytopes with n vertices for $n = d + 1, d + 2, \ldots$, or, more generally, with n k-faces.*

While trivial for $d = 2$, these questions are difficult for $d \geq 3$.

For $d = 3$ it follows from Steinitz' representation theorem that each combinatorial type of a convex polytope is rather simple. For $d > 3$ the universality theorem of Richter-Gebert says that combinatorial types may be arbitrarily complicated. For $d = 3$ there are asymptotic formulae due to Bender, Richmond and Wormald for the number of combinatorial types of convex polytopes with n vertices, edges, or facets, respectively, as $n \to \infty$. For $d > 3$ upper and lower estimates for the number of combinatorial types of convex polytopes with n vertices are due to the joint efforts of Shemer, Goodman and Pollack and Alon.

In the following, we describe these results but give no proofs. All polytopes considered in this section are proper convex polytopes in $\mathbb{E}^d$.

For more information we refer to Grünbaum [453] and the surveys of Bender [92], Bayer and Lee [83] and Klee and Kleinschmidt [595].

Combinatorial Types and Realization Spaces

Two (proper) convex polytopes in $\mathbb{E}^d$ are *of the same combinatorial type* if they have isomorphic boundary complexes. The equivalence classes of convex polytopes of the same combinatorial type are the *combinatorial types of convex polytopes*.

A more geometric way to express the fact that two convex polytopes P and Q are of the same combinatorial type is the following. It is possible to represent P and Q in the form

$$P = \operatorname{conv}\{x_1, \dots, x_n\},\ Q = \operatorname{conv}\{y_1, \dots, y_n\},$$

where the x_i and the y_i are the vertices of P and Q, respectively, which correspond to each other and such that

$$\operatorname{conv}\{x_{i_1}, \dots, x_{i_m}\} \in \mathcal{C}(P) \Leftrightarrow \operatorname{conv}\{y_{i_1}, \dots, y_{i_m}\} \in \mathcal{C}(Q)$$
$$\text{for each set } \{i_1, \dots, i_m\} \subseteq \{1, \dots, n\}.$$

For the investigation of a combinatorial type it is sometimes more convenient to study a proper sub-space (which easily yields the whole combinatorial type). Fix affinely independent points $x_1, \dots, x_{d+1}$ in $\mathbb{E}^d$ and let $P = \operatorname{conv}\{x_1, \dots, x_n\}$ be a proper convex polytope with vertices x_i such that the vertices $x_2, \dots, x_{d+1}$ are adjacent to the vertex x_1. The *realization space* $\mathcal{R}(P)$ of P is the family of all $d \times n$ matrices $(y_1, \dots, y_n) \in \mathbb{E}^{dn}$ such that the convex polytopes $P = \operatorname{conv}\{x_1, \dots, x_n\}$ and $Q = \operatorname{conv}\{y_1, \dots, y_n\}$ are of the same combinatorial type with corresponding vertices x_i and y_i, where $x_1 = y_1, \dots, x_{d+1} = y_{d+1}$. See Richter-Gebert and Ziegler [834] and Richter-Gebert [833].

Primary Semi-Algebraic Sets

A *primary semi-algebraic set* $\mathcal{A}$ (over $\mathbb{Z}$) in $\mathbb{E}^N$ is a set of the form

$$\mathcal{A} = \big\{x \in \mathbb{E}^N : p_1(x) = \dots = p_k(x) = 0,\ q_1(x), \dots, q_l(x) > 0\big\},$$

where $N \in \mathbb{N}$ and the p_i and the q_j are real polynomials on $\mathbb{E}^N$ with integer coefficients. For example, $\{0, 1\}$ and $]0, 1[$ are primary semi-algebraic sets, but $(0, 1]$ and $[0, 1]$ are not.

On the family of all primary semi-algebraic sets, there is an equivalence relation, called *stable equivalence*, which preserves certain geometric properties, for example the homotopy type, see [834].

What do Realization Spaces Look Like?

It is easy to show that the realization space of a convex polytope P is a primary semi-algebraic set. (The polynomials p_i and q_j are formed with determinants of $d \times d$ minors of the $d \times n$ matrices $x = (x_1, \dots, x_n) \in \mathbb{E}^N$.)

The following result, for $d = 3$, follows from a close inspection of a proof of the Steinitz representation theorem, see Richter-Gebert [833].

Theorem 15.10. *The realization space of a proper convex polytope P in $\mathbb{E}^3$ is a smooth open ball in $\mathbb{E}^{f_1 - 6}$, where f_1 is the number of edges of P.*

For $d > 3$ the situation is much more involved. Improving upon a deep theorem of Mnëv [746], Richter-Gebert [833] proved the following *universality theorem.*

Theorem 15.11. *For every primary semi-algebraic set $\mathcal{A}$, there is a (not necessarily proper) convex polytope P in $\mathbb{E}^4$, such that $\mathcal{A}$ is stably equivalent to $\mathcal{R}(P)$.*

Tools for the proof are so-called Lawrence extensions and connected sums. These are elementary geometric operations on polytopes.

Remark. The universality theorem shows that the realization spaces of convex polytopes may be arbitrarily complicated. This has been interpreted to indicate that there might not exist a reasonable extension of the representation theorem of Steinitz to higher dimensions. The universality theorem has a series of consequences, for example the following: There is a convex polytope in $\mathbb{E}^4$ which admits no realization with rational vertices, compare the discussion in Sect. 15.3.

How Many Combinatorial Types are There?

For $d = 3$ results of Bender and Wormald [94] and Bender and Richmond [93] yield asymptotic formulae for the number of combinatorial types of convex polytopes with n vertices, edges, or faces, as $n \to \infty$. For a survey, see Bender [92].

For $d > 3$ results of Shemer [930], Goodman and Pollack [385] and Alon [25] together yield the following estimates.

Theorem 15.12. *Let $c_s(d, n)$ and $c(d, n)$ be the numbers of combinatorial types of proper simplicial and general convex polytopes in $\mathbb{E}^d$ with n vertices. Then*

$$\left(\frac{n-d}{d}\right)^{\frac{nd}{4}} \le c_s(d, n) \le c(d, n) \le \left(\frac{n}{d}\right)^{d^2 n(1+o(1))} \quad \text{as } n \to \infty.$$

16 Volume of Polytopes and Hilbert's Third Problem

By general agreement, a notion of volume of convex polytopes is to be at least simply additive, translation invariant, non-negative and such that the volume of a cube of edge-length one equals one. Surprisingly, the proof that there is such a notion seems to require a limiting argument, for example the exhaustion method. The proof of the uniqueness is difficult. Rigorous treatments in the context of polytope theory came forth rather late. We mention Schatunovsky [884], Süss [977] and, in particular, Hadwiger [468].

A different line of attack is to try to reduce the volume problem for convex polytopes to that for cubes in the following way, where the volume of a cube is the dth power of its edge-length. Dissect each polytope into polytopal pieces which, when rearranged properly, form a cube. The volume of the polytope then is defined to be the volume of the cube. This is possible for $d = 2$ but not for $d \geq 3$. The latter was shown by Dehn [250], thereby solving Hilbert's third problem. While the volume problem thus cannot be solved by dissections, a rich theory developed around the question as to when two convex polytopes are $\mathcal{G}$-equidissectable, where $\mathcal{G}$ is a group of rigid motions. Important contributors are, amongst others, Hadwiger, Sydler, Jessen and Thorup, Sah, Schneider, and McMullen.

In this section, we first show that there is a unique notion of volume for convex polytopes. Secondly, the equidissectability result for polygons of Bolyai and Gerwien and the non-equidissectability result of Dehn for regular tetrahedra and cubes are presented.

For general information the reader is referred to Hadwiger [468], more special references will be given later.

16.1 Elementary Volume of Convex Polytopes

In Sect. 7.2, we defined the notions of the elementary volume of axis parallel boxes and of the Jordan measure of convex bodies. Both turned out to be valuations with special properties. Conversely, it was shown in Sect. 7.3 that valuations with these properties on the spaces of boxes and convex bodies are, up to multiplicative constants, the elementary volume and the Jordan measure, respectively.

Here, this program is extended to the elementary volume of convex polytopes, yet in a strange order. First it will be shown that there is at most one candidate for the notion of elementary volume. In this part of the proof a simple limiting argument is needed. Then we present a candidate and show that it has the required properties. While related to Hadwiger's [468] proof, the subsequent proof is slightly simpler. An alternative approach to the elementary volume is mentioned. We also make a remark, why we don't simply use Lebesgue or Jordan measure instead of the elementary volume.

The analogous result for spherical spaces is due to Schneider [903], see also Böhm and Hertel [135]. A general approach which also treats hyperbolic spaces, was outlined in McMullen and Schneider [716]. For discussions, see McMullen [714]. We refer also to the simple presentation of Rokhlin [855].

Dissections

A polytope $P \in \mathcal{P}_p$ is *dissected* into the polytopes $P_1, \ldots, P_n \in \mathcal{P}_p$, in symbols,

$$P = P_1 \dot\cup \cdots \dot\cup P_m,$$

if $P = P_1 \cup \cdots \cup P_m$ and the polytopes P_i have pairwise disjoint interiors. $\{P_1, \ldots, P_m\}$ then is said to be a *dissection* of P.

Uniqueness of Elementary Volume

The elementary volume on the space $\mathcal{P}$ of convex polytopes is a valuation with special properties. We first show that there is at most one such valuation.

Theorem 16.1. *Let Φ, Ψ be simple, translation invariant, monotone valuations on $\mathcal{P}$ with $\Phi([0,1]^d) = \Psi([0,1]^d) = 1$. Then $\Phi = \Psi$.*

Proof (by induction on d). If $d = 1$, then it is easy to see that $\Phi([\alpha, \beta]) = |\alpha - \beta| = \Psi([\alpha, \beta])$ for all intervals $[\alpha, \beta] \subseteq \mathbb{R}$, see the corresponding argument for boxes in the proof of Theorem 7.6.

Assume now that $d > 1$ and that the theorem holds for dimension $d - 1$. Since Φ and Ψ both are valuations on $\mathcal{P}$, they satisfy the inclusion–exclusion principle by Volland's extension theorem 7.2. The assumption that Φ and Ψ are simple then implies that they are *simply additive*:

(1) Let $P = P_1 \dot\cup \cdots \dot\cup P_m$, where $P, P_1, \ldots, P_m \in \mathcal{P}_p$.
Then $\Phi(P) = \Phi(P_1) + \cdots + \Phi(P_m)$ and similarly for Ψ.

Thus the translation invariance of Φ and Ψ together with $\Phi([0,1]^d) = \Psi([0,1]^d) = 1$ shows that

(2) $\Phi(K) = \Psi(K) = 1$ for any cube K in $\mathbb{E}^d$ of edge-length 1,

see the proof of statement (11) in the proof of Theorem 7.5.

Next, the following will be shown.

(3) $\Phi(Z) = \Psi(Z)$ for each right cylinder $Z \in \mathcal{P}$ of height 1.

Let H be a hyperplane and u a normal unit vector of H. Consider the functions $\varphi, \psi : \mathcal{P}(H) \to \mathbb{R}$ defined by

$$\varphi(Q) = \Phi(Q + [o, u]), \quad \psi(Q) = \Psi(Q + [o, u]) \text{ for } Q \in \mathcal{P}(H).$$

It is easy to see that φ and ψ are simple, translation invariant, monotone valuations on $\mathcal{P}(H)$. If L is a cube of edge-length 1 in H, then $L + [o, u]$ is a cube in $\mathbb{E}^d$ of edge-length 1, and thus

$$\varphi(L) = \Phi(L + [o, u]) = 1 = \Psi(L + [o, u]) = \psi(L)$$

by (2). Hence $\varphi = \psi$ by induction and therefore

$$\Phi(Q + [o, u]) = \varphi(Q) = \psi(Q) = \Psi(Q + [o, u]) \text{ for } Q \in \mathcal{P}(H).$$

Since this holds for any hyperplane in $\mathbb{E}^d$, the proof of (3) is complete.

For each hyperplane H the valuation φ is simple. Hence any two such valuations on different hyperplanes are 0 in the intersection of these hyperplanes and thus coincide for polytopes in the intersection. Thus all these valuations together yield a valuation ϕ, say, on the space of all polytopes $Q \in \mathcal{P}$ with $\dim Q \leq d - 1$.

The next proposition refines (3).

(4) $\Phi(Z) = \Psi(Z) = h\,\phi(Q)$
for each right cylinder $Z \in \mathcal{P}$ with base Q and height h.

For cylinders of height 1 this holds by (3) and its proof. For cylinders of height $1/n$ this holds by dissection, Proposition (1) and translation invariance. For cylinders of height l/n this then holds by dissection, Proposition (1) and translation invariance. For arbitrary real height, it finally follows by monotony.

After these preliminaries we will prove that

(5) $\Phi(P) = \Psi(P)$ for each $P \in \mathcal{P}$.

If P is improper, then $\Phi(P) = 0 = \Psi(P)$ since both Φ and Ψ are simple. Suppose now that P is proper. P can be dissected into proper simplices. Since Φ and Ψ are simply additive by (1), it suffices for the proof of (5) to show that $\Phi(S) = \Psi(S)$ for each proper simplex S. By considering the centre c of the inball of S of maximum radius and for any facet F of S its convex hull with c, we see that S can be dissected into simplices $T = \text{conv}(\{c\} \cup F)$, where the orthogonal projection b of c into the hyperplane $H = \text{aff}\, F$ is contained in F. It is thus sufficient to prove that $\Phi(T) = \Psi(T)$ for such simplices T. Let $n \in \mathbb{N}$. The n hyperplanes

$$H_i = H + \frac{i}{n}(c - b),\ i = 0, \dots, n-1,$$

dissect T into n polytopes $T_1, \dots, T_n \in \mathcal{P}_p$, say,

$$T = T_1 \dot{\cup} \cdots \dot{\cup} T_n. \tag{6}$$

Consider the cylinders $Y_1, \dots, Y_n,\ Z_1, \dots, Z_{n-1}$ where

$$Y_i = H_{i-1} \cap T + \left[o, \frac{1}{n}(c-b)\right],\ Z_i = H_i \cap T + \left[o, -\frac{1}{n}(c-b)\right].$$

Then $Y_i \supseteq T_i \supseteq Z_i$. This, the monotonicity of Φ, (6) and (1) together yield that

(7) $\Phi(Z_1) + \cdots + \Phi(Z_{n-1}) \leq \Phi(T_1) + \cdots + \Phi(T_n) = \Phi(T)$
$\leq \Phi(Y_1) + \cdots + \Phi(Y_n).$

Similar inequalities hold for Ψ. Noting (4), it then follows that

$$\Phi(Z_1) + \cdots + \Phi(Z_{n-1}) \leq \Psi(T) \leq \Phi(Y_1) + \cdots + \Phi(Y_n),$$

i.e.

$$(8)\quad -\Phi(Y_1) - \cdots - \Phi(Y_n) \le -\Psi(T) \le -\Phi(Z_1) - \cdots - \Phi(Z_{n-1}).$$

Adding (7) and (8) and taking into account that Y_{i+1} is a translate of Z_i, the translation invariance of Φ implies that

$$-\Phi(Y_1) \le \Phi(T) - \Psi(T) \le \Phi(Y_1),$$

i.e.

$$|\Phi(T) - \Psi(T)| \le \Phi(Y_1) = \frac{\|c-b\|}{n}\,\phi(F) \text{ for } n \in \mathbb{N}$$

by (4). Since this holds for all n, it follows that $\Phi(T) = \Psi(T)$, concluding the proof of (5). The induction is complete. □

Existence of Elementary Volume

The elementary volume of convex polytopes is to be a simple, translation invariant, non-negative (equivalently, a monotone) valuation on $\mathcal{P}$, such that for the unit cube $[0,1]^d$ its value is 1. If it exists, it is unique by what was shown earlier. A *candidate* for the elementary volume is the function $V : \mathcal{P} \to \mathbb{R}$, defined inductively as follows. For $d = 1$, let $V([\alpha, \beta]) = |\alpha - \beta|$ for each interval $[a, b] \subseteq \mathbb{R}$. Assume now that $d > 1$ and that the volume has been defined in dimension $d-1$. Then it is defined on each hyperplane and since it is simple, it is 0 on intersections of hyperplanes. This leads to a notion of elementary volume or area for all $(d-1)$-dimensional convex polytopes in $\mathbb{E}^d$. Denote it by v. Then V is defined by:

$$(9)\quad V(P) = \frac{1}{d}\sum_{i=1}^{m} h_P(u_i)\,v(F_i) \text{ for } P \in \mathcal{P}(\mathbb{E}^d),$$

where $F_1, \dots, F_m$ are the facets of P and $u_1, \dots, u_m$ the corresponding exterior normal unit vectors of P. If $\dim P < d-1$, the definition (9) is to be understood as $V(P) = 0$ (empty sum) and in case $\dim P = d-1$ the polytope P has two facets, both coinciding with P but with opposite exterior normal unit vectors and (9) yields $V(P) = 0$.

The following result shows that V, as defined in (9), has the required properties. By the uniqueness theorem 16.1 it is the unique such function and thus is legitimately called *the elementary volume* on $\mathcal{P}$.

Theorem 16.2. *V is a simple, translation invariant, monotone valuation on $\mathcal{P}$ with $V([0,1]^d) = 1$.*

Proof. The following assertions will be shown by induction on d:

(10) V is simply additive, that is, $V(P) = V(P_1) + \cdots + V(P_m)$ for $P, P_1, \dots, P_m \in \mathcal{P}_p$ such that $P = P_1 \dot\cup \cdots \dot\cup P_m$.

(11) V is a simple valuation on $\mathcal{P}$.

(12) The volume of any box in $\mathbb{E}^d$ equals the product of its edge-lengths. In particular, $V([0,1]^d) = 1$.

(13) Let L be a line through o and Z an unbounded cylinder parallel to L with polytopal cross-section and let $l > 0$. Then for all parallel slabs S for which $L \cap S$ is a line segment of length l, the volume $V(S \cap Z)$ is the same.

(14) V is translation invariant.

(15) V is non-negative.

(16) V is monotone.

For $d = 1$ statements (10)–(16) are trivial. Assume now that $d > 1$ and that (10)–(16) hold in (all hyperplanes of) $\mathbb{E}^d$ for v in place of V.

The proof of (10) for d is by induction on m. It is trivial for $m = 1$. Assume next that $m = 2$. Let P_1 and P_2 have the facets $F_1, \dots, F_j$ and $G_1, \dots, G_k$, respectively, with corresponding exterior normal unit vectors $u_1, \dots, u_j$ and $v_1, \dots, v_k$, say. Since $P = P_1 \dot{\cup} P_2 \in \mathcal{P}_p$, we may assume that

$F_1 = G_1,\ u_1 = -v_1$.
$F_2 \dot{\cup} G_2, \dots, F_l \dot{\cup} G_l$ are facets of P with exterior normal unit vectors $u_2 = v_2, \dots, u_l = v_l$.
$F_{l+1}, \dots, F_j,\ G_{l+1}, \dots, G_k$ are facets of P with exterior normal unit vectors $u_{l+1}, \dots, v_k$.

Since by the induction assumption on d, the assertion (10) is valid for v, it thus follows that

$$
\begin{aligned}
V(P) &= V(P_1 \dot{\cup} P_2)\\
&= \frac{1}{d}\sum_{i=2}^{l} h_P(u_i)\, v(F_i \dot{\cup} G_i) + \frac{1}{d}\sum_{i=l+1}^{j} h_P(u_i)\, v(F_i) + \sum_{i=l+1}^{k} h_P(v_i)\, v(G_i)\\
&= \frac{1}{d}\big(h_{P_1}(u_1)\, v(F_1) + h_{P_2}(v_1)\, v(G_1)\big) + \frac{1}{d}\sum_{i=2}^{l}\big(h_{P_1}(u_i)\, v(F_i) + h_{P_2}(v_i)\, v(G_i)\big)\\
&\quad + \frac{1}{d}\sum_{i=l+1}^{j} h_{P_1}(u_i)\, v(F_i) + \frac{1}{d}\sum_{i=l+1}^{k} h_{P_2}(v_i)\, v(G_i) = V(P_1) + V(P_2),
\end{aligned}
$$

concluding the proof of (10) for $m = 2$. Assume now that $m > 2$ and that (10) holds for $1, \dots, m-1$. Consider the case that $P = P_1 \dot{\cup} \cdots \dot{\cup} P_m$. The polytopes $P_1, P_m \in \mathcal{P}_p$ have disjoint interiors and thus can be separated by a hyperplane H, say. Let $H^{\pm}$ be the corresponding closed halfspaces where $P_1 \subseteq H^-$, $P_m \subseteq H^+$, say. We may assume that

$$
P_1, P_2, \dots, P_j \subseteq H^-,\ P_{j+1}, \dots, P_k \not\subseteq H^{\pm},\ P_{k+1}, \dots, P_m \subseteq H^+.
$$

Since we have already proved the assertion (10) for $m = 2$, the induction assumption on m then shows that

$$\begin{aligned} V(P) &= V\big((P \cap H^-)\dot{\cup}(P \cap H^+)\big) = V(P \cap H^-) + V(P \cap H^+) \\ &= V\big(P_1\dot{\cup}\cdots\dot{\cup}P_j\dot{\cup}(P_{j+1} \cap H^-)\dot{\cup}\cdots\dot{\cup}(P_k \cap H^-)\big) \\ &\quad + V\big((P_{j+1} \cap H^+)\dot{\cup}\cdots\dot{\cup}(P_k \cap H^+)\dot{\cup}P_{k+1}\dot{\cup}\cdots\dot{\cup}P_m\big) \\ &= V(P_1) + \cdots + V(P_j) + V(P_{j+1} \cap H^-) + \cdots + V(P_k \cap H^-) \\ &\quad + V(P_{j+1} \cap H^+) + \cdots + V(P_k \cap H^+) + V(P_{k+1}) + \cdots + V(P_m) \\ &= V(P_1) + \cdots + V(P_j) + V(P_{j+1}) + \cdots + V(P_k) + V(P_{k+1}) + \cdots + V(P_m). \end{aligned}$$

The induction on m is thus complete. Hence (10) holds for d and all m. This concludes the induction on d for (10).

For the proof of the assertion (11) note first that $V(P) = 0$ for $P \in \mathcal{P}$ with $\dim P \le d - 1$. Hence V is simple. To see that it is a valuation, let $P, Q \in \mathcal{P}$ such that $P \cup Q \in \mathcal{P}$. The cases where at least one of $P, Q, P \cap Q$ is improper are easily dealt with by the simplicity of V and (10). Assume now that $P, Q, P \cap Q$ all are proper. Consider all hyperplanes through facets of $P \cap Q$. These hyperplanes dissect P and Q into proper convex polytopes

$$P \cap Q,\ P_1, \dots, P_m \text{ and } P \cap Q,\ Q_1, \dots, Q_n,$$

say. Then

$$P \cap Q,\ P_1, \dots, P_m,\ Q_1, \dots, Q_n$$

form a dissection of $P \cup Q$ and it follows from (10) that

$$\begin{aligned} &V(P) + V(Q) \\ &\quad = V(P \cap Q) + V(P_1) + \cdots + V(P_m) + V(Q_1) + \cdots + V(Q_n) + V(P \cap Q) \\ &\quad = V(P \cup Q) + V(P \cap Q). \end{aligned}$$

Thus V is a valuation, concluding the induction for (11).

Assertion (12) easily follows from the definition of V by induction on d.

If $\dim Z < d$, the assertion (13) holds trivially. Assume now that $\dim Z = d$. To speak more easily, call L vertical for the proof of (13). Let S be a parallel slab as in (13) and T a horizontal slab of width l. For the proof of (13) it is sufficient to show that

(17) $V(S \cap Z) = V(T \cap Z).$

If S and T are parallel, this is easy to see. Assume now that S and T are not parallel. By induction,

(18) The contributions of the vertical facets of $S \cap Z$ and $T \cap Z$ to $V(S \cap Z)$ and $V(T \cap Z)$, respectively, are the same.

Next, the non-vertical facets will be considered. Let v be a unit vector parallel to L such that $F, F - lv$ and $G, G - lv$ are the top and bottom facets of $S \cap Z$ and

$T \cap Z$, respectively. Let $\pm u$ be the exterior normal unit vectors of $S \cap Z$ at the facets $F, F - lv$. Clearly, $\pm v$ are the exterior normal unit vectors of $T \cap Z$ at the facets $G, G - lv$. Choose $p \in F$ and $q \in G$ arbitrarily. Then

(19) The contributions of the facets $F, F - lv$ of $S \cap Z$ to $V(S \cap Z)$ and of the facets $G, G - lv$ of $T \cap Z$ to $V(T \cap Z)$ are

$$\frac{1}{d}\big(p \cdot u\ v(F) + (p - lv) \cdot (-u)\ v(F - lv)\big) = \frac{l}{d}\, u \cdot v\ v(F),$$
$$\frac{1}{d}\big(q \cdot v\ v(G) + (q - lv) \cdot (-v)\ v(G - lv)\big) = \frac{l}{d}\, v(G).$$

Here, the translation invariance of v was used. Clearly, $G = F'$, where " $'$ " is the orthogonal projection of the hyperplane $H_F = \operatorname{aff} F$ onto the hyperplane $H_G = \operatorname{aff} G$. Since S, T are not parallel, H_F, H_G are also not parallel. Let $w : \mathcal{P}(H_F) \to \mathbb{R}$ be defined by:

$$w(Q) = v(Q') \text{ for } Q \in \mathcal{P}(H_F).$$

Since, by induction, v is a simple translation invariant, monotone valuation on $\mathcal{P}(H_G)$, it is easy to see, that this also holds for w on $\mathcal{P}(H_F)$. Thus, there are two simple, translation invariant valuations v and w on $\mathcal{P}(H_F)$. The uniqueness theorem 16.1 then yields

$$w(Q) = \alpha\, v(Q) \text{ for } Q \in \mathcal{P}(H_F),$$

where $\alpha \geq 0$ is a suitable constant. To determine its value, choose a $(d-1)$-dimensional cube K in H_F of edge-length 1 such that a $(d-2)$-dimensional facet of K is contained in $H_F \cap H_G$. Then K' is a box in H_G of edge-lengths $1, \dots, 1, u \cdot v$. Thus

$$w(K) = v(K') = u \cdot v = \alpha\, v(K) = \alpha,$$

i.e. $\alpha = u \cdot v$. Here we have used property (12) for v. Hence

$$v(G) = v(F') = \alpha\, v(F) = u \cdot v\ v(F).$$

By (19), this shows that

(20) The contributions of the facets $F, F - lv$ of $S \cap Z$ to $V(S \cap Z)$ and of the facets $G, G - lv$ of $T \cap Z$ to $V(T \cap Z)$ are the same.

Having proved (18) and (20), the proof of (17) is complete, concluding the induction for (13).

For the proof of (14), let $P \in \mathcal{P}$ and $t \in \mathbb{E}^d$. If P is improper, $V(P) = 0 = V(P + t)$ since V is simple by (11), and we are done. Assume now that P is proper. Let L be a line through o with direction t and call it vertical. Let H be the $(d-1)$-dimensional subspace of $\mathbb{E}^d$ orthogonal to L. Let $F_1, \dots, F_m$ be the facets of P on the upper side and $G_1, \dots, G_n$ the facets on the lower side of P. The translate $P + t$ of P may be obtained as follows. Dissect each facet F_i into pieces $F_{i1}, \dots, F_{in}$ and each facet G_j into pieces $G_{1j}, \dots, G_{mj}$ such that

$$F'_{ij} = G'_{ij} = F'_i \cap G'_j \text{ for } i = 1, \dots, m, \ j = 1, \dots, n.$$

(Some pieces may be empty.) Now add to P those cylinders of the form $F_{ij} + [o, t]$ which are proper. This gives a dissection of $P + [o, t]$. Next remove from $P + [o, t]$ those cylinders of the form $G_{ij} + [o, t]$ which are proper. Then $P + t$ remains. Since by (13) $V(F_{ij} + [o, t]) = V(G_{ij} + [o, t])$, Proposition (10) shows that $V(P) = V(P + t)$. The induction for (14) is thus complete.

To prove (15), let $P \in \mathcal{P}$ and choose $p \in P$. Then $o \in P - p$. By induction, $v \geq 0$. The definition of V in (9) then yields $V(P - p) \geq 0$ and thus $V(P) = V(P - p) \geq 0$ by (14), concluding the induction for (15).

For the proof of (16), let $P, Q \in \mathcal{P}$ with $P \subseteq Q$. If P is improper, $V(P) = 0 \leq V(Q)$ by (13) and (15). Assume now that P is proper. The hyperplanes through the facets of P dissect Q into polytopes $P, Q_1, \dots, Q_m \in \mathcal{P}_p$, say. (15) and (10) then imply that

$$V(P) \leq V(P) + V(Q_1) + \dots + V(Q_m) = V(P \dot{\cup} Q_1 \dot{\cup} \cdots \dot{\cup} Q_m) = V(Q)$$

and the induction for (16) is complete.

Having proved (10)–(16) for d, the induction is complete. Thus (10)–(16) and, in particular, the theorem hold generally. □

Simple Consequences

An important property of V is the following.

Corollary 16.1. *V is rigid motion invariant.*

Proof. Let $r : \mathbb{E}^d \to \mathbb{E}^d$ be a rigid motion. Define a mapping $W : \mathcal{P} \to \mathbb{R}$ by $W(P) = V(rP)$ for $P \in \mathcal{P}$. Since $V(\cdot)$ is a simple, translation invariant monotone valuation on $\mathcal{P}$ by Theorem 16.2, it is immediate that this also holds for $W(\cdot)$. The proof of statement (11) in the proof of Theorem 7.5 then yields $W([0, 1]^d) = V([0, 1]^d) = 1$. An application of the uniqueness theorem 16.1 now shows that $V = W$, or $V(P) = V(rP)$ for $P \in \mathcal{P}$. □

The definition of $V(\cdot)$ and an easy induction argument yields the following property.

Proposition 16.1. *V is positively homogeneous of degree d.*

Volume and Elementary Volume

By Theorems 7.5 and 16.2 the volume (or Jordan measure) and the elementary volume are both simple, monotone and translation invariant valuations on $\mathcal{P}$ and their values for the unit cube both are 1. Thus Theorem 16.1 yields the following result:

Corollary 16.2. *On $\mathcal{P}$ volume, i.e. Jordan measure, and elementary volume coincide. Similarly, Minkowski surface area and elementary surface area, i.e. the sum of the elementary volumes of the facets, coincide on $\mathcal{P}$.*

An Alternative Approach to the Existence of the Elementary Volume on Convex Polytopes

Using a well known determinant formula from analytic geometry, define the elementary volume of simplices. It is easy to show that this is a valuation on the space of all simplices in $\mathbb{E}^d$. The latter can be extended, by the extension result of Ludwig and Reitzner [667] mentioned in Sect. 7.1, to a unique, simple, translation invariant, monotone valuation on the space $\mathcal{P}$ of all convex polytopes in $\mathbb{E}^d$. Since it is easy to see that this valuation is 1 for the unit cube, it coincides by Theorem 16.1 with the earlier notion of elementary volume.

Why not Simply use Lebesgue Measure?

There is general agreement, that a notion of volume or measure on the space of convex polytopes should be at least a translation invariant, simple and monotone valuation which assumes the value 1 for the unit cube. Lebesgue measure has these properties and is unique (on the space of Lebesgue measurable sets). So, why not simply use Lebesgue measure instead of the elementary volume, the introduction and uniqueness of which are so complicated to show. The reason is that it is by no means clear that the restriction of Lebesgue measure to the small subspace of convex polytopes is the only valuation having the mentioned properties. (Note that the requirements of being translation invariant, etc. on the space of convex polytopes is a much weaker property than the analogous requirement on the large space of measurable sets.) A similar remark applies to Jordan measure.

16.2 Hilbert's Third Problem

Let $\mathcal{G}$ be a group of rigid motions in $\mathbb{E}^d$. Two proper convex polytopes P, Q are *$\mathcal{G}$-equidissectable* if there are dissections $\{P_1, \dots, P_m\}$ of P and $\{Q_1, \dots, Q_m\}$ of Q such that

$$P_i = m_i Q_i \text{ with suitable } m_i \in \mathcal{G} \text{ for } i = 1, \dots, m.$$

By *equidissectability* we mean $\mathcal{G}$-equidissectability, where $\mathcal{G}$ is the group of all rigid motions.

If two convex polytopes are equidissectable, then they have equal volume. Does the converse hold? For $d = 2$ the answer is yes. While this can easily be shown with the geometric tools already known in antiquity, the first rigorous proofs are due to Bolyai [146] and Gerwien [372]. Farkas Bolyai published his proof in a book for high schools. In an appendix of this book Farkas's son János published his famous result on non-Euclidean geometry. Gerwien was a Prussian officer and amateur mathematician. Gauss, perhaps, was in doubt whether for $d = 3$ the answer still is yes, see his letters [365] to Gerling. In the late nineteenth century there were several attempts to prove that the answer was no, for example by Bricard [166], unfortunately with a gap. This seems to have been the motive for Hilbert [501] to state, in the third problem of his famous list of 23 problems, the following.

> In two letters to Gerling, Gauss expresses his regret that certain theorems of solid geometry depend upon the method of exhaustion, i.e. in modern phraseology, upon the axiom of continuity (or upon the axiom of Archimedes). Gauss mentions in particular the theorem of Euclid, that triangular pyramids of equal altitudes are to each other as their bases. Now the analogous problem in the plane has been solved. Gerling also succeeded in proving the equality of volume of symmetrical polyhedra by dividing them into congruent parts. Nevertheless, it seems to me probable that a general proof of this kind for the theorem of Euclid just mentioned is impossible, and it should be our task to give a rigorous proof of its impossibility. This would be obtained, as soon as we succeeded in *specifying two tetrahedra of equal bases and equal altitudes which can in no way be split up into congruent tetrahedra, and which cannot be combined with congruent tetrahedra to form two polyhedra which themselves could be split up into congruent tetrahedra.*

This problem was solved by Hilbert's student Dehn [250], even before Hilbert's list appeared in print.

The problem of necessary and sufficient conditions for $\mathcal{G}$-equidissectability was studied throughout the whole twentieth century for various groups $\mathcal{G}$ of rigid motions. Hadwiger [465] extended Dehn's necessary conditions for equidissectability to all dimensions and Sydler [979] and Jessen [547] showed their sufficiency for $d = 3$ and $d = 4$, respectively. For $d \geq 5$, the problem is open. For the group of translations necessary and sufficient conditions were given by Hadwiger and Glur [469] ($d = 2$) and Jessen and Thorup [548] (general d). Hadwiger [468], p. 58, showed that two convex polytopes $P, Q \in \mathcal{P}_p$ are $\mathcal{G}$-equidissectable if and only if $\phi(P) = \phi(Q)$ for all $\mathcal{G}$-invariant valuations ϕ on $\mathcal{P}_p$. The case $d = 3$ of this result is due to Jessen [546].

In this section we first prove the simple result of Bolyai and Gerwien by presenting several figures. Then Boltyanskiĭ's [144] concise proof of Dehn's result is presented in which he avoids Hamel functions and thus the axiom of choice. See also the expositions by Boltyanskiĭ [143] and in the nice collection of Aigner and Ziegler [6].

For more information we refer to the books of Hadwiger [468], Boltyanskiĭ [143] and Sah [873] and the surveys of McMullen and Schneider [716], Cartier [192], McMullen [714], Neumann [769], Kellerhals [571] and Dupont [279]. For a popular presentation of Hilbert's third problem, see Gray [393].

Equidissectability of Polygons

The following result is due independently to Bolyai [146] and Gerwien [372].

Theorem 16.3. *Let $P, Q \in \mathcal{P}_p(\mathbb{E}^2)$ such that $A(P) = A(Q)$. Then P and Q are equidissectable.*

Proof. Since equidissectability is a symmetric and transitive relation, it is sufficient to show that

(1) P is equidissectable to a rectangle of edge-lengths 1 and $A(P)$.

P clearly can be dissected into triangles. For the proof of (1) it is thus sufficient to show that

(2) Each triangle T is equidissectable to a rectangle of edge-lengths 1 and $A(T)$.

The following figure shows how one can obtain (2) and thus (1) (Fig. 16.1). □

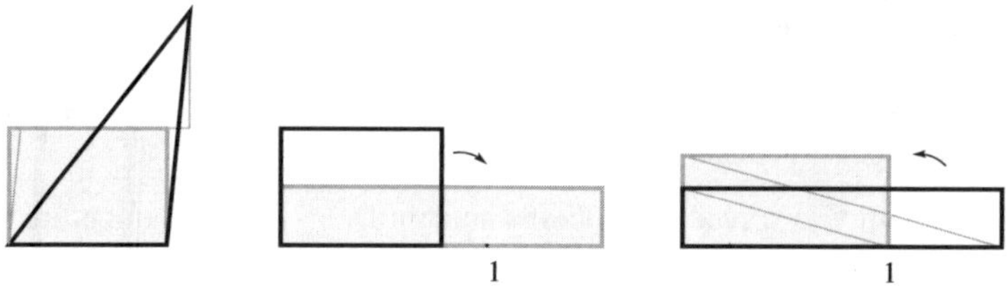

Fig. 16.1. Equidissectability of a triangle and a rectangle with one edge-length equal to 1

Hilbert's Third Problem

Let $\alpha_1, \dots, \alpha_m \in \mathbb{R}$ be rationally independent and let $\beta_1, \dots, \beta_m \in \mathbb{R}$. The corresponding *Hamel quasi-function* f is then defined by:

$$f(r_1\alpha_1 + \cdots + r_m\alpha_m) = r_1\beta_1 + \cdots + r_m\beta_m \text{ for rational } r_i.$$

f is *compatible* with a proper convex polytope P in $\mathbb{E}^3$ if the dihedral angles $\vartheta_1, \dots, \vartheta_m$ of P at its edges are all rational linear combinations of $\alpha_1, \dots, \alpha_m$. The *$f$-Dehn invariant* of P is

$$D_f(P) = \sum_{i=1}^{m} l_i f(\vartheta_i),$$

where l_i is the length of the edge of P corresponding to ϑ_i. Note, if $\alpha_1, \dots, \alpha_m$ and $\beta_1, \dots, \beta_m$ are extended to rationally independent $\alpha_1, \dots, \alpha_m, \dots, \alpha_k \in \mathbb{R}$ and $\beta_1, \dots, \beta_m, \dots, \beta_k \in \mathbb{R}$ and, correspondingly, f to a function g, then

$$D_f(P) = D_g(P).$$

Dehn's theorem is the following.

Theorem 16.4. *Let P, Q be equidissectable, proper convex polytopes in $\mathbb{E}^3$. Then $D_f(P) = D_f(Q)$ for each Hamel quasi-function f which is compatible with P and Q and such that $f(\pi)$ is defined and equal to* 0.

Proof. Let f be such a Hamel quasi-function. Since D_f is rigid motion invariant, it is sufficient to prove the following proposition.

(3) Let $P = P_1 \dot{\cup} \cdots \dot{\cup} P_k$, where $P_1, \dots, P_k \in \mathcal{P}_p$, and extend f to a Hamel quasi-function g compatible with $P_1, \dots, P_k$. Then

$$D_f(P) = D_g(P) = D_g(P_1) + \cdots + D_g(P_k).$$

The set of all vertices of $P, P_1, \dots, P_k$ dissect the edges of these polytopes into smaller line segments which we call *links*. The dihedral angle of one of the polytopes $P, P_1, \dots, P_k$ at a link is the dihedral angle of this polytope at the edge containing this link.

If a link is contained in an edge of P, then, for the dihedral angles $\vartheta, \vartheta_1, \dots, \vartheta_k$ of $P, P_1, \dots, P_k$ at this link, we have, $\vartheta = \vartheta_1 + \cdots + \vartheta_k$. Thus

$$f(\vartheta) = g(\vartheta) = g(\vartheta_1) + \cdots + g(\vartheta_k).$$

If a link is contained in the relative interior of a facet of P, then, for the dihedral angles $\pi, \vartheta_1, \dots, \vartheta_k$ of $P, P_1, \dots, P_k$ at this link, the equality $\pi = \vartheta_1 + \cdots + \vartheta_k$ holds. Thus

$$f(\pi) = 0 = g(\pi) = g(\vartheta_1) + \cdots + g(\vartheta_k).$$

If a link is in the interior of P, then, for the dihedral angles $2\pi, \vartheta_1, \dots, \vartheta_k$ of $P, P_1, \dots, P_k$ at this link, we have $2\pi = \vartheta_1 + \cdots + \vartheta_k$ and thus

$$f(2\pi) = 0 = g(2\pi) = g(\vartheta_1) + \cdots + g(\vartheta_k).$$

Multiplying these equalities by the lengths of the corresponding links and summing over all links yields the equality in (3). □

Corollary 16.3. *Let S be a regular simplex and K a cube in $\mathbb{E}^3$ with $V(S) = V(K)$. Then S and K are not equidissectable.*

Proof. The dihedral angle ϑ of S at any of its edges satisfies the equation $\cos\vartheta = 1/3$. Apply the formula $\cos(\alpha + \beta) = \cos\alpha\cos\beta - \sin\alpha\sin\beta$ with $\alpha = n\vartheta$ and $\beta = \pm\vartheta$ to express $\cos(n+1)\vartheta$ and $\cos(n-1)\vartheta$. Addition then yields

$$\cos(n+1)\vartheta = 2\cos\vartheta\cos n\vartheta - \cos(n-1)\vartheta \text{ for } n \in \mathbb{N}.$$

This, in turn, implies, by a simple induction argument, that

$$\cos n\vartheta = \frac{a_n}{3^n} \text{ for } n \in \mathbb{N}, \text{ where } a_n \in \mathbb{Z},\ 3 \not| a_n.$$

Since $a_n/3^n \neq \pm 1$, it follows that $n\vartheta$ is not an integer multiple of π for any $n \in \mathbb{N}$. Hence ϑ and π are rationally independent. Let f be a Hamel quasi-function such that $f(\vartheta) = 1$, $f(\pi) = 0$ and let l be the edge-length of S. Then

$$D_f(S) = 6l \neq 0 = D_f(K).$$

Thus S and K are not equidissectable by Dehn's theorem 16.4. □

17 Rigidity

Rigidity in the context of convex geometry can be traced back to Book XI of Euclid [310], but the first proper result seems to be Cauchy's [197] rigidity theorem for convex polytopal surfaces in $\mathbb{E}^3$. Cauchy's seminal result gave rise to a series of developments.

First, to flexibility results for general closed polytopal surfaces, including the result of Bricard [167] on flexible immersed octahedra, the examples of flexible embedded polytopal spheres of Connelly [217] and Steffen [954] and the more recent results of Sabitov [871]. Second, to rigidity in the context of differential geometry and, much later, to rigidity in Alexandrov's intrinsic geometry of closed convex surfaces, the latter culminating in the rigidity theorem of Pogorelov [803] for closed convex surfaces in $\mathbb{E}^3$, see Sect. 10.2. Third, to rigidity and infinitesimal rigidity for frameworks starting with Maxwell [700] and with contributions by Dehn [252], Alexandrov [16], Gluck [380] and Asimow and Roth [41].

In this section we present Cauchy's rigidity theorem for convex polytopal surfaces and a result of Asimov and Roth on the flexibility, resp. rigidity of convex frameworks.

References to the voluminous literature may be found in the following books and surveys: Alexandrov [16], Efimov [288, 289], Pogorelov [805], Ivanova-Karatopraklieva and Sabitov [538, 539] (differential geometry, intrinsic geometry of convex surfaces), Ivanova-Karatopraklieva and Sabitov [539], Connelly [218] (non-convex polytopal surfaces), Roth [859], Graver, Servatius and Servatius [391], Maehara [677], Graver [392] (convex and non-convex frameworks).

17.1 Cauchy's Rigidity Theorem for Convex Polytopal Surfaces

In Book XI of the Elements of Euclid [310] Definition 10 is as follows.

> Equal and similar solid figures are those contained by similar planes equal in multitude and magnitude.

The intensive study of Euclid in modern times led to the question whether this was a definition or, rather, a theorem saying that two polytopal surfaces are congruent if their corresponding facets are congruent. Legendre [639] definitely thought that it was a theorem, see the comment of Heath [310], but he was also aware that this theorem could not hold without additional assumption. This is shown by the two polytopal surfaces in Fig. 17.1. Legendre thought that convexity might be such an additional assumption. He drew the attention of young Cauchy to this problem and Cauchy [197] gave a positive answer. Minor errors in his ingenious proof were corrected later on, see [218].

Cauchy won high recognition with this result. He submitted it for publication to the Institute, as the Académie des sciences was called then. The referees Legendre, Carnot, and Biot gave an enthusiastic report which concluded with the following words:

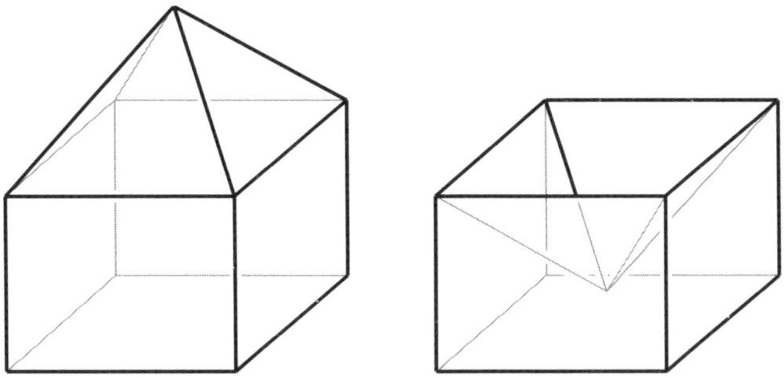

Fig. 17.1. Isometric but not congruent polytopal surfaces

> We wanted to give only an idea of M. Cauchy's proof, but have reproduced the argument almost completely. We have thus furnished further evidence of the brilliance with which this young geometer came to grips with a problem that had resisted even the efforts of the masters of the art, a problem whose solution was utterly essential if the theory of solids was to be perfected.

See Belhoste [91], p.28.

For a nice presentation of Cauchy's proof and a survey of related problems, we refer to Dolbilin [276]. See also the survey of Schlenker [889].

Cauchy's Rigidity Theorem for Convex Polytopal Surfaces

can be stated as follows:

Theorem 17.1. *Let P, Q be proper convex polytopes in $\mathbb{E}^3$. If there is a homeomorphism of* bd P *onto* bd Q *which maps each facet of P isometrically onto a facet of Q, then P and Q are congruent.*

There are many versions of Cauchy's proof in the literature, see, for example the proof in Aigner and Ziegler [6] which includes the arm lemma. The proof consists of two parts, a geometric and a combinatorial one. In the geometric part the following proposition is proved. Mark an edge of P by $+$, $-$, or leave it unmarked, if the dihedral angle of P at this edge is greater than, less than or equal to the dihedral angle of Q at the corresponding edge of Q. If no edge of P is marked, P and Q are congruent as can be seen by building up bd Q beginning with one facet and successively adding adjacent facets. If at least one edge of P is marked, then for any vertex of P on a marked edge the following can be shown. On circling around the vertex, there are at least four changes of sign of the marks encountered (omitting the edges not marked). This is ruled out in the combinatorial part of the proof by an argument based on the Euler polytope formula for planar connected graphs.

The first tool for the proof is the so-called *arm lemma of Cauchy*. We state it without proof. Proofs are elementary, yet complicated. For a proof, see Danzer [240]. For references to other proofs compare [218].

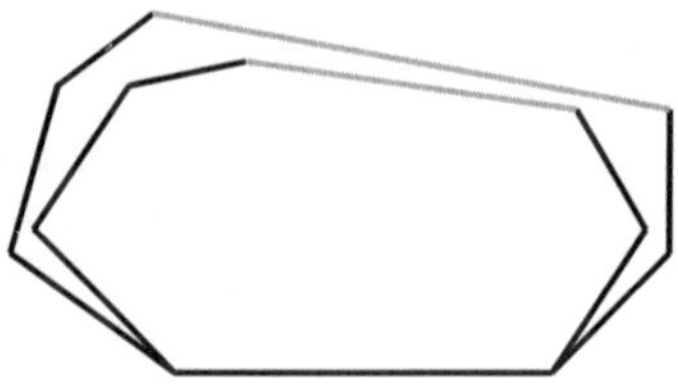

Fig. 17.2. Cauchy's arm lemma

Lemma 17.1. *Let S, T be convex spherical polygons on S^2. Let $s_1, \dots, s_n$ and $t_1, \dots, t_n$ be the vertices of S and T, respectively, in, say, counter clockwise order. Let $\sigma_1, \dots, \sigma_n$ and $\tau_1, \dots, \tau_n$ be the corresponding interior angles. Assume that for the spherical lengths of the edges of S and T,*

$$s_1 s_2 = t_1 t_2, \dots, s_{n-1} s_n = t_{n-1} t_n$$

and for the angles

(1) $\sigma_2 \leq \tau_2, \dots, \sigma_{n-1} \leq \tau_{n-1}$.

Then

(2) $s_n s_1 \leq t_n t_1$.

If, in (1)*, there is strict inequality at least once, then there is strict inequality in* (2) *(Fig. 17.2).*

Corollary 17.1. *Let $S, T, s_1, t_1, \dots, \sigma_1, \tau_1, \dots,$ be as in the lemma and assume that*

(3) $s_1 s_2 = t_1 t_2, \dots, s_{n-1} s_n = t_{n-1} t_n, s_n s_1 = t_n t_1$.

Mark the vertex s_i by $+, -,$ or leave it unmarked, if $\sigma_i > \tau_i$, $\sigma_i < \tau_i$, or $\sigma_i = \tau_i$. If at least one vertex of S is marked, then, on circling bd S*, we have at least four changes of sign of the marks encountered (omitting the unmarked vertices).*

Proof. It is not possible that there is a vertex marked by $-$ but no vertex marked by $+$. If this were the case, we may assume that s_i with $1 < i < n$ is marked by $-$. Then $s_n s_1 < t_n t_1$ by the arm lemma, contrary to (3). Similarly, it is not possible that there is a vertex marked by $+$ but no vertex has mark $-$. Thus there is at least one vertex marked by $+$ and at least one marked by $-$. Hence there are at least two changes of sign.

If there were precisely two changes of sign, then, by re-indexing, if necessary, we may suppose that at least one of the vertices $s_1, \dots, s_k$ is marked by $+$ while the others are also marked by $+$ or are unmarked, and similarly for the vertices $s_{k+1}, \dots, s_n$ with $+$ and $-$ exchanged. We then have the situation as shown in Fig. 17.3. Choose points a, b in the relative interiors of the edges s_k, s_{k+1} and s_n, s_1 of S and points c, d in the relative interiors of the edges t_k, t_{k+1} and t_n, t_1 of T such that $s_k a = t_k c$ and $s_1 b = t_1 d$. Applying the arm lemma to the polygons with vertices $b, s_1, \dots, s_k, a$ and $d, t_1, \dots, t_k, c$ and to the polygons with

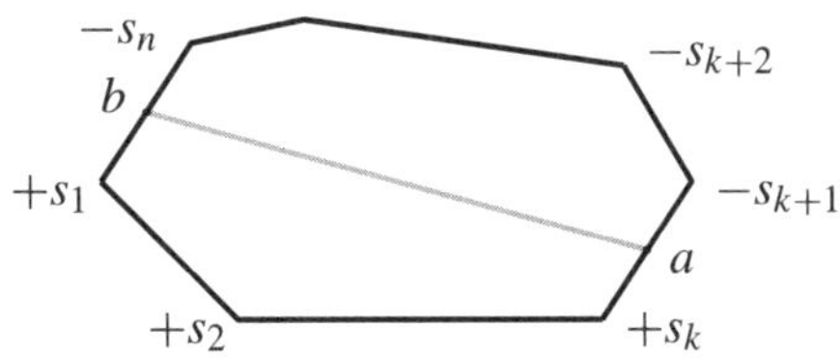

Fig. 17.3. Proof of Cauchy's rigidity theorem

vertices $a, s_{k+1}, \dots, s_n, b$ and $c, t_{k+1}, \dots, t_n, d$, we obtain the contradiction that $ab > cd > ab$. Thus there are more than two changes of sign.

Since the number of changes of sign is even, it follows that there are at least four changes of sign. □

The next step in our proof is *Cauchy's combinatorial lemma*:

Lemma 17.2. *Let P be a proper convex polytope in $\mathbb{E}^3$. Assume that the edges of P are marked by $+$, by $-$, or are unmarked in such a way that the following statement holds. On circling in* bd P *around a vertex of P which is an endpoint of a marked edge, there are at least four changes of sign of the marks encountered (omitting the unmarked edges). Then there are no marked edges.*

Proof. Assume that there are marked edges. Let $\mathcal{G}$ be a graph consisting of a maximal connected set of marked edges and their endpoints. Omitting from bd P the vertices and edges of $\mathcal{G}$ leaves a family of open connected sets in bd P, the countries of $\mathcal{G}$. Since $\mathcal{G}$ is connected, the countries are simply connected. Roughly speaking, each of these countries is a union of certain facets of P. Let v, e, f be the numbers of vertices, edges and countries, respectively, determined by the connected planar graph $\mathcal{G}$. Euler's formula for graphs in $\mathbb{E}^2$, see Theorem 15.2, then shows that

$$v - e + f = 2.$$

For $i = 2, 3, \dots,$ let f_i be the number of countries with a boundary consisting of i edges, where an edge is counted twice if, on circling the boundary of the country, it appears twice. The case $f_2 \neq 0$ cannot hold. Then,

$$f = f_3 + f_4 + \cdots$$

Since an edge of $\mathcal{G}$ is on the boundary of two countries or is counted twice if it is on the boundary of only one country, we see that

$$2e = 3f_3 + 4f_4 + \cdots$$

Hence

$$4v = 8 + 4e - 4f = 8 + 2f_3 + 4f_4 + 6f_5 + 8f_6 + 10f_7 + \cdots$$

Call the *index* of a vertex of $\mathcal{G}$ the number of changes of sign of the marks encountered on circling the vertex once. Let I be the sum of all indices. By the assumption of the lemma,

$$4v \leq I.$$

We now count the indices in a different way. If the boundary of a country determined by $\mathcal{G}$ consists of i edges, circling it there are at most i changes of sign if i is even and at most $i - 1$ if i is odd. Thus

$$I \leq 2f_3 + 4f_4 + 4f_5 + 6f_6 + 6f_7 + \cdots$$

and we obtain the contradiction that

$$8 + 2f_5 + 2f_6 + \cdots \leq 0. \quad \square$$

Proof of Cauchy's rigidity theorem. The given homeomorphism maps the facets, edges and vertices of P onto the facets, edges and vertices of Q, respectively, where corresponding facets are congruent and corresponding edges have equal length. For the proof of the congruence of P and Q it is sufficient to show that the dihedral angles of P and Q at corresponding edges are equal.

Mark an edge of P by $+$, $-$, or leave it unmarked, if the dihedral angle of P at this edge is greater than, less than, or equal to the dihedral angle of Q at the corresponding edge. We have to show that no edge of P is marked. To see the latter, we first show the following:

(4) Let p be a vertex of P which is incident with at least one marked edge. Then, circling around p, we encounter at least four changes of sign.

Consider spheres with centres at p and q, where q is the vertex of Q corresponding to p, and radius $\varrho > 0$. Choose ϱ so small that the sphere with centre p meets only those facets and edges of P which contain p and similarly for q. Intersecting these spheres with P and Q, gives convex spherical polygons S and T, respectively. Call vertices of S and T corresponding if they are determined by corresponding edges of P and Q. The interior angles of S and T at their vertices are simply the dihedral angles of P and Q at their edges with endpoints p and q. Mark a vertex of S by $+$, $-$, or leave it unmarked if this holds for the edge of P which determines this vertex. If a vertex of S is marked by $+$, $-$, or is left unmarked, then the interior angle of S at this vertex is greater than, less than, or equal to the interior angle of T at the corresponding vertex. An application of Corollary 17.1 then shows that circling around S we have at least four changes of sign. Translating this back to P, we thus see that circling around the vertex p, there are at least four changes of sign of the marks at the edges of P with endpoint p. The proof of (4) is complete.

Cauchy's combinatorial lemma and Proposition (4) finally show that there are no marked edges at all. $\square$

Higher Dimensions

Assume that one can prove Cauchy's theorem in $\mathbb{E}^k$. Then it holds also in S^k by the same arguments as for $\mathbb{E}^k$. By intersecting a closed convex polyhedral surface in $\mathbb{E}^{k+1}$ with small spheres centred at the vertices and applying the result in S^k, the polytopal surface turns out to be rigid at each vertex. This immediately yields Cauchy's theorem in $\mathbb{E}^{k+1}$. Since the rigidity theorem holds in $\mathbb{E}^3$ it thus holds in $\mathbb{E}^d$ for all $d \geq 3$. See Alexandrov [16] and Pogorelov [804]. For more information compare Connelly [218].

Flexible Non-convex Polytopal Spheres and the Bellows Conjecture

Cauchy's rigidity theorem implies the following. A closed convex polytopal surface in $\mathbb{E}^3$ with rigid facets and hinges along the edges cannot be flexed. Also the non-convex example of Legendre of Fig. 17.1 does not admit a flexing. More generally, Euler [311] conjectured that

> a closed spatial figure allows no changes as long as it is not ripped apart.

Bricard [167] gave an example of a flexible octahedral surface in $\mathbb{E}^3$. Unfortunately, it suffers from the defect that the surface is *not* embedded, that is, it has self-intersections. The next step of this story is due to Gluck [380] who showed that the closed, simply connected polytopal embedded surfaces in $\mathbb{E}^3$ are generically rigid, that is, at least the large majority is rigid. Finally, Connelly [217], surprisingly, specified a flexible, embedded polytopal sphere in $\mathbb{E}^3$. A simpler example is due to Steffen [954].

The so-called *bellows conjecture* asserts that the volume of a flexible polytopal sphere does not change while flexing. Sabitov's [871] affirmative answer is based on an interesting formula in which the volume is expressed as a polynomial in terms of edge-lengths. This formula may be considered as a far-reaching extension of formulae of Heron for the area of a triangle and Euler for the volume of a tetrahedron. For a survey see Schlenker [889].

17.2 Rigidity of Frameworks

A framework is a system of rods in $\mathbb{E}^d$ with joints at common endpoints such that the rods can rotate freely. A basic question is to decide whether a given framework is rigid or flexible.

Early results on frameworks in the nineteenth century are due to Maxwell [700], Peaucellier [787], Kempe [573], Bricard [167]. Throughout the twentieth century and, in particular, in the last quarter of it, a multitude of results on frameworks were given. Amongst others, these results deal with rigidity and infinitesimal rigidity, with stresses and self-stresses.

In this section a result of Asimow and Roth [41] will be presented, showing that a framework consisting of the edges and vertices of a proper convex polytope in $\mathbb{E}^3$ is rigid if and only if all facets are triangular. We follow Roth [859].

Definitions

An *abstract framework* $\mathcal{F}$ consists of two sets, the *set of vertices* $\mathcal{V} = \mathcal{V}(\mathcal{F}) = \{1, \dots, v\}$ and the *set of edges* $\mathcal{E} = \mathcal{E}(\mathcal{F})$, the latter consisting of two-element subsets of $\mathcal{V}$. For $i \in \mathcal{V}$ let $\mathfrak{a}(i) = \{j \in \mathcal{V} : \{i, j\} \in \mathcal{E}\}$ be the set of vertices *adjacent* to the vertex i. A *framework* $\mathcal{F}(\mathfrak{p})$ in $\mathbb{E}^d$ is an abstract framework $\mathcal{F} = \langle \mathcal{V}, \mathcal{E} \rangle$ together with a point $\mathfrak{p} = (p_1, \dots, p_v) \in \mathbb{E}^d \times \dots \times \mathbb{E}^d = \mathbb{E}^{dv}$. A point p_i, $i \in \mathcal{V}$, is a *vertex* and a line segment $[p_i, p_j]$, $\{i, j\} \in \mathcal{E}$, is an *edge* of $\mathcal{F}(\mathfrak{p})$. The framework $\mathcal{F}(\mathfrak{p})$ is called a *realization* of the abstract framework $\mathcal{F}$. Let e be the number of edges of $\mathcal{F}$. Consider the e functions $f_{\{ij\}} : \mathbb{E}^{dv} \to \mathbb{R}$, $\{i, j\} \in \mathcal{E}$, defined by:

$$f_{\{ij\}}(\mathfrak{x}) = ||x_i - x_j||^2 \text{ for } \mathfrak{x} = (x_1, \dots, x_v) \in \mathbb{E}^{dv}.$$

Let

$$\begin{aligned} \mathcal{R}(\mathfrak{p}) &= \{\mathfrak{x} : f_{\{ij\}}(\mathfrak{x}) = f_{\{ij\}}(\mathfrak{p}) \text{ for all } \{i, j\} \in \mathcal{E}\} \\ &= \bigcap_{\{i,j\} \in \mathcal{E}} \{\mathfrak{x} : f_{\{ij\}}(\mathfrak{x}) = f_{\{ij\}}(\mathfrak{p})\} \subseteq \mathbb{E}^{dv}. \end{aligned}$$

Then

$$\{\mathcal{F}(\mathfrak{x}) : \mathfrak{x} \in \mathcal{R}(\mathfrak{p})\}$$

is the set of all realizations $\mathcal{F}(\mathfrak{x})$ of $\mathcal{F}$ with edge-lengths equal to the corresponding edge-lengths of $\mathcal{F}(\mathfrak{p})$. Call $\mathfrak{x} = (x_1, \dots, x_v)$ *congruent* to $\mathfrak{p} = (p_1, \dots, p_v)$ if there is a rigid motion $m : \mathbb{E}^d \to \mathbb{E}^d$ such that $x_1 = m\, p_1, \dots, x_v = m\, p_v$. Let

$$\mathcal{C}(\mathfrak{p}) = \{\mathfrak{x} : \mathfrak{x} \text{ congruent to } \mathfrak{p}\} \subseteq \mathcal{R}(\mathfrak{p}).$$

If $\text{aff}\{p_1, \dots, p_v\} = \mathbb{E}^d$, then it is not too difficult to show that $\mathcal{C}(\mathfrak{p})$ is a smooth manifold of dimension $\frac{1}{2}d(d+1)$ in $\mathbb{E}^{dv}$, where by *smooth* we mean of class $\mathcal{C}^\infty$. Here $\frac{1}{2}d(d-1)$ dimensions come from the rotations and d from the translations. Then $\{\mathcal{F}(\mathfrak{x}) : \mathfrak{x} \in \mathcal{C}(\mathfrak{p})\}$ is the set of all realizations $\mathcal{F}(\mathfrak{x})$ of $\mathcal{F}$ which are congruent to $\mathcal{F}(\mathfrak{p})$ in the ordinary sense. The framework $\mathcal{F}(\mathfrak{p})$ is *flexible* if there is a continuous function $\mathfrak{x} : [0, 1] \to \mathbb{E}^{dv}$ such that

(1) $$\mathfrak{x}(0) = \mathfrak{p} \in \mathcal{C}(\mathfrak{p}) \text{ and } \mathfrak{x}(t) \in \mathcal{R}(\mathfrak{p}) \backslash \mathcal{C}(\mathfrak{p}) \text{ for } 0 < t \leq 1.$$

$\mathcal{F}(\mathfrak{p})$ is *rigid* if it is not flexible.

A Rigidity Criterion

The above makes it clear that flexibility of a framework $\mathcal{F}(\mathfrak{p})$ is $\mathbb{E}^d$ is determined by the way in which $\mathcal{C}(\mathfrak{p})$ is included in $\mathcal{R}(\mathfrak{p})$ in a neighbourhood of $\mathfrak{p}$. The following result is a simple version of the *rigidity predictor theorem* of Gluck [381].

Theorem 17.2. *Let $\mathcal{F}(\mathfrak{p})$ be a framework in $\mathbb{E}^d$, where $\mathfrak{p} = (p_1, \dots, p_v)$ and* $\text{aff}\{p_1, \dots, p_d\} = \mathbb{E}^d$. *Assume that the e vectors*

$$\text{grad}\, f_{\{ij\}}(\mathfrak{p}), \{i, j\} \in \mathcal{E},$$

are linearly independent in $\mathbb{E}^{dv}$. Then $e \leq dv - \frac{1}{2}d(d+1)$ and $\mathcal{F}(\mathfrak{p})$ is rigid if and only if equality holds.

Proof. Clearly,

$$\mathcal{R}(\mathfrak{p}) = \bigcap_{\{i,j\}\in\mathcal{E}} \mathcal{R}_{\{ij\}} \text{ where } \mathcal{R}_{\{ij\}} = \big\{\mathfrak{x} : f_{\{ij\}}(\mathfrak{x}) = f_{\{ij\}}(\mathfrak{p})\big\}.$$

Since grad $f_{\{ij\}}(\mathfrak{p}) \neq \mathfrak{o}$, the implicit function theorem shows that, in a neighbourhood of $\mathfrak{p}$ in $\mathbb{E}^d$, the set $\mathcal{R}_{\{ij\}}$ is a smooth hypersurface in $\mathbb{E}^{dv}$ with normal vector grad $f_{\{ij\}}(\mathfrak{p})$ at $\mathfrak{p}$. Since, by assumption, the e vectors grad $f_{\{ij\}}(\mathfrak{p})$ are linearly independent, in a suitable neighbourhood of $\mathfrak{p}$ in $\mathbb{E}^d$, the set $\mathcal{R}(\mathfrak{p})$ is a smooth manifold of dimension $dv - e$. See, e.g. Auslander-MacKenzie [43], p. 32. Since, in this neighbourhood, $\mathcal{C}(\mathfrak{p})$ is a sub-manifold of $\mathcal{R}(\mathfrak{p})$, we have $\frac{1}{2}d(d+1) \leq dv - e$ or $e \leq dv - \frac{1}{2}d(d+1)$.

If $e < dv - \frac{1}{2}(d+1)$ then $\mathcal{C}(\mathfrak{p})$ is a proper sub-manifold of $\mathcal{R}(\mathfrak{p})$ and we can choose a continuous function $\mathfrak{x} : [0,1] \to \mathbb{E}^{dv}$ such that (1) holds, i.e. $\mathcal{F}(\mathfrak{p})$ is flexible. If $e = dv - \frac{1}{2}(d+1)$ then $\mathcal{C}(\mathfrak{p})$ has the same dimension as $\mathcal{R}(\mathfrak{p})$ and thus coincides with $\mathcal{R}(\mathfrak{p})$ in a neighbourhood of $\mathfrak{p}$. In this case there is no such continuous function $\mathfrak{x}(t)$ which satisfies (1). Hence $\mathcal{F}(\mathfrak{p})$ is rigid. □

Rigidity of Convex Frameworks in $\mathbb{E}^3$

The rigidity predictor theorem and arguments from the proof of Cauchy's rigidity theorem yield the following result of Asimow and Roth [41].

Theorem 17.3. *Let $\mathcal{F}(\mathfrak{p})$ be the framework consisting of the vertices and edges of a proper convex polytope P in $\mathbb{E}^3$. Then the following statements are equivalent:*
(i) *All facets of P are triangles.*
(ii) *$\mathcal{F}(\mathfrak{p})$ is rigid.*

The implication (i)⇒(ii) is an immediate consequence of Cauchy's rigidity theorem.

Proof. Let $\mathfrak{p} = (p_1, \dots, p_v) \in \mathbb{E}^{3v}$ where $p_1, \dots, p_v$ are the vertices of P. Let $\mathcal{E} = \{\{i,j\} : [p_i, p_j] \text{ is an edge of } P\}$ and define $f_{\{ij\}}(\mathfrak{x}) = \|x_i - x_j\|^2$ for $\mathfrak{x} = (x_1, \dots, x_v) \in \mathbb{E}^{3v}$ and $\{i,j\} \in \mathcal{E}$. Let $\mathfrak{a}(i) = \{j : \{i,j\} \in \mathcal{E}\}$, $i = 1, \dots, v$.

The main step of the proof is to show that

(2) The vectors grad $f_{\{ij\}}(\mathfrak{p})$, $\{i,j\} \in \mathcal{E}$, are linearly independent.

To see this, assume the contrary. Then there are real numbers $\omega_{\{ij\}}$, not all 0, such that

$$(3) \quad \sum_{\{i,j\}\in\mathcal{E}} \omega_{\{ij\}} \text{ grad } f_{\{ij\}}(\mathfrak{p}) = \mathfrak{o}.$$

Note that

$$\text{grad } f_{\{ij\}}(\mathfrak{p}) = 2(o, \dots o, p_i - p_j, o, \dots, o, p_j - p_i, o, \dots, o) \in \mathbb{E}^{3v},$$
for $\{i,j\} \in \mathcal{E}$, or, equivalently, for $i \in \{1, \dots, v\}$ and $j \in \mathfrak{a}(i)$.

Thus (3) implies that

$$(4)\quad \sum_{j\in\mathfrak{a}(i)} \omega_{\{ij\}}(p_i - p_j) = o \text{ for } i = 1, \dots, v.$$

If $\omega_{\{ij\}} > 0$ mark the edge $[p_i, p_j]$ of P by $+$, if $\omega_{\{ij\}} < 0$ mark it by $-$, and leave it unmarked if $\omega_{\{ij\}} = 0$. If p_i is a vertex which is the endpoint of a marked edge, the *index* of p_i is the number of changes of sign of the marks encountered on circling p_i once in bd P. We show the following.

(5) Let p_i be a vertex of P which is the endpoint of at least one marked edge. Then p_i has index at least four.

The index cannot be 0 since, then, all $\omega_{\{ij\}}$ are non-negative and at least one is positive, or all are non-positive and at least one is negative. Let u be an exterior normal vector of a support plane of P which meets P only at p_i. Then $u \cdot (p_i - p_j) > 0$ for all $j \in \mathfrak{a}(i)$. Thus

$$\sum_{j\in\mathfrak{a}(i)} \omega_{\{ij\}}\, u \cdot (p_i - p_j) > 0,$$

contrary to (4). The index cannot be 2 since, then, we may separate the edges of P with endpoint p_i which are marked $+$ from the edges marked $-$ by a hyperplane through p_i. Let u be a normal vector of this hyperplane. Then

$$\sum_{j\in\mathfrak{a}(i)} \omega_{\{ij\}}\, u \cdot (p_i - p_j) = \sum_{\substack{j\in\mathfrak{a}(i)\\ \omega_{\{ij\}}>0}} \omega_{\{ij\}}\, u \cdot (p_i - p_j) + \sum_{\substack{j\in\mathfrak{a}(i)\\ \omega_{\{ij\}}<0}} \omega_{\{ij\}}\, u \cdot (p_i - p_j) \neq 0,$$

again contrary to (4). Since the index is even, it is thus at least 4, concluding the proof of (5).

Combining (5) with the combinatorial lemma 17.2 of Cauchy, we see that there are no marked edges at all, i.e. all $\omega_{\{ij\}}$ are 0. This contradicts the assumption in the proof of (2). The proof of (2) is complete.

Having proved (2), we may apply Theorem 17.2 to see that

(6) $\mathcal{F}(\mathfrak{p})$ is rigid if and only if $e = 3v - 6$.

By Euler's polytope formula 15.1

$$(7)\quad 3v - 6 = 3(v - 2) = 3(e - f) = e + (2e - 3f),$$

where f is the number of facets of P. Let f_i, $i = 3, 4, \dots$, be the number of facets of P with i edges. Since

$$3f = 3\sum_{i\geq 3} f_i \leq \sum_{i\geq 3} i f_i = 2e,$$

where equality holds if and only if $f = f_3$, we conclude from (7) that $3v - 6 \geq e$, where equality holds if and only if all facets of P are triangular. Together with (6), this shows that $\mathcal{F}(\mathfrak{p})$ is rigid if and only if all facets of P are triangular. □

Remark. To Günter Ziegler [1046] we owe the following comment. The proof of the theorem of Asimow and Roth is via the rigidity matrix, which characterizes infinitesimal rigidity. Thus it shows that, in the simplicial case, the polytope actually is infinitesimally rigid, which implies rigidity, but is a stronger property in general.

18 Theorems of Alexandrov, Minkowski and Lindelöf

In convex geometry there are results of geometric interest, but not (yet?) of a systematic character. This does not exclude that such a result is useful as a tool or that there are related results. Of course, this may change over the years. In the nineteenth century Steiner's formula for the volume of parallel bodies was an interesting curiosity, through the work of Minkowski it is now an essential part of the Brunn–Minkowski theory.

In the following we present results of this type on convex polytopes due to Alexandrov [16], Minkowski [739] and Lindelöf [658]. Besides the geometric interest of these results, it is the methods of proof which contribute to their appeal.

18.1 Alexandrov's Uniqueness Theorem for Convex Polytopes

Given two convex polytopes, what conditions ensure that one is a translate of the other? There are several such results in the literature, dealing with projections, sections or properties of faces.

This section contains Alexandrov's sufficient condition for the congruence of convex polytopes in $\mathbb{E}^3$, see [16]. The proof makes use of Cauchy's combinatorial lemma.

Alexandrov's Uniqueness Theorem

In the proof, unconventional terminology is used. A *side* of a (possibly improper) convex polygon or polytope is a vertex or an edge. If F and G are convex polygons, then by *parallel sides* we mean sides defined by support lines of F and G, respectively, with the same exterior normal vector. Let $H = F + G$. Then each edge of H is the sum of *corresponding* (unique, parallel) *sides* of F and G. We say that F can be *embedded into* G if $F + t \subsetneq G$ for a suitable vector t. These definitions may easily be extended to vertices, edges and facets, i.e. to faces of convex polytopes in $\mathbb{E}^3$. *Alexandrov's uniqueness theorem* for convex polytopes in $\mathbb{E}^3$ can now be stated as follows:

Theorem 18.1. *Let P, Q be proper convex polytopes in $\mathbb{E}^3$. Then the following statements are equivalent:*

(i) *For each pair of parallel faces of P and Q, at least one of which is a facet, neither can be embedded in the other.*
(ii) *P and Q coincide up to translation.*

Proof. (i)$\Rightarrow$(ii) The proof of the following simple elementary proposition is left to the reader; see also Alexandrov [16].

(1) Let $H = F + G$, where F and G are proper convex polygons. Mark an edge of H by $+$, $-$, or leave it unmarked, if its corresponding side in F has length greater than, less than, or equal to its corresponding side in G. If no edge of H is marked, then F and G coincide up to translation. If at least one edge of H is marked then, omitting the unmarked edges, there are at least four changes of sign on each circuit of H.

Consider the polytope

$$R = P + Q.$$

Each edge of R is the sum of corresponding sides in P and Q, respectively. At least one of these sides is an edge, and if both are edges, they are parallel in the ordinary sense. Mark an edge of R by $+$, $-$, or leave it unmarked, if the length of its corresponding side in P is greater than, less than or equal to the length of its corresponding side in Q.

The main part of the proof is to show the following proposition.

(2) Let $H = F + G$ be a facet of R, where F and G are the corresponding faces in P and Q, respectively. If no edge of H is marked, then F and G both are facets which coincide up to translation. If at least one edge of H is marked then, omitting the unmarked edges, there are at least four changes of sign on each circuit of H.

The following simple remark will be useful in the proof of (2).

(3) For each edge of H the corresponding sides in P and Q actually are corresponding sides in F and G, respectively. Thus we may define the signs of the edges of H by means of F and G.

To show (2), we distinguish four cases, according to the different possibilities for F and G.

First, F and G are facets. Then take into account the assumption in statement (i) and apply (3) and (1).

Second, F is a facet and G an edge (or vice versa). Since, by the assumption in statement (i), G cannot be embedded into F, the sides of F parallel to G have shorter length than G. These sides of F are separated by edges of F which are not parallel to G. Clearly, for these edges of F, the parallel sides in G are the endpoints of G and thus have length 0. Noting (3), we see then that there are at least four changes of sign on each circuit of H.

Third, F and G are non-parallel edges. Then H is a parallelogram and, noting (3), the edges of H have alternating sign.

Fourth, F is a facet and G a vertex (or vice versa). Since then G can be embedded into F, this possibility is ruled out by the assumption in statement (i). The proof of (2) is complete.

Next, define a graph $\mathcal{G}$ on bd R as follows: in each facet of R choose a relative interior point, a *knot*. Two knots are connected by an *edge*, that is a Jordan curve in bd R if the facets containing the knots have a common (ordinary) edge (of R). Clearly, the edges of $\mathcal{G}$ may be chosen such that they meet only at their endpoints, if at all. Mark an edge of $\mathcal{G}$ by $+$, $-$, or leave it unmarked, if the corresponding edge of R is so marked. Proposition (2) now reads as follows.

> If a knot of $\mathcal{G}$ is the endpoint of a marked edge of $\mathcal{G}$, then, on a circuit of the knot, there are at least four changes of sign of the edges of $\mathcal{G}$, omitting the unmarked edges.

Cauchy's combinatorial lemma 17.2 then shows that there is no marked edge of $\mathcal{G}$. This, in turn, implies that none of the edges of R is marked. Hence (2) implies the following.

> For each facet $H = F+G$ of R the faces F and G of P and Q, respectively, are both facets and they coincide up to translation.

Since this exhausts all facets of P and Q, we see that

> The facets of P and Q appear in parallel pairs and the facets of any such pair coincide up to translation.

Finally, building up bd P and bd Q starting with a pair of parallel facets and adding the adjacent facets, etc., we see that bd P and bd Q and thus P and Q coincide up to translation.

(ii)$\Rightarrow$(i) Trivial. □

Remark. This result does *not* extend to $\mathbb{E}^d$, $d \geq 4$ in a straightforward way. To see this, consider a box in $\mathbb{E}^d$ with edge-lengths $1, 1, 3, \dots, 3$ and a cube of edge-length 2 with edges parallel to those of the box.

Problem 18.1. *Find a version of Alexandrov's theorem which holds in every dimension.*

18.2 Minkowski's Existence Theorem and Symmetry Condition

It is a natural question to ask whether a convex polytope is determined by the areas of its facets or the curvatures at its vertices. The first pertinent result of this type seems to be Minkowski's [739] existence and uniqueness theorem for convex polytopes with given exterior normal vectors and areas of the facets. This result is the first step in the proof of the existence of a convex body with given surface area measure, see Sect. 10.1. A different result is Alexandrov's theorem [16] on the existence and uniqueness of convex polytopes with vertices on fixed rays and given corresponding curvatures.

Here, we present Minkowski's existence and uniqueness theorem for convex polytopes. As an application, we show Minkowski's condition for the central symmetry of a convex polytope which will be used in the proof of the Venkov–McMullen theorem on tilings in Sect. 32.2.

For other pertinent results, see Alexandrov [16]. See also the report in Schneider [907].

Minkowski's Existence and Uniqueness Theorem for Convex Polytopes

Minkowski [739] proved the following result; the proof is taken from Alexandrov [16]. It makes use of the Lagrange multiplier theorem from calculus.

Theorem 18.2. *Let $u_1, \dots, u_n \in S^{d-1}$ and $\alpha_1, \dots, \alpha_n > 0$. Then the following statements are equivalent:*

(i) *$u_1, \dots, u_n$ and $\alpha_1, \dots, \alpha_n$ are the exterior normal unit vectors and the areas of the corresponding facets of a proper convex polytope P in $\mathbb{E}^d$ which is unique up to translation.*

(ii) *$u_1, \dots, u_n$ are not contained in a halfspace whose boundary hyperplane contains o and*

$$\sum_{i=1}^{n} \alpha_i u_i = o.$$

Proof. (i)⇒(ii) Clearly, $P = \{x : u_i \cdot x \le h_P(u_i),\ i = 1, \dots, n\}$. Let $x \in P$. If there were a vector $u \neq o$ such that $u_i \cdot u \le 0$ for $i = 1, \dots, n$, then $x + \lambda u \in P$ for all $\lambda \ge 0$. This contradicts the boundedness of P and thus proves the first assertion in (ii). To see the second assertion, note that

$$\left(\sum_{i=1}^{n} \alpha_i u_i\right) \cdot u = \sum_{u_i \cdot u > 0} \alpha_i\, u_i \cdot u - \sum_{u_i \cdot u < 0} \alpha_i\, u_i \cdot (-u) \text{ for each } u \in S^{d-1}.$$

Since both sums on the right hand side are equal to the area of the orthogonal projection of P into the hyperplane $u^{\perp}$, it follows that

$$\left(\sum_{i=1}^{n} \alpha_i u_i\right) \cdot u = 0 \text{ for each } u \in S^{d-1}.$$

This readily implies the second assertion in (ii).

(ii)⇒(i) First, the existence of P will be shown. The first step is to prove that

(1) $P\big(s = (s_1, \dots, s_n)\big) = \{x : u_i \cdot x \le s_i,\ i = 1, \dots, n\}$
is a convex polytope with $o \in P(s)$ for $s_1, \dots, s_n \ge 0$.

Clearly, $P(s)$ is a convex polyhedron with $o \in P(s)$. We have to show that $P(s)$ is bounded. If not, then $P(s)$ contains a ray starting at o, say $\{\lambda u : \lambda \ge 0\}$, where $u \neq o$. Then

$$u_i \cdot u \le \frac{s_i}{\lambda} \text{ for all } \lambda > 0 \text{ and thus } u_i \cdot u \le 0 \text{ for } i = 1, \dots, n,$$

in contradiction to (ii), concluding the proof of (1).

For $s_1, \dots, s_n \ge 0$, let $A_i\big(s = (s_1, \dots, s_n)\big)$ be the area of the (possibly empty) face $F_i(s) = P(s) \cap \{x : u_i \cdot x = s_i\}$ of $P(s)$. (If $F_i(s) = \emptyset$, put $A_i(s) = 0$.) Then

(2) $V(s) = V\big(P(s)\big)$ is differentiable for $s_1, \dots, s_n > 0$, and

$$\frac{\partial V(s)}{\partial s_i} = A_i(s) \text{ for } i = 1, \dots, n.$$

The proof of the formula for the partial derivatives of $V(s)$, in the case where $A_i(s) = 0$, is left to the reader. If $A_i(s) > 0$, then, for sufficiently small $|h|$, the polytopes $P(s_1, \dots, s_i + h, \dots, s_n)$ and $P(s_1, \dots, s_n)$ differ by the convex hull of $F_i(s_1, \dots, s_i + h, \dots, s_n)$ and $F_i(s_1, \dots, s_n)$. Since $F_i(s_1, \dots, s_i + h, \dots, s_n)$ tends to $F(s_1, \dots, s_n)$ as $h \to 0$ (with respect to the Hausdorff metric), the volume of the convex hull is $A_i(s_1, \dots, s_n)|h| + o(|h|)$. Hence

$$V(s_1, \dots, s_i + h, \dots, s_n) - V(s_1, \dots, s_n) = A(s_1, \dots, s_n)h + o(h) \text{ as } h \to 0.$$

Now divide by h and let h tend to 0 to get the formula for the partial derivatives of $V(s)$ in (2). From the formula and the continuity of A_i, the differentiability of V follows.

The simplex

$$(3)\quad S = \left\{ s : s_i \geq 0, \sum_{i=1}^{n} \alpha_i s_i = 1 \right\} \subseteq \mathbb{E}^n$$

is compact. Since V is continuous on S, it attains its maximum on S at a point of S, say a. If $a \notin \operatorname{relint} S$, at least one a_i is 0 and thus $o \in \operatorname{bd} P(a)$ by (1). Choose a vector $v \neq o$ such that $o \in \operatorname{int}\big(P(a) + v\big)$. Then

$$\begin{aligned} P(a) + v &= \big\{x + v : u_i \cdot x \leq a_i,\ i = 1, \dots, n\big\} \\ &= \big\{y : u_i \cdot y \leq a_i + u_i \cdot v,\ i = 1, \dots, n\big\} \\ &= P(b), \text{ where } b = a + (u_1 \cdot v, \dots, u_n \cdot v). \end{aligned}$$

Since $o \in \operatorname{int}\big(P(a) + v\big) = \operatorname{int} P(b)$, we thus see that $b_1, \dots, b_n > 0$. Since $a \in S$, the equality in statement (ii) implies that

$$\sum_{i=1}^{n} \alpha_i b_i = \sum_{i=1}^{n} \alpha_i a_i + \sum_{i=1}^{n} \alpha_i\, u_i \cdot v = 1.$$

Hence $b \in \operatorname{relint} S$. We thus have shown that

$$V \text{ attains its maximum on } S \text{ at the point } b \in \operatorname{relint} S.$$

Propositions (2), (3) and the Lagrange multiplier theorem then yield

$$\frac{\partial}{\partial s_i}\left(V(s) - \lambda \sum_{i=1}^{n} \alpha_i s_i \right)\Bigg|_{s=b} = A_i(b) - \lambda \alpha_i = 0 \text{ for } i = 1, \dots, n,$$

where λ is a suitable constant. Since $b_1, \dots, b_n > 0$, the convex polytope $P(b)$ is proper and thus $A_i(b) > 0$ for certain i. Hence $\lambda \neq 0$ and we obtain

$$\alpha_i = \frac{1}{\lambda} A_i(b) \text{ for } i = 1, \dots, n.$$

The polytope $P = \lambda^{-\frac{1}{d-1}} P(b)$ then has the desired properties.

Secondly, we show that P is unique up to translation. For $d = 3$, this is an immediate consequence of Alexandrov's uniqueness theorem 18.1. For general d, we argue as follows:

Let $P, Q \in \mathcal{P}_p$ be two convex polytopes with $u_1, \dots, u_n \in S^{d-1}$ and $\alpha_1, \dots, \alpha_n > 0$ as exterior normals and areas of their facets. By Lemma 6.4, we have

$$V(P, Q, \dots, Q) = \frac{1}{d} \sum_{i=1}^{n} h_P(u_i)\, \alpha_i = V(P, \dots, P) = V(P).$$

Thus

$$V(P)^d = V(P, Q, \dots, Q)^d \geq V(P)V(Q)^{d-1}, \text{ or } V(P) \geq V(Q)$$

by Minkowski's first inequality, see Theorem 6.11. Similarly, $V(Q) \geq V(P)$ and therefore,

$$V(P) = V(Q) = V(P, Q, \dots, Q).$$

Hence there is equality in Minkowski's first inequality. Together with $V(P) = V(Q)$ this shows that P and Q coincide up to translation. □

Remark. Minkowski's theorem is an existence and uniqueness theorem. A first algorithm to *construct* P, given the exterior normal vectors and the areas of the facets, is due to Little [662]. Complexity questions are studied by Gritzmann and Hufnagel [395].

Minkowski's Symmetry Condition

In some contexts, it is important to know whether a convex body or a polytope is centrally symmetric, for example in problems dealing with packing and tiling, compare Sects. 30.1–30.3 and 32.2. Thus symmetry criteria are of interest. As an immediate consequence of the uniqueness statement in Minkowski's existence and uniqueness theorem we have the following result.

Corollary 18.1. *Let $P \in \mathcal{P}_p$. Then the following statements are equivalent:*

(i) *P is centrally symmetric.*

(ii) *For any facet of P there is a (unique) facet of P with equal area and opposite exterior unit normal vector.*

Proof. Since the implication (i)⇒(ii) is trivial, it is sufficient to show that

(ii)⇒(i) Let $u_1, -u_1, \dots, u_m, -u_m$ be the exterior normal unit vectors of the facets of P and $\alpha_1, \alpha_1, \dots, \alpha_m, \alpha_m$ the corresponding areas. By (ii), $u_1, -u_1, \dots, u_m, -u_m$ and $\alpha_1, \alpha_1, \dots, \alpha_m, \alpha_m$ are the exterior normal unit vectors and the areas of the facets of $-P$. Hence the uniqueness statement in the existence and uniqueness theorem implies that P is a translate of $-P$. This, in turn, shows that P is centrally symmetric. □

A useful result is the following *symmetry condition of Minkowski* [736].

Theorem 18.3. *Let $P = P_1 \dot{\cup} \cdots \dot{\cup} P_m$, where $P, P_1, \dots, P_m \in \mathcal{P}_p$. If each of the convex polytopes P_i is centrally symmetric, then P is also centrally symmetric (Fig. 18.1).*

Proof. By the above corollary, it is sufficient to show the following statement.

(4) Let F be a facet of P with exterior normal unit vector u, say, and G the face of P with exterior normal unit vector $-u$. Then $v(F) = v(G)$ and, in particular, G is a facet of P.

Orient $\mathbb{E}^d$ by the vector u. Consider the facets of the polytopes P_i which are parallel to aff F. If such a facet H is on the upper side of a polytope P_i, let its *signed area* $w(H)$ be equal to $v(H)$, and equal to $-v(H)$ otherwise. Then

$$\sum_H w(H) = 0$$

since all P_i are centrally symmetric. Since the signed areas of the facets of the P_i, which are strictly between the supporting hyperplanes of P parallel to aff F, cancel out,

$$\sum_H w(H) = \sum w(H^k) + \sum w(H_l),$$

where the H^k are the facets contained in F and the H_l the facets contained in G. Clearly, the facets in F form a dissection of F and similarly for G. Hence,

$$v(F) - v(G) = \sum v(H^k) - \sum v(H_l) = \sum w(H) = 0,$$

concluding the proof of (4) and thus of the theorem. □

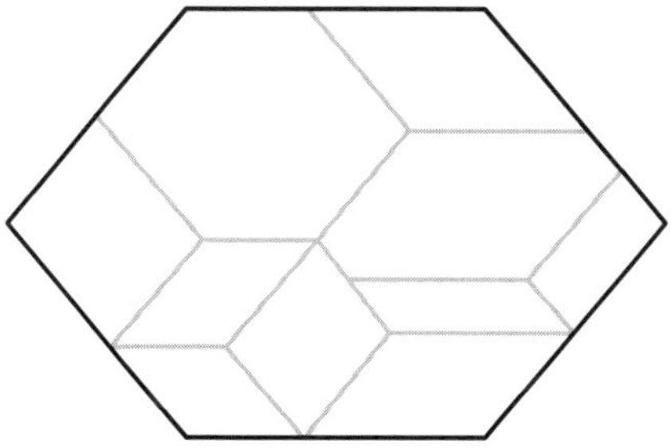

Fig. 18.1. Minkowski's symmetry condition

18.3 The Isoperimetric Problem for Convex Polytopes and Lindelöf's Theorem

Considering convex polytopes, the following natural isoperimetric problem arises. Determine, among all proper convex polytopes in $\mathbb{E}^d$ with n facets, those with minimum isoperimetric quotient and specify the value of the latter.

This section contains Lindelöf's necessary condition for polytopes with minimum isoperimetric quotient amongst all convex polytopes in $\mathbb{E}^d$ with n facets.

Pertinent results in $\mathbb{E}^3$, for small values of n, are reviewed by Fejes Tóth [330] and Florian [337]. This includes characterizations of regular polytopes. For asymptotic results about the minimum isoperimetric quotient and the form of the minimizing polytopes as $n \to \infty$, see Gruber [439–441, 443]. Compare also Sects. 8.3, 8.6 and 9.2.

Lindelöf's Necessary Condition for Minimum Isoperimetric Quotient

The following result of Lindelöf [658] was vaguely anticipated by Steiner [960]. The proof given below is modeled along the lines of the proof of the classical isoperimetric theorem 8.7.

Theorem 18.4. *Among all proper convex polytopes in $\mathbb{E}^d$ with given exterior normals of the facets, it is precisely the polytopes circumscribed to a ball that have minimum isoperimetric quotient (Fig. 18.2).*

The proof which we present in the following makes use of Minkowski's theorem on mixed volumes and the Brunn–Minkowski theorem. A different proof, which is left to the reader, is based on Minkowski's first inequality for mixed volumes. These proofs are very similar to the second and third proof of the isoperimetric inequality, see Theorem 8.7.

Proof. Since homotheties do not change the isoperimetric quotient of a convex body, it is sufficient to show the following:

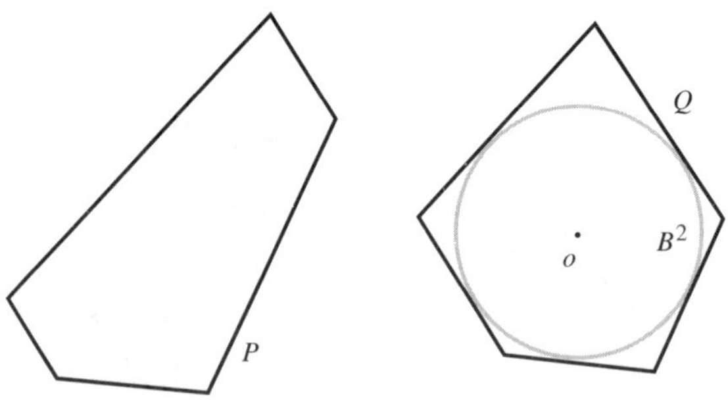

$$\frac{S(P)^d}{V(P)^{d-1}} \geq \frac{S(Q)^d}{V(P)^{d-1}}$$

Fig. 18.2. Lindelöf's isoperimetric theorem for polytopes

(1) Let $P, Q \in \mathcal{P}_p$ have the same set of exterior normal vectors of their facets and such that Q is circumscribed to B^d. Then

$$\frac{S(P)^d}{V(P)^{d-1}} \geq \frac{S(Q)^d}{V(Q)^{d-1}},$$

where equality holds if and only if P is homothetic to Q (and thus also circumscribed to a ball).

Since Q is circumscribed to the unit ball B^d, dissecting Q into pyramids, all with apex o, the formula for the volume of a pyramid then shows that

(2) $S(Q) = d\, V(Q)$.

Minkowski's theorem on mixed volumes 6.5 yields

(3) $V\big((1-\lambda)P + \lambda Q\big)$ is a polynomial in λ for $0 \leq \lambda \leq 1$.

Q is circumscribed to B^d. Choose $\varrho > 0$ such that $Q \subseteq \varrho B^d$. For $0 \leq \lambda \leq 1$ the polytope $(1-\lambda)P + \lambda Q$ can thus be dissected into $(1-\lambda)P$, right prisms of height λ with the facets of $(1-\lambda)P$ as bases, and a set in the $\lambda\varrho$-neighbourhood of the union of the $(d-2)$-faces of $(1-\lambda)P$. Thus

(4) $V\big((1-\lambda)P + \lambda Q\big) = (1-\lambda)^d V(P) + (1-\lambda)^{d-1}\lambda S(P) + O(\lambda^2)$ as $\lambda \to +0$.

The Brunn–Minkowski theorem 8.3 shows that

(5) the function $f(\lambda) = V\big((1-\lambda)P + \lambda Q\big)^{\frac{1}{d}} - (1-\lambda)V(P)^{\frac{1}{d}} - \lambda V(Q)^{\frac{1}{d}}$ for $0 \leq \lambda \leq 1$ with $f(0) = f(1) = 0$ is strictly concave, unless P is homothetic to Q, in which case it is identically 0.

By (3) f is differentiable. Thus (5) shows that

(6) $f'(0) \geq 0$ where equality holds if and only if P is homothetic to Q.

Using the definition of f in (5), (4), and (2), a calculation which is almost identical to that in the proof of the isoperimetric theorem 8.7, yields (1). □

Corollary 18.2. *Among all proper convex polytopes in* $\mathbb{E}^d$ *with a given number of facets, there are polytopes with minimum isoperimetric quotient and these polytopes are circumscribed to a ball.*

In the proof, the first step is to show that there is a polytope with minimum isoperimetric quotient. By Lindelöf's theorem this polytope then is circumscribed to a ball.

Remark. Diskant [274] extended Lindelöf's theorem to the case where ordinary surface area is replaced by generalized surface area, see Sect. 8.3. The above corollary and its generalization by Diskant are used by Gruber [439,443] to obtain information about the geometric form of convex polytopes with minimum isoperimetric quotient.

19 Lattice Polytopes

A *convex lattice polytope* in $\mathbb{E}^d$ is the convex hull of a finite subset of the integer lattice $\mathbb{Z}^d$. Equivalently, it is a convex polytope, all vertices of which are in $\mathbb{Z}^d$. Let $\mathcal{P}_{\mathbb{Z}^d}$ and $\mathcal{P}_{\mathbb{Z}^d\,p}$ denote the spaces of all convex, resp. proper convex lattice polytopes in $\mathbb{E}^d$. Lattice polytopes play a prominent role in convexity and several other branches of mathematics, including the following:

Algebraic geometry (toric varieties, Newton polytopes)

Integer optimization

Tiling (Delone triangulations)

Crystallography

Combinatorial geometry (counting problems)

In this section we first present results of Ehrhart on lattice point enumerators and study the relation of lattice point enumerators to the volume of lattice polytopes. In particular, results of Pick, Reeve and Macdonald are proved. Next, we give a version of Minkowski's theorem on mixed volumes for integer linear combinations of convex lattice polytopes due to McMullen and Bernstein. Finally, we present the theorem of Betke–Kneser on valuations on the space of convex lattice polytopes which is analogous to Hadwiger's functional theorem on valuations on the space of convex bodies. This result leads to short proofs of the lattice point enumeration theorems of Ehrhart. For a different approach to enumeration problems for lattice polytopes based on generating functions, we refer to Barvinok [80, 81]. Applications of lattice polytopes deal with the irreducibility of polynomials in several variables and the Minding–Kouchnirenko–Bernstein theorem on the number of zeros of a generic system of polynomial equations.

Some of the proofs are rather complicated. The reader should not be misled by a first look at the (much shorter) original proofs. In this section we use simple properties of lattices and the Euler characteristic which are not specified in the pertinent sections, but are easy to prove using the tools developed there. In addition, we use simple material on Abelian groups. Tools on polynomials are proved as the proofs are not easily available elsewhere. Let $\mathcal{U}$ denote the family of all *integer unimodular $d \times d$ matrices*, i.e. $d \times d$ matrices with integer entries and determinant ± 1.

For general information on lattice polytopes and lattice polyhedra, see the books of Schrijver [915], Erdös, Gruber and Hammer [307], Gruber and Lekkerkerker [447], Handelman [478], Barvinok [80] and Beck and Robins [86]. Schrijver, in particular, studies convex lattice polyhedra in the context of integer optimization. See also the surveys of McMullen and Schneider [716], Gritzmann and Wills [397], Lagarias [625], Barvinok [81] and DeLoera [253] and the pertinent articles in the collection on Integer Points in Polyhedra – Geometry, Number Theory, Algebra, Optimization [536].

19.1 Ehrhart's Results on Lattice Point Enumerators

The *lattice point enumerators* $L, L^o, L^b : \mathcal{P}_{\mathbb{Z}^d} \to \mathbb{Z}$ are defined as follows, where # denotes the counting function:

$$L(P) = \#(P \cap \mathbb{Z}^d)$$
$$L^o(P) = (-1)^{d-\dim P}\#(\text{relint } P \cap \mathbb{Z}^d)$$
$$L^b(P) = L(P) - L^o(P) \text{ for } P \in \mathcal{P}_{\mathbb{Z}^d}$$

relint P is the interior of P relative to the affine hull aff P. Some authors use slightly different definitions.

The systematic study of lattice point enumerators started with the work of Reeve [825, 826], Macdonald [675] and the polynomiality results of Ehrhart [292–294] and is now part of the theory of valuations on $\mathcal{P}_{\mathbb{Z}^d}$. The results of Ehrhart are related to the Riemann–Roch theorem, see Brion [169]. Ehrhart polynomials and their coefficients play an essential role in combinatorics and the geometry of numbers. For references see the book of Stanley [952], the article of Henk, Schürmann and Wills [492] and the survey of Henk and Wills [493]. A relation between the roots of Ehrhart polynomials and successive minima due to Henk, Schürmann and Wills will be stated in Sect. 23.1.

This section contains the proofs of two results of Ehrhart, following in part the line of Macdonald [675]. Different proofs will be given in Sect. 19.4, using the theorem of Betke and Kneser.

For more information the reader is referred to Ehrhart's monograph [294], to the books cited in the introduction of Sect. 19 and to the surveys of McMullen and Schneider [716], Gritzmann and Wills [397], Brion [170] and Simion [940]. Beck, De Loera, Develin, Pfeifle and Stanley [85] investigated the coefficients and roots of the Ehrhart polynomials.

Ehrhart's Polynomiality and Reciprocity Results for L and L^o

The following results are due to Ehrhart [292–294]. The first result is called *Ehrhart's polynomiality theorem*, the second *Ehrhart's reciprocity theorem.*

Theorem 19.1. *Let P be a proper convex lattice polytope in $\mathbb{E}^d$. Then the following claims hold:*

(i) $L(nP) = p_P(n)$ *for* $n \in \mathbb{N}$, *where* p_P *is a polynomial of degree* d, *with leading coefficient* $V(P)$ *and constant term* 1.
(ii) $L^o(nP) = (-1)^d p_P(-n)$ *for* $n \in \mathbb{N}$.

These results yield the Reeve–Macdonald formulae for the volume of lattice polytopes, see Sect. 19.2. In our proof of the theorem, the notion of lattice and simple related concepts are used, for which the reader may wish to consult Sect. 21.

Proof. (i) First, the following will be shown.

(1) Let $S \in \mathcal{P}_{\mathbb{Z}^d p}$ be a simplex. Then

$$\sum_{n=0}^{\infty} L(nS)t^n = q(t)\Big(1 + \binom{d+1}{1}t + \binom{d+2}{2}t^2 + \cdots\Big) \text{ for } |t| < 1,$$

where q is a polynomial of degree at most d with integer coefficients.

Embed $\mathbb{E}^d$ into $\mathbb{E}^{d+1}$ as usual (first d coordinates) and let $q_1, \dots, q_{d+1} \in \mathbb{Z}^{d+1}$ be the vertices of the lattice simplex $T = S + (0, \dots, 0, 1)$ in the hyperplane $\{x : x_{d+1} = 1\}$. The lattice $\mathbb{Z}^{d+1}$ has determinant $d(\mathbb{Z}^{d+1}) = 1$. Let L be the sublattice of $\mathbb{Z}^{d+1}$ with basis $\{q_1, \dots, q_{d+1}\}$. The set $F = \{\alpha_1 q_1 + \dots + \alpha_{d+1} q_{d+1} : 0 \leq \alpha_i < 1\}$ is a fundamental parallelotope of L. Thus, for the determinant of L we have $d(L) = V_{d+1}(F)$, where V_{d+1} stands for the volume in $\mathbb{E}^{d+1}$. Then the following hold.

(2) For each $u \in \mathbb{Z}^{d+1}$, there is precisely one point $v \in F \cap \mathbb{Z}^{d+1}$ such that $u \in v + L$. Further, $V_{d+1}(F) = \#(F \cap \mathbb{Z}^{d+1}) = h$, say.

If $u = \beta_1 q_1 + \dots + \beta_{d+1} q_{d+1} \in \mathbb{Z}^{d+1}$, then $v = (\beta_1 - \lfloor \beta_1 \rfloor) q_1 + \dots + (\beta_{d+1} - \lfloor \beta_{d+1} \rfloor) q_{d+1}$ is the unique point $v \in F \cap \mathbb{Z}^{d+1}$ with $u \in v + L$. The index of the sublattice (subgroup) L in the lattice (group) $\mathbb{Z}^{d+1}$ is thus $\#(F \cap \mathbb{Z}^{d+1})$. Since the index is also equal to $d(L)/d(\mathbb{Z}^{d+1}) = V_{d+1}(F)$, we see that $V_{d+1}(F) = \#(F \cap \mathbb{Z}^{d+1})$, concluding the proof of (2). Compare Sect. 21.3. For each $u \in nT \cap \mathbb{Z}^{d+1}$, according to (2), we have

(3) $u = v + \sum_{i=1}^{d+1} m_i q_i$ with suitable integers $m_i \geq 0$.

To see this, note that $u = \beta_1 q_1 + \dots + \beta_{d+1} q_{d+1}$ with $\beta_i \geq 0$ and put $m_i = \lfloor \beta_i \rfloor$. Considering the last coordinate, (3) shows that

(4) $n = v_{d+1} + \sum_{i=1}^{d+1} m_i$ with integers $m_i \geq 0$.

Conversely, given $v \in F \cap \mathbb{Z}^{d+1}$, any solution of (4) in integers $m_i \geq 0$ gives rise to a unique point $u \in nT \cap \mathbb{Z}^{d+1}$ with $u \in v + L$. The number of such solutions of (4) equals the coefficient of t^n in the power series

$$\begin{aligned} t^{v_{d+1}}(1 + t + t^2 + \cdots)^{d+1} &= \frac{t^{v_{d+1}}}{(1-t)^{d+1}} \\ &= t^{v_{d+1}}\left(1 + \binom{-d-1}{1}(-t) + \binom{-d-1}{2}(-t)^2 + \cdots\right) \\ &= t^{v_{d+1}}\left(1 + \binom{d+1}{1} t + \binom{d+2}{2} t^2 + \cdots\right), \end{aligned}$$

where we have applied Newton's binomial series. Hence

(5) $L(nS)(= L(nT))$ equals the coefficient of t^n in the power series

$$q(t)\left(1 + \binom{d+1}{1} t + \binom{d+2}{2} t^2 + \cdots\right), \text{ where } q(t) = \sum_{v \in F \cap \mathbb{Z}^{d+1}} t^{v_{d+1}}.$$

To determine the degree of $q(\cdot)$, note that each $v \in F \cap \mathbb{Z}^{d+1}$ can be represented in the form $v = \alpha_1 q_1 + \cdots \alpha_{d+1} q_{d+1}$ where $0 \leq \alpha_i < 1$. Each q_i has last coordinate

1. Hence $v_{d+1} = \alpha_1 + \cdots + \alpha_{d+1} < d+1$. Thus, being int shows that q has degree at most d. The proof of (1) is comple

Second, we show Proposition (i) for simplices.

(6) Let $S \in \mathcal{P}_{\mathbb{Z}^d p}$ be a simplex. Then $L(nS) = p_S(n)$ for $n \in \mathbb{N}$, w polynomial of degree d, with leading coefficient $V(S)$ and constant

To show (6), note that according to (1), $q(t) = a_0 + a_1 t + \cdots + a_d t^d$. By the definitio of q in (5), the coefficient a_i is the number of points $v \in F \cap \mathbb{Z}^{d+1}$ with $v_{d+1} = i$. Thus, in particular,

(7) $a_0 = 1$ and $a_0 + a_1 + \cdots + a_d = h$,

see (2). Denote the k-dimensional volume by V_k. We have,

$$\begin{aligned} V_{d+1}(F) &= (d+1)!V_{d+1}\big(\operatorname{conv}(T \cup \{o\})\big) \\ &= \frac{(d+1)!}{d+1} V_d(T) = d\,!V(S). \end{aligned}$$

Since $h = V_{d+1}(F)$ by (2), it follows that

(8) $V(S) = \dfrac{h}{d\,!}$.

Inserting

$$q(t) = a_0 + a_1 t + \cdots + a_d t^d$$

into (1) and comparing the coefficients of t^n in the two power series, yields

$$\begin{aligned} L(nS) &= \binom{d+n}{d} a_0 + \binom{d+n-1}{d} a_1 + \cdots + \binom{n}{d} a_d \\ &= a_0 + \cdots + \frac{a_0 + \cdots + a_d}{d\,!} n^d = 1 + \cdots + V(S)\, n^d = p_S(n), \text{ say,} \\ &\quad \text{for } n \in \mathbb{N} \end{aligned}$$

by (7) and (8). The proof of (6) is complete.

Using (6), we now show the following related statement.

(9) Let $R \in \mathcal{P}_{\mathbb{Z}^d}$ be a simplex with $c = \dim R < d$. Then $L(nR) = p_R(n)$ for $n \in \mathbb{N}$, where p_R is a polynomial of degree less than d with constant term 1.

Embed $\mathbb{E}^c$ into $\mathbb{E}^d$ as usual (first c coordinates). There is an integer unimodular $d \times d$ matrix $U \in \mathcal{U}$ such that UR is a simplex in $\mathcal{P}_{\mathbb{Z}^c p}$. Now apply (6) to UR and note that $L(nR) = L(nUR)$ to get (9).

Third, the following proposition will be shown.

(10) Let $S \in \mathcal{P}_{\mathbb{Z}^d p}$ be a simplex. Then $L(n \operatorname{int} S) = q_S(n)$ for $n \in \mathbb{N}$, where q_S is a polynomial of degree d with leading coefficient $V(S)$ and constant term $(-1)^d$.

see this, note that

$$L(n\,\mathrm{int}\,S) = L(nS) - \sum_R L(nR) + \sum_Q L(nQ) - + \cdots,$$

where R ranges over all facets of S, Q over all $(d-2)$-dimensional faces, etc. Now take into account (6) and (9) to see that $L(n\,\mathrm{int}\,S) = q_S(n)$ where q_S is a polynomial of degree d with leading coefficient $V(S)$ and constant term

$$1 - \binom{d+1}{1} + \binom{d+1}{2} - + \cdots + (-1)^d \binom{d+1}{d} = (-1)^d,$$

concluding the proof of (10).

Analogous to the derivation of (9) from (6), the following proposition is a consequence of (10).

(11) Let $R \in \mathcal{P}_{\mathbb{Z}^d}$ be a simplex with $c = \dim R < d$. Then $L(n\,\mathrm{relint}\,R) = q_R(n)$ for $n \in \mathbb{N}$, where q_R is a polynomial of degree $< d$ with constant term $(-1)^c$.

In the fourth, and last, step of the proof of (i), we extend (6) from simplices to convex polytopes.

(12) Let $P \in \mathcal{P}_{\mathbb{Z}^d\,p}$. Then $L(nP) = p_P(n)$ for $n \in \mathbb{N}$, where p_P is a polynomial of degree d with leading coefficient $V(P)$ and constant term 1.

By Theorem 14.9, the proper lattice polytope P is the union of all simplices of a suitable simplicial complex where all simplices are lattice simplices. Hence P is the disjoint union of the relative interiors of these simplices. See, e.g. Alexandroff and Hopf [9], p. 128. Propositions (10) and (11) thus yield (12), except for the statement that the constant term is 1. To see this note that the Euler characteristic of a simplicial complex is the number of its vertices minus the number of its edges plus the number of its 2-dimensional simplices, etc. This together with (10) and (11) shows that the constant term in p_P is just the Euler characteristic of the simplicial complex; but this equals the Euler characteristic of P and is thus 1. The proof of (12) and thus of Proposition (i) is complete.

(ii) Since the proof of (ii) is similar to that of (i), some details are omitted. The first step is to show the following analogue of (1).

(13) Let $S \in \mathcal{P}_{\mathbb{Z}^d\,p}$ be a simplex. Then

$$\sum_{n=0}^{\infty} L^o(nS)t^n = r(t)\left(1 + \binom{d+1}{1}t + \binom{d+2}{2}t^2 + \cdots\right) \text{ for } |t| < 1,$$

where r is a polynomial with integer coefficients of degree $\le d+1$.

Instead of the fundamental parallelotope F in the proof of (1), here the fundamental parallelotope $G = \{x = \alpha_1 q_1 + \cdots + \alpha_{d+1} q_{d+1} : 0 < \alpha_i \le 1\}$ is used.

In (13), $r(t) = b_1 t + \cdots + b_{d+1} t^{d+1}$, where b_k equals the number of points $v \in G \cap \mathbb{Z}^{d+1}$ with $v_{d+1} = k$. By symmetry, $a_i = b_{d+1-i}$ for $i = 0, \ldots, d$. Hence

$$r(t) = a_d t + \cdots + a_1 t^d + a_0 t^{d+1}$$

and we obtain, from (13)

$$\begin{aligned} L^o(nS) &= \binom{d+n-1}{d} a_d + \binom{d+n-2}{d} a_{d-1} + \cdots + \binom{n-1}{d} a_0 \\ &= \frac{1}{d\,!}\big((d+n-1)\cdots(n+1)na_d + \cdots + (n-1)\cdots(n-1-d+1)a_0\big) \\ &= \frac{(-1)^d}{d\,!}\big((-n+d)\cdots(-n+1)a_0 + \cdots + (-n)\cdots(-n-d+1)a_d\big) \\ &= (-1)^d\left(\binom{-n+d}{d} a_0 + \cdots + \binom{-n}{d} a_d\right) = (-1)^d p_S(-n). \end{aligned}$$

This, together with the corresponding results in lower dimensions, yields the following counterpart of (6) and (9).

(14) Let $S \in \mathcal{P}_{\mathbb{Z}^d}$ be a simplex with $c = \dim S \le d$. Then $L^o(nS) = (-1)^c p_S(-n)$ for $n \in \mathbb{N}$.

In the final step we extend (14) to proper convex lattice polytopes.

(15) Let $P \in \mathcal{P}_{\mathbb{Z}^d p}$. Then $L^o(nP) = (-1)^d p_P(-n)$ for $n \in \mathbb{N}$.

Represent P as the disjoint union of the relative interiors of the simplices of a complex of lattice simplices as in the proof of (12). Then

$$\begin{aligned} (16)\quad p_P(n) = L(nP) &= \sum_S L^o(nS) + \sum_R L^o(nR) + \cdots \\ &= \sum_S (-1)^d p_S(-n) + \sum_R (-1)^{d-1} p_R(-n) + \cdots \text{ for } n \in \mathbb{N}, \end{aligned}$$

where S ranges over all proper simplices of this complex, R over all $(d-1)$-dimensional simplices, etc. Since $p_P, p_S, \dots,$ all are polynomials and since (16) holds for all $n \in \mathbb{N}$, it holds for all $t \in \mathbb{R}$ in place of $n \in \mathbb{N}$. Thus it holds in particular for all $-n$ where $n \in \mathbb{N}$. Hence

$$\begin{aligned} (17)\quad (-1)^d p_P(-n) &= \sum_S p_S(n) - \sum_R p_R(n) + \cdots \\ &= \sum_S L(nS) - \sum_R L(nR) + \cdots \text{ for } n \in \mathbb{N}, \end{aligned}$$

by (i). To finish the proof of (15,) it suffices to show that the last line in (17) is equal to $L^o(nP)$. Let $l \in \mathbb{Z}^d$. The contribution of l to the last line in (17) equals the expression

$$\#\big(\{S : l \in nS\}\big) - \#\big(\{R : l \in nR\}\big) + - \dots$$

If $l \notin nP$, this expression is 0. If $l \in \operatorname{bd}(nP)$, this expression may be interpreted (up to the sign) as the Euler characteristic of (a complex of convex polyhedral cones with apex l and union equal to) the polyhedral support cone of P at l. Thus it is 0. If $l \in \operatorname{int} nP$, this expression is (up to the sign) equal to the Euler characteristic of (a complex of convex polyhedral cones with apex l and union equal to) $\mathbb{E}^d$. Thus it is 1. The last line in (17) thus equals $L^o(nP)$, concluding the proof of (15) and thus of (ii). □

The Coefficients of the Ehrhart Polynomial

Based on Barvinok's [78, 79] method for counting lattice points, DeLoera, Hemmecke, Tauzer and Yoshida [254] developed an efficient algorithm to calculate the coefficients of Ehrhart polynomials.

19.2 Theorems of Pick, Reeve and Macdonald on Volume and Lattice Point Enumerators

Can the volume of a lattice polytope P be calculated in terms of the number of lattice points in P, in the relative interior of P and on the relative boundary of P, that is, in terms of $L(P)$, $L^o(P)$ and $L^b(P)$?

A nice result of Pick [801] says that, for a Jordan lattice polygon P in $\mathbb{E}^2$,

(1) $A(P) = L(P) - \frac{1}{2}L^b(P) - 1,$

where A stands for area. For this result there are many proofs, variants and extensions known, as a look into the American Mathematical Monthly, the Mathematics Teacher and similar journals shows. We mention the far-reaching generalizations of Hadwiger and Wills [470] and Grünbaum and Shephard [456].

There is no direct generalization of the formula (1) to higher dimension as can be seen from the simplices $S_n = \text{conv}\{o, (1,0,0), (0,1,0), (1,1,n)\}$ in $\mathbb{E}^3$.

$$V(S_n) = \frac{n}{3!}, \text{ while } L(S_n) = L^b(S_n) = 4,\ L^o(S_n) = 0 \text{ for } n \in \mathbb{N}.$$

Reeve [825, 826] ($d = 3$) and Macdonald [675] (general d) were able to express the volume of a (general) lattice polytope P in terms of $L(P)$, $L^b(P)$, $L(2P)$, $L^b(2P), \dots, L\big((d-1)P\big), L^b\big((d-1)P\big)$.

This section contains the proofs of Pick's theorem and of results of Reeve and Macdonald in the special case of convex lattice polytopes.

For more information, see the references given earlier and the references at the beginning of Sect. 19.

Pick's Lattice Point Theorem in $\mathbb{E}^2$

A *Jordan lattice polygon* is a solid polygon in $\mathbb{E}^2$ bounded by a closed Jordan polygonal curve in $\mathbb{E}^2$, all vertices of which are points of $\mathbb{Z}^2$. Let $\mathfrak{J}_{\mathbb{Z}^2}$ be the family of all Jordan lattice polygons. Clearly, the lattice point enumerators extend to $\mathfrak{J}_{\mathbb{Z}^2}$. The following result is due to Pick [801].

Theorem 19.2. *Let $P \in \mathfrak{J}_{\mathbb{Z}^2}$. Then (Fig. 19.1)*

(1) $A(P) = L(P) - \frac{1}{2}L^b(P) - 1.$

Proof. We first show that

(2) The expression $M = L - \frac{1}{2}L^b - 1$ is *simply additive* on $\mathfrak{J}_{\mathbb{Z}^2}$.

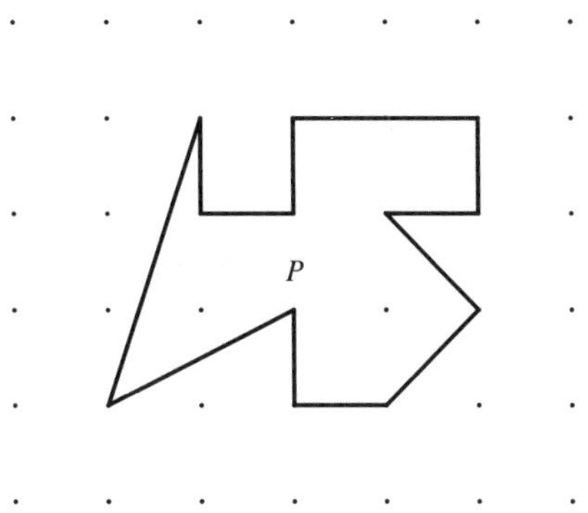

$A(P) = 7\frac{1}{2},\ L(P) = 15,\ L^b(P) = 13$

Fig. 19.1. Pick's theorem

By the latter we mean the following: If $P, Q \in \mathfrak{J}_{\mathbb{Z}^2}$ are such that $P \cup Q \in \mathfrak{J}_{\mathbb{Z}^2}$ and $P \cap Q$ is a polygonal Jordan arc, then

$$M(P \cup Q) = M(P) + M(Q).$$

To see (2), assume that the common arc $P \cap Q$ of P and Q contains m points of $\mathbb{Z}^2$. Then

$$\begin{aligned} L(P \cup Q) &= L(P) + L(Q) - m, \\ L^b(P \cup Q) &= L^b(P) + L^b(Q) - 2m + 2. \end{aligned}$$

This, in turn, implies (2):

$$\begin{aligned} M(P \cup Q) &= L(P \cup Q) - \frac{1}{2}L^b(P \cup Q) - 1 \\ &= L(P) - \frac{1}{2}L^b(P) + L(Q) - \frac{1}{2}L^b(Q) - m + m - 1 - 1 \\ &= M(P) + M(Q). \end{aligned}$$

To prove the theorem, it is sufficient to show the following:

(3) Let $n = 3, 4, \ldots$ Then (1) holds for each $P \in \mathfrak{J}_{\mathbb{Z}^2}$ with $L(P) = n$.

The proof is by induction on n. If $n = 3$, then P is a lattice triangle which contains no point of $\mathbb{Z}^2$, except its vertices. We may assume that $P = \operatorname{conv}\{o, b_1, b_2\}$. The triangle $-P + b_1 + b_2$ is also a lattice triangle which contains no point of $\mathbb{Z}^2$, except its vertices. Hence o is the only point of $\mathbb{Z}^2$ in the parallelogram $\{\alpha_1 b_1 + \alpha_2 b_2 : 0 \le \alpha_i < 1\} \subseteq P \cup (-P + b_1 + b_2)$. Thus $\{b_1, b_2\}$ is a basis of $\mathbb{Z}^2$ and therefore,

$$A(P) = \frac{1}{2}|\det(b_1, b_2)| = \frac{1}{2} = 3 - \frac{1}{2}3 - 1 = L(P) - \frac{1}{2}L^b(P) - 1.$$

Assume now that $n > 3$ and that (3) holds for $3, 4, \ldots, n - 1$. Let q be a vertex of $\operatorname{conv} P$ and let p and r be the points of $\mathbb{Z}^2$ on $\operatorname{bd} P$ just before and after q.

If the triangle $S = \mathrm{conv}\{p, q, r\}$ contains no further point of $\mathbb{Z}^2$, dissect P along the line segment $[p, r]$. This gives two Jordan lattice polygons, S and T, say, where $L(S), L(T) < n$. Hence (3) holds for S and T by the induction assumption and we obtain that

$$\begin{aligned} A(P) &= A(S \cup T) = A(S) + A(T) \\ &= L(S) - \frac{1}{2}L^b(S) - 1 + L(T) - \frac{1}{2}L^b(T) - 1 \\ &= L(S \cup T) - \frac{1}{2}L^b(S \cup T) - 1 = L(P) - \frac{1}{2}L^b(P) - 1 \end{aligned}$$

by (2). Hence (3) holds for P, too. If $S = \mathrm{conv}\{p, q, r\}$ contains points of $\mathbb{Z}^2$ different from p, q, r, choose one which is closest to q, say s. Now dissect P along $[q, s]$ and proceed as before to show that (3) also holds for P in the present case. The induction is thus complete. This concludes the proof of (3) and thus of the theorem. □

Remark. Since the perimeter $P(P)$ of a polygon $P \in \mathcal{J}_{\mathbb{Z}^2}$ is at least $L^b(P)$, Pick's equality (1) yields the following estimate.

$$A(P) \geq L(P) - \frac{1}{2}P(P) - 1.$$

If C is a planar convex body such that $P = \mathrm{conv}(C \cap \mathbb{Z}^2)$ is a proper convex lattice polygon, then

$$A(C) \geq A(P) \geq L(P) - \frac{1}{2}P(P) - 1 \geq L(C) - \frac{1}{2}P(C) - 1.$$

This inequality was first proved by Nosarzewska [773].

The Reeve–Macdonald Lattice Point Results in $\mathbb{E}^d$

A very satisfying extension of Pick's theorem to higher dimensions is due to Reeve [825,826] ($d = 3$) and Macdonald [675] (general d). We present it in the case where the lattice polytope is convex.

Theorem 19.3. *Let $P \in \mathcal{P}_{\mathbb{Z}^d p}$. Then*

(i) $d!V(P) = L(dP) - \binom{d}{1} L\big((d-1)P\big) + \cdots + (-1)^{d-1}\binom{d}{d-1} L(P) + (-1)^d$, *and*

(ii) $$\frac{(d-1)d!}{2}V(P) = M\big((d-1)P\big) - \binom{d-1}{1} M\big((d-2)P\big) + \cdots \cdots + (-1)^{d-2}\binom{d-1}{d-2} M(P) + \frac{1}{2} + \frac{1}{2}(-1)^d,$$

where $M(P) = L(P) - \frac{1}{2}L^b(P) = \frac{1}{2}\big(L(P) + L^o(P)\big)$.

The original proof of (ii) by Macdonald was rather complicated. Using Ehrhart's reciprocity theorem 19.1(ii), the following proof of (ii) is along the lines of the proof of Proposition (i) and much shorter.

Proof. (i) By Proposition (i) in Ehrhart's theorem 19.1,

(4) $L(nP) = p_P(n)$ for $n \in \mathbb{N}$, where p_P is a polynomial of degree d with leading coefficient $V(P)$ and constant term 1.

Since $p_P(i) = L(iP) \neq 0$ for $i = 0, \dots, d$, Lagrange's theorem on partial fractions shows that

$$(5)\quad \frac{p_P(t)}{\prod_{i=0}^{d}(t-i)} = \sum_{j=0}^{d} \frac{a_j}{t-j}$$

with suitable coefficients a_j. Then

$$p_P(t) = \sum_{j=0}^{d} a_j \prod_{\substack{i=0\\ i\neq j}}^{d} (t-i).$$

Put $t = j$ to see that

$$L(jP) = p_P(j) = a_j \prod_{\substack{i=0\\ i\neq j}}^{d} (j-i), \text{ or } a_j = \frac{(-1)^{d-j} L(jP)}{(d-j)!j!}.$$

Multiplying (5) by $\prod_{i=0}^{d}(t-i)$ and comparing the coefficient of t^d on both sides shows that

$$d!V(P) = L(dP) - \binom{d}{1} L\big((d-1)P\big) + - \cdots + (-1)^d$$

by (4).

(ii) Propositions (i) and (ii) in Ehrhart's theorem 19.1 together imply that

(6) $M(nP) = \frac{1}{2}\big(L(nP) + L^o(nP)\big) = \frac{1}{2}\big(p_P(n) + (-1)^d p_P(-n)\big) = q_P(n)$, say, for $n \in \mathbb{N}$, where q_P is a polynomial of degree d with leading coefficient $V(P)$, second coefficient 0, and constant term $\frac{1}{2} + \frac{1}{2}(-1)^d$.

Since $q_P(i) = \frac{1}{2}\big(L(iP) + L^o(iP)\big) \neq 0$ for $i = 0, \dots, d-1$, Lagrange's theorem yields

$$\frac{q_P(t)}{\prod_{i=0}^{d-1}(t-i)} = \sum_{j=0}^{d-1} \frac{b_j}{t-j} + V(P)$$

with suitable coefficients b_j. Then

$$q_P(t) = \sum_{j=0}^{d-1} b_j \prod_{\substack{i=0\\i\neq j}}^{d-1} (t-i) + V(P) \prod_{i=0}^{d-1} (t-i).$$

Now put $t = j$. This gives

$$M(jP) = q_P(j) = b_j \prod_{\substack{i=0\\i\neq j}}^{d-1} (j-i), \text{ or } b_j = \frac{(-1)^{d-1-j} M(jP)}{(d-1-j)!j!},$$

where by $M(0P)$ we mean $\frac{1}{2} + \frac{1}{2}(-1)^d$. Hence

$$(d-1)!\, q_P(t) = \sum_{j=0}^{d-1} (-1)^{d-1-j} \binom{d-1}{j} M(jP) \prod_{\substack{i=0\\i\neq j}}^{d-1} (t-i) + (d-1)!\, V(P) \prod_{i=0}^{d-1} (t-i).$$

Note (6) and compare the coefficient of t^{d-1} on both sides. This yields the following:

$$0 = \sum_{j=0}^{d-1} (-1)^{d-1-j} \binom{d-1}{j} M(jP) + (d-1)! \sum_{j=0}^{d-1} (-j) V(P),$$

or

$$\frac{1}{2}(d-1)d\,!V(P) = \sum_{j=0}^{d-1} (-1)^{d-1-j} \binom{d-1}{j} M(jP),$$

concluding the proof of (ii). □

19.3 The McMullen–Bernstein Theorem on Sums of Lattice Polytopes

One of the fascinating results of early convex geometry is Minkowski's theorem on mixed volumes 6.5. It says that the volume of a linear combination of convex bodies with non-negative coefficients is a homogeneous polynomial in the coefficients. McMullen [708,709] and Bernstein [101] proved an analogous result for linear combinations of lattice polytopes with integer coefficients where, instead of the volume, the number of lattice points is considered. Both the volume and the lattice point enumerators are valuations and the results of Minkowski and McMullen–Bernstein are now part of the theory of valuations. While this is classical for the volume, it is due to the efforts of McMullen [709] for lattice point enumerators, see the survey [714].

This section contains Bernstein's proof of the McMullen–Bernstein theorem. We start with some properties of polynomials which, surprisingly, we could not find in the literature in the required form.

For more information on pertinent material, see McMullen [708, 709] and the surveys of McMullen and Schneider [716], Gritzmann and Wills [397] and McMullen [714].

Sufficient Conditions for Polynomiality

We begin with two *criteria for polynomiality*. Both seem to be known. A result similar to the first criterion is due to Carroll [191]. I am grateful to Iskander Aliev [22] for this reference. Here $\mathbb{N} = \{0, 1, \dots\}$.

Lemma 19.1. *Let $k \in \mathbb{N}$ and $p : \mathbb{N}^d \to \mathbb{R}$ such that for any $d-1$ variables fixed, p is (the restriction of) a real polynomial of degree at most k in the remaining variable. Then p is a real polynomial in all d variables of degree at most k.*

Proof. We prove the lemma in case $d = 2$. The same idea, together with a simple induction argument, yields the general case.

By assumption,

$$(1)\quad p(m,n) = \sum_{i=0}^{k} a_i(m)n^i \text{ for } m, n \in \mathbb{N},$$

where the $a_i(m)$ are suitable coefficients. Thus, in particular,

$$p(m,j) = \sum_{i=0}^{k} a_i(m)j^i \text{ for } m \in \mathbb{N},\ j = 0, \dots, k.$$

Cramer's rule and the Gram determinant then imply that

$$(2)\quad a_i(m) = \sum_{j=0}^{k} b_{ij}\, p(m,j) \text{ for } m \in \mathbb{N},\ i = 0, \dots, k,$$

with suitable real coefficients b_{ij}. Thus

$$p(m,n) = \sum_{i,j=0}^{k} b_{ij}\, p(m,j)n^i \text{ for } m, n \in \mathbb{N},$$

by (1) and (2). Since, by the assumption of the lemma, $p(m, j)$ is a polynomial in m $(\in \mathbb{N})$ for each $j = 0, 1, \dots, k$, $p(m, n)$ is also a polynomial in m, n $(\in \mathbb{N})$. □

Lemma 19.2. *Let $q : \mathbb{N} \to \mathbb{R}$ be such that $q(n+1) - q(n)$ is (the restriction of) a real polynomial for $n \in \mathbb{N}$. Then q is a real polynomial in one variable.*

Proof. Let

$$q(n+1) - q(n) = a_0 + a_1 n + \cdots + a_k n^k \text{ for } n \in \mathbb{N},$$

with suitable coefficients a_i. Then

$$\begin{aligned} q(n+1) - q(1) &= q(n+1) - q(n) + q(n) - q(n-1) + \cdots + q(2) - q(1) \\ &= a_0 + a_1 n + \cdots + a_k n^k + a_0 + a_1(n-1) + \cdots + a_k(n-1)^k + \cdots \\ &= a_0(1 + 1 + \cdots + 1) + a_1(1 + 2 + \cdots + n) + \cdots + a_k(1 + 2^k + \cdots + n^k). \end{aligned}$$

Now, take into account that the sum of the ith-order arithmetic series $1+2^i+\cdots+n^i$ is a polynomial in n of degree $i+1$, where the coefficients depend on the Bernoulli numbers. Hence $q(n+1)-q(1)$ and thus $q(n+1)$ and then also $q(n)$ is a polynomial in n of degree $k+1$. □

The McMullen–Bernstein Theorem on Lattice Points in Sums of Lattice Polytopes

is as follows.

Theorem 19.4. *Let $P_1, \dots, P_m \in \mathcal{P}_{\mathbb{Z}^d\,p}$. Then $L(n_1 P_1+\cdots+n_m P_m)$ is a polynomial in $n_1, \dots, n_m \in \mathbb{N}$ of degree d.*

Proof. We begin with some simple preparations. The *inclusion–exclusion formula for finite sets* is as follows.

(3) Let $A_1, \dots, A_m$ be finite sets. Then

$$\#\bigcup_i A_i = \sum_i \#A_i - \sum_{i<j} \#(A_i \cap A_j) + \sum_{i<j<k} \#(A_i \cap A_j \cap A_k) - \cdots$$

For $S \in \mathcal{P}_{\mathbb{Z}^d}$ and $u \in \mathbb{E}^d$, $u \neq o$, the support set of S with exterior normal u is the set

$$S(u) = \big\{x \in S : u \cdot x = h_S(u)\big\}.$$

For a finite set $U \subseteq \mathbb{E}^d$ with $o \notin U$, let

$$S(U) = \big\{x \in S : u \cdot x = h_S(u) \text{ for each } u \in U\big\} = \bigcap_{u \in U} S(u).$$

An easy argument yields the following.

(4) Let $S, T \in \mathcal{P}_{\mathbb{Z}^d}$ and let $U \subseteq \mathbb{E}^d$ be a finite set with $o \notin U$. Then the support sets $S(U), T(U), (S+T)(U)$ are in $\mathcal{P}_{\mathbb{Z}^d}$, are faces of $S, T, S+T$ and

$$(S+T)(U) = \bigcap_{u \in U} \big(S(u)+T(u)\big) = S(U)+T(U).$$

The main step of the proof of the theorem is to show the following proposition.

(5) Let $P, Q \in \mathcal{P}_{\mathbb{Z}^d}$. Then $L(P+nQ)$ is the restriction of a polynomial of degree $\leq d$ to $n \in \mathbb{N}$.

The proof is by induction on $l = \dim Q$. If $l = 0$, then Q is a point of $\mathbb{Z}^d$ and for each $n \in \mathbb{N}$, the polytope $P+nQ$ is a translate of P by a vector of $\mathbb{Z}^d$. Hence $L(P+nQ)$ is a constant.

Assume now that $l > 0$ and that (5) holds in case where $\dim Q = 0, 1, \dots, l-1$. We have to prove it for $\dim Q = l$. Since translations of P and Q by vectors of $\mathbb{Z}^d$ do not affect (5), we may assume that

$$o \in P, Q.$$

As a consequence of Corollary 21.2 and Theorem 21.1 the following holds. There is an integer unimodular $d \times d$ matrix which maps the lattice $\operatorname{lin}(P+Q) \cap \mathbb{Z}^d$ onto a lattice of the form $\mathbb{Z}^c$ in $\mathbb{E}^d$, where $c = \dim \operatorname{lin}(P+Q)$ and $\mathbb{E}^c$ is embedded into $\mathbb{E}^d$ as usual (first c coordinates). Then P, Q are mapped onto convex lattice polytopes with respect to $\mathbb{Z}^c$ the sum of which is proper. Thus we may suppose that already $P+Q \in \mathcal{P}_{\mathbb{Z}^d p}$. Embed $\mathbb{E}^d$ into $\mathbb{E}^{d+1}$ as usual, denote by "$'$" the orthogonal projection from $\mathbb{E}^{d+1}$ onto $\mathbb{E}^d$, and let $u_1 = (0, \dots, 0, 1) \in \mathbb{E}^{d+1}$. When speaking of upper side, etc. of a convex polytope in $\mathbb{E}^{d+1}$, this is meant with respect to the last coordinate. Let $R = \operatorname{conv}\big(Q \cup \{u_1\}\big) \in \mathcal{P}_{\mathbb{Z}^{d+1}}$.

After these preparations we show that

(6) $L\big(P+(n+1)Q\big) - L(P+nQ)$ is a polynomial in $n \in \mathbb{N}$.

By the above, $P+nQ+R \in \mathcal{P}_{\mathbb{Z}^{d+1} p}$ and its upper side contains the horizontal facet $F_1 = P + nQ + u_1$ with exterior normal vector u_1 and certain "non-horizontal" facets, say $F_2, \dots, F_m$. Let $u_2, \dots, u_m$ be exterior normal vectors of these. Then

(7) $F_1 = P + nQ + u_1,$
$F_i = P(u_i) + nQ(u_i) + R(u_i)$ for $i = 2, \dots, m$

by (4). Next, it will be shown that

(8) $u_2, \dots, u_m$ are not orthogonal to $\operatorname{lin} Q$.

For, assume that $u_i \perp \operatorname{lin} Q \; (\subseteq \mathbb{E}^d)$. Then $P(u_i) \subseteq P$, $Q(u_i) \subseteq Q$. Since $R = \operatorname{conv}(Q \cup \{u_1\})$, $u_i \perp \operatorname{lin} Q$, $u_1 = (0, \dots, 0, 1)$ and u_i has last coordinate greater than 0, we have $R(u_i) = \{u_1\}$. Thus $F_i = (P+nQ+R)(u_i) = P(u_i)+nQ(u_i)+R(u_i) \subseteq P+nQ+\{u_1\} = F_1$ by (4) and (7). This is impossible, concluding the proof of (8). From $P, Q \in \mathcal{P}_{\mathbb{Z}^d}$, $R \in \mathcal{P}_{\mathbb{Z}^{d+1}}$ and (8) it follows that

(9) $P(u_i), Q(u_i), \big(R(u_i) \in \mathcal{P}_{\mathbb{Z}^{d+1}}$ and thus$\big)$ $R(u_i)' \in \mathcal{P}_{\mathbb{Z}^d}$ for $i = 1, \dots, m$,
$\dim Q(u_i) < \dim Q = l$ for $i = 2, \dots, m$.

Hence (4) yields

(10) $P\big(\{u_i, u_j\}\big), Q\big(\{u_i, u_j\}\big), R\big(\{u_i, u_j\}\big)' \in \mathcal{P}_{\mathbb{Z}^d}$ for $1 \le i < j \le m$,
$\dim Q\big(\{u_i, u_j\}\big) \; (\le \dim Q(u_j)) < \dim Q = l$ for $1 \le i < j \le m$.
Analogous statements hold for $P\big(\{u_i, u_j, u_k\}\big), \dots$

Since $F_1', \dots, F_m'$ tile $(P+nQ+R)' = P+nQ+R' = P+nQ+Q = P+(n+1)Q$ and $F_1' = P+nQ$, propositions (3), (7) and (4) imply that

$$
\begin{aligned}
L\big(P+(n+1)Q\big) &= \sum_{i} L(F_i') - \sum_{i<j} L(F_i' \cap F_j') + \cdots \\
&= L(P+nQ) + \sum_{i=2}^{m} L(F_i') - \sum_{i<j} L\big((F_i \cap F_j)'\big) + \cdots \\
&= L(P+nQ) + \sum_{i=2}^{m} L\big(P(u_i) + nQ(u_i) + R(u_i)'\big) \\
&\quad - \sum_{i<j} L\big(P(\{u_i, u_j\}) + nQ(\{u_i, u_j\}) + R(\{u_i, u_j\})'\big) + \cdots .
\end{aligned}
$$

This, combined with (9), (10) and the induction assumption, shows that $L(P + (n+1)Q) - L(P+nQ)$ is a polynomial in $n \in \mathbb{N}$, concluding the proof of (6).

Having proved (6), Lemma 19.2 shows that $L(P+nQ)$ is a polynomial in $n \in \mathbb{N}$. If K is a lattice cube such that $P, Q \subseteq K$, then $P + nQ \subseteq (n+1)K$ and thus $L(P+nQ) \le L((n+1)K) = O(n^d)$. Hence the polynomial $L(P+nQ)$ has degree $\le d$. Thus (5) holds for $Q \in \mathcal{P}_{\mathbb{Z}^d}$ with $\dim Q = l$. The induction is thus complete and (5) holds generally.

The theorem finally follows from (5) and Lemma 19.1. □

19.4 The Betke–Kneser Theorem on Valuations

Many of the functions ϕ on the space $\mathcal{P}_{\mathbb{Z}^d}$ of convex lattice polytopes which have been studied are *valuations* with values in $\mathbb{R}$ or in some Abelian group $\mathcal{A}$. This means that

$$
\phi(P \cup Q) + \phi(P \cap Q) = \phi(P) + \phi(Q)
$$
$$
\text{whenever } P, Q, P \cup Q, P \cap Q \in \mathcal{P}_{\mathbb{Z}^d}, \text{ and } \phi(\emptyset) = 0.
$$

Examples are the volume and the lattice point enumerators. Among such valuations many are *integer unimodular invariant*. By this we mean that

$$
\phi(P) = \phi(UP + u) \text{ for } P \in \mathcal{P}_{\mathbb{Z}^d},\ U \in \mathcal{U},\ u \in \mathbb{Z}^d,
$$

where $\mathcal{U}$ is the family of all integer unimodular $d \times d$ matrices. A central result in this context due to Betke [105] and Betke and Kneser [108] shows that the structure of the linear space of the real, integer unimodular invariant valuations on $\mathcal{P}_{\mathbb{Z}^d}$ is surprisingly simple. It parallels Hadwiger's functional theorem 7.9 for real, continuous and motion invariant valuations on the space $\mathcal{C}$ of convex bodies. The theorem of Betke and Kneser yields simple proofs of the results of Ehrhart [292, 293] on lattice point enumerators.

In the following we state without proof an unpublished result of Stein and Betke on the inclusion–exclusion principle, give a proof of the Betke–Kneser theorem and show how it implies the results of Ehrhart. We will make use of some algebraic tools.

For more information, see the survey of McMullen [714] and the references in the introduction of Sect. 19, to which we add Kantor [564].

The Inclusion–Exclusion Formula

R. Stein [955] and Betke [106] proved the following result but, unfortunately, did not publish their proofs.

Theorem 19.5. *Let $\phi : \mathcal{P}_{\mathbb{Z}^d} \to \mathcal{A}$ be an integer unimodular invariant valuation, where $\mathcal{A}$ is an Abelian group. Then ϕ satisfies the following inclusion–exclusion formula for lattice polytopes:*

$$\phi(P_1 \cup \dots \cup P_m) = \sum_i \phi(P_i) - \sum_{i<j} \phi(P_i \cap P_j) + \dots + (-1)^{m-1} \phi(P_1 \cap \dots \cap P_m),$$

whenever $P_i, P_i \cap P_j, \dots, P_1 \cap \dots \cap P_m, P_1 \cup \dots \cup P_m \in \mathcal{P}_{\mathbb{Z}^d}$.

Algebraic Preparations

Let $\mathcal{G}_p^d$ be the free Abelian group generated by the proper convex lattice polytopes $P \in \mathcal{P}_{\mathbb{Z}^d p}$ and let $\mathcal{H}_p^d$ be its subgroup generated by the following elements of $\mathcal{G}_p^d$:

$$P - UP - u : P \in \mathcal{P}_{\mathbb{Z}^d p},\ U \in \mathcal{U},\ u \in \mathbb{Z}^d$$
$$P - \sum_i P_i : P = P_1 \dot{\cup} \cdots \dot{\cup} P_m,\ P_i \in \mathcal{P}_{\mathbb{Z}^d p},\ P_i \cap P_j, P_i \cap P_j \cap P_k, \dots \in \mathcal{P}_{\mathbb{Z}^d}$$

Let $S_0 = \{o\}$ and denote by S_i the simplex $\operatorname{conv}\{o, b_1, \dots, b_i\}$ for $i = 1, \dots, d$, where $\{b_1, \dots, b_d\}$ is the standard basis of $\mathbb{E}^d$.

Proposition 19.1. *$\mathcal{G}_p^d / \mathcal{H}_p^d$ is an infinite cyclic group generated by the coset $S_d + \mathcal{H}_p^d$.*

Proof. Let $P \in \mathcal{P}_{\mathbb{Z}^d p}$. By Theorem 14.9, $P = T_1 \dot{\cup} \cdots \dot{\cup} T_m$, where $T_i \in \mathcal{P}_{\mathbb{Z}^d p}$ and $T_i \cap T_j, T_i \cap T_j \cap T_k, \dots \in \mathcal{P}_{\mathbb{Z}^d}$ are simplices. The definition of $\mathcal{H}_p^d$ then shows that

$$P + \mathcal{H}_p^d = T_1 + \mathcal{H}_p^d + \dots + T_m + \mathcal{H}_p^d.$$

For the proof of the proposition it is thus sufficient to show the following:

(1) Let $S \in \mathcal{P}_{\mathbb{Z}^d p}$ be a simplex of volume V (V is an integer multiple of $1/d!$). Then

$$S + \mathcal{H}_p^d = (d!V) S_d + \mathcal{H}_p^d.$$

This will be proved by induction on $d!V$. If $d!V = 1$, there are $U \in \mathcal{U}$, and $u \in \mathbb{Z}^d$, such that $S = US_d + u$, which implies (1), on noting that

$$\begin{aligned} S + \mathcal{H}_p^d &= S - US_d - u + US_d + u + \mathcal{H}_p^d = US_d + u + \mathcal{H}_p^d \\ &= US_d + u - S_d + S_d + \mathcal{H}_p^d = S_d + \mathcal{H}_p^d \end{aligned}$$

by the definition of $\mathcal{H}_p^d$.

Assume now that $d!V > 1$ and that (1) holds for all proper convex lattice simplices of volume less than V. Let $S = \operatorname{conv}\{p_0, \dots, p_d\} \in \mathcal{P}_{\mathbb{Z}^d p}$ be a simplex with

$V(S) = V$. Let j be the largest index such that $p_1 - p_0, \ldots, p_{j-1} - p_0 \in \mathbb{Z}^d$ are such that the parallelotope spanned by these vectors contains no point of $\mathbb{Z}^d$ except for its vertices. An application of Theorem 21.3 shows that there is a basis $\{c_1, \ldots, c_d\}$ of $\mathbb{Z}^d$ such that

$$p_i - p_0 = u_{i1}c_1 + \cdots + u_{ii}c_i \text{ for } i = 1, \ldots, d$$

with suitable $u_{ik} \in \mathbb{Z}^d$. The assumption on j shows that o is the only point of $\mathbb{Z}^d$ in the parallelotope $\{\alpha_1(p_1 - p_0) + \cdots + \alpha_{j-1}(p_{j-1} - p_0) : 0 \le \alpha_i < 1\}$ and that $|u_{jj}| > 1$. The former can be used to show that $|u_{11}| = \cdots = |u_{j-1,j-1}| = 1$. By replacing c_i by $-c_i$, if necessary, and renaming, we may assume that $u_{11} = \cdots = u_{j-1,j-1} = 1$ and $u_{jj} > 1$. Putting $d_1 = c_1, d_2 = u_{21}c_1 + c_2, \ldots, d_{j-1} = u_{j-1,1}c_1 + \cdots + u_{j-1,j-2}c_{j-2} + c_{j-1}$ and $d_j = u_1 d_1 + \cdots + u_j d_{j-1} + c_j$ with suitable $u_i \in \mathbb{Z}^d$, and $d_{j+1} = c_{j+1}, \ldots, d_d = c_d$, we obtain a basis $\{d_1, \ldots, d_d\}$ of $\mathbb{Z}^d$ such that

$$\begin{aligned}
p_1 - p_0 &= d_1 \\
p_2 - p_0 &= \quad d_2 \\
&\cdots\cdots\cdots \\
p_{j-1} - p_0 &= \qquad\qquad d_{j-1} \\
p_j - p_0 &= v_{j1}d_1 + \cdots + v_{jj}d_j \\
&\cdots\cdots\cdots\cdots\cdots\cdots \\
p_d - p_0 &= v_{d1}d_1 + \cdots\cdots\cdots + v_{dd}d_d
\end{aligned}$$

with suitable $v_{ik} \in \mathbb{Z}$, where $0 \le v_{j1}, \ldots, v_{j,j-1} \le v_{jj}, v_{jj} > 1$. Finally, permuting $d_1, \ldots, d_{j-1}$ suitably, if necessary, and retaining $d_j, \ldots, d_d$, we obtain a basis $e_1, \ldots, e_d$ of $\mathbb{Z}^d$ such that

$$\begin{aligned}
V(p_1 - p_0) &= e_1 \\
V(p_2 - p_0) &= \quad e_2 \\
&\cdots\cdots\cdots \\
V(p_{j-1} - p_0) &= \qquad\qquad e_{j-1} \\
V(p_j - p_0) &= w_{j1}e_1 + \cdots + w_{jj}e_j \\
&\cdots\cdots\cdots\cdots\cdots\cdots \\
V(p_d - p_0) &= w_{d1}e_1 + \cdots\cdots\cdots + w_{dd}e_d
\end{aligned}$$

with a suitable integer unimodular $d \times d$ matrix V and integers $w_{ik} \in \mathbb{Z}$ such that $0 \le w_{j1} \le w_{j2} \le \cdots \le w_{jj}, w_{jj} > 1$. Now choose an integer unimodular $d \times d$ matrix W such that $We_i = b_i$ and put $U = WV$. Then

$$\begin{aligned}
&U(p_i - p_0) = b_i \text{ for } i = 1, \ldots, j-1, \\
&U(p_j - p_0) = w_{j1}b_1 + \cdots + w_{jj}b_j, \\
&\qquad\qquad \text{where } w_{jk} \in \mathbb{Z},\ 0 \le w_{j1} \le \cdots \le w_{jj},\ w_{jj} > 1.
\end{aligned}$$

Since $S + \mathcal{H}_p^d = US - Up_0 + \mathcal{H}_p^d$, by the definition of $\mathcal{H}_p^d$, we may assume that S already has this form. The facets of S are

$$F_i = \operatorname{conv}\{p_0, \ldots, p_{i-1}, p_{i+1}, \ldots, p_d\},\ i = 0, \ldots d.$$

If $p \in \mathbb{Z}^d$, represent $\operatorname{conv}(S \cup \{p\})$ as follows:

$$(2)\ \operatorname{conv}\big(S\cup\{p\}\big) = S\dot{\cup}\operatorname{conv}\big(F_{i_1}\cup\{p\}\big)\dot{\cup}\cdots\dot{\cup}\operatorname{conv}\big(F_{i_k}\cup\{p\}\big) = \operatorname{conv}\big(F_{j_1}\cup\{p\}\big)\dot{\cup}\cdots\dot{\cup}\operatorname{conv}\big(F_{j_l}\cup\{p\}\big).$$

Here, for a facet F_{i_n}, the point p is in the interior of the halfspace aff $F_{i_n}^+$ determined by the support hyperplane aff F_{i_n} of S, and, for a facet F_{j_n}, the point p is in the interior of the support halfspace aff $F_{j_n}^-$ of S. Then

$$(3)\ S+\operatorname{conv}\big(F_{i_1}\cup\{p\}\big)+\cdots+\operatorname{conv}\big(F_{i_k}\cup\{p\}\big)+\mathcal{H}_p^d = \operatorname{conv}\big(F_{j_1}\cup\{p\}\big)+\cdots+\operatorname{conv}\big(F_{j_l}\cup\{p\}\big)+\mathcal{H}_p^d$$

by the definition of $\mathcal{H}_p^d$. We next show that

(4) There is a point $p\in\mathbb{Z}^d$ such that
$V\big(\operatorname{conv}\big(F_i\cup\{p\}\big)\big) < V = V(S)$ for $i=0,\dots,d$.

The proof of (4) is elementary and will only be outlined. Choose m such that

$$ma_j+1\le a_1+\cdots+a_j\le (m+1)a_j$$

and put $p=(0,\dots,0,1,\dots,1,0,\dots,0)\in\mathbb{Z}^d$ with $j-m-1$ zeros, $m+1$ ones and $d-j$ zeros (in this order). Considering the equations of the hyperplanes aff F_i for $i=0, i=1,\dots,j-1, i=j, i=j+1,\dots,d$, it turns out that $p\in$ aff F_i for $i=j+1,\dots,d$. For all other i the point p is closer to aff F_i than p_i. This concludes the proof of (4). Propositions (4), (3) and the induction assumption show that S satisfies (1). The induction is thus complete and (1) holds generally, concluding the proof of the proposition. □

Let $\mathcal{G}^d$ be the free Abelian group generated by the convex lattice polytopes $P\in\mathcal{P}_{\mathbb{Z}^d}$ and let $\mathcal{H}^d$ be the subgroup generated by the following elements of the group $\mathcal{G}^d$:

$$P-UP-u,\ P\in\mathcal{P}_{\mathbb{Z}^d},\ U\in\mathcal{U},\ u\in\mathbb{Z}^d$$

$$P-\sum_i P_i+\sum_{i<j}P_i\cap P_j-\cdots,\ P=P_1\cup\cdots\cup P_m,\ P_i,\ P_i\cap P_j,\dots\in\mathcal{P}_{\mathbb{Z}^d}$$

Let $\mathcal{A}$ be an Abelian group.

Proposition 19.2. *Between the integer unimodular invariant valuations $\phi:\mathcal{P}_{\mathbb{Z}^d}\to\mathcal{A}$ and the homomorphisms $\psi:\mathcal{G}^d/\mathcal{H}^d\to\mathcal{A}$ there is a one-to-one correspondence such that*

(5) $\phi(P)=\psi(P+\mathcal{H}^d)$ *for* $P\in\mathcal{P}_{\mathbb{Z}^d}$.

Proof. The first step is to show the following statement:

(6) Let $\phi:\mathcal{P}_{\mathbb{Z}^d}\to\mathcal{A}$ be an integer unimodular invariant valuation. Then there is a unique homomorphism $\psi:\mathcal{G}^d/\mathcal{H}^d\to\mathcal{A}$ such that (5) is satisfied.

First, define a mapping $\psi:\mathcal{G}^d\to\mathcal{A}$ as follows: for $P\in\mathcal{P}_{\mathbb{Z}^d}$ let $\psi(P)=\phi(P)$. Then extend ψ to a mapping $\psi:\mathcal{G}^d\to\mathcal{A}$ by linearity (over $\mathbb{Z}^d$). This is possible since $\mathcal{G}^d$ is the free Abelian group generated by the polytopes in $\mathcal{P}_{\mathbb{Z}^d}$. Clearly ψ is a homomorphism. Next,

(7) $\psi|\mathcal{H}^d = 0.$

To see this, note that for the generating elements of $\mathcal{H}^d$,

$$\psi(P - UP - u) = \psi(P) - \psi(UP + u) = \phi(P) - \phi(UP + u) = 0,$$

$$\begin{aligned}\psi\left(P - \sum_i P_i + \sum_{i<j} P_i \cap P_j - \cdots\right) \\ &= \psi(P) - \sum_i \psi(P_i) + \sum_{i<j} \psi(P_i \cap P_j) - \cdots \\ &= \phi(P) - \sum_i \phi(P_i) + \sum_{i<j} \phi(P_i \cap P_j) - \cdots = 0,\end{aligned}$$

where we have used the facts that ϕ is integer unimodular invariant by assumption and satisfies the inclusion–exclusion formula by Theorem 19.5. Since this holds for the generating elements of $\mathcal{H}^d$, we obtain (7). Since $\psi : \mathcal{G}^d \to \mathcal{A}$ is a homomorphism, (7) shows that it gives rise to a homomorphism of $\mathcal{G}^d/\mathcal{H}^d \to \mathcal{A}$. We also denote it by ψ. Clearly $\psi(P + \mathcal{H}^d) = \psi(P) = \phi(P)$ for $P \in \mathcal{P}_{\mathbb{Z}^d}$. To conclude the proof of (6) we have to show that ψ is unique. Let μ be another homomorphism of $\mathcal{G}^d/\mathcal{H}^d \to \mathcal{A}$ satisfying (5). Then

$$\psi(P + \mathcal{H}^d) = \phi(P) = \mu(P + \mathcal{H}^d) \text{ for } P \in \mathcal{P}_{\mathbb{Z}^d}.$$

Since the cosets $P + \mathcal{H}^d : P \in \mathcal{P}_{\mathbb{Z}^d}$ generate $\mathcal{G}^d/\mathcal{H}^d$ and ψ, μ are both homomorphisms, $\psi = \mu$, concluding the proof of (6).

The second step is to prove the following reverse statement:

(8) Let $\psi : \mathcal{G}^d/\mathcal{H}^d \to \mathcal{A}$ be a homomorphism. Then the mapping $\phi : \mathcal{P}_{\mathbb{Z}^d} \to \mathcal{A}$, defined by (5), is a integer unimodular invariant valuation.

We first show that ϕ is a valuation. Let $P, Q \in \mathcal{P}_{\mathbb{Z}^d}$, where $P \cup Q, P \cap Q \in \mathcal{P}_{\mathbb{Z}^d}$. Then

$$\begin{aligned}\phi(P \cup Q) &= \psi(P \cup Q + \mathcal{H}^d) \\ &= \psi\big(P \cup Q - (P \cup Q - P - Q + P \cap Q) + \mathcal{H}^d\big) \\ &= \psi(P + Q - P \cap Q + \mathcal{H}^d) \\ &= \psi(P + \mathcal{H}^d) + \psi(Q + \mathcal{H}^d) - \psi(P \cap Q + \mathcal{H}^d) \\ &= \phi(P) + \phi(Q) - \phi(P \cap Q)\end{aligned}$$

by the definitions of ϕ and $\mathcal{H}^d$, since ψ is a homomorphism. To see that ϕ is integer unimodular invariant, let $P \in \mathcal{P}_{\mathbb{Z}^d}$, $U \in \mathcal{U}$, $u \in \mathbb{Z}^d$. Then

$$\begin{aligned}\phi(UP + u) &= \psi(UP + u + \mathcal{H}^d) = \psi(UP + u + (P - UP - u) + \mathcal{H}^d) \\ &= \psi(P + \mathcal{H}^d) = \phi(P)\end{aligned}$$

by the definitions of ϕ and $\mathcal{H}^d$, concluding the proof of (8).

Having shown (6) and (8), the proof of the proposition is complete. □

Proposition 19.3. *$\mathcal{G}^d/\mathcal{H}^d$ is the free Abelian group generated by the cosets $S_0 + \mathcal{H}^d, \dots, S_d + \mathcal{H}^d$.*

Proof. By induction on d we prove that

(9) $S_0 + \mathcal{H}^d, \dots, S_d + \mathcal{H}^d$ generate $\mathcal{G}^d/\mathcal{H}^d$.

For $d = 0$, (9) is trivial. Assume now that $d > 0$ and that (9) holds for $d - 1$. Let $\mathbb{E}^{d-1}$ be embedded into $\mathbb{E}^d$ as usual (first $d - 1$ coordinates). Then $\mathcal{G}^d$ is generated by (the elements of) $\mathcal{G}^d_p$ and $\mathcal{G}^{d-1}$. According to Proposition 19.1, $\mathcal{G}^d_p$ is generated by S_d and $\mathcal{H}^d_p$ and $\mathcal{G}^{d-1}$ is generated by $S_0, \dots, S_{d-1}$ and $\mathcal{H}^{d-1} (\subseteq \mathcal{H}^d)$ by the induction assumption. Thus $\mathcal{G}^d$ is generated by $S_0, \dots, S_d$ and $\mathcal{H}^d$. This concludes the induction and thus yields (9).

It remains to show that

(10) $S_0 + \mathcal{H}^d, \dots, S_d + \mathcal{H}^d$ are linearly independent (with respect to $\mathbb{Z}$).

A simple induction argument shows that

$$L(nS_i) = \binom{n+i}{i} \text{ for } n \in \mathbb{N},\ i = 0, \dots, d.$$

For $n \in \mathbb{N}$ define $\phi_n : \mathcal{P}_{\mathbb{Z}^d} \to \mathbb{R}$ by

$$\phi_n(P) = L(nP) \text{ for } P \in \mathcal{P}_{\mathbb{Z}^d}.$$

Since the lattice point enumerator $L : \mathcal{P}_{\mathbb{Z}^d} \to \mathbb{R}$ is a valuation, each ϕ_n is also a valuation. Let $\mu_n : \mathcal{G}^d/\mathcal{H}^d \to \mathbb{R}$ be the corresponding homomorphism, see Proposition 19.2. If (10) did not hold, then

$$a_0 S_0 + \cdots + a_d S_d + \mathcal{H}^d = \mathcal{H}^d$$

for suitable integers a_i, not all 0. Then

$$\begin{aligned} 0 &= \mu_n(a_0 S_0 + \cdots + a_d S_d + \mathcal{H}^d) \\ &= a_0 \mu_n(S_0 + \mathcal{H}^d) + \cdots + a_d \mu_n(S_d + \mathcal{H}^d) \\ &= a_0 \phi_n(S_0) + \cdots + a_d \phi_n(S_d) = a_0 L(nS_0) + \cdots + a_d L(nS_d) \\ &= a_0 \binom{n+0}{0} + a_1 \binom{n+1}{1} + \cdots + a_d \binom{n+d}{d} \text{ for } n \in \mathbb{N}. \end{aligned}$$

Since $\binom{n+i}{i}$ is a polynomial in n of degree i, this can hold only if $a_d = \cdots = a_0 = 0$. This contradiction concludes the proof of (10).

Claims (9) and (10) together yield Proposition 19.3. □

The Theorem of Betke and Kneser

As a corollary of the above results, Betke and Kneser proved the following result.

Theorem 19.6. *The integer unimodular invariant valuations $\phi : \mathcal{P}_{\mathbb{Z}^d} \to \mathbb{R}$ (with ordinary addition and multiplication by real numbers) form a real vector space of dimension $d+1$. This space has a basis $\{L_0, \dots, L_d\}$ such that*

$$L_i(nP) = n^i L_i(P) \text{ and } L(nP) = L_0(P) + L_1(P)n + \cdots + L_d(P)n^d$$

for $P \in \mathcal{P}_{\mathbb{Z}^d}$, $n \in \mathbb{N}$, $i = 0, \dots, d$.

Proof. We first show that

(11) The homomorphisms $\psi : \mathcal{G}^d/\mathcal{H}^d \to \mathbb{R}$ with ordinary addition and multiplication with real numbers form a real vector space of dimension $d+1$.

These homomorphisms clearly form a real vector space. For the proof that it is of dimension $d+1$, use Proposition 19.3 to define $d+1$ homomorphisms $\psi_i : \mathcal{G}^d/\mathcal{H}^d \to \mathbb{R}$, $i = 0, \dots, d$, by:

$$\psi_i(a_0 S_0 + \cdots + a_d S_d + \mathcal{H}^d) = a_i \text{ for } a_0 S_0 + \cdots + a_d S_d + \mathcal{H}^d \in \mathcal{G}^d/\mathcal{H}^d.$$

To show that $\psi_0, \dots, \psi_d$ are linearly independent, let $\alpha_0, \dots, \alpha_d \in \mathbb{R}$ be such that $\alpha_0\psi_0 + \cdots + \alpha_d\psi_d = 0$. Then

$$\begin{aligned} 0 &= \alpha_0\psi_0(a_0 S_0 + \cdots + a_d S_d + \mathcal{H}^d) + \cdots + \alpha_d\psi_d(a_0 S_0 + \cdots + a_d S_d + \mathcal{H}^d) \\ &= \alpha_0 a_0 + \cdots + \alpha_d a_d \text{ for all } a_0, \dots, a_d \in \mathbb{Z}, \end{aligned}$$

which implies that $\alpha_0 = \cdots = \alpha_d = 0$ and thus shows the linear independence of $\psi_0, \dots, \psi_d$. To show that $\psi_0, \dots, \psi_d$ form a basis, let $\psi : \mathcal{G}^d/\mathcal{H}^d \to \mathbb{R}$ be a homomorphism. Then

$$\begin{aligned} \psi(a_0 S_0 + \cdots + a_d S_d + \mathcal{H}^d) &= a_0\psi(S_0 + \mathcal{H}^d) + \cdots + a_d\psi(S_d + \mathcal{H}^d) \\ &= a_0\beta_0 + \cdots + a_d\beta_d \text{ say, where } \beta_i = \psi(S_i + \mathcal{H}^d) \\ &= \beta_0\psi_0(a_0 S_0 + \cdots + a_d S_d + \mathcal{H}^d) + \cdots + \beta_d\psi_d(a_0 S_0 + \cdots + a_d S_d + \mathcal{H}^d) \\ &\text{for all } a_0 S_0 + \cdots + a_d S_d + \mathcal{H}^d \in \mathcal{G}^d/\mathcal{H}^d. \end{aligned}$$

Hence $\psi = \beta_0\psi_0 + \cdots + \beta_d\psi_d$. Thus $\{\psi_0, \dots, \psi_d\}$ is a basis, concluding the proof of (11).

By Proposition 19.2, the integer unimodular invariant valuations $\phi : \mathcal{P}_{\mathbb{Z}^d} \to \mathbb{R}$ coincide with the restrictions of the homomorphisms $\psi : \mathcal{G}^d/\mathcal{H}^d \to \mathbb{R}$ to a certain subset of $\mathcal{G}^d/\mathcal{H}^d$. Thus (11) implies that

(12) The integer unimodular invariant valuations $\phi : \mathcal{P}_{\mathbb{Z}^d} \to \mathbb{R}$ form a real vector space of dimension at most $d+1$.

We next show that

(13) There are $d+1$ linearly independent valuations $L_0, \dots, L_d : \mathcal{P}_{\mathbb{Z}^d} \to \mathbb{R}$ such that

$$L_i(nP) = n^i L_i(P),\ L(nP) = L_0(P) + L_1(P)n + \cdots + L_d(P)n^d$$

for $P \in \mathcal{P}_{\mathbb{Z}^d}$, $n \in \mathbb{N}$, $i = 0, \dots, d$.

Let ϕ_n, μ_n be as in the proof of Proposition 19.3. Then

$$\begin{aligned}
L(nP) &= \phi_n(P) = \mu_n(P + \mathcal{H}^d) = \mu_n(a_0 S_0 + \cdots + a_d S_d + \mathcal{H}^d) \\
&= a_0 \mu_n(S_0 + \mathcal{H}^d) + \cdots + a_d \mu_n(S_d + \mathcal{H}^d) \\
&= a_0 \phi_n(S_0) + \cdots + a_d \phi_n(S_d) = a_0 \binom{n+0}{0} + \cdots + a_d \binom{n+d}{d} \\
&= b_0 + b_1 n + \cdots + b_d n^d \text{ for } P \in \mathcal{P}_{\mathbb{Z}^d},\ n \in \mathbb{N},
\end{aligned}$$

by Proposition 19.3, where $a_0, \dots, a_d$ and thus $b_0, \dots, b_d$ are suitable integers, depending only on P. Thus we may write,

$$(14)\quad L(nP) = L_0(P) + L_1(P)n + \cdots + L_d(P)n^d \text{ for } P \in \mathcal{P}_{\mathbb{Z}^d},\ n \in \mathbb{N}.$$

It is easy to check that the mappings $L_i : \mathcal{P}_{\mathbb{Z}^d} \to \mathbb{R}$ are integer unimodular invariant valuations: let $P, Q \in \mathcal{P}_{\mathbb{Z}^d}$ be such that $P \cup Q,\ P \cap Q \in \mathcal{P}_{\mathbb{Z}^d}$. Then $L\big(n(P \cup Q)\big) + L\big(n(P \cap Q)\big) = L(nP) + L(nQ)$. Represent each of these four expressions as a polynomial in n according to (14) and compare coefficients. By (14),

$$\begin{aligned}
L(mnP) &= L_0(nP) + L_1(nP)m + \cdots + L_d(nP)m^d \\
&= L_0(P) + L_1(P)mn + \cdots + L_d(P)m^d n^d \\
&\text{for } P \in \mathcal{P}_{\mathbb{Z}^d},\ m, n \in \mathbb{N}.
\end{aligned}$$

Fixing $n \in \mathbb{N}$, this shows that

$$L_i(nP) = n^i L_i(P) \text{ for } P \in \mathcal{P}_{\mathbb{Z}^d},\ n \in \mathbb{N},\ i = 0, \dots, d.$$

Since $L(nS_i) = \binom{n+i}{i}$ is a polynomial in n of degree i, we obtain

$$L_i(S_i) \neq 0,\ L_{i+1}(S_i) = \cdots = L_d(S_i) = 0 \text{ for } i = 0, \dots, d.$$

This readily implies that $L_0, \dots, L_d$ are linearly independent. The proof of (13) is complete.

The theorem finally follows from (12) and (13). □

Ehrhart's Lattice Point Enumerators Theorem

As simple corollaries of the theorem of Betke and Kneser we get again the theorem 19.1 of Ehrhart [292, 293] on lattice point enumerators, but without the information that the constant term in the polynomial is 1.

Corollary 19.1. *Let $P \in \mathcal{P}_{\mathbb{Z}^d\, p}$. Then the following hold:*

(i) $L(nP) = p_P(n)$ *for $n \in \mathbb{N}$, where p_P is a polynomial of degree d with leading coefficient $V(P)$.*

(ii) $L^o(nP) = (-1)^d p_P(-n)$ *for $n \in \mathbb{N}$.*

Proof. (i) By the Betke–Kneser theorem,

(15) $L(nP) = L_0(P) + \cdots + L_d(P)n^d = p_P(n)$ for $n \in \mathbb{N}$,

where p_P is a polynomial of degree $\leq d$. We have to show that $L_d(P) = V(P)$. Using the formula for the calculation of Jordan measure in Sect. 7.2, it follows that

$$\begin{aligned} V(P) &= \lim_{n\to\infty} \frac{1}{n^d} \# \left(P \cap \frac{1}{n}\mathbb{Z}^d \right) = \lim_{n\to\infty} \frac{1}{n^d} \#(nP \cap \mathbb{Z}^d) \\ &= \lim_{n\to\infty} \frac{L(nP)}{n^d} = L_d(P). \end{aligned}$$

(ii) We first show that L^o is a valuation on $\mathcal{P}_{\mathbb{Z}^d}$. Let $Q, R \in \mathcal{P}_{\mathbb{Z}^d}$ be such that $Q \cup R, Q \cap R \in \mathcal{P}_{\mathbb{Z}^d}$. Considering the cases $Q \subseteq R$, or $R \subseteq Q$; $\dim Q = \dim R = \dim(Q \cap R) + 1$; $\dim Q = \dim R = \dim(Q \cap R)$, it is easy to see that $L^o(Q \cup R) + L^o(Q \cap R) = L^o(Q) + L^o(R)$, i.e. L^o is a valuation. Clearly, L^o is integer unimodular invariant. Thus the theorem of Betke and Kneser implies that

(16) $L^o(nP) = \sum_{i=0}^{d} a_i L_i(nP) = \sum_{i=0}^{d} a_i n^i L_i(P)$ for $n \in \mathbb{N}$

with suitable coefficients a_i independent of P and n. In order to determine the coefficients a_i take the lattice cube $[0, 1]^d$ instead of P. Then

(17) $L\big(n[0, 1]^d\big) = (n + 1)^d = \sum_{i=0}^{d} n^i \binom{d}{i}$ for $n \in \mathbb{N}$,

(18) $L^o\big(n[0, 1]^d\big) = (n - 1)^d = \sum_{i=0}^{d} n^i (-1)^{d-i} \binom{d}{i}$ for $n \in \mathbb{N}$.

Now compare (15) and (17) and also (16) and (18) to see that $a_i = (-1)^{d-i}$. Thus

$$\begin{aligned} L^o(nP) &= \sum_{i=0}^{d} (-1)^{d-i} L_i(P) n^i = (-1)^d \sum_{i=0}^{d} L_i(P)(-n)^i \\ &= (-1)^d p_P(n) \text{ for } n \in \mathbb{N} \end{aligned}$$

by (16) and (15), concluding the proof of (ii). □

19.5 Newton Polytopes: Irreducibility of Polynomials and the Minding–Kouchnirenko–Bernstein Theorem

Let $p = p(x_1, \dots, x_d)$ be a polynomial in d variables over $\mathbb{R}$, $\mathbb{C}$, or some other field. Its *Newton polytope* N_p is the convex hull of all points $(u_1, \dots, u_d) \in \mathbb{Z}^d$, such that, in p, there is a monomial of the form

$$a x_1^{u_1} \cdots x_d^{u_d} \text{ where } a \neq 0.$$

The Newton polytope N_p conveys important properties of the polynomial p. Newton polytopes turned out to be of interest in algebra, algebraic geometry and numerical analysis. They constitute a bridge between convexity and algebraic geometry.

Irreducibility of polynomials has a long history and there is a large body of pertinent results. See, e.g. Schmidt [895], Lidl and Niederreiter [656], Mignotte and Stefanescu [723] and Schinzel [887]. It plays a crucial role in finite geometry, see Hirschfeld [506], combinatorics, see Szönyi [983], and coding, compare Stichtenoth [969]. Surprisingly, Newton polytopes provide an effective tool to describe large classes of (absolute) irreducible polynomials, see Gao [355] and the references there.

A remarkable result of Bernstein [100] says that the number of roots of a generic system of polynomial equations over $\mathbb{C}$ of the form

$$\begin{aligned} p_1(z_1, \dots, z_d) &= 0 \\ p_2(z_1, \dots, z_d) &= 0 \\ \dots\dots\dots&\dots\dots \\ p_d(z_1, \dots, z_d) &= 0 \end{aligned}$$

equals $d!\,V(N_{p_1}, \dots, N_{p_d})$. The case $d = 2$ was anticipated by Minding [731] in 1841 using different terminology, as pointed out by Khovanskiĭ [583]. The case where the Newton polytopes of $p_1, \dots, p_d$ all coincide was treated slightly before Bernstein by Kouchnirenko [612]. This result gave rise to several alternative proofs, generalizations and expositions, see Huber and Sturmfels [525, 526], Sturmfels [974], Rojas [853, 854] and the nice book by Sturmfels [975]. In the latter there are described applications to economics (Nash equilibria of n-person games), statistics (random walks on $\mathbb{Z}^d$, numerical algorithms for maximum likelihood equations), linear partial differential equations and other areas.

This section contains a simple result of Ostrowski [781] which yields an irreducibility criterion, see Gao [355]. Special cases include the Stepanov–Schmidt [895] criterion for irreducibility. Then the result of Minding–Bernstein is given without proof.

For more detailed information, see the articles, surveys and books cited earlier.

An Irreducibility Criterion

The first result is a simple observation of Ostrowski [781] on Newton polytopes, see also Gao [355].

Proposition 19.4. *Let $p, q : \mathbb{E}^d \to \mathbb{R}$ be real polynomials. Then*

$$N_{pq} = N_p + N_q.$$

Proof. Since each monomial of the polynomial pq is a sum of products of a monomial of p and a monomial of q, we have

$$N_{pq} \subseteq N_p + N_q.$$

To see the reverse inclusion, it is sufficient to show that each vertex u of the convex lattice polytope $N_p + N_q$ is contained in N_{pq}. By Lemma 6.1, u is the sum of a support set of N_p and a support set of N_q. Since u is a singleton, these support sets have to be singletons too and thus are vertices of N_p and N_q, respectively, say v, w. Then $u = v + w$ and this is the only possible representation of u as the sum of a point of N_p and a point of N_q. The product of the corresponding monomials $bx_1^{v_1} \cdots x_d^{v_d}$ of p and $cx_1^{w_1} \cdots x_d^{w_d}$ of q is thus a monomial of the form $ax_1^{u_1} \cdots x_d^{u_d}$ and no other product of a monomial of p and a monomial of q has this form. Hence $ax_1^{u_1} \cdots x_d^{u_d}$ is the monomial in pq corresponding to u, or $u \in N_{pq}$. Thus

$$N_p + N_q \subseteq N_{pq},$$

concluding the proof. □

A real polynomial on $\mathbb{E}^d$ is *irreducible* if it cannot be represented as a product of two real polynomials on $\mathbb{E}^d$, each consisting of more than one monomial. Call a convex lattice polytope in $\mathbb{E}^d$ *integer irreducible* if it cannot be represented as the sum of two convex lattice polytopes, each consisting of more than one point. As an immediate consequence of the earlier Proposition 19.4, we obtain the following *irreducibility criterion*, see Gao [355].

Theorem 19.7. *Let $p : \mathbb{E}^d \to \mathbb{R}$ be a real polynomial with Newton polytope N_p. If N_p is integer irreducible, then p is irreducible.*

Remark. Actually, this result holds for any polynomial p in d variables over any field $\mathcal{F}$. In addition, irreducibility of p may be sharpened to *absolute irreducibility*, i.e. irreducibility in the algebraic closure of $\mathcal{F}$. When N_p is not integer irreducible, then p may still be irreducible. An example is provided by the irreducible real polynomial $p(x, y) = 1 + y + xy + x^2 + y^2$, whose Newton polytope $N_p = \operatorname{conv}\big\{o, (0, 2), (2, 0)\big\}$ is integer reducible:

$$\operatorname{conv}\{o, (0, 2), (2, 0)\} = \operatorname{conv}\{o, (0, 1), (1, 0)\} + \operatorname{conv}\{o, (0, 1), (1, 0)\}.$$

Gao and Lauder [356] study the integer irreducibility of convex lattice polytopes. It turns out that the problem to decide whether a convex lattice polytope is integer irreducible is NP-complete.

The Stepanov–Schmidt Irreducibility Criterion

Clearly, a convex lattice polygon in $\mathbb{E}^2$, which has an edge of the form $[(0, m), (n, 0)]$ with m and n relatively prime and which is contained in the triangle $\operatorname{conv}\{o, (0, m), (n, 0)\}$, is integer irreducible. This leads to the following criterion of Stepanov and Schmidt, see W. Schmidt [895], p. 92, and Gao [355].

Corollary 19.2. *Let p be a real polynomial in two variables, such that its Newton polygon N_p contains an edge of the form $[(0, m), (n, 0)]$, where m and n are relatively prime, and is contained in the triangle $\operatorname{conv}\big\{o, (0, m), (n, 0)\big\}$. Then p is irreducible.*

For example, the polynomials $x^2 + y^3$ and $x^3 + y^7 + xy + xy^2 + x^2y^2$ are irreducible.

Zeros of Systems of Polynomial Equations, the Minding–Kouchnirenko–Bernstein Theorem

The problem to determine the number of (positive, negative, real or complex) roots of a polynomial equation or of a system of such equations has attracted interest for several centuries. Highlights of the study of this problem are Descartes's rule of signs, the fundamental theorem of algebra, and theorems of Sturm, Routh and Hurwitz. For systems of polynomial equations *Bezout's theorem* is as follows: If the system

$$(1) \qquad \begin{array}{c} p_1(z_1, \dots, z_d) = 0 \\ p_2(z_1, \dots, z_d) = 0 \\ \dots\dots\dots\dots\dots\dots \\ p_d(z_1, \dots, z_d) = 0 \end{array}$$

of d complex polynomial equations in d complex variables has only finitely many common complex zeros $(z_1, \dots, z_d)$, then the number of these zeros is at most the product of the degrees of the polynomials $p_1, \dots, p_d$. For an elementary proof, see the book by Cox, Little and O'Shea [228]. In general this upper estimate is far too large. In the generic case, the following *theorem of Minding–Kouchnirenko–Bernstein* gives the precise answer. We state it without proof and do not explain what is meant by generic. For detailed information, see the book of Sturmfels [975], or the original article of Bernstein [100].

Theorem 19.8. *For generic systems of polynomial equations over* $\mathbb{C}$ *of the form* (1) *the number of common solutions in* $(\mathbb{C}\backslash\{0\})^d$ *is finite and equals* $d!\,V(N_{p_1}, \dots, N_{p_d})$.

20 Linear Optimization

A *linear optimization* (or *linear programming*) *problem* entails minimizing or maximizing a linear form on a convex polytope or polyhedron. A typical form is the following:

$$(1) \quad \sup\left\{c \cdot x : x \in \mathbb{E}^d,\ Ax \le b\right\},$$

where A is a real $m \times d$ matrix, $c \in \mathbb{E}^d$, $b \in \mathbb{E}^m$ and the inequality is to be understood componentwise. When writing (1), we mean the problem is to determine the supremum and, if it is finite, to find a point at which it is attained. It turns out that all common linear optimization problems are polynomially equivalent to one of the form (1).

With early contributions dating back to the eighteenth century, a first vague version of linear programming was given by Fourier [342] at the beginning of the nineteenth century. Pertinent later results on systems of linear inequalities are due to Gordan, Farkas, Stiemke, Motzkin and others. Linear optimization, as it is used at present, started with the work of Kantorovich [565] which won him a Nobel prize in economics, von Neumann [768], Koopmans [608] and, in particular, Dantzig [237], who specified the simplex algorithm. While the Klee-Minty [597] cube shows that the common version of the simplex algorithm is *not* polynomial, Borgwardt [151] proved that, on the average, it *is* polynomial. See also a more recent result of Spielman and Teng [950]. In practice the simplex algorithm works very effectively. Khachiyan [580] indicated a proof that the ellipsoid algorithm of Shor [933] and Yudin and Nemirovskiĭ [1033] is polynomial. The ellipsoid algorithm did not replace the simplex algorithm in practice and was never stably implemented. There is no running code available, not even for small test problems. A different polynomial algorithm is that of Karmarkar [566]. While it was not put to practical use, it started the development of a very efficient, huge class of interior-point methods for linear programming, some of which are both polynomial and efficient in practice. For many types of problems the interior point methods are better than the simplex algorithm and are widely used in practice.

Integer linear optimization (or *integer linear programming*) is linear optimization with the variables restricted to the integers. A standard problem is the following,

$$\sup\{c \cdot x : x \in \mathbb{Z}^d,\ Ax \leq b\},$$

where A is a rational $m \times d$ matrix and $b \in \mathbb{E}^m$, $c \in \mathbb{E}^d$ are rational vectors. An integer linear optimization problem may be interpreted as the search of optimum points of the integer lattice $\mathbb{Z}^d$ contained in the convex polyhedron $\{x : Ax \leq b\}$. Integer linear optimization is essentially different from linear optimization: there is no duality, no polynomial algorithm is known and, presumably, does not exist.

In this section we first consider a classical duality result, then describe the simplex algorithm in geometric terms and explain how to find feasible solutions with the ellipsoid algorithm. In integer optimization we consider so-called totally dual integral systems for which integer optimization is easier than in the general case. Their relations to lattice polyhedra are touched and Hilbert bases are used to characterize totally dual integral systems. We have borrowed freely from Schrijver's book [915].

To simplify the presentation, we often consider row vectors as being contained in $\mathbb{E}^d$ or $\mathbb{E}^m$. Using the matrix product we write cx instead of $c^T \cdot x$ where c is a row vector and x a column vector in $\mathbb{E}^d$.

There exists a rich literature on linear and integer optimization, including the classic of Dantzig [238] and the monographs of Schrijver [915], Borgwardt [152], Grötschel, Lovász and Schrijver [409], Berkovitz [99], Dantzig and Thapa [239] and Schrijver [916]. See also the surveys of Shamir [928], Burkard [179], Gritzmann and Klee [396] and Bartels [75]. For information on the history of optimization, see [508] and [915].

20.1 Preliminaries and Duality

As remarked before, any of the standard linear optimization problems can be reduced in polynomial time to any of the others. It turns out that certain pairs of linear optimization problems, one a maximization, the other one a minimization problem, are particularly strongly related, in the sense that the values of their solutions coincide. Such results are called *duality theorems*.

After some preliminary results on normal cones we present the duality theorem of von Neumann, Gale, Kuhn and Tucker.

For more information, see the literature cited earlier.

Terminology and Normal Cones

Given a linear optimization problem, say

$$\sup\{cx : Ax \leq b\},$$

the function $x \to cx,\ x \in \mathbb{E}^d$ is its *objective function* and the convex polyhedron $\{x : Ax \leq b\}$ its *feasible set*. A point of the feasible set is a *feasible solution*. If the supremum is attained at a feasible solution, the latter is called an *optimum solution*. There is an analogous notation for the other linear optimization problems.

Let C be a closed convex cone in $\mathbb{E}^d$ with apex o. Its *normal cone* $N_C(o)$ of C at the apex o is the closed convex cone with apex o consisting of all exterior normal vectors of support hyperplanes of C at o, that is,

$$N_C(o) = \big\{u : ux \leq 0 \text{ for all } x \in C\big\},$$

see Sect. 14.2. Let P be a convex polyhedron. The *normal cone* $N_P(p)$ *of* P *at a point* $p \in \operatorname{bd} P$ is the closed convex cone of all exterior normal vectors of support hyperplanes of P at p. The *normal cone* $N_P(F)$ *of* P *at a face* $F \in \mathcal{F}(P)$ is the closed convex cone of all exterior normal vectors of support hyperplanes of P which contain F. If p is a relative interior point of F, then it is easy to see that

$$N_P(F) = N_P(p).$$

An immediate extension of Proposition 14.1 is the following result.

Proposition 20.1. *Let* $P = \{x : Ax \leq b\}$ *be a convex polyhedron and let* $p \in \operatorname{bd} P$, *resp.* $F \in \mathcal{F}(P)$. *If* $a_1 x \leq \beta_1, \dots, a_k x \leq \beta_k$ *are the inequalities among the inequalities* $Ax \leq b$ *which are satisfied with the equality sign by* p, *resp. by all* $x \in F$, *then*

$$N_P(p), \text{ resp. } N_P(F) = \operatorname{pos}\{a_1, \dots, a_k\}.$$

Let $F \neq \emptyset$ be a face of a convex polyhedron P. F is called a *minimum face* of P if there is no face of P properly contained in F, except the empty face. This means that the polyhedron F has no proper face except for the empty face. Thus the minimum faces are planes of dimension 0 (vertices), 1 (edges, unbounded in both directions), 2,...

Existence of Solutions

Since each convex polyhedron $P \neq \emptyset$ can be represented in the form

$$\{\lambda_1 p_1 + \cdots + \lambda_m p_m : \lambda_i \geq 0,\ \lambda_1 + \cdots + \lambda_m = 1\} + \{\mu_1 q_1 + \cdots + \mu_n q_n : \mu_j \geq 0\},$$

the following result is easy to see:

Proposition 20.2. *Assume that the linear optimization problem* $\sup\{cx : Ax \leq b\}$ *has non-empty feasible set* $P = \{x : Ax \leq b\}$*. Then the supremum is either* $+\infty$*, or it is finite and attained. An analogous result holds for the problem* $\inf\{yb : y \geq o,\ yA = c\}$.

The Duality Theorem

of von Neumann [768] and Gale, Kuhn and Tucker [353] is as follows.

Theorem 20.1. *Let* A *be a real* $m \times d$ *matrix,* $b \in \mathbb{E}^m$, $c \in \mathbb{E}^d$*. If at least one of the extreme* $\sup\{cx : Ax \leq b\}$ *and* $\inf\{yb : y \geq o,\ yA = c\}$ *is attained, then so is the other and*

(1) $\max\{cx : Ax \leq b\} = \min\{yb : y \geq o,\ yA = c\}$.

Proof. We first show the following inequality, where $P = \{x : Ax \leq b\}$ and $Q = \{y : y \geq o,\ yA = c\}$:

(2) If $P, Q \neq \emptyset$, then $\sup\{cx : Ax \leq b\}$ and $\inf\{yb : y \geq o,\ yA = c\}$ both are attained and

$$\max\{cx : Ax \leq b\} \leq \min\{yb : y \geq o, yA = c\}.$$

Let $x \in P, y \in Q$. Then $cx = yAx \leq yb$. Hence $\sup\{cx : Ax \leq b\} \leq \inf\{yb : y \geq o,\ yA \leq c\}$ and both are attained by Proposition 20.2, concluding the proof of (2).

Assume now that $P \neq \emptyset$ and that $\delta = \sup\{cx : Ax \leq b\}$ is attained at $p \in P = \{x : Ax \leq b\}$, say. Then p is a boundary point of P and the hyperplane through p with normal vector c supports P at p. Clearly, c is an exterior normal vector of this support hyperplane. Let $a_1x \leq \beta_1, \dots, a_kx \leq \beta_k$ be the inequalities among the m inequalities $Ax \leq b$, which are satisfied by p with the equality sign. We may assume that $a_1, \dots, a_k$ are the first k row vectors of A and $\beta_1, \dots, \beta_k$ the first k entries of b. Proposition 20.1 then implies that

(3) $c = \lambda_1 a_1 + \cdots + \lambda_k a_k$ with suitable $\lambda_i \geq 0$.

Thus,

$$\delta = cp = \lambda_1 a_1 p + \cdots + \lambda_k a_k p = \lambda_1 \beta_1 + \cdots + \lambda_k \beta_k$$

and therefore

$$\begin{aligned}\max\{cx : Ax \leq b\} &= \delta = cp = \lambda_1\beta_1 + \cdots + \lambda_k\beta_k \\ &= (\lambda_1, \dots, \lambda_k, 0, \dots, 0)b \geq \inf\{yb : y \geq o,\ yA = c\},\end{aligned}$$

on noting that $(\lambda_1, \dots, \lambda_k, 0, \dots, 0) \in Q = \{y : y \geq o,\ yA = c\}$ by (3). In particular, $Q \neq \emptyset$. (2) then shows that the infimum is also attained and that the maximum is less than or equal to the minimum. Thus (1) holds in case where $P \neq \emptyset$ and the supremum is attained.

Assume, finally, that $Q \neq \emptyset$ and that $\varepsilon = \inf\{yb : y \geq o,\ yA = c\}$ is attained at $q \in Q = \{y : yA = c, -y \leq o\}$, say. Then $-b$ is an exterior normal vector of the support hyperplane $\{y : yb = \varepsilon\}$ of Q at q. Proposition 20.1 then shows that

(4) $b = \mu_1 b_1 + \dots + \mu_d b_d + \nu_1 e_1 + \dots + \nu_l e_l$ with suitable $\mu_i \in \mathbb{R}, \nu_j \geq 0$,

where $b_1, \dots, b_d$ are the column vectors of A and $e_1, \dots, e_l$ are standard unit vectors of $\mathbb{E}^d$ such that e_j has entry 1 where q has entry 0. In particular,

(5) $qe_j = 0$ for $j = 1, \dots, l$.

(4) implies that

$$A(\mu_1, \dots, \mu_d)^T = \mu_1 b_1 + \dots + \mu_d b_d = b - \nu_1 e_1 - \dots - \nu_l e_l \leq b.$$

Thus $(\mu_1, \dots, \mu_d)^T \in P$ and therefore $P \neq \emptyset$. This, together with (4) and (5), shows that

$$\begin{aligned}\min\{yb : y \geq o, yA = c\} &= \varepsilon = qb = \mu_1 q b_1 + \dots + \mu_d q b_d \\ &= qA(\mu_1, \dots, \mu_d)^T = c(\mu_1, \dots, \mu_d)^T \\ &\leq \sup\{cx : Ax \leq b\}.\end{aligned}$$

Since $P, Q \neq \emptyset$, (2) shows that the supremum is also attained and that the maximum is less than or equal to the minimum. Thus (1) holds also in the case where $Q \neq \emptyset$ and the infimum is attained. The proof of the theorem is complete. □

20.2 The Simplex Algorithm

The simplex algorithm of Dantzig [237], or versions of it, are still the common method for linear optimization problems.

In this section we give a description of the simplex algorithm and show that it leads to a solution. The presentation follows Schrijver [915].

The Idea of the Simplex Algorithm

Consider a linear optimization problem of the form

(1) $\sup\{cx : Ax \leq b\}$

where a vertex v_0 of the feasible set $P = \{x : Ax \leq b\}$ is given. Check whether there is an edge of P starting at v_0 along which the objective function cx increases. If there is no such edge, v_0 is an optimum solution. Otherwise move along one of these edges. If this edge is a ray, the supremum is infinite. If not, let v_1 be the other vertex on this edge. Repeat this step with v_1 instead of v_0, etc. This leads to an optimum solution or shows that the supremum is infinite in finitely many steps.

Description of a Standard Version of the Simplex Algorithm, Given a Vertex

Let $a_1, \dots, a_m$ be the row vectors of A. The vertex v_0 of the feasible set $P = \{x : Ax \leq b\}$ is the intersection of a suitable subfamily of d hyperplanes among the m hyperplanes $Ax = b$, say $A_0 x = b_0$, where A_0 is a non-singular $d \times d$ submatrix of A and b_0 is obtained from b by deleting the entries corresponding to the rows of A not in A_0. Clearly, $A_0 v_0 = b_0$ and, since A_0 is non-singular, we may represent the row vector c as a linear combination of the rows of A_0, say

$$(2)\quad c = u_0 A.$$

Here u_0 is a row vector in $\mathbb{E}^m$ with entries outside (the row indices of) A_0 equal to 0. We distinguish two cases:

(i) $u_0 \geq o$. Then c is a non-negative linear combination of the rows of A_0. The hyperplane $\{x : cx = cv_0\}$ thus supports P at v_0 and has exterior normal vector c by Proposition 20.1. Hence v_0 is an optimum solution of (1).

(ii) $u_0 \ngeq o$. Let i_0 be the smallest index of a row in A_0 with $u_{0i_0} < 0$. Since A_0 is a non-singular $d \times d$ matrix, the following system of linear equations has a unique solution x_0:

$$(3)\quad a_i x_0 = 0 \text{ for each row } a_i \text{ of } A_0, \text{ except for } a_{i_0} \text{ for which } a_{i_0} x_0 = -1.$$

Then $v_0 + \lambda x_0$ is on a bounded edge of P, on an unbounded edge of P, or outside P for $\lambda > 0$. In addition,

$$(4)\quad cx_0 = u_0 A x_0 = -u_{0i_0} > 0.$$

Case (ii) splits into two subcases:

(iia) $ax_0 \leq 0$ for each row a of A. Then $v_0 + \lambda x_0 \in P$ for all $\lambda \geq 0$. Noting (4), it follows that the supremum in (1) is $+\infty$.

(iib) $ax_0 > 0$ for a suitable row a of A. Let $\lambda_0 \geq 0$ be the largest $\lambda \geq 0$ such that $v_0 + \lambda x_0 \in P$, i.e.

$$\begin{aligned}\lambda_0 &= \max\left\{\lambda \geq 0 : Av_0 + \lambda A x_0 \leq b\right\}\\ &= \max\left\{\lambda \geq 0 : \lambda a_j x_0 \leq \beta_j - a_j v_0 \text{ for all } j = 1, \dots, m \text{ with } a_j x_0 > 0\right\}.\end{aligned}$$

Let j_0 be the smallest index j of a row in A for which the maximum is attained. Let A_1 be the $d \times d$ matrix which is obtained from the non-singular $d \times d$ matrix A_0 by deleting the row a_{i_0} and inserting the row a_{j_0} at the appropriate position. Since $a_{j_0} x_0 > 0$ by our choice of j_0, (3) and the fact that A_0 is non-singular show that A_1 is also non-singular. Since $a_i(v_0 + \lambda_0 x_0) = \beta_i$ for each row of the non-singular matrix A_1, by (3) and the choice of A_0 and j_0, and since $a_i(v_0 + \lambda_0 x_0) \leq \beta_i$ for all other rows of A, by the definition of λ_0, it follows that $v_1 = v_0 + \lambda_0 x_0$ is a also vertex of P. We have $A_1 v_1 = b_1$, where b_1 is obtained from b_0 by deleting the entry β_{i_0} and inserting the entry β_{j_0} at the appropriate position (Fig. 20.1).

Repeat this step with v_1, A_1 instead of v_0, A_0.

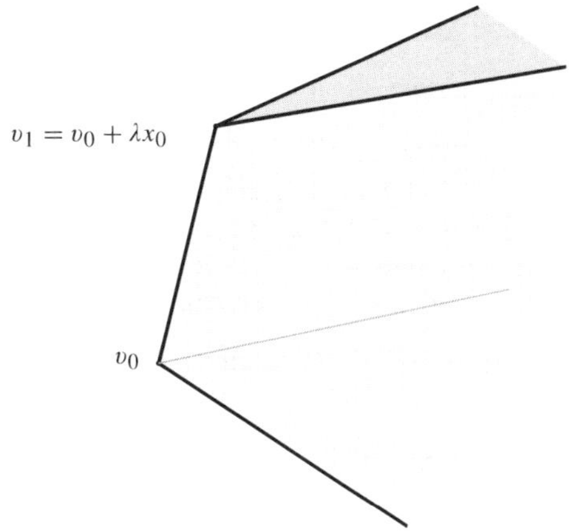

Fig. 20.1. Simplex algorithm

The Simplex Algorithm Terminates

We prove the following result.

Theorem 20.2. *The simplex algorithm either shows that the supremum in* (1) *is infinite or leads to an optimum solution in finitely many steps.*

Proof. Assume that, on the contrary, the simplex algorithm applied to (1) does not terminate. Then it does not show that the supremum in (1) is infinite and produces sequences of matrices, vectors and reals,

$$A_k, b_k, v_k, u_k, x_k, \lambda_k, \ k = 0, 1, \ldots$$

We note that no step can lead to an unbounded edge, and that P has only finitely many vertices. Moving from one vertex to the next one increases the objective function or leaves it the same if the vertices coincide, we therefore have the following: from a certain index on, each step leads to case (iib) and always to the same vertex. Since for the $d \times d$ sub-matrices of A there are only finitely many choices, there are indices $k < l$, such that

$$A_k = A_l, v_k = v_{k+1} = \cdots = v_l, \lambda_k = \lambda_{k+1} = \cdots = \lambda_l = 0.$$

Let n be the largest index of a row which is removed from one of $A_k, \ldots, A_l$ in some step, say from A_p. Then A_{p+1} no longer has this row. Since A_p contains this row, it must have been inserted into one of $A_{p+1}, \ldots, A_l = A_k, A_{k+1}, \ldots, A_{p-1}$. Let A_{q+1} be the first matrix in this sequence of matrices which again contains the nth row. By our choice of n,

(5) None of the rows a_i of A_p with $i > n$ is removed in any of the following steps.

By (2) for A_p and (4) for A_q instead of A_0, we have $u_p A x_q = c x_q > 0$. Thus

(6) $u_{pi} a_i x_q > 0$ for at least one row a_i in A_p.

We now distinguish several cases:

$i < n$: Then $u_{pi} \geq 0$ since n is the smallest index of a row in A_p with $u_{pn} < 0$. In addition, $a_i x_q \leq 0$ since n is the smallest index of a row in A with $a_n v_q = \beta_n$ and $a_n x_q > 0$ (note that $\lambda_q = 0$).
$i = n$: Then $u_{pn} < 0$ and $a_n x_q > 0$, see the case $i < n$.
$i > n$: Then $a_i x_q = 0$ since a_i is not deleted from A_q, see (3).

Each case is in contradiction to (6), concluding the proof of the theorem. □

What to do, if no Vertex is Known?

If no vertex of the feasible set $P = \{x : Ax \leq b\} \neq \emptyset$ of a linear optimization problem

(7) $\sup\{cx : Ax \leq b\}$

is known, we proceed as follows: Let S be the linear subspace $\mathrm{lin}\{a_1, \dots, a_m\}$ of $\mathbb{E}^d$. Then it is easy to see that

$$P = Q \oplus S^\perp,$$

where $Q = P \cap S$ is a convex polyhedron *with* vertices. If $c \notin S$, then cx assumes arbitrarily large values on $P = Q \oplus S^\perp$. Then the supremum is $+\infty$ and we are done. If $c \in S$, then

$$\sup\{cx : x \in P\} = \sup\{cx : x \in Q\}.$$

Thus we have reduced our problem (7) to a problem where it is clear that the feasible set *has* vertices, but we do *not know* them.

Changing notation, we assume that the new problem has the form

$$\sup\{cx : Ax \leq b\},$$

where the feasible set $P = \{x : Ax \leq b\}$ has vertices. Consider the following linear optimization problem with one more variable z (Fig. 20.2),

$$\sup\left\{z : Ax - bz \leq o, -z \leq 0, z \leq 1\right\} \tag{8}$$

with the feasible set

$$Q = \mathrm{cl\,conv}\left(\{(o, 0)\} \cup \left(P + (o, 1)\right)\right).$$

$(o, 0)$ is a vertex of Q. Let A_0 be a non-singular $d \times d$ sub-matrix of A. Then $(o, 0)$ is the intersection of the $d + 1$ hyperplanes

$$\begin{aligned} a_i x - \beta_i z &= 0 \ , \ a_i \text{ row of } A_0, \\ z &= 0. \end{aligned}$$

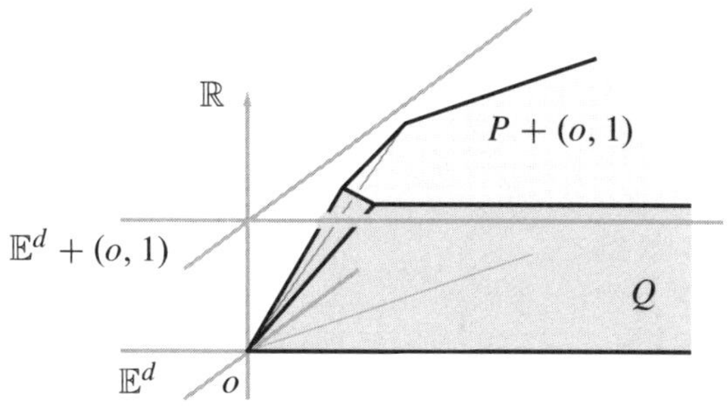

Fig. 20.2. How to find a vertex?

Start the simplex algorithm for the linear optimization problem (8) with this vertex. In finitely many steps this leads to a vertex of the form $(v, 1)$ of Q, i.e. to a vertex v of P.

Considering the earlier, the task remains to determine whether the feasible set P is empty or not. If $P = \emptyset$, the supremum in (8) is less than 1. So this determines feasibility.

20.3 The Ellipsoid Algorithm

The ellipsoid algorithm was originally developed for convex optimization problems by Shor [933] and Yudin and Nemirovskiĭ [1033]. Khachiyan [580] sketched a proof that an extension of it is polynomial for linear optimization. This sketch was made into a proof by Gács and Lovász [348] and then elaborated by Grötschel, Lovász and Schrijver [408]. The definitive treatment is in their book [409]. This does not mean that, in practical applications, it is superior to the simplex algorithm or versions of the latter.

In the following we present a simple version of the ellipsoid algorithm to find a feasible solution of a linear optimization problem (under additional assumptions).

Ellipsoids and Half Ellipsoids

We first prove a simple lemma on ellipsoids.

Lemma 20.1. *Let E be a solid ellipsoid in $\mathbb{E}^d$ and H^- a halfspace whose boundary hyperplane contains the centre of E. Then $E \cap H^-$ is contained in an ellipsoid F such that*

$$\frac{V(F)}{V(E)} \leq e^{-\frac{1}{2(d+1)}}.$$

Proof. We may assume that $E = B^d$ and $H^- = \{x : x_d \geq 0\}$. Let

$$F = \left\{ x : \frac{d^2-1}{d^2} x_1^2 + \cdots + \frac{d^2-1}{d^2} x_{d-1}^2 + \frac{(d+1)^2}{d^2} \left(x_d - \frac{1}{d+1} \right)^2 \leq 1 \right\}.$$

For $x \in B^d \cap H^-$ we have $x_d \geq 0$ and $x_1^2 + \cdots + x_{d-1}^2 \leq 1 - x_d^2$, or

$$\frac{d^2-1}{d^2}x_1^2 + \cdots + \frac{d^2-1}{d^2}x_{d-1}^2 \leq \frac{d^2-1}{d^2} - \frac{d^2-1}{d^2}x_d^2.$$

Clearly,

$$\frac{(d+1)^2}{d^2}\left(x_d - \frac{1}{d+1}\right)^2 = \frac{(d+1)^2}{d^2}x_d^2 - \frac{2(d+1)}{d^2}x_d + \frac{1}{d^2}.$$

Addition then gives

$$\frac{d^2-1}{d^2}x_1^2 + \cdots + \frac{d^2-1}{d^2}x_{d-1}^2 + \frac{(d+1)^2}{d^2}\left(x_d - \frac{1}{d+1}\right)^2 \leq 1 + \frac{2d+2}{d^2}(x_d^2 - x_d) \leq 1$$

on noting that $0 \leq x_d \leq 1$ implies that $x_d^2 - x_d \leq 0$. Thus $x \in F$, concluding the proof that $B^d \cap H^- \subseteq F$. Finally,

$$\frac{V(F)}{V(B^d)} = \left(\frac{d^2}{d^2-1}\right)^{\frac{d-1}{2}} \frac{d}{d+1} \leq e^{\frac{1}{d^2-1}\frac{d-1}{2}} e^{-\frac{1}{d+1}} = e^{-\frac{1}{2(d+1)}},$$

where we have used the fact that $1 + x \leq e^x$ for $x = \frac{1}{d^2-1}, \frac{-1}{d+1}$. □

How to Find a Feasible Solution by the Ellipsoid Algorithm?

We consider the feasible set $P = \{x : Ax \leq b\}$ of a linear optimization problem. Assuming that there are $\varrho, \delta > 0$ such that $P \subseteq \varrho B^d$ and $V(P) \geq \delta$, we describe how to find a feasible solution, i.e. a point of P.

Let E_0 be the ellipsoid ϱB^d with centre $c_0 = o$. Then $P \subseteq E_0$. If $c_0 \in P$, then c_0 is the required feasible solution. If $c_0 \notin P$, then c_0 is not contained in at least one of the defining halfspaces of P, say $c_0 \notin \{x : a_{i_0}x \leq \beta_{i_0}\}$. Clearly, $P \subseteq E_0 \cap H_0^-$ where H_0^- is the halfspace $\{x : a_{i_0}x \leq a_{i_0}c_0\}$ which contains c_0 on its boundary hyperplane. By Lemma 20.1, there is an ellipsoid E_1 with centre c_1 such that $P \subseteq E_0 \cap H_0^- \subseteq E_1$ and

$$\frac{V(E_1)}{V(E_0)} \leq e^{-\frac{1}{2(d+1)}}.$$

If $c_1 \in P$, then c_1 is the required feasible solution, otherwise repeat this step with E_1, c_1 instead of E_0, c_0.

In this way we either get a feasible solution $c_n \in P$ in finitely many steps, or there is a sequence of ellipsoids $E_0, E_1, \cdots \supseteq P$ with

$$V(E_n) = \frac{V(E_n)}{V(E_{n-1})} \cdots \frac{V(E_1)}{V(E_0)} V(E_0) \leq e^{-\frac{n}{2(d+1)}} V(E_o) \text{ for } n = 1, 2, \ldots$$

Thus $V(E_n) < \delta \leq V(P)$ for all sufficiently large n, in contradiction to $E_n \supseteq P$. The latter alternative is thus ruled out.

Remark. There is no need to give the feasible set P in an explicit form as earlier. For the ellipsoid algorithm it is sufficient to *know*, at the nth step, whether $c_n \in P$ or to specify a halfspace which contains P but not c_n. This can be achieved by a separation oracle for P.

Complexity of the Ellipsoid Algorithm

The result of Khachiyan [580] shows that one can find a feasible solution of a rational system of linear inequalities

$$\{x : Ax \leq b\}$$

in polynomial time by a refined version of the ellipsoid algorithm. Since this is polynomially equivalent to the solution of the linear optimization problem

$$\sup\{cx : Ax \leq b\}$$

with rational c, there is a polynomial time algorithm for rational linear optimization problems. See Schrijver [915].

20.4 Lattice Polyhedra and Totally Dual Integral Systems

Lattice polyhedra and polytopes play an important role in several branches of mathematics, including convex geometry and the geometry of numbers and in applied fields such as crystallography. See Sects. 8.4, 19 and 32.1. Here we study lattice polyhedra and polytopes in the context of integer linear optimization. Basic results on lattice polyhedra in optimization are due to Gordan and many living mathematicians, including Gomory, Lenstra, Chvátal, Grötschel, Lovász and Schrijver, Papadimitriou and Edmonds and Giles.

Integer linear optimization problems, for example the problem to determine

$$\sup\{cx : Ax \leq b, x \in \mathbb{Z}^d\},$$

behave much worse than corresponding linear optimization problems, but in the special case, where the inequality system

$$Ax \leq b$$

is a so-called totally dual integral system, they behave quite well. The study of such systems was initiated by Edmonds and Giles.

In this section we first prove some simple yet important results on lattice polyhedra and then introduce the notion of totally dual integral systems of linear inequalities.

For more information, see Schrijver [915] and the references cited there.

Rational Polyhedra and Lattice Polyhedra

A convex polyhedron P in $\mathbb{E}^d$ is *rational*, if it has a representation of the form

$$P = \{x : Ax \leq b\},$$

where A is a rational matrix and b a rational vector. The convex polyhedron P is called *integer, integral* or a *lattice polyhedron* if it is the convex hull of the points of the integer lattice $\mathbb{Z}^d$ contained in it. It is not difficult to show the following equivalences, where a *minimum face* of a convex polyhedron is a face which does not contain a proper subface. Thus a minimum face is a vertex, a line or a plane of dimension ≥ 2.

Proposition 20.3. *Let P be a rational convex polyhedron in $\mathbb{E}^d$. Then the following statements are equivalent:*

(i) *P is a lattice polyhedron.*

(ii) *Each face of P contains a point of $\mathbb{Z}^d$.*

(iii) *Each minimum face of P contains a point of $\mathbb{Z}^d$.*

(iv) *If in a linear optimization problem of the form* $\sup\{cx : x \in P\}$ *the supremum is finite, it is attained at a point of $\mathbb{Z}^d$.*

(v) *Each support hyperplane of P contains a point of $\mathbb{Z}^d$.*

Some Properties of Lattice Polyhedra

The following result of Edmonds and Giles [287] contains a series of earlier results as special cases.

Theorem 20.3. *Let P be a rational convex polyhedron in $\mathbb{E}^d$. Then the following statements are equivalent:*

(i) *P is a lattice polyhedron.*

(ii) *Each rational support hyperplane of P contains a point of $\mathbb{Z}^d$.*

Proof. (i)⇒(ii) Clear.

(ii)⇒(i) We may assume that $P = \{x : Ax \leq b\}$, where A, b are integer. It is sufficient to show that each minimum face of P contains a point of $\mathbb{Z}^d$. To see this, assume that, on the contrary, there is a minimum face F of P which contains no point of $\mathbb{Z}^d$. Being a minimum face, F is a plane and thus can be represented in the form

$$F = \{x : A'x = b'\},$$

where the matrix A' consists of, say k ($\leq m$) rows of A and the column b' of the corresponding entries of b. We now construct a rational hyperplane H which supports P but contains no point of $\mathbb{Z}^d$ in contradiction to (ii). For this we need the definition and simple properties of polar lattices in Sect. 21.4. The columns $b'_1, \dots, b'_d$ of A' are integer vectors in $\mathbb{E}^k$ and are contained in the subspace $S = \text{lin}\{b'_1, \dots, b'_d\}$

of $\mathbb{E}^k$. They generate a lattice L in S. We show that $b' \in S \backslash L$. Let $x \in F$. Then $b' = A'x = x_1 b'_1 + \cdots + x_d b'_d \in S$. If $b' \in L$, then $b' = u_1 b'_1 + \cdots + u_d b'_d = A'u$ for suitable $u \in \mathbb{Z}^d$. Hence F contains the integer vector u, which is excluded by assumption. Let L^* be the polar lattice of L in S. Since $b' \notin L$ there is a rational row $y' \in L^*$ with

$$y'b' \notin \mathbb{Z}, \text{ while } y'A' = (y'b'_1, \dots, y'b'_d) \in \mathbb{Z}^d$$

by the definition of L^*. By adding suitable positive integers to the entries of y', if necessary, we may suppose that

$$y' \geq o \text{ while still } y'b' \notin \mathbb{Z},\ y'A' \in \mathbb{Z}^d.$$

If $a_1, \dots, a_k$ are the rows of A', then

$$\begin{aligned} c &= y'A' = y'_1 a_1 + \cdots + y'_k a_k \ \in \text{pos}\{a_1, \dots, a_k\} \cap \mathbb{Z}^d = N_P(F) \cap \mathbb{Z}^d, \\ \beta &= y'b' \in \mathbb{Q} \setminus \mathbb{Z} \end{aligned}$$

by Proposition 20.1. The hyperplane

$$H = \{x : cx = \beta\} = \{x : y'A'x = y'b'\}$$

is rational, contains $F = \{x : A'x = b'\}$ and its normal vector c is in $N_P(F)$. Thus H supports P. H contains no $u \in \mathbb{Z}^d$ since otherwise $y'b' = y'A'u \in \mathbb{Z}$ while $y'b' \notin \mathbb{Z}$. This contradicts (ii) and thus concludes the proof. □

A consequence of this result is the following:

Corollary 20.1. *Let $Ax \leq b$ be a rational system of linear inequalities. Then the following statements are equivalent:*

(i) *$\sup\{cx : Ax \leq b\}$ is attained by an integer vector x for each rational row c for which the supremum is finite.*

(ii) *$\sup\{cx : Ax \leq b\}$ is an integer for each integer row c for which the supremum is finite.*

(iii) *$P = \{x : Ax \leq b\}$ is a lattice polyhedron.*

Proof. (i)⇒(ii) Let c be an integer row and such that $\sup\{cx : Ax \leq b\}$ is finite. By (i), the supremum is attained at an integer vector x and thus is an integer.

(ii)⇒(iii) We first show the following:

(2) Let $H = \{x : cx = \delta\}$ be a rational support hyperplane of P. Then H contains a point of $\mathbb{Z}^d$.

By multiplying the equation $cx = \delta$ by a suitable positive rational number and changing notation, if necessary, we may suppose that c is an integer row vector with relatively prime entries. Since H is a support hyperplane of P, $\sup\{cx : Ax \leq b\} = \delta$ is finite. By (ii), δ is then an integer. Since c has relatively prime integer entries, there is a $u \in \mathbb{Z}^d$ such that $cu = \delta$. Hence $u \in H$, concluding the proof of (2). Having proved (2), Theorem 20.3 implies statement (iii).

(iii)⇒(i) Trivial. □

Totally Dual Integral Systems of Linear Inequalities

The implication (i)⇒(ii) in this corollary says the following. Let $Ax \leq b$ be a rational system of linear inequalities. If $\sup\{c\,x : Ax \leq b\}$ is attained by an integer vector x for each rational row vector c for which the supremum is finite, then it is an integer for each integer row vector c for which the supremum is finite. The duality equality

$$\sup\{c\,x : Ax \leq b\} = \inf\{y\,b : y \geq o,\ yA = c\}$$

then led Edmonds and Giles [287] to define the following: a rational system of linear inequalities $Ax \leq b$ is *totally dual integral* if $\inf\{y\,b : y \geq o,\ yA = c\}$ is attained by an integer row vector y for each integer row vector c for which the infimum is finite. This implies, in particular, that Proposition (ii) of the above corollary holds, which, in turn, shows that $P = \{x : Ax \leq b\}$ is a lattice polyhedron.

Note that there are rational systems of linear inequalities which define the same lattice polyhedron and such that one system is totally dual integral while the other is not.

Complexity of Integer Linear Optimization for Lattice Polyhedra and Totally Dual Integral Systems

We have stated earlier that, presumably, there is no polynomial time algorithm for general integer linear optimization problems. Fortunately, for lattice polyhedra and totally dual integral systems the situation is better:

There is a polynomial algorithm by Lenstra [647] which finds for a fixed number of variables an optimum solution of the integer linear optimization problem

$$\sup\{c\,x : Ax \leq b,\ x \in \mathbb{Z}^d\}.$$

Similarly, there is a polynomial algorithm which finds an integral optimum solution for the linear optimization problem

$$\inf\{y\,b : y \geq o, yA = c\}$$

if $Ax \leq b$ is a totally dual integral system with A integer and c an integer row vector. For proofs and references, see Schrijver [915], pp. 232, 331.

Considering these remarks, it is of interest to find out whether a given rational system $Ax \leq b$ defines a lattice polyhedron or is totally dual integral. See [915].

20.5 Hilbert Bases and Totally Dual Integral Systems

For a better understanding of totally dual integral systems of linear inequalities, Giles and Pulleyblank [378] introduced the notion of a Hilbert basis of a polyhedral convex cone.

In this section we present the geometric Hilbert basis theorem of Gordan and van der Corput and show how one uses Hilbert bases to characterize totally dual integral systems of linear inequalities. An algorithm for geometric Hilbert bases is due to Hirzebruch [507] and Jung [556].

For references and detailed information, compare Schrijver [915] and Bertsimas and Weismantel [104].

Geometric Hilbert Bases

A *(geometric) Hilbert basis* of a polyhedral convex cone C in $\mathbb{E}^d$ with apex o is a set of vectors $\{a_1, \dots, a_m\}$ in C such that each integer vector in C is an integer linear combination of $a_1, \dots, a_m$. Of particular interest are *integer (geometric) Hilbert bases*, that is, Hilbert bases consisting of integer vectors.

The Geometric Hilbert Basis Theorem

Old results of Gordan [387] and van der Corput [225] on systems of linear equations can be formulated as follows.

Theorem 20.4. *Let C be a pointed rational polyhedral convex cone in $\mathbb{E}^d$ with apex o. Then C has an integer Hilbert basis. If C is pointed, it has a unique minimal (with respect to inclusion) integer Hilbert basis.*

Proof. Existence: Let $C = \text{pos}\{q_1, \dots, q_n\}$ where the q_i are rational vectors. We may suppose that $q_i \in \mathbb{Z}^d$. We prove the following:

(1) Let $\{a_1, \dots, a_m\} = \{\lambda_1 q_1 + \dots + \lambda_n q_n : 0 \le \lambda_i \le 1\} \cap \mathbb{Z}^d$. Then $\{a_1, \dots, a_m\}$ is an integer Hilbert basis of C and $C = \text{pos}\{a_1, \dots, a_m\}$.

Since $C = \text{pos}\{q_1, \dots, q_n\}$ and $\{q_1, \dots, q_n\} \subseteq \{a_1, \dots, a_m\} \subseteq C$, clearly $C = \text{pos}\{a_1, \dots, a_m\}$. To see that $\{a_1, \dots, a_m\}$ is an integer Hilbert basis of C, let $u \in C \cap \mathbb{Z}^d$. Since $C = \text{pos}\{q_1, \dots, q_n\}$, there are $\mu_1, \dots, \mu_n \ge 0$ such that $u = \mu_1 q_1 + \dots + \mu_n q_n$. Then

$$u - \lfloor \mu_1 \rfloor q_1 - \dots - \lfloor \mu_n \rfloor q_n = (\mu_1 - \lfloor \mu_1 \rfloor) q_1 + \dots + (\mu_n - \lfloor \mu_n \rfloor) q_n \in C \cap \mathbb{Z}^d.$$

The vector $(\mu_1 - \lfloor \mu_1 \rfloor) q_1 + \dots + (\mu_n - \lfloor \mu_n \rfloor) q_n$ thus occurs among $a_1, \dots, a_m$. Since the $q_1, \dots, q_n$ also occur among the $a_1, \dots, a_m$, we see that u is a non-negative integer linear combination of $a_1, \dots, a_m$, concluding the proof of (1).

Uniqueness: Let C be pointed. We will show that

(2) $B = \{a \in C \cap \mathbb{Z}^d \setminus \{o\} : a \text{ is not a sum of vectors in } C \cap \mathbb{Z}^d \setminus \{o\}\}$ is the unique minimal Hilbert basis of C.

Clearly, B is contained in any Hilbert basis of C. Since the Hilbert basis in (1) is finite, B is also finite. B is integer. Thus, to finish the proof of (2), we have to show that B is a Hilbert basis of C. Since C is pointed, o is an extreme point and thus a vertex of C. Thus there is a support hyperplane $\{x : c\,x = 0\}$ of C at o which meets C only at o. We may assume that $c\,x > 0$ for each $x \in C \setminus \{o\}$. Suppose that there are vectors in $C \cap \mathbb{Z}^d \setminus \{o\}$, which are not non-negative integer linear combinations of the vectors in B. Let u be such a vector with $c\,u$ (> 0) as close as possible to 0. Then, in particular, $u \notin B$ and we may choose vectors $v, w \in C \cap \mathbb{Z}^d \setminus \{o\}$, with $u = v + w$. Then $c\,u = c\,v + c\,w$. Since $v, w \in C \setminus \{o\}$ we thus have that $0 < c\,v, c\,w < c\,u$ which contradicts our choice of u. Thus every vector in $C \cap \mathbb{Z}^d$, including o, is a non-negative integer linear combination of vectors in B. The proof of (2) is complete. □

The notion of geometric Hilbert basis and the geometric Hilbert basis theorem remind one of the Hilbert [500] basis theorem for ideals in polynomial rings over Noetherian rings. There are actually relations between these two topics, see [915], p. 376 for references.

Characterization of Totally dual Integral Systems by Hilbert Bases

We prove the following result, where a row a_i of A is *active* on F if $a_i x = \beta_i$ for each $x \in F$:

Theorem 20.5. *Let $Ax \le b$ be a rational system of linear inequalities. Then the following propositions are equivalent:*

(i) *$Ax \le b$ is totally dual integral.*

(ii) *For each face F of the convex polyhedron $P = \{x : Ax \le b\}$ the rows of A which are active on F form a Hilbert basis of the cone generated by these rows.*

Proof. (i)⇒(ii) Let F be a face of P and let $a_1, \dots, a_k$ be the rows of A which are active on F. Then, clearly, the following hold:

(3) Let x be a relatively interior point of F. Let a be a row of A and β the corresponding entry of b. Then

$$a\,x = \beta \text{ if } a \text{ is one of } a_1, \dots, a_k \text{ and } ax < \beta \text{ otherwise.}$$

To show that $\{a_1, \dots, a_k\}$ is a Hilbert basis of $\operatorname{pos}\{a_1, \dots, a_k\} = N_P(F)$, let

$$c \in N_P(F) \cap \mathbb{Z}^d \setminus \{o\}.$$

Then the supremum in the equality

(4) $\sup\{c\,x : Ax \le b\} = \inf\{yb : y \ge o,\ yA = c\}$

is attained by any point of F, in particular at a relatively interior point x, say. The infimum in (4) is thus finite. Then (i) shows that it is attained by an integer row $y \ge o$, say. Then $c\,x = yAx \le yb = cx$ and therefore $yAx = yb$, or $y(b - Ax) = 0$. Noting (3), this shows that an entry of y is 0 if the corresponding row of A is not active on F. Hence $c = yA$ is an integer linear combination just of $a_1, \dots, a_k$. Hence $\{a_1, \dots, a_k\}$ is a Hilbert basis of $\operatorname{pos}\{a_1, \dots, a_k\}$.

(ii)⇒(i) Let $c \in \mathbb{Z}^d$ be such that the infimum in (4) and thus also the supremum are attained. Let F be a minimum face of P such that the supremum in (4) is attained at each point of F, in particular at a relatively interior point x, say. Let $a_1, \dots, a_k$ be the row vectors of A which are active on F. Then

$$a_i x = \beta_i \text{ for } i = 1, \dots, k.$$

Since c is the exterior normal vector of a hyperplane which supports P at F, $c \in N_P(x) = \operatorname{pos}\{a_1, \dots, a_k\}$ by Proposition 20.1. By (ii), $c = \lambda_1 a_1 + \dots + \lambda_k a_k$ with suitable integers $\lambda_i \geq 0$. Enlarge the integer row $(\lambda_1, \dots, \lambda_k)$ by appropriately inserting 0s to get an integer row $y \geq o$ with

$$yA = \lambda_1 a_1 + \dots + \lambda_k a_k = c,$$
$$yb = \lambda_1 \beta_1 + \dots + \lambda_k \beta_k = (\lambda_1 a_1 + \dots + \lambda_k a_k)x = cx.$$

Thus the infimum in (4) is attained at the integer row y. As this is true for each c for which the infimum in (4) is finite, the system $Ax \leq b$ is totally dual integral. □

Geometry of Numbers and Aspects of Discrete Geometry

The roots of discrete geometry and geometry of numbers date back to the seventeenth or eighteenth century. We mention the ball packing problem, first treated by Kepler. The well known discussion between Newton and Gregory on the problem of the 13 balls involved the following question: given a unit ball, is it possible to arrange 13 non-overlapping unit balls such that each touches the given ball, or not? Gregory said yes, Newton no. Sporadic results in the late eighteenth and the nineteenth century are due to Lagrange, Gauss, Dirichlet, Korkin and Zolotarev (packing of balls and positive definite quadratic forms), Fedorov (tiling), and Thue (irregular packing of circular discs). Both areas became well-established branches of mathematics only at the turn of the nineteenth and during the twentieth century. The major figures at the beginning of the systematic era were Minkowski (fundamental theorems, applications to Diophantine approximation) and Voronoĭ (geometric theory of quadratic forms) in the geometry of numbers and, 50 years later, Fejes Tóth (packing and covering) in discrete geometry. Other contributors to both areas in the twentieth century were Delone, Siegel, Mahler, Davenport, Kneser, Rogers, Ryshkov and many living mathematicians, including Hlawka, Bambah and Schmidt. Important topics are lattice and non-lattice packing, covering and tiling. Both areas have strong ties to other parts of mathematics and the applied sciences, for example to crystallography, coding and data transmission, modular functions, computational and algorithmic geometry, graph theory, number theory and algebraic geometry. There are applications to numerical integration and the Riemann mapping theorem.

At the heart of the geometry of numbers is the interplay of the group-theoretic notion of lattice and the geometric concept of convex set, the lattices representing periodicity, the convex sets geometry. In discrete geometry similar problems are considered as in the geometry of numbers, but relaxing periodicity.

While the important problems of the geometry of numbers and of discrete geometry are easy to state, their solution, in general, is difficult. Thus progress is slow. Major results in recent years are at the boundary of the classical theory, dealing, for example, with positive and indefinite quadratic forms and computational and algorithmic aspects. It seems that fundamental advance in the future will require new ideas and additional tools from other areas.

The aim of this chapter is to present basic results from these two areas. We start with regular, i.e. lattice results and topics from the geometry of numbers such as the fundamental theorems of Minkowski, the Minkowski–Hlawka theorem, the geometric theory of positive definite quadratic forms and reduction. Minkowski's fundamental theorem and the Minkowski–Hlawka theorem are opposite cornerstones of the geometry of numbers, the first yielding a simple upper bound for the density of lattice packings, the second a lower bound. To show the reader the flavour of the geometry of numbers of the English school we discuss some classical particular arithmetic–geometric problems. Computational and algorithmic aspects are touched. Then irregular, i.e. non-lattice results of a systematic character of discrete geometry are presented, dealing with packing, covering, tiling, optimum quantization and Koebe's representation theorem for planar graphs. Besides quantization and Koebe's theorem, a further result of an instrumental character in classical discrete geometry is Euler's polytope formula, see Sect. 15.1. For corresponding problems the irregular case for obvious reasons is more difficult than the regular one. Thus it is not surprising that lattice results are, in general, much farther reaching than corresponding non-lattice results. In our presentation the emphasis is on the geometry. The given applications and relations to other areas deal with Diophantine approximation, polynomials, error correcting codes, data transmission, numerical integration, graphs and the Riemann mapping theorem. Lattice polytopes and some of their applications were treated in Sect. 19 in the chapter on convex polytopes, but would also fit well into the present chapter.

The reader who wants to get more detailed information is referred to the books and surveys of Fejes Tóth [329, 330], Rogers [851], Gruber [416], Conway and Sloane [220], Pach and Agarwal [783], Erdös, Gruber and Hammer [307], Cassels [195], Gruber and Lekkerkerker [447], Ryshkov and Baranovskiĭ [867], Kannan [563], Grötschel, Lovász and Schrijver [409], Siegel [937], Gruber [430], Zong [1048, 1049], Lagarias [625], Olds, Lax and Davidoff [778], Coppel [223], Ryshkov, Barykinskiĭ and Kucherinenko [868], Matoušek [695], Böröczky [155], Ryshkov [866], Bombieri and Gubler [148], to the collected or selected works of Minkowski [745], Voronoĭ [1014], Davenport [246] and Hlawka [516] and to the pertinent articles in the Handbooks of Convex Geometry [475] and of Discrete and Computational Geometry [476] and in Discrete and Computational Geometry [273]. A large collection of research problems in discrete geometry is due to Brass, Moser and Pach [164].

Finite packing and covering problems, Erdös type problems, arrangements and matroids will not be considered in the following. For these, see the book of Pach and Agarwal [783], the monograph of Böröczky [155] on finite packing and covering, the book of Matoušek [695] and the monograph of Bokowski [137] on oriented matroids. Similarly, we consider lattice points in large convex bodies in the sense of the circle problem of Gauss only in passing and instead refer to Gruber and Lekkerkerker [447], Sect. iii, and articles in the Proceedings on Fourier Analysis and Convexity [343], together with the references cited there.

Extension to Crystallographic Groups?

Most results in the geometry of numbers and part of the results in discrete geometry rest on the notion of lattices, that is, discrete groups of translations in $\mathbb{E}^d$. Considering this, it is surprising that there is no equally elaborate theory for other crystallographic groups, although there are some pertinent results. We mention a result of Delone [258] on the number of facets of a stereohedron, i.e. a space filler by means of a crystallographic group, the negative solution of Hilbert's 18th problem, a version of Blichfeldt's theorem due to Schmidt [896], p. 30 based on a crystallographic structure, and a result on ball packings with crystallographic groups of Horváth and Molnár [522]. A starting point for research in this direction could be Engel's article [297]. See also the short chapter on crystallography in Erdös, Gruber and Hammer [307] and the books of Senechal [925] and Engel, Michel and Senechal [301].

The crystallographer Peter Engel [300] has reservations about a substantial parallel theory for crystallographic groups – in spite of his important pertinent contributions. His argument is that the context of general crystallographic groups is so complicated that one may not expect a lot of non-trivial results.

All convex bodies in this chapter are proper.

21 Lattices

The notion of lattice already appeared implicitly in the work of Kepler [576, 577], who used it in the context of packing of balls. Crystallographers such as Haüy [483] in the eighteenth century and many crystallographers in the nineteenth century based their investigations on lattices, although experimental proof that lattices are underlying crystals was given only in the early twentieth century by von Laue and father and son Bragg, for which all three got the Nobel Prize. A different source for lattices is number theory. Here the classical reference is Gauss [364], who seems to have first seen the relation between positive definite quadratic forms and lattice packing of balls. Lattices and convex bodies are the main ingredients of the geometry of numbers. Results dealing with lattices are often the starting point for more general investigations in discrete geometry.

In this section, basic notions related to lattices and some of the fundamental properties of lattices are presented, as needed in the context of the geometry of numbers.

For additional information on lattices, mainly from the viewpoint of the geometry of numbers, see Cassels [195], Gruber and Lekkerkerker [447] and Lagarias [625]. While we study relations between lattices in general, special lattices, theta series and codes are treated in the book of Gruber and Lekkerkerker. More information on these topics is presented by Ebeling [282] and Conway and Sloane [220]. For relations to crystallography, see Erdös, Gruber and Hammer [307], Engel [296, 297] and Lagarias [625]. Engel, Michel and Senechal [301] treat lattices from a crystallographic viewpoint. For algorithmic problems and results on lattices compare the references in Sect. 28.

21.1 Basic Concepts and Properties and a Linear Diophantine Equation

Discrete sets of various sorts in $\mathbb{E}^d$ play an important role in numerous branches of mathematics and other fields, including discrete geometry and the geometry of numbers. It is thus a natural problem to study interesting classes of discrete sets. In many cases such sets are uniformly distributed over $\mathbb{E}^d$, or they have periodicity properties. We mention Delone sets and refer for these to Sect. 32.1, orbits of crystallographic groups, periodic sets as considered by Zassenhaus, see Sects. 30.4 and 31.4, and lattices.

In the following we define the notion of lattice, basis, fundamental parallelotope and determinant, and state the relations between different bases of a lattice. An application deals with a simple Diophantine equation.

Lattices and Lattice Bases

A *(geometric) lattice* L in $\mathbb{E}^d$ is the system of all integer linear combinations of d linearly independent vectors $b_1, \dots, b_d \in \mathbb{E}^d$,

$$L = \{u_1 b_1 + \cdots + u_d b_d : u_i \in \mathbb{Z}\}.$$

The d-tuple $\{b_1, \dots, b_d\}$ is called a *basis* of L.

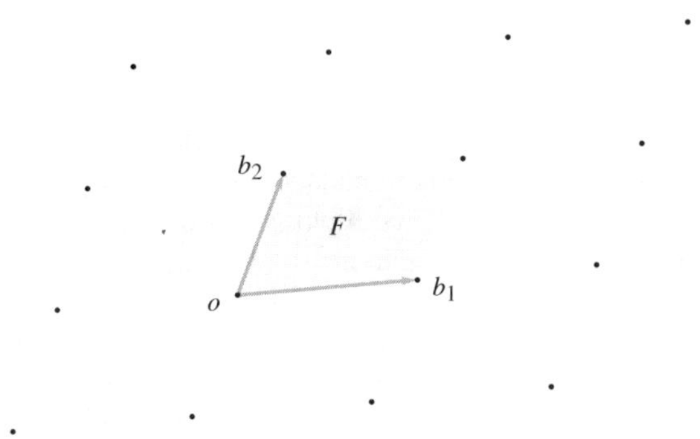

Fig. 21.1. Lattice

An example of a lattice is the *integer lattice*

$$\mathbb{Z}^d = \{(u_1, \dots, u_d) : u_i \in \mathbb{Z}\}.$$

The vectors

$$(1, 0, \dots, 0), (0, 1, \dots, 0), \dots, (0, 0, \dots, 1)$$

form the *standard basis* of $\mathbb{Z}^d$.

Lattices appear in many different branches of mathematics, including Diophantine approximation, algebraic number theory and algebraic geometry, complex analysis (periods of doubly periodic analytic functions), numerical analysis (nodes for numerical integration), integer programming, coding, and crystallography (Fig. 21.1).

Relations Between Different Bases

Different bases of a given lattice are related in a rather simple way. Recall, an integer unimodular $d \times d$ matrix is a $d \times d$ matrix U with integer entries and $\det U = \pm 1$.

Theorem 21.1. *Let $\{b_1, \dots, b_d\}$ be a basis of a lattice L in $\mathbb{E}^d$. Then the following statements hold:*

(i) *Let $\{c_1, \dots, c_d\}$ be another basis of L. Then*

$$(1)\qquad \begin{array}{l} c_1 = u_{11}b_1 + \cdots + u_{1d}b_d \\ c_2 = u_{21}b_1 + \cdots + u_{2d}b_d \\ \dots\dots\dots\dots\dots\dots\dots \\ c_d = u_{d1}b_1 + \cdots + u_{dd}b_d \end{array} \quad \textit{or} \quad (c_1, \dots, c_d) = (b_1, \dots, b_d)U^T,$$

where $U = (u_{ik})$ is a suitable integer unimodular $d \times d$ matrix.

(ii) *Let $U = (u_{ik})$ be an integer unimodular $d \times d$ matrix and let $\{c_1, \dots, c_d\}$ be defined by* (1)*. Then $\{c_1, \dots, c_d\}$ is a basis of L.*

Proof. (i) Since $\{b_1, \dots, b_d\}$ is a basis of L, each vector of L is an integer linear combination of $b_1, \dots, b_d$. This implies (1), where $U = (u_{ik})$ is an integer $d \times d$ matrix. Noting that $\{c_1, \dots, c_d\}$ is also a basis of L, it follows that, conversely,

$$(2)\qquad \begin{array}{l} b_1 = v_{11}c_1 + \cdots + v_{1d}c_d \\ b_2 = v_{21}c_1 + \cdots + v_{2d}c_d \\ \dots\dots\dots\dots\dots\dots\dots \\ b_d = v_{d1}c_1 + \cdots + v_{dd}c_d \end{array} \quad \text{or} \quad (b_1, \dots, b_d) = (c_1, \dots, c_d)V^T,$$

where $V = (v_{ik})$ is a suitable integer $d \times d$ matrix. From (1) and (2) we conclude that

$$(3)\quad (c_1, \dots, c_d) = (b_1, \dots, b_d)U^T = (c_1, \dots, c_d)V^T U^T = (c_1, \dots, c_d)(UV)^T$$

Since $c_1, \dots, c_d$ are linearly independent and thus $(c_1, \dots, c_d)$ a non-singular $d \times d$ matrix, it follows from (3) that $\det(UV)^T = 1$ or $\det U \det V = 1$. Since U and V are integer matrices, their determinants are also integers. This then shows that $\det U = \pm 1$, concluding the proof of (i).

(ii) Since U is an integer unimodular $d \times d$ matrix, (1) implies that $c_1, \dots, c_d$ are in L, are linearly independent and

$$(b_1, \dots, b_d) = (c_1, \dots, c_d)V^T, \text{ where } V = U^{-1}.$$

Being the inverse of the integer unimodular $d \times d$ matrix U, the matrix V is also an integer unimodular matrix. Thus each b_i is an integer linear combination of the vectors $c_1, \dots, c_d$. Since each vector of L is an integer linear combination of the vectors $b_1, \dots, b_d$, it follows that each vector of L is an integer linear combination of the vectors $c_1, \dots, c_d$. Since $c_1, \dots, c_d$ are linearly independent, $\{c_1, \dots, c_d\}$ is a basis. □

Particular Bases of a Lattice; Reduction

Since, for $d \geq 2$, there are infinitely many integer unimodular $d \times d$-matrices, any lattice in $\mathbb{E}^d$, $d \geq 2$, has infinitely many different bases. A major problem of reduction theory is to single out bases which have particularly nice geometric properties, e.g. bases consisting of short vectors or bases where the vectors are almost orthogonal. For more information, see Sect. 28.

The Bravais Classification of Lattices

A square lattice and a hexagonal lattice in $\mathbb{E}^2$, clearly, are different, but what makes them different? One way to distinguish lattices is to classify them by means of their groups of isometries, keeping the origin fixed. This is the *Bravais classification* of lattices from crystallography, see Engel [296,297], Erdös, Gruber and Hammer [307] and [295] and Engel, Michel and Senechal [301].

Fundamental Parallelotope and Determinant of a Lattice

Given a basis $\{b_1, \dots, b_d\}$ of a lattice L, the corresponding *fundamental parallelotope* F is defined by:

$$F = \{\alpha_1 b_1 + \cdots + \alpha_d b_d : 0 \leq \alpha_i < 1\}.$$

The *determinant* $d(L)$ of the lattice L is the volume of F,

$$d(L) = V(F) = |\det(b_1, \dots, b_d)|.$$

It follows from Theorem 21.1 that $d(L)$ is independent of the particular choice of a basis of $\mathbb{E}^d$.

A Linear Diophantine Equation

There are many known proofs of the following result, in particular proofs based on the Euclidean algorithm. See, e.g. Mordell [754]. The first indication of such a proof is due to Āryabhata about 500 AD. A later contributor is Brahmagupta in the seventh century. The proof presented later resulted from a discussion with Keith Ball [55].

Proposition 21.1. *Let u, v be positive integers with greatest common divisor 1. Then there are integers x, y such that*

$$uy - vx = 1.$$

Proof. The point (u, v) is a *primitive point* of the integer lattice $\mathbb{Z}^2$, i.e. there is no lattice point on the line segment $[o, (u, v)]$ except for $o, (u, v)$. Consider the line segment $[o, (u, v)]$ and move it parallel to itself to the left until it first hits a point of $\mathbb{Z}^2$, say (x, y). We assert that $(u, v), (x, y)$ form a basis of $\mathbb{Z}^2$. By construction, the triangle conv$\{o, (u, v), (x, y)\}$ contains no point of $\mathbb{Z}^2$, except its vertices. Considering its mirror image in o and the translation of the latter by the vector $(u, v) + (x, y)$, we see that the parallelogram generated by (u, v) and (x, y) contains no point of $\mathbb{Z}^2$, except its vertices. The parallelogram $F = \{\alpha(u, v) + \beta(x, y) : 0 \leq \alpha, \beta < 1\}$ then contains only the point o of $\mathbb{Z}^2$. Given a point $l \in \mathbb{Z}^2$, we may subtract integer multiples of (u, v) and (x, y) from it such that the resulting lattice point is contained in F and thus must coincide with o. Hence l is an integer linear combination of (u, v) and (x, y). Thus $(u, v), (x, y)$ form a basis of the lattice $\mathbb{Z}^2$. By Theorem 21.1 their determinant is ± 1 and by our choice of (x, y) it is positive. Hence $uy - vx = 1$. □

21.2 Characterization of Lattices

Since the notion of lattice is important for many purposes, it is sometimes useful to have at hand an alternative description.

This section contains a simple characterization of lattices. This result or, more precisely, its proof will be used in Proof of Theorem 21.3. It is also a tool for the Venkov–McMullen theorem 32.3 on characterization of parallelohedra.

A Characterization of Lattices

A subset of $\mathbb{E}^d$ is called *discrete* if any bounded set contains only finitely many of its points or, equivalently, if it has no point of accumulation. A characterization of lattices based on the notions of group and discrete set now is as follows.

Theorem 21.2. *Let $L \subseteq \mathbb{E}^d$. Then the following statements are equivalent:*

(i) *L is a lattice.*

(ii) *L is a discrete sub-group of $\mathbb{E}^d$ which is not contained in a hyperplane.*

Proof. (i)⇒(ii) Let $\{b_1, \dots, b_d\}$ be a basis of L. If l, m are integer linear combinations of $b_1, \dots, b_d$, then so is $l - m$. Hence L is a sub-group of $\mathbb{E}^d$. For the proof that L is discrete, note that

(1) $\big\{\alpha_1 b_1 + \dots + \alpha_d b_d : -1 < \alpha_i < 1\big\} \cap L = \{o\}.$

Let $\varrho > 0$ be the radius of a ball with centre at o which is contained in the open parallelotope in (1). Then the distance from o to any point of $L \backslash \{o\}$ is at least ϱ. Therefore, we have $\|l - m\| \geq \varrho$ for $l, m \in L$, $l \neq m$. If L is not discrete, it contains a bounded infinite subset. This subset then has at least one accumulation point. Any two distinct points of this subset, which are sufficiently close to the accumulation point, have distance less than ϱ. This contradiction concludes the proof that L is discrete. L is not contained in a hyperplane since it contains the points $o, b_1, \dots, b_d$.

(ii)⇒(i) It is sufficient to show the following:

(2) There are d linearly independent vectors $b_1, \dots, b_d \in L$ such that $L = \big\{u_1 b_1 + \dots + u_d b_d : u_i \in \mathbb{Z}\big\}$.

This is an immediate consequence of the case $i = d$ of the following proposition:

(3) Let $c_1, \dots, c_d \in L$ be linearly independent. Then, for $i = 1, \dots, d$, we have the following: There are i linearly independent vectors $b_1, \dots, b_i \in L$ such that

$$\begin{aligned} c_1 &= u_{11}b_1 \\ c_2 &= u_{21}b_1 + u_{22}b_2 \\ &\dots\dots\dots\dots\dots\dots \\ c_i &= u_{i1}b_1 + \cdots\cdots + u_{ii}b_i \end{aligned} \qquad \text{where} \quad u_{jk} \in \mathbb{Z}, u_{jj} \neq 0$$

and

$$\text{lin}\{b_1, \dots, b_i\} \cap L = \{u_1 b_1 + \cdots + u_i b_i : u_j \in \mathbb{Z}\}.$$

Proof by induction: For $i = 1$, choose, among all points of L on the line $\text{lin}\{c_1\}$, one with minimum positive distance from o, say b_1. Since L is discrete, this is possible. Then the assertion in (3), for $i = 1$, holds with this b_1.

Next, let $i < d$ and assume that the assertion in (3) holds for i. Consider the unbounded parallelotope

$$P = \big\{\alpha_1 b_1 + \cdots + \alpha_i b_i + \alpha c_{i+1} : 0 \leq \alpha_j < 1, \ \alpha \in \mathbb{R}\big\}.$$

All points of P, which are sufficiently far from o, have arbitrarily large distance from $\text{lin}\{b_1, \dots, b_i\}$. Since $(P \cap L) \setminus \text{lin}\{b_1, \dots, b_i\} \supseteq \{c_{i+1}\} \neq \{o\}$ and since L is discrete, we thus may choose a point $b_{i+1} \in P \cap L$ which is not contained in $\text{lin}\{b_1, \dots, b_i\}$ and has minimum distance from $\text{lin}\{b_1, \dots, b_i\}$. Then

(4) $b_1, \dots, b_i, b_{i+1} \in L$ are linearly independent.

Next, note that, for any point of L in $\text{lin}\{b_1, \dots, b_i, b_{i+1}\}$, we obtain a point of P by adding a suitable integer linear combination of $b_1, \dots, b_i$. These two points then have the same distance from $\text{lin}\{b_1, \dots, b_i\}$. Thus b_{i+1} has minimum distance from $\text{lin}\{b_1, \dots, b_i\}$, not only among all points of $(P \cap L) \setminus \text{lin}\{b_1, \dots, b_i\}$, but also among all points of $\big(\text{lin}\{b_1, \dots, b_i, b_{i+1}\} \cap L\big) \setminus \text{lin}\{b_1, \dots, b_i\}$. This yields, in particular,

(5) $\big\{\alpha_1 b_1 + \cdots + \alpha_i b_i + \alpha_{i+1} b_{i+1} : 0 \leq \alpha_j < 1\big\} \cap L = \{o\}$.

We now show that

(6) $\text{lin}\{b_1, \dots, b_{i+1}\} \cap L = \big\{u_1 b_1 + \cdots + u_i b_i + u_{i+1} b_{i+1} : u_j \in \mathbb{Z}\big\}$.

Let $x \in \text{lin}\{b_1, \dots, b_{i+1}\} \cap L$. Hence $x = u_1 b_1 + \cdots + u_{i+1} b_{i+1}$ with suitable $u_i \in \mathbb{R}$. Then

$$x - \lfloor u_1 \rfloor b_1 - \cdots - \lfloor u_{i+1} \rfloor b_{i+1} \in \big\{\alpha_1 b_1 + \cdots + \alpha_{i+1} b_{i+1} : 0 \leq \alpha_j < 1\big\} \cap L = \{o\}$$

by (5) and thus $x = \lfloor u_1 \rfloor b_1 + \cdots + \lfloor u_{i+1} \rfloor b_{i+1}$. Comparing the two representations of x and taking into account the fact that $b_1, \dots, b_{i+1}$ are linearly independent by (4), it follows that $u_j = \lfloor u_j \rfloor \in \mathbb{Z}$, for $j = 1, \dots, i + 1$. Thus, the left-hand side in (6) is contained in the right-hand side. Since the converse is obvious, the proof of (6) is complete.

By definition of b_{i+1},

$$c_{i+1} \in \big(\operatorname{lin}\{b_1, \dots, b_{i+1}\} \cap L\big) \backslash \operatorname{lin}\{b_1, \dots, b_i\}.$$

Thus (6) yields

(7) $c_{i+1} = u_{i+1\,1} b_1 + \cdots + u_{i+1\,i+1} b_{i+1}$ where $u_{i+1\,j} \in \mathbb{Z}, u_{i+1\,i+1} \neq 0$.

Considering (4), (3) and (7), and (6), the induction is complete, concluding the proof of (3). Since (3) implies (2), the proof of the implication (ii)⇒(i) is complete. □

The Sub-Groups of $\mathbb{E}^d$

From a general mathematical viewpoint, it is of interest to describe all sub-groups of $\mathbb{E}^d$. It turns out that the closed sub-groups of $\mathbb{E}^d$ are the direct sums of the form $L \oplus S$ where L is a lattice in a linear sub-space R of $\mathbb{E}^d$ and S is a linear sub-space of $\mathbb{E}^d$ such that $R \cap S = \{o\}$. The general sub-groups of $\mathbb{E}^d$ are the direct sums of the form $L \oplus D$ where L is a lattice in a linear sub-space R of $\mathbb{E}^d$ and D a dense sub-group of a linear sub-space S of $\mathbb{E}^d$ where $R \cap S = \{o\}$. See Siegel [937].

21.3 Sub-Lattices

Given a mathematical structure, it is a basic problem to describe its sub-structures and their properties. We study sub-lattices of a given lattice. In general, a lattice is given by specifying one of its bases. Since Minkowski [743], Sect. 14, it is known that, for each basis of a sub-lattice, there is a basis of the lattice such that these bases are related in a particularly simple way and vice versa, for each basis of the lattice there is such a basis of the sub-lattice. Max Köcher [604] pointed out that one may select bases of the lattice and the sub-lattice which are related in an even simpler way. He did not communicate a proof, but seems to have had in mind a proof based on the theory of elementary divisors.

In this section we present these results. Our proof of Köcher's result is elementary.

For additional information on sub-lattices compare Cassels [195] and the author and Lekkerkerker [447].

Relations Between the Bases of a Lattice and its Sub-Lattices

If a lattice is contained in a lattice L, it is a *sub-lattice* of L.

Theorem 21.3. *Let M be a sub-lattice of a lattice L in $\mathbb{E}^d$. Then the following hold:*

(i) *Given a basis $\{c_1, \dots, c_d\}$ of M, there is a basis $\{b_1, \dots, b_d\}$ of L such that*

$$(1)\qquad \begin{aligned} c_1 &= u_{11} b_1 \\ c_2 &= u_{21} b_1 + u_{22} b_2 \\ &\cdots\cdots\cdots\cdots\cdots \\ c_d &= u_{d1} b_1 + \cdots\cdots + u_{dd} b_d \end{aligned} \qquad \text{where} \quad u_{ik} \in \mathbb{Z}, u_{ii} \neq 0.$$

(ii) *Given a basis* $\{b_1, \dots, b_d\}$ *of* L*, there is a basis* $\{c_1, \dots, c_d\}$ *of* M *such that* (1) *holds.*

Proof. (i) This is an immediate consequence of the case $i = d$ of Proposition (3) in the proof of Theorem 21.2.

(ii) We first show that

(2) $uL \subseteq M$ with a suitable positive $u \in \mathbb{Z}$.

Consider bases $\{b_1, \dots, b_d\}$ of L and $\{c_1, \dots, c_d\}$ of M such that (1) holds. The inverse of the integer lower triangular matrix (u_{ik}), with determinant $u = u_{11} \cdots u_{dd} \neq 0$, is a lower triangular matrix, the entries of which are of the form v_{ik}/u, where $v_{ik} \in \mathbb{Z}$. Hence

$$\begin{aligned} ub_1 &= v_{11}c_1 \\ ub_2 &= v_{21}c_1 + v_{22}c_2 \\ &\cdots\cdots\cdots\cdots\cdots\cdots \\ ub_d &= v_{d1}c_1 + \cdots\cdots + v_{dd}c_d. \end{aligned}$$

From this it follows that uL is a sub-lattice of M, concluding the proof of (2).

To prove (ii), let $\{b_1, \dots, b_d\}$ be a basis of L. Choose u as in (2). Then $\{ub_1, \dots, ub_d\}$ is a basis of uL. An application of (i) to the sub-lattice uL of M thus yields a basis $\{d_1, \dots, d_d\}$ of M such that

$$(3)\quad \begin{aligned} ub_1 &= w_{11}d_1 \\ ub_2 &= w_{21}d_1 + w_{22}d_2 \\ &\cdots\cdots\cdots\cdots\cdots\cdots \\ ub_d &= w_{d1}d_1 + \cdots\cdots + w_{dd}d_d \end{aligned} \qquad \text{where} \quad w_{ik} \in \mathbb{Z},\ w_{ii} \neq 0.$$

Solving (3) for $d_1, \dots, d_d$, we see that we may express $d_1, \dots, d_d$ in the form (1) with $d_1, \dots, d_d$ instead of $c_1, \dots, c_d$, but where the u_{ik} are rational. Since $\{b_1, \dots, b_d\}$ is a basis of L and M a sub-lattice, the u_{ik} are, in fact, integers. □

Primitive Points and Bases

If $\{b_1, \dots, b_d\}$ is a basis of L and $b = u_1b_1 + \cdots + u_db_d$, then b is primitive if and only if 1 is the greatest common divisor of $u_1, \dots, u_d$.

Corollary 21.1. *Let* b *be a primitive point of a lattice* L *in* $\mathbb{E}^d$*. Then there are points* $b_2, \dots, b_d \in L$*, such that* $\{b, b_2, \dots, b_d\}$ *is a basis of* L*.*

Proof. Choose $c_2, \dots, c_d \in L$, such that $b, c_2, \dots, c_d$ are linearly independent. By Proposition (i) of the above theorem, there is a basis $\{b_1, \dots, b_d\}$ of L, such that

$$\begin{aligned} b\ &= u_{11}b_1 \\ c_2 &= u_{21}b_1 + u_{22}b_2 \\ &\cdots\cdots\cdots\cdots\cdots\cdots \\ c_d &= u_{d1}b_1 + \cdots\cdots + u_{dd}b_d \end{aligned} \qquad \text{where} \quad u_{ik} \in \mathbb{Z},\ u_{ii} \neq 0.$$

Since b is primitive, this can hold only if $u_{11} = \pm 1$ or $b = \pm b_1$. Hence $\{b, b_2, \dots, b_d\}$ is also a basis of L. □

A similar, slightly more complicated proof leads to the following result:

Corollary 21.2. *Let $b_1, \dots, b_k$ be k linearly independent points of a lattice L such that*

$$\{\alpha_1 b_1 + \cdots + \alpha_k b_k : 0 \le \alpha_i < 1\} \cap L = \{o\}.$$

Then, there are points $b_{k+1}, \dots, b_d \in L$ such that $\{b_1, \dots, b_d\}$ is a basis of L.

Closely Related Bases of a Lattice and a Sub-Lattice

The following result shows that, given a sub-lattice M of a lattice L, there are bases of M and L which are related in a particularly simple way. We could ascertain whether this result is a consequence of a more general result on Abelian groups.

Theorem 21.4. *Let M be a sub-lattice of a lattice L in $\mathbb{E}^d$. Then there are bases $\{c_1, \dots, c_d\}$ of M and $\{b_1, \dots, b_d\}$ of L, such that*

$$c_1 = u_1 b_1, \dots, c_d = u_d b_d, \text{ where } u_i \in \mathbb{Z} \setminus \{0\}.$$

Proof (by induction on d). If $d = 1$, the theorem is easy to see. Assume now that $d > 1$ and that the theorem holds for $d - 1$. The proof for d is divided into two steps.

In the first step we treat a special case:

(4) Let M be a sub-lattice of a lattice L in $\mathbb{E}^d$ which contains a primitive point b_1 of L. Then the theorem holds.

In the following, we consider lower dimensional lattices in $\mathbb{E}^d$, but this should not cause difficulties. Since b_1 is a primitive point of L, there is a basis of L of the form $\{b_1, b_2, \dots, b_d\}$ by Corollary 21.1. Then

$$L = \{u b_1 + l : u \in \mathbb{Z},\ l \in \operatorname{lin}\{b_2, \dots, b_d\} \cap L = L'\}.$$

L' is a lattice of dimension $d - 1$ in $\mathbb{E}^d$. Let a point $m \in M \subseteq L$ be given. Then, since $b_1 \in M$, we have $m - u b_1 \in \operatorname{lin}\{b_2, \dots, b_d\} \cap M$ for suitable $u \in \mathbb{Z}$. Hence

$$M = \{u b_1 + n : u \in \mathbb{Z},\ n \in \operatorname{lin}\{b_2, \dots, b_d\} \cap M = M'\}.$$

M' is a $(d - 1)$-dimensional sub-lattice of the $(d - 1)$-dimensional lattice L'. By induction, there are bases $\{b_2, \dots, b_d\}$ of L' and $\{c_2, \dots, c_d\}$ of M', such that

$$c_2 = u_2 b_2, \dots, c_d = u_d b_d \text{ for suitable } u_i \in \mathbb{Z}.$$

The bases $\{b_1, b_2, \dots, b_d\}$ of L and $\{b_1, c_2, \dots, c_d\}$ of M are then of the desired form, concluding the proof for d in the special case (4).

In the second step, we consider a general sub-lattice M of L. For each point $m \in M \setminus \{o\}$, let j be the unique positive integer such that $m = j\, l$, where $l \in L$ is primitive. Call j the *index* of m. Choose $c \in M \setminus \{o\}$, such that c has minimum index, say i, and let $c = ib_1$, where $b_1 \in L$ is primitive. Now the following will be shown:

(5) Let $m \in M \setminus \{o\}$ and let j be the index of m. Then $i | j$.

If m is linearly dependent on c, Proposition (5) is easy to see. Assume now that c and m are linearly independent. The 2-dimensional lattice $M' = \text{lin}\{c, m\} \cap M$ is a sub-lattice of the 2-dimensional lattice $L' = \text{lin}\{c, m\} \cap L$. Since b_1 is a primitive point of L, and thus of L', Corollary 21.1 shows that there is a basis of the form $\{b_1, b_2\}$ of L'. Let $m = j(ub_1 + vb_2)$, where u, v are relatively prime integers. Note that $c = ib_1$ and choose $h \in \mathbb{Z}$, such that

$$n = m - hc = (ju - ih)b_1 + (jv)b_2 = k(wb_1 + zb_2), \text{ where}$$
$$0 \le ju - ih = kw < i \text{ and } w, z \text{ are relatively prime integers.}$$

Clearly, k is the index of n. The inequality $0 \le kw < i$ and the inequality $i \le k$, which follows from the definition of i, are compatible only if $w = 0$. Thus $ju - ih = 0$, or

(6) $i | ju$,

and $n = jvb_2$. Next, consider the point $c + n = ib_1 + jvb_2 \in M \setminus \{o\}$. Since i is the minimum index of all points of $M \setminus \{o\}$, we have

(7) $i | jv$.

Since the integers u, v are relatively prime, there are integers x, y such that $uy - vx = 1$, see Corollary 21.1. Propositions (6) and (7) then imply that $i | j(uy - vx) = j$, concluding the proof of (5).

The lattice $(1/i)M$ is a sub-lattice of L by (5) and $b_1 = (1/i)c \in (1/i)M$ is a primitive point of L. Thus, we may apply (4) to see that there are bases of L and $(1/i)M$, and thus of L and M, of the desired type. The induction, and thus the proof of the theorem, is complete. □

The Index of a Sub-Lattice

Let M be a sub-lattice of a lattice L and choose bases $\{b_1, \dots, b_d\}$ of L and $\{c_1, \dots, c_d\}$ of M, as in Theorem 21.4. The fundamental parallelotope,

$$\{\alpha_1 c_1 + \dots + \alpha_d c_d : 0 \le \alpha_i < 1\},$$

of M, and thus every fundamental parallelotope of M, contains precisely $|u_1 \cdots u_d|$ points of L, where $u_1, \dots, u_d$ are as in the last theorem. Thus the number of translates of M by vectors of L, which are needed to make up L, is precisely $|u_1 \cdots u_d|$. In other words, $|u_1 \cdots u_d|$ is the index of the sub-group $\langle M, + \rangle$ in the group $\langle L, + \rangle$. Call this the *index of the sub-lattice M in the lattice L*. It follows, from $c_1 = u_1 b_1, \dots, c_d = u_d b_d$, that the index of the sub-lattice M in the lattice L equals

$$|u_1 \cdots u_d| = \frac{d(M)}{d(L)}.$$

21.4 Polar Lattices

For each lattice L in $\mathbb{E}^d$, there exists a sort of dual lattice, called the polar lattice of L, which is a useful tool in various contexts. Polar lattices seem to have appeared first in the work of Bravais in crystallography, see the discussion by Rigault [838]. There is a sort of weak duality between a convex body and a lattice, on the one hand, and the polar body and the polar lattice, on the other hand. For an example of this duality, see Theorem 23.2.

In this section, we introduce the notion of the polar lattice of a given lattice and show how their bases are related.

The Polar Lattice of a Given Lattice

Our aim here is to show the following simple result, where B^{-T} is the transposed of the inverse of the $d \times d$ matrix B.

Theorem 21.5. *Let L be a lattice in* $\mathbb{E}^d$. *Then*

$$L^* = \{m \in \mathbb{E}^d : l \cdot m \in \mathbb{Z} \text{ for each } l \in L\}$$

is a lattice, called the dual *or* polar lattice *of L. If* $\{b_1, \dots, b_d\}$ *is a basis of L and B the (non-singular) matrix with columns* $b_1, \dots, b_d$, *then the columns of the matrix* $B^* = B^{-T}$ *form a basis* $\{b_1^*, \dots, b_d^*\}$ *of* L^*, *the* dual *or* polar basis *of the given basis.*

Proof. Let $\{b_1, \dots, b_d\}$ be a basis of L. Then $b_1^*, \dots, b_d^*$ are linearly independent. To see that

$$(1)\quad b_i \cdot b_k^* = \begin{cases} 1 \text{ for } i = k, \\ 0 \text{ for } i \neq k, \end{cases}$$

note that, for the $d \times d$ matrix $(b_i \cdot b_k^*)$, we have,

$$(b_i \cdot b_k^*) = B^T B^{-T} = (B^{-1}B)^T = I^T = I,$$

where I is the $d \times d$ unit matrix. Since $b_1^*, \dots, b_d^*$ are linearly independent, it is sufficient, for the proof of the theorem, to show that

$$(2)\quad L^* = \{v_1 b_1^* + \dots + v_d b_d^* : v_i \in \mathbb{Z}\}.$$

First, let $m \in L^*$. Since $b_1^*, \dots, b_d^*$ are linearly independent, $m = w_1 b_1^* + \dots + w_d b_d^*$ for suitable $w_i \in \mathbb{R}$. Then

$$w_i = b_i \cdot (w_1 b_1^* + \dots + w_d b_d^*) = b_i \cdot m \in \mathbb{Z}$$

by (1) and the definition of L^*. Hence $m \in \{v_1 b_1^* + \dots + v_d b_d^* : v_i \in \mathbb{Z}\}$. Second, let $m \in \{v_1 b_1^* + \dots + v_d b_d^* : v_i \in \mathbb{Z}\}$, say $m = v_1 b_1^* + \dots + v_d b_d^*$, where $v_i \in \mathbb{Z}$. Then we have

$$l \cdot m = (u_1 b_1 + \dots + u_d b_d) \cdot (v_1 b_1^* + \dots + v_d b_d^*) = u_1 v_1 + \dots + u_d v_d \in \mathbb{Z},$$

for any $l = u_1 b_1 + \dots + u_d b_d \in L$ by (1), and thus $m \in L^*$. This concludes proof of (2) and thus of the theorem. □

22 Minkowski's First Fundamental Theorem

We cite Cassels [195], prologue:

> We owe to MINKOWSKI the fertile observation that certain results which can be made almost intuitive by consideration of figures in n-dimensional Euclidean space have far-reaching consequences in diverse branches of number theory. For example, he simplified the theory of units in algebraic number fields and both simplified and extended the theory of approximation of irrational numbers by rational ones (Diophantine Approximation). This new branch of number theory, which MINKOWSKI christened "The Geometry of Numbers", has developed into an independent branch of number theory which, indeed, has many applications elsewhere but which is well worth studying for its own sake.

Minkowski's fundamental theorem is one among a small number of basic results of the geometry of numbers alluded to by Cassels. It relates the basic notions of lattices and convex bodies. The fundamental theorem is simple, almost trivial and, at the same time, deep. There exist numerous generalizations and arithmetic consequences of it. The following remark of Hilbert well describes the situation, see Rose [856]:

> The art of doing mathematics consists in finding that special case which contains all the germs of generality.

In this section we present the fundamental theorem together with some of its classical applications dealing with Diophantine approximation, representation of integers as sums of squares, and estimates for discriminants of polynomials.

For a wealth of different versions, related results and further applications, see Cassels [195], Kannan [563], Gruber and Lekkerkerker [447] and Erdös, Gruber and Hammer [307].

22.1 The First Fundamental Theorem

Let $f : \mathbb{E}^d \to \mathbb{R}$. The arithmetic problem of finding a solution $u = (u_1, \dots, u_d)$ of the inequality

$$f(x_1, \dots, x_d) \leq 1,$$

where the u_i are integers, not all equal to 0, is equivalent to the geometric problem of finding a point of the integer lattice $\mathbb{Z}^d$ different from o and contained in the set

$$\{x : f(x) \leq 1\}.$$

This led Minkowski to look for conditions which guarantee that a set, in particular a convex body, contains points of a lattice different from the origin. The result was his first fundamental theorem.

In this section we present Minkowski's theorem together with three elegant and convincing proofs, each based on a different idea.

The First Fundamental Theorem

A very satisfying answer to the above problem is the following *first fundamental theorem of Minkowski* [734] or *Minkowski's first theorem*:

Theorem 22.1. *Let C be an o-symmetric convex body and L a lattice in $\mathbb{E}^d$ such that $V(C) \geq 2^d d(L)$. Then C contains a pair of points $\pm l \in L \setminus \{o\}$.*

The first proof is based on Dirichlet's pigeon hole principle and a formula to calculate the volume of a convex body. The idea to use the pigeon hole principle in this context seems to be due to Scherrer [886]. The second proof is due to Siegel [935] and makes use of Fourier series and Parseval's theorem. The third one utilizes a close relation between the fundamental theorem and the notion of density of a lattice packing of convex bodies (Fig. 22.1).

Proof (using the pigeon hole principle). Since C is compact and L discrete, it is sufficient to prove the theorem under the stronger assumption that

(1) $V(C) > 2^d d(L)$.

The convex body $\frac{1}{2}C$ is Jordan measurable by Theorem 7.4. Using the substitution rule for multiple integrals, the asymptotic formula (3) in Sect. 7.2 to calculate the Jordan measure then implies that

$$V\Big(\frac{1}{2}C\Big) \sim \#\Big(\frac{1}{2}C \cap \frac{1}{n}L\Big) \cdot d\Big(\frac{1}{n}L\Big) \text{ as } n \to \infty.$$

Since by (1) $V(\frac{1}{2}C) > d(L)$, it follows that

(2) $\#\Big(\frac{1}{2}C \cap \frac{1}{n}L\Big) > n^d$ for all sufficiently large n.

Keep such an n fixed and choose a basis $\{b_1, \dots, b_d\}$ of L. To each point

$$\frac{u_1}{n}b_1 + \cdots + \frac{u_d}{n}b_d \in \frac{1}{2}C \cap \frac{1}{n}L$$

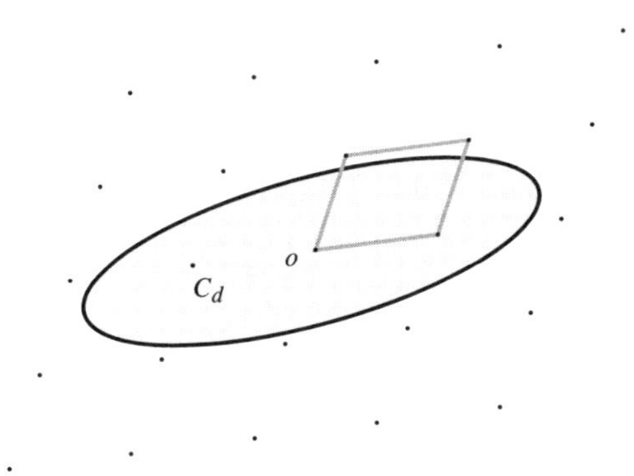

Fig. 22.1. Fundamental theorem

we associate the point

$$\frac{w_1}{n}b_1 + \cdots + \frac{w_d}{n}b_d \text{ where } w_i \equiv u_i \bmod n \text{ and } w_i \in \{0, \ldots, n-1\}.$$

There are precisely n^d points of the latter form. Since, by (2), there are more than n^d points in $\frac{1}{2}C \cap \frac{1}{n}L$, the Dirichlet pigeon hole principle implies that there are two distinct points, say

$$\frac{u_1}{n}b_1 + \cdots + \frac{u_d}{n}b_d,\ \frac{v_1}{n}b_1 + \cdots + \frac{v_d}{n}b_d \in \frac{1}{2}C \cap \frac{1}{n}L,$$

for which the associated points coincide. Then $u_i \equiv v_i \bmod n$, or $n|(u_i - v_i)$ for $i = 1, \ldots, d$, and we obtain

$$o \neq \underbrace{\frac{u_1 - v_1}{n}}_{\in\mathbb{Z}} b_1 + \cdots + \underbrace{\frac{u_d - v_d}{n}}_{\in\mathbb{Z}} b_d \in \Big(\frac{1}{2}C - \frac{1}{2}C\Big) \cap L = C \cap L.$$

This concludes the first proof of the fundamental theorem. □

Proof (of Siegel with Fourier series). It is sufficient to show the following proposition:

(3) Let $C \cap L = \{o\}$. Then $V(C) \leq 2^d d(L)$.

For the proof of (3) we assume that $C \cap L = \{o\}$ and first show the following:

(4) The convex bodies $\frac{1}{2}C + l$, $l \in L$, are pairwise disjoint.

If (4) did not hold, then $\frac{1}{2}x + l = \frac{1}{2}y + m$ for suitable $x, y \in C$ and $l, m \in L$, $l \neq m$. Hence $o \neq l - m = \frac{1}{2}y - \frac{1}{2}x \in (\frac{1}{2}C - \frac{1}{2}C) \cap L = C \cap L$. This contradicts the assumption in (3) and thus concludes the proof of (4).

Let $\mathbb{1}$ be the characteristic function of $\frac{1}{2}C$. Clearly, the function $\psi : \mathbb{E}^d \to \mathbb{R}$ defined by:

(5) $\psi(x) = \sum_{l \in L} \mathbb{1}(x + l)$ is L-periodic.

Because of (4), we have $\psi(x) = 0$ or 1 for each $x \in \mathbb{E}^d$ and thus

(6) $\psi^2 = \psi$.

To ψ corresponds the Fourier series

$$\sum_{m \in L^*} c(m) e^{2\pi i\, m \cdot x},$$

where, for the Fourier coefficients $c(m)$, we have the following representations:

$$(7)\qquad c(m)=\frac{1}{d(L)}\int\limits_F \psi(x)e^{-2\pi i\, m\cdot x}dx=\frac{1}{d(L)}\int\limits_F\sum_{l\in L}\mathbb{1}(x+l)e^{-2\pi i\, m\cdot x}dx$$
$$=\frac{1}{d(L)}\sum_{l\in L}\int\limits_F \mathbb{1}(x+l)e^{-2\pi i\, m\cdot x}dx$$
$$=\frac{1}{d(L)}\sum_{l\in L}\int\limits_{F+l}\mathbb{1}(y)e^{-2\pi i\, m\cdot(y-l)}dy=\frac{1}{d(L)}\int\limits_{\mathbb{E}^d}\mathbb{1}(y)e^{-2\pi i\, m\cdot y}dy,$$

by (5). Here $i=\sqrt{-1}$ and F is a fundamental parallelotope of L. Note that the sums are all finite and thus integration and summation may be interchanged. A similar calculation shows that:

$$(8)\qquad \int\limits_F \psi(x)^2dx=\int\limits_F \psi(x)\,dx=\int\limits_F\Big(\sum_{l\in L}\mathbb{1}(x+l)\Big)\,dx$$
$$=\sum_{l\in L}\int\limits_F \mathbb{1}(x+l)\,dx=\sum_{l\in L}\int\limits_{F+l}\mathbb{1}(y)\,dy=\int\limits_{\mathbb{E}^d}\mathbb{1}(y)\,dy$$
$$=V\Big(\frac{1}{2}C\Big),$$

where we have used (6), (5) and the definition of $\mathbb{1}$. Finally, (8), Parseval's theorem for Fourier series, (7) and the definition of $\mathbb{1}$ show that:

$$V\Big(\frac{1}{2}C\Big)=\int\limits_F \psi(x)^2dx=d(L)\sum_{m\in L^*}|c(m)|^2$$
$$=d(L)|c(o)|^2+d(L)\sum_{m\in L^*\setminus\{o\}}|c(m)|^2$$
$$=d(L)\frac{1}{d(L)^2}V\Big(\frac{1}{2}C\Big)^2+d(L)\sum_{m\in L^*\setminus\{o\}}|c(m)|^2,$$

i.e.

$$2^d d(L)=V(C)+\frac{4^d d(L)^2}{V(C)}\sum_{m\in L^*\setminus\{o\}}|c(m)|^2\geq V(C).\qquad\square$$

Proof (by means of lattice packing). The reader who is not familiar with lattice packing of convex bodies and the notion of density may wish to consult Sect. 30.1 first.

As before, it is sufficient to show (3). Statement (4) says that $\{\frac{1}{2}C+l:l\in L\}$ is a lattice packing. The density of a lattice packing is at most 1. Hence

$$\frac{V(\frac{1}{2}C)}{d(L)}\leq 1,\ \text{ or } V(C)\leq 2^d d(L).\qquad\square$$

The Background of the Fundamental Theorem

The principle underlying the fundamental theorem and some of its refinements or extensions, such as Blichfeldt's [130] theorem or more modern generalizations of it, is Dirichlet's pigeon hole principle or, put analytically, the following simple observation: If, on a measure space with total measure 1, the integral of a function is greater than 1, then the function assumes values greater than 1.

Effective Methods for Finding Lattice Points

The fundamental theorem guarantees the *existence* of lattice points different from o in C, but it does not tell us how to *find* such a point. In the context of algorithmic geometry of numbers, a polynomial time algorithm has been specified which finds such points, supposing that the volume of C is substantially larger than $2^d d(L)$. See the remarks on the shortest lattice vector problem in Sect. 28.2 and, for more information, Grötschel, Lovász and Schrijver [409].

22.2 Diophantine Approximation and Discriminants of Polynomials

The first fundamental theorem has numerous classical applications. These include Minkowski's applications to positive definite quadratic forms, Diophantine approximation, his linear form theorem and discriminants of polynomials.

In the following, we present a selection of applications due to Minkowski and a result of Lagrange on the representation of integers as sums of squares. For further applications of the fundamental theorem. See, e.g. Gruber and Lekkerkerker [447] and Schmidt [896].

Homogeneous Minimum of a Positive Definite Quadratic Form

Using a geometric argument on lattices and balls, Minkowski [732] was able to improve a theorem of Hermite [495] on positive definite quadratic forms. At the time when he published his result, Minkowski did not yet have the fundamental theorem in its general form, but the argument used led him to the fundamental theorem shortly afterwards. Minkowski's result is as follows:

Corollary 22.1. *Let*

$$q(x) = \sum_{i,k=1}^{d} a_{ik}\, x_i x_k \text{ for } x = (x_1, \dots, x_d) \in \mathbb{E}^d$$

be a positive definite quadratic form ($a_{ik} = a_{ki}$). Then the following inequality has a non-trivial integer solution u, i.e. $u \in \mathbb{Z}^d \setminus \{o\}$:

$$q(u) \le 4\Big(\frac{\det(a_{ik})}{V(B^d)^2}\Big)^{\frac{1}{d}}.$$

Proof. To see this estimate, note that for $\varrho > 0$ the set $E_\varrho = \{x : q(x) \le \varrho\}$ is a solid ellipsoid in $\mathbb{E}^d$ with centre o and volume

$$V(E_\varrho) = \frac{\varrho^{\frac{d}{2}} V(B^d)}{\det(a_{ik})^{\frac{1}{2}}}.$$

If

$$\varrho = 4\Big(\frac{\det(a_{ik})}{V(B^d)^2}\Big)^{\frac{1}{d}},$$

then $V(E_\varrho) = 2^d$. Now apply the fundamental theorem to the convex body E_ϱ and the lattice $\mathbb{Z}^d$. □

For smaller upper estimates, but in the terminology of density of packings of balls, see Sect. 29.2 and the articles of Blichfeldt, Sidel'nikov, Levenstein and Kabat'janskiĭ cited there. The best known upper bound is that of Levenstein and Kabat'janskiĭ.

Minkowski's Linear Form Theorem

This result can be formulated as follows, see Minkowski [735]:

Corollary 22.2. *Let $l_1, \dots, l_d$ be d real linear forms in d real variables, such that the absolute value δ of their determinant is positive. Assume further that $\tau_1, \dots, \tau_d > 0$ are such that $\tau_1 \cdots \tau_d \ge \delta$. Then the following system of inequalities has a non-trivial integer solution*

$$|l_1(x)| \le \tau_1, \dots, |l_d(x)| \le \tau_d.$$

Proof. Apply the fundamental theorem to the parallelotope P and the lattice $\mathbb{Z}^d$, where

$$P = \big\{x : |l_1(x)| \le \tau_1, \dots, |l_d(x)| \le \tau_d\big\}, V(P) = \frac{2^d \tau_1 \cdots \tau_d}{\delta} \ge 2^d. \quad \square$$

The estimate in the linear form theorem cannot be improved for all systems of linear forms.

Simultaneous Diophantine Approximation

As a consequence of the linear form theorem, we obtain a classical approximation result due to Kronecker [618]. He proved it by means of Dirichlet's pigeon hole principle. Here, we follow Minkowski [735] who used his linear form theorem.

Corollary 22.3. *Let $\vartheta_1, \dots, \vartheta_d \in \mathbb{R}$. Then the following system of inequalities has infinitely many integer solutions $(u_0, u_1, \dots, u_d)$, where $u_0 \neq 0$:*

$$\Big|\vartheta_1 - \frac{u_1}{u_0}\Big| \le \frac{1}{u_0^{1+\frac{1}{d}}}, \dots, \Big|\vartheta_d - \frac{u_d}{u_0}\Big| \le \frac{1}{u_0^{1+\frac{1}{d}}}.$$

Proof. If $\vartheta_1, \dots, \vartheta_d$ all are rational, choose $(u_0, \dots, u_d) \in \mathbb{Z}^{d+1}$, $u_0 \neq 0$, such that

$$\vartheta_1 = \frac{u_1}{u_0}, \dots, \vartheta_d = \frac{u_d}{u_0}.$$

Then the corollary holds trivially by considering all integer multiples of the integer vector $(u_0, u_1, \dots, u_d)$. Assume, now, that, among $\vartheta_1, \dots, \vartheta_d$, not all are rational, say ϑ_1 is irrational. Let $0 < \varepsilon_1 < 1$. Then the linear form theorem, applied in $\mathbb{E}^{d+1}$, shows that the following system of $d+1$ inequalities has a non-trivial integer solution $(u_0, u_1, \dots, u_d)$:

$$|u_0\vartheta_1 - u_1| \leq \varepsilon_1, \dots, |u_0\vartheta_d - u_d| \leq \varepsilon_1, |u_0| \leq \frac{1}{\varepsilon_1^d}.$$

Here $u_0 \neq 0$ since otherwise $u_1 = \cdots = u_d = 0$ (note that $0 < \varepsilon_1 < 1$). This gives a first solution. Clearly, $0 < |u_0\vartheta_1 - u_1|$. Next, let $0 < \varepsilon_2 < \varepsilon_1$ such that

$$\varepsilon_2 < |u_0\vartheta_1 - u_1|$$

and argue as before. This, again, gives a solution, different from the first one. Continuing in this way, one gets an infinite set of solutions as required. □

Approximation of Linear Forms

Similar arguments lead to the following counterpart of Corollary 22.3, proved earlier by Dirichlet [271]:

Corollary 22.4. *Let $\vartheta_1, \dots, \vartheta_d \in \mathbb{R}$. Then the following inequality has infinitely many integer solutions $(u_0, u_1, \dots, u_d)$, where $u_1, \dots, u_d$ are not all* 0*:*

$$|u_1\vartheta_1 + \cdots + u_d\vartheta_d - u_0| \leq \frac{1}{\max\{|u_1|, \dots, |u_d|\}^d}.$$

The Four Squares Theorem of Lagrange

Slightly more complicated is the proof of the following result of Lagrange, known already to Diophantus and conjectured by Bachet.

Corollary 22.5. *Let u be a positive integer. Then there are integers u_1, u_2, u_3, u_4 such that*

$$u = u_1^2 + \cdots + u_4^2.$$

Proof (following Davenport [245]). For the proof, we may assume that u is square free and $u > 1$. Thus $u = p_1 \cdots p_n$, where the p_i are different primes. First, the following will be shown:

(1) For each p_i there are integers a_i, b_i such that $a_i^2 + b_i^2 + 1 \equiv 0 \bmod p_i$.

If $p_i = 2$ put $a_i = 1$, $b_i = 0$. If p_i is odd, each of the sets $\{a^2 : 0 \le a < p_i/2\}$ and $\{-1 - b^2 : 0 \le b < p_i/2\}$ consists of $\frac{1}{2}(p_i + 1)$ integers which are pairwise incongruent modulo p_i. Since there are only p_i residue classes modulo p_i, these sets must contain elements a_i^2 and $-1 - b_i^2$, say, which are congruent modulo p_i. Then $a_i^2 + b_i^2 + 1 \equiv 0 \bmod p_i$, concluding the proof of (1).

As a consequence of (1), we shall prove that:

(2) There are integers a, b such that $a^2 + b^2 + 1 \equiv 0 \bmod u$.

For each i, choose an integer P_i such that $P_i p_i = u$. Since P_i and p_i are relatively prime, there is an integer q_i such that $P_i q_i \equiv 1 \bmod p_i$, see Proposition 21.1. Now let

$$a = P_1 q_1 a_1 + \cdots + P_n q_n a_n,\ b = P_1 q_1 b_1 + \cdots + P_n q_n b_n.$$

Then

$$\begin{aligned} a^2 + b^2 + 1 &\equiv P_1^2 q_1^2 a_1^2 + \cdots + P_n^2 q_n^2 a_n^2 + P_1^2 q_1^2 b_1^2 + \cdots + P_n^2 q_n^2 b_n^2 + 1 \\ &\equiv P_i^2 q_i^2 (a_i^2 + b_i^2) + 1 \equiv a_i^2 + b_i^2 + 1 \equiv 0 \bmod p_i, \end{aligned}$$

by (1) and our choice of P_i, q_i. Thus $p_i | a^2 + b^2 + 1$ for each i. This implies that $u = p_1 \cdots p_n | a^2 + b^2 + 1$, concluding the proof of (2).

Next consider the lattice L in $\mathbb{E}^4$ with basis

$$(1, 0, a, -b), (0, 1, b, a), (0, 0, u, 0), (0, 0, 0, u).$$

Clearly $d(L) = u^2$. Let

$$\varrho = \frac{2^{\frac{5}{4}} u^{\frac{1}{2}}}{\pi^{\frac{1}{2}}}.$$

Then

$$V(\varrho B^4) = \varrho^4 V(B^4) = \frac{2^5 u^2}{\pi^2} \frac{\pi^{\frac{4}{2}}}{\Gamma(1 + \frac{4}{2})} = 2^4 u^2,$$

where B^4 is the solid Euclidean unit ball in $\mathbb{E}^4$. An application of the fundamental theorem, with $C = \varrho B^4$ and the lattice L just defined, yields a point $\ne o$ of L in ϱB^4. Thus, there are integers $v_1, \dots, v_4$, not all 0, such that

$$0 < v_1^2 + v_2^2 + (av_1 + bv_2 + uv_3)^2 + (-bv_1 + av_2 + uv_4)^2 \le \varrho^2 < 2u.$$

Since

$$\begin{aligned} &v_1^2 + v_2^2 + (av_1 + bv_2 + uv_3)^2 + (-bv_1 + av_2 + uv_4)^2 \\ &\quad \equiv v_1^2 + v_2^2 + (av_1 + bv_2)^2 + (-bv_1 + av_2)^2 \\ &\quad \equiv v_1^2(a^2 + b^2 + 1) + v_2^2(a^2 + b^2 + 1) \equiv 0 \bmod u, \end{aligned}$$

by (2), we finally obtain the representation of u we were looking for:

$$u = v_1^2 + v_2^2 + (av_1 + bv_2 + uv_3)^2 + (-bv_1 + av_2 + uv_4)^2. \quad \square$$

Remark. There are versions of Lagrange's theorem for algebraic integers in certain number fields. See, e.g. the article of Deutsch [264] and the references there.

Remark. Note that all the earlier applications of the fundamental theorem guarantee the *existence* of solutions, but do not help us to *determine* such efficiently. We return to this question in Sect. 28.2.

Lower Estimate of the Discriminant of an Irreducible Polynomial

We will present a special case of a result of Minkowski [743] on the discriminant of irreducible polynomials. See also Siegel [937].

A polynomial $p(t) = a_0 + a_1 t + \cdots + a_{d-1} t^{d-1} + t^d$ with rational coefficients a_i is *irreducible* over $\mathbb{Q}$ if it cannot be represented as the product of two non-constant polynomials with rational coefficients. This implies that a root of p is not a root of a polynomial of lower degree which is not identically zero and has rational coefficients. Also, every root of p is simple, for otherwise it would be a root of p' which is of lower degree and does not vanish identically. For irreducibility criteria based on the notion of Newton polytopes, see Sect. 19.5.

If $t_1, \dots, t_d$ denote the roots of p, then the *discriminant* D of p is defined by:

$$D = \prod_{1 \leq i < j \leq d} (t_i - t_j)^2.$$

By Vandermonde's theorem on determinants,

$$D = \det \begin{pmatrix} 1 \, t_1 \, \dots \, t_1^{d-1} \\ 1 \, t_2 \, \dots \, t_2^{d-1} \\ \\ 1 \, t_d \, \dots \, t_d^{d-1} \end{pmatrix}^2 .$$

The theorem on elementary symmetric functions implies the following: a symmetric polynomial q in $t_1, \dots, t_d$, with integer coefficients, can be expressed as a polynomial in $a_0, \dots, a_{d-1}$, with integer coefficients. If, in particular, $a_0, \dots, a_{d-1}$ are integers, then $q(t_1, \dots, t_d)$ is an integer. A lattice is *admissible* for a set if it contains no interior point of the set, except, possibly, the origin.

Corollary 22.6. *Let $p(t) = a_0 + a_1 t + \cdots + a_{d-1} t^{d-1} + t^d$ be an irreducible polynomial with integer coefficients $a_0, \dots, a_{d-1}$. If all roots of p are real, then the discriminant D of p satisfies*

$$D \geq \left(\frac{d^d}{d!}\right)^2.$$

Proof. Consider the linear forms

$$\begin{aligned} l_1(u) &= u_1 + t_1 u_2 + \cdots + t_1^{d-1} u_d \\ l_2(u) &= u_1 + t_2 u_2 + \cdots + t_2^{d-1} u_d \\ &\cdots\cdots\cdots\cdots\cdots\cdots \\ l_d(u) &= u_1 + t_d u_2 + \cdots + t_d^{d-1} u_d, \end{aligned}$$

where $t_1, \dots, t_d$ are the roots of p. For $u \in \mathbb{Z}^d \setminus \{o\}$, the linear forms $l_i(u)$, $i = 1, \dots, d$, all are different from 0, for, if $l_i(u) = 0$, then t_i is a root of the polynomial $l_i(u)$ of degree at most $d - 1$ in t_i with integer coefficients. This contradicts the irreducibility of p. Therefore $l_1(u) \cdots l_d(u) \neq 0$ and, since this product is a symmetric polynomial in $t_1, \dots, t_d$ with integer coefficients, it must be an integer. Thus

$$|l_1(u) \cdots l_d(u)| \geq 1 \text{ for } u \in \mathbb{Z}^d \setminus \{o\}.$$

The lattice $L = \{l = (l_1(u), \dots, l_d(u)) : u \in \mathbb{Z}^d\}$ is thus admissible for the star set

$$\{x : |x_1 \cdots x_d| \leq 1\}.$$

By the inequality of the geometric and arithmetic mean, this star set contains the o-symmetric cross-polytope

$$O = \Big\{x : \frac{1}{d}\big(|x_1| + \cdots + |x_d|\big) \leq 1\Big\},$$

of volume $2^d d^d / d\,!$. Thus L is also admissible for O. The fundamental theorem then shows that $d(L) \geq d^d / d\,!$. Noting that $D = d(L)^2$, the proof is complete. □

23 Successive Minima

Successive minima of star bodies or convex bodies with respect to lattices were first defined and investigated by Minkowski in the context of the geometry of numbers. Minkowski put them to use in algebraic number theory. A hundred years later, successive minima still play a role in the geometry of numbers and in algebraic and transcendental number theory. See, e.g. Bertrand [102], Chen [204] and Matveev [697], but they are also important in Diophantine Approximation, see, e.g. Schmidt [896], and in computational geometry, compare Lagarias, Lenstra and Schnorr [626], Schnorr [913] and Blömer [134]. There is a surprising link to Nevanlinna's value distribution theory, a branch of complex analysis, see Wong [1028] and Hyuga [534]. Relations to lattice polytopes and roots of Ehrhart polynomials were studied by Stanley, Henk, Schürmann and Wills, see the references in Sect. 19.1 and the survey of Henk and Wills [493].

In this section, we first present Minkowski's theorem on successive minima and prove a result of Mahler relating successive minima of a convex body with respect to a lattice and successive minima of the polar body with respect to the polar lattice. Then Jarník's transference theorem is proved. It connects lattice packing and lattice covering of a given convex body. Finally, we give a result of Perron and Khintchine on Diophantine approximation.

23.1 Successive Minima and Minkowski's Second Fundamental Theorem

There are many extensions and refinements of the first fundamental theorem, including Blichfeldt's [130] theorem and its relatives. See, e.g. Gruber and Lekkerkerker [447] and Lagarias [625]. A particularly refined result is the theorem on successive minima or second fundamental theorem of Minkowski [735].

Henk, Schürmann and Wills [492] discovered an interesting connection between the successive minima of an o-symmetric lattice polytope with respect to the integer lattice $\mathbb{Z}^d$ and the roots of the corresponding Ehrhart polynomial.

This section contains Minkowski's proof of the second fundamental theorem, streamlined by Henk, and Mahler's theorem relating successive minima and polarity. In addition, we state the result of Henk, Schürmann and Wills.

Successive Minima

Let C be an o-symmetric convex body and L a lattice in $\mathbb{E}^d$. The *successive minima* $\lambda_i = \lambda_i(C, L),\ i = 1, \dots, d$, of C with respect to L are defined by:

$$\lambda_i = \min\big\{\lambda > 0 : \lambda C \text{ contains } i \text{ linearly independent points of } L\big\}.$$

Clearly,

$$(1)\quad 0 < \lambda_1 \le \lambda_2 \le \cdots \le \lambda_d < +\infty.$$

The Second Fundamental Theorem

It is easy to see that Minkowski's first fundamental theorem is equivalent to the inequality

$$V(\lambda_1 C) \le 2^d d(L) \text{ or } \lambda_1^d V(C) \le 2^d d(L).$$

Thus the following *second fundamental theorem* or *theorem on successive minima of Minkowski* is a refinement of the first fundamental theorem.

Theorem 23.1. *Let C be an o-symmetric convex body and L a lattice in $\mathbb{E}^d$. Let $\lambda_i = \lambda_i(C, L), i = 1, \dots, d$, be the successive minima of C with respect to L. Then*

$$\frac{2^d}{d\,!}\, d(L) \le \lambda_1 \cdots \lambda_d V(C) \le 2^d d(L).$$

The difficult part of the proof is to show the right-hand inequality. For a long time the original proof of Minkowski [745] was considered to be rather obscure. Alternative proofs were given by Bambah, Woods and Zassenhaus [63] and others. It was a great surprise when a careful scrutiny of Minkowski's proof by Henk [491] revealed that, after eliminating the proof of a Fubini-type result from Minkowski's proof and streamlining the rest, the proof was perfectly correct and elegant. It is Henk's version of Minkowski's proof which is reproduced below.

Proof. It is sufficient to consider the case where $L = \mathbb{Z}^d$.

Right-hand inequality: Let

$$C_i = \frac{\lambda_i}{2}\, C \text{ for } i = 1, \dots, d.$$

It follows from the definition of successive minima that there are d linearly independent points $l_1, \dots, l_d \in \mathbb{Z}^d$, such that

$$l_i \in \lambda_i C \text{ for } i = 1, \dots, d.$$

After assigning to $\{l_1, \dots, l_d\}$ a basis of $\mathbb{Z}^d$ by Theorem 21.3 (i) and then transforming this basis into the standard basis of $\mathbb{Z}^d$ by an integer unimodular $d \times d$ matrix according to Theorem 21.1, we may assume that

$$l_i \in E_i = \{x = (x_1, \dots, x_i, 0, \dots, 0) \in \mathbb{E}^d\} \text{ for } i = 1, \dots, d.$$

Next, let

$$L_{dk} = \{l = (u_1, \dots, u_d) \in \mathbb{Z}^d : |u_i| \le k\},\ L_{ik} = E_i \cap L_{dk} \text{ for } k = 1, 2, \dots$$

Since C is bounded, there is a constant $\alpha > 0$, which depends only on C, such that

(2) $V(C_d + L_{dk}) \le (2k + 1 + \alpha)^d.$

By the definition of λ_1 we have, for $C_1 = \frac{\lambda_1}{2} C$, $(\operatorname{int} C_1 + l) \cap (\operatorname{int} C_1 + m) = \emptyset$ for $l, m \in \mathbb{Z}^d, l \neq m$. Thus

(3) $V(C_1 + L_{dk}) = (2k+1)^d\, V(C_1) = (2k+1)^d\, \dfrac{\lambda_1^d}{2^d}\, V(C).$

The main step of the proof is to show the following estimate:

(4) $V(C_{i+1} + L_{dk}) \ge \left(\dfrac{\lambda_{i+1}}{\lambda_i}\right)^{d-i} V(C_i + L_{dk}) \text{ for } i = 1, \dots, d-1.$

If $\lambda_{i+1} = \lambda_i$, the inequality (4) is trivial. We may thus assume that $\lambda_{i+1} > \lambda_i$. Then the following statement holds:

(5) $(\operatorname{int} C_{i+1} + l) \cap (\operatorname{int} C_{i+1} + m) = \emptyset$
for $l = (u_1, \dots, u_d),\ m = (v_1, \dots, v_d) \in \mathbb{Z}^d$,
where $(u_{i+1}, \dots, u_d) \neq (v_{i+1}, \dots, v_d)$.

Otherwise the $i+1$ linearly independent lattice points $l_1, \dots, l_i,\ l - m$ would be contained in the interior of $\lambda_{i+1} C = C_{i+1} - C_{i+1}$, in contradiction to the definition of λ_{i+1}. Proposition (5) implies that

$$(6)\quad \begin{aligned} V(C_i + L_{dk}) &= (2k+1)^{d-i}\, V(C_i + L_{ik}),\\ V(C_{i+1} + L_{dk}) &= (2k+1)^{d-i}\, V(C_{i+1} + L_{ik}). \end{aligned}$$

Let

$$E_i^{\perp} = \{x = (0, \dots, 0, x_{i+1}, \dots, x_d) \in \mathbb{E}^d\}$$

and define linear maps $f,\ g : \mathbb{E}^d \to \mathbb{E}^d$ by:

$$f(x) = \Big(\frac{\lambda_{i+1}}{\lambda_i}x_1, \dots, \frac{\lambda_{i+1}}{\lambda_i}x_i, x_{i+1}, \dots, x_d\Big) \text{ for } x \in \mathbb{E}^d,$$
$$g(x) = \Big(x_1, \dots, x_i, \frac{\lambda_{i+1}}{\lambda_i}x_{i+1}, \dots, \frac{\lambda_{i+1}}{\lambda_i}x_d\Big) \text{ for } x \in \mathbb{E}^d.$$

For every $x \in E_i^{\perp}$, there is a point $y \in E_i$ with $C_i \cap (x+E_i) \subseteq f(C_i) \cap (x+E_i) + y$, and so

$$\begin{aligned} V(C_i + L_{ik}) &= \int\limits_{E_i^{\perp}} v\big((C_i + L_{ik}) \cap (x + E_i)\big)\, dx \\ &\le \int\limits_{E_i^{\perp}} v\big((f(C_i) + L_{ik}) \cap (x + E_i)\big)\, dx = V\big(f(C_i) + L_{ik}\big), \end{aligned}$$

by Fubini's theorem, where $v(\cdot)$ stands for i-dimensional volume. Since $g\big(f(C_i)\big) + L_{ik} = C_{i+1} + L_{ik}$, we conclude that

$$\begin{aligned} V(C_{i+1} + L_{ik}) &= V\big(g\big(f(C_i)\big) + L_{ik}\big) \\ &= \Big(\frac{\lambda_{i+1}}{\lambda_i}\Big)^{d-i} V\big(f(C_i) + L_{ik}\big) \ge \Big(\frac{\lambda_{i+1}}{\lambda_i}\Big)^{d-i} V(C_i + L_{ik}). \end{aligned}$$

Now, multiply both sides of this inequality by $(2k+1)^{d-i}$ and use (6) to get the estimate (4).

Finally, (3), (4) and (2) together imply the following:

$$\begin{aligned} &(2k+1)^d \Big(\frac{\lambda_1}{2}\Big)^d V(C) \\ &= V(C_1 + L_{dk}) \\ &\le \Big(\frac{\lambda_1}{\lambda_2}\Big)^{d-1} V(C_2 + L_{dk}) \le \cdots \le \Big(\frac{\lambda_1}{\lambda_2}\Big)^{d-1} \cdots \Big(\frac{\lambda_{d-1}}{\lambda_d}\Big)^1 V(C_d + L_{dk}) \\ &\le \frac{\lambda_1^d (2k+1+\alpha)^d}{\lambda_1 \cdots \lambda_d} \text{ for } k = 1, 2, \dots \end{aligned}$$

This yields the right-hand inequality.

Left-hand inequality: Again, consider d linearly independent points $l_1, \dots, l_d \in \mathbb{Z}^d$ such that

$$l_i \in \lambda_i C \text{ or } \frac{l_i}{\lambda_i} \in C \text{ for } i = 1, \dots, d.$$

Since C is o-symmetric and convex, it contains the cross-polytope

$$O = \operatorname{conv}\Big\{\pm\frac{l_1}{\lambda_1}, \dots, \pm\frac{l_d}{\lambda_d}\Big\}.$$

Thus

$$V(C) \geq V(O) = \frac{2^d}{d\,!}\left|\det\left(\frac{l_1}{\lambda_1},\dots,\frac{l_d}{\lambda_d}\right)\right|$$
$$= \frac{2^d|\det(l_1,\dots,l_d)|}{d\,!\,\lambda_1\cdots\lambda_d} \geq \frac{2^d}{d\,!\,\lambda_1\cdots\lambda_d}. \qquad \square$$

Polar Lattices and Polar Bodies

Given a convex body C in $\mathbb{E}^d$ with $o \in \operatorname{int} C$, its *polar body* C^* is defined by:

$$C^* = \{y : x \cdot y \leq 1 \text{ for all } x \in C\},$$

compare Sect. 9.1. The following theorem is due to Mahler [679]:

Theorem 23.2. *Let C be an o-symmetric convex body and L a lattice in $\mathbb{E}^d$ and let $\lambda_i = \lambda_i(C, L)$, $\lambda_j^* = \lambda_j(C^*, L^*)$ for $i, j = 1, \dots, d$. Then*

$$(7)\quad 1 \leq \lambda_{d-k+1}\lambda_k^* \leq \frac{4^d}{V(C)V(C^*)} \leq (d\,!)^2 \ \textit{for}\ k = 1,\dots,d.$$

Proof. We need the following inequality of Mahler, see Theorem 9.6:

$$(8)\quad V(C)V(C^*) \geq \frac{4^d}{(d\,!)^2}.$$

The definition of polar lattice implies that

$$(9)\quad d(L)\,d(L^*) = 1.$$

For the proof of the left-hand inequality, choose linearly independent points $l_1, \dots, l_d \in L$ and $m_1, \dots, m_d \in L^*$ such that

$$l_i \in \lambda_i \operatorname{bd} C,\ m_j \in \lambda_j^* \operatorname{bd} C^* \text{ for } i, j = 1, \dots, d.$$

Then $\pm\frac{1}{\lambda_i} l_i \in \operatorname{bd} C$ and $\pm\frac{1}{\lambda_j^*} m_j \in \operatorname{bd} C^*$ and the definition of C^* implies that

$$\pm\frac{l_i}{\lambda_i} \cdot \frac{m_j}{\lambda_j^*} \leq 1 \text{ or } \lambda_i\lambda_j^* \geq \pm l_i \cdot m_j.$$

Taking into account the definition of L^*, it follows that

$$(10)\quad \lambda_i\lambda_j^* \geq 1 \text{ or } l_i \cdot m_j = 0 \text{ for } i, j = 1, \dots, d.$$

Let $k \in \{1, \dots, d\}$. Since $m_1, \dots, m_k$ are linearly independent, the set $\{x : x \cdot m_1 = \dots = x \cdot m_k = 0\}$ is a subspace of $\mathbb{E}^d$ of dimension $d - k$. Thus, at least one of the $d-k+1$ linearly independent points $l_1, \dots, l_{d-k+1}$ is not contained in this subspace.

Hence $l_i \cdot m_j \neq 0$ for suitable i, j, where $1 \leq i \leq d-k+1$, and $1 \leq j \leq k$. Then, Propositions (1) and (10) yield the left inequality:

$$\lambda_{d-k+1}\lambda_k^* \geq \lambda_i\lambda_j^* \geq 1.$$

We now prove the right-hand inequality. The second fundamental theorem shows that

$$\lambda_1 \cdots \lambda_d V(C) \leq 2^d d(L),\ \lambda_1^* \cdots \lambda_d^* V(C^*) \leq 2^d d(L^*).$$

Thus

$$(\lambda_1\lambda_d^*) \cdots (\lambda_{d-k+1}\lambda_k^*) \cdots (\lambda_d\lambda_1^*) V(C) V(C^*) \leq 4^d d(L) d(L^*).$$

Combining this and the inequalities

$$1 \leq \lambda_1\lambda_d^*, \dots, \lambda_{d-k+1}\lambda_k^*, \dots, \lambda_d\lambda_1^*,$$

which follow from the left inequality in (7), Propositions (8) and (9), we obtain the right inequality. □

Remark. The right inequality has been refined substantially. The case where $C = B^d$ has attracted particular attention. For a discussion and references, see Gruber [430].

A Relation Between Successive Minima and the Roots of the Ehrhart Polynomial

Given a proper lattice polytope $P \in \mathcal{P}_{\mathbb{Z}^d}$, Ehrhart's polynomiality theorem for the lattice point enumerator L shows that

$$L(nP) = \#(nP \cap \mathbb{Z}^d) = p_P(n),\ n \in \mathbb{N},$$

where p_P is a polynomial of degree d, called the Ehrhart polynomial of P. See Sect. 19.1. The result of Henk, Schürmann and Wills is as follows.

Theorem 23.3. *Let $P \in \mathcal{P}_{\mathbb{Z}^d}$ be a proper, o-symmetric lattice polytope in $\mathbb{E}^d$. Let $\lambda_i = \lambda_i(P, \mathbb{Z}^d)$ be its successive minima with respect to the integer lattice $\mathbb{Z}^d$ and let $-\gamma_i = -\gamma_i(P, \mathbb{Z}^d)$ be the roots of its Ehrhart polynomial. Then*

$$\gamma_1 + \cdots + \gamma_d \leq \frac{1}{2}(\lambda_1 + \cdots + \lambda_d).$$

Equality is attained for the cube $P = \{x : -1 \leq x_i \leq 1\}$.

23.2 Jarník's Transference Theorem and a Theorem of Perron and Khintchine

Let C be an o-symmetric convex body and L a lattice in $\mathbb{E}^d$. The family $\{C + l : l \in L\}$ of translates of C by the vectors of L is called a *set lattice*. If any two distinct translates have disjoint interiors, the set lattice is a *lattice packing* of C with *packing lattice* L. If the translates cover $\mathbb{E}^d$, the set lattice is a *lattice covering* of C with *covering lattice* L.

Given C and L, the problem arises to determine the numbers

$$\begin{aligned}\varrho(C,L)&=\max\bigl\{\varrho>0:\{\varrho C+l:l\in L\}\text{ is a lattice packing}\bigr\},\\ \mu(C,L)&=\min\bigl\{\mu>0:\{\mu C+l:l\in L\}\text{ is a lattice covering}\bigr\},\end{aligned}$$

the *packing radius* and the *covering radius* of C with respect to L. From a more arithmetic viewpoint, $2\varrho(C,L)=\lambda_1(C,L)$ and $\mu(C,L)$ are called the *homogeneous* and the *inhomogeneous minimum* of C with respect to L.

There are several inequalities between the quantities $\varrho(C,L)$, $\mu(C,L)$, $\lambda_1(C,L)$, $\dots$, $\lambda_d(C,L)$, respectively, called transference theorems since they transfer information from one situation to another situation, for example from packing to covering and vice versa. A first result of this nature is the following transference theorem of Jarník [543]. For a different transference theorem due to Kneser [601], see Theorem 26.2.

A basic problem in Diophantine approximation is the simultaneous approximation of real numbers by rationals. Equally important is the related problem of approximation of linear forms. Compare Corollaries 22.3 and 22.4 to get an idea of such results.

In the following we present Jarník's transference theorem and a deep result of Perron and Khintchine, which relates the approximation of d real numbers by rationals with common denominator and the approximation of the linear form with these reals as coefficients.

Jarník's Transference Theorem

Our aim is to show the following estimates:

Theorem 23.4. *Let C be an o-symmetric convex body and L a lattice in $\mathbb{E}^d$. Further, let $\lambda_i=\lambda_i(C,L)$ for $i=1,\dots,d$ and $\mu=\mu(C,L)$. Then*

$$\frac{1}{2}\,\lambda_d\le\mu\le\frac{1}{2}\,(\lambda_1+\dots+\lambda_d).$$

Proof. Left-hand inequality: Let $\{b_1,\dots,b_d\}$ be a basis of L. The definition of μ shows that there are vectors $l_i\in L$ such that $\frac{1}{2}\,b_i-l_i\in\mu\,C$ or $b_i-2l_i\in 2\mu C$ for $i=1,\dots,d$. The vectors $b_1-2l_1,\dots,b_d-2l_d$ are linearly independent. (For otherwise, there is a linear combination of these vectors with integer coefficients not all 0 which is equal to o. Hence there are integers $u_1,\dots,u_d$, with greatest common divisor 1, such that we have the equality $\sum u_ib_i=2\sum u_il_i$. Since $b_1,\dots,b_d$ form a basis, the point $\sum u_ib_i$ is primitive, and this equality cannot hold.) The definition of λ_d then shows that $\lambda_d\le 2\mu$, concluding the proof of the left-hand inequality.

Right-hand inequality: It is sufficient to show that

$$(1)\quad\Bigl\{\frac{1}{2}\,(\lambda_1+\dots+\lambda_d)C+l:l\in L\Bigr\}\text{ is a lattice covering.}$$

For this, it is sufficient to prove the following implication:

(2) Let $x \in \mathbb{E}^d$. Then $x \in \frac{1}{2}(\lambda_1 + \cdots + \lambda_d)C + l$ for suitable $l \in L$.

To see this, choose d linearly independent points $l_1, \dots, l_d \in L$ such that $l_i \in \lambda_i C$. Represent x in the form $x = x_1 l_1 + \cdots + x_d l_d$ with $x_i \in \mathbb{R}$. Choose $u_1, \dots, u_d \in \mathbb{Z}$ such that $|x_i - u_i| \le \frac{1}{2}$ and let $l = u_1 l_1 + \cdots + u_d l_d$. Then

$$\begin{aligned} x - l = (x_1 - u_1) l_1 + \cdots + (x_d - u_d) l_d &\in \frac{\lambda_1}{2} C + \cdots + \frac{\lambda_d}{2} C \\ &= \frac{1}{2}(\lambda_1 + \cdots + \lambda_d) C. \end{aligned}$$

The proof of (2), and thus of (1), is complete, concluding the proof of the right-hand inequality. □

A Result of Perron and Khintchine on Diophantine Approximation

We first state a result due to Mahler [1939].

Lemma 23.1. *Let*

$$\begin{aligned} P &= \big\{x : |a_{i1}x_1 + \cdots + a_{id}x_d| \le 1,\ i = 1, \dots, d\big\}, \\ Q &= \big\{x : |b_{i1}x_1 + \cdots + b_{id}x_d| \le 1,\ i = 1, \dots, d\big\} \end{aligned}$$

be two parallelotopes in $\mathbb{E}^d$ *such that* $A^* = (a_{ik})^{-T} = (b_{ik}) = B$. *Let* $\lambda_P = \lambda_1(P, \mathbb{Z}^d)$ *and* $\lambda_Q = \lambda_1(Q, \mathbb{Z}^d)$. *Then*

$$\begin{aligned} \lambda_P &\le \big(d\, \lambda_Q |\det A|\big)^{\frac{1}{d-1}}, \\ \lambda_Q &\le \big(d\, \lambda_P |\det B|\big)^{\frac{1}{d-1}}. \end{aligned}$$

Proof. To see that

$$P^* = \big\{y : |b_{11}y_1 + \cdots + b_{1d}y_d| + \cdots + |b_{d1}y_1 + \cdots + b_{dd}y_d| \le 1\big\},$$

note that

$$\{x : |x_i| \le 1\}^* = \{y : x \cdot y \le 1 \text{ for all } x \text{ with } |x_i| \le 1\} = \{y : |y_1| + \cdots + |y_d| \le 1\}$$

and, for an o-symmetric convex body C,

$$\begin{aligned} (A^{-1}C)^* &= \{z : z \cdot A^{-1}x \le \text{ for all } x \in C\} \\ &= \{A^T A^{-T} z : z^T A^{-1} x = (A^{-T}z)^T x = A^{-T} z \cdot x \le 1 \text{ for all } x \in C\} \\ &= \{A^T y : y \cdot x \le 1 \text{ for all } x \in C\} = A^T C^*. \end{aligned}$$

Thus,

$$\begin{aligned} P^* &= \big(A^{-1}\{x : |x_i| \le 1\}\big)^* = A^T\{x : |x_i| \le 1\}^* \\ &= B^{-1}\{y : |y_1| + \cdots + |y_d| \le 1\} \\ &= \{z : |b_{11}z_1 + \cdots + b_{1d}z_d| + \cdots + |b_{d1}z_1 + \cdots + b_{dd}z_d|\}, \end{aligned}$$

as required. This representation of P^* immediately yields the inclusions

$$\frac{1}{d}\,Q \subseteq P^* \subseteq Q$$

which, in turn, show that

(3) $\lambda_Q \le \lambda_1(P^*, \mathbb{Z}^d) \le d\,\lambda_Q.$

The theorem on successive minima implies the following:

$$\lambda_P^{d-1} \le \lambda_1(P, \mathbb{Z}^d) \cdots \lambda_{d-1}(P, \mathbb{Z}^d) \le \frac{2^d}{\lambda_d(P, \mathbb{Z}^d) V(P)} = \frac{|\det A|}{\lambda_d(P, \mathbb{Z}^d)}.$$

This, together with Theorem 23.2 and Proposition (3), shows that

$$\lambda_P^{d-1} \le \lambda_1(P^*, \mathbb{Z}^d) |\det A| \le d\,\lambda_Q |\det A|.$$

This yields the first inequality. The second follows by symmetry. □

The following result goes back to Perron [793] and Khintchine [581]. It shows that the results on simultaneous Diophantine approximation and on approximation of linear forms in Sect. 22.2 are closely related.

Theorem 23.5. *Let $\vartheta_1, \dots, \vartheta_d \in \mathbb{R}$. Then the following propositions are equivalent:*

(i) *There is a constant $\alpha > 0$ such that the following system of inequalities has no integer solution $(u_0, \dots, u_d)$ where $u_0 \ne 0$:*

$$\left|\vartheta_1 - \frac{u_1}{u_0}\right| \le \frac{\alpha}{u_0^{1+\frac{1}{d}}}, \dots, \left|\vartheta_d - \frac{u_d}{u_0}\right| \le \frac{\alpha}{u_0^{1+\frac{1}{d}}}.$$

(ii) *There is a constant $\beta > 0$ such that the following inequality has no integer solution $(u_0, \dots, u_d)$ where $(u_1, \dots, u_d) \ne o$:*

$$|u_1\vartheta_1 + \dots + u_d\vartheta_d - u_0| \le \frac{\beta}{\max\{|u_1|, \dots, |u_d|\}^d}.$$

The following interpretation may help in understanding the meaning of this result: $\vartheta_1, \dots, \vartheta_d$ cannot be simultaneously approximated well by rationals with the same denominator if and only if the linear form $u_1\vartheta_1 + \dots + u_d\vartheta_d$, for integers $u_1, \dots, u_d$ not all 0, cannot be approximated well by integers.

Proof. (i)⇒(ii) If (i) holds, then, for each $\tau > \alpha^d$, the following system of inequalities has only the trivial integer solution:

$$|u_0\vartheta_1 - u_1| \le \frac{\alpha}{\tau^{\frac{1}{d}}}, \dots, |u_0\vartheta_d - u_d| \le \frac{\alpha}{\tau^{\frac{1}{d}}}, |u_0| \le \tau.$$

(If there were a non-trivial integer solution, then $u_0 \neq 0$, and we obtain a contradiction to (i).) Thus the parallelotope

$$P = \Big\{x \in \mathbb{E}^{d+1} : \Big|\frac{\tau^{\frac{1}{d}}}{\alpha}(x_1 - \vartheta_1 x_0)\Big| \le 1, \dots, \Big|\frac{\tau^{\frac{1}{d}}}{\alpha}(x_d - \vartheta_d x_0)\Big| \le 1, \Big|\frac{1}{\tau}x_0\Big| \le 1\Big\}$$

in $\mathbb{E}^{d+1}$ contains only the point o of $\mathbb{Z}^{d+1}$, and therefore,

(4) $\lambda_P = \lambda_1(P, \mathbb{Z}^{d+1}) > 1.$

Let A be the coefficient matrix of the linear forms which determine P. Then

$$A^{-T} = \begin{pmatrix} \frac{1}{\tau} & 0 & \dots & 0 \\ -\frac{\tau^{\frac{1}{d}}\vartheta_1}{\alpha} & \frac{\tau^{\frac{1}{d}}}{\alpha} & \dots & 0 \\ \dots & \dots & \dots & \dots \\ -\frac{\tau^{\frac{1}{d}}\vartheta_d}{\alpha} & 0 & \dots & \frac{\tau^{\frac{1}{d}}}{\alpha} \end{pmatrix}^{-T} = \begin{pmatrix} \tau & \tau\vartheta_1 & \dots & \tau\vartheta_d \\ 0 & \frac{\alpha}{\tau^{\frac{1}{d}}} & \dots & 0 \\ \dots & \dots & \dots & \dots \\ 0 & 0 & \dots & \frac{\alpha}{\tau^{\frac{1}{d}}} \end{pmatrix}.$$

Consider the parallelotope

$$Q = \Big\{x \in \mathbb{E}^{d+1} : \Big|\frac{\alpha}{\tau^{\frac{1}{d}}}x_1\Big| \le 1, \dots, \Big|\frac{\alpha}{\tau^{\frac{1}{d}}}x_d\Big| \le 1,\ \tau|\vartheta_1 x_1 + \dots + \vartheta_d x_d - x_0| \le 1\Big\}$$

$$= \Big\{x \in \mathbb{E}^{d+1} : |x_1| \le \frac{\tau^{\frac{1}{d}}}{\alpha}, \dots, |x_d| \le \frac{\tau^{\frac{1}{d}}}{\alpha}, |\vartheta_1 x_1 + \dots + \vartheta_d x_d - x_0| \le \frac{1}{\tau}\Big\}.$$

Note that A^{-T} is the coefficient matrix of the linear forms which determine Q. Let $\lambda_Q = \lambda_1(Q, \mathbb{Z}^{d+1})$. An application of (4) and Lemma 23.1 then shows that

$$1 < \lambda_P \le \big((d+1)\lambda_Q|\det A|\big)^{\frac{1}{d}}, \text{ or } \lambda_Q \ge \frac{\lambda_P^d}{(d+1)|\det A|} > \frac{\alpha^d}{d+1}.$$

The latter implies that

$$\frac{\alpha^d}{d+1} Q \cap \mathbb{Z}^{d+1} = \{o\}$$

which, in turn, shows that the system of inequalities

$$|x_1| \le \frac{\alpha^{d-1}\tau^{\frac{1}{d}}}{d+1}, \dots, |x_d| \le \frac{\alpha^{d-1}\tau^{\frac{1}{d}}}{d+1}, |\vartheta_1 x_1 + \dots + \vartheta_d x_d - x_0| \le \frac{\alpha^d}{\tau(d+1)}$$

has no non-trivial integer solution and thus, a fortiori, no integer solution $(u_0, \dots, u_d)$ where $(u_1, \dots, u_d) \neq o$. Put $\sigma = \alpha^{d-1}\tau^{\frac{1}{d}}/(d+1)$ and $\beta = \alpha^{d^2}/(d+1)^{d+1}$. Then the system of inequalities

$$|x_1| \le \sigma, \dots, |x_d| \le \sigma, |\vartheta_1 x_1 + \dots + \vartheta_d x_d - x_0| \le \frac{\beta}{\sigma^d}$$

has no integer solution $(u_0, \dots, u_d)$ where $(u_1, \dots, u_d) \neq o$. This implies (ii).

(ii)⇒(i) This implication is shown similarly. □

24 The Minkowski–Hlawka Theorem

Given an o-symmetric convex body C, Minkowski's fundamental theorem says that any lattice L with sufficiently small determinant contains a point of C other than o. A theorem of Hlawka [509] which verifies a conjecture of Minkowski that the latter stated in slightly different forms at various places, for example in [733], yields the following counterpart of this statement: There is a lattice, the determinant of which is not too large, which contains no point of C other than o.

The Minkowski–Hlawka theorem has attracted interest ever since it was proved by Hlawka in 1944. In the first decades, emphasis was on alternative proofs, refinements and generalizations. We mention Siegel, Rogers, Macbeath, Cassels and Schmidt. More recently, its relations to error correcting codes have been studied by Sloane and Rush amongst others. A Minkowski–Hlawka theorem, in the adelic setting, is due to Thunder [998].

In this section we present a basic version of the Minkowski–Hlawka theorem and state a beautiful generalization of it, the mean value theorem of Siegel. Applications to lattice packing will be given in Theorem 30.4.

For further pertinent results and numerous references, see [447].

24.1 The Minkowski–Hlawka Theorem

In the following we prove a classical version of the Minkowski–Hlawka theorem.

A Version of the Minkowski–Hlawka Theorem

Theorem 24.1. *Let J be a Jordan measurable set in $\mathbb{E}^d$ with $V(J) < 1$. Then there is a lattice L in $\mathbb{E}^d$ with $d(L) = 1$ which contains no point of J, with the possible exception of o.*

We give two short and elegant proofs, one more discrete, the other more analytic. The first proof is due to Rogers [845], the second is a version of a proof of Davenport and Rogers [247] and Cassels [193].

Proof (by Rogers). Let p be a prime. For each point $u \in \mathbb{Z}^d$ with $0 \leq u_1, \dots, u_{d-1} < p$ and $u_d = 1$, let $L(p, u)$ be the lattice with basis

$$\{(p, 0, \dots, 0), (0, p, \dots, 0), \dots, (0, 0, \dots, 0, p, 0), u\}.$$

Then the following hold:

(1) For each $v = (v_1, \dots, v_d) \in \mathbb{Z}^d$ with $v_d \neq 0, \pm p, \pm 2p, \dots,$ there is a unique lattice $L(p, u)$ with $v \in L(p, u)$.

To show this, it is sufficient to show that there are unique integers $u_1, \dots, u_{d-1}$, with $0 \le u_1, \dots, u_{d-1} < p$, such that, for suitable integers $k_1, \dots, k_{d-1}, k$, we have

$$\begin{aligned} v_1 &= ku_1 + k_1 p \\ &\dots\dots \\ v_{d-1} &= ku_{d-1} + k_{d-1} p \\ v_d &= k \end{aligned} \quad \text{or} \quad \begin{aligned} v_1 &\equiv ku_1 \mod p \\ &\dots\dots \\ v_{d-1} &\equiv ku_{d-1} \mod p \\ v_d &= k. \end{aligned}$$

Since $v_d = k$ is not an integer multiple of p, these congruences uniquely determine integers $u_1, \dots, u_{d-1}$ with $0 \le u_1, \dots, u_{d-1} < p$, concluding the proof of (1). An immediate consequence of (1) is as follows:

(2) The set $Y = \{v \in \mathbb{Z}^d : v_d \neq 0, \pm p, \pm 2p, \dots\} \subseteq \mathbb{Z}^d$ is the disjoint union of the p^{d-1} sets $L(p, u) \cap Y : 0 \le u_1, \dots, u_{d-1} < p$.

Since J is Jordan measurable and thus bounded, and since $V(J) < 1$, we can choose a prime p which is so large that

(3) $J \subseteq \{x : |x_i| < p^{\frac{1}{d}}\}$

and

(4) $\dfrac{1}{p^{d-1}} \#\Big(J \cap \dfrac{1}{p^{\frac{d-1}{d}}} Y\Big) \le \dfrac{1}{p^{d-1}} \#\Big(J \cap \dfrac{1}{p^{\frac{d-1}{d}}} \mathbb{Z}^d\Big) < 1.$

For the latter, note that $(1/p^{(d-1)/d})\mathbb{Z}^d$ is a lattice with determinant $1/p^{d-1}$.

Propositions (2) and (4) imply that there is a lattice $L(p, u)$ such that

$$\#\Big(J \cap \frac{1}{p^{\frac{d-1}{d}}} \big(L(p, u) \cap Y\big)\Big) < 1,$$

or

(5) $p^{\frac{d-1}{d}} J \cap L(p, u) \cap Y = \emptyset.$

Since

$$L(p, u) \cap \{x : |x_1|, \dots, |x_{d-1}| < p,\ x_d = 0\} = \{o\},$$

(3) and (5) imply that

$$p^{\frac{d-1}{d}} J \cap L(p, u) \subseteq \{o\} \text{ or } J \cap \frac{1}{p^{\frac{d-1}{d}}} L(p, u) \subseteq \{o\}.$$

Since $(1/p^{(d-1)/d})L(p, u)$ is a lattice of determinant 1, the proof is complete. □

Proof (by Davenport, Rogers and Cassels). By replacing J by its closure, if necessary, we may assume that J is compact. For $\lambda > 0$ consider the hyperplanes $H_n = \{x : x_d = n/\lambda^{d-1}\}$, $n = 0, \pm 1, \dots$, parallel to $\mathbb{E}^{d-1} = H_0$. Let $v(\cdot)$ denote $(d-1)$-dimensional measure. Since J is Jordan measurable and compact, Fubini's theorem for Riemann integrals and the definition of 1-dimensional Riemann integrals show that we may choose λ so large that the following statements hold:

(6) $(V(J)\approx)\ \sum_{n\neq 0} v(H_n\cap J)\frac{1}{\lambda^{d-1}} < 1,$

(7) $J\subseteq\{x : |x_i| < \lambda\}$.

Let L be the lattice $\lambda\mathbb{Z}^{d-1}$ in $\mathbb{E}^{d-1}$. The determinant $d(L)$ of L equals λ^{d-1}. Let $F=\{x\in\mathbb{E}^{d-1} : 0\le x_i < \lambda\}$. Noting that $\mathbb{E}^{d-1}$ is the disjoint union of the sets $F+l$, $l\in L$, that $L-l=L$ for $l\in L$ and that F contains precisely one point of each translate $L+x$ of L, we obtain the following formula:

$$
\begin{aligned}
(8)\quad \int_F \#\big(K\cap(L+x)\big)\,dx &= \sum_{l\in L}\int_F \#\big((F+l)\cap K\cap(L+x)\big)\,dx\\
&= \sum_{l\in L}\int_F \#\big(F\cap(K-l)\cap(L+x)\big)\,dx = \sum_{l\in L} v\big(F\cap(K-l)\big)\\
&= \sum_{l\in L} v\big((F+l)\cap K\big) = v(K) \text{ for any measurable } K\subseteq\mathbb{E}^{d-1}.
\end{aligned}
$$

For the proof of the next formula, dissect nF into n^{d-1} disjoint sets of the form $F+l$, where $l\in L$ and note that $L+l=L$ for $l\in L$. This shows that

$$
\begin{aligned}
(9)\quad \int_F \#\big(K\cap(L+nx)\big)\,dx &= \frac{1}{n^{d-1}}\int_{nF} \#\big(K\cap(L+y)\big)\,dy\\
&= \int_F \#\big(K\cap(L+x)\big)\,dx = v(K)\\
&\text{for any measurable } K\subseteq\mathbb{E}^{d-1} \text{ and } n=\pm1,\dots
\end{aligned}
$$

by (8).

Finally, for $x\in F$, let $L(x)$ be the lattice

$$L+\mathbb{Z}\big(x_1,\dots,x_{d-1},\frac{1}{\lambda^{d-1}}\big) \text{ in } \mathbb{E}^d.$$

Then $d\big(L(x)\big)=d(L)/\lambda^{d-1}=1$. The definitions of H_n, L and $L(x)$, together with (7), show that $H_0\cap J\cap L(x)\subseteq\{o\}$. Then (9) and (6) yield the following, where #* counts points different from o:

$$
\begin{aligned}
\int_F \#^*\big(J\cap L(x)\big)\,dx &= \int_F \sum_{n\neq 0}\#\big(H_n\cap J\cap L(x)\big)\,dx\\
&= \sum_{n\neq 0}\int_F \#\big(J_n\cap(L+nx)\big)\,dx, \text{ where } J_n=H_n\cap J-\big(0,\dots,0,\frac{n}{\lambda^{d-1}}\big),\\
&= \sum_{n\neq 0} v(J_n) = \sum_{n\neq 0} v(H_n\cap J) < \lambda^{d-1} = v(F).
\end{aligned}
$$

The integral of the non-negative function $\#^*\big(J\cap L(x)\big)$ over the set F is thus less than the measure of F. Therefore, there is a point $x\in F$ such that $\#^*\big(J\cap L(x)\big) < 1$. $L(x)$ is then the desired lattice. □

Hlawka [514] described the idea of his proof of the Minkowski–Hlawka theorem as follows:

> This theorem requires one to pick out from the infinite set of lattices, lattices which have a particular property. Since it is difficult to find such lattices by chance, it is plausible to make the following comparison: consider the problem of catching fish of given length from a pond. Making one haul, one may catch such a fish only by chance. For this reason it makes sense to catch many fish, hoping that a fish of the desired length is among them. In probability theory this is called a random sample. For this sample one considers the mean value, in our case the mean value of the length, to get information on the sample. This was the idea which I applied in 1942 ...

The two proofs given earlier – as all other known proofs, including Hlawka's original proof – are based on mean value arguments dealing with huge sets of lattices. This fact prevents the effective construction of lattices as specified in the theorem. In particular, this explains why, so far, there is no effective algorithm available to construct lattice packings of o-symmetric convex bodies of density at least 2^{-d}, although such packings exist by the Minkowski–Hlawka theorem, compare Theorem 30.4. For balls, the situation is slightly better but by no means satisfactory. There are constructions of rather dense lattice packings of balls using codes. In many cases the codes can be given effectively. Unfortunately the codes used by Rush [862] to reach the Minkowski–Hlawka bound $2^{-d+o(d)}$ are not of this type. See Sect. 29.3.

Refinements

There are several refinements of the Minkowski–Hlawka theorem, see Gruber and Lekkerkerker [447]. The best known estimate is due to Schmidt [893]: there is an absolute constant $\alpha > 0$ such that, for each Borel set B in $\mathbb{E}^d$ with $V(B) \leq d \log \sqrt{2} - \alpha$, there is a lattice in $\mathbb{E}^d$ of determinant 1 which contains no point of B except, possibly, o. At present, it is the belief of many people working in the geometry of numbers that, in essence, the theorem of Minkowski–Hlawka cannot be refined. In the past, this was not always so, but Edmund Hlawka [517] told the author that he was always convinced that no essential refinement was possible.

24.2 Siegel's Mean Value Theorem and the Variance Theorem of Rogers–Schmidt

A natural question to ask in the geometry of numbers is the following: Given a property which a lattice may or may not have, is the set of lattices which have this property, large or small? Tools which sometimes help to give an answer are measure and Baire categories. Measure has turned out to be a versatile tool which applies to many such questions, ever since Siegel [936] defined and put to use a natural measure on the space of all lattices of determinant 1. Later contributions are due to Rogers, Macbeath and Schmidt. Because of great technical difficulties, in particular in the work of Schmidt, the development seems to have reached a deadlock. Despite this, we believe that many important measure results in the geometry of numbers still await discovery. Category on the space of lattices has rarely been used, a minor exception is the article of Aliev and the author [23].

In this section we describe the natural measure on the space of lattices of determinant 1 and state an elegant generalization of the Minkowski–Hlawka theorem, Siegel's mean value theorem. Then, an important adjunct result is specified which, perhaps, is best called the variance theorem of Rogers and Schmidt.

The reader who wants to get more information on measure theory on spaces of lattices is referred to the books of Rogers [851], Gruber and Lekkerkerker [447] and Siegel [937].

Measure on the Space of Lattices of Determinant 1

Let $\mathcal{L}(1)$ be the space of all lattices of determinant 1 in $\mathbb{E}^d$. Since, by Mahler's selection theorem 25.1, respectively, its Corollary 25.1, this space is locally compact, it carries many measures. A measure on this space, which has proved extremely useful for the geometry of numbers, can be defined as follows.

Let $\mathcal{SL}(d)$ be the locally compact multiplicative group of all real $d \times d$ matrices with determinant 1, the *special linear group*. We consider the Haar measure on $\mathcal{SL}(d)$. Up to normalization, this measure can be described as follows: Representing a $d \times d$ matrix as a point in $\mathbb{E}^{d^2}$, the space $\mathcal{SL}(d)$ is a surface in $\mathbb{E}^{d^2}$. Given a Borel set $\mathcal{B} \subseteq \mathcal{SL}(d)$, consider the cone with basis $\mathcal{B}$ and apex at the origin, that is the set $\{\lambda B : 0 \le \lambda \le 1, B \in \mathcal{B}\} \subseteq \mathbb{E}^{d^2}$. This set is again Borel and its Lebesgue measure is the Haar measure of $\mathcal{B}$.

Let $\mathcal{U}$ be the sub-group of $\mathcal{SL}(d)$, consisting of all integer unimodular matrices. Given a lattice $L \in \mathcal{L}(1)$, any basis of L with positive determinant can be identified with a matrix $B \in \mathcal{SL}(d)$. The family of all bases of L, with positive determinant, is then the set of all matrices of the form BU where $U \in \mathcal{U}$, that is, a left coset of $\mathcal{U}$. There exists a fundamental domain $\mathcal{F}$ of $\mathcal{SL}(1)$ with respect to $\mathcal{U}$, that is a set which contains precisely one matrix from each coset of $\mathcal{U}$, which is Borel and has positive finite Haar measure. Clearly, the Haar measure can be normalized such that $\mathcal{F}$ has measure 1. Since $\mathcal{F}$ contains precisely one basis of each lattice in $\mathcal{L}(1)$, this gives a measure μ on the space of all lattices of determinant 1 in $\mathbb{E}^d$. This measure was defined by Siegel [937] using reduction theory of positive definite quadratic forms, thus following an idea of Minkowski.

Siegel's Mean Value Theorem

Using this measure, Siegel [936] proved the following elegant result.

Theorem 24.2. *Let $f : \mathbb{E}^d \to \mathbb{R}$ be Riemann integrable. Then*

$$\int\limits_{\mathcal{L}(1)} \sum \big\{f(l) : l \in L \setminus \{o\}\big\}\, d\mu(L) = \int\limits_{\mathbb{E}^d} f(x)\, dx.$$

If f is the characteristic function of a Jordan measurable set J then, in particular,

$$\int\limits_{\mathcal{L}(1)} \#^*(L \cap J)\, d\mu(L) = V(J).$$

This says that the mean value of the number of points $\neq o$ of a lattice L of determinant 1 in J equals $V(J)$. A stronger version of the Minkowski–Hlawka theorem is an immediate consequence: If $V(J) < 1$ then all lattices of $\mathcal{L}(1)$ are disjoint from $J \setminus \{o\}$, except for a set of lattices of measure at most $V(J)$.

We shall not give a proof of Siegel's theorem, but refer the reader to Siegel [936, 937] and Gruber and Lekkerkerker [447].

Heuristic Observations

Comparing the first proof of the Minkowski–Hlawka theorem and Siegel's mean value theorem, we see that the average over a certain set of lattices is the same as the integral over the space of all lattices of determinant 1. It thus seems plausible that the set of lattices employed in the proof of the Minkowski–Hlawka theorem is not only dense in $\mathcal{L}(1)$, as remarked in Sect. 25.1, but even uniformly distributed in the sense of uniform distribution theory, see Hlawka [515]. Compare a pertinent remark by Schmidt [894].

The following is the result of a discussion with Hendrik Lenstra [649]. It has been conjectured that the Minkowski–Hlawka theorem, in essence, is best possible, even for balls. Assume that this is true. Then there is a positive function $\varphi(d)$ where $\varphi(d) \to 0$, $\varphi(d)d \to \infty$ as $d \to \infty$ and such that the following statement holds. Let B be the ball with centre at the origin and volume 1. Then, each lattice $L \in \mathcal{L}(1)$ contains a point different from o in the ball $2^{\varphi(d)}B$ of volume $2^{\varphi(d)d}$. In contrast, Siegel's mean value theorem implies that all lattices in $\mathcal{L}(1)$, up to a set of lattices of measure $2^{-\varphi(d)d} = o(1)$, are admissible for the ball $2^{-\varphi(d)}B$ of volume $2^{-\varphi(d)d} = o(1)$. This may be expressed in terms of $\lambda_1(B, L)$ as follows:

$\lambda_1(2^{\varphi(d)}B, L) \leq 1$ for all $L \in \mathcal{L}(1)$ and
$\lambda_1(2^{-\varphi(d)}B, L) \geq 1$ for all $L \in \mathcal{L}(1)$, up to a set of lattices of measure at most $o(1)$.

Equivalently,

$\lambda_1(B^d, L) \leq V(B^d)^{\frac{1}{d}} 2^{\varphi(d)} = V(B^d)^{\frac{1}{d}}(1 + o(1))$ for all $L \in \mathcal{L}(1)$ and
$\lambda_1(B^d, L) \geq V(B^d)^{\frac{1}{d}} 2^{-\varphi(d)} = V(B^d)^{\frac{1}{d}}(1 - o(1))$ for all $L \in \mathcal{L}(1)$, up to a set of lattices of measure at most $o(1)$.

Hence, assuming that the Minkowski–Hlawka theorem, in essence, is best possible, the maximum value of $\lambda_1(B^d, L)$ for a lattice of determinant 1 is about $V(B^d)^{1/d}$ and, for a majority of lattices of determinant 1, the homogeneous minimum $\lambda_1(B^d, L)$ is close to $V(B^d)^{1/d}$. The following variance theorem shows that, in fact, the latter statement is true. This phenomenon is an instance of the heuristic remark in Sect. 11.2, which says that in many complicated situations, the average configuration attains almost the extremal value. For other examples of this phenomenon, see Sects. 8.6 and 11.2.

The Variance Theorem of Rogers and Schmidt

An interesting complement of Siegel's mean value theorem is the following result of Rogers [847] and Schmidt [892]:

Theorem 24.3. *There is an absolute constant $\alpha > 0$ such that the following statement holds: Let $d \geq 3$ and let B be a Borel set in $\mathbb{E}^d$, then*

$$\int\limits_{\mathcal{L}(1)} \Big(\#^*(L \cap B) - V(B)\Big)^2 \, d\mu(L) < \alpha \, V(B).$$

Remark. For $d = 2$, a slightly weaker result is due to Schmidt [892]. A consequence of these results is the following: Let B be a Borel set in $\mathbb{E}^d$ with $V(B) = \infty$. Then, almost every lattice $L \in \mathcal{L}(1)$ has infinitely many primitive points in common with B. A refinement of the latter result was given by Aliev and the author [23]. It says that, for almost every lattice $L \in \mathcal{L}(1)$, the set B contains infinitely many pairwise disjoint d-tuples of linearly independent primitive points of L.

25 Mahler's Selection Theorem

Several problems in the geometry of numbers amount to the question as to whether there exist lattices with a given extremum property. For example, given a convex body C, do there exist lattice packings and lattice coverings of C with maximum, respectively, minimum density? In several such situations, Mahler's selection theorem yields the existence of extremal lattices.

Related results of a similar character in other areas of mathematics are the Bolzano–Weierstrass theorem from calculus, the Arzelà-Ascoli theorem in analysis and Blaschke's selection theorem 6.3 in convex geometry. As will be seen later, the Blaschke selection theorem can be used to prove Mahler's theorem.

In this section, we define a topology on the space of all lattices in $\mathbb{E}^d$ and present Mahler's selection theorem. For applications to lattice packing and covering, see Theorems 30.1 and 31.1.

25.1 Topology on the Space of Lattices

There is a natural topology on the space of lattices. Endowed with this topology, the space $\mathcal{L}$ of all lattices, and thus also the closed subspace $\mathcal{L}(1)$ of all lattices with determinant 1, in $\mathbb{E}^d$ are locally compact.

In this section, we define this topology on the space $\mathcal{L}$ and state some results due to Rogers, Woods and Schmidt on dense subsets. These results are useful for measure results.

For more information, see Cassels [195] and [447].

Topology on the Space of Lattices

We define the natural topology on the space $\mathcal{L}$ of all lattices in $\mathbb{E}^d$ by specifying a basis. A *basis of the topology* on $\mathcal{L}$ is given by the following sets of lattices:

$$\big\{M \in \mathcal{L} : M \text{ has a basis } \{c_1, \dots, c_d\} \text{ with } \|c_i - b_i\| < \varepsilon,\ i = 1, \dots, d\big\}$$
$$\text{where } L \in \mathcal{L},\ \{b_1, \dots, b_d\} \text{ is a basis of } L, \text{ and } \varepsilon > 0.$$

The corresponding notion of convergence can be described as follows. A sequence (L_n) of lattices in $\mathbb{E}^d$ is *convergent*, if there is a lattice L in $\mathbb{E}^d$ such that, for suitable bases $\{b_{n1}, \dots, b_{nd}\}$ of L_n, $n = 1, 2, \dots$, and $\{b_1, \dots, b_d\}$ of L, respectively, we have

$$b_{n1} \to b_1, \dots, b_{nd} \to b_d \text{ as } n \to \infty.$$

From now on, we assume that $\mathcal{L}$ is endowed with this topology.

It is not difficult to see that this topology is induced by a suitable metric on $\mathcal{L}$. See [447].

Dense Sets of Lattices

Confirming a statement of Rogers [851], Woods [1030] proved that the set of lattices, where the basis vectors are the columns of the following matrices, is dense in the space $\mathcal{L}(1)$:

$$\begin{pmatrix} \alpha & 0 & \dots & 0 & \alpha_1 \\ 0 & \alpha & \dots & 0 & \alpha_2 \\ \dots & \dots & \dots & \dots & \dots \\ 0 & 0 & \dots & \alpha & \alpha_{d-1} \\ 0 & 0 & \dots & 0 & \alpha^{1-d} \end{pmatrix}, \text{ where } \alpha_1, \dots, \alpha_{d-1} \in \mathbb{R} \text{ and } \alpha > 0.$$

This set of lattices can be used to prove the Minkowski–Hlawka theorem, see Sect. 24.1.

A refinement of Wood's result is the following result due to Schmidt [894]. For almost all $(d-1)$-tuples $(\beta_1, \dots, \beta_{d-1}) \in \mathbb{E}^{d-1}$, the set of lattices, where the basis vectors are the columns of the matrices

$$\begin{pmatrix} k & 0 & \dots & 0 & \beta_1 k \\ 0 & k & \dots & 0 & \beta_2 k \\ \dots & \dots & \dots & \dots & \dots \\ 0 & 0 & \dots & k & \beta_{d-1} k \\ 0 & 0 & \dots & 0 & k^{1-d} \end{pmatrix}, \quad k = 1, 2, \dots,$$

is dense in $\mathcal{L}(1)$.

25.2 Mahler's Selection Theorem

Mahler's selection theorem is the main topological result for the space of lattices. It provides a firm basis for several results which before were clear only intuitively.

In the following we present a standard version of it.

The Selection Theorem of Mahler [680]

Theorem 25.1. *Let* (L_n) *be a sequence of lattices in* $\mathbb{E}^d$ *such that, for suitable constants* $\alpha, \varrho > 0$, *the following hold for* $n = 1, 2, \dots$:

(i) $d(L_n) \le \alpha$.

(ii) L_n *is admissible for* ϱB^d.

Then the sequence of lattices contains a convergent subsequence.

We first present a familiar proof of Mahler's theorem and then outline a beautiful geometric proof due to Groemer [401], based on the notion of Dirichlet–Voronoĭ cells and Blaschke's selection theorem.

Proof. In the first step, the following will be shown:

(1) Let $l_1, \dots, l_d$ be d linearly independent points of a lattice in $\mathbb{E}^d$. Then there is a basis $\{b_1, \dots, b_d\}$ of this lattice such that:

$$\|b_i\| \le \|l_1\| + \dots + \|l_i\| \text{ for } i = 1, \dots, d.$$

By Theorem 21.3, there is a basis $\{c_1, \dots, c_d\}$ such that:

$$\begin{aligned} l_1 &= u_{11}c_1 \\ l_2 &= u_{21}c_1 + u_{22}c_2 \\ &\dots\dots\dots\dots\dots\dots \\ l_d &= u_{d1}c_1 + \dots\dots + u_{dd}c_d \end{aligned} \qquad \text{where } u_{ik} \in \mathbb{Z},\ u_{ii} \ne 0.$$

Then

$$\begin{aligned} c_1 &= u_{11}^{-1} l_1, \\ c_2 &= u_{22}^{-1} l_2 + t_{21} l_1 \\ &\dots\dots\dots\dots\dots\dots\dots\dots \\ c_d &= u_{dd}^{-1} l_d + t_{dd-1} l_{d-1} + \dots + t_{d1} l_1 \end{aligned} \qquad \text{where } t_{ik} \in \mathbb{R}.$$

Since $\{c_1, \dots, c_d\}$ is a basis of L, the d vectors

$$\begin{aligned} b_1 &= c_1 & &= c_1 \\ b_2 &= c_2 - \lfloor t_{21} \rfloor l_1 & &= c_2 + v_{21} c_1 \\ &\dots\dots\dots\dots\dots\dots\dots\dots\dots\dots\dots\dots\dots\dots\dots\dots \\ b_d &= c_d - \lfloor t_{d1} \rfloor l_1 - \dots - \lfloor t_{dd-1} \rfloor l_{d-1} & &= c_d + v_{dd-1} c_{d-1} + \dots + v_{d1} c_1 \\ &\quad \text{where } v_{ik} \in \mathbb{Z} \end{aligned}$$

also form a basis and

$$\begin{aligned} \|b_i\| &= \|u_{ii}^{-1} l_i + (t_{ii-1} - \lfloor t_{ii-1} \rfloor) l_{i-1} + \dots + (t_{i1} - \lfloor t_{i1} \rfloor) l_1\| \\ &\le |u_{ii}^{-1}| \|l_i\| + \|l_{i-1}\| + \dots + \|l_1\| \le \|l_i\| + \|l_{i-1}\| + \dots + \|l_1\| \\ &\quad \text{for } i = 1, \dots, d, \end{aligned}$$

concluding the proof of (1).

The second step is to show the following, where α, ϱ are as in (i) and (ii):

(2) Let L be a lattice in $\mathbb{E}^d$ such that $d(L) \le \alpha$ and which is admissible for ϱB^d. Then there is a basis $\{b_1, \dots, b_d\}$ of L with

$$\|b_i\| \le \beta, \text{ where } \beta = \frac{d\, 2^d \alpha}{\varrho^{d-1} V(B^d)},$$
$$d(L) \ge \gamma, \text{ where } \gamma = \frac{\varrho^d V(B^d)}{2^d}.$$

To see this, choose d linearly independent points $l_1, \dots, l_d \in L$ such that:

$$l_i \in \lambda_i \operatorname{bd} B^d \text{ or } \|l_i\| = \lambda_i,$$

where $\lambda_i = \lambda_i(B^d, L)$. The assumptions in (2), together with Minkowski's theorem on successive minima, yield the following:

$$(3)\quad \varrho \le \|l_i\| \le \frac{2^d d(L)}{\|l_1\| \cdots \|l_{i-1}\| \, \|l_{i+1}\| \cdots \|l_d\| V(B^d)} \le \frac{2^d \alpha}{\varrho^{d-1} V(B^d)}.$$

Proposition (1) now shows that there is a basis $\{b_1, \dots, b_d\}$ of L such that:

$$\|b_i\| \le \frac{d\, 2^d \alpha}{\varrho^{d-1} V(B^d)} = \beta.$$

This proves the first assertion in (2). The second assertion also follows from (3), since $\|l_1\|, \dots, \|l_d\| \ge \varrho$ by the assumptions in (2). The proof of (2) is complete.

In the final step of our proof, note that, by the assumptions of the theorem and (2), there are bases $\{b_{n1}, \dots, b_{nd}\}$ of L_n for $n = 1, 2, \dots$, such that the following hold:

$$\varrho \le \|b_{n1}\|, \dots, \|b_{nd}\| \le \beta \text{ and } |\det\{b_{n1}, \dots, b_{nd}\}| = d(L_n) \ge \gamma.$$

Now apply a Bolzano–Weierstrass type argument to the first basis vectors, then to the second basis vectors, etc., to get the selection theorem. □

Proof (outline, following Groemer). Let $\alpha, \varrho > 0$ be as in the theorem. Consider, for $n = 1, 2, \dots$, the *Dirichlet–Voronoĭ cell* $D_n = D(L_n, o)$ of o with respect to L_n,

$$D_n = \big\{x : \|x\| \le \|x - l\| \text{ for all } l \in L_n\big\}.$$

It consists of all points of $\mathbb{E}^d$ which are at least as close to o as to any other point of L_n. Since L_n is admissible for ϱB^d, we have,

$$(4)\quad \frac{\varrho}{2} B^d \subseteq D_n.$$

Since $\{D_n + l : l \in L_n\}$ is a *tiling* of $\mathbb{E}^d$, i.e. both a packing and a covering, we conclude that $V(D_n) = d(L_n) \le \alpha$. Noting that D_n is convex, (4) then implies that

$$(5)\quad D_n \subseteq \beta B^d.$$

where $\beta > 0$ is a suitable constant. Propositions (4) and (5), together with Blaschke's selection theorem 6.3, imply that a suitable subsequence of the sequence (D_n) converges to a convex body D, where

$$\frac{\varrho}{2} B^d \subseteq D \subseteq \beta B^d.$$

It turns out that D is the Dirichlet–Voronoĭ cell of a lattice L. The subsequence of (L_n) which corresponds to the convergent subsequence of the sequence of Dirichlet–Voronoĭ cells, then converges to L. □

An Alternative Version

In some cases the following version of Mahler's selection theorem is useful:

Corollary 25.1. *The space $\mathfrak{L}$ of lattices in $\mathbb{E}^d$ is locally compact.*

26 The Torus Group $\mathbb{E}^d/L$

A lattice is a sub-group of the additive group of $\mathbb{E}^d$. Given a lattice L, a natural object to investigate is the quotient or torus group $\mathbb{E}^d/L$. Since this group is compact and Abelian, there is a Haar measure m defined on it. In essence it is the ordinary Lebesgue measure on a fundamental parallelotope. The question arises, to estimate the measure

$$m(\mathfrak{U}+\mathfrak{V})$$

of the sum $\mathfrak{U}+\mathfrak{V} = \{\mathfrak{u}+\mathfrak{v} : \mathfrak{u} \in \mathfrak{U},\ \mathfrak{v} \in \mathfrak{V}\}$ for measurable sets $\mathfrak{U}, \mathfrak{V}$ in $\mathbb{E}^d/L$ for which $\mathfrak{U}+\mathfrak{V}$ is also measurable. A satisfying answer to this question is the sum theorem of Macbeath and Kneser. It was used by Kneser to prove a strong transference theorem.

In this section, we first study the quotient group $\langle \mathbb{E}^d/L, +\rangle$, then present two proofs of the sum theorem and, finally, use it to show Kneser's transference theorem.

For information on topological groups, in particular for measure theory on topological groups, see Nachbin [760].

26.1 Definitions and Simple Properties of $\mathbb{E}^d/L$

Let L be a lattice in $\mathbb{E}^d$. Then $\mathbb{E}^d/L$ is a group which is endowed with a natural topology and a natural measure. The topology on $\mathbb{E}^d/L$ can be defined with a particular notion of distance. $\mathbb{E}^d/L$ thus carries a considerable lot of structure. Hence it is plausible to expect interesting results.

In this section we define the group $\mathbb{E}^d/L$, a notion of distance which makes it a topological group, and a measure. Both the distance and the measure are inherited from $\mathbb{E}^d$ in a simple way.

The Group $\mathbb{E}^d/L$

With respect to vector addition $+$, the space $\mathbb{E}^d$ is an Abelian group and L a subgroup of it. The *torus* or *quotient group* $\mathbb{E}^d/L$ is the Abelian group consisting of all cosets of L, that is translates $L+x = \{l+x : l \in L\}$ of L, where $x \in \mathbb{E}^d$. For $L+x$ we also write $\mathfrak{x}$. Addition $+$ on $\mathbb{E}^d/L$ is defined by:

$$\mathfrak{x}+\mathfrak{y} = L+x+L+y = L+x+y \text{ for } \mathfrak{x} = L+x,\ \mathfrak{y} = L+y \in \mathbb{E}^d/L.$$

Clearly, addition on $\mathbb{E}^d/L$ is independent of the particular choice of x, y. A coset $\mathfrak{x} = L+x$ is also called an *inhomogeneous lattice*. A *fundamental domain* of $\mathbb{E}^d/L$ is a subset of $\mathbb{E}^d$ which contains precisely one point of each coset. If $\{b_1, \dots, b_d\}$ is a basis of L, then the corresponding fundamental parallelotope

$$F = \{\alpha_1 b_1 + \dots + \alpha_d b_d : 0 \le \alpha_i < 1\}$$

is a fundamental domain. Addition in $\mathbb{E}^d/L$ corresponds to *addition modulo L* in F, symbolized by $+_L$ (Fig. 26.1):

$$x +_L y = (L+x+y) \cap F \text{ for } x, y \in F.$$

Proposition 26.1. *The groups $\langle \mathbb{E}^d/L, + \rangle$ and $\langle F, +_L \rangle$ are isomorphic. An isomorphism is given by the mapping*

$$\mathfrak{x} \to \mathfrak{x} \cap F \text{ for } \mathfrak{x} \in \mathbb{E}^d/L.$$

Proof. Left to the reader. □

Topology on $\mathbb{E}^d/L$

Define a *distance modulo L*, $\|\cdot\|_L$, on $\mathbb{E}^d/L$ by:

$$\|\mathfrak{x}\|_L = \inf\{\|l+x\| : l \in L\} \text{ for } \mathfrak{x} = L+x \in \mathbb{E}^d/L.$$

Since L is discrete, $\mathfrak{x} = L+x$ is also discrete. Hence the infimum is attained for a suitable $l \in L$. The distance $\|\cdot\|_L$ yields a metric and thus induces a topology on $\mathbb{E}^d/L$. We assume, from now on, that $\mathbb{E}^d/L$ is endowed with this topology.

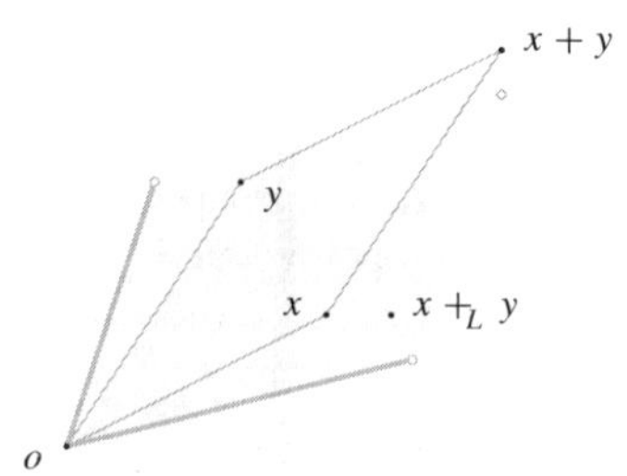

Fig. 26.1. Addition modulo L

Proposition 26.2. *$\mathbb{E}^d/L$ is a compact Abelian topological group.*

Proof. Clearly, $\mathbb{E}^d/L$ is Abelian. For the proof that $\mathbb{E}^d/L$ is a topological group, we have to show that the mapping

(1) $(\mathfrak{x}, \mathfrak{y}) \to \mathfrak{x} - \mathfrak{y}$ for $\mathfrak{x}, \mathfrak{y} \in \mathbb{E}^d/L$ is continuous.

Let $\mathfrak{a}, \mathfrak{b} \in \mathbb{E}^d/L$ and $\varepsilon > 0$. Let $\mathfrak{x}, \mathfrak{y} \in \mathbb{E}^d/L$ be contained in the $\varepsilon/2$-neighbourhoods of $\mathfrak{a}$ and $\mathfrak{b}$, respectively, i.e.

$$\|\mathfrak{x} - \mathfrak{a}\|_L, \|\mathfrak{y} - \mathfrak{b}\|_L < \frac{\varepsilon}{2}.$$

Choose $l, m \in L$ such that:

$$\|\mathfrak{x} - \mathfrak{a}\|_L = \|x - a - l\|, \|\mathfrak{y} - \mathfrak{b}\|_L = \|y - b - m\|.$$

The definition of $\|\cdot\|_L$ then shows that

$$\begin{aligned}&\|\mathfrak{x} - \mathfrak{y} - (\mathfrak{a} - \mathfrak{b})\|_L \le \|x - y - a + b - l + m\| \\ &\quad \le \|x - a - l\| + \|y - b - m\| = \|\mathfrak{x} - \mathfrak{a}\|_L + \|\mathfrak{y} - \mathfrak{b}\|_L < \frac{\varepsilon}{2} + \frac{\varepsilon}{2} = \varepsilon.\end{aligned}$$

Thus $\mathfrak{x} - \mathfrak{y}$ is contained in the ε-neighbourhood of $\mathfrak{a} - \mathfrak{b}$, concluding the proof of (1).

$\mathbb{E}^d/L$ is a metric space. For the proof that it is compact, it is thus sufficient to show that each sequence in $\mathbb{E}^d/L$ has a convergent subsequence. Let $(\mathfrak{x}_n)$ be a sequence in $\mathbb{E}^d/L$. Let F be a fundamental parallelotope. Let $x_n \in F$ be such that $\mathfrak{x}_n \cap F = \{x_n\}$. This gives a sequence (x_n) in the bounded set $F \subseteq \mathbb{E}^d$. The Bolzano–Weierstrass theorem then shows that the sequence (x_n) contains a subsequence (x_{n_k}) converging to a point $x \in \mathbb{E}^d$, say. Let $\mathfrak{x} = L + x$. The definition of $\|\cdot\|_L$ now implies that

$$\|\mathfrak{x}_{n_k} - \mathfrak{x}\|_L \le \|x_{n_k} - x\| \to 0 \text{ as } n \to \infty.$$

Thus $(\mathfrak{x}_{n_k})$ converges to $\mathfrak{x}$. □

Given a fundamental parallelotope F of L, define the *distance modulo* L on F by:

$$\|x\|_L = \inf\left\{\|x + l\| : l \in L\right\} = \|\mathfrak{x}\|_L \text{ for } x \in F.$$

$\|\cdot\|_L$ yields a metric, and thus a topology, on F. Let F be endowed with this topology (Fig. 26.2).

Proposition 26.3. *The metric spaces $\langle \mathbb{E}^d/L, \|\cdot\|_L\rangle$ and $\langle F, \|\cdot\|_L\rangle$ are isometric. An isometry is given by the mapping*

$$\mathfrak{x} \to \mathfrak{x} \cap F \text{ for } x \in \mathbb{E}^d/L.$$

Proof. Left to the reader. □

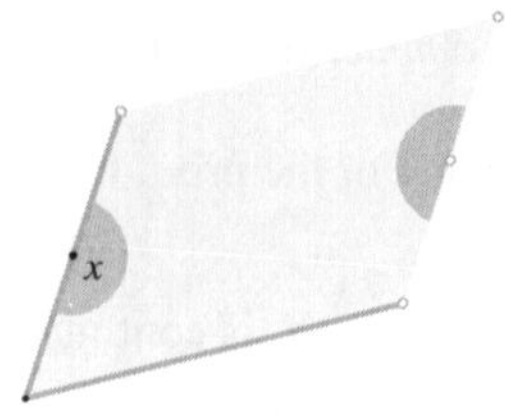

Fig. 26.2. Neighbourhood of a point

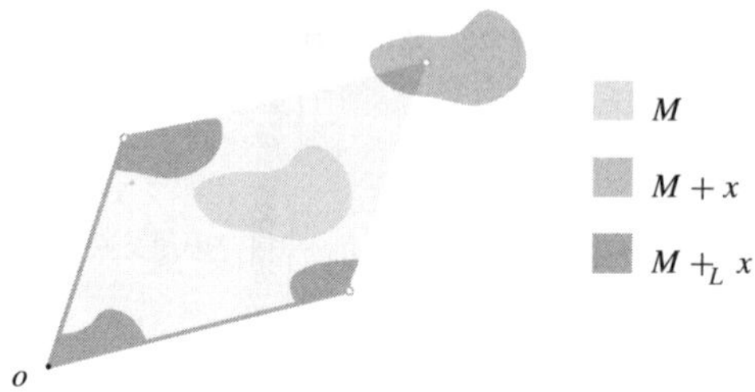

Fig. 26.3. Translation of a set modulo L

Measure on $\mathbb{E}^d/L$

$\mathbb{E}^d/L$ is a compact Abelian group. Hence there is a unique (complete) Haar measure m on it which is normalized such that $m(\mathbb{E}^d/L) = d(L)$.

Let F be a fundamental parallelotope of L. For Lebesgue measurable sets S in F, let $V(S)$ denote the Lebesgue measure of S. This makes F into a measurable space with measure $V(\cdot)$. It is not difficult to see that $V(\cdot)$ is invariant with respect to addition modulo L in F (Fig. 26.3).

The mapping $\mathfrak{x} \to \mathfrak{x} \cap F$ of $\mathbb{E}^d/L$ onto F is both an isomorphism and an isometry. The Haar measures on $\mathbb{E}^d/L$ and on F are both unique and normalized such that $\mathbb{E}^d/L$ and F both have measure $d(L)$. This leads to the following result.

Proposition 26.4. *A set $\mathfrak{S} \subseteq \mathbb{E}^d/L$ is measurable in $\mathbb{E}^d/L$ if and only if the set $S = F \cap \mathfrak{S}$ is measurable in F. In the case of measurability $m(\mathfrak{S}) = V(S)$.*

Proof. Left to the reader. □

26.2 The Sum Theorem of Macbeath–Kneser

Given an additive group and two of its subsets, it is a natural question to investigate properties of their sum. For example, if the group is finite, one may ask for estimates of the number of elements of the sum-set in terms of the numbers of elements of the given sets. Or, if the group is endowed with a measure, one may try to relate the measure of the sum-set to the measures of the given sets. Such problems of $\alpha+\beta$ type were considered in additive number theory, in the theory of congruences, in group theory, see Mann [687, 688] and Danilov [235], and in other areas of mathematics. Minkowski's theorem on mixed volumes 6.5 and the Brunn–Minkowski theorem 8.1 are also $\alpha+\beta$ type results.

The sum theorem of Macbeath and Kneser, which will be proved later, gives a satisfying answer to this question for the torus group. Let L be a lattice in $\mathbb{E}^d$.

The Sum Theorem of Kneser–Macbeath

Macbeath [673] and Kneser [601, 602] proved the following result.

Theorem 26.1. *Let $\mathfrak{S}, \mathfrak{T}$ be non-empty sets in $\mathbb{E}^d/L$ such that $\mathfrak{S}, \mathfrak{T}$ and $\mathfrak{S}+\mathfrak{T}$ are measurable. Then the following claims hold:*

(i) *If $m(\mathfrak{S})+m(\mathfrak{T})>d(L)$, then $\mathfrak{S}+\mathfrak{T}=\mathbb{E}^d/L$.*
(ii) *If $m(\mathfrak{S})+m(\mathfrak{T})\le d(L)$, then $m(\mathfrak{S}+\mathfrak{T})\ge m(\mathfrak{S})+m(\mathfrak{T})$.*

Note that the Lebesgue measurability of $\mathfrak{S}$ and $\mathfrak{T}$ does *not* necessarily imply the Lebesgue measurability of $\mathfrak{S}+\mathfrak{T}$. If $\mathfrak{S}$ and $\mathfrak{T}$ are Borel, their sum is Lebesgue measurable, but need not be Borel. For references, see the Remark in Sect. 8.2.

If $\mathfrak{S}+\mathfrak{T}$ is not measurable, a result similar to the sum theorem of Kneser and Macbeath holds, where $m(\mathfrak{S}+\mathfrak{T})$ is replaced by the corresponding inner measure of $\mathfrak{S}+\mathfrak{T}$.

We give two proofs. The first one is due to Kneser [602], see also Cassels [195]. The second proof is due to Macbeath [673] and Kneser [601]. In the second proof, we consider only the case where $\mathfrak{S}, \mathfrak{T}$ and $\mathfrak{S}+\mathfrak{T}$ all are Jordan measurable sets.

Proof of Kneser

As a preparation for the first proof, we show the following result; versions of it are useful tools in integral geometry. See, e.g. Hadwiger [468], Santaló [881] or Schneider and Weil [911].

Lemma 26.1. *Let $\mathfrak{U}, \mathfrak{V}\subseteq\mathbb{E}^d/L$ be measurable. Then there is $\mathfrak{z}\in\mathbb{E}^d/L$ such that:*

$$m\big((\mathfrak{U}+\mathfrak{z})\cap\mathfrak{V}\big)\,d(L)=m(\mathfrak{U})\,m(\mathfrak{V}).$$

Proof. Let $\mathbb{1}_{\mathfrak{U}}$ and $\mathbb{1}_{\mathfrak{V}}$ be the characteristic functions of $\mathfrak{U}$ and $\mathfrak{V}$, respectively. Then

$$\begin{aligned} m\big((\mathfrak{U}+\mathfrak{y})\cap\mathfrak{V}\big) &= \int\limits_{\mathbb{E}^d/L} \mathbb{1}_{\mathfrak{U}+\mathfrak{y}}(\mathfrak{x})\mathbb{1}_{\mathfrak{V}}(\mathfrak{x})\,dm(\mathfrak{x}) \\ &= \int\limits_{\mathbb{E}^d/L} \mathbb{1}_{\mathfrak{U}}(\mathfrak{x}-\mathfrak{y})\mathbb{1}_{\mathfrak{V}}(\mathfrak{x})\,dm(\mathfrak{x}). \end{aligned}$$

It is well known that convolutions are continuous. This implies that the expression $m\big((\mathfrak{U}+\mathfrak{y})\cap\mathfrak{V}\big)$ is continuous as a function of $\mathfrak{y}$. Taking into account the fact that $\mathbb{E}^d/L$ is compact, this expression is integrable. Then

$$\begin{aligned}\int\limits_{\mathbb{E}^d/L} & m\big((\mathfrak{U}+\mathfrak{y})\cap\mathfrak{V}\big)\,dm(\mathfrak{y}) \\ &= \int\limits_{\mathbb{E}^d/L}\int\limits_{\mathbb{E}^d/L} \mathbb{1}_{\mathfrak{U}}(\mathfrak{x}-\mathfrak{y})\mathbb{1}_{\mathfrak{V}}(\mathfrak{x})\,dm(\mathfrak{x})\,dm(\mathfrak{y}) \\ &= \int\limits_{\mathbb{E}^d/L}\int\limits_{\mathbb{E}^d/L} \mathbb{1}_{\mathfrak{U}}(\mathfrak{x}-\mathfrak{y})\mathbb{1}_{\mathfrak{V}}(\mathfrak{x})\,dm(\mathfrak{y})\,dm(\mathfrak{x}) \\ &= \int\limits_{\mathbb{E}^d/L} m(\mathfrak{U})\mathbb{1}_{\mathfrak{V}}(\mathfrak{x})\,dm(\mathfrak{x}) = m(\mathfrak{U})\,m(\mathfrak{V}),\end{aligned}$$

by Fubini's theorem, since $\mathbb{E}^d/L$ is Abelian and thus unimodular. Since $m\big((\mathfrak{U}+\mathfrak{y})\cap\mathfrak{V}\big)$ is continuous, as a function of $\mathfrak{y}$, on the compact connected space $\mathbb{E}^d/L$ of measure $d(L)$, there is a coset $\mathfrak{z}\in\mathbb{E}^d/L$ such that:

$$\int\limits_{\mathbb{E}^d/L} m\big((\mathfrak{U}+\mathfrak{y})\cap\mathfrak{V}\big)\,dm(\mathfrak{y}) = m\big((\mathfrak{U}+z)\cap\mathfrak{V}\big)\,d(L). \ \square$$

Proof (of the sum theorem). (i) If Proposition (i) does not hold, choose $\mathfrak{x}\in\mathbb{E}^d/L$ such that $\mathfrak{x}\notin\mathfrak{S}+\mathfrak{T}$. Then $\mathfrak{x}-\mathfrak{s}\notin\mathfrak{T}$ for $\mathfrak{s}\in\mathfrak{S}$, or

$$(\mathfrak{x}-\mathfrak{S})\cap\mathfrak{T}=\emptyset.$$

This yields a contradiction and thus concludes the proof of (i):

$$m(\mathfrak{S})+m(\mathfrak{T}) = m(\mathfrak{x}-\mathfrak{S})+m(\mathfrak{T}) = m\big((\mathfrak{x}-\mathfrak{S})\cup\mathfrak{T}\big)\le m(\mathbb{E}^d/L)=d(L).$$

(ii) First, we show the following.

(1) Let $\mathfrak{U},\mathfrak{V}\subseteq\mathbb{E}^d/L$ where $\mathfrak{U}\cap\mathfrak{V}\neq\emptyset$. Then $\mathfrak{U}+\mathfrak{V}\supseteq\mathfrak{U}\cap\mathfrak{V}+\mathfrak{U}\cup\mathfrak{V}$.

Let $\mathfrak{x}\in\mathfrak{U}\cap\mathfrak{V}$ and $\mathfrak{y}\in\mathfrak{U}\cup\mathfrak{V}$. If $\mathfrak{y}$ belongs to $\mathfrak{U}$, then we may regard $\mathfrak{x}$ as belonging to $\mathfrak{V}$, since it belongs to both $\mathfrak{U}$ and $\mathfrak{V}$. Hence $\mathfrak{x}+\mathfrak{y}=\mathfrak{y}+\mathfrak{x}\in\mathfrak{U}+\mathfrak{V}$. Similarly, if $\mathfrak{y}$ belongs to $\mathfrak{V}$, we regard $\mathfrak{x}$ as belonging to $\mathfrak{U}$.

The crucial task in the proof of (ii) is to show it for compact sets $\mathfrak{S}$ and $\mathfrak{T}$. Let

(2) $m(\mathfrak{S})=\sigma d(L),\ m(\mathfrak{T})=\tau d(L)$.

By assumption, $\sigma+\tau\le 1$. If $\sigma=0$ or $\tau=0$, then (ii) holds trivially. After exchanging $\mathfrak{S}$ and $\mathfrak{T}$ and renaming, if necessary, we may thus assume that

(3) $0<\sigma\le\tau$ and $\sigma+\tau\le 1$.

The main step of the proof is to show the following statement by induction:

(4) For $n = 1, 2, \dots$, there are non-empty compact sets $\mathfrak{S}_n, \mathfrak{T}_n$ in $\mathbb{E}^d/L$ and real numbers σ_n, τ_n, where

$$\mathfrak{S}_1 = \mathfrak{S},\ \mathfrak{T}_1 = \mathfrak{T},\ \sigma_1 = \sigma,\ \tau_1 = \tau,$$

and such that:

$$\begin{aligned}&\mathfrak{S}_n, \mathfrak{T}_n \text{ are compact, } \mathfrak{S}_n + \mathfrak{T}_n \subseteq \mathfrak{S} + \mathfrak{T},\\ &m(\mathfrak{S}_n) = \sigma_n d(L),\ m(\mathfrak{T}_n) = \tau_n d(L),\\ &\sigma_n + \tau_n = \sigma + \tau,\ \sigma_n = \sigma_{n-1}\tau_{n-1}\ (n \geq 2).\end{aligned}$$

The case $n = 1$ is trivial. Since the proof for $n = 2$ is the same as that for $n + 1$, assuming that (4) holds for $n \geq 2$, only the latter will be presented. An application of Lemma 26.1, with $\mathfrak{U} = \mathfrak{S}_n$ and $\mathfrak{V} = \mathfrak{T}_n$, shows that there is a $\mathfrak{z}_n \in \mathbb{E}^d/L$ such that:

(5) $\mathfrak{S}_{n+1} = (\mathfrak{S}_n + \mathfrak{z}_n) \cap \mathfrak{T}_n \neq \emptyset$,
$m(\mathfrak{S}_{n+1})d(L) = m(\mathfrak{S}_n)m(\mathfrak{T}_n) = \sigma_n\tau_n d(L)^2$.

Let

(6) $\mathfrak{T}_{n+1} + \mathfrak{z}_n = (\mathfrak{S}_n + \mathfrak{z}_n) \cup \mathfrak{T}_n$.

Since $\mathfrak{S}_n, \mathfrak{T}_n$ are compact,

(7) $\mathfrak{S}_{n+1}, \mathfrak{T}_{n+1}$ are also compact.

An application of (1) with $\mathfrak{U} = \mathfrak{S}_n + \mathfrak{z}_n, \mathfrak{V} = \mathfrak{T}_n$ shows that

$$\mathfrak{S}_n + \mathfrak{z}_n + \mathfrak{T}_n \supseteq (\mathfrak{S}_n + \mathfrak{z}_n) \cap \mathfrak{T}_n + (\mathfrak{S}_n + \mathfrak{z}_n) \cup \mathfrak{T}_n = \mathfrak{S}_{n+1} + \mathfrak{T}_{n+1} + \mathfrak{z}_n,$$

i.e.

(8) $\mathfrak{S}_{n+1} + \mathfrak{T}_{n+1} \subseteq \mathfrak{S}_n + \mathfrak{T}_n \subseteq \mathfrak{S} + \mathfrak{T}$.

Put

(9) $m(\mathfrak{S}_{n+1}) = \sigma_{n+1}d(L), m(\mathfrak{T}_{n+1}) = \tau_{n+1}d(L)$.

Since

$$\begin{aligned}m(\mathfrak{S}_n) + m(\mathfrak{T}_n) &= m(\mathfrak{S}_n + \mathfrak{z}_n) + m(\mathfrak{T}_n)\\ &= m\big((\mathfrak{S}_n + \mathfrak{z}_n) \cap \mathfrak{T}_n\big) + m\big((\mathfrak{S}_n + \mathfrak{z}_n) \cup \mathfrak{T}_n\big)\\ &= m(\mathfrak{S}_{n+1}) + m(\mathfrak{T}_{n+1} + \mathfrak{z}_n) = m(\mathfrak{S}_{n+1}) + m(\mathfrak{T}_{n+1}),\end{aligned}$$

the induction hypothesis and Propositions (9), (5) and (2) show that

(10) $\sigma_{n+1} + \tau_{n+1} = \sigma_n + \tau_n = \sigma + \tau,\ \sigma_{n+1} = \sigma_n\tau_n$.

Having shown (7)–(10), the induction is complete, concluding the proof of the statement (4).

Clearly, as a sum of compact sets in $\mathbb{E}^d/L$,

(11) $\mathfrak{S}_n + \mathfrak{T}_n$ is compact and thus measurable and
$m(\mathfrak{S}_n + \mathfrak{T}_n) \geq m(\mathfrak{T}_n) = \tau_n d(L)$ for $n = 1, 2, \dots$

It follows from (3) and (4) that

$$0 \leq \sigma_n \leq 1,\ \sigma_n \leq \sigma_{n-1}(1 - \sigma_{n-1}) \text{ for } n = 2, 3, \dots$$

and therefore, $\sigma_n \to 0$ as $n \to \infty$. Hence (4) yields $\tau_n \to \sigma + \tau$ as $n \to \infty$. Combine this with (11), (4) and (2) to get

$$m(\mathfrak{S} + \mathfrak{T}) \geq (\sigma + \tau)\, d(L) = m(\mathfrak{S}) + m(\mathfrak{T}),$$

as required. The proof of (ii) for compact sets $\mathfrak{S}, \mathfrak{T}$ is complete.

If $\mathfrak{S}, \mathfrak{T}$ and $\mathfrak{S} + \mathfrak{T}$ are measurable, then there are non-decreasing sequences $\mathfrak{S}_1 \subseteq \mathfrak{S}_2 \subseteq \dots \subseteq \mathfrak{S}$, $\mathfrak{T}_1 \subseteq \mathfrak{T}_2 \subseteq \dots \subseteq \mathfrak{T}$ of non-empty compact sets the measures of which tend to $m(\mathfrak{S})$ and $m(\mathfrak{T})$, respectively. Applying (ii) to the non-empty compact sets $\mathfrak{S}_n, \mathfrak{T}_n$ and letting $n \to \infty$, yields (ii) for $\mathfrak{S}, \mathfrak{T}$. □

Proof of Macbeath and Kneser

As an essential tool for the second proof, we state, without proof, the following result on finite Abelian groups. For a proof see Kneser [601].

Lemma 26.2. *Let A, B be non-empty subsets of a finite Abelian group G. Then the following claims hold:*

(i) *If $\#A + \#B > \#G$ then $A + B = G$.*
(ii) *If $\#A + \#B \leq \#G$ then $\#(A + B) \geq \#A + \#B - \#H$,*
where H is a proper sub-group of G.

Proof (of the sum theorem for $\mathfrak{S}, \mathfrak{T}, \mathfrak{S} + \mathfrak{T}$ Jordan measurable).

(i) The simple proof of statement (i) is as earlier.

(ii) For the proof of Statement (ii) of the sum theorem, it is sufficient to show the following proposition:

(12) Let $S, T \subseteq [0, 1)^d$ be Jordan measurable such that $V(S) + V(T) < 1$ and $S +_{\mathbb{Z}^d} T$ is Jordan measurable. Then $V(S +_{\mathbb{Z}^d} T) \geq V(S) + V(T)$.

Let p be a prime number and consider the group

$$G_p = \Big\{\Big(\frac{u_1}{p}, \dots, \frac{u_d}{p}\Big) : u_i \in \{0, 1, \dots, p - 1\}\Big\}$$

with componentwise addition modulo 1. This group has p^d elements. Thus, each proper sub-group has at most p^{d-1} elements. If we consider G_p as a subset of $[0, 1)^d$, addition in G_p and in $[0, 1)^d$ coincide. Clearly,

$$(S +_{\mathbb{Z}^d} T) \cap G_p \supseteq S \cap G_p +_{\mathbb{Z}^d} T \cap G_p.$$

The formula to calculate the Jordan measure in Sect. 7.2 shows that

$$\lim_{\substack{p\to\infty\\ p \text{ prime}}} \frac{\#G_p}{p^d} = 1 > V(S) + V(T) = \lim_{\substack{p\to\infty\\ p \text{ prime}}} \frac{\#(S\cap G_p)}{p^d} + \lim_{\substack{p\to\infty\\ p \text{ prime}}} \frac{\#(T\cap G_p)}{p^d}.$$

This, in turn, implies that $\#(S\cap G_p) + \#(T\cap G_p) \le \#G_p$ for all sufficiently large primes p. The lemma then shows that

$\#\big((S+_{\mathbb{Z}^d} T)\cap G_p\big) \ge \#(S\cap G_p) + \#(T\cap G_p) - \#H_p$ for all sufficiently large primes p, with H_p a proper sub-group of G_p.

Thus,

$$\begin{aligned} V(S+_{\mathbb{Z}^d} T) &= \lim_{\substack{p\to\infty\\ p \text{ prime}}} \frac{\#\big((S+_{\mathbb{Z}^d} T)\cap G_p\big)}{p^d} \\ &\ge \lim_{\substack{p\to\infty\\ p \text{ prime}}} \frac{\#(S\cap G_p)}{p^d} + \lim_{\substack{p\to\infty\\ p \text{ prime}}} \frac{\#(T\cap G_p)}{p^d} - \lim_{\substack{p\to\infty\\ p \text{ prime}}} \frac{p^{d-1}}{p^d} \\ &= V(S) + V(T), \end{aligned}$$

concluding the proof of Proposition (12) and thus of Statement (ii). □

26.3 Kneser's Transference Theorem

It is to Jarník's credit that he first proved a transference theorem. In the subsequent development more refined tools and ideas led to further transference theorems. We mention, in particular, Hlawka's [513] transference theorem which he proved using the method of the additional variable and Kneser's [601] transference theorem which is based on the sum theorem. Both relate the packing and the covering radius of an o-symmetric convex body with respect to a given lattice.

In this section Kneser's transference theorem is proved. We use notions and simple properties of lattice packing and covering; see Sects. 30.1 and 31.1.

For more information the reader is referred to [447].

The Transference Theorem of Kneser

As a consequence of the sum theorem, Kneser [601] proved the following result, where for the definitions of the packing radius $\varrho(C,L)$ and the covering radius $\mu(C,L)$ see Sect. 23.2. The transference theorem relates lattice packing and covering.

Theorem 26.2. *Given an o-symmetric convex body C and a lattice L in $\mathbb{E}^d$ and let*

(1) $q = \left\lfloor \dfrac{d(L)}{\varrho(C,L)^d V(C)} \right\rfloor$ *and* $r = \dfrac{d(L)}{\varrho(C,L)^d V(C)} - q.$

Then

(2) $\mu(C,L) \le \varrho(C,L)(q + r^{\frac{1}{d}}).$

Proof. Put $\varrho = \varrho(C, L)$ and $\mu = \mu(C, L)$, and let F be a fundamental parallelotope of L. By $t\mathfrak{C}$ for $t \geq 0$ we mean the set $tC + L$. It follows from Proposition 26.4 that

(3) $m(t\mathfrak{C}) = V(F \cap t\mathfrak{C})$ for $t \geq 0$.

Using (3), we prove the following two assertions.

First,

(4) $m(t\mathfrak{C}) = t^d V(C)$ for $0 \leq t \leq \varrho$.

For $0 \leq t \leq \varrho$ the set lattice $\{tC + l : l \in L\}$ is a packing. Hence its density $V(tC)/d(L) = t^d V(C)/d(L)$ is the proportion of $\mathbb{E}^d$ covered by the sets $tC + l$, see Sect. 30.1. The latter equals the proportion of F covered by the sets $tC + l$ and thus is equal to $V\big(F \cap (tC + L)\big)/d(L) = V(F \cap t\mathfrak{C})/d(L)$. Hence (3) yields (4).

Second,

(5) $\mu = \inf\big\{t > 0 : m(t\mathfrak{C}) = d(L)\big\}$.

If $t > \mu$, then $\{tC + l : l \in L\}$ is a covering. Then $t\mathfrak{C} = tC + L = \mathbb{E}^d$ and thus $m(t\mathfrak{C}) = V(F) = d(L)$ by (3). If $t < \mu$, then $\{tC + l : l \in L\}$ is not a covering. In particular, the family $tC + l : l \in L$ of translates of the (compact) convex body by vectors of the lattice L do not cover F. Hence $m(t\mathfrak{C}) = V(F \cap t\mathfrak{C}) = V\big(F \cap (tC + L)\big) < V(F) = d(L)$ by (3). The proof of (5) is now complete.

Noting that $(s + t)\mathfrak{C} \supseteq s\mathfrak{C} + t\mathfrak{C}$, the sum theorem shows that

(6) $m\big((s + t)\mathfrak{C}\big) \geq \min\big\{m(s\mathfrak{C}) + m(t\mathfrak{C}), d(L)\big\}$ for $s, t \geq 0$.

Having shown (4)–(6), the proof of the theorem is easy: By (4),

(7) $m(\varrho\mathfrak{C}) = \varrho^d V(C)$, $m(r^{\frac{1}{d}}\varrho\mathfrak{C}) = r\varrho^d V(C)$.

Applying (6) several times, (7) and (1), it follows that

$$\begin{aligned} m\big(\varrho(q + r^{\frac{1}{d}})\mathfrak{C}\big) &\geq \min\big\{q\, m(\varrho\mathfrak{C}) + m(r^{\frac{1}{d}}\varrho\mathfrak{C}), d(L)\big\} \\ &= \min\big\{\varrho^d (q + r)V(C), d(L)\big\} = d(L). \end{aligned}$$

This, together with (5), finally yields (2). □

27 Special Problems in the Geometry of Numbers

In this chapter we have so far outlined systematic features of the geometry of numbers. What about problems? In contrast to other areas of mathematics, in the geometry of numbers there is only a rather small number of basic particular problems. Roughly speaking, these are of two types. First, special problems of a more arithmetic character, including problems on forms of various types. A selection of these will be considered here. Second, sets of problems involving reduction, packing, covering and tiling. While reduction has arithmetic and geometric aspects, the other three sets of problems of the second group definitely are geometric. Reduction, packing, covering and tiling will be studied in the subsequent sections. In this section we consider the following special problems:

The conjecture on the product of inhomogeneous linear forms
$DOTU$-matrices
Mordell's inverse problem for the linear form theorem of Minkowski
Minima of the Epstein zeta function
Lattice points in large convex bodies

A further group of special problems deals with homogeneous and inhomogeneous minima of indefinite quadratic forms. These problems have been solved, mainly through the efforts of the Indian school of the geometry of numbers of Bambah, Dumir and Hans-Gill and their students, see the survey by Bambah, Dumir and Hans-Gill [58]. While, in recent decades, much progress has been achieved for indefinite quadratic forms, there is not much advance visible on the other problems. Exceptions are the results of Narzullaev and Ramharter [765] on $DOTU$-matrices and the proof of McMullen [704] of the conjecture on the product of inhomogeneous linear forms for $d = 6$.

For more detailed information, see the book of the author and Lekkerkerker [447] and the reports of Malyshev [682], Bambah [56], Bambah, Dumir and Hans-Gill [58] and Bayer-Fluckiger and Nebe [84]

27.1 The Product of Inhomogeneous Linear Forms and DOTU Matrices

A special problem, which has attracted interest since Minkowski first studied the 2-dimensional case, is the conjecture on the product of inhomogeneous linear forms. In spite of numerous contributions, the general case is still open and there are doubts whether the conjecture is true generally. The conjecture is one of those seminal problems which, over a century, has generated numerous notions, problems and results of different types in number theory and, in particular, in the geometry of numbers. One such notion is that of $DOTU$-matrices. Tools used in this context and, in particular, tools to attack the conjecture range from algebraic topology to measure and algebraic number theory.

In this section we state the conjecture, describe the main lines of attack, additional results and problems and make some remarks on $DOTU$-matrices.

The reader who is interested in more precise information may wish to consult [447], Malyshev [682], Bambah [56], Bambah, Dumir and Hans-Gill [58] and Narzullaev and Ramharter [765] and the references in these sources.

The Conjecture on the Product of Inhomogeneous Linear Forms

The following conjecture has been attributed to Minkowski, but according to Dyson [281] it is not contained in his written work.

Conjecture 27.1. *Let $l_1, \dots, l_d$ be d real linear forms in d variables such that the absolute value δ of their determinant is positive. Then, for any $\alpha_1, \dots, \alpha_d \in \mathbb{R}$, there is a point $u \in \mathbb{Z}^d$ such that:*

$$\big|\big(l_1(u) - \alpha_1\big) \cdots \big(l_d(u) - \alpha_d\big)\big| \le \frac{\delta}{2^d}.$$

Here equality occurs precisely in case when there is an integer unimodular transformation of the variables, such that the linear forms assume the form $\beta_1 v_1, \dots, \beta_d v_d$ *and* $\alpha_1 \equiv \frac{1}{2}\beta_1 \mod \beta_1, \dots, \alpha_d \equiv \frac{1}{2}\beta_d \mod \beta_d$.

The conjecture may also be expressed as follows:

Conjecture 27.2. *For any lattice L in* $\mathbb{E}^d$ *the family of all translates of the set*

(1) $|x_1 \cdots x_d| \le \dfrac{d(L)}{2^d}$

by vectors of L covers $\mathbb{E}^d$*. Precisely for the lattices L not of the form* $D\mathbb{Z}^d$*, where D is a* $d \times d$ *diagonal matrix, the set in* (1) *can be replaced by the set*

$$|x_1 \cdots x_d| < \frac{d(L)}{2^d}.$$

The Cases d=2,...,6

The conjecture is true for:

$d = 2$: Minkowski [737, 743], a multitude of further proofs
$d = 3$: Remak [830], Davenport [243], who simplified Remak's proof, Birch and Swinnerton-Dyer [117], Narzullaev [762]
$d = 4$: Dyson [281], Skubenko [941], Bambah and Woods [60]
$d = 5$: Skubenko [941]. Bambah and Woods [62] gave a proof along similar lines, clarifying Skubenko's arguments
$d = 6$: McMullen [704]

Most of the proofs mentioned so far follow the line of proof of Remak-Davenport: For the proof of the conjecture (except for the equality case) it is sufficient to show the following two assertions:

(i) For each lattice L in $\mathbb{E}^d$ there is an ellipsoid E of the form:

$$a_1 x_1^2 + \cdots + a_d x_d^2 \le 1$$

which contains d linearly independent points of L on its boundary but no point of L in its interior, except o.

(ii) Let L be a lattice of determinant 1 and ρB^d a ball which contains d linearly independent points of L on its boundary but no point of L in its interior, except o. Then the following family of balls covers $\mathbb{E}^d$:

$$\left\{\frac{\sqrt{d}}{2} B^d + l : l \in L\right\}.$$

Let L be a lattice of determinant 1 and let E be the corresponding ellipsoid. After applying to L and E a suitable diagonal transformation of determinant 1, we may assume that E is a Euclidean ball as in (ii). Then the translates of the ball $(\sqrt{d}/2)B^d$ by the vectors of L cover $\mathbb{E}^d$. Since the inequality of the arithmetic and geometric mean shows that the ball $(\sqrt{d}/2)B^d$ is contained in the set $|x_1 \cdots x_d| \le 1/2^d$, the translates of this set by the vectors of L cover $\mathbb{E}^d$.

Asymptotic Estimates

Chebotarev [202] proved that, for each lattice L, the family of all translates of the set

$$|x_1 \cdots x_d| \le \frac{d(L)}{2^{\frac{d}{2}}}$$

by vectors of L covers E^d. The proof uses a clever application of Minkowski's first fundamental theorem. Using refinements of Minkowski's theorem, many successive improvements of this result were given, where $2^{d/2}$ is replaced by $\nu_d 2^{d/2}$ with $\nu_d > 1$. We mention the following; for others, see the author and Lekkerkerker [447] and Bambah, Dumir and Hans-Gill [58]:

$\nu_d \sim 2e - 1$: Davenport [244]
$\nu_d \sim 2(2e - 1)$: Woods [1029]
$\nu_d \sim 3.0001(2e - 1)$: Bombieri [147]
$\nu_d \sim 3(2e - 1)$: Gruber [410]
$\nu_d \sim e^{-\frac{2}{3}} d^{\frac{1}{3}} \log^{-\frac{2}{3}} d$ as $d \to \infty$ Skubenko [942]

See also Narzullaev and Skubenko [764], Mukhsinov [759] and Andriyasyan, Il'in and Malyshev [34].

Sets of Lattices for which the Conjecture is true and $DOTU$-Matrices

A result of Birch and Swinnerton-Dyer [117] is as follows: If the conjecture holds for dimensions $2, \dots, d-1$, then it holds in dimension d for all lattices L for which the homogeneous minimum

(2) $\lambda(L) = \inf\left\{|l_1 \cdots l_d| : l \in L \setminus \{o\}\right\}$

is 0. Since it is easy to prove that $\lambda(L) = 0$ for almost all lattices $L \in \mathcal{L}(1)$ in the sense of the measure introduced by Siegel on the space of all lattices of determinant 1, we see that, if the conjecture holds in dimensions $2, \dots, d-1$, then it holds in dimension d for almost all lattices of determinant 1, see [411]. In the same article, it was shown that, in all dimensions, the measure of the set of lattices of determinant 1 which cover $\mathbb{E}^d$ by the set

$$|x_1 \cdots x_d| \le \frac{d\,d!}{d^d} \sim \frac{\sqrt{2\pi}}{\sqrt{d}e^d} = e^{-d+o(d)} \text{ as } d \to \infty,$$

is at least $1 - e^{-0.279\,d}$.

A real $d \times d$-matrix A is a *$DOTU$-matrix* if it can be written in the form:

$$A = DOTU,$$

where D is a diagonal, O an orthogonal, T an upper triangular matrix with 1's in the diagonal and U an integer unimodular $d \times d$ matrix. This notion was introduced by

Macbeath [674]. An easy proof led Macbeath to the following result: If $L = A\mathbb{Z}^d$, where A is a $DOTU$-matrix, then the conjecture holds for the lattice L. It is thus of interest to find out whether all $d \times d$-matrices are $DOTU$-matrices or not. It is easy to see that each real 2×2 matrix is $DOTU$. After some unsuccessful attempts, Narzulaev [763] established this for $d = 3$. Narzullaev and Ramharter [765] proved that all non-singular 4×4 matrices A, for which the homogeneous minimum (2) of the lattice $L = A\mathbb{Z}^4$ is sufficiently small, are $DOTU$-matrices. It is easy to see that for all d the rational $d \times d$ matrices are $DOTU$ and Macbeath showed that the set of $DOTU$-matrices is open. Hence, the $d \times d$ $DOTU$-matrices form a dense open set in the space of all non-singular $d \times d$ matrices.

Unfortunately, Gruber [413] and Ahmedov [4] showed for infinitely many, respectively, all sufficiently large d, the existence of $d \times d$ matrices which are not $DOTU$. Skubenko [943] gave an example of such a matrix for $d = 2880$ and Hendrik Lenstra [648] communicated an example with $d = 64$. These examples make use of algebraic number theory and, in particular, of class field theory.

A Related Conjecture

which has attracted some interest is the following; we state it in geometric form:

Conjecture 27.3. *For $k = 0, \dots, d$ and any lattice L in $\mathbb{E}^d$, the family of all translates of the set*

$$\left\{x : x_1, \dots, x_k \geq 0,\ |x_1 \cdots x_d| \leq \frac{d(L)}{2^{d-k}}\right\}$$

by vectors of L covers $\mathbb{E}^d$.

This conjecture was proved by Chalk [200] for $k = d$ and all d, by Cole [212] for $k = d - 1$ and all d and by Bambah and Woods [61] for $k = 2$ and $d = 3$. All other cases seem to be open.

27.2 Mordell's Inverse Problem and the Epstein Zeta-Function

A geometric version of the Minkowski linear form theorem is as follows, see Corollary 22.2: Let L be a lattice in $\mathbb{E}^d$ with $d(L) = 1$. Given numbers $\tau_1, \dots, \tau_d > 0$ such that $\tau_1 \cdots \tau_d \geq 1$, the lattice L contains a point $\neq o$ in the box

(1) $|x_1| \leq \tau_1, \dots, |x_d| \leq \tau_d.$

This led Mordell [753] to consider the problem to choose $\tau_1, \dots, \tau_d > 0$ such that $\tau_1 \cdots \tau_d$ is as large as possible and there is no point of $L \setminus \{o\}$ in the box (1).

Given a lattice L in $\mathbb{E}^d$, the *Epstein zeta-function* $\zeta(L, \cdot)$ is defined as follows:

$$\zeta(L, s) = \sum_{l \in L \setminus \{o\}} \frac{1}{\|l\|^{2s}} \text{ for } s > \frac{d}{2}.$$

This function was first studied by Epstein [305] and re-discovered by Sobolev [945]. It is important for the determination of potentials of crystal lattices, for the lattice energy, for the dynamics of viscous fluids, and for the numerical integration of functions belonging to Sobolev classes. Its study, in the context of the geometry of numbers, was cultivated in the 1950s and 1960s and later in Great Britain and in Russia. The motive for the study in Great Britain was purely number-theoretic, while the Russian school of the geometry of numbers studied it following a suggestion of Sobolev in the context of the problem of the optimal choice of nodes for numerical integration formulae.

In the following we describe some of the developments induced by Mordell's problem. For more information and references the reader is referred to [447] and Narzullaev and Ramharter [765]. In addition, we state some results on local minima of the Epstein zeta-function. See also [447] and the report [438].

Mordell's Inverse Problem

Given a lattice L in $\mathbb{E}^d$, let

$$\kappa(L) = \sup\left\{\tau_1 \cdots \tau_d : \tau_i > 0,\ \{x : |x_j| < \tau_j\} \cap L = \{o\}\right\}.$$

Using the quantity

$$\kappa_d = \inf\left\{\frac{\kappa(L)}{d(L)} : L \in \mathcal{L}\right\}.$$

Mordell's problem can be formulated as follows:

Problem 27.1. *Determine κ_d for $d = 2, 3, \ldots$*

The problem is solved for $d = 2, 3$:

$$\kappa_2 = \frac{1}{2} + \frac{1}{2\sqrt{5}} = 0.7236\ldots : \quad \text{Szekeres [982]}$$

$$\kappa_3 = \frac{8}{7}\cos^2\left(\frac{\pi}{7}\right)\cos\left(\frac{2\pi}{7}\right) = 0.578416\ldots : \text{Ramharter [822]}$$

Hlawka [512] (lower bound) and Gruber and Ramharter [450] (upper bound) gave the estimates:

$$2^{-\frac{d^2}{2}(1+o(1))} = \frac{d}{d!^2 2^{\frac{d(d+1)}{2}}} < \kappa_d \le \kappa_2^{\frac{d-1}{2}} = (0.7236\ldots)^{\frac{d-1}{2}}.$$

Minima of the Epstein Zeta-Function

The main problem on the Epstein zeta-function in the context of the geometry of numbers is the following:

Problem 27.2. *Given $s > \frac{d}{2}$, determine among all lattices L in $\mathbb{E}^d$ with $d(L) = 1$ those for which $\zeta(L, s)$ has a global or a local minimum.*

Collecting the work of Rankin [823], Cassels [194], Ennola [303], Diananda [266] and Montgomery [752] (global minima) and Ryshkov [864] (local minima) we state the following result.

Theorem 27.1. *The following propositions hold:*

(i) *Let $s > 1$. Then $\zeta(\cdot, s)$ attains its global minimum among all lattices in $\mathbb{E}^2$ of determinant 1 precisely for the regular hexagonal lattices.*

(ii) *Let $s \geq 3$. Then the only local minimum of $\zeta(\cdot, s)$ among all lattices in $\mathbb{E}^2$ of determinant 1 is the global minimum.*

The 3-dimensional case was studied by Ennola [304] and Sandakova [876].

27.3 Lattice Points in Large Convex Bodies

A classical problem of analytic number theory, the *Gauss circle problem* [366], is to estimate the number of points of the integer lattice $\mathbb{Z}^2$ in the circular disc ρB^2 for large $\rho > 0$. Clearly,

$$A(\rho) = \#(\mathbb{Z}^2 \cap \rho B^2) = \rho^2 \pi + O(\rho) \text{ as } \rho \to \infty.$$

$O(\rho)$ may be replaced by $O\big(\rho^{\frac{2}{3}}\big)$ as shown by Sierpiński [938]. The best upper estimate, at present, is due to Huxley [533], who showed that instead of $O(\rho)$ one may put $O\big(\rho^{\frac{131}{208}} (\log \rho)^{\frac{18637}{4160}}\big)$. It was proved by Hardy [479] that $O(\rho)$ may not be replaced by $o\big(\rho^{\frac{1}{2}}\big)$. The best pertinent result known was given by Soundararajan [949]. In an article which is not yet published, Cappell and Shaneson [189] seem to have proved the bound $\mathrm{O}\,(\rho^{\frac{1}{2}+\epsilon})$.

The circle problem was the starting point of a voluminous literature in the context of analytic number theory, see the survey of Ivić, Krätzel, Kühleitner and Nowak [540], the books of Fricker [344] and Krätzel [616] and the short report in Gruber and Lekkerkerker [447].

In this section we present a result of Hlawka [511] which is representative of this area. Since all available proofs are technically involved and require sophisticated analytic tools, no proof is given.

Hlawka's Lattice Point Theorem

Theorem 27.2. *Let C be an o-symmetric convex body, the boundary of which is of class $\mathcal{C}^2$ and has positive Gauss curvature. Then*

$$A(\rho) = \#(\mathbb{Z}^d \cap \rho C) = \rho^d V(C) + O\big(\rho^{d-2+\frac{2}{d+1}}\big) \textit{ as } \rho \to \infty.$$

Here $O\big(\rho^{d-2+\frac{2}{d+1}}\big)$ cannot be replaced by $o\big(\rho^{\frac{d-1}{2}}\big)$.

For various refinements due to Krätzel, Nowak, Müller and others, see the survey of Ivić, Krätzel, Kühleitner and Nowak [540].

28 Basis Reduction and Polynomial Algorithms

The basic problem of reduction theory can be stated in two equivalent ways. First, given a lattice, determine a (not necessarily unique) basis having nice geometric properties, a reduced basis. Here, nice may mean that the basis vectors are short or almost orthogonal. Second, given a positive definite quadratic form, find a (not necessarily unique) equivalent form having nice arithmetic properties, a reduced form.

There are several classical reduction methods for positive definite quadratic forms, going back at least to Lagrange [628]. Rather geometric are the reduction methods of Korkin and Zolotarev [610] and Lenstra, Lenstra and Lovász [646]. The latter provides polynomial time algorithms for geometric problems dealing with the shortest lattice vector problem and thus with Minkowski's fundamental theorem and with the nearest lattice point problem. In addition, it yields polynomial time algorithms for a multitude of other problems. These include the factoring of polynomials, Diophantine approximation, integer programming and cryptography. It was used for the disproof of Mertens's conjecture on the Riemann zeta-function and had a strong impact on complexity problems in algorithmic geometry. The Lenstra, Lenstra and Lovász algorithm was inspired by the 2-dimensional reduction method of Gauss.

In the following we first present the LLL-basis reduction algorithm and then describe its applications to Diophantine approximation and the nearest lattice point problem.

For additional information and references on reduction and geometric algorithms, see Gruber and Lekkerkerker [447], Kannan [563], Grötschel, Lovász and Schrijver [409], Gruber [430], Koy and Schnorr [613] and Micciancio and Goldwasser [721]. We are not aware of a comprehensive and unified exposition of the various reduction methods.

28.1 LLL-Basis Reduction

We will describe the Lenstra, Lenstra and Lovász [646] basis reduction algorithm and show that it is polynomial in input size. Our presentation follows the exposition of Grötschel, Lovász and Schrijver [409].

Definition of LLL-Reduced Bases

Let $\{b_1, \dots, b_d\}$ be an (ordered) basis of a lattice L in $\mathbb{E}^d$. Using the Gram–Schmidt orthogonalization method, we assign to $\{b_1, \dots, b_d\}$ an (ordered) system $\{\hat{b}_1, \dots, \hat{b}_d\}$ of d orthogonal vectors such that:

$$(1)\quad \begin{aligned} b_1 &= \hat{b}_1 \\ b_2 &= \mu_{21}\hat{b}_1 + \hat{b}_2 \\ &\cdots\cdots\cdots\cdots \\ b_d &= \mu_{d1}\hat{b}_1 + \dots + \mu_{d\,d-1}\hat{b}_{d-1} + \hat{b}_d \end{aligned} \qquad \text{where} \quad \mu_{ji} = \frac{b_j \cdot \hat{b}_i}{\|\hat{b}_j\|^2}.$$

$\hat{b}_{j+1}$ is the orthogonal projection of b_{j+1} onto the subspace $\text{lin}\{b_1, \dots, b_j\}^\perp = \text{lin}\{\hat{b}_1, \dots, \hat{b}_j\}^\perp$. The following result exhibits a connection between the orthogonal system $\{\hat{b}_1, \dots, \hat{b}_d\}$ and the shortest (non-zero) vector problem in L.

Lemma 28.1. *Let $\{b_1, \dots, b_d\}$ be a basis of a lattice L and $\{\hat{b}_1, \dots, \hat{b}_d\}$ its Gram–Schmidt orthogonalization. Then*

$$\|l\| \geq \min\{\|\hat{b}_1\|, \dots, \|\hat{b}_d\|\} \text{ for any } l \in L \setminus \{o\}.$$

Proof. Let $l \in L$. Then

$$l = \sum_i u_i b_i \text{ where } u_i \in \mathbb{Z}.$$

Let k be the largest index with $u_k \neq 0$. Replace $b_1, \dots, b_k$ by their expressions from (1). This gives

$$l = \sum_{i=1}^{k} \alpha_i \hat{b}_i \text{ where } \alpha_k = u_k \in \mathbb{Z} \setminus \{0\}.$$

Since $\hat{b}_1, \dots, \hat{b}_k$ are pairwise orthogonal, this yields the desired inequality:

$$\|l\|^2 = \sum_{i=1}^{k} \alpha_i^2 \|\hat{b}_i\|^2 \geq \alpha_k^2 \|\hat{b}_k\|^2 \geq \|\hat{b}_k\|^2. \quad \square$$

After these preparations, the definition of an LLL-reduced basis is as follows: Let $\{b_1, \dots, b_d\}$ be a basis of a lattice L, $\{\hat{b}_1, \dots, \hat{b}_d\}$ its Gram–Schmidt orthogonalization and the numbers μ_{ji} as in (1). Then $\{b_1, \dots, b_d\}$ is an *LLL-reduced basis* of L if it satisfies the following conditions:

(2) $|\mu_{ji}| \leq \dfrac{1}{2}$ for $i, j = 1, \dots, d, i < j$,

(3) $\|\hat{b}_{j+1} + \mu_{j+1\,j} \hat{b}_j\|^2 \geq \dfrac{3}{4} \|\hat{b}_j\|^2$ for $j = 1, \dots, d-1$.

Roughly speaking, the first condition means that the basis vectors $\{b_1, \dots, b_d\}$ are pairwise almost orthogonal. The vectors $\hat{b}_j$ and $\hat{b}_{j+1} + \mu_{j+1\,j} \hat{b}_j$ are the projections of b_j and b_{j+1} onto $\text{lin}\{\hat{b}_1, \dots, \hat{b}_{j-1}\}^\perp$. Thus, the second condition says that the length of the projection of b_{j+1} cannot be much smaller than the length of the projection of b_j.

Properties of LLL-Reduced Bases

We collect some basic properties of LLL-reduced bases which will be needed later:

Theorem 28.1. *Let $\{b_1, \dots, b_d\}$ be an LLL-reduced basis of a lattice L in $\mathbb{E}^d$. Then the following hold:*

(i) $\|b_1\| \leq 2^{\frac{1}{4}(d-1)} d(L)^{\frac{1}{d}}$

(ii) $\|b_1\| \le 2^{\frac{1}{2}(d-1)} \min\{\|l\| : l \in L \setminus \{o\}\}$

(iii) $\|b_1\| \cdots \|b_d\| \le 2^{\frac{1}{4}d(d-1)} d(L)$

Proof. (i) Since the vectors $\hat{b}_j$ are pairwise orthogonal, condition (3) shows that

$$\frac{3}{4}\|\hat{b}_j\|^2 \le \|\hat{b}_{j+1}\|^2 + \mu_{j+1\,j}^2\|\hat{b}_j\|^2 \text{ for } j = 1, \dots, d-1.$$

Since $\mu_{j+1\,j}^2 \le \frac{1}{4}$ by (2),

$$\|\hat{b}_{j+1}\|^2 \ge \frac{1}{2}\|\hat{b}_j\|^2 \text{ for } j = 1, \dots, d-1$$

follows. Then induction implies that

(4) $\|\hat{b}_j\|^2 \ge 2^{i-j}\|\hat{b}_i\|^2$ for $i, j = 1, \dots, d,\ i < j$.

In particular,

(5) $\|\hat{b}_j\|^2 \ge 2^{1-j}\|\hat{b}_1\|^2 = 2^{1-j}\|b_1\|^2$ for $j = 1, \dots, d$.

Multiplying the latter inequalities for $j = 1, \dots, d$, and taking into account the orthogonality of the system $\{\hat{b}_1, \dots, \hat{b}_d\}$ and (1), Statement (i) is obtained as follows:

$$\begin{aligned} 2^{-\frac{1}{2}d(d-1)}\|b_1\|^{2d} &\le \|\hat{b}_1\|^2 \cdots \|\hat{b}_d\|^2 = \det(\hat{b}_1, \dots, \hat{b}_d)^2 \\ &= \det(b_1, \dots, b_d)^2 = d(L)^2. \end{aligned}$$

(ii) Proposition (5) yields

$$\min\{\|\hat{b}_1\|, \dots, \|\hat{b}_d\|\} \ge 2^{-\frac{1}{2}(d-1)}\|b_1\|.$$

Now apply Lemma 28.1.

(iii) Using (1) and (2), we see that:

$$\|b_j\|^2 = \|\hat{b}_j\|^2 + \sum_{i=1}^{j-1} \mu_{ji}^2\|\hat{b}_i\|^2 \le \|\hat{b}_j\|^2 + \sum_{i=1}^{j-1} \frac{1}{4}\|\hat{b}_i\|^2.$$

So, by (4),

$$\|b_j\|^2 \le \Big(1 + \sum_{i=1}^{j-1} \frac{1}{2} 2^{j-i}\Big)\|\hat{b}_j\|^2 \le 2^{j-1}\|\hat{b}_j\|^2.$$

Multiplying this for $j = 1, \dots, d$, yields (iii)

$$\|b_1\|^2 \cdots \|b_d\|^2 \le 2^{\frac{1}{2}d(d-1)}\|\hat{b}_1\|^2 \cdots \|\hat{b}_d\|^2 = 2^{\frac{1}{2}d(d-1)}d(L). \quad \square$$

LLL-Reduction is Polynomial

We now come to the main result, where a *lattice* in $\mathbb{E}^d$ is *rational* if all its points have rational coordinates.

Theorem 28.2. *There is a polynomial time algorithm that finds, for any given basis* $\{a_1, \dots, a_d\}$ *of a rational lattice* L *in* $\mathbb{E}^d$, *an* LLL*-reduced basis.*

Proof. Without loss of generality, we may assume that $L \subseteq \mathbb{Z}^d$, and let $\{a_1, \dots, a_d\}$ be a basis of L. We shall transform this basis until the conditions (2) and (3) are satisfied. The algorithm starts with the basis $\{b_1, \dots, b_d\} = \{a_1, \dots, a_d\}$. Let $\{\hat{b}_1, \dots, \hat{b}_d\}$ be the corresponding Gram–Schmidt orthogonalization and let the coefficients μ_{ji} be as in (1).

Step I. *For* $j = 1, \dots, d$, *and, given* j, *for* $i = 1, \dots, j-1$, *replace* b_j *by* $b_j - \lceil \mu_{ji} \rfloor b_i$, *where* $\lceil \mu_{ji} \rfloor$ *is the integer nearest to* μ_{ji}. *(In case of ambiguity choose the smaller integer.) Then go to Step II.*

Step I does not change the Gram–Schmidt orthogonalization of the basis. Hence, after executing (the substeps of) Step I for a certain j, the new μ_{ji} will satisfy $|\mu_{ji}| \le \frac{1}{2}$ for $i = 1, \dots, j-1$. Executing (the substeps of) Step I for a larger j will not spoil this property. Thus executing (all substeps of) Step I yields a basis which satisfies condition (2).

Step II. *If there is an index* j *violating condition* (3), *interchange* b_j *and* b_{j+1} *and return to Step I.*

We analyze this step. Let

$$c_1 = b_1, \dots, c_{j-1} = b_{j-1}, c_j = b_{j+1}, c_{j+1} = b_j, c_{j+2} = b_{j+2}, \dots, c_d = b_d.$$

Then $\hat{c}_i = \hat{b}_i$ for $i \neq j, j+1$. Further, $\hat{c}_j = \hat{b}_{j+1} + \mu_{j+1\,j}\hat{b}_j$ and since condition (3) is violated for j, we have that:

$$(6)\quad \|\hat{c}_j\|^2 < \frac{3}{4}\|\hat{b}_j\|^2.$$

The formula for $\hat{c}_{j+1}$ is more complicated, but it is not needed. For our purpose it is sufficient to note that from

$$\|\hat{c}_1\|^2 \cdots \|\hat{c}_d\|^2 = \|\hat{b}_1\|^2 \cdots \|\hat{b}_d\|^2 = d(L)^2$$

it follows that $\|\hat{c}_j\|^2 \|\hat{c}_{j+1}\|^2 = \|\hat{b}_j\|^2 \|\hat{b}_{j+1}\|^2$ and thus by (6)

$$(7)\quad \|\hat{c}_j\|^{2(d-j)+2} \|\hat{c}_{j+1}\|^{2(d-j)} < \frac{3}{4}\|\hat{b}_j\|^{2(d-j)+2} \|\hat{b}_{j+1}\|^{2(d-j)}.$$

After having described the algorithm we come to the proof of the Theorem. First, the following will be shown:

(8) The total number of arithmetic operations in the algorithm is polynomial in the encoding length of the input.

To estimate the number of times Steps I and II are executed before a reduced basis is obtained (if at all), we study the quantity

$$D = \|\hat{b}_1\|^{2d}\|\hat{b}_2\|^{2(d-1)}\cdots\|\hat{b}_d\|^2.$$

By the earlier remarks and (7), Step I does not change the value of D, while Step II decreases it by a factor less than $\frac{3}{4}$. It follows from Gram's determinant theorem that

$$\|\hat{b}_1\|^2\cdots\|\hat{b}_j\|^2 = \det\big((b_k\cdot b_l)_{k,l\leq j}\big) > 0 \text{ for } j = 1,\dots,d,$$

see [295]. Thus

$$(9)\quad D = \prod_{j=1}^{d} D_j, \text{ where } D_j = \det\big((b_k\cdot b_l)_{k,l\leq j}\big) > 0.$$

Since $L \subseteq \mathbb{Z}^d$, all bases obtained by the execution of Steps I and II are integer. Hence all D_js and thus also D are integers at least 1. At the beginning of the algorithms we have,

$$D = D_0 = \|\hat{a}_1\|^{2d}\cdots\|\hat{a}_d\|^2 \leq \|a_1\|^{2d}\cdots\|a_d\|^2 \leq (\|a_1\|\cdots\|a_d\|)^{2d}.$$

Since Step I leaves D unchanged and Step II decreases it by a factor less than $3/4$ and since D is always at least 1, Step II is executed at most

$$\frac{\log D_0}{\log\frac{4}{3}} \leq \frac{2d}{\log\frac{4}{3}}(\log\|a_1\| + \cdots + \log\|a_d\|)$$

times. This is polynomial in the encoding length of the input. Between two executions of Step II there is at most one execution of Step I which has $O(d^3)$ arithmetic operations. Together this yields (8).

In the following parts of the proof, it will be shown that the denominators and the numerators of the (rational) numbers occurring in the algorithm are not too large. This is done first for the bases and their orthogonalizations resulting from Steps I and II and then for the bases and the numbers μ_{ji} resulting from the substeps of Step I.

In the second part of the proof, the following will be shown:

(10) For every basis $\{b_1,\dots,b_d\}$ resulting from Steps I or II in the algorithm every coordinate of any vector $\hat{b}_j$ is a rational whose denominator is bounded by D_0 and whose numerator is bounded by A_0D_0, where $A_0 = \max\{\|a_1\|,\dots,\|a_d\|\}$.

We begin by showing that:

(11 $D\hat{b}_j \in \mathbb{Z}^d$ for $j = 1,\dots,d$.

To see this, note that (1) implies that

$$(12)\quad \hat{b}_j = b_j - \lambda_{j1}b_1 - \cdots - \lambda_{j\,j-1}b_{j-1}$$

with suitable $\lambda_{ji} \in \mathbb{R}$. Multiplying this representation of $\hat{b}_j$ by b_i for $i = 1, \dots, j-1$, gives the following system of linear equations for $\lambda_{j1}, \dots, \lambda_{j\,j-1}$:

$$b_j \cdot b_i = \lambda_{j1} b_1 \cdot b_i + \dots + \lambda_{j\,j-1} b_{j-1} \cdot b_i, \; i = 1, \dots, j-1.$$

(Note that $\hat{b}_j \cdot b_i = 0$.) This system has determinant $D_j > 0$, see (9). Since $b_1, \dots, b_d \in L \subseteq \mathbb{Z}^d$ and thus all inner products $b_i \cdot b_k$ are integer, it follows that $D_j \lambda_{ji} \in \mathbb{Z}$ for $i = 1, \dots, j-1$. Since, by (9), the integer $D_j > 0$ divides the integer $D > 0$, (12) implies (11). Next we show that

(13) $\max\{\|\hat{b}_1\|, \dots, \|\hat{b}_d\|\} \le \max\{\|a_1\|, \dots, \|a_d\|\} = A_0$,

say. Step I does not change this maximum. In Step II, $\hat{c}_i = \hat{b}_i$ for $i \neq j, j+1$. Further $\|\hat{c}_j\| < \|\hat{b}_j\|$, by the condition of Step II, see also (6), and $\|\hat{c}_{j+1}\| \le \|\hat{b}_j\|$ since $\hat{c}_{j+1}$ is a suitable orthogonal projection of $\hat{b}_j$. Hence Step II never increases the maximum. This concludes the proof of (13). Since $D \le D_0$, (11) and (13) together yield (10).

Third, we prove the following statement:

(14) For every basis $\{b_1, \dots, b_d\}$ resulting from Steps I or II in the algorithm every coordinates of any vector b_j is an integer bounded by $\sqrt{d} A_0$.

For the proof it is sufficient to show that:

(15) $\max\{\|b_1\|^2, \dots, \|b_d\|^2\} \le d A_0^2$.

After Step I has been executed, $|\mu_{ji}| \le \frac{1}{2}$. Then (1), together with (13), implies that:

$$\|b_j\|^2 = \sum_{i=1}^{j} \mu_{ij}^2 \|\hat{b}_i\|^2 \le \sum_{i=1}^{j} \|\hat{b}_i\|^2 \le d A_0^2.$$

Hence (15) holds after an execution of Step I. Since Step II does not change the maximum, (15) holds generally, concluding the proof of (14).

The fourth step is to show that:

(16) For every basis $\{b_1, \dots, b_d\}$ resulting from a substep of Step I in the algorithm, every coordinate of any b_j is an integer bounded by $(2D_0A_0)^d \sqrt{d} A_0$.

Note that the Gram–Schmidt orthogonalization does not change during an execution of Step I. Thus, for the bases appearing during Step I, statement (10) is valid. If, in a substep of Step I, the basis vector b_j is replaced by $b_j - \lceil \mu_{ji} \rfloor b_i$, then

$$|\mu_{ji}| = \frac{|b_j \cdot \hat{b}_i|}{\|\hat{b}_i\|^2} \le \frac{\|b_j\|}{\|\hat{b}_i\|} \le D_0 \|b_j\|$$

by the Cauchy–Schwarz inequality and (10). Thus

$$|\lceil \mu_{ji} \rfloor| \le 2|\mu_{ji}| \le 2D_0 \|b_j\|$$

and so

$$\|b_j - \lceil\mu_{ji}\rfloor b_i\| \leq \|b_j\| + |\lceil\mu_{ji}\rfloor|\|b_i\| \leq \|b_j\| + 2D_0\|b_i\|\|b_j\|.$$

Since $i < j$ the b_i which appears here will not change any more through executions of sub-steps of Step I and thus will appear in the basis resulting from Step I. Hence $\|b_i\| \leq \sqrt{d}A_0$ by (15) and we conclude that:

$$\|b_j - \lceil\mu_{ji}\rfloor b_i\| \leq (1 + 2\sqrt{d}A_0D_0)\|b_j\| \leq 2dA_0D_0\|b_j\|.$$

Since b_j is changed at most $j - 1 < d$ times, its length is increased at most by the factor $(2dA_0D_0)^d$. Since we start with a basis resulting from Steps I or II, an application of (14) then yields (16).

Fifth, we prove that:

(17) For every basis $\{b_1, \dots, b_d\}$ resulting from Steps I and II, respectively, from a sub-step of Step I, each number μ_{ji} is a rational where the denominator is bounded by $d^2A_0(2A_0D_0)^{d+3}$ and the numerator by $dA_0D_0^3$.

Note that

$$\mu_{ji} = \frac{b_j \cdot \hat{b}_i}{\|\hat{b}_i\|^2}$$

and apply (10) and (14), respectively, (16).

Having proved Propositions (8), (10), (14), (16) and (17), the theorem follows.

□

Remark. Actually, the results of Lenstra, Lenstra and Lovász [646] are more explicit than the above results. For refinements, see Grötschel, Lovász, Schrijver [409].

28.2 Diophantine Approximation, the Shortest and the Nearest Lattice Vector Problem

Typical classical results in number theory and, in particular, in Diophantine approximation, Diophantine equations and in the classical geometry of numbers say that a given problem, a given inequality, or a given equation *has* an integer solution. The problem, how to *find* all or, at least one solution by means of an algorithm, say, was left open. This situation was considered in the last few decades to be rather unsatisfactory and much effort has been spent to remedy it. The LLL-basis reduction algorithm was defined with such applications in mind and has turned out to provide polynomial time algorithms for the solution of numerous problems in algebra, number theory and computational geometry.

In the following, we present applications to Diophantine approximations and the shortest and the nearest lattice point problem.

For more information consult Kannan [563], Grötschel, Lovász and Schrijver [409], Koy and Schnorr [613] and Micciancio and Goldwasser [721].

Simultaneous Diophantine Approximation

The following is an approximate algorithmic version of a result of Kronecker [618] and Minkowski [735], see Corollary 22.3. It is due to Lenstra, Lenstra and Lovász [646].

Corollary 28.1. *There is a polynomial time algorithm which, given rational numbers* $\vartheta_1, \dots, \vartheta_d$ *and* $0 < \varepsilon < 1$, *finds integers* $u_0, u_1, \dots, u_d$ *where* $u_0 \neq 0$, *such that:*

$$\left|\vartheta_1 - \frac{u_1}{u_0}\right| \le \frac{\varepsilon}{u_0} \le \frac{2^{\frac{1}{4}(d+1)}}{u_0^{1+\frac{1}{d}}}, \dots, \left|\vartheta_d - \frac{u_d}{u_0}\right| \le \frac{\varepsilon}{u_0} \le \frac{2^{\frac{1}{4}(d+1)}}{u_0^{1+\frac{1}{d}}}, 1 \le u_0 \le \frac{2^{\frac{1}{4}d(d+1)}}{\varepsilon^d}.$$

Proof. Let L be the lattice in $\mathbb{E}^{d+1}$ with basis

$$a_0 = \left(\vartheta_1, \dots, \vartheta_d, \frac{\varepsilon^{d+1}}{2^{\frac{1}{4}d(d+1)}}\right), a_1 = (1, 0, \dots, 0), \dots, a_d = (0, \dots, 0, 1, 0).$$

Then

$$\det L = \frac{\varepsilon^{d+1}}{2^{\frac{1}{4}d(d+1)}}.$$

By the earlier theorem of Lenstra, Lenstra and Lovász, there is a polynomial time algorithm which finds a reduced basis of L. The first vector of this basis, say b_0, satisfies

$$(1) \quad \|b_0\| \le 2^{\frac{d}{4}} (\det L)^{\frac{1}{d+1}} = \varepsilon < 1$$

by Theorem 28.1. Write $b_0 = u_0 a_0 - u_1 a_1 - \dots - u_d a_d$, with $u_0, u_1, \dots, u_d \in \mathbb{Z}$. By (1) we have $u_0 \neq 0$. Thus we may assume that $u_0 \geq 1$. Since the last coordinate of b_0 is $\varepsilon^{d+1} u_0 2^{-d(d+1)/4}$ and since the ith coordinate is $u_0 \vartheta - u_i$ for $i = 1, \dots, d$, it follows from (1) that

$$|u_0 \vartheta - u_1| \le \varepsilon, \dots, |u_0 \vartheta - u_d| \le \varepsilon, 0 < u_0 \frac{\varepsilon^{d+1}}{2^{\frac{1}{4}d(d+1)}} < \varepsilon \text{ or } 1 \le u_0 \le \frac{2^{\frac{1}{4}d(d+1)}}{\varepsilon^d}.$$

This clearly implies the corollary. □

Approximation of Linear Forms

The approximate algorithmic version of the result of Dirichlet [271], see Corollary 22.4, can be proved in a similar way. It is as follows:

Corollary 28.2. *There is a polynomial time algorithm which, given rational numbers* $\vartheta_1, \dots, \vartheta_d$ *and* $0 < \varepsilon < 1$, *finds integers* u_0 *and* $u_1, \dots, u_d$, *not all* 0, *such that:*

$$|u_1 \vartheta_1 + \dots + u_d \vartheta_d - u_0| \le \varepsilon \le \frac{2^{\frac{1}{4}d(d+1)}}{\max\{u_1, \dots, u_d\}^d}, |u_1|, \dots, |u_d| \le \frac{2^{\frac{1}{4}(d+1)}}{\varepsilon^{\frac{1}{d}}}.$$

Shortest Lattice Vector Problem

The formal statement of this homogeneous problem is as follows:

Problem 28.1. *Given a lattice L in $\mathbb{E}^d$, find a point $l \in L \setminus \{o\}$, such that:*

$$\|l\| = \min\left\{\|m\| : m \in L \setminus \{o\}\right\}.$$

In the present chapter this problem appears in different versions:

Find a point $v \in \mathbb{Z}^d \setminus \{o\}$ such that $q(v) = \min\left\{q(u) : u \in \mathbb{Z}^d \setminus \{o\}\right\}$, where $q(\cdot)$ is a positive definite quadratic form, see Corollary 22.1.

Determine $\lambda_1(B^d, L) = 2\varrho(B^d, L)$ or, more generally, determine $\lambda_1(C, L) = 2\varrho(C, L)$ where C is a o-symmetric convex body. In essence, the latter means that the Euclidean norm is replaced by an arbitrary norm on $\mathbb{E}^d$. See Sects. 23.2 and 26.3.

The Minkowski fundamental theorem applied to B^d shows that:

$$\min\left\{\|m\| : m \in L \setminus \{o\}\right\} = \lambda_1(B^d, L) \le \left(\frac{2^d d(L)}{V(C)}\right)^{\frac{1}{d}},$$

but it does not explicitly provide a point $l \in L \setminus \{o\}$, such that $\|l\| = \lambda_1(B^d, L)$. As an immediate consequence of the Lenstra, Lenstra, Lovász theorem, we obtain the following approximate answer to the shortest lattice vector problem.

Corollary 28.3. *There is a polynomial time algorithm which, for any given basis $\{a_1, \dots, a_d\}$ of a rational lattice L in $\mathbb{E}^d$, finds a vector $l \in L \setminus \{o\}$, such that:*

$$\|l\| \le 2^{\frac{1}{2}(d-1)} \min\left\{\|m\| : m \in L \setminus \{o\}\right\}.$$

Remark. For any fixed $\varepsilon > 0$ there is a polynomial time algorithm due to Schnorr [912] with the factor $(1+\varepsilon)^{\frac{1}{2}(d-1)}$ instead of $2^{\frac{1}{2}(d-1)}$.

Remark. Using an approximate algorithmic version of John's theorem 11.2, this result can easily be extended to arbitrary norms instead of the Euclidean norm. This, then, is an approximate algorithmic version of Minkowski's fundamental theorem.

Nearest Lattice Point Problem

This is the following inhomogeneous problem.

Problem 28.2. *Given a lattice L in $\mathbb{E}^d$, find for any $x \in \mathbb{E}^d$ a point $l \in L$ such that:*

$$\|x - l\| = \min\left\{\|x - m\| : m \in L\right\}.$$

Van Emde Boaz [1007] has shown that this problem is NP-hard. Of course, it may be considered also for norms different from $\|\cdot\|$. Using the LLL-basis reduction, Babai [44] showed that an approximate answer to the nearest lattice point problem is possible in polynomial time:

Corollary 28.4. *There is a polynomial time algorithm which, for any given basis* $\{a_1, \dots, a_d\}$ *of a rational lattice* L *in* $\mathbb{E}^d$ *and any point* $x \in \mathbb{E}^d$ *with rational coordinates, finds a vector* $l \in L$ *such that:*

$$\|x - l\| \le 2^{\frac{d}{2}-1} \min\left\{\|x - m\| : m \in L\right\}.$$

Proof. First, the LLL-theorem shows that there is a polynomial time algorithm which finds an LLL-reduced basis $\{b_1, \dots, b_d\}$. Let $\{\hat{b}_1, \dots, \hat{b}_d\}$ be its Gram–Schmidt orthogonalization. We will use the following relations:

$$(2)\quad \begin{array}{ll} b_1 = \hat{b}_1 & \hat{b}_1 = b_1 \\ b_2 = \mu_{21}\hat{b}_1 + \hat{b}_2 & \hat{b}_2 = \nu_{21} b_1 + b_2 \\ \dots\dots\dots & \dots\dots\dots \\ b_d = \mu_{d1}\hat{b}_1 + \dots + \mu_{dd-1}\hat{b}_{d-1} + \hat{b}_d & \hat{b}_d = \nu_{d1} b_1 + \dots + \nu_{dd-1} b_{d-1} + b_d. \end{array}$$

Second,

(3) There is a polynomial time algorithm which finds for given rational $x \in \mathbb{E}^d$ a point $l \in L$ such that:

$$x - l = \sum_{i=1}^{d} \lambda_i \hat{b}_i, \text{ where } |\lambda_i| \le \frac{1}{2} \text{ for } i = 1, \dots, d.$$

Start with a representation of x of the form

$$x = \sum_i \lambda_{0i} \hat{b}_i.$$

Subtract $\lceil \lambda_{0d} \rfloor b_d$ to get a representation of the form

$$x - \lceil \lambda_{0d} \rfloor b_d = \sum_i \lambda_{1i} \hat{b}_i \text{ where } |\lambda_{1d}| \le \frac{1}{2}.$$

Next subtract $\lceil \lambda_{1d-1} \rfloor b_{d-1}$, etc. Taking into account (2) we arrive, after d steps, at (3).

Third, it will be shown that:

(4) $\|x - l\| \le 2^{\frac{d}{2}-1} \|x - m\|$ for any $m \in L$, where l is as in (3).

Let $m \in L$ and write

$$x - m = \sum_i \mu_i \hat{b}_i.$$

Let k be the largest index such that $\lambda_k \ne \mu_k$. Then

$$l - m = \sum_{i=1}^{k} (\lambda_i - \mu_i)\hat{b}_i = (\lambda_k - \mu_k) b_k + \text{ lin. comb. of } b_1, \dots, b_{k-1} \in L$$

by (2). Thus $\lambda_k - \mu_k$ is a non-zero integer. In particular $|\lambda_k - \mu_k| \ge 1$. Since $|\lambda_k| \le \frac{1}{2}$, it follows that $|\mu_k| \ge \frac{1}{2}$. So,

$$\|x - m\|^2 = \sum_i \mu_i^2 \|\hat{b}_i\|^2 \geq \frac{1}{4}\|\hat{b}_k\|^2 + \sum_{i=k+1}^{d} \mu_i^2 \|\hat{b}_i\|^2$$
$$\geq \frac{1}{4}\|\hat{b}_k\|^2 + \sum_{i=k+1}^{d} \lambda_i^2 \|\hat{b}_i\|^2$$

and therefore

$$\|x - l\|^2 = \|\sum_i \lambda_i \hat{b}_i\|^2 \leq \frac{1}{4}\sum_{i=1}^{k} \|\hat{b}_i\|^2 + \sum_{i=k+1}^{d} \lambda_i^2 \|\hat{b}_i\|^2$$
$$\leq \frac{1}{4}\sum_{i=1}^{k} 2^{k-i} \|\hat{b}_k\|^2 + \sum_{i=k+1}^{d} \lambda_i^2 \|\hat{b}_i\|^2$$
$$\leq 2^{k-2}\|x - m\|^2 \leq 2^{d-2}\|x - m\|^2,$$

where we have used the inequality (4) in the proof of Theorem 28.1. The proof of (4) and thus of the corollary is complete. □

29 Packing of Balls and Positive Quadratic Forms

Packing of balls is a story with many chapters starting with Kepler [576, 577]. The corresponding theory of positive definite quadratic forms dates back to Lagrange [628] and Gauss [364]. It seems that the main reason for the intensive geometric work on lattice packing of balls in the nineteenth and twentieth century was the arithmetic background. Rogers [851], p.2, expressed this in his little classic on Packing and Covering as follows:

> Largely because of its connection with the arithmetic minimum of a positive definite quadratic form, much effort has been devoted to the study of $\delta_L(K_n)$, where K_n is the unit sphere in n-dimensional space.

For the notions of packing, upper density, maximum density $\delta_L(B^d)$ of lattice packings of B^d and maximum density $\delta_T(B^d)$ of packings of translates of B^d, see Sect. 30.1.

In this section we outline a selection of classical and modern results and methods. We begin with densest lattice packing of balls in 2 and 3 dimensions and the density bounds of Blichfeldt and Minkowski–Hlawka. Then a chapter on the geometry of positive definite quadratic forms is presented, which goes back at least to Korkin and Zolotarev and to Voronoĭ. Finally, relations between ball packing and error-correcting codes are discussed.

For more information, compare the monographs of Thompson [995], Conway and Sloane [220], Leppmeier [650], Zong [1049] and Martinet [691] and the surveys of Bambah [56] and Pfender and Ziegler [799].

29.1 Densest Lattice Packing of Balls in Dimensions 2 and 3

The densest lattice packings of balls are known in dimensions:

$d = 2$: Lagrange [628] (in arithmetic form)
$d = 3$: Gauss [364]
$d = 4, 5$: Korkin and Zolotarev [609, 611]
$d = 6, 7, 8$: Blichfeldt [132]
$d = 24$: Cohn and Kumar [211]

The densest lattice packing of balls in dimension $d = 24$ is provided by the Leech lattice.

The proofs for $d = 3$, up to the year 2000, were either arithmetic, or geometric but rather clumsy. Hales [472] gave a simple and elegant geometric argument which is reproduced below. We make some remarks on the result of Hales that $\delta_L(B^3) = \delta_T(B^3)$ and add heuristic considerations for higher dimensions.

Densest Lattice Packing in Dimensions 2 and 3

We prove the following results of Lagrange and Gauss.

Theorem 29.1. $\delta_L(B^2) = \dfrac{\pi}{2\sqrt{3}} = 0.906\,899..,\ \ \delta_L(B^3) = \dfrac{\pi}{3\sqrt{2}} = 0.740\,480...$

Proof. The case $d = 2$ is left to the reader (Fig. 29.1).

For the proof in case $d = 3$, note first that by Theorem 30.1 there is a packing lattice L of B^3 which yields a packing of B^3 of maximum density. If the ball B^3 did not touch any other ball in this packing then contracting L slightly would yield a lattice packing of higher density, which is impossible. Thus, by periodicity, the balls in the packing are arranged as beads in parallel strings such that neighbouring balls in each string touch. If balls in different strings did not touch then, again, by pushing the strings suitably together, we obtain a lattice packing of B^3 of higher density, which is impossible. Thus we may arrange the strings in parallel layers where, in

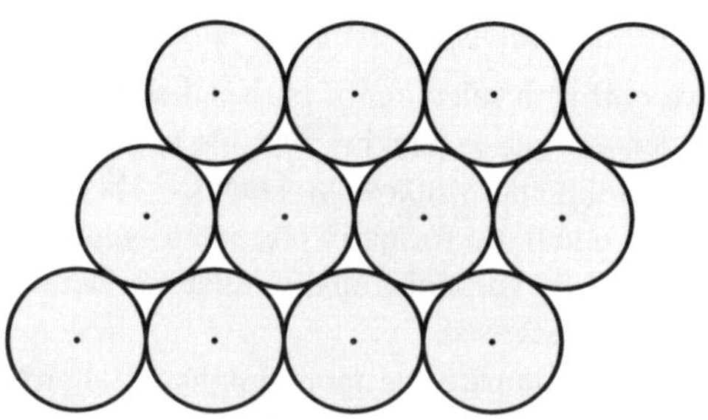

Fig. 29.1. Densest lattice packing of circular discs in $\mathbb{E}^2$

each layer, neighbouring strings touch. Given a layer, the balls in each neighbouring layer must rest in the pockets formed by three balls in our layer and touch all three balls; otherwise we can form a lattice packing of higher density, which is impossible. (Of these three balls two are in one string and the third one in a neighbouring string.) We thus have obtained three balls, touching pairwise. (Of these three balls two are neighbours in a string in our layer and one is from a parallel neighbouring layer.) We now change our point of view. By periodicity we may suppose that these three balls have centres at $o, a, b \in L$, where o, a, b form an equilateral triangle. The balls with centres in the 2-dimensional sub-lattice of L generated by a, b form a layer in which the balls are arranged hexagonally. The balls in the parallel neighbouring layers must rest in the pockets formed by three pairwise touching balls in our layer and touch all three balls; otherwise we can obtain a lattice packing of higher density. This shows that L is generated by points a, b, c, such that o, a, b, c form the vertices of a regular tetrahedron. That is, L is a *face-centred cubic lattice* and an elementary calculation shows that the density is as required. □

Densest Packing of Translates in Dimension 2 and 3

For $d = 2$ Thue [996, 997] showed that

$$\delta_T(B^2) = \delta_L(B^2).$$

A gap in Thue's proof was filled by Fejes Tóth [327]. For a proof of a more general result due to Fejes Tóth and Rogers, see Sect. 30.4. For a proof of Thue's theorem, based on Fejes Tóth's moment theorem 33.1, compare Sect. 33.4.

In a recent breakthrough Hales has proved the so-called *Kepler conjecture*, which says that

$$\delta_L(B^3) = \delta_T(B^3).$$

See Hales [472–474]. Hales' proof is long and computationally involved. In [474] an outline if given. Many mathematicians had previously worked on this problem, including Fejes Tóth, Zassenhaus and Hsiang. Fejes Tóth, in particular, described in his book [329] a plan of a proof in which the problem is reduced to an optimization problem in finitely many variables over a compact set which, possibly, can be solved in the future on a computer. This would yield a proof with a transparent mathematical part.

Heuristic Observation

Consider densest lattice packings of the unit ball in $\mathbb{E}^d$ for $d = 2, 3, \ldots$. The space between the balls of the packings seems to become bigger and bigger as the dimension increases. Thus, for sufficiently large d, a suitable translate of B^d should fit between the balls of a densest lattice packing. Then, by periodicity, a suitable translate of the whole lattice packing also fits into the space left uncovered by the original lattice packing. Together with the original lattice packing of B^d, it forms a packing of translates of B^d with density $2\delta_L(B^d)$. In this dimension we then have,

$$\delta_L(B^d) \le \frac{1}{2}\,\delta_T(B^d).$$

29.2 Density Bounds of Blichfeldt and Minkowski–Hlawka

While the problem of finding tight lower and upper bounds or asymptotic formulae as $d \to \infty$ for the maximum density of lattice and non-lattice packings of balls remains unsolved, there are substantial pertinent results.

In this section we give Blichfeldt's upper estimate for $\delta_T(B^d)$ and the lower estimate for $\delta_L(B^d)$ which follows from the Minkowski–Hlawka theorem. Then more precise upper estimates for $\delta_T(B^d)$ due to Sidel'nikov, Kabat'janski and Levenstein and lower estimates for $\delta_L(B^d)$ of Schmidt and Ball are described.

Upper Estimate for $\delta_T(B^d)$; Blichfeldt's Enlargement Method

The following result of Blichfeldt [131] was the first substantial improvement of the trivial estimate $\delta_T(B^d) \le 1$.

Theorem 29.2. $\delta_T(B^d) \le \dfrac{d+2}{2}\,2^{-\frac{d}{2}} = 2^{-\frac{d}{2}+o(d)}$ *as* $d \to \infty$.

Proof. The first step is to show the *inequality of Blichfeldt*:

$$(1)\quad \sum_{j,k=1}^{n} \|t_j - t_k\|^2 \le 2n \sum_{j=1}^{n} \|s - t_j\|^2 \text{ for all } t_1, \dots, t_n, s \in \mathbb{E}^d.$$

Clearly,

$$\begin{aligned}\sum_{j,k=1}^{n} (a_j - a_k)^2 &= \sum_{j,k} (a_j^2 + a_k^2 - 2a_j a_k) = \sum_{j,k} (a_j^2 + a_k^2) - 2\Big(\sum_j a_j\Big)^2 \\ &\le \sum_{j,k} (a_j^2 + a_k^2) = 2n \sum_j a_j^2 \text{ for all } a_1, \dots, a_n \in \mathbb{R}.\end{aligned}$$

Then,

$$\begin{aligned}\sum_{j,k} \|t_j - t_k\|^2 &= \sum_{j,k} \|(s - t_j) - (s - t_k)\|^2 = \sum_{j,k} \sum_i \big((s_i - t_{ji}) - (s_i - t_{ki})\big)^2 \\ &\le \sum_i \sum_{j,k} \big((s_i - t_{ji}) - (s_i - t_{ki})\big)^2 \le 2n \sum_i \sum_j (s_i - t_{ji})^2 \\ &= 2n \sum_j \sum_i (s_i - t_{ji})^2 = 2n \sum_j \|s - t_j\|^2,\end{aligned}$$

concluding the proof of Blichfeldt's inequality (1).

In the second step, we prove the following estimate, where the function f is defined by $f(r) = \max\big\{0, 1 - \frac{1}{2}r^2\big\}$ for $r \ge 0$.

(2) Let $\{B^d + t : t \in T\}$ be a packing of the unit ball B^d. Then

$$\sum_{t \in T} f(\|s - t\|) \leq 1 \text{ for each } s \in \mathbb{E}^d.$$

Since $\{B^d + t : t \in T\}$ is a packing, any two distinct vectors $t \in T$ have distance at least 2. Hence

$$\sum_{j,k} \|t_j - t_k\|^2 \geq 4n(n-1) \text{ for any distinct } t_1, \dots, t_n \in T.$$

An application of Blichfeldt's inequality (1) implies that:

$$\sum_{j} \|s - t_j\|^2 \geq 2(n-1) \text{ for any distinct } t_1, \dots, t_n \in T \text{ and } s \in \mathbb{E}^d.$$

Thus

$$\begin{aligned}\sum_{t \in T} f(\|s - t\|) &= \sum_{j} f(\|s - t_j\|) = \sum_{j} \Big(1 - \frac{1}{2}\|s - t_j\|^2\Big)\\ &\leq n - (n-1) = 1 \text{ for } s \in \mathbb{E}^d,\end{aligned}$$

where, for given s, the points $t_1, \dots, t_n$ are precisely the points t of T with $\|s - t\| < \sqrt{2}$, i.e. those points t of T with $f(\|s - t\|) > 0$. The proof of (2) is complete.

In the third step we show the following:

(3) Let $\{B^d + t : t \in T\}$ be a packing of B^d. Then its upper density is at most $\dfrac{d+2}{2} 2^{-\frac{d}{2}}$.

Let K be the cube $\{x : |x_i| \leq 1\}$. Proposition (2) and the definition of f then yield the following:

$$\begin{aligned}V\big((\tau + 2\sqrt{2})K\big) &\geq \int\limits_{(\tau+2\sqrt{2})K} \Big\{\sum_{t \in T} f(\|s - t\|)\Big\} ds\\
&\geq \sum_{t \in T \cap (\tau+\sqrt{2})K} \int\limits_{\sqrt{2}B^d + t} f(\|s - t\|)\, ds = \sum_{t \in T \cap (\tau+\sqrt{2})K} \int\limits_{\sqrt{2}B^d} f(\|s\|)\, ds\\
&= \sum_{t \in T \cap (\tau+\sqrt{2})K} \int_0^{\sqrt{2}} S(rB^d) f(r)\, dr = \sum_{t \in T \cap (\tau+\sqrt{2})K} d\, V(B^d) \int_0^{\sqrt{2}} r^{d-1}\Big(1 - \frac{r^2}{2}\Big) dr\\
&= \frac{2}{d+2} 2^{\frac{d}{2}} \sum_{t \in T \cap (\tau+\sqrt{2})K} V(B^d) \geq \frac{2^{\frac{d}{2}+1}}{d+2} \sum_{t \in T} V\big((B^d + t) \cap \tau K\big),\end{aligned}$$

where $S(\cdot)$ stands for ordinary surface area in $\mathbb{E}^d$. Thus,

$$\frac{1}{(2\tau)^d}\sum_{t\in T} V\big((B^d+t)\cap \tau K\big) \le \frac{d+2}{2}\,2^{-\frac{d}{2}}\frac{\big(2(\tau+2\sqrt{2})\big)^d}{(2\tau)^d}.$$

Now let $\tau \to \infty$ to get Proposition (3).

Having proved (3), the definition of $\delta_T(B^d)$ readily yields the theorem. □

More Precise Upper Estimates for $\delta_T(B^d)$

A slight improvement of Blichfeldt's result is due to Rogers [849]:

$$\delta_T(B^d) \le (d/e)2^{-0.5d}(1+o(1)).$$

A refinement of Rogers' upper estimate for small dimensions is due to Bezdek [111]. For many decades, it was a widespread belief among number theorists that, in essence, Blichfeldt's upper estimate for $\delta_T(B^d)$ was best possible. Thus it was a great surprise when, in the 1970s, essential improvements were achieved using spherical harmonics:

$\delta_T(B^d) \le 2^{-0.5096d+o(d)}$: Sidel'nikov [934]
$\delta_T(B^d) \le 2^{-0.5237d+o(d)}$: Levenštein [651]
$\delta_T(B^d) \le 2^{-0.599d+o(d)}$: Kabat'janski and Levenštein [557]

For an outline of the proof of the last estimate, see Fejes Tóth and Kuperberg [325].

Lower Estimate for $\delta_L(B^d)$

In Sect. 30.3 we shall see, in the more general context of lattice packing of convex bodies, that, as a consequence of the Minkowski–Hlawka theorem 24.1, the following result holds.

Theorem 29.3. $\delta_L(B^d) \ge 2^{-d}$.

Actually, slightly more is true, where $\zeta(\cdot)$ denotes the Riemann zeta function:

$\delta_L(B^d) \ge 2\zeta(d)\,2^{-d}$. This follows from Hlawka's [509] version of the Minkowski–Hlawka theorem for star bodies.

$\delta_L(B^d) \ge 2dc_d\,2^{-d}$, where $c_d \to \log\sqrt{2}$ as $d\to\infty$. This follows from Schmidt's [893] refinement of the Minkowski–Hlawka theorem.

$\delta_L(B^d) \ge 2(d-1)\zeta(d)\,2^{-d}$. This is the best known lower bound. It is due to Ball [52].

It is believed that no essential improvement of these estimates is possible in the sense that the best estimate is of the form

$$\delta_L(B^d) \ge 2^{-d+o(d)} \text{ as } d\to\infty.$$

29.3 Error Correcting Codes and Ball Packing

The Minkowski–Hlawka theorem guarantees the existence of (comparatively) dense lattice packings of balls and, more generally, of centrally symmetric convex bodies, but does not provide constructions of such.

Starting with the work of Leech [637], error correcting codes turned out to be a powerful tool for the construction of dense lattice and non-lattice packings of balls. Work in this direction culminated in the construction of Rush [862] of lattice packings of balls of density $2^{-d+o(d)}$, thus reaching the Minkowski–Hlawka bound. An explicit construction of dense lattice or non-lattice packings based on error correcting codes requires codes which can be given explicitly. Unfortunately this is not the case for the codes used by Rush.

A different way to construct dense packings, in dimensions which are not too large, is to pack congruent layers of ball packings in stacks to get a layer of 1-dimension more, and use induction.

In this section we first give the necessary definitions from coding theory and describe two constructions of ball packings based on codes due to Leech and Sloane [638]. Then the layer construction is outlined.

For more information, see the monographs of Conway and Sloane [220] and Zong [1049] and the survey of Rush [863] on sphere packing.

Binary Error Correcting Codes

In a data transmission system it is the task of error correcting codes to correct errors which might have occurred during the transmission in the channel. See Sect. 33.4 for some information.

A *(binary error correcting) code* C of *length* d consists of a set of ordered d-tuples of 0s and 1s, the so-called *code-words*. Clearly, C can be identified with a subset of the set of vertices of the unit cube $\{x : 0 \leq x_i \leq 1\}$ in $\mathbb{E}^d$ or with a subset of F^d, where F is the Galois field consisting of 0 and 1. The 0s and 1s in a code-word are called its *letters*. Given two code-words of C, the number of letters in which they differ is their *Hamming distance*. The minimum of the Hamming distances of any two distinct code-words of C is the *minimum distance* of C. A code C of length d consisting of M code-words and of minimum distance m is called a *(d, M, m)-code*. A (d, M, m)-code C is *linear of dimension* k if it is a k-dimensional linear subspace of the d-dimensional vector space F^d over F, i.e. F^d with coordinatewise addition modulo 2 and scalar multiplication. In this case C is called a *$[d, k, m]$-code*. If the result of the transmission in the channel of a code-word is a *word*, i.e. a d-tuple of 0s and 1s, we assign to it (one of) the closest code-word(s). This is the original code-word if the number of errors is at most $\lceil \frac{m}{2} \rceil - 1$. Thus, given d, a code is good if, for given number of code-words, its minimum distance is large, or for given minimum distance the number of code-words is large.

Examples of Codes

We present three codes which are important for packing of balls:

Our first example is the $[d, d-1, 2]$-*even weight code* consisting of all words with an even number of 1s.

A *Hadamard matrix* H is a $d \times d$ matrix with entries ± 1 such that $HH^T = dI$, where I is the $d \times d$ unit matrix. It is known that Hadamard matrices can exist only for $d = 1, 2$ and multiples of 4, but whether there are Hadamard matrices for all multiples of 4 is an open question. The following inductive construction yields special Hadamard matrices for powers of 2:

$$H_1 = (1),\ H_2 = \begin{pmatrix} H_1 & H_1 \\ H_1 & -H_1 \end{pmatrix}, \dots,\ H_{2k} = \begin{pmatrix} H_k & H_k \\ H_k & -H_k \end{pmatrix}, \dots$$

Observe that the first row and column of H_8 contain only 1s. Now consider the matrix $\widetilde{H}_8$ by replacing the 1s by 0s and the -1s by 1s. Delete the first column of $\widetilde{H}_8$. The rows of the remaining matrix then form a $(7, 8, 4)$-code. The code-words form the vertices of a regular simplex inscribed into the unit cube in $\mathbb{E}^7$.

The *Golay code* G_{24} is a $(24, 2^{12}, 8)$-code which is defined as follows: Let G_{12} be the 12×12 matrix (c_{ik}), where

$$c_{ik} = \begin{cases} 0 \text{ if } i = k = 1 \text{ or } i, k \geq 2 \text{ and } i + k - 4 \text{ is a} \\ \quad \text{quadratic residue } \bmod 11, \\ 1 \text{ otherwise,} \end{cases}$$

and let I_{12} be the 12×12 unit matrix. Then G_{24} is the 12-dimensional subspace of F^{24} spanned by the rows of the matrix (I_{12}, G_{12}).

Construction of Ball Packings by Means of Codes

Leech and Sloane [638] specified three basic constructions of packings of balls using error correcting binary codes. We describe two of these, constructions A and B.

Theorem 29.4. *Construction A: Let C be a (d, M, m)-code. We consider C as a subset of the set of vertices of the unit cube in $\mathbb{E}^d$. Let*

$$T = C + 2\mathbb{Z}^d,\ \rho = \min\left\{1, \frac{\sqrt{m}}{2}\right\}.$$

Then $\{\rho B^d + t : t \in T\}$ is a packing of density

$$\frac{M\rho^d\, V(B^d)}{2^d}.$$

If C is linear, then T is a lattice.

Proof. In order to show that $\{\rho B^d + t : t \in T\}$ is a packing it is sufficient to prove the following:

(1) $s, t \in T,\ s \neq t \Rightarrow \|s - t\| \geq 2\rho.$

Let $s = c + 2u, t = d + 2v$. If $c \neq d$, then $c - d$ has at least m coordinates equal to ± 1. Thus $s - t = c - d + 2(u - v)$ has at least m odd coordinates. Hence $\|s - t\| \geq \sqrt{m} \geq 2\rho$. If $c = d$, then $s - t = 2(u - v) \neq o$ and thus $\|s - t\| \geq 2 \geq 2\rho$. This concludes the proof of (1) and thus shows that $\{\rho B^d + t : t \in T\}$ is a packing. The other assertions are obvious. □

An application of the Construction A to the $[d, d - 1, 2]$-even weight code for $d = 3, 4, 5$ yields the densest lattice packing of balls in these dimensions. If this construction is applied to the $(7, 8, 4)$-code constructed earlier by means of the Hadamard matrix H_8, it yields the densest lattice packing of balls for $d = 7$.

Using a refined version of construction A and codes with large alphabets, Rush [862] was able to construct lattice packings of B^d of density $2^{-d+o(d)}$, thus reaching the Minkowski–Hlawka bound, see Theorem 29.3.

Theorem 29.5. *Construction B: Let C be a (d, M, m)-code such that the weight of each code-word is even. Let*

$$T = \left\{t \in C + 2\mathbb{Z}^d : 4 \mid t_1 + \cdots + t_d\right\},\ \rho = \min\left\{\sqrt{2}, \frac{\sqrt{m}}{2}\right\}.$$

Then $\{\rho B^d + t : t \in T\}$ is a packing of density

$$\frac{M\rho^d\, V(B^d)}{2^{d+1}}.$$

If C is linear, then T is a lattice.

The proof of this result is similar to that of Theorem 29.4 and thus is omitted.

The Leech Lattice

If Construction B is applied to the Golay code G_{24}, we obtain a lattice packing $\{B^{24} + (1/\sqrt{2})\, l : l \in L\}$. It turns out that a suitable translate of this packing fits into the space left uncovered. This gives again a lattice packing $\{B^{24} + (1/\sqrt{2})\, m : m \in M\}$ where M is the lattice $L \cup (L + a), a = (1/2)\,(1, 1, \ldots, 1, -3)$, the *Leech lattice*. The Leech lattice goes back to the Göttingen thesis of Niemeier [771]. Cohn and Kumar [211] proved that it provides the densest lattice packing of balls in $\mathbb{E}^{24}$. The number of neighbours of B^{24} in the Leech lattice packing of B^{24} is 196560. This is the maximum number of neighbours of B^{24} in any lattice packing and, moreover, in any packing of B^{24} as shown by Levenštein [652] and Odlyzko and Sloane [775]. See also Zong [1049].

The Layer Construction

A different construction for dense lattice packings of balls can be described as follows: For $d = 1$ consider the packing $\{B^1 + u : u \in 2\mathbb{Z}\}$. Assume now that $d > 1$ and that we have constructed a lattice packing of B^d, say $\{B^d + l : l \in L^d\}$ where L^d is a lattice in $\mathbb{E}^d$. For $d + 1$ proceed as follows: Consider $\mathbb{E}^d$ as being embedded in $\mathbb{E}^{d+1}$ as usual (first d coordinates). Clearly, $\{B^{d+1} + l : l \in L^d\}$ is a layer of non-overlapping balls in $\mathbb{E}^{d+1}$. Consider a translate $\{B^{d+1} + l + b : l \in L^d\}$ of this layer where $b = (b_1, \dots, b_{d+1})$ is chosen such that $b_{d+1} > 0$ is minimal and such that $\{B^{d+1} + l + ub : l \in L^d, u \in \mathbb{Z}\}$ is a packing of B^{d+1} with packing lattice $L^{d+1} = \{l + ub : l \in L^d, u \in \mathbb{Z}\}$.

This construction yields sequences of lattices (not necessarily unique, not necessarily infinite). In this way one can obtain the densest lattice packing in dimensions $d = 2, \dots, 8$ and 24.

29.4 Geometry of Positive Definite Quadratic Forms and Ball Packing

The geometric theory of positive definite quadratic forms, which is related to lattice packing and covering with balls, was developed since the nineteenth century mainly by the Russian school of the geometry of numbers. Among the contributors are Korkin, Zolotarev, Voronoĭ, Delone, Ryshkov and their students. Voronoĭ deserves particular mention. We list also Minkowski, Blichfeldt, Watson and Barnes. Recent contributions are due to the French school of quadratic forms of Martinet, see [691].

Incidentally, note that the study of general, not necessarily positive definite quadratic forms is a quite different thing. In the context of the geometry of numbers it was cultivated, amongst others, by Bambah, Dumir, Hans-Gill, Raka and their disciples at Chandigarh, see the report of Bambah, Dumir and Hans-Gill [57].

The main problem of the geometric theory of positive definite quadratic form is to determine the extreme and the absolute extreme forms. This is equivalent to the determination of the lattice packings of balls which have maximum density either locally, i.e. among all sufficiently close lattice packings, or globally, i.e. among all lattice packings. Other problems deal with minimum points, covering and reduction.

A quadratic form on $\mathbb{E}^d$ may be represented by the vector of its coefficients in $\mathbb{E}^{\frac{1}{2}d(d+1)}$. This allows us to transform certain problems on positive definite quadratic forms and lattice packing of balls in $\mathbb{E}^d$ into geometric problems about subsets of $\mathbb{E}^{\frac{1}{2}d(d+1)}$ which, in many cases, are more accessible. The solution of the geometric problem, finally, is translated back into the language of positive forms. The systematic use of this idea is due to Voronoĭ. A different application of it is our proof of John's characterization of the ellipsoid of maximum volume inscribed into a convex body, see Theorem 11.2.

In the following we describe pertinent results of Korkin and Zolotarev, Voronoĭ, Delone and Ryshkov, using the approach of Ryshkov [865].

In the book of Martinet [691], the theory of positive definite quadratic forms is treated from an arithmetic point of view. A comprehensive modern geometric exposition has yet to be completed, although parts of it can be found in the literature, for example in Delone [257] and Delone and Ryshkov [259]. See also [1015].

The Cone of Positive Definite Quadratic Forms

Let q be a real quadratic form on $\mathbb{E}^d$,

$$q(x) = x^T A x = \sum_{i,k=1}^{d} a_{ik}\, x_i x_k \text{ for } x \in \mathbb{E}^d,$$

where

$$A = (a_{ik}),\ a_{ik} = a_{ki},$$

is the symmetric $d \times d$ *coefficient matrix* of q. The *coefficient vector* of q is the vector

$$(a_{11}, a_{12}, \dots, a_{1d}, a_{22}, a_{23}, \dots, a_{dd}) \in \mathbb{E}^{\frac{1}{2}d(d+1)}.$$

The determinant $\det A$ is the *discriminant* of q. In the following we will not distinguish between a quadratic form, its coefficient matrix and its coefficient vector. A quadratic form q on $\mathbb{E}^d$ is *positive definite* if $q(x) > 0$ for all $x \in \mathbb{E}^d \setminus \{o\}$.

Let $\mathcal{P}$ be the set of all (coefficient vectors of) positive definite quadratic forms on $\mathbb{E}^d$. If $q, r \in \mathcal{P}$, then $\lambda q + \mu r \in \mathcal{P}$ for all $\lambda, \mu > 0$. Thus $\mathcal{P}$ is a convex cone in $\mathbb{E}^{\frac{1}{2}d(d+1)}$ with apex at the origin. A result from linear algebra says that a quadratic form on $\mathbb{E}^d$ is positive definite if and only if all principal minors of its coefficient matrix are positive. Thus, if $q \in \mathcal{P}$, all quadratic forms on $\mathbb{E}^d$, the coefficients of which are sufficiently close to that of q, are also positive definite and thus are in $\mathcal{P}$. This means that $\mathcal{P}$ is open. $\mathcal{P}$ is called the *(open convex) cone of (coefficient vectors of) positive definite quadratic forms* on $\mathbb{E}^d$. $\mathcal{P}$ is not a polyhedral cone although it shares several properties with polyhedral cones. For more information on the algebraic and geometric properties of $\mathcal{P}$ see the articles of Bertraneu and Fichet [103] and Gruber [444].

Let U be an integer $d \times d$ matrix. It yields the linear transformation

$$x \to Ux$$

of $\mathbb{E}^d$ onto itself. In turn, this linear transformation induces a transformation

$$q(x) = x^T A x \to q(Ux) = x^T U^T A U x$$

of the space of all quadratic forms on $\mathbb{E}^d$ onto itself, or in terms of coefficient vectors, a transformation

$$\mathcal{U} : (a_{ik}) \to \Big(\sum_{l,m} u_{li} u_{mk} a_{lm}\Big)$$

of the space of coefficient vectors onto itself. $\mathcal{U}$ may be considered as a linear transformation of $\mathbb{E}^{\frac{1}{2}d(d+1)}$ onto itself which maps $\mathcal{P}$ onto itself.

Two forms $q, r \in \mathcal{P}$ are *equivalent*, if there is an integer unimodular $d \times d$ matrix U such that $r(x) = q(Ux)$. Since $U\mathbb{Z}^d = \mathbb{Z}^d$, equivalent forms assume the same values for integer values of the variables, that is, on $\mathbb{Z}^d$.

Lattice Packing of Balls and Positive Definite Quadratic Forms

First, let B be a non-singular $d \times d$ matrix, not necessarily symmetric. Then $L = B\mathbb{Z}^d$ is a lattice in $\mathbb{E}^d$. The columns $b_1, \dots, b_d$ of B form a basis of L. To L or, more precisely, to the basis B we associate the positive definite quadratic form q on $\mathbb{E}^d$ defined by:

$$q(x) = \|Bx\|^2 = Bx \cdot Bx = x^T B^T Bx = x^T Ax \text{ for } x \in \mathbb{E}^d,$$

where $A = (a_{ik}) = (b_i \cdot b_k) = B^T B$ is a symmetric $d \times d$ matrix. q is called the *metric form of L associated to the basis* $\{b_1, \dots, b_d\}$. If $\{c_1, \dots, c_d\}$ is a different basis of L and C the matrix with columns $c_1, \dots, c_d$, then there is an integer unimodular $d \times d$ matrix U such that $C = BU$. The metric form r of L associated to the basis $\{c_1, \dots, c_n\}$ then is

$$r(x) = x^T C^T Cx = x^T U^T B^T BUx = x^T U^T AUx = q(Ux) \text{ for } x \in \mathbb{E}^d$$

and thus is equivalent to q. If, on the other hand, r is a positive definite quadratic form equivalent to q, then there is a basis of L such that r is the metric form of L associated to this basis.

Second, let $q(x) = x^T Ax$ be a positive definite quadratic form on $\mathbb{E}^d$. Then by a result from linear algebra, there is a $d \times d$ matrix B such that $A = B^T B$. Thus q is the metric form of the lattice $L = B\mathbb{Z}^d$ associated to the basis $\{b_1, \dots, b_d\}$ which consists of the columns of B. Besides B it is precisely the matrices of the form $C = SB$, where S is any orthogonal $d \times d$ matrix, for which $A = C^T C$. Thus L is unique up to (proper and improper) rotations.

Given a lattice L, we have, for the packing radius of B^d with respect to L,

$$\varrho(B^d, L) = \frac{1}{2} \min\left\{\|l\| : l \in L \setminus \{o\}\right\} = \frac{1}{2} \lambda_1(B^d, L).$$

The density of the lattice packing $\left\{\varrho(B^d, L)B^d + l : l \in L\right\}$ is given by:

$$\frac{V\left(\varrho(B^d, L)B^d\right)}{d(L)} = \frac{\varrho(B^d, L)^d\, V(B^d)}{d(L)}.$$

We say that L provides a *locally densest lattice packing* of balls if

$$\frac{\varrho(B^d, M)^d\, V(B^d)}{d(M)} \le \frac{\varrho(B^d, L)^d\, V(B^d)}{d(L)}$$

for all lattices M in a suitable neighbourhood of L. If q is a positive definite quadratic form on $\mathbb{E}^d$, its *arithmetic minimum* $\min\{q\}$ is defined by:

$$\min\{q\} = \min\left\{q(u) : u \in \mathbb{Z}^d \setminus \{o\}\right\}.$$

The points $u \in \mathbb{Z}^d \setminus \{o\}$, for which $\min\{q\}$ is attained are the *minimum points* or *vectors* of q. We say that q is an *extreme form* if

$$\frac{\min\{r\}^d}{\det r} \le \frac{\min\{q\}^d}{\det q}$$

for all positive definite quadratic forms r in a suitable neighbourhood of q, i.e. if the coefficient vector of r is in a suitable neighbourhood in $\mathcal{P}$ of the coefficient vector of q.

If L and q are such that q is the metric form of L associated to the basis $\{b_1, \dots, b_d\}$ of L and B is the $d \times d$ matrix with columns $b_1, \dots, b_d$, then $L = B\mathbb{Z}^d$ and

$$\begin{aligned}\min\{q\} &= \min\left\{q(u) = u^T B^T Bu = Bu \cdot Bu = \|Bu\|^2 : u \in \mathbb{Z}^d \setminus \{o\}\right\}\\ &= \min\left\{\|l\|^2 : l \in L \setminus \{o\}\right\} = 4\varrho(L, B^d)^2,\\ \det q &= \det(B^T B) = \det(B)^2 = d(L)^2.\end{aligned}$$

Hence,

$$\frac{\min\{q\}^d}{\det q} = \frac{4^d}{V(B^d)^2}\left(\frac{\varrho(L, B^d)^d\, V(B^d)}{d(L)}\right)^2.$$

From this it follows that the problem to determine the locally densest lattice packings of balls in $\mathbb{E}^d$ and the problem to determine the extreme positive definite quadratic forms on $\mathbb{E}^d$ are equivalent. In the following we consider only the latter.

The Set of Minimum Points

The following result contains some information on the set of minimum points. It is due to Delone and Ryshkov [259]. Proposition (i) refines an old theorem of Korkin and Zolotarev [611].

Theorem 29.6. *Let q be a positive definite quadratic form on $\mathbb{E}^d$. Then the following hold:*

(i) *The absolute value of the determinant of any d-tuple of minimum vectors is bounded above by $d\,!$.*

(ii) *Assume that there are d linearly independent minimum vectors. Then there is a basis of $\mathbb{Z}^d$ such that for any minimum vector the absolute value of its coordinates with respect to this basis is bounded above by $d\,(d\,!)^2$.*

Proof. (i) The *Dirichlet–Voronoĭ cell D* of the origin o with respect to the lattice $\mathbb{Z}^d$ and the norm $q(\cdot)^{1/2}$ is defined by:

$$D = \left\{x : q(x) \le q(x-u) \text{ for all } u \in \mathbb{Z}^d\right\}.$$

D is a convex o-symmetric polytope and $\{D + u : u \in \mathbb{Z}^d\}$ is a tiling of $\mathbb{E}^d$. In particular, $V(D) = 1$. Next,

(1) $\frac{1}{2}m \in D$ for each minimum vector m.

To see this note that, since m is a minimum vector, $q(m) \leq q(m-u)$ for all $u \in \mathbb{Z}^d \setminus \{m\}$. Note that $m \neq 2u$ for each $u \in \mathbb{Z}^d$. Hence $q(m) \leq q(m-2u)$ and thus $q\left(\frac{1}{2}m\right) \leq q\left(\frac{1}{2}m - u\right)$ for all $u \in \mathbb{Z}^d$.

Now, let $m_1, \dots, m_d \in \mathbb{Z}^d$ be d minimum vectors. If these are linearly dependent, we are done. Otherwise, there is a basis $\{b_1, \dots, b_d\}$ of $\mathbb{Z}^d$ such that:

$$\text{(2)} \quad \begin{aligned} m_1 &= u_{11}b_1 \\ m_2 &= u_{21}b_1 + u_{22}b_2 \\ &\cdots\cdots\cdots\cdots \\ m_d &= u_{d1}b_1 + \cdots\cdots + u_{dd}b_d \end{aligned} \qquad \text{where} \quad u_{ik} \in \mathbb{Z},\ |u_{ik}| \leq |u_{ii}|,\ u_{ii} \neq 0.$$

To see this, apply Theorem 21.3. This gives a basis satisfying (2), except, possibly, for the condition that $|u_{ik}| \leq |u_{ii}|$, but this condition can easily be obtained by adding suitable integer multiples of $b_1, \dots, b_{i-1}$ to b_i for $i = 2, \dots, d$ and taking the vectors thus obtained as the required basis. Since $\pm\frac{1}{2}m_1, \dots, \pm\frac{1}{2}m_d \in D$ by (1) and D is convex, D contains the cross-polytope $\operatorname{conv}\{\pm\frac{1}{2}m_1, \dots, \pm\frac{1}{2}m_d\}$. Thus

$$\begin{aligned} 1 = V(D) &\geq V\left(\operatorname{conv}\left\{\pm\frac{1}{2}m_1, \dots, \pm\frac{1}{2}m_d\right\}\right) \\ &= \frac{2^d}{d\,!}\left|\det\left(\frac{1}{2}m_1, \dots, \frac{1}{2}m_d\right)\right| = \frac{1}{d\,!}|\det(m_1, \dots, m_d)| \\ &= \frac{1}{d\,!}|u_{11}\cdots u_{dd}|\,|\det(b_1, \dots, b_d)| = \frac{1}{d\,!}|u_{11}\cdots u_{dd}|. \end{aligned}$$

by (2). Hence $|\det(m_1, \dots, m_d)| \leq d\,!$, concluding the proof of (i). Note also that $|u_{ii}| \leq d\,!$ and therefore

(3) $|u_{ik}| \leq d\,!$.

(ii) Let $m_1, \dots, m_d$ be d linearly independent minimum vectors and let $\{b_1, \dots, b_d\}$ be a basis of $\mathbb{Z}^d$ such that (2) and (3) hold. We then show the following:

(4) Let m be a minimum vector. Then $m = a_1m_1 + \cdots + a_dm_d$ where $|a_i| \leq d\,!$ for $i = 1, \dots, d$.

$m_1, \dots, m_{i-1}, m, m_{i+1}, \dots, m_d$ are d minimum vectors. Thus (i) implies that

$$\begin{aligned} d\,! &\geq |\det(m_1, \dots, m_{i-1}, m, m_{i+1}, \dots, m_d)| \\ &= |a_i|\,|\det(m_1, \dots, m_{i-1}, m_i, m_{i+1}, \dots, m_d)| \geq |a_i|, \end{aligned}$$

concluding the proof of (4). Finally, (4), (2) and (3) together yield (ii). □

Perfect Forms

A positive definite quadratic form on $\mathbb{E}^d$ is *perfect* if it is uniquely determined by its minimum vectors and the value of the minimum. In Proposition 29.7 it will be shown that the perfect forms correspond precisely to the vertices of the so-called Ryshkov polyhedron.

Later we will see that perfect forms – besides being of interest per se – play an important role for the determination of extreme forms or, more geometrically, for the determination of lattice packings of B^d of locally maximum density. Minkowski [742] and Voronoĭ [1012] both proposed methods for finding all perfect forms in d variables. Unfortunately, these methods are not efficient and it is still a task to determine all perfect forms even for moderately large d.

If the minimum vectors of a positive definite quadratic form on $\mathbb{E}^d$ do not span $\mathbb{E}^d$, the form cannot be perfect. As a consequence of the above theorem and Theorem 21.1 we thus obtain the following result:

Theorem 29.7. *There are only finitely many non-equivalent perfect positive definite quadratic forms on* $\mathbb{E}^d$ *or, in other words, only finitely many equivalence classes of perfect positive definite quadratic forms on* $\mathbb{E}^d$*, with the same minimum.*

The numbers of non-equivalent perfect positive definite quadratic forms with given minimum have been evaluated for dimensions $d = 2, \dots, 8$:

$d = 2$:	1 :	Lagrange [628]
$d = 3$:	1 :	Gauss [364]
$d = 4$:	2 :	Korkin and Zolotarev [611]
$d = 5$:	3 :	Korkin and Zolotarev [611]
$d = 6$:	7 :	Barnes [71]
$d = 7$:	33 :	Jaquet-Chiffelle [542]
$d = 8$:	10916 :	Dutour Sikiric and Schürmann [280]

For more information see Martinet [691] and Nebe [767].

Ryshkov's Approach to Results of Korkin–Zolotarev and Voronoĭ on Perfect and Eutactic Forms

In the following we describe an elegant path to the fundamental theorems of Korkin–Zolotarev and Voronoĭ, where as a tool for the determination of extreme forms a generalized convex polyhedron $\mathcal{R}(m) \subseteq \mathcal{P}$, $m > 0$, is used, dual to the so-called Voronoĭ polyhedron. (For the definition of generalized convex polyhedra, see Sect. 14.2.) It was introduced by Ryshkov [865], and we call it the *Ryshkov's polyhedron*:

$$\mathcal{R}(m) = \bigcap_{\substack{u \in \mathbb{Z}^d \setminus \{o\} \\ \text{primitive}}} \Big\{ (a_{11}, \dots, a_{1d}, a_{22}, \dots, a_{dd}) \in \mathbb{E}^{\frac{1}{2}d(d+1)} : \sum_{i,k} a_{ik}\, u_i u_k \ge m \Big\}.$$

$\mathcal{R}(m)$ is the intersection of closed halfspaces in $\mathbb{E}^{\frac{1}{2}d(d+1)}$ with (interior) normal vectors $(u_1^2, 2u_1u_2, \dots, 2u_1u_d, u_2^2, 2u_2u_3, \dots, u_d^2)$. We list several properties of $\mathcal{R}(m)$:

Proposition 29.1. $\mathcal{R}(m)$ *is closed, convex and has non-empty interior. Each ray in* $\mathcal{P}$ *starting at the origin meets* $\operatorname{bd} \mathcal{R}(m)$ *in precisely one point and, from that point on it, is contained in* $\operatorname{int} \mathcal{R}(m)$.

Proof. As an intersection of closed halfspaces, $\mathcal{R}(m)$ is closed and convex. The definition of $\mathcal{R}(m)$ implies that each ray $\mathcal{R}$ in the open convex cone $\mathcal{P}$ starting at the origin intersects $\mathcal{R}(m)$ in a half-line. Let $\mathcal{R}$ be such a ray and consider the first point of the halfline $\mathcal{R} \cap \mathcal{R}(m)$, say A. Next choose $\frac{1}{2}d(d+1)$ further such rays, say $\mathcal{R}_i, i = 1, \dots, \frac{1}{2}d(d+1)$, in $\mathcal{P}$ which determine a simplicial cone which contains the ray $\mathcal{R}$ in its interior. For any $B \in \mathcal{R} \cap \mathcal{R}(m), B \neq A$, we may choose points $A_i \in \mathcal{R}_i \cap \mathcal{R}(m)$ such that B is an interior point of the simplex $\operatorname{conv}\{A, A_1, \dots, A_{\frac{1}{2}d(d+1)}\}$ which, in turn, is contained in the convex set $\mathcal{R}(m)$. This implies that $\operatorname{int}\mathcal{R}(m) \neq \emptyset$ and thus concludes the proof of the proposition. □

Proposition 29.2. $\mathcal{R}(m) \subseteq \mathcal{P}$.

Proof. Consider a point in $\mathcal{R}(m)$ and let q be the corresponding quadratic form. We have to show that q is positive definite. By the definition of $\mathcal{R}(m)$ we have $q(u) \geq m > 0$ for each $u \in \mathbb{Z}^d \setminus \{o\}$. Thus $q(r) > 0$ for each point $r \in \mathbb{E}^d \setminus \{o\}$, with rational coordinates and therefore $q(x) \geq 0$ for each $x \in \mathbb{E}^d$ by continuity. Thus q is positive semi-definite. If q were not positive definite, then a well known arithmetic result says that for any positive number and thus in particular for m, there is a point $u \in \mathbb{Z}^d \setminus \{o\}$, with $q(u)$ less than this number. This is the required contradiction. □

Proposition 29.3. *The points of* $\mathcal{R}(m)$*, respectively, of* $\operatorname{bd}\mathcal{R}(m)$ *correspond precisely to the positive definite quadratic forms with arithmetic minimum at least* m*, respectively, equal to* m*.*

Proof. This is an immediate consequence of the definition of $\mathcal{R}(m)$ and the earlier two propositions. □

Proposition 29.4. *Let* U *be an integer unimodular* $d \times d$ *matrix and let* $\mathcal{U}$ *be the corresponding transformation of* $\mathcal{P}$*. Then* $\mathcal{U}(\mathcal{R}(m)) = \mathcal{R}(m)$ *and* $\mathcal{U}(\operatorname{bd}\mathcal{R}(m)) = \operatorname{bd}\mathcal{R}(m)$.

Proof. This follows from Proposition 29.3 by taking into account that with any positive definite quadratic form having arithmetic minimum at least m or equal to m, any equivalent form is also positive definite with arithmetic minimum greater or equal to m, or equal to m, respectively. □

Proposition 29.5. $\mathcal{R}(m)$ *is a generalized convex polyhedron.*

Proof. Considering the definition of generalized convex polyhedra in Sect. 14.2, it is sufficient to prove that, for any $q \in \operatorname{bd}\mathcal{R}(m)$, there is a cube $\mathcal{K}$ with centre q in $\mathbb{E}^{\frac{1}{2}d(d+1)}$ such that $\mathcal{K} \cap \mathcal{R}(m)$ is a convex polytope. Let $q \in \operatorname{bd}\mathcal{R}(m)$. The positive quadratic form q has at most $2^d - 1$ pairs of minimum points, see Theorem 30.2. If a cubic neighbourhood $\mathcal{K}$ of q is chosen sufficiently small, then $\mathcal{K} \subseteq \mathcal{P}$ and the minimum points of any form in $\mathcal{K}$ are amongst those of q. This means that

$$\mathcal{K} \cap \mathcal{R}(m) = \mathcal{K} \cap \bigcap_{\substack{u \in \mathbb{Z}^d \\ u \text{ minimum} \\ \text{point of } q}} \Big\{(a_{11}, \dots, a_{dd}) \in \mathbb{E}^{\frac{1}{2}d(d+1)} : \sum_{i,k} a_{ik}\, u_i u_k \geq m\Big\}$$

is a polytope, concluding the proof. □

Proposition 29.6. *Any two facets of $\mathcal{R}(m)$ are equivalent via a transformation of the form $\mathcal{U}$ where the corresponding $d \times d$ matrix U is integer and unimodular.*

Proof. Let $\mathcal{F}$ and $\mathcal{G}$ be facets of $\mathcal{R}(m)$. They are determined by two primitive integer minimum vectors $u_{\mathcal{F}}, u_{\mathcal{G}} \in \mathbb{Z}^d$, see the proof of Proposition 29.5. Now take for U any integer unimodular $d \times d$ matrix U with $U u_{\mathcal{F}} = u_{\mathcal{G}}$. □

Proposition 29.7. *The perfect quadratic forms on $\mathbb{E}^d$ with arithmetic minimum m correspond precisely to the vertices of $\mathcal{R}(m)$.*

Proof. This result is clear. □

Since at least $\frac{1}{2}d(d+1)$ facets meet at each vertex of the polyhedron $\mathcal{R}(m)$ in $\mathbb{E}^{\frac{1}{2}d(d+1)}$, the following result of Korkin and Zolotarev [611] is an immediate consequence of this proposition:

Theorem 29.8. *Each perfect positive definite quadratic form on $\mathbb{E}^d$ has at least $\frac{1}{2}d(d+1)$ pairs $\pm u \neq o$ of minimum points. Among the minimum points are d linearly independent ones.*

An extension of this result to lattice packings of an o-symmetric convex body of locally maximum density is due to Swinnerton-Dyer [978], see Theorem 30.3.

The following property of Ryshkov's polyhedron is an immediate consequence of Proposition 29.7 and Theorem 29.7.

Proposition 29.8. *There are only finitely many vertices of $\mathcal{R}(m)$ which are pairwise non-equivalent via transformations of the form $\mathcal{U}$ where U is an integer unimodular $d \times d$ matrix.*

The *(equi-)discriminant surface* $\mathcal{D}(\delta)$, where $\delta > 0$, consists of all (points of $\mathcal{P}$ corresponding to) positive definite quadratic forms on $\mathbb{E}^d$ with discriminant δ. We establish the following properties of $\mathcal{D}(\delta)$.

Proposition 29.9. *The following statements hold:*

(i) *Each ray in $\mathcal{P}$ starting at the origin meets $\mathcal{D}(\delta)$ at precisely one point.*
(ii) *$\mathcal{D}(\delta)$ is strictly convex and smooth.*

The strict convexity of the discriminant surface plays an important role in our proof of John's theorem 11.2.

Proof. (i) is trivial.

(ii) We first show the strict convexity: let $A, B \in \mathcal{D}(\delta)$ be such that $A \neq B$. Since A and B are not proportional to each other, Minkowski's determinant inequality for symmetric, positive semi-definite matrices then implies that $\det\big((1-\lambda)A+\lambda B\big) > \delta$ for $0 < \lambda < 1$. It remains to show that $\mathcal{D}(\delta)$ is a smooth surface: Since the gradient of the determinant does not vanish on $\mathcal{P}$ and thus on $\mathcal{D}(\delta)$, a version of the implicit function theorem from calculus shows that the discriminant surface $\mathcal{D}(\delta) = \{A \in \mathcal{P} : \det A = \delta\}$ is smooth at each if its points. □

Let $q(x) = \sum a_{ik}\, x_i x_k$ be a positive definite quadratic form q on $\mathbb{E}^d$, with arithmetic minimum m and discriminant $\delta = \det(a_{ik})$. q is *eutactic* if the following statement holds: let $(b_{ik}) = (a_{ik})^{-1}$. Then

$$(b_{11}, 2b_{12}, \dots, 2b_{1d}, b_{22}, 2b_{23}, \dots, b_{dd}),$$

the normal vector of the discriminant surface $\mathcal{D}(\delta)$ at its point q, is a linear combination with positive coefficients of the vectors

$$(u_1^2, 2u_1u_2, \dots, 2u_1u_d, u_2^2, 2u_2u_3, \dots, u_d^2)$$

where $\pm(u_1, \dots, u_d) \in \mathbb{Z}^d$ ranges over the minimum vectors of q. These vectors are the normal vectors of the facets of the Ryshkov polyhedron $\mathcal{R}(m)$ which contain the boundary point q of $\mathcal{R}(m)$ and thus generate the normal cone of $\mathcal{R}(m)$ at q. This seemingly strange definition is, in fact, perfectly natural, as will be clear from the following basic *theorem of Voronoĭ* [1012] and its proof.

It seems that Coxeter [229] was the first to use the word eutactic in the present context. Presumably, he wanted to express the fact that (b_{ik}) is well (=eu in Greek)-determined (=tactic) by the vectors $(u_1^2, \dots, 2u_1u_d, \dots, u_d^2)$. See also Martinet [690].

Theorem 29.9. *A positive definite quadratic form on $\mathbb{E}^d$ is extreme if and only if it is perfect and eutactic.*

Proof (by means of the Ryshkov polyhedron). Let q be a positive definite quadratic form on $\mathbb{E}^d$ with coefficients a_{ik}, arithmetic minimum m and discriminant δ. For the proof of the theorem, it is sufficient to show that the following statements are equivalent:

(i) q is extreme.

(ii) A suitable neighbourhood of $(a_{11}, a_{12}, \dots, a_{dd})$ in the polyhedron $\mathcal{R}(m)$ is contained in the unbounded convex body determined by the smooth and strictly convex surface $\mathcal{D}(\delta)$ through $(a_{11}, a_{12}, \dots, a_{dd})$.

(iii) $(a_{11}, a_{12}, \dots, a_{dd})$ is the only point of $\mathcal{R}(m)$ in the tangent hyperplane of $\mathcal{D}(\delta)$ at $(a_{11}, a_{12}, \dots, a_{dd})$, that is the hyperplane

$$\Big\{v = (v_{11}, \dots, v_{dd}) \in \mathbb{E}^{\frac{1}{2}d(d+1)} : \sum_{i,k} b_{ik}\, v_{ik} = d,\ v_{ik} = v_{ki}\Big\},$$

where $(a_{ik})^{-1} = (b_{ik})$.

(iv) q is perfect and eutactic.

Only (iii)⇔(iv) needs justification: Consider a point of a convex polyhedron and a hyperplane through it. Then the following are equivalent (a) the hyperplane meets the polyhedron only at this point and (b) this point is a vertex of the polyhedron and thus the unique point contained in all facets through it, and the exterior normal vector of the hyperplane is a linear combination with positive coefficients of the exterior normal vectors of these facets, see Proposition 14.1. □

The following is a list of the numbers of extreme forms for dimensions $d = 2, \dots, 8$:

$d = 2$: 1 : Korkin and Zolotarev [611]
$d = 3$: 1 : Korkin and Zolotarev [611]
$d = 4$: 2 : Korkin and Zolotarev [611]
$d = 5$: 3 : Korkin and Zolotarev [611]
$d = 6$: 6 : Hofreiter [518], Barnes [71, 72]
$d = 7$: 30 : Conway and Sloane [221]
$d = 8$: 2408 : Riener [836, 837]

For more information see Martinet [691].

Remark. Using Voronoĭ's theorem, one can decide, at least in principle, whether a given positive quadratic form is extreme or, equivalently, whether a given lattice provides a locally densest packing of balls.

In order to find all extreme forms it is sufficient to consider a maximal set of inequivalent vertices of $\mathcal{R}(m)$ and to take those vertices q for which the tangent hyperplane of the discriminant surface through q meets $\mathcal{R}(m)$ only at q. Unfortunately, no effective algorithm to determine the extreme forms is yet known.

30 Packing of Convex Bodies

The problem of packing of convex bodies and, in particular, of Euclidean balls has attracted interest ever since Kepler first considered such questions. One reason for this is that packing results can have interesting arithmetic interpretations. Among the eminent contributors we mention Kelvin, Minkowski, Thue, Voronoĭ and Fejes Tóth. The first investigations of non-lattice packings are due to Thue [996, 997], the later development was strongly influenced by the seminal work of Fejes Tóth. As a nice historical curiosity we cite the following observation of Reynolds [832] on the distortion of a dense packing of grains of sand, see Coxeter [231]:

> As the foot presses upon the sand when the falling tide leaves it firm, the portion of it immediately surrounding the foot becomes momentarily dry. .... The pressure of the foot causes dilatation of the sand, and so more water is [drawn] through the interstices of the surrounding sand ..., leaving it dry until a sufficient supply has been obtained from below, when it again becomes wet. On raising the foot we generally see that the sand under and around it becomes wet for a little time. This is because the sand contracts when the distorting forces are removed, and the excess of water escapes at the surface.

At the beginning the sand grains form a dense packing. Then the pressure of the foot distorts the packing which consequently becomes less dense and thus provides more space for the water. By capillary forces the water then is drawn into the interior of the sand. On raising the foot with the water as lubricant the sand grains again glide back into a dense packing which provides less space for the water.

How seriously Hilbert [501] took the packing problem can be seen from the following question in his 18th problem:

> How can one arrange most densely in space an infinite number of equal solids of given form, e.g. spheres with given radii or regular tetrahedra with given edges (or in prescribed position), that is, how can one so fit them together that the ratio of the filled to the unfilled space may be as great as possible?

In this section we consider lattice packings and packings of translates of a given convex body. After some definitions and simple remarks, we show that densest lattice packings of convex bodies always exist. Then bounds for the number of neighbours in lattice packings of convex bodies are given. Next, we mention an algorithm of Betke and Henk, which permits us to determine the densest lattice packing of convex polytopes in $\mathbb{E}^3$, and present a lower bound for the maximum density of lattice packings as a consequence of the Minkowski–Hlawka theorem. Finally, we consider, in the planar case, the relation between the lattice and the non-lattice case.

For more detailed expositions, see the books of Fejes Tóth [329,330] and Rogers [851], and the surveys of Fejes Tóth [322], Fejes Tóth and Kuperberg [325], Gruber [438] and Bambah [56].

30.1 Definitions and the Existence of Densest Lattice Packings

This section contains the definitions of packing and density and some simple, yet important, properties of packings. In particular, we study the relation between packings of convex bodies and packings of their central symmetrizations, and show the existence of lattice packings of convex bodies of maximum density.

Packing of Convex Bodies and the Notion of Density

A family of convex bodies in $\mathbb{E}^d$ is a *packing* if any two distinct bodies have disjoint interior. We will consider *packings of translates* and *lattice packings* of a given proper convex body C, i.e. packings of the form $\{C+t : t \in T\}$ and $\{C+l : l \in L\}$, where T a discrete set and L a lattice in $\mathbb{E}^d$, respectively. If $\{C+l : l \in L\}$ is a lattice packing of C then L is called a *packing lattice* of C. Let K be the cube $\{x : |x_i| \le 1\}$.

If T is a discrete set in $\mathbb{E}^d$ then its *upper* and *lower density* are

$$\limsup_{\tau\to+\infty} \frac{\#(T \cap \tau K)}{(2\tau)^d}, \quad \liminf_{\tau\to+\infty} \frac{\#(T \cap \tau K)}{(2\tau)^d}.$$

If the upper and the lower density of T coincide, their common value is the *density* $\delta(T)$ of the discrete set T. If the density of T exists, it may be interpreted as the number of points of T per unit volume or, roughly speaking, as the number of points in T divided by the volume of $\mathbb{E}^d$.

Next, given a convex body C and a discrete set T, consider the family $\{C + t : t \in T\}$ of translates of C by the vectors of T. Its *upper* and *lower density* are

$$\limsup_{\tau \to +\infty} \frac{1}{(2\tau)^d} \sum_{t \in T} V\big((C + t) \cap \tau K\big), \ \liminf_{\tau \to +\infty} \frac{1}{(2\tau)^d} \sum_{t \in T} V\big((C + t) \cap \tau K\big).$$

If they coincide, their common value is called the *density* $\delta(C, T)$ of the given family. In other sources different definitions are used. For packings and coverings with translates of a convex body, these amount to the same values for the upper and lower densities. For more general families of convex bodies our definitions are still close to the intuitive notion of density and avoid strange occurrences such as packings with upper density $+\infty$.

Roughly speaking, the density of a family $\{C + t : t \in T\}$ of translates of C by the vectors of T is the total volume of the bodies divided by the total volume of $\mathbb{E}^d$. In the case where this family is a packing, the density may be considered as the proportion of $\mathbb{E}^d$ which is covered by the bodies of the packing, or as the probability that a 'random point' of $\mathbb{E}^d$ is contained in one of the bodies of the packing.

Given a convex body C, let $\delta_T(C)$ and $\delta_L(C)$ denote the supremum of the upper densities of all packings of translates of C, and all lattice packings of C, respectively. $\delta_T(C)$ and $\delta_L(C)$ are called the *(maximum) translative packing density* and the *(maximum) lattice packing density* of C, respectively. Clearly,

$$0 < \delta_L(C) \le \delta_T(C) \le 1.$$

Density of Lattices and of Families of Translates of Convex Bodies

We start with the density of lattices:

Proposition 30.1. *Let L be a lattice in $\mathbb{E}^d$. Then its density $\delta(L)$ exists and equals $1/d(L)$.*

Proof. Let F be a fundamental parallelotope of L and choose $\sigma > 0$ such that:

(1) $F \subseteq \sigma K$.

For the proof of the proposition, it is sufficient to show that

$$(2) \quad \frac{2^d(\tau - \sigma)^d}{d(L)} \le \#(L \cap \tau K) \le \frac{2^d(\tau + \sigma)^d}{d(L)} \text{ for } \tau > \sigma.$$

The parallelotopes $\{F + l : l \in L\}$ are pairwise disjoint and cover $\mathbb{E}^d$. Thus the parallelotopes in this family which intersect τK, in fact, cover τK. By (1) these parallelotopes are all contained in $(\tau + \sigma)K$. Thus if m is their number, we have

$$\#(L \cap \tau K) \le m = \frac{mV(F)}{V(F)} \le \frac{V\big((\tau + \sigma)K\big)}{V(F)} = \frac{2^d(\tau + \sigma)^d}{d(L)}.$$

This proves the right-hand inequality in (2). Next, consider the parallelotopes from our family which intersect $(\tau - \sigma)K$. These cover $(\tau - \sigma)K$ and by (1) are all contained in τK. If n is their number, it thus follows that

$$\#(L \cap \tau K) \geq n = \frac{nV(F)}{V(F)} \geq \frac{V\big((\tau - \sigma)K\big)}{V(F)} = \frac{2^d(\tau - \sigma)^d}{d(L)},$$

concluding the proof of the left-hand inequality in (2). □

Remark. Using the Möbius inversion formula from number theory, it can be shown that the density of the set of primitive points of a lattice L in $\mathbb{E}^d$ is $1/(\zeta(d)d(L))$, where $\zeta(\cdot)$ is the Riemann zeta function. In particular, this shows that the probability that a point of a lattice is primitive is $1/\zeta(d)$.

Next, the density of a discrete set T and the density of the family of translates of a given convex body by the vectors of T will be related:

Proposition 30.2. *Let C be a proper convex body and T a discrete set in $\mathbb{E}^d$. Then the upper density of the family $\{C + t : t \in T\}$ is equal to $V(C)$ times the upper density of T. Analogous statements hold for the lower density and the density, if the latter exists.*

Proof. Choose $\sigma > 0$ such that $C \subseteq \sigma K$. Clearly,

$$(C + t) \cap \tau K \neq \emptyset \Rightarrow t \in (\tau + \sigma)K,$$
$$t \in (\tau + \sigma)K \Rightarrow (C + t) \subseteq (\tau + 2\sigma)K.$$

Then

$$\frac{1}{(2\tau)^d} \sum_{t \in T} V\big((C + t) \cap \tau K\big) \leq \frac{1}{(2\tau)^d} \#\big(T \cap (\tau + \sigma)K\big)V(C)$$
$$\leq \frac{1}{\big(2(\tau + 2\sigma)\big)^d} \sum_{t \in T} V\big((C + t) \cap (\tau + 2\sigma)K\big) \frac{\big(2(\tau + 2\sigma)\big)^d}{(2\tau)^d}.$$

Now let $\tau \to \infty$ to get the equalities for the upper and the lower density and the density. □

Corollary 30.1. *Let C be a proper convex body and L a lattice in $\mathbb{E}^d$. Then the family $\{C + l : l \in L\}$ of translates of C by the vectors of L has density*

$$\delta(C, L) = \frac{V(C)}{d(L)}.$$

The family $\{C + l : l \in L\}$ of translates of the body C by the vectors of the lattice L is sometimes called a *set lattice* with set C and lattice L and $\delta(C, L)$ is its *density*.

Packing and Central Symmetrization

Given a convex body C, its *central symmetrization* is the convex body

$$D = \frac{1}{2}(C - C) \left(= \left\{ \frac{1}{2}(x - y) : x, y \in C \right\} \right).$$

(The convex body $2D = C - C$ is called the *difference body* of C.) Minkowski [740] discovered that lattice packings of a convex body and its difference body are closely related. His proof yields the following, slightly more general result.

Proposition 30.3. *Let C be a proper convex body, $D = \frac{1}{2}(C - C)$ its central symmetrization and T a discrete set in $\mathbb{E}^d$. Then the following statements are equivalent:*

(i) $\{C + t : t \in T\}$ *is a packing.*
(ii) $\{D + t : t \in T\}$ *is a packing.*

Proof. It is sufficient to show the following equivalence:

(3) Let $s, t \in \mathbb{E}^d$. Then

$$(\operatorname{int} C + s) \cap (\operatorname{int} C + t) \neq \emptyset \Leftrightarrow (\operatorname{int} D + s) \cap (\operatorname{int} D + t) \neq \emptyset.$$

We shall use the equalities

(4) $\operatorname{int} D = \frac{1}{2}(\operatorname{int} C - \operatorname{int} C)$,

(5) $\operatorname{int} C = \frac{1}{2}(\operatorname{int} C + \operatorname{int} C)$.

Then

$$\begin{aligned}
&(\operatorname{int} C + s) \cap (\operatorname{int} C + t) \neq \emptyset \\
&\Rightarrow x + s = y + t \text{ for suitable } x, y \in \operatorname{int} C \\
&\Rightarrow \frac{1}{2}(x - y) + s = \frac{1}{2}(y - x) + t \\
&\Rightarrow (\operatorname{int} D + s) \cap (\operatorname{int} D + t) \neq \emptyset \text{ by (4)} \\
&\Rightarrow \frac{1}{2}(u - v) + s = \frac{1}{2}(w - z) + t \text{ for suitable } u, v, w, z \in \operatorname{int} C \text{ by (4)} \\
&\Rightarrow \frac{1}{2}(u + z) + s = \frac{1}{2}(w + v) + t \\
&\Rightarrow (\operatorname{int} C + s) \cap (\operatorname{int} C + t) \neq \emptyset \text{ by (5)},
\end{aligned}$$

concluding the proof of (3) and thus of the proposition. □

Remark. If L is a packing lattice of C or, equivalently, of $D = \frac{1}{2}(C - C)$, the corresponding densities are

$$\delta(C, L) = \frac{V(C)}{d(L)} \text{ and } \delta(D, L) = \frac{V(D)}{d(L)},$$

respectively. By the Brunn–Minkowski theorem 8.1 and the inequality of Rogers and Shephard 9.10, these densities are related as follows:

$$\delta(C,L) \le \delta(D,L) \le \frac{1}{2^d}\binom{2d}{d}\delta(C,L) \sim \frac{2^d}{\sqrt{\pi d}}\,\delta(C,L).$$

There is equality in the left inequality if and only if C is centrally symmetric and in the right inequality if and only if C is a simplex.

Admissible and Critical Lattices

Let C be a convex body in $\mathbb{E}^d$ with $o \in \operatorname{int} C$. A lattice L is *admissible* for C if o is the only point of L in $\operatorname{int} C$. L is *critical* for C if it is admissible and has minimum determinant among all admissible lattices. This determinant is denoted $\Delta(C)$ and is called the *critical determinant* of C. The lattice L is *locally critical* for C if it is admissible and has minimum determinant among all admissible lattices in a suitable neighbourhood of it.

Proposition 30.4. *Let C be a proper convex body, $D = \frac{1}{2}(C - C)$ its central symmetrization and L a lattice in $\mathbb{E}^d$. Then the following statements are equivalent:*

(i) $\{C + l : l \in L\}$ *is a packing.*
(ii) $\{D + l : l \in L\}$ *is a packing.*
(iii) L *is admissible for* $2D$.

Proof. Note Proposition 30.3 and its proof. Then

$$\begin{aligned} &\{C+l : l \in L\} \text{ is a packing}\\ \Leftrightarrow\ &\{D+l : l \in L\} \text{ is a packing}\\ \Leftrightarrow\ &\operatorname{int} D \cap (\operatorname{int} D + l) = \emptyset \text{ for each } l \in L\setminus\{o\},\\ \Leftrightarrow\ &l \notin \operatorname{int} D - \operatorname{int} D \text{ for each } l \in L\setminus\{o\},\\ \Leftrightarrow\ &l \notin \operatorname{int} 2D \text{ for each } l \in L\setminus\{o\}\\ \Leftrightarrow\ &L \text{ is admissible for } 2D. \end{aligned}$$ □

Corollary 30.2. *Let C be a proper convex body, $D = \frac{1}{2}(C - C)$ its central symmetrization and L a lattice in $\mathbb{E}^d$. Then the following statements are equivalent:*

(i) $\{C + l : l \in L\}$ *is a packing of maximum density.*
(ii) L *is a critical lattice of* $2D = C - C$.

Note that

$$\delta_L(C) = \frac{V(C)}{\Delta(2D)}.$$

Existence of Densest Lattice Packings

As a consequence of Mahler's selection theorem 25.1 we prove that, for any convex body, there exist lattice packings of maximum density.

Theorem 30.1. *Let C be a proper convex body in $\mathbb{E}^d$ with $o \in \operatorname{int} C$. Then there is a packing lattice L of C such that $\delta(C, L) = \delta_L(C)$.*

Proof. Let (L_n) be a sequence of packing lattices of C such that:

$$(6)\quad 0 < \delta(C, L_1) = \frac{V(C)}{d(L_1)} \le \delta(C, L_2) = \frac{V(C)}{d(L_2)} \le \dots \to \delta_L(C).$$

Then L_n is admissible for $2D = C - C$ by Proposition 30.4 and $d(L_n) \le d(L_1)$ for all n. By Mahler's selection theorem, the sequence (L_n) has a convergent subsequence. After suitable cancellation and renumbering, if necessary, we may assume that $L_1, L_2, \dots \to L$, where L is a suitable lattice. The definition of convergence then implies that

(7) $d(L_1), d(L_2), \dots \to d(L)$, and

(8) for each $l \in L$ there are points $l_n \in L_n$, $n = 1, 2, \dots,$ such that $l_1, l_2, \dots \to l$.

Then,

(9) L is a packing lattice of C.

Otherwise, there is a vector $l \in L \setminus \{o\}$, such that $\operatorname{int} C \cap (\operatorname{int} C + l) \ne \emptyset$. By (8) we may choose points $l_n \in L_n$ converging to l. Then $\operatorname{int} C \cap (\operatorname{int} C + l_n) \ne \emptyset$ and $l_n \ne o$ for all sufficiently large n. But, since L_n is a packing lattice of C by assumption, we have $\operatorname{int} C \cap (\operatorname{int} C + l_n) = \emptyset$ for all n with $l_n \ne o$. This contradiction concludes the proof of (9).

Finally, (6), (7) and (9) together yield the equality $\delta(C, L) = \delta_L(C)$. □

Remark. Similarly, $\delta_T(C)$ is attained for suitable configurations, see Hlawka [510] and Groemer [399] for this and more general results.

30.2 Neighbours

Let C be a convex body and L a lattice in $\mathbb{E}^d$. If $\{C + l : l \in L\}$ is a packing, two distinct bodies of it are *neighbours* if they intersect. The question arises, as to how many neighbours can C have? Well-known pertinent results are due to Minkowski and Swinnerton-Dyer. More recent are results of the author [419] and Engel [299]. One reason, why the notion of neighbour has attracted interest seems to be the fact that neighbours correspond to minimum points in Diophantine inequalities. Compare Theorems 29.6 and 29.8, where minimum points of positive definite quadratic forms are studied.

Versions of the above question play an important role in the context of finite packing. See, e.g. the books of Zong [1049] and Böröczky [155], in particular the results dealing with the notions of kissing and Hadwiger numbers.

In this section upper and lower estimates for the number of neighbours in lattice packings are given.

Upper Estimates for the Number of Neighbours

The following are classical estimates due to Minkowski [743].

Theorem 30.2. *Let C be a proper convex body and L a lattice in $\mathbb{E}^d$ such that $\{C+l : l \in L\}$ is a packing. Then C has at most $3^d - 1$ neighbours. If C is strictly convex, it has at most $2^{d+1} - 2$ neighbours.*

Proof. Let $D = \frac{1}{2}(C - C)$. Since we have the following,

$$\{C + l : l \in L\} \text{ packing} \Leftrightarrow \{D + l : l \in L\} \text{ packing},$$
$$C + l \text{ neighbour of } C \Leftrightarrow D + l \text{ neighbour of } D,$$
$$C \text{ strictly convex} \Leftrightarrow D \text{ strictly convex},$$

it is sufficient to prove the result for D instead of C. Since

$$\{D + l : l \in L\} \text{ packing} \Leftrightarrow L \text{ is admissible for } 2D,$$
$$D + l \text{ neighbour of } D \Leftrightarrow l \in \operatorname{bd} 2D,$$

it is sufficient to show the following:

(1) the lattice L, which is admissible for $2D$, has at most $\frac{1}{2}(3^d - 1)$ pairs of points $\pm l$ on $\operatorname{bd} 2D$. Here $\frac{1}{2}(3^d - 1)$ can be replaced by $2^d - 1$ if $2D$ is strictly convex.

First, let C and thus $2D$ be strictly convex and let $\{b_1, \dots, b_d\}$ be a basis of L. If $\pm l = \pm(u_1 b_1 + \dots + u_d b_d) \in L \cap \operatorname{bd} 2D$, then we cannot have $u_i \equiv 0 \bmod 2$ for $i = 1, \dots, d$; otherwise all u_i are even and $\frac{1}{2} l \in (L \cap \operatorname{int} 2D) \setminus \{o\}$, which contradicts the admissibility of L for $2D$. If $\pm l = \pm(u_1 b_1 + \dots + u_d b_d)$, $\pm m = \pm(v_1 b_1 + \dots + v_d b_d) \in (L \cap \operatorname{bd} 2D)$ where $\pm l \neq \pm m$, then we cannot have $u_i \equiv v_i \bmod 2$ for $i = 1, \dots, d$; otherwise $\frac{1}{2}(l - m) \in L \setminus \{o\}$ and by the strict convexity of $2D$ we have $\frac{1}{2}(l - m) \in \operatorname{int} 2D$, which again contradicts the admissibility of L for $2D$. Since there are $2^d - 1$ residue classes for $(u_1, \dots, u_d)$ modulo 2, excluding $(0, \dots, 0)$, there can be at most $2^d - 1$ different pairs of points $\pm l \in L \cap \operatorname{bd} 2D$, as required.

If, second, C and thus $2D$ is not strictly convex, similar arguments with congruences modulo 3 lead to the bound $\frac{1}{2}(3^d - 1)$. □

Remark. Helmut Groemer [406] pointed out that the following simple geometric argument shows that, in any packing of translates of a convex body C, the number of neighbours of a fixed translate is at most $3^d - 1$: It is sufficient to prove this for the difference body D instead of C. Since D is symmetric in o and convex, a translate of D is a neighbour of D if and only if it is contained in $3D$. Considering volumes, we see that in $3D$ there is space for at most 3^d non-overlapping translates of D, including D.

Remark. The bound $\frac{1}{2}(3^d - 1)$ is attained if $C = \{\alpha_1 b_1 + \cdots + \alpha_d b_d : |\alpha_i| \le \frac{1}{2}\}$ where $\{b_1, \ldots, b_d\}$ is a basis of L, and Groemer [398] showed that this is the only case.

Remark. A result of the author [419] says that for a typical convex body C (in the sense of Baire categories) for any lattice packing which (locally) has maximum density, the number of neighbours of C is at most $2d^2$. We think that this can still be improved and state the following conjecture, which is also in accordance with the next result.

Conjecture 30.1. *For a typical proper convex body C, for any lattice packing which has locally maximum density, the number of neighbours of C is precisely $d(d+1)$.*

Lower Estimate for the Number of Neighbours

The following result of Swinnerton-Dyer [978] extends the corresponding estimate of Korkin and Zolotarev [609–611] for Euclidean balls, see Theorem 29.8. We present the particularly elegant proof of Swinnerton-Dyer.

Theorem 30.3. *Let C be a proper convex body and L a lattice in $\mathbb{E}^d$ such that $\{C+l : l \in L\}$ is a packing which has locally maximum density. Then C has at least $d(d+1)$ neighbours.*

Proof. Again, it is sufficient to show that the lattice L which is locally critical for $2D = C - C$, has at least $\frac{1}{2}d(d+1)$ pairs of points $\pm l$ on $\operatorname{bd} 2D$.

Assume that there are only $n < \frac{1}{2}d(d+1)$ such pairs of points of L on $\operatorname{bd} 2D$, say $\pm l_j$, $j = 1, \ldots, n$. Consider supporting hyperplanes of $2D$ at these points and denote their exterior normal unit vectors by $\pm u_j$. Determine a real $d \times d$ matrix $A = (a_{ik})$ different from the $d \times d$ zero matrix O by the conditions

$$\begin{aligned} a_{ik} - a_{ki} &= 0 \text{ for } i < k, \\ l_j^T A u_j &= 0 \text{ for } j = 1, \ldots, n. \end{aligned}$$

There is such a matrix $A \neq O$ since these conditions form a homogeneous system of $n + \frac{1}{2}d(d-1) < d^2$ linear equations for the d^2 variables a_{ik}. For any real s, the linear transformation

$$x \to (I + sA)x$$

maps each of the $2n$ points $\pm l_j$ onto a point of the corresponding supporting hyperplane and thus onto a point not in $\operatorname{int} 2D$. If $|s|$ is sufficiently small, the linear transformation has determinant $\neq 0$ and maps any point $l \in L\}$ different from $o, \pm l_1, \dots, \pm l_n$, onto a point outside $2D$. Thus the lattice $L(s) = (I + sA)L$ is admissible for $2D$. Its determinant is given by:

$$d\big(L(s)\big) = |\det(I + sA)|\, d(L)$$

$$= \det \begin{pmatrix} 1 + s\,a_{11} & s\,a_{12} & \dots & s\,a_{1d} \\ s\,a_{21} & 1 + s\,a_{22} & \dots & s\,a_{2d} \\ \dots & \dots & \dots & \dots \\ s\,a_{d1} & s\,a_{d2} & \dots & 1 + s\,a_{dd} \end{pmatrix} d(L)$$

$$= (1 + s\,a_1 + s^2 a_2 + \dots + s^d a_d) d(L),$$

$$\text{where } a_1 = \sum_i a_{ii},\ a_2 = \sum_{i<k} (a_{ii}a_{kk} - a_{ik}a_{ki}).$$

(The expressions for $a_3, \dots, a_d$ are not needed.) Since L is locally critical, $d\big(L(s)\big) \geq d(L)$ for all sufficiently small $|s|$. Hence

$$1 + s\,a_1 + s^2 a_2 + \dots \geq 1 \text{ for all sufficiently small } |s|.$$

Clearly, this is possible only if $a_1 = 0, a_2 \geq 0$. Thus

$$0 \leq 2a_2 - a_1^2 = -\sum_{i,k} a_{ik}^2 \text{ and thus } a_{ik} = 0 \text{ for all } i, k.$$

This contradicts $A \neq O$, and thus concludes the proof of the theorem. □

30.3 The Betke–Henk Algorithm and the Lower Bound for $\delta_L(C)$ of Minkowski–Hlawka

In this section we discuss general results on lattice packings of maximum density.

How to Find Lattice Packings of Convex Polytopes in $\mathbb{E}^3$ of Maximum Density?

The problem of finding lattice packings of maximum density for a given convex body is difficult. While there are many pertinent results in the planar case, until recently this problem has been solved in $\mathbb{E}^3$ only for balls, frustrums of balls, double cylinders, cubes, truncated cubes, tetrahedra, cubo-octahedra, cylinders with convex base and, trivially, space fillers. See the references in Erdös, Gruber and Hammer [307]. Minkowski [743] stated conditions for lattice packings of maximum density of a convex body in $\mathbb{E}^3$, but these conditions are difficult to apply in concrete cases. It thus was a breakthrough when Betke and Henk [107] presented an *efficient* algorithm for computing the density of a densest lattice packing of an arbitrary convex polytope in $\mathbb{E}^3$. As an application they calculated the densest lattice packings of all regular and Archimedean polytopes. Let alone the trivial case of space fillers, all explicit results in dimensions $d \geq 4$ deal with balls, see Sect. 29.1.

A Lower Estimate for the Maximum Lattice Packing Density

As a consequence of the Minkowski-Hlawka theorem and the inequality of Rogers and Shephard on difference bodies, we obtain the following estimates:

Theorem 30.4. *Let C be a proper convex body in $\mathbb{E}^d$. Then, as $d \to \infty$,*

(i) $\delta_L(C) \geq 2^{-d}$ *if C is centrally symmetric.*
(ii) $\delta_L(C) \geq 4^{-d+o(d)}$ *for general C.*

Proof. (i) Let C be an o-symmetric proper convex body. The Minkowski–Hlawka theorem 24.1 then shows that there are lattices L which contain no point of $2C$, except o and with determinant $d(L)$ greater than but arbitrarily close to $V(2C)$. These lattices provide packings of C where the densities $V(C)/d(L)$ are less than, but arbitrarily close to, 2^{-d}. This clearly yields $\delta_L(C) \geq 2^{-d}$.

(ii) Let C be a proper convex body. Its central symmetrization $D = \frac{1}{2}(C - C)$ then is a proper, o-symmetric convex body. By (i), there are lattices L which provide packings of D with density $V(D)/d(L)$ arbitrarily close to 2^{-d}. By Proposition 30.3, each such lattice provides a packing of C where for the density we have

$$\frac{V(C)}{d(L)} = \frac{V(C)}{V(D)} \frac{V(D)}{d(L)} \geq \frac{2^d}{\binom{2d}{d}} \frac{V(D)}{d(L)}$$

by the Rogers–Shephard inequality. Since this is arbitrarily close to

$$\frac{2^d}{\binom{2d}{d}} 2^{-d} = 4^{-d+o(d)},$$

we obtain $\delta_L(C) \geq 4^{-d+o(d)}$. □

Heuristic Observations

We now extend the heuristic observations in Sect. 24.2. There is reason to believe that the bound in (i) cannot be improved essentially for certain o-symmetric convex bodies, perhaps even for Euclidean balls. If this is true, there is a function $\psi : \mathbb{N} \to \mathbb{R}$ where

$$\psi(d) = o(d), \ \psi(d) \to \infty \text{ as } d \to \infty,$$

such that the following hold: for each d there is an o-symmetric convex body C in $\mathbb{E}^d$ with $V(2C) = 2^{-\psi(d)}$ and each lattice $L \in \mathfrak{L}(1)$ contains a point $\neq o$ in the interior of the convex body $2^{1+2\psi(d)/d}C$ of volume $2^{\psi(d)}$. Thus, no lattice $L \in \mathfrak{L}(1)$ provides a packing of the convex body $2^{2\psi(d)/d}C$ of volume $2^{-d+\psi(d)}$. This implies that $\delta_L(2^{2\psi(d)/d}C) \leq 2^{-d+\psi(d)}$. Since the maximum lattice packing density is invariant with respect to dilatations we have,

$$\delta_L(C) \leq 2^{-d+\psi(d)}.$$

Let $\mathcal{A}$ be the set of all lattices in $\mathcal{L}(1)$ which contain a point $\neq o$ in the convex body $2C$ of volume $2^{-\psi(d)} = o(1)$. Then Siegel's mean value theorem shows that

$$\mu(\mathcal{A}) \leq \int\limits_{\mathcal{L}(1)} \#^*(L \cap 2C)\, d\mu(L) = V(2C) = o(1).$$

Thus, we have the following:

> All lattices $L \in \mathcal{L}(1)$, with a set of exceptions of measure $o(1)$, provide a packing of the convex body C with density $2^{-d-\psi(d)}$.

Supposing that, for the Minkowski–Hlawka theorem, there is no essential improvement possible for certain o-symmetric convex bodies, this shows the following: a large majority of lattices of determinant 1 provide lattice packings of such bodies with density close to the maximum lattice packing density.

30.4 Lattice Packing Versus Packing of Translates

Roughly speaking, the geometry of numbers deals with regular configurations, in particular with lattice packing, covering and tiling of convex and, possibly, non-convex bodies. Discrete geometry investigates the irregular case, in particular packing, covering and tiling of translates and congruent copies of convex and non-convex bodies.

Classical results in $\mathbb{E}^2$ say that, in several cases, general extremal configurations are no better than the corresponding extremal lattice configurations. An example is a result of Fejes Tóth and Rogers. It says that densest lattice packings of convex discs have maximum density among all packings by translates. For more information see the books of Fejes Tóth [327, 329] and Pach and Agarwal [783]. A stability result of the author [436] gives information on the geometric appearance of general packings of circular discs of maximum density. Such packings are asymptotically regular hexagonal. For other results of this type see Sects. 31.4 and 33.4 and the author's survey [438].

For $d \geq 3$ the only pertinent result for convex bodies is due to Hales. It says that $\delta_L(B^3) = \delta_T(B^3)$, see Sect. 29.2. (An earlier result of Bezdek and Kuperberg [109] deals with unbounded circular cylinders.) So far there is no example known of a convex body in $\mathbb{E}^d$, $d \geq 3$, with the property that the maximum lattice packing density is smaller than the maximum density of a packing by translates. If non-convex bodies are admitted, examples for this phenomenon are known, see Szabó [980]. Compare also the discussion in Sect. 32.3. Bezdek and Kuperberg [110] specified, for each $d \geq 3$, packings of congruent ellipsoids in $\mathbb{E}^d$ which have density larger than $\delta_L(B^d)$.

In the following we give a proof of the result of Fejes Tóth [328,329] and Rogers [846]. The special case of solid circular discs is due to Thue [996,997] and Fejes Tóth [327]. For a direct proof of the latter, see also Sect. 33.4.

Lattice Packing and Packing of Translates in $\mathbb{E}^2$

We show the following result of Fejes Tóth and Rogers:

Theorem 30.5. *Let C be a proper convex disc in $\mathbb{E}^2$. Then $\delta_T(C) = \delta_L(C)$.*

The following proof was given by Fejes Tóth [331]. It seems to be the shortest proof known.

Proof. Since the strictly convex discs are dense among all convex discs, it is sufficient to prove this result for strictly convex discs. The result is first proved for *triangular discs*, that is strictly convex discs C contained in a convex hexagon with vertices a, b, c, d, e, f, parallel opposite edges and such that $a, c, e \in C$. Call the line through a and b *vertical*, a and e the *left* and *right* and c the *bottom vertex* of C (Fig. 30.1).

The first step in the proof of the theorem is to show the following:

(1) Let $\{C + t : t \in T\}$ be a packing of translates of a triangular disc C. Then for each translate $C + t$ there is an associated set A_t such that:

- (i) $\text{diam}\, A_t \leq 3\, \text{diam}\, C$.
- (ii) The sets $A_t \cup (C + t),\ t \in T$, do not overlap.
- (iii) The *density* of $C + t$ in $A_t \cup (C + t)$, i.e. the quotient of the areas of these sets, equals the density of a certain lattice packing of C.

Given $C + t$, consider translates $C + r, C + s$, not necessarily belonging to the given packing, such that the bottom vertices of $C + s$ and $C + r$ coincide with the left and right vertices of $C + t$. The translates $C + r, C + s, C + t$ enclose a region A. We distinguish two cases (Fig. 30.2):

First, no translate $C + u,\ u \in T$, overlaps A. Then put $A_t = A$. The region A_t is contained in a triangle, a congruent copy of which is contained in $C + t$. Hence

(2) $\text{diam}\, A_t \leq \text{diam}\, C$.

Further,

(3) each point of A_t is connected to a point of $C + t$ by a vertical line segment and this line segment does not meet any translate $C + u,\ u \in T \setminus \{t\}$.

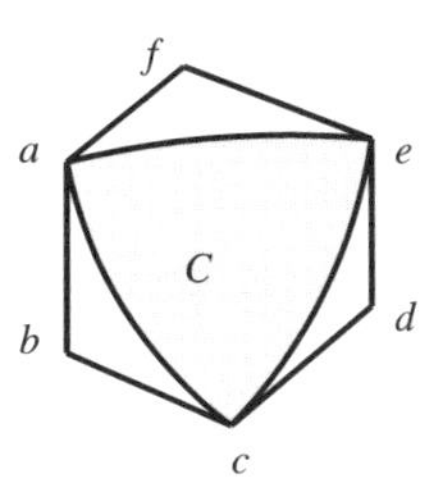

Fig. 30.1. Triangular disc

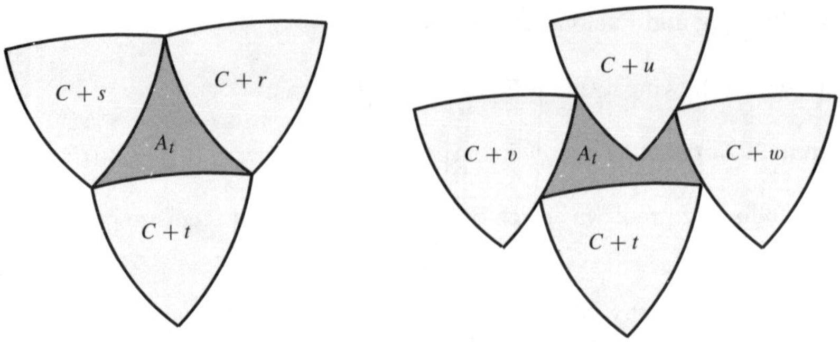

Fig. 30.2. Packing with triangular discs

Let L be the lattice with basis $r-t, s-t$. Then $\{C+t+l : l \in L\}$ is a lattice packing of $C+t$. It contains $C+r, C+s, C+t$. Since $\{A_t \cup (C+t)+l : l \in L\}$ is a lattice tiling, $d(L) = A(A_t \cup (C+t)) = A(A_t) + A(C)$. Thus,

(4) the density of $C+t$ in $A_t \cup (C+t)$, that is $A(C)/(A(A_t)+A(C))$, equals the density of the lattice packing $\{C+t+l : l \in L\}$.

Second, there is a translate $C+u$, $u \in T$ which overlaps A.

Choose two translates $C+v, C+w$, not necessarily from the given packing, such that $C+v, C+w$ touch $C+t$ at the left and right vertex of $C+t$, respectively, and $C+u$ at the right vertex of $C+v$ and the left vertex of $C+w$, respectively. Since C is strictly convex, the discs $C+v, C+w$ are unique. The translates $C+v, C+u, C+w, C+t$ enclose a region A_t, say. Since $\operatorname{diam} A_t = \operatorname{diam} \operatorname{bd} A_t$,

(5) $\operatorname{diam} A_t \leq 3 \operatorname{diam} C$.

Since no translate $C+p$, $p \in T$, can overlap A_t, we see that

(6) each point of A_t is connected to a point of $C+t$ by a vertical line segment and this line segment does not meet any translate $C+p$, $p \in T \setminus \{t\}$.

Let L be the lattice with basis $v-t, w-t$. Then $\{C+t+l : l \in L\}$ is a lattice packing of C containing $C+v, C+w, C+t, C+u$. (The lattice translation $v-t$ maps $C+w$ onto $C+v+w-t$ and the latter meets $C+w$ at its left vertex. Similarly, $C+v+w-t$ meets $C+v$ at its right vertex. Since C is strictly convex, $C+u$ is the unique translate which meets $C+v$ at its right and $C+w$ at its left vertex, we see that $C+u = C+v+w-t$.) Since $\{A_t \cup (C+t)+l : l \in L\}$ is a tiling, $d(L) = A(A_t \cup (C+t)) = A(A_t) + A(C)$. Thus

(7) the density of $C+t$ in $A_t \cup (C+t)$, that is $A(C)/(A(A_t)+A(C))$ equals the density of the lattice packing $\{C+t+l : l \in L\}$.

Propositions (2), (5); (3), (6); and (4), (7) imply (i), (ii) and (iii) in (1), respectively, concluding the proof of (1).

An immediate consequence of (1) is the following:

(8) Let C be a triangular disc. Then the upper density of a packing of translates of C never exceeds the density of the densest lattice packing of C.

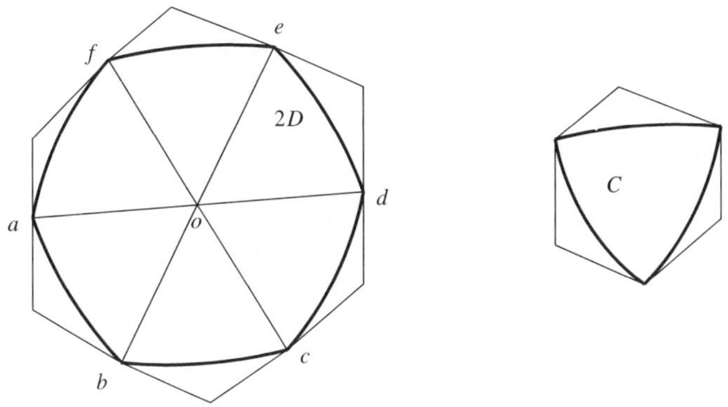

Fig. 30.3. Disc and triangular disc

In the second step of the proof, the following will be shown:

(9) Let D be an o-symmetric strictly convex disc. Then there is a triangular disc C such that $D = \frac{1}{2}(C - C)$ (Fig. 30.3).

To prove this, we first show that there is an affine regular convex hexagon H inscribed in $2D$. For $a \in \operatorname{bd} 2D$ let H be the unique convex hexagon H inscribed in $2D$ with opposite vertices a and $-a$ such that its edges parallel to the line segment $[a, -a]$ both have length $\|a\|$. Since $2D$ is strictly convex and symmetric in o, these edges are also symmetric in o. Thus H is affine regular.

Choose an affine regular hexagon with consecutive vertices a, b, c, d, e, f and centre at o which is inscribed in $2D$. Translate the triangles with vertices a, b, o; o, c, d; f, o, e along with the adjacent lunae cut off from $2D$ by the line segments $[a, b]$, $[c, d]$ and $[f, e]$, such that the triangles coincide. Let C be the union of the translated triangles and lunae. Considering an o-symmetric convex hexagon circumscribed to $2D$ the edges of which touch $2D$ at the points $a, \dots, f$, it follows that C is triangular. Clearly, $2D = C - C$, concluding the proof of (9).

In the third step of the proof, we show the following proposition:

(10) Let D be an o-symmetric strictly convex disc. Then the upper density of a packing of translates of D never exceeds the density of the densest lattice packing of D.

Let $\{D + t : t \in T\}$ be a packing. By Proposition 30.2, its upper density is $A(D)\delta$, where δ is the upper density of T. Choose a triangular disc C such that $D = \frac{1}{2}(C - C)$. This is possible by (9). By Proposition 30.4, $\{C + t : t \in T\}$ is also a packing. Its upper density is $A(C)\delta$. By (8), there is a packing lattice L of C and $A(C)\delta \le A(C)/d(L)$ by Corollary 30.1. Thus $\delta \le 1/d(L)$. Proposition 30.4 then shows that L is also a packing lattice of D. Thus $\{D + l : l \in L\}$ is a lattice packing of D of density $A(D)/d(L) \ge A(D)\delta$. Now note that $A(D)\delta$ is the upper density of the packing $\{D + t : t \in T\}$.

The fourth step is to show the following:

(11) Let E be a strictly convex disc. Then the upper density of a packing of translates of E never exceeds the density of the densest lattice packing of E.

Let $D = \frac{1}{2}(E - E)$. Similar arguments as in the proof of (10) but with E, D instead of D, C yield (11), concluding the proof of the theorem for strictly convex discs.

For convex discs which are not strictly convex consider approximation with strictly convex discs and use (11). The details are tedious. □

Remark. This result does not extend to packing by congruent copies, as the example of a triangle shows. More precisely, for a typical convex disc C (in the sense of Baire categories), Fejes Tóth [320] showed that there are packings of congruent copies of C which have density larger than $\delta_L(C)$. It has been conjectured that this holds in all dimensions $d \geq 2$. For certain ellipsoids this was shown by Bezdek and Kuperberg [110] and Rogers [851] conjectured it for Euclidean balls for all sufficiently large d.

In the special case of packings by congruent copies of centrally symmetric convex discs, the theorem continues to hold according to a result of Fejes Tóth [327], [329], p. 86, which is a consequence of one of his more general results.

A Conjecture of Zassenhaus for Densest Packings in $\mathbb{E}^d$

A *periodic packing* of a proper convex body C is a packing by translates, where the set T of translation vectors is of the form

$$T = L \cup (L + t_1) \cup \cdots \cup (L + t_m)$$

with a lattice L and vectors $t_1, \ldots, t_m \in \mathbb{E}^d$. It is easy to see that, for every proper convex body C, there are periodic packings with density arbitrarily close to $\delta_T(C)$. An interesting open *conjecture of Zassenhaus* [1042] asserts even more:

Conjecture 30.2. *Let C be a proper convex body in $\mathbb{E}^d$. Then $\delta_T(C)$ is attained by a suitable periodic packing of C.*

31 Covering with Convex Bodies

The theory of covering is less rich than the theory of packing with convex bodies and appeared much later in the literature, the first landmark being a result of Kershner [578]. It shows that the minimum density of coverings of $\mathbb{E}^2$ by congruent circular discs is attained by lattice coverings. One reason for the fact that covering results have attracted less interest is that the arithmetic and number theoretic interpretations of covering results seem to have attracted less attention than corresponding interpretations of packing results. Yet, in the last two or three decades, covering with convex bodies has become important in the local theory of normed spaces, see the report of Giannopoulos and Milman [375], and Schneider [905] discovered that coverings with congruent balls play an essential role for Hausdorff approximation of convex bodies by polytopes.

This section deals with lattice coverings and coverings with translates of convex bodies. We begin with definitions and elementary remarks. Then star numbers of coverings are considered. Next, the upper bound for minimum densities, due to Rogers, is given. Finally we touch the relation between coverings with translates and lattice coverings.

For more information, see the references cited at the beginning of Sect. 30 to which we add a survey of Fejes Tóth [324].

31.1 Definitions, Existence of Thinnest Lattice Coverings and the Covering Criterion of Wills

In this section we give definitions of coverings and covering density, state several simple properties of coverings and show the existence of lattice coverings with minimum density. In addition, we state the covering criterion of Wills.

Covering with Convex Bodies and the Notion of Density

A family of convex bodies in $\mathbb{E}^d$ is a *covering* if their union equals $\mathbb{E}^d$. We consider only *coverings by translates* and *lattice coverings* of a given convex body C, i.e. coverings of the form $\{C + t : t \in T\}$ and $\{C + l : l \in L\}$ where T is a discrete set and L a lattice in $\mathbb{E}^d$, respectively. If $\{C + l : l \in L\}$ is a lattice covering of C, then L is a *covering lattice* of C. The upper and lower density and the density of a discrete set or of a family of translates of a convex body are defined in Sect. 30.1. Corollary 30.1 says that the density $\delta(C, L)$ of a family $\{C + l : l \in L\}$ of translates of a proper convex body by the vectors of a lattice L equals

$$\frac{V(C)}{d(L)}.$$

Consider a covering by translates of a convex body. Cum grano salis, its density may be interpreted as the total volume of the bodies divided by the volume of $\mathbb{E}^d$, or as the expectation of the number of bodies of the covering in which a random point of $\mathbb{E}^d$ is contained.

Given a convex body C, let $\vartheta_T(C)$ and $\vartheta_L(C)$ denote the infima of the lower densities of all coverings by translates of C and of all lattice coverings of C, respectively. $\vartheta_T(C)$ and $\vartheta_L(C)$ are called the *(minimum) translative covering density* and the *(minimum) lattice covering density* of C, respectively. Clearly,

$$1 \leq \vartheta_T(C) \leq \vartheta_L(C) < \infty.$$

Existence of Thinnest Coverings

Using Jarník's transference theorem, the lower estimate in the theorem on successive minima of Minkowski and Mahler's selection theorem, we prove the following result.

Theorem 31.1. *Let C be a proper convex body in $\mathbb{E}^d$ with $o \in \operatorname{int} C$. Then there is a covering lattice L of C such that $\delta(C, L) = \vartheta_L(C)$.*

Proof. Let (L_n) be a sequence of covering lattices of C such that:

$$(1) \qquad \delta(C, L_1) = \frac{V(C)}{d(L_1)} \geq \delta(C, L_2) = \frac{V(C)}{d(L_2)} \geq \cdots \to \vartheta_L(C) \geq 1.$$

Then

$$(2) \qquad V(C) \geq d(L_n).$$

Let B be a solid Euclidean ball with centre o such that $B \supseteq C$. Since L_n is a covering lattice of C and thus a fortiori of B, it follows that for the covering radius $\mu(B, L_n)$ of B with respect to L_n that $\mu(B, L_n) \leq 1$. Thus Jarník's transference theorem 23.4 implies that, for the dth successive minimum of B with respect to L_n, we have

$$(3) \qquad \lambda_d(B, L_n) \leq 2\mu(B, L_n) \leq 2.$$

Considering the inequality

$$\lambda_1(B, L_n) \cdots \lambda_d(B, L_n) V(B) \geq \frac{2^d}{d\,!}\, d(L_n),$$

which holds by the theorem 23.1 on successive minima, it follows from (1) and (3) that

$$\lambda_1(B, L_n) \geq \frac{2^d\, d(L_n)}{d\,!\,\lambda_d(B, L_n)^{d-1} V(B)} \geq \frac{2^d\, V(C)}{d\,!\,2^{d-1} V(B)\,\delta(C, L_1)} = \alpha,$$

say. Thus

(4) L_n is admissible for the ball αB with centre o.

Noting (2) and (4), Mahler's selection theorem 25.1 implies that the sequence (L_n) has a convergent subsequence. After appropriate cancellation and renumbering of indices, if necessary, we may assume that $L_1, L_2, \cdots \to L$, where L is a suitable lattice. The definition of convergence of lattices implies the following propositions:

$$(5) \qquad d(L_1), d(L_2), \cdots \to d(L).$$

(6) If $l_n \in L_n$ are such that $l_1, l_2, \cdots \to l \in \mathbb{E}^d$, then $l \in L$.

For the proof that

(7) L is a covering lattice of C,

let $x \in \mathbb{E}^d$. We have to show that $x \in C + l$ for suitable $l \in L$. Since, by assumption, the lattices L_n are covering lattices of C, there are vectors $l_n \in L_n$ such that

$x \in C + l_n$. Hence (l_n) is a bounded sequence in $\mathbb{E}^d$. By cancellation and renumbering, if necessary, we may assume that $l_1, l_2, \dots \to l \in \mathbb{E}^d$, say. Then $l \in L$ by (6). The inclusion $x \in C + l_n$ and the fact that C is closed yield $x \in C + l$. The proof of (7) is complete.

Concluding, L is a covering lattice of C, by (7), and Propositions (5) and (1) imply that

$$\delta(C, L) = \frac{V(C)}{d(L)} = \lim_{n\to\infty} \frac{V(C)}{d(L_n)} = \vartheta_L(C). \quad \square$$

Remark. A different proof shows that $\vartheta_T(C)$ is attained for suitable discrete sets T, see Hlawka [510] and Groemer [399].

The Covering Criterion of Wills

Given a convex body C in $\mathbb{E}^d$, when does a lattice provide a covering of C? There is a small number of pertinent results. We cite the following interesting theorem of Wills [1026].

Theorem 31.2. *Let C be a proper convex body in $\mathbb{E}^d$ with $V(C) > \frac{1}{2}S(C)$. Then the integer lattice $\mathbb{Z}^d$ is a covering lattice of C.*

Remark. A generalization of this result which deals with multiple coverings is due to Bokowski, Hadwiger and Wills [138]. For other covering criteria, see [447].

Considering the covering criterion, the following problem arises.

Problem 31.1. *Extend the covering criterion to general lattices.*

31.2 Star Numbers

Given a lattice covering of a convex body C in $\mathbb{E}^d$, its *star number* is the number of the translates of C by lattice vectors, including C, which intersect the body C.

In the following we give a lower bound for star numbers due to Erdös and Rogers. An upper bound in the spirit of the lower estimate of Swinnerton-Dyer for neighbours in packings still seems to be missing. We formulate this as a problem:

Problem 31.2. *Find a tight upper bound for the star number of lattice coverings of (symmetric or general) proper convex bodies which (locally) have minimum density.*

Lower Estimate for the Star Number

The precise lower bound for star numbers of lattice coverings of o-symmetric convex bodies is due to Erdös and Rogers [308]:

Theorem 31.3. *Let C be a proper, o-symmetric convex body and L a covering lattice of C in $\mathbb{E}^d$. Then the star number of the covering $\{C + l : l \in L\}$ is at least $2^{d+1} - 1$.*

Proof. Call $x, y \in \mathbb{E}^d$ *congruent modulo L* if $x - y \in L$. The points of the form $\frac{1}{2} l$ with $l \in L$ then fall into 2^d congruence classes modulo L. (To see this, let $\{b_1, \dots, b_d\}$ be a basis of L. Then each point of the form $\frac{1}{2} l$ with $l \in L$ is congruent to one of the 2^d points

$$\frac{u_1}{2} b_1 + \dots + \frac{u_d}{2} b_d \text{ where } u_i \in \{0, 1\}$$

and these points are pairwise incongruent.) Let $o, \frac{1}{2} l_1, \dots, \frac{1}{2} l_n$, with $l_i \in L$ and $n = 2^d - 1$ be representatives of these congruence classes. Then, as the translates of C by the vectors of L cover $\mathbb{E}^d$, there are $m_1, \dots, m_n \in L$ such that $\frac{1}{2} l_i \in C + m_i$, or

$$\frac{1}{2} l_i - m_i \in C \text{ for } i = 1, \dots, n.$$

These points lie in different congruence classes modulo L and are not congruent to o. Since $-(\frac{1}{2} l_i - m_i) \equiv \frac{1}{2} l_i - m_i$ modulo L, we see that the points $-(\frac{1}{2} l_i - m_i)$ and $\frac{1}{2} l_i - m_i$ lie in the same congruence class but clearly are different. Since C is o-symmetric it follows that

> the $2n = 2^{d+1} - 2$ points $\pm(\frac{1}{2} l_i - m_i)$, $i = 1, \dots, n$,
> are pairwise different, different from o and all are contained in C.

Thus C meets each of the $2n + 1 = 2^{d+1} - 1$ distinct bodies C, $C \pm (l_i - 2m_i)$, $i = 1, \dots, n$, of the covering. □

31.3 Rogers's Upper Bound for $\vartheta_T(C)$

Numerous early attempts to construct or to establish the existence of dense lattice and non-lattice coverings of $\mathbb{E}^d$ with convex bodies were relatively unsuccessful. The upper bounds which were obtained all were of the form c^d with suitable constants $c > 1$. The breakthrough finally was achieved by Rogers who introduced averaging methods to this problem which led to surprisingly small upper estimates for the minimum density of a covering of $\mathbb{E}^d$ by translates of a convex body and for lattice coverings.

In this section we prove Rogers' upper estimate for the minimum density of a covering of $\mathbb{E}^d$ by translates of a convex body. His proof is based on periodic sets and a mean value argument.

For an exposition of the work of Rogers on coverings, the reader may wish to consult Rogers' classical Cambridge tract [851].

An Inequality Between $\delta_T(C)$ and $\vartheta_T(C)$

Before embarking on Rogers' result, we present a result which relates the packing and the covering case and, as a consequence, yields a (large) upper bound for $\vartheta_T(C)$. We point out the nice idea of proof, which might have been a starting point for Rogers' ingenious proof, see later.

Proposition 31.1. *Let C be a proper, o-symmetric convex body in $\mathbb{E}^d$. Then* $(1 \leq)\ \vartheta_T(C) \leq 2^d \delta_T(C)\ (\leq 2^d)$. *In particular,* $\delta_T(C) \geq 2^{-d}$.

Proof. Consider a packing of translates of C in $\mathbb{E}^d$. By successively inserting additional translates of C into the interstitial space of the packing, we finally arrive at a packing of translates of C, say $\{C + t : t \in T\}$, such that, for any $x \in \mathbb{E}^d$, the translate $C + x$ meets at least one translate of the packing, say $C + t$. Thus there are $p, q \in C$ such that $p + x = q + t$, and therefore $x = q - p + t \in C - C + t = 2C + t$ since C is o-symmetric. In other words, $\{2C + t : t \in T\}$ is a covering of translates of C. □

If C is an o-symmetric convex body for which the Minkowski–Hlawka bound $2^{-d+o(d)}$ for $\delta_L(C)$ is best possible, then $\vartheta_L(C) \leq 2^{o(d)}$. The upper estimates of Rogers for $\vartheta_T(C)$ and $\vartheta_L(C)$ are more explicit and hold for all o-symmetric convex bodies C.

Rogers's Upper Bound for $\vartheta_T(C)$

Rogers [848, 851] proved the following upper estimate for $\vartheta_T(C)$. See Füredi and Kang [347] for an elegant proof of a slightly weaker result.

Theorem 31.4. *Let C be a proper convex body in $\mathbb{E}^d$. Then* $\vartheta_T(C) \leq d \log d \,\big(1 + o(1)\big)$ *as* $d \to \infty$.

The following rough outline of the proof may help the reader:

It is enough to cover a large box and to continue by periodicity.
Randomly placed translates of C cover almost the whole box
in an economic way.
With an additional trick the small holes are covered.

Proof. We may suppose that

$$V(C) = 1 \tag{1}$$

and that o is the centroid of C. Then a well-known result, which can easily be proved, says that

$$-\frac{1}{d}\, C \subseteq C. \tag{2}$$

Choose $s > 0$ so large that, for the lattice $L = s\,\mathbb{Z}^d$,

(3) any two distinct bodies of the family $\{C + l : l \in L\}$ are disjoint.

Let $F = \{x : 0 \leq x_i < s\}$ be a fundamental parallelotope of L and let m be a large positive integer.

After these preparations the following will be shown first:

(4) There is a set $X = \{x_1, \dots, x_m\} \subseteq F$ such that the proportion of space left uncovered by the set $C + L + X$, that is by the family $\{C + l + x : l \in L,\ x \in X\}$, is at most

$$\Bigl(1 - \frac{1}{s^d}\Bigr)^m.$$

Let $\mathbb{1}_C$ be the characteristic function of C. It follows from (3) that the characteristic function of the set $C+L$, that is the union of the family $\{C+l : l \in L\}$, is $\sum\{\mathbb{1}_C(x-l) : l \in L\}$. Thus the set $C + L + x_i$ where $x_i \in F$ has characteristic function $\sum\{\mathbb{1}_C(x - l - x_i) : l \in L\}$. This shows that, for $X = \{x_1, \dots, x_m\} \subseteq F$, the characteristic function of the set $S = \mathbb{E}^d \backslash (C + l + X)$ left uncovered by the set $C + L + X$, that is by the family $\{C + l + x_i : l \in L,\ x_i \in X\}$, is given by:

$$\mathbb{1}_S(x) = \prod_{i=1}^m \Bigl(1 - \sum_{l\in L} \mathbb{1}_C(x - l - x_i)\Bigr).$$

$\mathbb{1}_S$ is periodic with respect to L. Thus the proportion of space left uncovered by the set $C + L + X$ equals

$$\frac{V(S \cap F)}{V(F)} = \frac{1}{V(F)} \int_F \mathbb{1}_S(x)\,dx.$$

The mean value of this proportion extended over all choices $X = \{x_1, \dots, x_m\}$ of m points in F is thus

$$\begin{aligned}
&\frac{1}{V(F)^m} \int_F \cdots \int_F \Bigl(\frac{1}{V(F)} \int_F \mathbb{1}_S(x)\ dx\Bigr) dx_1 \cdots dx_m \\
&= \frac{1}{V(F)^{m+1}} \int_F \Bigl(\int_F \cdots \int_F \prod_{i=1}^m \Bigl(1 - \sum_{l\in L} \mathbb{1}_C(x - l - x_i)\Bigr) dx_1 \cdots dx_m\Bigr) dx \\
&= \frac{1}{V(F)^{m+1}} \int_F \Bigl(\prod_{i=1}^m \int_F \Bigl(1 - \sum_{l\in L} \mathbb{1}_C(x - l - x_i)\Bigr) dx_i\Bigr) dx \\
&= \frac{1}{V(F)^{m+1}} \int_F \prod_{i=1}^m \Bigl(V(F) - \sum_{l\in L} \int_F \mathbb{1}_C(x - l - x_i) dx_i\Bigr) dx \\
&= \frac{1}{V(F)^{m+1}} \int_F \prod_{i=1}^m \Bigl(V(F) - \sum_{l\in L} \int_{F-x+l} \mathbb{1}_C(-y) dy\Bigr) dx \\
&= \frac{1}{V(F)^{m+1}} \int_F \prod_{i=1}^m \Bigl(V(F) - \int_{\mathbb{E}^d} \mathbb{1}_C(-y) dy\Bigr) dx \\
&= \frac{1}{V(F)^{m+1}} V(F)\bigl(V(F) - V(C)\bigr)^m = \Bigl(1 - \frac{V(C)}{V(F)}\Bigr)^m = \Bigl(1 - \frac{1}{s^d}\Bigr)^m
\end{aligned}$$

by (1). Since the mean value extended over all choices of m points in F of the proportion in question is $(1 - 1/s^d)^m$, there is at least one choice $X = \{x_1, \dots, x_m\}$ of m points in F for which this proportion is at most $(1 - 1/s^d)^m$, concluding the proof of (4).

Second, we prove the following proposition where the set X is as in (4).

(5) Let $0 < t \le 1/d$ and $Y = \{y_1, \dots, y_n\} \subseteq F$ with n maximum such that the family $\{-tC + l + y : l \in L,\ y \in Y\}$ is a packing contained in $\mathbb{E}^d \backslash (C + L + X)$. Then

$$n \le \frac{s^d}{t^d}\left(1 - \frac{1}{s^d}\right)^m.$$

This is easy. First, the proportion of space covered by the packing $\{-tC+l+y : l \in L, y \in Y\}$, that is its density, is $n\,t^d\,V(C)/V(F) = n\,t^d/s^d$ by (1) and the choice of $L = s\,\mathbb{Z}^d$. Second, this packing is contained in $\mathbb{E}^d \backslash (C + L + X)$ and the proportion of space covered by the latter set is at most $(1 - 1/s^d)^m$ by (4). The proof of (5) is complete.

Third, note (4) and (5). In order to prove that

$$\{(1+t)C + l + z : l \in L,\ z \in X \cup Y\} \text{ is a covering,} \tag{6}$$

let $w \in \mathbb{E}^d$. We have to show that w is contained in at least one of the bodies in (6). By (2), (3) and since by (5) $0 < t \le \frac{1}{d}$, the bodies $-tC+m+w,\ m \in L$, are pairwise disjoint. We distinguish two cases. First, $(-tC + L + w) \cap (C + L + X) \ne \emptyset$. Then there are $p, q \in C,\ l, m \in L,\ x_i \in X$ such that $-tp + l + w = q + m + x_i$, or $w = tp + q - l + m + x_i \in (1+t)C + L + X$. Second, if the first alternative does not hold, then $-tC + L + w \subseteq \mathbb{E}^d \backslash (C + L + X)$. By the maximality of the family $\{-tC+l+y : l \in L, y \in Y\}$ we then have $(-tC+L+w) \cap (-tC+L+Y) \ne \emptyset$. Thus there are $p, q \in C,\ l, m \in L,\ y_j \in Y$ such that $-tp+l+w = -td+m+y_j$, or $w = tp + tq - l + m + y_j \in (1+t)C + L + Y$ by (2) and since $0 < t \le 1/d$ by (5). This concludes the proof of (6).

Fourth, noting (1), the definition of $L = s\,\mathbb{Z}^d$, Propositions (4) and (5) show that

(7) the density of the covering in (6) is

$$\frac{(1+t)^d(m+n)}{s^d} \le (1+t)^d\left(\frac{m}{s^d} + \frac{1}{t^d}\left(1 - \frac{1}{s^d}\right)^m\right).$$

By choosing s, m, t suitably this finally yields the desired upper bound for a covering of $\mathbb{E}^d$ by translates of $(1+t)C$, and thus for a corresponding covering of $\mathbb{E}^d$ by translates of C: Let

$$\frac{m}{s^d} = d\ \log\frac{1}{t},\quad t = \frac{1}{d\log d}.$$

Then

$$\begin{aligned}
&(1+t)^d\left(\frac{m}{s^d}+\frac{1}{t^d}\left(1-\frac{1}{s^d}\right)^m\right)\\
&\le e^{dt}\left(d\log\frac{1}{t}+\frac{1}{t^d}e^{-\frac{m}{s^d}}\right)\\
&\le e^{\frac{1}{\log d}}\left(d\log(d\log d)+\frac{1}{t^d}e^{-d\log\frac{1}{t}}\right)\\
&\le \left(1+\frac{2}{\log d}\right)(d\log d+d\log\log d+1)\\
&\le d\log d+d\log\log d+5d=d\log d\,\big(1+o(1)\big). \quad \square
\end{aligned}$$

Remark. The following lower estimate of Coxeter, Few and Rogers [233] shows that Theorem 31.4 cannot be improved much, if at all:

$$\vartheta_T(B^d) \gtrsim \frac{d}{e\sqrt{e}} \quad \text{as } d \to \infty.$$

Rogers's Upper Bound for $\vartheta_L(C)$

A more refined proof led Rogers [850] to a corresponding result for lattice coverings which is stated without proof:

Theorem 31.5. *Let C be a proper convex body in $\mathbb{E}^d$, where $d \ge 3$. Then*

$$\vartheta_L(C) \le d^{\log_2 d\,(1+o(1))} \textit{as } d \to \infty.$$

31.4 Lattice Covering Versus Covering with Translates

In Sect. 30.4 we encountered the phenomenon that certain general extremal configurations are not better than corresponding extremal lattice configurations. In addition, general extremal configurations may have lattice characteristics. A planar example of this, dealing with packing of convex discs was presented and a case where general extremal configurations exhibit regular hexagonality mentioned.

Thinnest lattice coverings in $\mathbb{E}^2$ with o-symmetric convex discs have minimum density among all coverings by translates as shown by Kershner. A stability result of Gruber [436] says that thinnest coverings of $\mathbb{E}^2$ by circular discs are arranged asymptotically in the form of a regular hexagonal pattern. For more information compare the survey [438].

A Result of Fejes Tóth and Bambah and Rogers

Without proof, we state the following result of Fejes Tóth [328] and Bambah and Rogers [59]. The special case for circular discs is due to Kershner [578]. For an alternative proof of Kershner's result, see Sect. 33.4. It is an open problem, whether the symmetry assumption can be omitted.

Theorem 31.6. *Let C be a proper o-symmetric convex disc in $\mathbb{E}^2$. Then $\vartheta_T(C) = \vartheta_L(C)$.*

Remark. It has been conjectured that the equality in this theorem holds for all convex discs. A corresponding result in $\mathbb{E}^2$ for coverings with congruent copies of a centrally symmetric convex disc has been conjectured but, so far, this was proved only under additional restrictions, see Fejes Tóth [323]. For a typical convex (not necessarily centrally symmetric) disc C the thinnest covering with congruent copies has density smaller than $\theta_L(C)$, as shown by Fejes Tóth and Zamfirescu [326].

A Conjecture of Zassenhaus on Thinnest Coverings in $\mathbb{E}^d$

Considering the packing case in Sect. 30.4, it is clear what is meant by a *periodic covering* with a convex body C. Zassenhaus [1042] stated a conjecture for coverings analogous to his Conjecture 30.1 on packings as follows:

Conjecture 31.1. *Let C be a proper convex body in $\mathbb{E}^d$. Then $\vartheta_T(C)$ is attained by a suitable periodic covering of C.*

32 Tiling with Convex Polytopes

Tiling problems date back to antiquity, see the historical remarks on honeycombs by Hales [472]. We mention also Kepler, MacLaurin and Kelvin. In his vortex theory of planetary motion, Descartes [262] used figures which are almost Dirichlet–Voronoĭ tilings, compare Gaukroger [362]. I owe this information to Rolf Klein [598]. Problems on quadratic forms led Dirichlet [272] and later Voronoĭ [1014] and Delone [255], his disciple Ryshkov and their school to study lattice and non-lattice tiling, while Fedorov's [319] research on tiling was the outgrowth of his seminal work in crystallography. Hilbert's [501] 18th problem made tiling problems more popular. Since then there has been a continuous stream of tiling results.

In this section we first present Dirichlet–Voronoĭ and Delone tilings. Then the basic theorem of Venkov–McMullen will be given. Finally, we discuss a conjecture of Voronoĭ and Hilbert's 18th problem.

For detailed information on tiling the reader is referred to the surveys and books of Heesch and Kienzle [486], Rogers [851], Fejes Tóth [327], Delone and Ryshkov [259], Grünbaum and Shephard [454], Engel [296, 297], Gruber and Lekkerkerker [447], Erdös, Gruber and Hammer [307], Schulte [918], Senechal [925], Johnson [550], Schattschneider and Senechal [883] and Engel, Michel and Senechal [301].

32.1 Dirichlet–Voronoĭ and Delone Tilings and Polyhedral Complexes

Among the most important examples of tilings in $\mathbb{E}^d$ are the Dirichlet–Voronoĭ tilings and the Delone triangulations. The former are important for quadratic forms and in computational geometry, discrete geometry and convexity, and are of use in econometry, geography, sociology, biology, microbiology, metallurgy, crystallography, data transmission and other fields. This is expressed, for example, by Tanemura [988]:

> ... usefulness of the concept of Voronoĭ tessellation is three-fold. Firstly, the Voronoĭ tessellation can be one of the ways of describing the manner of a spatial distribution of particles. ... Secondly, the Voronoĭ tessellation is useful for modeling tiling patterns which are observed in nature. ... Thirdly, the Voronoĭ tessellation can be used as a tool for reducing the load of computation.

After stating basic definitions, we describe, in this section, Dirichlet–Voronoĭ tilings and Delone triangulations and show that locally finite facet-to-facet tilings give rise to polyhedral complexes. Finally, a bound, due to Minkowski, for the number of facets of a lattice tile is given.

The tilings that will be considered are all *locally finite*, i.e. any bounded set in $\mathbb{E}^d$ meets only finitely many tiles. Tilings which are not locally finite and, in particular, such tilings in infinite dimensional normed spaces may have completely unexpected properties. For some results and references to the literature, see, e.g. Klee, Maluta and Zanco [596], Klee [594] and Gruber [432].

The reader who wants to know more on tilings in the context of Dirichlet–Voronoĭ and Delone tilings may consult the books of Møller [749], Engel and Syta [302] and Okabe, Boots, Sugihara and Chiu [777] and the surveys of Fortune [341] and Aurenhammer and Klein [42].

Tiling with Convex Bodies

A family of proper convex or unbounded proper convex bodies in $\mathbb{E}^d$ is a *(convex) tiling* if it is both a packing and a covering. The bodies are called *tiles*. We will consider mainly, but not exclusively, tilings which consist of lattice translates, translates, or congruent copies of a given convex body P. These tilings are called *lattice tilings, translative tilings* and *tilings of congruent copies* of P. The convex body P then is a convex polytope as will be shown in this section. It is called the *prototile* of the tiling. The prototile of a lattice tiling is called a *parallelohedron*. Particular tilings of congruent copies of P are those where the tiles are the images of P under the rigid motions of a crystallographic group. If this is the case, P is a *stereohedron*. A tiling is *facet-to-facet* if, for any two tiles with $(d-1)$-dimensional intersection, the intersection is a facet of both of them. It is *face-to-face*, if the intersection of any two tiles is a face of both of them.

Dirichlet–Voronoĭ Tilings

Dirichlet [272] first introduced tilings of the following form, where L is a lattice in $\mathbb{E}^d$:

$$\{P+l : l \in L\},\ \ P = \big\{x : \|x\| \le \|x-m\| \text{ for all } m \in L\big\}.$$

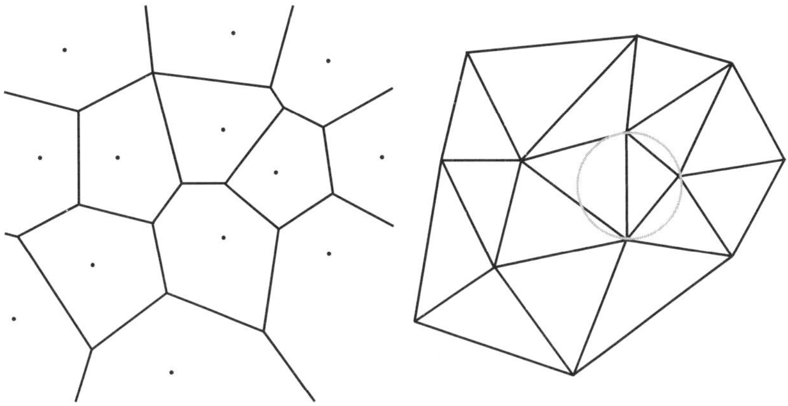

Fig. 32.1. Dirichlet–Voronoĭ tiling and Delone triangulation

The systematic study of such tilings started with Voronoĭ [1014]. More generally, let D be a discrete set in $\mathbb{E}^d$. Then the sets

$$\{D(p) : p \in D\},\ D(p) = \big\{x : \|x - p\| \leq \|x - q\| \text{ for all } q \in D\big\}$$

are called *Dirichlet–Voronoĭ cells* of D. Clearly, any cell $D(p)$ consists of all points x which are at least as close to p as to any other point $q \in D$. Dirichlet–Voronoĭ cells are known under very different names, for example honeycombs, domains of action, Brillouin, or Wigner-Seitz zones. The corresponding tilings are called *Dirichlet–Voronoĭ tilings* (Fig. 32.1).

Proposition 32.1. *Let D be a discrete set. Then the corresponding Dirichlet–Voronoĭ cells are proper, generalized convex polyhedra and form a facet-to-facet tiling of $\mathbb{E}^d$.*

Proof. First, the following will be shown:

(1) Each Dirichlet–Voronoĭ cell $D(p)$, $p \in D$, is a proper generalized convex polyhedron.

As the intersection of closed halfspaces, each cell is a closed convex set in $\mathbb{E}^d$. Since D is discrete, each cell has non-empty interior and thus is proper. To show that it is a generalized polyhedron, it is sufficient to prove the following: Let K be a cube. Then the intersection $D(p) \cap K$ is a convex polytope for each $p \in D$. Clearly, $D(p) \cap K$ is the intersection of the cube K with the halfspaces $\{x : \|x - p\| \leq \|x - q\|\}$, where $q \in D$, $q \neq p$. Since D is discrete, all but finitely many of these halfspaces contain K in their interior. Thus $D(p) \cap K$ is the intersection of K with a finite family of these halfspaces and thus is a convex polytope, concluding the proof of (1).

The second step is to show the following statement:

(2) The Dirichlet–Voronoĭ cells $\{D(p) : p \in D\}$ form a locally-finite tiling of $\mathbb{E}^d$.

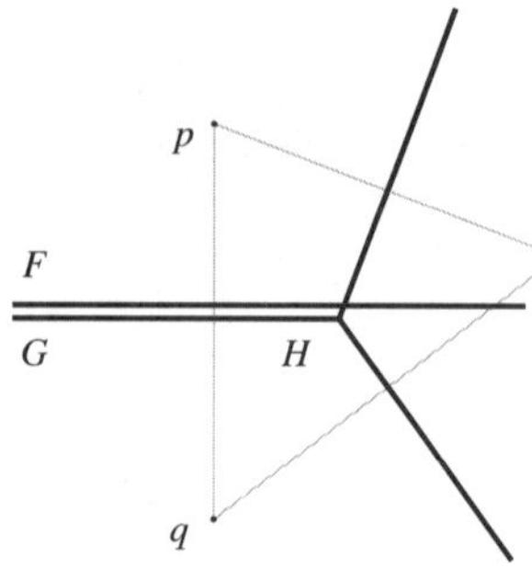

Fig. 32.2. Dirichlet–Voronoĭ tilings are facet-to-facet

For the proof of (2) it is sufficient to show the following: if K is a cube, then only finitely many of the Dirichlet–Voronoĭ cells $D(p)$ meet K and these form a tiling of K. Let $q \in D$. Since D is discrete, all but finitely many of the halfspaces $\{x : \|x - p\| \leq \|x - q\|\}$ are disjoint from K. Thus, only finitely many Dirichlet–Voronoĭ cells $D(p)$ meet K, say $D(p_1), \dots, D(p_k)$. Any two of these cells are separated by a hyperplane and thus have disjoint interiors. Since D is discrete, for each point of $\mathbb{E}^d$ there is a least one nearest point in D. Thus $\mathbb{E}^d$ is the union of all cells. Together, this implies that $\{D(p_1) \cap K, \dots, D(p_k) \cap K\}$ form a tiling of K.

The third and final step is to show the following:

(3) The Dirichlet–Voronoĭ cells $\{D(p) : p \in D\}$ form a facet-to-facet tiling of $\mathbb{E}^d$.

By (1) and (2), each facet of a cell $D(p),\ p \in D$, is covered by facets of other such cells. Thus, if (3) did not hold, there are points $p, q \in D$ such that the tiles $D(p)$ and $D(q)$ have facets F and G, respectively, which overlap, but do not coincide. Then there is a $(d-2)$-dimensional face H of G, say, which meets the relative interior of F. Project F, G, H onto a 2-dimensional plane orthogonal to H. Then we have the configuration of Fig. 32.2.

Since H is the intersection of two facets of $D(q)$, one of which is G, there is a point $r \in D$ which, together with q, determines this facet. Then the hyperplane $\{x : \|x - p\| = \|x - r\|\}$ cuts off the part of F outside H. Thus this part is not contained in $D(p)$. This contradiction concludes the proof of (3).

Having shown (1)–(3), the proof of the proposition is complete. □

Delone Sets, the Empty Sphere Method and Delone Triangulations

We follow Delone [255], see also Delone and Ryshkov [259]. Let D be an (r, R)-*system* or *Delone set* in $\mathbb{E}^d$ where $r, R > 0$. That is, a discrete set D in $\mathbb{E}^d$ such that any two distinct points have distance at least r and for any point in $\mathbb{E}^d$ there is a point in D at distance at most R. Lattices and periodic sets as considered by Zassenhaus, are Delone sets but not vice versa. Delone sets still have a certain uniformity and regularity. Now the aim is to construct a tiling of $\mathbb{E}^d$ with convex polytopes such that the vertices of these polytopes are precisely the points of D. Particular such tilings are called *Delone triangulations* or *Delone tilings*. These can be defined by Delone's *empty sphere method*: Consider all Euclidean spheres such that D contains no point in the interior but $d+1$ or more of its points are on the sphere and such that these points are not contained in a hyperplane. For each such *empty sphere* take the convex hull of the points of D on it. This gives a proper convex polytope.

Proposition 32.2. *Let D be a Delone set. Then the proper convex polytopes obtained by the empty sphere method form a facet-to-facet tiling of* $\mathbb{E}^d$.

Proof. We first show the following:

(4) Let P, Q be two distinct convex polytopes obtained by the empty sphere method. Then $\operatorname{int} P \cap \operatorname{int} Q = \emptyset$.

Let S, T be the corresponding empty spheres. None of these can contain the other one. If they are disjoint or touch, we are done. If not, they intersect along a common $(d-2)$-sphere. The hyperplane H through the latter cuts off from each of the spheres S, T a spherical cap contained in the interior of the other sphere. Thus none of these caps can contain a point of D. This shows that H separates the points of $D \cap S$ from the points of $D \cap T$, which, in turn, implies (4).

Next, we prove the following:

(5) Let P be a polytope obtained by the empty sphere method from D and F a facet of P. Then there is another such polytope which meets P in F.

Let S be the empty sphere corresponding to P. Move the centre of S in the direction of the exterior normal of F while keeping the vertices of F on the sphere. By the definition of D, this will eventually lead to an empty sphere T which contains the vertices of F and one or more points of D on the far side of F. The polytope corresponding to T is the desired polytope which meets P in F.

The last step is to show that the following holds:

(6) The polytopes obtained by the empty sphere method from D cover $\mathbb{E}^d$.

Since the centre of an empty sphere has distance at most R from the nearest point of D, its radius is at most R. Hence each of the polytopes obtained by the empty sphere method has diameter at most $2R$. The vertices of each of these polytopes are contained in D. Since D is discrete, each bounded set meets only finitely many such polytopes. If (6) did not hold, connect a point of the complement with a line segment which avoids all faces of our polytopes of dimension at most $d-2$ to a point in the interior of one of the polytopes. The first point where this line segment meets a polytope, say P, is an interior point of a facet F of P and there is no polytope which meets P in F. Since this contradicts (5), the proof of (6) is complete.

The proposition now follows from (4) to (6). □

An Alternative way to Construct Dirichlet–Voronoi and Delone Tilings

Let D be a general discrete set or a Delone set. Embed $\mathbb{E}^d$ into $\mathbb{E}^{d+1}$ as usual (first d coordinates). Consider, in $\mathbb{E}^{d+1}$, the solid paraboloid

$$S = \{(x, z) : x \in \mathbb{E}^d,\ z \geq \|x\|^2\}$$

and choose a subset $E \subseteq \operatorname{bd} S$ such that its orthogonal projection into $\mathbb{E}^d$ equals D. Define generalized polyhedra P, Q as follows: $P = \operatorname{conv} E$ and Q is the intersection of the support halfspaces of S at the points of E. Then $P \subseteq S \subseteq Q$, the orthogonal projections of the facets of P into $\mathbb{E}^d$ give the Delone triangulation and the projections of the facets of Q into $\mathbb{E}^d$ the Dirichlet–Voronoi tiling corresponding to D.

Facet-to-Facet Tilings give Rise to Polyhedral Complexes

A *(generalized) polyhedral complex* $\mathcal{C}$ is a family of (generalized) convex polyhedra in $\mathbb{E}^d$ with the following property: The intersection of any two of its polyhedra is a face of each of these and any face of any of its polyhedra is contained in the family. The polyhedra of $\mathcal{C}$ are called the *cells* of $\mathcal{C}$.

The proof of the following result is easy and left to the reader.

Proposition 32.3. *The tiles of a locally finite, convex facet-to-facet tiling of* $\mathbb{E}^d$ *are generalized convex polyhedra.*

Our next aim is to prove the following result of the Gruber and Ryshkov [451]:

Theorem 32.1. *Let* $\mathcal{T}$ *be a locally finite, convex facet-to-facet tiling of* $\mathbb{E}^d$*. Then* $\mathcal{T}$ gives rise to a (generalized) polyhedral complex, *that is, the family of all tiles and all faces of tiles, including the empty face, is a (generalized) polyhedral complex.*

Proof. We have to show that the intersection of any two faces of tiles is again a face of a tile.

To prove this, we first show the following:

(7) Let $S, T \in \mathcal{T}$. Then $S \cap T$ is a face of both S and T.

The main tool for the proof of (7) is the following proposition, the proof of which is left to the reader.

(8) If a face of S contains a relative interior point of a convex subset of S (in its relative interior), then it contains (the relative interior of) this subset (in its relative interior).

Let $f \in \operatorname{relint}(S \cap T)$. Since $\operatorname{bd} S$ is the disjoint union of the sets $\operatorname{relint} F$, F a face of S, there is a face F of S with

(9) $f \in \operatorname{relint} F$.

Similarly, there is a face G of T, such that

(9) $f \in \operatorname{relint} G$.

For the proof of (7) it is sufficient to show that

(10) $F = G (\text{andthus} F = G = S \cap T)$.

Let N be a bounded convex neighbourhood of f. Since $\mathcal{T}$ is locally finite, N intersects only finitely many tiles. By choosing N even smaller, if necessary, we may suppose that N meets only those faces of a tile which contain f. Let $s, t \in N$ be interior points of S and T, respectively, such that the line segment $[s, t]$ intersects the boundaries of tiles (all of which meet N) only in interior points of facets. Order the tiles which intersect $[s, t]$, say $S = T_1, \dots, T_m = T$, so that successive tiles have common facets. Let F_{12} be the common facet of $S = T_1$ and T_2. By our choice of N, the facet F_{12} contains the relative interior point f of the face F of $S = T_1$. Hence $F \subseteq F_{12}$. Since F_{12} is a facet of T_2 too, we deduce that F is contained in and is a face of T_2. Let F_{23} be the common facet of T_2 and T_3. By our choice of N, the facet F_{23} contains the point f in the relative interior of the face F of T_2. Hence $F \subseteq F_{23}$. Continuing in this way, we finally see that F is contained in and is a facet of $T_m = T$. Thus F, G are both faces of T. Since f is a relative interior point of each of them, this can hold only if $F = G$, concluding the proof of (10) and thus of (7).

Next, the following will be shown:

(11) Let $S, T \in \mathcal{T}$, F a face of S and G a face of T. Then $F \cap G$ is a face of both S and T.

If $F = S$ or $G = T$, the statement (11) is an immediate consequence of (7). Otherwise, each of F, G is the intersection of finitely many facets of S, respectively, T. Since $\mathcal{T}$ is a facet-to-facet tiling, we then have

$$F = S \cap S_1 \cap \cdots \cap S_m, \; G = T_1 \cap \cdots \cap T_n \cap T,$$

say, where $S_1, \dots, T_n \in \mathcal{T}$. Hence

$$F \cap G = (S \cap S_1) \cap \cdots \cap (S \cap S_m) \cap (S \cap T_1) \cap \cdots \cap (S \cap T_n).$$

is an intersection of faces of S by (7), and thus itself is a face of S. Similarly, $F \cap G$ is a face of T, concluding the proof of (11).

Having proved (11), the proof of the theorem is complete. □

An immediate consequence of Propositions 32.1, 32.2 and of Theorem 32.1, is the following result.

Corollary 32.1. *Dirichlet–Voronoĭ tilings and Delone triangulations give rise to polyhedral complexes.*

Dirichlet–Voronoĭ and Delone Tilings are Dual

Two (generalized) polyhedral complexes $\mathcal{C}, \mathcal{D}$ in $\mathbb{E}^d$ are *dual* polyhedral complexes if there is a one-to one mapping of the cells of $\mathcal{C}$ onto the cells of $\mathcal{D}$ such that the following statements hold:

(i) Each cell $F \in \mathcal{C}$ of dimension i is mapped onto a cell $G \in \mathcal{D}$ of dimension $d - i$ such that these cells have precisely one point in common which is in the relative interior of both F and G. F is disjoint from any other cell $H \in \mathcal{D}$ of dimension $d - i$ and similarly for G. The cells F and G are then said to be *dual* to each other.

(ii) If cells $F \in \mathcal{C}$ and $G \in \mathcal{D}$ have non-empty intersection, then F is dual to a suitable face of G and G is dual to a suitable face of F.

Given a Delone set, consider the corresponding Dirichlet–Voronoĭ and Delone tilings. It can be shown that the complexes which arise from these tilings are dual to each other. For a proof, in the case of simplicial complexes, see Rogers [846] Chap. 8.

The Number of Facets of a Parallelohedron

There is not much information available on parallelohedra. The following is an immediate consequence of Minkowski's theorem 30.2 on the number of neighbours of a convex body in a lattice packing. For a few other properties of parallelohedra, see the Venkov–McMullen characterization of translative tiles and thus of parallelohedra in Sect. 32.2. See also the short report on the conjecture of Voronoĭ in Sect. 32.3 and the result of Ryshkov and Bol'shakova [869] cited in Sect. 32.2 later.

Proposition 32.4. *The number of facets of a parallelohedron in* $\mathbb{E}^d$ *is at most* $2^{d+1} - 2$.

This result was extended by Delone [258] to an estimate for the number of facets of a stereohedron. For further information on stereohedra, see Delone and Sandakova [260]. Compare also the report of Schulte [918].

32.2 The Venkov–Alexandrov–McMullen Characterization of Translative Tiles

From the viewpoint of geometry and crystallography it is of interest to study tilings by translates and congruent copies of a given convex polytope, the prototile. A major problem in this context is to characterize the prototiles of translative tilings and tilings by congruent copies. A different problem is to find out whether a prototile which tiles by translation or by congruent copies, actually is a parallelohedron, i.e. it is a lattice tile, or a stereohedron, that is, it tiles by means of a crystallographic group. For translative tilings both problems were solved in a satisfying way by Venkov, Alexandrov and McMullen. For tilings by congruent copies the first problem remains unsolved while the second problem, which goes back to Hilbert's 18th problem, in general has a negative answer. This was proved by Reinhardt.

Alexandrov [16], p. 349 (English edition), described the problem of characterizing parallelohedra as follows:

> The problem consists in finding all possible parallelohedra. This first implies finding all possible types of structures of parallelohedra and, second, describing the metric characteristics for such types which, if enjoyed by a polyhedron, ensure that it is a parallelohedron.
>
> This problem is interesting not only in itself but also by its connections with crystallography and number theory. It was first solved by the great Russian crystallographer E.S. Fedorov in 1890 ...

This sub-section contains a proof of a characterization of translative convex tiles due to Venkov [1008], Alexandrov [17] and, independently but much later, McMullen [710]. The proof is difficult, a slightly different version of the proof is due to Zong [1048] and in [712] McMullen has given an outline. The Venkov–Alexandrov–McMullen theorem shows that, in particular, a translative convex tile necessarily is a lattice tile or parallelohedron. Thus it is an example of the phenomenon that an object which has a certain property, in some cases has an even stronger such property, see the heuristic remark in Sect. 2.1.

For extensions and auxilliary results we refer to Alexandrov [17] and Groemer [400]. For remarks on earlier results consult [710, 711].

A conjecture of Voronoĭ asserts, that parallelohedra are affine images of Dirichlet–Voronoi cells of lattices. For the conjecture and some pertinent results see Sect. 32.3. For Dirichlet–Voronoĭ cells Ryshkov and Bol'shakova [869] proved an interesting decomposition theorem.

Necessity of the Conditions

First, some notation is introduced. Let P be a (proper) centrally symmetric convex polytope in $\mathbb{E}^d$, each facet of which is also centrally symmetric. A *sub-facet*, or *ridge* of P is a $(d-2)$-dimensional face of P. Given a sub-facet G of P, the *belt* of P corresponding to it consists of all facets of P which contain translates of G and $-G$ as faces.

Theorem 32.2. *Let P be a proper convex body in $\mathbb{E}^d$ which is a translative tile. Then the following claims hold:*

(i) *P is a convex polytope.*

(ii) *P is centrally symmetric.*

(iii) *Each facet of P is centrally symmetric.*

(iv) *Each belt of P consists of* 4 *or* 6 *facets.*

Proof. Let $\{P+t : t \in T\}$ be a translative tiling of $\mathbb{E}^d$ with $o \in T$.

(i) Let $s, t \in T$, $s \neq t$. Then $\operatorname{int}(P+s) \cap \operatorname{int}(P+t) = \emptyset$. Thus $s - t \notin \operatorname{int} P - \operatorname{int} P = \operatorname{int}(P-P)$. Since $o \in \operatorname{int}(P-P)$, it follows that $\|s-t\|$ is bounded below by a positive constant. Hence T is a discrete set in $\mathbb{E}^d$. This, and the compactness of P, show that only finitely many translates touch P, say the translates $P+t_i$, $t_i \in T\backslash\{o\}$, $i=1,\dots,k$, and all other translates $P+t$, $t \in T\backslash\{t_1,\dots,t_k\}$, have distance from P bounded below by a positive constant. Hence each point of bd P is contained in one of the finitely many touching sets $P\cap(P+t_i)$. The separation theorem 4.4 for convex bodies then shows that each point in bd P is contained in one of finitely many supporting hyperplanes. This finally implies that the convex body P is actually a convex polytope.

(ii) Clearly, $\{-P-t : -t \in -T\}$ is also a tiling of $\mathbb{E}^d$. Since T and thus $-T$ are discrete as shown earlier, and since $-P$ is compact, there are only finitely many translates $-P-t_i$, $t_i \in T$, $i=1,\dots,k$, say, which meet P. Thus P is the union of the k non-overlapping centrally symmetric convex polytopes $P\cap(-P-t_i)$,

$i = 1, \dots, k$. An application of Minkowski's symmetry theorem 18.3 then shows that P itself is centrally symmetric.

A different proof of (ii) can be obtained along the following lines: the tiling $\{P + t : t \in T\}$ clearly has density 1. This density is $V(P)\,\delta(T)$ where $\delta(T)$ is the density of the discrete set T. $\{P + t : t \in T\}$ is a tiling and thus a packing. Then $\bigl\{\frac{1}{2}(P - P) + t : t \in T\bigr\}$ is also a packing by Proposition 30.4. It thus has density $V\bigl(\frac{1}{2}(P - P)\bigr)\delta(T) \le 1$. To prove (ii), assume that P is not centrally symmetric. Then P and $-P$ are not homothetic and the Brunn–Minkowski theorem 8.1 yields the contradiction:

$$1 \ge V\left(\frac{1}{2}(P - P)\right)\delta(T) > V(P)\,\delta(T) = 1.$$

(iii) By (ii) we may assume that o is the centre of P. Let F be a facet of P and $-F$ its opposite facet. Each point in the relative interior of F is contained in the facet $-F + t$ of a suitable translate $P + t,\ t \in T\backslash\{o\}$, of the given tiling (Fig. 32.3). This implies that the relative interior of F, and thus also F, is the union of finitely many non-overlapping, centrally symmetric $(d - 1)$-dimensional convex polytopes of the form $F \cap (-F + t)$. Now apply Minkowski's symmetry theorem 18.3 (Fig. 32.4).

(iv) Note that we have already proved statements (i)–(iii). Let G be a sub-facet of P and consider the corresponding belt. It consists of, say, k pairs of opposite facets. For the proof of (iv) assume that, on the contrary, $k \ge 4$. Consider the orthogonal

Fig. 32.3. Lattice tiles in $\mathbb{E}^2$

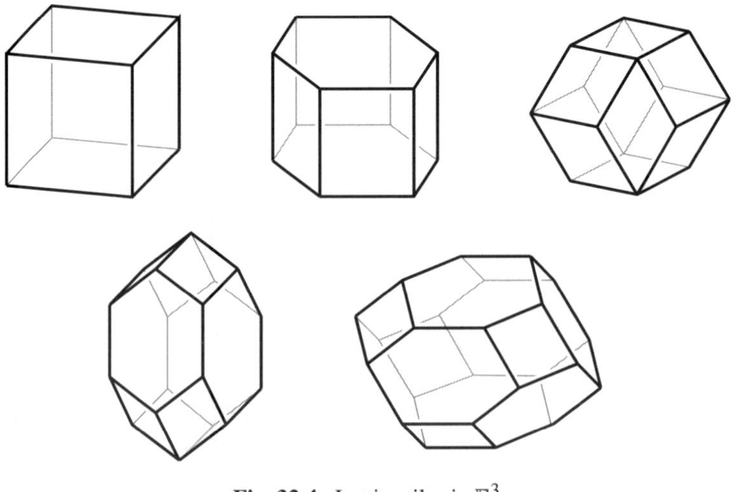

Fig. 32.4. Lattice tiles in $\mathbb{E}^3$

projection " $'$" of $\mathbb{E}^d$ onto the 2-dimensional subspace orthogonal to G. Then P' is a centrally symmetric convex $2k$-gon. The edges of P' are the projections of the facets of the belt corresponding to G and the vertices of P' are the projections of the sub-facets parallel to G, more precisely, which are translates of G or $-G$. The dihedral angle of P at two adjacent facets of the belt corresponding to G equals the internal angle of the corresponding edges of the centrally symmetric convex $2k$-gon P'. Since $k \geq 4$, the following statements hold:

(1) The dihedral angle of P at a sub-facet parallel to G is less than π.

(2) The sum of the dihedral angles of P at two non-opposite sub-facets parallel to G is greater than π and less than 2π.

(3) The sum of the dihedral angles of P at three pairwise non-opposite sub-facets parallel to G is greater than 2π.

Choose a point $g \in \operatorname{relint} G$ which is not contained in a face of any tile $P + t$, $t \in T$ of dimension less than $d - 2$ or of dimension $d - 2$ but not parallel to G. If $g \in \operatorname{relint} F$ where F is a facet of a tile $P+s$, $s \in T \setminus \{o\}$, Proposition (1) applied to G, shows that there is a tile $P+t$, $t \in T\setminus\{o, s\}$, which contains g and is contained in the wedge determined by the facet F of $P + s$ and a facet E of P where G is a sub-facet of E. Considering P', $P' + s'$, $P' + t'$, we see that P has dihedral angles at two non-opposite sub-facets parallel to G with sum at most π, in contradiction to (2). By our choice of g and the case just settled, we see, if g is contained in a tile $P + s$, $s \in T \setminus \{o\}$, then g is contained in the relative interior of a sub-facet of $P + s$ which is parallel to G. The sum of the dihedral angles at g of such tiles is 2π. By (1)–(3) we see that g belongs to at least 3 sub-facets parallel to G. We distinguish the following two cases: first, g belongs to 2 opposite sub-facets. Then g belongs to 2 pairs of non-opposite sub-facets which is impossible by (2). Second, g belongs to 3 pairwise non-opposite sub-facets which is impossible by (3). This shows that our assumption that $k \geq 4$ is wrong and thus concludes the proof of (iv). □

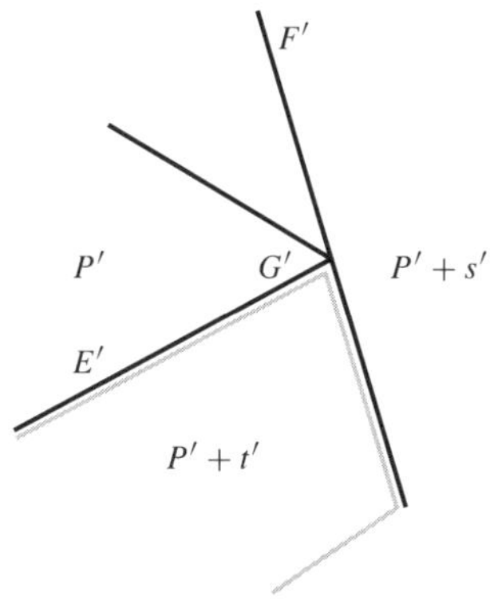

Fig. 32.5. On the proof of the Venkov–McMullen theorem

Remark. Groemer [400] proved that a convex polytope which admits a tiling of $\mathbb{E}^d$ by (positive) homothetic copies, is already a translative tile and thus a parallelohedron by Theorems 32.2 and 32.3.

Sufficiency of the Conditions

More difficult is the proof of the following converse of Theorem 32.2:

Theorem 32.3. *Let P be a proper convex body in $\mathbb{E}^d$ which satisfies Properties* (i)–(iv) *in Theorem* 32.2. *Then P is a translative tile, more precisely even a lattice tile; i.e. it is a parallelohedron.*

Proof. In the following int is the interior relative to $\mathbb{E}^d$ or a sphere S and by relint we mean the interior relative to the affine or the spherical hull. We may assume that o is the centre of P.

In the first step we describe a family of translates of P which is a candidate for a facet-to-facet lattice tiling of P. If F is a facet of P, then the facet $-F$ is a translate of F (note (ii) and (iii) and that o is the centre of P), say $F = -F + t_F$. Clearly, $P \cap (P + t_F) = F = -F + t_F$. Now consider the following family of translates of P:

(4) $\{P + l : l \in L\}$, where $L = \left\{\sum u_F t_F : F \text{ facet of } P, u_F \in \mathbb{Z}\right\}$.

We will show that this is the desired facet-to-facet lattice tiling of P. The following notation will be needed. Given $l \in L$ and a face F of the translate $P + l$, let $L(l, F)$ be the subset of L defined recursively as follows: $l \in L(l, F)$ and $n \in L(l, F)$ if there is a point $m \in L(l, F)$ such that $P + m$ and $P + n$ have a facet in common which contains F as a face. Clearly, $L(o, \emptyset) = L$ and $L(o, P) = \{o\}$. Since L is an additive sub-group of $\mathbb{E}^d$ and thus $L(l, F) = L(o, F - l) + l$, it is sufficient to study $L(o, F)$ where F is a face of P. Since P contains only finitely many faces which are translates of F,

(5) $L(l, F)$ is finite for each $l \in L$ and each face $F \neq \emptyset$ of $P + l$.

In the second step we will show that

(6) $\{P + l : l \in L\}$ is a covering of $\mathbb{E}^d$.

To see this, it will be proved first that

(7) for $k = d, d - 1, \dots, 0$, one has the following inclusion:
relint $F \subseteq \operatorname{int} \bigcup \{P + l : l \in L(o, F)\}$ for each face F of P with $\dim F = k$.

Obviously, (7) holds for $k = d$ (where $L(o, P) = \{o\}$) and $k = d - 1$ (where $L(o, F) = \{o, t_F\}$). Assume now that $k < d - 1$ and (7) holds for all faces F of P with $\dim F = k + 1$. Let G be a face of P with $\dim G = k$. Let S be a sphere of dimension $d - k - 1$ centred at a point of relint G, orthogonal to aff G, and so small that it meets only those faces of $P + l, l \in L(o, G)$, which contain G. For the proof that (7) holds for the face G, it is sufficient to show that

(8) $S \subseteq \bigcup \{P + l : l \in L(o, G)\}$.

Define

$$Q(l, G) = \bigcup\Big\{\bigcup\{P + m : m \in L(l, F)\} : F \text{ face of } P + l, G \subsetneq F\Big\}$$
for $l \in L(o, G)$.

Let $l \in L(o, G)$. Each point of $(P + l) \cap S$ lies in the relative interior of a suitable face F of $P + l$ with $\dim F > \dim G = k$. The induction assumption thus shows that each point of $(\text{relint } F) \cap S$ has a neighbourhood in S which is contained in

$$\bigcup\{P + m : m \in L(l, F)\} \subseteq Q(l, G).$$

Since $(P + l) \cap S$ is compact, there is a $\delta > 0$ such that the δ-neighbourhood of $(P + l) \cap S$ in S is contained in $Q(l, G)$. As $L(o, G)$ is finite by (5), we may take the same δ for each $l \in L(o, G)$. Hence the δ-neighbourhood in S of each point of the set

$$R = \bigcup\{P + l : l \in L(o, G)\} \cap S \subseteq S$$

is contained in

$$\bigcup\{Q(l, G) : l \in L(o, G)\} = \bigcup\{P + l : l \in L(o, G)\}.$$

Since S is arcwise connected and R compact and non-empty, this can hold only if $R = S$, concluding the proof of (8). This concludes the induction and thus proves (7).

P is the disjoint union of the sets relint F where F ranges over the $0, 1, \dots, d$-dimensional faces of P. It thus follows from (7) that $P \subseteq \text{int} \bigcup\{P + l : l \in L\}$. Taking into account the fact that P is compact, a suitable δ-neighbourhood of P is contained in $\text{int} \bigcup\{P + l : l \in L\}$ too. By periodicity the δ-neighbourhood of $\bigcup\{P + l : l \in L\}$ is also contained in $\text{int} \bigcup\{P + l : l \in L\}$. This is possible only if $\bigcup\{P + l : l \in L\} = \mathbb{E}^d$. The proof of (6) is complete.

The third step is to show that

(9) $\{P + l : l \in L\}$ is a packing of $\mathbb{E}^d$.

In order to show this, we first prove that

(10) for $k = d, d - 1, \dots, 0$, we have the equality $\text{int}(P + l) \cap \text{int}(P + m) = \emptyset$ for $l, m \in L(o, F)$, $l \neq m$, and any face F of P with $\dim F = k$.

Clearly, (10) holds for $k = d$ (where $L(o, P) = \{o\}$) and $k = d-1$ (where $L(o, F) = \{o, t_F\}$). If $k = d - 2$, Theorem 32.2 implies that $L(o, F) = \{o, l, m\}$ with suitable $l, m \in L$ (when the belt corresponding to F consists of six facets) or $L(o, F) = \{o, l, m, l + m\}$ (when the belt corresponding to F consists of four facets) and the translates $P + l$, $l \in L(o, F)$, have pairwise disjoint interiors. Hence (10) holds for $k = d - 2$ too. Assume now that $k < d - 2$ and that (10) holds for all faces F of P with $\dim F = k + 1$. Let G be a face of P with $\dim G = k$. For the proof that (10) holds for G, assume the contrary. Then there are $l, m \in L(o, G)$, $l \neq m$, such that $\text{int}(P + l) \cap \text{int}(P + m) \neq \emptyset$. By the definition of $L(o, G)$ we can find points $l = l_1, l_2, \dots, l_n = m \in L(o, G)$, $l_i \neq l_j$, for $i \neq j$, such that each set

$(P+l_i)\cap(P+l_{i+1})$ is a common facet of $P+l_i$ and $P+l_{i+1}$ which contains G as a face. The sequence of polytopes $P+l = P+l_1, P+l_2, \dots, P+l_n = P+m$ then is called an $\{l,m\}$*-chain*. Next, let the small sphere S of dimension $d-k-1 \geq 2$ be chosen as in the proof of (7). Our $\{l,m\}$-chain then gives rise to a sequence of $(d-k-1)$-dimensional spherically convex polytopes $R_1 = (P+l_1)\cap S, \dots, R_n = (P+l_n)\cap S$ on S, called a *spherical* $\{l,m\}$*-chain*. From the proof of (7), we see that the spherical polytopes $(P+l)\cap S$, $l \in L(o,G)$, cover S. By assumption, $\operatorname{int} R_1 \cap \operatorname{int} R_n = \operatorname{int}(P+l)\cap \operatorname{int}(P+m)\cap S \neq \emptyset$.

Choose a point $p \in \operatorname{int} R_1 \cap \operatorname{int} R_n$ such that its antipode in S is not contained in the spherical hull of any face of any of the spherical polytopes $(P+l)\cap S$, $l \in L(o,G)$. By a *loop based at* p *associated with the spherical* $\{l,m\}$*-chain* $R_1,\dots,R_n$, we mean a closed spherical polygonal curve C on S starting at p which can be dissected into the (great circular) arcs $C_1 \subseteq R_1, \dots, C_n \subseteq R_n$ such that C_i and C_{i+1} meet in $R_i \cap R_{i+1}$. Of course, C_n and C_1 meet at p. The loop C is an *interior loop* if each arc C_i is contained in $\operatorname{relint}(R_{i-1}\cap R_i)\cup \operatorname{int} R_i \cup \operatorname{relint}(R_i \cap R_{i+1})$. Denote the *length* of C by $|C|$. Let λ denote the infimum of the lengths of the interior loops based at p associated with all possible spherical $\{l,m\}$-chains.

To conclude the induction, we will show that both $\lambda = 0$ and $\lambda > 0$ lead to a contradiction. Clearly, $\lambda = 0$ contradicts the definition of loops based at p. It remains to consider the case $\lambda > 0$. By (5) and the definition of $\{l,m\}$-chains, there are only finitely many $\{l,m\}$-chains. A compactness argument then implies that there is a loop D based at p with $|D| = \lambda$ and associated with the spherical $\{l,m\}$-chain $R_1 = (P+l_1)\cap S, \dots, R_m = (P+l_m)\cap S$, say. An interior loop based at p associated with a spherical $\{l,m\}$-chain can always be deformed into such a loop of smaller length. Thus D cannot be an interior loop. By our choice of p, most great circles on S through p avoid all spherical faces of dimension at most $d-k-3$ of all spherical polytopes $(P+l)\cap S$, $l \in L(o,G)$. (To see this project these spherical faces from p, respectively, its antipode into the equator of S corresponding to the north pole p.) Each such great circle is an interior loop based at p of length 2π. Since an interior loop cannot have minimal length, $\lambda < 2\pi$. Hence D cannot be a great circle. Thus D has a non-straight angle at a relative interior point q of a spherical face $F\cap S$ of the spherical polygon $(P+l_i)\cap S$, say, where F is a face of $P+l_i$ with $G \subsetneq F$. (Note that a spherical convex polytope is the disjoint union of the relative interiors of its faces.) Clearly, $\dim F \geq k+1$. By the proof of (7), there is a δ-neighbourhood N of q in S with

$$N \subseteq \bigcup \{P+g : g \in L(l_i,F)\} \cap S.$$

Choose $r,s \in D\cap N$ such that q is strictly between r and s. Construct a new loop E by replacing the part of D between r and s by the arc $\widehat{rs} \subseteq N$ of the great circle through r,s. Then

$$\widehat{rs} \subseteq N \subseteq \operatorname{int}\Big(\bigcup \{P+g : g \in L(l_i,F)\} \cap S\Big).$$

Clearly, $|E| < |D|$ and so a contradiction will be obtained if it can be shown that the loop E based at p is associated with some $\{l,m\}$-chain.

Let $u \in \{1, \ldots, n\}$ be the smallest index such that $r \in R_u$ and $v \in \{1, \ldots, n\}$ the largest index such that $s \in R_v$. Clearly, $u \neq v$. Since $L(l_i, F)$ is finite, we can find two sequences (r_w) in $(\text{int } R_u) \cap N$ and (s_w) in $(\text{int } R_v) \cap N$ with $r_w \to r$ and $s_w \to r$ as $w \to \infty$, such that the arcs $\widehat{r_w s_w}$ meet no face of dimension at most $d-k-3$ of any polytope $(P+g) \cap S$, $g \in L(l_i, F)$, (and thus meet no face of dimension at most $d-2$ of any polytope $P+g$, $g \in L(l_i, F)$,) and such that the arcs $\widehat{r_w s_w}$ pass through the same sequence $S_1 = R_u, S_2, \ldots, S_a = R_v$ where $S_j = (P+m_j) \cap S, m_j \in L(l_i, F)$ for $j = 1, \ldots, a$. Then, since $\dim F \geq k+1$, the inductive assumption shows that $(P+m_j) \cap (P+m_{j+1})$ is a common facet of both $P+m_j$ and $P+m_{j+1}$ containing F and G for $j = 1, \ldots, a-1$. Finally, since $\widehat{r_w s_w} \to \widehat{rs}$, we see that $P+l = P+l_1, \ldots, P+l_u, P+m_1, \ldots, P+m_a, P+l_v, \ldots, P+l_n = P+m$ is a new $\{l, m\}$-chain with which the loop E is associated. Since $|E| < |D| = \lambda$, this is impossible. The proof of the induction and thus of (10) is complete.

To conclude the proof of (9) we now show that

(11) $\text{int}(P+l) \cap \text{int}(P+m) = \emptyset$ for $l, m \in L$, $l \neq m$.

It follows, from the second step, that there is a number $\delta > 0$ such that the δ-neighbourhood of $P+l$, $l \in L$, is contained in the set

$$\bigcup \Big\{ \bigcup \{P+m : m \in L(l, G\} : G \text{ face of } P+l \Big\}.$$

Let i be the number of translates $P+m$ in this set. Clearly, this set is contained in the union of all sequences of translates of P starting with $P+l$ which can be obtained by fitting together facet-to-facet at most i translates of P. This union obviously contains the δ-neighbourhood of $P+l$. Applying the same process to each translate of this first union, we obtain a second union of translates of P, which contains the 2δ-neighbourhood of $P+l$, etc. Continuing in this way, we arrive at the following statement:

(12) Let $\lambda > 0$ and $l \in L$. Then there is a number j such that the union of all sequences of translates of P starting with $P+l$ which can be obtained by fitting together facet-to-facet at most j translates of P contains the 2λ-neighbourhood of $P+l$.

To prove (11), assume the contrary. Then there are $l, m \in L$, $l \neq m$, such that $\text{int}(P+l) \cap \text{int}(P+m) \neq \emptyset$. Call a sequence $P+l = P+l_1, P+l_2, \ldots, P+l_n = P+m$, $l_i \in L$, $l_i \neq l_j$, for $i \neq j$, an *$\{l, m\}$-chain* if $(P+l_i) \cap (P+l_{i+1})$ is a common facet of both $P+l_i$ and $P+l_{i+1}$ for $i = 1, \ldots, n-1$. Choose $p \in \text{int}(P+l) \cap \text{int}(P+m)$. Define a *loop based at p associated with the $\{l, m\}$-chain* $P+l_1, \ldots, P+l_n$ to be a closed polygon C starting at p which can be dissected into line segments $C_1 \subseteq P+l_1, \ldots, C_n \subseteq P+l_n$, such that C_i and C_{i+1} meet in $(P+l_i) \cap (P+l_{i+1})$ for $i = 1, \ldots, n$ and $l_{n+1} = l_1$. C is an *interior loop* if each line segment C_i is contained in relint $\big((P+l_{i-1}) \cap (P+l_i)\big) \cup \text{int}(P+l_i) \cup \text{relint}\big((P+l_i) \cap (P+l_{i+1})\big)$ for $i = 1, \ldots, n$ and $l_{n+1} = L_1$. Let λ be the infimum of the lengths of interior loops. Consider all loops of length at most 2λ. Then using (8) and a similar compactness argument to that in the third step, we see that there is a loop of length λ. Now arguing as in the third step we arrive at a contradiction, concluding the proof of (11) and thus of (9).

Finally, the definition of L in (4) together with Proposition (9) implies by Theorem 21.2 that L is a lattice. Then (6) and (9) show that L provides a tiling of $\mathbb{E}^d$ with prototile P. □

Parallelohedra and Zonotopes

A zonotope is a finite sum of line segments. It thus has the property that its faces of all dimensions are centrally symmetric. If, conversely, a convex polytope has the property that for $k = 2$ in case $d = 2, 3$ and for a given $k \in \{2, \dots, d-2\}$ in case $d \geq 4$, its k-dimensional faces are all centrally symmetric, then it is already a zonotope as shown by McMullen [706]. Hence, all 2– and 3-dimensional parallelohedra are zonotopes. The example of the so-called 24-*cell* in $\mathbb{E}^4$, which is a parallelohedron, shows that a parallelohedron may have faces which are not centrally symmetric and thus is not a zonotope. For information on the 24-cell, see Coxeter [230].

Non-Convex Tilings are Different

The convexity assumption in Theorem 32.3 cannot be omitted. Stein [956] specified simple star bodies in $\mathbb{E}^5$ and $\mathbb{E}^{10}$ which tile by translations but do not admit lattice tilings. See Stein and Szabó [957].

32.3 Conjectures and Problems of Voronoĭ, Hilbert, Minkowski, and Keller

There is a small series of tiling problems with marked impact on the literature in discrete geometry and convexity in the twentieth century. We mention the conjecture of Voronoĭ, Hilbert's 18th problem, the cube conjecture of Minkowski and its stronger version due to Keller.

In this section these problems are described and some references to the literature given.

Conjecture of Voronoĭ

A characterization of convex lattice tiles different from the earlier characterization of Venkov, Alexandrov and McMullen, is indicated by the following *conjecture of Voronoĭ on parallelohedra.*

Conjecture 32.1. *Let P be a proper convex polytope which admits a lattice tiling of $\mathbb{E}^d$, i.e. P is a parallelohedron. Then there is a lattice L such that P is a suitable linear image of the Dirichlet–Voronoĭ cell*

$$\big\{x : \|x\| \leq \|x - l\| \text{ for all } l \in L\big\}.$$

This conjecture has been proved for $d = 2, 3, 4$ by Delone [256] and Engel [298] established for $d = 5$ the slightly weaker result that each parallelohedron is combinatorially equivalent to a Dirichlet–Voronoĭ cell of a lattice. For general d Voronoĭ [1013] proved the conjecture if P is the prototile of a *primitive lattice tiling*, that is, the tiling is facet-to-facet and at each vertex of a tile there meet precisely $d + 1$ tiles. Žitomirskiĭ [1052] observes that Voronoĭ needs in his proof only that at each sub-facet of a tile there meet precisely three tiles and gives a far-reaching generalization of Voronoĭ's result, but his proof is difficult to understand. If P is a zonotope, then the conjecture is true according to Erdahl [306]. Unfortunately there are convex polytopes which are lattice tiles but not zonotopes, see the remark on parallelohedra and zonotopes in the last section. There are several statements which are equivalent to Voronoĭ's conjecture, see Deza and Grishukhin [265].

On Hilbert's 18th Problem

It follows from the earlier results of Venkov–McMullen that a convex polytope which provides a translative tiling of $\mathbb{E}^d$, necessarily is a parallelohedron. Considering this result, the following conjecture is highly plausible:

Conjecture 32.2. *Let P be a convex polytope which provides a tiling of $\mathbb{E}^d$ by congruent copies. Then P is a stereohedron that is, there is a crystallographic group $\mathcal{G}$ such that $\{gP : g \in \mathcal{G}\}$ is a tiling of $\mathbb{E}^d$.*

Hilbert seems to have had doubts whether this was true since, in his 18th problem, he asked the following:

> The question arises: whether polyhedra also exist which do not appear as fundamental regions of groups of motions, by means of which nevertheless by a suitable juxtaposition of congruent copies a complete filling up to all space is possible.

See [501]. Reinhardt [828] produced the first example of a (non-convex) polytope P in $\mathbb{E}^3$ which yields a tiling with congruent copies but no tiling by means of the copies of P under the rigid motions of a crystallographic group. The first (non-convex) planar examples are due to Heesch [485] and the first convex example in $\mathbb{E}^2$ was given by Kershner [579]. In all these examples the congruent copies of P are not just translates. It is thus of particular interest that Szabó [980] exhibited an example of a (non-convex) polytope in $\mathbb{E}^3$ which tiles by translation but not by means of a crystallographic group.

The answer to Hilbert's problem is in the negative and thus the Conjecture false. Hence the following problem arises.

Problem 32.1. *Specify mild additional conditions which imply that a convex polytope in $\mathbb{E}^d$ which tiles $\mathbb{E}^d$ with congruent copies, is a stereohedron.*

Conjectures of Minkowski and Keller

Minkowski [735] conjectured that, in any lattice tiling of a cube, there are translates of this cube which meet facet-to-facet and proved this for $d = 2, 3$. The general proof, surprisingly, turned out to be extremely difficult. It was finally given by Hajós [471] who reduced it to a problem on Abelian groups. Keller [570] conjectured that, more generally, in any tiling of $\mathbb{E}^d$ with translates of a cube, there are two cubes which meet facet-to-facet. For $d \leq 6$ this was confirmed by Perron [794]. Making essential use of a reformulation of Keller's conjecture by Corrádi and Szabó [226], Lagarias and Shor [627] disproved Keller's conjecture for $d \geq 10$ and Mackey [676] for $d \geq 8$. Thus Keller's conjecture is open only in case $d = 7$. For more information, see Zong [1051].

33 Optimum Quantization

In the sequel we investigate integrals of the following type:

$$\int_J \min_{s \in S}\{\|x - s\|^2\} dx,$$

where $J \subseteq \mathbb{E}^d$ is Jordan measurable and $S \subseteq J$ or $S \subseteq \mathbb{E}^d$ a finite set consisting of n points, say. This integral may be interpreted as the volume above sea level of a mountain landscape over J with n valleys, each a piece of a paraboloid of revolution. The deepest points of the valleys are at sea-level and are the points of S. Given J and n, the problem is to determine the minimum volume above sea level of such mountain landscapes and to describe the minimizing configurations S. While precise solutions are out of reach, asymptotic results as $n \to \infty$ are possible.

Since the late 1940s this seemingly unspectacular problem has appeared in several rather different areas, including the following:

Data transmission, see Gray and Neuhoff [394]

Discrete geometry and location theory, see Fejes Tóth [329] and Matérn and Persson [693]

Numerical integration, see Chernaya [206]

Probability theory, see Graf and Luschgy [389]

Convex geometry, see Gruber [439, 443]

In this section we prove Fejes Tóth's inequality for sums of moments in $\mathbb{E}^2$ and an asymptotic formula for integrals of this type due to Zador. The structure of the minimizing arrangements will also be discussed. These results are then applied to packing and covering problems for circular discs, to problems of data transmission and to numerical integration. For an application to the volume approximation of convex bodies by circumscribed convex polytopes and the isoperimetric problem for convex polytopes compare Sect. 11.2.

For additional information compare Du, Faber and Gunzburger [278] and the author [442].

33.1 Fejes Tóth's Inequality for Sums of Moments

The 2-dimensional case of the problem has attracted the interest of the Hungarian school of discrete geometry, ever since Fejes Tóth published his inequality on sums of moments. By now, there are more than a dozen proofs of it known. The inequality has applications to packing and covering problems for circular discs, to volume approximation of convex bodies in $\mathbb{E}^3$ by circumscribed convex polytopes and to problems in areas outside of mathematics such as economics and geography. For a planar generalization of it, to which Fejes Tóth, Fejes Tóth, Imre and Florian contributed, see the report of Florian [337]. The inequality on sums of moments indicates that, for certain geometric and analytic problems, regular hexagonal configurations are close to optimal, and are possibly optimal.

In this section the sum theorem of Fejes Tóth will be presented.

For more information, see the book of Fejes Tóth [329] and the surveys of Florian [337] and Gruber [438, 442].

Sums of Moments

Let $f : [0, +\infty) \to [0, +\infty)$ be non-decreasing, where $f(0) = 0$, and let H be a convex 3, 4, 5, or 6-gon in $\mathbb{E}^2$. Then, given a dissection $C_1, \dots, C_n$ of H and a set $S = \{s_1, \dots, s_n\}$ of n points in H or in $\mathbb{E}^d$, the sum

$$\sum_i \int_{C_i} f(\|x - s_i\|)dx$$

is called a *sum of moments*. If $f(t) = t^2$, this is a sum of moments of inertia. If for the sets C_i we take the Dirichlet–Voronoĭ cells

$$D_i = \big\{x \in H : \|x - s_i\| \le \|x - s_j\| \text{ for } j = 1, \dots, n\big\},\ i = 1, \dots, n,$$

in H corresponding to S, then the sum of moments decreases and, in addition,

$$\sum_i \int_{D_i} f(\|x - s_i\|)dx = \int_H \min_{s \in S}\{f(\|x - s\|)\}dx.$$

Fejes Tóth's Inequality for Sums of Moments

Our aim is to prove the following result of Fejes Tóth [329].

Theorem 33.1. *Let $f : [0, +\infty) \to [0, +\infty)$ be non-decreasing where $f(0) = 0$ and let $H \subseteq \mathbb{E}^2$ be a convex 3, 4, 5, or 6-gon. Then,*

$$(1)\quad I(H, f, n) = \inf_{\substack{S \subseteq \mathbb{E}^2 \\ \#S = n}} \int_H \min_{s \in S}\{f(\|x - s\|)\}dx \ge n \int_{H_n} f(\|x\|)dx,$$

where H_n is a regular hexagon in $\mathbb{E}^2$ of area $A(H_n) = A(H)/n$ and centre at the origin o.

The following analytic proof is due to the author [437]. It uses the moment lemma of Fejes Tóth [329], p. 198.

Proof. It is sufficient to prove the theorem for functions f with positive continuous derivative on $(0, +\infty)$. Let $S \subseteq \mathbb{E}^2$, $\#S = n$. Then

$$(2) \quad \int\limits_H \min_{s \in S}\{f(\|x - s\|)\}dx = \sum_i \int\limits_{D_i} f(\|x - s_i\|)dx,$$

where the sets D_i, $i = 1, \dots, n$, are the Dirichlet–Voronoĭ cells in H corresponding to S. D_i is a convex polygon of area a_i with v_i vertices, say. The *moment lemma* of Fejes Tóth says that

$$(3) \quad \int\limits_{D_i} f(\|x - s_i\|)dx \geq \int\limits_{R_i} f(\|x\|)dx = M(a_i, v_i),$$

say, where R_i is a regular v_i-gon with area a_i and centre o. Let g be defined by $g(r^2) = f(r)$ for $r \geq 0$. Then $g(0) = 0$ and g has positive continuous derivative on $(0, +\infty)$. Let G be such that $G(0) = 0$ and $G' = g$. Finally, let $h(a, v) = a/(v \tan(\pi/v))$ for $a > 0, v \geq 3$. Clearly, the following hold:

If R is a regular polygon with centre o, area a, and v vertices,
then $h^{\frac{1}{2}}$ is its inradius, and

$$(4) \quad M(a, v) = \int\limits_R f(\|x\|)dx = 2v \int\limits_0^{\frac{\pi}{v}} \int\limits_0^{\frac{h^{1/2}}{\cos\psi}} g(r^2) r\, dr\, d\psi = v \int\limits_0^{\frac{\pi}{v}} G\Big(\frac{h}{\cos^2\psi}\Big) d\psi.$$

Define $M(a, v)$ for $a > 0, v \geq 3$ by the latter integral.

After these preparations the main step of the proof of the theorem is to show that the moment

(5) $M(a, v)$ is convex for $a > 0, v \geq 3$.

Let

$$I = \int\limits_0^{\frac{\pi}{v}} g\Big(\frac{h}{\cos^2\psi}\Big)\frac{d\psi}{\cos^2\psi}, \quad J = \int\limits_0^{\frac{\pi}{v}} g'\Big(\frac{h}{\cos^2\psi}\Big)\frac{d\psi}{\cos^4\psi}, \quad K = g\Big(\frac{h}{\cos^2\frac{\pi}{v}}\Big).$$

Elementary calculus yields for the second order partial derivatives of M,

$$M_{aa} = vh_a^2 J \ (> 0), \ M_{av} = (h_a + vh_{av})I + vh_a h_v J - \frac{\pi h_a}{v\cos^2\frac{\pi}{v}} K,$$
$$M_{vv} = (2h_v + vh_{vv})I + vh_v^2 J + \frac{2\pi a}{v\cos^2\frac{\pi}{v}}\Big(\frac{\pi}{v^3} - h_{av}\Big)K.$$

Noting that

$$2h_v^2 - hh_{vv} = \frac{2\pi^2 a^2}{v^6 \sin^2 \frac{\pi}{v}}, \quad h + v h_v = \frac{\pi a}{v^2 \sin^2 \frac{\pi}{v}},$$

$$K - \frac{\cos \frac{\pi}{v}}{\sin \frac{\pi}{v}} I = \frac{2a \cos^2 \frac{\pi}{v}}{v \sin^2 \frac{\pi}{v}} \int_0^{\frac{\pi}{v}} g'\Big(\frac{h}{\cos^2 \psi}\Big) \frac{\sin^2 \psi}{\cos^4 \psi} d\psi \ (> 0),$$

a simple, yet lengthy calculation then shows that

$$\begin{aligned}
M_{aa} M_{vv} - M_{av}^2 \\
&= -v^2 h_a (2h_v h_{av} - h_a h_{vv}) IJ + \frac{2\pi^2 a h_a^2}{v^3 \cos^2 \frac{\pi}{v}} JK - \Big((h_a + v h_{av}) I - \frac{\pi h_a}{v \cos^2 \frac{\pi}{v}} K\Big)^2 \\
&= -\frac{2\pi^2 a \cos \frac{\pi}{v}}{v^5 \sin^3 \frac{\pi}{v}} IJ + \frac{2\pi^2 a}{v^5 \sin^2 \frac{\pi}{v}} JK - \frac{\pi^2}{v^4 \sin^2 \frac{\pi}{v} \cos^2 \frac{\pi}{v}} \Big(K - \frac{\cos \frac{\pi}{v}}{\sin \frac{\pi}{v}} I\Big)^2 \\
&= \frac{2\pi^2 a}{v^5 \sin^2 \frac{\pi}{v}} J \Big(K - \frac{\cos \frac{\pi}{v}}{\sin \frac{\pi}{v}} I\Big) - \frac{2\pi^2 a}{v^5 \sin^2 \frac{\pi}{v}} \frac{v}{2a \cos^2 \frac{\pi}{v}} \Big(K - \frac{\cos \frac{\pi}{v}}{\sin \frac{\pi}{v}} I\Big)^2 \\
&= \frac{2\pi^2 a}{v^5 \sin^2 \frac{\pi}{v}} \Big(J - \frac{v}{2a \cos^2 \frac{\pi}{v}} \Big(K - \frac{\cos \frac{\pi}{v}}{\sin \frac{\pi}{v}} I\Big)\Big) \Big(K - \frac{\cos \frac{\pi}{v}}{\sin \frac{\pi}{v}} I\Big) \\
&= \frac{2\pi^2 a}{v^5 \sin^2 \frac{\pi}{v}} \int_0^{\frac{\pi}{v}} g'\Big(\frac{h}{\cos^2 \psi}\Big) \Big(1 - \frac{\sin^2 \psi}{\sin^2 \frac{\pi}{v}}\Big) \frac{d\psi}{\cos^4 \psi} \ \frac{2a \cos^2 \frac{\pi}{v}}{v \sin^2 \frac{\pi}{v}} \\
&\quad \times \int_0^{\frac{\pi}{v}} g'\Big(\frac{h}{\cos^2 \psi}\Big) \frac{\sin^2 \psi}{\cos^4 \psi} d\psi \ > 0.
\end{aligned}$$

Having proved that M_{aa} and $M_{aa} M_{vv} - M_{av}^2$ are positive for $a > 0, v \geq 3$, it follows that the Hessian matrix of M is positive definite, which in turn implies (5), see the convexity criterion 2.10 of Brunn and Hadamard.

Our next tool is the following simple consequence of Euler's polytope formula. See, e.g.
aiFejes Tóth, L. [329], p. 16:

(6) $v_1 + \cdots + v_n \leq 6n$.

Since, for fixed a, the function $M(a, v)$ is convex in v by (5) and has a limit as $v \to +\infty$ (the moment of the circular disc with centre o and area a), we see the following:

(7) For a fixed, $M(a, v)$ is decreasing in v.

Now, combine (1), (2), (4), apply Jensen's inequality 2.1 for convex functions, and use (6) and (7) to see that

$$\begin{aligned} I(H, f, n) &\geq \sum_{i=1}^{n} M(a_i, v_i) \geq n\, M\Big(\frac{a_1 + \cdots + a_n}{n}, \frac{v_1 + \cdots + v_n}{n}\Big) \\ &\geq n\, M\Big(\frac{|H|}{n}, 6\Big). \end{aligned}$$

The proof of the theorem is complete. □

33.2 Zador's Asymptotic Formula for Minimum Distortion

In higher dimensions, the problem raised in the introduction of Sect. 33 appeared first in the work of Zador [1035] in the context of data transmission, more precisely, in the problem to evaluate the quality of certain encoders as the number of code-words of the code-books used tends to infinity. Later appearances are in error estimates for numerical integration, approximation of probability measures by discrete probability measures, approximation of convex bodies by polytopes and approximation of functions by step functions. Far-reaching refinements and generalizations of Zador's results are due to Chernaya [206] and Gruber [443].

In the following, we give an asymptotic formula due to Zador for general d, which is sufficient for many applications. For applications see Sect. 33.4.

Minimum Distortion

Let J be a Jordan measurable set in $\mathbb{E}^d$ with $V(J) > 0$ and let $\alpha > 0$. The expression

$$I(J, \alpha, n) = \inf_{\substack{S \subseteq J \\ \#S = n}} \int_J \min_{s \in S}\{\|x - s\|^\alpha\}\, dx$$

is called *minimum distortion.* This notion first appeared in the context of data transmission, compare Sect. 33.4. In some cases the following slightly different quantity is of interest:

$$K(J, \alpha, n) = \inf_{\substack{S \subseteq \mathbb{E}^d \\ \#S = n}} \int_J \min_{s \in S}\{\|x - s\|^\alpha\}\, dx.$$

Zador's Asymptotic Formula

The following result goes back to Zador [1035]. The proof is a simplified version of the author's [443] proof of a more general result. A *body* is a compact set which is equal to the closure of its interior.

Theorem 33.2. *Let $\alpha > 0$. Then there is a constant $\delta = \delta_{\alpha,d} > 0$, depending only on α and d, such that, for any Jordan measurable body $J \subseteq \mathbb{E}^d$ with $V(J) > 0$,*

$$I(J, \alpha, n),\ K(J, \alpha, n) \sim \delta\, \frac{V(J)^{\frac{\alpha+d}{d}}}{n^{\frac{\alpha}{d}}} \quad \textit{as } n \to \infty.$$

The constant $\delta = \delta_{2,d-1}$ has appeared already in the context of volume approximation of convex bodies by circumscribed convex polytopes, see Theorem 11.4. The formula

$$\delta_{\alpha,d} \sim \frac{d^{\frac{\alpha}{2}}}{(2\pi e)^{\frac{\alpha}{2}}} \text{ as } d \to \infty$$

will be proved later, see Proposition 33.1.

We prove only the asymptotic formula for $I(J, \alpha, n)$. The proof for $K(J, \alpha, n)$ is very similar, but in one detail slightly more complicated.

Proof. In the following const stands for a positive constant which may depend on α and d. If const appears several times in the same context, this does not mean that it is always the same constant.

In the first step of the proof we show that

$$(1) \quad \frac{\text{const}}{n^{\frac{\alpha}{d}}} \le I\big(K, \alpha, n\big) \le \frac{\text{const}}{n^{\frac{\alpha}{d}}}, \text{ where } K = \{x : 0 \le x_i \le 1\}.$$

For the proof of the lower estimate some preparations are needed. First, the following will be shown:

(2) Let $I \subseteq \mathbb{E}^d$ be a convex body and let $\varrho > 0$ and $s \in \mathbb{E}^d$ be such that $V(I) = V(\varrho B^d + s)$. Then

$$\int_I \|x - s\|^\alpha dx \ge \int_{\varrho B^d + s} \|x - s\|^\alpha dx = \int_{\varrho B^d} \|x\|^\alpha dx.$$

To see this, note that the sets $I \backslash (\varrho B^d + s)$ and $(\varrho B^d + s) \backslash I$ have the same volume and $\|x - s\|^\alpha$ is greater on the first set. Second, dissecting ϱB^d into infinitesimal shells of the form $(t + dt)B^d \backslash t B^d$ of volume $d\, V(B^d) t^{d-1} dt$ and adding, we obtain,

$$(3) \quad \int_{\varrho B^d} \|x\|^\alpha dx = d\, V(B^d) \int_0^\varrho t^{d-1} t^\alpha dt = \frac{d}{\alpha + d} V(B^d) \varrho^{\alpha+d}.$$

After these preparations, choose minimizing configurations $S_n = \{s_{n1}, \dots, s_{nn}\} \subseteq K$ for $I\big(K, \alpha, n\big)$ and consider the corresponding Dirichlet–Voronoĭ cells in K,

$$D_{ni} = \big\{x \in K : \|x - s_{ni}\| \le \|x - s_{nj}\| \text{ for } j = 1, \dots, n\big\},\ i = 1, \dots, n.$$

The cells D_{ni} are convex polytopes which tile K and are such that, for $x \in D_{ni}$, one has $\|x - s_{ni}\|^\alpha \le \|x - s_{nj}\|^\alpha$ for $j = 1, \dots, n$. This together with (2), (3) and Jensen's inequality, applied to the convex function $t \to t^{(\alpha+d)/d}$, yields the lower estimate in (1):

$$\begin{aligned}
I(K,\alpha,n) &= \int\limits_K \min_{s\in S_n}\{\|x-s_{ni}\|^\alpha\}\,dx = \sum_i \int\limits_{D_{ni}} \|x-s_{ni}\|^\alpha\,dx\\
&\geq \sum_i \int\limits_{\varrho_{ni} B^d} \|x\|^\alpha\,dx \text{ where } \varrho_{ni} = \Big(\frac{V(D_{ni})}{V(B^d)}\Big)^{\frac{1}{d}}\\
&= \frac{d\,V(B^d)}{\alpha+d}\sum_i \varrho_{ni}^{\alpha+d} = \frac{d\,V(B^d)}{(\alpha+d)\,V(B^d)^{\frac{\alpha+d}{d}}}\sum_i V(D_{ni})^{\frac{\alpha+d}{d}}\\
&\geq \frac{d}{(\alpha+d)V(B^d)^{\frac{\alpha}{d}}}\,n\Big(\frac{1}{n}\sum_i V(D_{ni})\Big)^{\frac{\alpha+d}{d}} = \frac{d}{(\alpha+d)\,V(B^d)^{\frac{\alpha}{d}}}\,\frac{1}{n^{\frac{\alpha}{d}}}.
\end{aligned}$$

For the proof of the upper estimate in (1), note that we can cover the cube K with n balls of radius ϱ_n, say, such that the total volume of these balls is bounded above by a constant independent of n. (For $n = l^d$ take as set of centres a square grid of edge-length $\frac{1}{l}$ in K. For general n choose l such that $l^d \leq n < (l+1)^d$ and as set of centres choose the same as before plus $n - l^d$ arbitrary points in K.) Thus

$$n\,\varrho_n^d V(B^d) \leq \text{const}, \text{ or } \varrho_n \leq \frac{\text{const}}{n^{\frac{1}{d}}}.$$

If $T_n = \{t_{n1},\dots,t_{nn}\}$ is the set of centres of the balls of the nth covering, we obtain the upper estimate in (1) as follows:

$$\begin{aligned}
I(K,\alpha,n) &\leq \int\limits_K \min_{i=1,\dots,n}\{\|x-t_{ni}\|^\alpha\}\,dx \leq \sum_i \int\limits_{\varrho_n B^d + t_{ni}} \|x-t_{ni}\|^\alpha\,dx\\
&= \sum_i \int\limits_{\varrho_n B^d} \|x\|^\alpha\,dx \leq n\,V(\varrho_n B^d)\varrho_n^\alpha \leq \frac{\text{const}}{n^{\frac{\alpha}{d}}}.
\end{aligned}$$

In the second step of the proof, it will be shown that we have the following:

(4) There is a constant $\delta > 0$, depending only on α and d such that:

$$I(K,\alpha,n) \sim \frac{\delta}{n^{\frac{\alpha}{d}}} \text{ as } n \to \infty.$$

Clearly, (1) implies,

(5) $\delta := \liminf\limits_{n\to\infty}\big(n^{\frac{\alpha}{d}} I(K,\alpha,n)\big) \in \mathbb{R}^+.$

We show that one may replace lim inf in (5) by lim. Let $\lambda > 1$ and choose k such that:

(6) $I(K,\alpha,k) < \dfrac{\lambda\,\delta}{k^{\frac{\alpha}{d}}}.$

First consider n of the form $n = kl^d$, $l = 1, 2, \dots$ Clearly, $I(K,\alpha,k)$ is the volume of the mountain landscape

(7) $\big\{(x,y) : x \in K,\ 0 \le y \le \min_{s \in S_k}\{\|x-s\|^\alpha\}\big\} \subseteq \mathbb{E}^{d+1}$

over the cube K with k valleys. Here S_k is a minimizing configuration in K, consisting of k points. Each of the l^d affinities

$$x_1 \to \frac{x_1+a_1}{l}, \dots, x_d \to \frac{x_d+a_d}{l}, y \to \frac{y}{l^\alpha},$$

$$a_1, \dots, a_d \in \{0, 1, \dots, l-1\}$$

maps the landscape (7) onto a small landscape over a small cube of edge-length $\frac{1}{l}$, where the small cubes tile the unit cube K. Thus, these small landscapes put together, form a landscape over K with at most kl^d valleys. This landscape contains the following landscape:

$$\Big\{(x,y) : x \in K,\ 0 \le y \le \min\Big\{\|x-s\|^\alpha : s \in \frac{1}{l}(S_k + a),\ a_i \in \{0, 1 \dots, l-1\}\Big\}\Big\}.$$

Since this landscape is among those over which we form the infimum in the definition of $I(K, \alpha, n = kl^d)$ it follows that

(8) $I\big(K, \alpha, n = kl^d\big) \le l^d \dfrac{1}{l^d l^\alpha} I\big(K, \alpha, k\big) \le \dfrac{\lambda\,\delta}{(kl^d)^{\frac{\alpha}{d}}}$

by (6). Secondly, we consider general n. Choose l_0 so large that

(9) $\left(\dfrac{l_0+1}{l_0}\right) \le \lambda.$

Then,

(10) $I\big(K, \alpha, n\big) \le \dfrac{\lambda^2\delta}{n^{\frac{\alpha}{d}}}$ for $n \ge kl_0^d$.

To see this, let $n \ge kl_0^d$, and choose $l \ge l_0$ such that $kl^d \le n < k(l+1)^d$. Then the definition of I, (8) and (9) yield (10):

$$I\big(K, \alpha, n\big) n^{\frac{\alpha}{d}} \le I\big(K, \alpha, kl^d\big)\big(k(l+1)^d\big)^{\frac{\alpha}{d}} \le \lambda\,\delta\Big(\frac{l+1}{l}\Big)^\alpha \le \lambda^2\delta.$$

Since $\lambda > 1$ was arbitrary, (5) and (10) together yield (4).

By applying a suitable affine transformation, we see that (4) implies the following asymptotic formula:

(11) Let $C \subseteq \mathbb{E}^d$ be a cube. Then $I(C, \alpha, n) \sim \delta \dfrac{V(C)^{\frac{\alpha+d}{d}}}{n^{\frac{\alpha}{d}}}$ as $n \to \infty$.

The third step of the proof is to show that

(12) $I(J, \alpha, n) \gtrsim \delta \dfrac{V(J)^{\frac{\alpha+d}{d}}}{n^{\frac{\alpha}{d}}}$ as $n \to \infty$.

Arguments similar to those which led to (1) yield the inequalities

(13) $\frac{\text{const}}{n^{\frac{\alpha}{d}}} \le I(J, \alpha, n) \le \frac{\text{const}}{n^{\frac{\alpha}{d}}}.$

Choose minimizing configurations $S_n = \{s_{n1}, \dots, s_{nn}\} \subseteq J$ for $I(J, \alpha, n)$ and consider the corresponding Dirichlet–Voronoĭ cells in J,

$$D_{ni} = \big\{x \in J : \|x - s_{ni}\| \le \|x - s_{nj}\| \text{ for } j = 1, \dots, n\big\},\ i = 1, \dots, n.$$

Since J is a body and $I(J, \alpha, n) \to 0$ as $n \to \infty$ by (13), a simple indirect compactness argument implies that

(14) $\max_i \{\operatorname{diam} D_{ni}\} \to 0$ as $n \to \infty$.

Next, we introduce some necessary notation. If C is a cube, then

$$n(C) = \#\{i : C \cap D_{ni} \neq \emptyset\},\ S_n(C) = \{s_{ni} : C \cap D_{ni} \neq \emptyset\}.$$

Now $n(C)$ will be estimated below and above:

(15) Let $C \subseteq J$ be a cube. Then $\text{const}\, n \le n(C) \le n$.

Clearly,

$$\begin{aligned} I\big(C, \alpha, n(C)\big) &\le \int_C \min_{s \in S_n(C)} \{\|x - s\|^\alpha\}\, dx \\ &= \int_C \min_{s \in S_n} \{\|x - s\|^\alpha\}\, dx \le I(J, \alpha, n) \le \frac{\text{const}}{n^{\frac{\alpha}{d}}} \end{aligned}$$

by (13). Hence $I\big(C, \alpha, n(C)\big) \to 0$ and thus $n(C) \to \infty$ as $n \to \infty$. This, in turn, shows by (11) that

$$I\big(C, \alpha, n(C)\big) \sim \delta \frac{V(C)^{\frac{\alpha+d}{d}}}{n(C)^{\frac{\alpha}{d}}} \text{ as } n \to \infty.$$

Hence $n(C) \ge \text{const}\, n$, concluding the proof of (15).

For the proof of (12) let $\lambda > 1$. Since J is compact and Jordan measurable, the following holds:

(16) There are pairwise disjoint cubes $C_1, \dots, C_k \subseteq J$ such that:

$$\sum_i V(C_i) \ge \frac{1}{\lambda} V(J).$$

Since the cubes $C_1, \dots, C_k$ are pairwise disjoint, it follows from (14) that the sets $S_n(C_i)$, $i = 1, \dots, k$, are also pairwise disjoint and thus

(17) $\sum_i n(C_i) \le n$ for sufficiently large n.

Next, we prove that

$$(18)\quad \liminf_{n\to\infty} I(J,\alpha,n)\, n^{\frac{\alpha}{d}} \overset{>}{\sim} \frac{\delta\, V(J)^{\frac{\alpha+d}{d}}}{\lambda^{1+\frac{\alpha}{d}}}.$$

Clearly, the following hold:

(19) There are a subsequence of $1, 2, \dots,$ and constants $\sigma_i > 0,\ i = 1, \dots, k$, such that:

$$I(J,\alpha,n)\, n^{\frac{\alpha}{d}} \to \liminf_{n\to\infty} I(J,\alpha,n)\, n^{\frac{\alpha}{d}},$$
$$n(C_i) \sim \sigma_i\, n \text{ for } i = 1, \dots, k,$$
$$\sum_i \sigma_i \le 1$$

as $n \to \infty$ in this subsequence.

Here we have applied (15) and (17). Finally, taking into account the fact that the cubes $C_i,\ i = 1, \dots, k$, are pairwise disjoint by (16), the definitions of $S_n(C_i)$, I, (11), (15), (19), Jensen's inequality (Theorem 1.9) applied to the convex function $t \to t^{-\alpha/d}, t > 0$, and (16) together show the following:

$$\begin{aligned}
I(J,\alpha,n)\, n^{\frac{\alpha}{d}} &= \int_J \min_{s\in S_n}\{\|x-s\|^\alpha\}\, dx\, n^{\frac{\alpha}{d}} \\
&\ge \sum_i \int_{C_i} \min_{s\in S_n(C_i)}\{\|x-s\|^\alpha\}\, dx\, n^{\frac{\alpha}{d}} \ge \sum_i I\big(C_i,\alpha,n(C_i)\big)\, n^{\frac{\alpha}{d}} \\
&\sim \delta \sum_i V(C_i)^{\frac{\alpha+d}{d}} \frac{n^{\frac{\alpha}{d}}}{n(C_i)^{\frac{\alpha}{d}}} \sim \delta \sum_i V(C_i)^{\frac{\alpha+d}{d}} \sigma_i^{-\frac{\alpha}{d}} \\
&= \delta \sum_j V(C_j) \sum_i \frac{V(C_i)}{\sum_j V(C_j)} \Big(\frac{\sigma_i}{V(C_i)}\Big)^{-\frac{\alpha}{d}} \\
&\ge \delta \sum_j V(C_j) \Big(\sum_i \frac{V(C_i)}{\sum_j V(C_j)} \frac{\sigma_i}{V(C_i)}\Big)^{-\frac{\alpha}{d}} \\
&= \delta \Big(\sum_j V(C_j)\Big)^{\frac{\alpha+d}{d}} \Big(\sum_i \sigma_i\Big)^{-\frac{\alpha}{d}} \ge \frac{\delta\, V(J)^{\frac{\alpha+d}{d}}}{\lambda^{\frac{\alpha+d}{d}}},
\end{aligned}$$

as $n \to \infty$ in the subsequence from Proposition (19). The proof of (18) is complete. Since $\lambda > 1$ was arbitrary, (18) immediately yields (12).

In the last step of the proof it will be shown that

$$(20)\quad I(J,\alpha,n) \overset{<}{\sim} \delta V(J)^{\frac{\alpha+d}{d}} \frac{1}{n^{\frac{\alpha}{d}}} \text{ as } n \to \infty.$$

Before proving (20), note that the definition of I yields the following inequality:

(21) Let C be a cube. Then $I(C \cap J, \alpha, l) \le I(C, \alpha, l)$.

For the proof of (20) let $\lambda > 1$. Since J is Jordan measurable, the following statement holds:

(22) There are cubes $C_1, \ldots, C_k$ such that:

$$J \subseteq C_1 \cup \cdots \cup C_k,\ C_i \cap J \neq \emptyset \text{ for } i = 1, \ldots, k,$$

$$V(C_i) \text{ all are equal and } \leq \frac{\lambda V(J)}{k},$$

$$\sum_{C_i \not\subseteq J} V(C_i)^{\frac{\alpha+d}{d}} \leq (\lambda - 1) \sum_i V(C_i)^{\frac{\alpha+d}{d}}.$$

Using this, we prove that

(23) $I(J, \alpha, n) \lesssim \lambda^{3+\frac{\alpha}{d}} \delta V(J)^{\frac{\alpha+d}{d}} \dfrac{1}{n^{\frac{\alpha}{d}}}$ as $n \to \infty$.

At first, the case where $n = kl$, $l = 1, 2, \ldots$, will be considered. For $i = 1, \ldots, k$ and $l = 1, 2, \ldots$, choose minimizing configurations S_{il} for $I(C \cap J, \alpha, l)$ with $\# S_{il} = l$. Let $S_n = \bigcup_i S_{il}$. Then $\# S_n \leq n = kl$. The definition of I together with (22), (21), (11) and (22) shows that

$$\begin{aligned}
I(J, \alpha, n = kl)\, n^{\frac{\alpha}{d}} &\leq \int_J \min_{s \in S_n} \{\|x - s\|^\alpha\}\, dx\, n^{\frac{\alpha}{d}} \\
&\leq \sum_i \int_{C_i \cap J} \min_{s \in S_{il}} \{\|x - s\|^\alpha\}\, dx\, n^{\frac{\alpha}{d}} \\
&\leq \Big(\sum_{C_i \subseteq J} I(C_i, \alpha, l) + \sum_{C_i \not\subseteq J} I(C_i \cap J, \alpha, l) \Big) n^{\frac{\alpha}{d}} \\
&\leq \Big(\sum_{C_i \subseteq J} I(C_i, \alpha, l) + \sum_{C_i \not\subseteq J} I(C_i, \alpha, l) \Big) n^{\frac{\alpha}{d}} \\
&\sim \delta \Big(\sum_{C_i \subseteq J} V(C_i)^{\frac{\alpha+d}{d}} + \sum_{C_i \not\subseteq J} V(C_i)^{\frac{\alpha+d}{d}} \Big) k^{\frac{\alpha}{d}} \\
&\leq \delta \Big(\sum_i V(C_i)^{\frac{\alpha+d}{d}} + (\lambda - 1) \sum_i V(C_i)^{\frac{\alpha+d}{d}} \Big) k^{\frac{\alpha}{d}} \\
&\leq \lambda\, \delta \sum_i V(C_i)^{\frac{\alpha+d}{d}} k^{\frac{\alpha}{d}} = \lambda\, \delta\, k \Big(\frac{\lambda V(J)}{k} \Big)^{\frac{\alpha+d}{d}} k^{\frac{\alpha}{d}} \\
&= \lambda^{2+\frac{\alpha}{d}} \delta V(J)^{\frac{\alpha+d}{d}} \text{ as } l \to \infty.
\end{aligned}$$

Thus (23) holds for n of the form $n = kl$, $l = 1, 2, \ldots$ with $\lambda^{2+\alpha/d}$ instead of $\lambda^{3+\alpha/d}$. A similar argument as the one that led to (10), then yields (23).

Since $\lambda > 1$ was arbitrary, (23) implies (20). The desired asymptotic formula for $I(J, \alpha, n)$ finally follows from (12) and (20). □

Generalizations

From the point of view of applications, it is of interest to extend Zador's asymptotic formula to asymptotic formulae for expressions of the following form, where f is a non-decreasing function on $[0, +\infty)$, $J \subseteq \mathbb{E}^d$ Jordan measurable, $\|\cdot\|$ a norm on $\mathbb{E}^d$, possibly different from the standard Euclidean norm, and $w : J \to \mathbb{R}^+$ a continuous weight function:

$$(24)\quad \inf_{\substack{S\subseteq J\\ \#S=n}} \int_J \min_{s\in S}\{f(\|x-s\|)\}w(x)\,dx,$$

or of the form

$$(25)\quad \inf_{\substack{S\subseteq J\\ \#S=n}} \int_J \min_{s\in S}\{f(\rho_M(x,s))\}w(x)\,d\omega_M(x),$$

where $\langle M, \rho_M, \omega_M\rangle$ is a d-dimensional Riemannian manifold with Riemannian metric ρ_M and measure ω_M, $J \subseteq M$ Jordan measurable and $w : J \to \mathbb{R}^+$ a weight function as before. Similar problems arise if the sets S are not necessarily contained in the set J. For many functions f corresponding asymptotic formulae exist. For pertinent results and more information compare Gruber [442, 443].

The Constant δ

The proof of the following asymptotic formula was communicated by Karoly Böröczky [157].

Proposition 33.1. Let $\alpha > 0$. Then $\delta_{\alpha,d} \sim \dfrac{d^{\frac{\alpha}{2}}}{(2\pi e)^{\frac{\alpha}{2}}}$ as $d \to \infty$.

Proof. We make use of the material in the first two steps of the proof of the theorem of Zador. Note that

$$\delta_{\alpha,d} = \lim_{n\to\infty} \big(n^{\frac{\alpha}{d}} I(K,\alpha,n)\big).$$

Estimate below: From the proof of (1) we have,

$$(26)\quad n^{\frac{\alpha}{d}} I(K,\alpha,n) \geq \frac{d}{(\alpha+d)\,V(B^d)^{\frac{\alpha}{d}}}.$$

Estimate above: By Rogers's covering theorem 31.4, for all sufficiently large n there are points $\{x_{n1},\dots,x_{nn}\}$ in the cube K and $\rho_n > 0$ such that the balls $\rho_n B^d + x_{ni}$ cover the cube and their total volume is less than $2d\log d$. Then $n\,\rho_n^d\,V(B^d) \leq 2d\log d$, or

$$n^{\frac{\alpha}{d}}\rho_n^{\alpha} \leq \frac{(2d\log d)^{\frac{\alpha}{d}}}{V(B^d)^{\frac{\alpha}{d}}}.$$

Let D_{ni} be the Dirichlet–Voronoĭ cell of x_{ni} in K with respect to the set $\{x_{n1}, \dots, x_{nn}\}$. Then $D_{ni} \subseteq \rho_n B^d + x_{ni}$ and the cells D_{ni} form a tiling of the unit cube K. This yields the following rough estimate:

$$n^{\frac{\alpha}{d}} I(K, \alpha, n) \le n^{\frac{\alpha}{d}} \int_K \min_{x_{ni}} \{\|x - x_{ni}\|^\alpha\} \, dx$$
$$= n^{\frac{\alpha}{d}} \sum_i \int_{D_{ni}} \|x - x_{ni}\|^\alpha \, dx \le n^{\frac{\alpha}{d}} \sum_i V(D_{ni}) \rho_n^\alpha = n^{\frac{\alpha}{d}} \rho_n^\alpha .$$

Thus,

$$(27) \quad n^{\frac{\alpha}{d}} I(K, \alpha, n) \le \frac{(2d \log d)^{\frac{\alpha}{d}}}{V(B^d)^{\frac{\alpha}{d}}} .$$

Since $V(B^d) = \pi^{\frac{d}{2}} / \Gamma(1 + \frac{d}{2})$, Stirling's formula for the gamma function and the inequalities (26) and (27) yield the asymptotic formula for $\delta_{\alpha,d}$, as required. □

33.3 Structure of Minimizing Configurations, and a Heuristic Principle

While it is out of reach to give a precise description of the configurations S_n which minimize the expressions (24) and (25) for all n, information is available as $n \to \infty$. For $d = 2$ the minimizing configurations S_n are asymptotically regular hexagonal and for $d \ge 3$, they are still distributed rather regularly over J.

Without giving details, we roughly outline what is known. For precise information, see the author [439, 442, 443]. In addition, a conjecture of Gersho [371] on the distribution of S_n over J will be discussed.

The Case $d = 2$

For a wide class of strictly increasing functions $f : [0, +\infty) \to [0, +\infty)$ a weak stability result of Gruber [439] says the following: For J in $\mathbb{E}^2$ or on a Riemannian 2-manifold $\langle M, \rho_M, \omega_M \rangle$ and $w =$ const the minimizing configurations S_n for the expressions in (24) for the Euclidean norm and for (25) are asymptotically regular hexagonal patterns in J as $n \to \infty$. If w is not constant then a result of this type still holds, but its formulation is slightly more complicated. A related weak stability result for the Euclidean norm for a wider class of functions f is due to Fejes Tóth [321].

The Case $d \ge 2$

A rather precise description of the minimizing sets S_n is indicated by the following *conjecture of Gersho* [371] which we state without giving precise definitions.

Conjecture 33.1. *There is a convex polytope P with $V(P) = 1$ which tiles $\mathbb{E}^d$ with congruent copies such that the following holds: Let $J \subseteq \mathbb{E}^d$ be a Jordan measurable* *set with $V(J) > 0$ and let $S_n = \{s_{n1}, \dots, s_{nn}\} \subseteq J$, $n = 1, 2, \dots$, be minimizing configurations for $I(J, 2, n)$. Then the corresponding Dirichlet–Voronoĭ cells*

$$D_{ni} = \{x : \|x - s_{ni}\| \leq \|x - s_{nj}\| \text{ for } j = 1, \dots, n\}$$

are asymptotically congruent to $\big(V(J)/n\big)^{1/d} P$ as $n \to \infty$.

This conjecture has so far been proved only for $d = 2$, where it is an immediate consequence of the weak stability result mentioned earlier. The convex polytope P then is a regular hexagon. The conjecture also follows from the result of Fejes Tóth [321]. Assertions in the literature that it is a consequence of earlier results are not justified.

While for dimensions $d > 2$ Gersho's conjecture is open, weaker results on what S_n looks like have been proved by Gruber [443]: Let J, f, w be as earlier, such that f is from a certain class of non-decreasing functions, the boundary of J is not too fuzzy and w has a positive lower bound. Let S_n, $n = 1, 2, \dots$, be minimizing configurations for the expression in (24) for the Euclidean norm or in (25). Then:

(i) There is a $\beta > 1$ such that S_n is a $\big(1/\beta n^{1/d}, \beta/n^{1/d}\big)$-Delone set in J.
(ii) S_n is *uniformly distributed* in J.

For the definition of a Delone set in J, compare Sect. 32.1. The sequence of sets (S_n) is *uniformly distributed* in J, if

$$\#(K \cap S_n) \sim \frac{V(K)}{V(J)} n \text{ as } n \to \infty \text{ for each Jordan measurable set } K \subseteq J.$$

That is, each K contains the appropriate share of points of S_n if n is sufficiently large. For more on uniform distribution see Hlawka [515].

Heuristic Observations

If a convex body is optimal or almost optimal with respect to an inequality, then, in many cases, it is particularly regular or symmetric in a certain sense. See Groemer's survey [403] on geometric stability results.

The weak stability result for point configurations and its applications are also examples of this phenomenon, see the surveys [439, 442, 443]. Results on packing and covering of circular discs with maximum, respectively, minimum density and Gersho's conjecture are further examples of this, see [436, 438] and Sects. 30.4 and 31.4. In all these cases the extremal configurations are point configurations which are distributed over subsets of $\mathbb{E}^d$ or Riemannian manifolds in a rather regular way. More generally, we express this as follows:

Heuristic Principle. *In many simple situations, for example in low dimensions or depending on few parameters, the extremal configurations are rather regular, possibly in an asymptotic sense.*

33.4 Packing and Covering of Circles, Data Transmission and Numerical Integration

Fejes Tóth's theorem on sums of moments and Zador's theorem on minimum distortion, the refinements of the latter by the author [443] and the results on the structure of the minimizing configurations are tools for a series of applications, see the references cited in the introduction of Sect. 33.

In the following, applications to packing and covering with circular discs, to data transmission and to numerical integration are presented. For an application of Zador's theorem to the approximation of convex bodies by circumscribed convex polytopes see the approximation theorem 11.4.

Packing and Covering of B^2

Using Fejes Tóth's theorem on sums of moments, we prove a result of Thue [996, 997] and Fejes Tóth [327] which says that the maximum density of a packing in $\mathbb{E}^2$ with circular discs equals the maximum lattice packing density, compare Sects. 29.1 and 31.4.

Corollary 33.1. $\delta_T(B^2) = \delta_L(B^2) = \dfrac{\pi}{\sqrt{12}} = 0.906\,899\ldots$

Proof. We first show a weaker statement:

(1) Let $\tau > 0$ and consider a packing of n translates of B^2 in the square τK, where $K = \{x : |x_i| \leq 1\}$. Then

$$\frac{n\pi}{4\tau^2} \leq \frac{\pi}{\sqrt{12}}.$$

Let S_n be the set of centres of this packing. Fejes Tóth's inequality on sums of moments then shows that

$$(2)\quad \int\limits_{\tau K} \min_{s\in S_n}\{f(\|x-s\|)\}\,dx \geq n\int\limits_{H_n} f(\|x\|)\,dx \text{ where } f(t) = \begin{cases} 0 \text{ for } 0 \leq t \leq 1, \\ 1 \text{ for } t > 1, \end{cases}$$

and H_n is a regular hexagon of area $A(H_n) = A(H)/n = 4\tau^2/n$ with centre o. Note that

$$(3)\quad \int\limits_{\tau K} \min_{s\in S_n}\{f(\|x-s\|)\}\,dx = 4\tau^2 - n\pi$$

and

$$(4)\quad n\int\limits_{H_n} f(\|x\|)\,dx = nA(H_n\backslash B^2) \geq nA(H_n) - nA(B^2) = 4\tau^2 - n\pi.$$

In (4) equality holds precisely in case where H_n contains B^2. Since inequality is excluded by Propositions (2)–(4), the hexagon H_n contains the circular disc B^2. Then

$$A(H_n) \geq \frac{2\sqrt{3}}{\pi} A(B^2) \text{ and thus } \frac{n\pi}{4\tau^2} \leq \frac{\pi}{\sqrt{12}},$$

concluding the proof of statement (1).

Second, the following stronger statement will be shown:

(5) Let $\{B^2 + s : s \in S\}$ be a packing. Then

$$\frac{1}{4\tau^2} \sum_{s \in S} A\big((B^2 + s) \cap \tau K\big) \leq \frac{\pi}{\sqrt{12}} + O\Big(\frac{1}{\tau}\Big).$$

To see this, rewrite the left side of this inequality in the form

$$\frac{1}{4\tau^2} \sum_{\substack{s \in S \\ B^2 + s \subseteq \tau K}} A(B^2 + s) + \frac{1}{4\tau^2} \sum_{\substack{s \in S \\ B^2 + s \not\subseteq \tau K \\ (B^2+s) \cap \tau K \neq \emptyset}} A\big((B^2 + s) \cap \tau K\big).$$

Apply (1) to the first expression and note that in the second expression only discs are considered which intersect $\operatorname{bd} \tau K$. Since we consider a packing, these discs do not overlap. The total area of these discs is thus $O(\tau)$. This gives the second term on the right side of the inequality in (5).

Proposition (5) implies that $\delta_T(B^2) \leq \pi/\sqrt{12}$. Since there is a lattice packing of B^2 of density $\pi/\sqrt{12}$, the equality $\delta_L(B^2) = \delta_L(B^2) = \pi/\sqrt{12}$ follows. □

Similar arguments lead to the following counterpart of the result of Thue and Fejes Tóth due to Kershner [578]. It says that the minimum density of a covering of $\mathbb{E}^2$ with circular discs equals the minimum lattice covering density.

Corollary 33.2. $\vartheta_T(B^2) = \vartheta_L(B^2) = \dfrac{2\pi}{\sqrt{27}} = 1.209\,199\ldots$

Proof. First, we show the following:

(6) Let $\tau > 0$ and consider a covering of τK with n translates of B^2. Then,

$$\frac{n\pi}{4\tau^2} \geq \frac{2\pi}{\sqrt{27}}.$$

We may assume that all centres of these translates are in τK. Let S_n be the set of centres. Fejes Tóth's inequality then yields

$$\int_{\tau K} \min_{s \in S_n}\{f(\|x - s\|)\}\, dx \geq n \int_{H_n} f(\|x\|)\, dx$$

with f and H_n as before. Since the translates of B^2 cover τK, the integrand of the first integral is 0. Being non-negative, the integrand of the second integral thus is 0 too. Taking into account the definition of f, this implies that $H_n \subseteq B^2$ which, in turn, yields (6).

Second, the following statement holds:

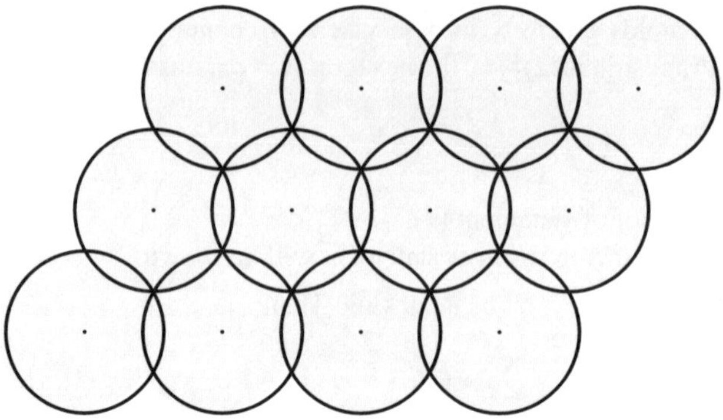

Fig. 33.1. Densest lattice covering with circular discs in $\mathbb{E}^2$

(7) Let $\{B^2 + s : s \in S\}$ be a covering (of $\mathbb{E}^2$). Then

$$\begin{aligned}\frac{1}{4\tau^2}&\sum_{s\in S} A\big((B^2+s)\cap \tau K\big)\\ &\geq \frac{4(\tau-2)^2}{4\tau^2}\,\frac{1}{4(\tau-2)^2}\sum_{\substack{s\in S\\ (B^2+s)\cap(\tau-2)K\neq\emptyset}} A(B^2+s)\\ &\geq \frac{4(\tau-2)^2}{4\tau^2}\,\frac{2\pi}{\sqrt{27}} \quad \text{for } \tau > 2\end{aligned}$$

by (6), applied to $(\tau-2)K$ instead of τK.

Proposition (7) shows that $\theta_T(B^2) \geq 2\pi/\sqrt{27}$. Since there are lattice coverings of B^2 of density $2\pi/\sqrt{27}$, the equality $\theta_T(B^2) = \theta_L(B^2) = 2\pi/\sqrt{27}$ follows (Fig. 33.1). □

Minimum Distortion of Vector Quantization

The scheme of a *data transmission system* is as follows:

$$x \longrightarrow \boxed{\text{encoder}} \xrightarrow{C} \boxed{\text{channel}} \xrightarrow{C'} \boxed{\text{decoder}} \longrightarrow x'.$$

A source produces signals. The signals are the input of the encoder. The encoder assigns to each incoming signal x a code-word c which is taken from a code-book consisting of finitely many code-words. The code-word c then is transmitted in a channel. The output of the channel is a word c', possibly different from c, from which the decoder produces a signal x'. We study the quality of the encoder in a particularly important special case, compare Gray and Neuhoff [394].

The *signals* are the points of a Jordan measurable set C in $\mathbb{E}^d$ such that $V(C) > 0$. The *encoder* consists of a dissection of C into n Jordan measurable sets, say $C_1, \dots, C_n$, the *cells*, and a set $\{c_1, \dots, c_n\}$ of n points in $\mathbb{E}^d$, the *code-book*. The encoder now works as follows: If x is a signal, it finds out to which set of the dissection x belongs. If it belongs to C_i, the encoder assigns the *code-word* c_i, here also called *code-vector*, to x. In case of ambiguity, which can occur only for x in a set of Jordan measure 0, choose any code-vector c_i with $x \in C_i$. Common measures for the quality of the thus defined *encoder* or *(vector-)quantizer* on C can be described as follows: Let $\alpha > 0$. The corresponding *(average) distortion* of the encoder is defined to be

$$\sum_i \int_{C_i} \|x - c_i\|^\alpha \, dx.$$

How should the cells C_i and the code-vectors c_i be chosen in order to minimize the distortion? Given the code-book $\{c_1, \dots, c_n\}$, the distortion is minimized if $C_1, \dots, C_n$ are the Dirichlet–Voronoĭ cells in C corresponding to the code-book $\{c_1, \dots, c_n\}$. Then the distortion is easily seen to be

$$\int_C \min_{i \in \{1,\dots,n\}} \{\|x - c_i\|^\alpha\} dx.$$

Thus the *minimum distortion* is given by

$$\inf_{\substack{S \subseteq \mathbb{E}^d \\ \#S = n}} \int_C \min_{c \in S} \{\|x - c\|^\alpha\} dx.$$

The literature on vector quantization began with fundamental papers of Shannon [929] and Zador [1034, 1035]. Shannon considered the minimum distortion as $d \to \infty$, while in Zador's high resolution theory the case $n \to \infty$ is investigated.

An immediate consequence of Zador's theorem proved earlier, actually a reformulation of it, is the following result of Zador [1034, 1035].

Corollary 33.3. *Let $\alpha > 0$. Then there is a constant $\delta = \delta_{\alpha,d} > 0$, depending only on α and d, such that the following holds: let $C \subseteq \mathbb{E}^d$ be a compact, Jordan measurable body. Then the minimum distortion of a vector-quantizer on C with n code-words is asymptotically equal to*

$$\delta \frac{V(C)^{\frac{\alpha+d}{\alpha}}}{n^{\frac{\alpha}{d}}} \text{ as } n \to \infty.$$

Remark. See Gruber [443] for more precise results.

Minimum Error of Numerical Integration Formulae

Let $J \subseteq \mathbb{E}^d$ be a compact Jordan measurable set with $V(J) > 0$ and $\mathcal{F}$ a class of Riemann integrable functions on J. For given sets of n *nodes* $N = \{p_1, \dots, p_n\} \subseteq J$ and n *weights* $W = \{w_1, \dots, w_n\} \subseteq \mathbb{R}$ the *error* of the numerical integration formula

$$\int_J f(x)\, dx \approx \sum_i f(p_i) w_i \text{ for } f \in \mathcal{F}$$

is defined to be

$$E(J,\mathcal{F},N,W) = \sup_{f\in\mathcal{F}} \Big\{ \Big| \int_J f(x)\,dx - \sum_i f(p_i)w_i \Big| \Big\}.$$

The *minimum error* then is

$$E(J,\mathcal{F},n) = \inf_{\substack{N\subseteq J,\#N=n\\ W\subseteq\mathbb{R},\#W=n}} \big\{E(\mathcal{F},N,W)\big\}.$$

Precise solutions of the problems to determine, for all n, the minimum error and to describe the corresponding sets of nodes and weights are out of reach. Upper estimates and asymptotic formulae as $n\to\infty$ of the minimum error for several classes $\mathcal{F}$ of functions have been given by Koksma and Hlawka, see Hlawka [515], by Sobolev [946] and his school and by the Dnepropetrovsk school of numerical analysis, see Chernaya [206, 207].

As a consequence of Zador's theorem 33.2, we obtain the following result of Chernaya [207]:

Corollary 33.4. *Let* $0<\alpha\le 1$. *Then there is a constant* $\delta=\delta_{\alpha,d}>0$, *depending only on* α *and* d, *such that the following statement holds: Let* $J\subseteq\mathbb{E}^d$ *be a compact Jordan measurable body and consider the following class* $\mathcal{H}^\alpha$ *of Hölder continuous real functions on* J:

$$\mathcal{H}^\alpha = \big\{f : J\to\mathbb{R} : |f(x)-f(y)|\le\|x-y\|^\alpha \text{ for } x,y\in J\big\}.$$

Then

$$E(J,\mathcal{H}^\alpha,n)\sim\delta\,\frac{V(J)^{\frac{\alpha+d}{d}}}{n^{\frac{\alpha}{d}}} \text{ as } n\to\infty.$$

Proof. Taking into account Zador's theorem, it is sufficient to prove that

$$(1)\quad E(J,\mathcal{H}^\alpha,n) = \inf_{\substack{N\subseteq J\\ \#N=n}} \left\{ \int_J \min_{p\in N}\{\|x-p\|^\alpha\}dx \right\}.$$

To see this, we proceed as follows

First, the following will be shown:

(2) Let $N=\{p_1,\dots,p_n\}\subseteq J$ and let $h: J\to\mathbb{R}$ be defined by $h(x)=\min_{p\in N}\{\|x-p\|^\alpha\}$. Then $h\in\mathcal{H}^\alpha$.

Let $x,y\in J$. By exchanging x and y, if necessary, we may assume that $h(x)\ge h(y)$. Then

$$\begin{aligned}
0 &\le h(x)-h(y) = \min_{p\in N}\{\|x-p\|^\alpha\} - \min_{q\in N}\{\|y-q\|^\alpha\}\\
&= \|x-p\|^\alpha - \|y-q\|^\alpha \text{ for suitable } p,q\in N\\
&\le \|x-q\|^\alpha - \|y-q\|^\alpha\\
&= \|x-y+y-q\|^\alpha - \|y-q\|^\alpha \le (\|x-y\|+\|y-q\|)^\alpha - \|y-q\|^\alpha\\
&\le \|x-y\|^\alpha + \|y-q\|^\alpha - \|y-q\|^\alpha = \|x-y\|^\alpha,
\end{aligned}$$

where we have used that $(s+t)^\alpha \leq s^\alpha + t^\alpha$ for $s, t \geq 0$ (note that $0 < \alpha \leq 1$). This concludes the proof of (2).

Second,

(3) let $N = \{p_1, \dots, p_n\} \subseteq J$ and $W = \{w_1, \dots, w_n\} \subseteq \mathbb{R}$. Define $D_i = \big\{x \in J : \|x - p_i\| \leq \|x - p_j\| \text{ for } j = 1, \dots, n\big\}$, $i = 1, \dots, n$. Let $\overline{w}_i = V(D_i),\ i = 1, \dots, n,\ \overline{W} = \{\overline{w}_1, \dots, \overline{w}_n\}$. Then

$$E(J, \mathcal{H}^\alpha, N, W) \geq E(J, \mathcal{H}^\alpha, N, \overline{W}) = \int_J \min_{p \in N}\{\|x - p\|^\alpha\} dx.$$

This can be seen as follows:

$$\begin{aligned}
E(J, \mathcal{H}^\alpha, N, \overline{W}) &= \sup_{f \in \mathcal{H}^\alpha} \Big\{\Big| \int_J f(x)\,dx - \sum_i f(p_i) \int_{D_i} dx \Big|\Big\} \\
&\leq \sup_{f \in \mathcal{H}^\alpha} \Big\{\sum_i \int_{D_i} |f(x) - f(p_i)|\,dx\Big\} \\
&\leq \sum_i \int_{D_i} \|x - p_i\|^\alpha dx = \int_J \min_{p \in N}\{\|x - p\|^\alpha\}\,dx \\
&= \int_J h(x)\,dx = \int_J h(x)\,dx - \sum_i h(p_i) w_i \leq E(J, \mathcal{H}^\alpha, N, W)
\end{aligned}$$

by the assumption in (3), (2) and since $h(p_i) = 0$.

Third, it follows from (3) that

$$\begin{aligned}
E(J, \mathcal{H}^\alpha, n) &= \inf_{\substack{N \subseteq J, \#N = n \\ W \subseteq \mathbb{R}, \#W = n}} \{E(J, \mathcal{H}^\alpha, N, W)\} \\
&= \inf_{\substack{N \subseteq J \\ \#N = n}} \{E(J, \mathcal{H}^\alpha, N, \overline{W})\} = \inf_{\substack{N \subseteq J \\ \#N = n}} \Big\{\int_J \min_{p \in N}\{\|x - p\|^\alpha\}\,dx\Big\},
\end{aligned}$$

concluding the proof of (1) and thus of the theorem. □

Remark. See Chernaya [206] and Gruber [443] for more general results.

34 Koebe's Representation Theorem for Planar Graphs

Let $\mathcal{G}$ be a finite, 3-connected planar graph. Koebe [605] showed that one may assign to each vertex of $\mathcal{G}$ a (circular) disc such that these discs form a packing in $\mathbb{E}^2$ where two discs touch precisely in case when the corresponding vertices of $\mathcal{G}$ are connected by an edge. This result was rediscovered by Andreev [29, 30] and Thurston [999]. Thurston [1000] also specified a procedure, Thurston's algorithm, for finding such packings. Its convergence was proved by Rodin and Sullivan [844] and Colin de

Verdière [214,215]. An extension of Koebe's theorem to a representation of $\mathcal{G}$ and its dual $\mathcal{G}^*$ by two related packings of discs is due to Brightwell and Scheinerman [168].

These results have a series of important consequences: Miller and Thurston (unpublished) showed that Koebe's theorem yields the basic theorem of Lipton and Tarjan [661] on separation of graphs, see Miller, Teng, Thurston and Vavaies [724] and Pach and Agarwal [783]. Koebe's theorem and the corresponding algorithms of Thurston [999], Rodin and Sullivan [844] and Mohar [747] yield constructive approximations to the analytic functions as in the Riemann mapping theorem. The extension of Brightwell and Scheinerman proves a conjecture of Tutte [1002] on simultaneous straight line representation of $\mathcal{G}$ and its dual $\mathcal{G}^*$. For us, it is important that the result of Brightwell and Scheinerman readily yields a refined version of the representation theorem 15.6 of Steinitz for convex polytopes in $\mathbb{E}^3$.

In this section we first present the theorem of Brightwell and Scheinerman, then discuss the algorithm of Thurston which yields a construction of the circle packing corresponding to a graph with triangular countries that is, for Koebe's theorem, and outline how this can be used to obtain approximations for the Riemann mapping theorem.

For more information and references to the literature, see the books of Pach and Agarwal [783] on combinatorial geometry, of Mohar and Thomassen [748] and Felsner [332] on graphs and of Stephenson [967] on packings of circular discs in the context of discrete analytic functions. See also Sachs [872] and Stephenson [966].

34.1 The Extension of Koebe's Theorem by Brightwell and Scheinerman

After introducing needed terminology on planar graphs, we present a proof of the theorem of Brightwell and Scheinerman.

Graph Terminology

Recall the definitions and notation in Sect. 15.1. Let $\mathcal{G}$ be a 3-connected planar graph in $\mathbb{C} \cup \{\infty\}$. We assume that ∞ is not a vertex of $\mathcal{G}$. Since $\mathcal{G}$ is 3-connected, any two countries of $\mathcal{G}$ are either disjoint, or have one common vertex, or one common edge. The *exterior country* of $\mathcal{G}$ is the country containing the point ∞ in its interior. Up to isomorphisms the *dual graph* $\mathcal{G}^*$ of $\mathcal{G}$ is defined as follows: In each country of $\mathcal{G}$, including the exterior country, choose a point. These points are the vertices of $\mathcal{G}^*$. Distinct vertices of $\mathcal{G}^*$ are connected by an edge in $\mathcal{G}^*$ if the corresponding countries of $\mathcal{G}$ have an edge of $\mathcal{G}$ in common. Clearly, $\mathcal{G}^*$ can be drawn in $\mathbb{C} \cup \{\infty\}$. It can be shown that $\mathcal{G}^*$ is also 3-connected and planar. Next, the *vertex-country incidence graph* $\mathcal{G}^\wedge$ of $\mathcal{G}$ will be defined: Its vertices are the vertices of $\mathcal{G}$ and $\mathcal{G}^*$. An edge of $\mathcal{G}^\wedge$ connects a vertex v of $\mathcal{G}$ with a vertex w of $\mathcal{G}^*$ if v is a vertex of the country of $\mathcal{G}$ corresponding to w. There are no other edges. By applying a suitable Möbius transformation, if necessary, we may assume that ∞ is the vertex of $\mathcal{G}^*$ which corresponds to the exterior country of $\mathcal{G}$. Let $\mathcal{G}'$ denote the graph obtained from $\mathcal{G}^\wedge$ by deleting the vertex ∞ and all edges incident with it. See Bollobás [142] and Mohar and Thomassen [748] (Fig. 34.1).

Fig. 34.1. Graph, dual graph and vertex-country incidence graph

By a *primal-dual circle packing* of $\mathcal{G}$, we mean a system of (circular) discs in $\mathbb{C} \cup \{\infty\}$, each disc corresponding to a vertex of $\mathcal{G}^\wedge$, such that the following properties hold:

(i) The discs corresponding to the vertices of $\mathcal{G}$ form a packing such that two discs touch precisely in case where the corresponding vertices are connected by an edge of $\mathcal{G}$.

(ii) A disc with centre ∞ is simply the complement in $\mathbb{C} \cup \{\infty\}$ of an open disc with centre 0. Its radius is $-\rho$ where ρ is the radius of the open disc. The discs corresponding to the vertices of $\mathcal{G}^*$ or, equivalently, to the countries of $\mathcal{G}$, form a packing such that two discs touch precisely in case where the corresponding countries have an edge of $\mathcal{G}$ in common.

(iii) Let D_v, D_w be discs corresponding to vertices v, w of $\mathcal{G}$ which are connected by an edge of $\mathcal{G}$ and D_x, D_y the discs corresponding to the countries of $\mathcal{G}$ adjacent to the edge vw. Then the four discs D_v, D_w, D_x, D_y have a common boundary point at which the boundary circles of D_v, D_w intersect the boundary circles of D_x, D_y orthogonally.

A *weak primal-dual circle packing* of $\mathcal{G}$ is defined similarly with the only exception that there is no disc corresponding to the exterior country of $\mathcal{G}$.

The Extension of Koebe's Theorem by Brightwell and Scheinerman

Following Brightwell and Scheinerman [168] and Mohar and Thomassen [748], we prove the following refinement of Koebe's theorem.

Theorem 34.1. *Let $\mathcal{G}$ be a 3-connected planar graph. Then $\mathcal{G}$ admits a primal-dual circle packing.*

Proof. The proof is split into several steps.

In the first step we derive necessary conditions for a primal-dual circle packing of $\mathcal{G}$.

(1) Let $\big(\rho_v : v \in \mathcal{V}(\mathcal{G}')\big)$ be the radii of a primal-dual circle packing of $\mathcal{G}$. Then

(i) $$\sum_{vw \in \mathcal{E}(\mathcal{G}')} \arctan \frac{\rho_w}{\rho_v} = \pi \text{ for } v \in \mathcal{V}(\mathcal{G}'),\ v\,\infty \notin \mathcal{E}(\mathcal{G}^\wedge).$$

Let $v_1, \dots, v_k \in \mathcal{V}(\mathcal{G}')$ be such that $v_i \infty \in \mathcal{E}(\mathcal{G}^\wedge)$, i.e. $v_1 v_2 \cdots v_k$ is the outer facet cycle of $\mathcal{G}$, and let

$$\alpha_i = 2 \sum_{v_i w \in \mathcal{E}(\mathcal{G}')} \arctan \frac{\rho_w}{\rho_{v_i}} \text{ for } i = 1, \dots, k.$$

Then

(ii) $0 < \alpha_i < \pi$ for $i = 1, \dots, k$, and $\sum_{i=1}^{k} \alpha_i = (k-2)\pi$.

We identify $\mathcal{G}$ with the *nerve* of the packing $\{D_v : v \in \mathcal{V}(\mathcal{G})\}$, i.e. its vertices are the centres of the discs D_v and two vertices are connected by an edge if the corresponding discs touch, and similarly for $\mathcal{G}^*$. For the proof of (i) consider the case that $v \in \mathcal{V}(\mathcal{G})$, the proof being similar in case that $v \in \mathcal{V}(\mathcal{G}^*)$. The countries of $\mathcal{G}$ that contain v are all bounded. The corresponding vertices w in $\mathcal{G}^*$ are precisely the neighbours w of v in $\mathcal{G}^\wedge$. The cycle determined by the neighbours w is circumscribed to the disc D_v since the boundaries of D_v and D_w intersect orthogonally. This yields (1)(i) (Fig. 34.2).

To see (1)(ii), note that α_i is the internal angle of the outer country cycle of $\mathcal{G}$ at the vertex v_i. Since the outer country cycle is circumscribed to D_∞, it is a convex polygon which implies (1)(ii) (Fig. 34.3).

The proof that the conditions (1)(i),(ii) are sufficient for the existence of a weak primal-dual circle packing of $\mathcal{G}$ makes use of a topological result on graphs which will be given in steps four and five.

In step two a topological tool is presented. It will yield a topological result on graphs in step three. A map $f : \mathbb{C} \to \mathbb{C}$ is a *covering map* if it is continuous, onto and such that for each $x \in \mathbb{C}$ there are neighbourhoods U of x and W of $f(x)$, respectively, such that f maps U homeomorphically onto W.

(2) Let $f : \mathbb{C} \to \mathbb{C}$ be a covering map such that the set $\{x \in \mathbb{C} : \#f^{-1}(f(x)) \geq 2\}$ is bounded. Then f is a homeomorphism.

It is sufficient to show that $\{x : \#f^{-1}(f(x)) \geq 2\} = \emptyset$ or, equivalently, that $\{z : \#f^{-1}(z) \geq 2\} = \emptyset$. To see the latter, we first prove that the set $\{z : \#f^{-1}(z) \geq 2\}$

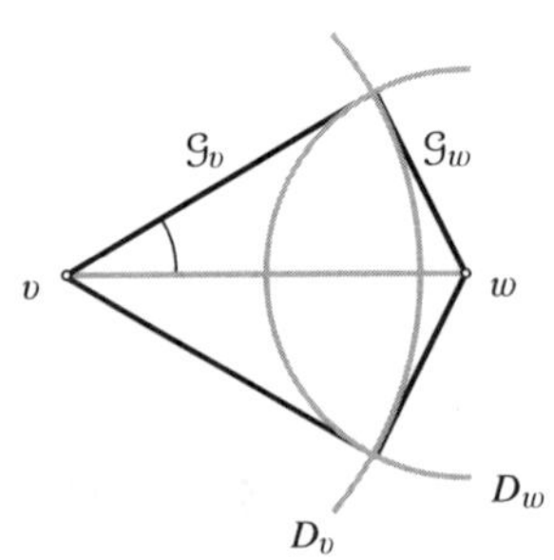

Fig. 34.2. Country circle and vertex circle

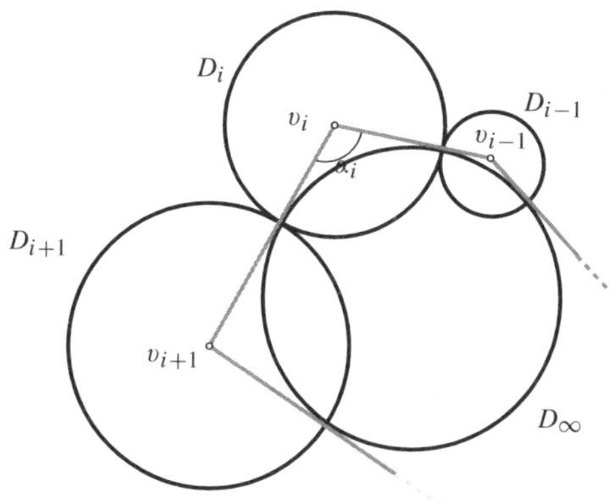

Fig. 34.3. Country circle and vertex circles

is open. Let w be one of its points. Then there are points $x, y \in \mathbb{C}^2$, $x \neq y$, with $f(x) = f(y) = w$. Choose disjoint neighbourhoods U of x and V of y and a neighbourhood W of w such that f maps each of U, V homeomorphically onto W. Then, clearly, $W \subseteq \{z : \#f^{-1}(z) \geq 2\}$. Secondly, it will be shown that the set $\{z : \#f^{-1}(z) = 1\}$ is also open. For suppose not. Then there are points $z, z_n, n = 1, 2, \ldots$, with $z_n \to z$ and $\#f^{-1}(z) = 1$, $\#f^{-1}(z_n) \geq 2$ for all n. Let $\{x\} = f^{-1}(z)$ and choose neighbourhoods U of x and W of z such that f maps U homeomorphically onto W. By omitting finitely many indices and renumbering, if necessary, we may assume that there are $x_n \in U$, $y_n \in \mathbb{C}\backslash U$ such that $f(x_n) = f(y_n) = z_n$ and $x_n \to x$. Since $\{y : \#f^{-1}(f(y)) \geq 2\}$ is bounded by assumption, by considering a suitable subsequence and renumbering, if necessary, we may assume that $y_n \to y \in \mathbb{C}\backslash U$, say. The continuity of f then implies that $f(x) = f(y) = z$, while $x \neq y$. This is a contradiction. $\mathbb{C}$ is thus the disjoint union of the two open sets $\{z : \#f^{-1}(z) \geq 2\}$ and $\{z : \#f^{-1}(z) = 1\}$. By assumption, $\{z : \#f^{-1}(z) \geq 2\}$ is bounded. Thus $\{z : \#f^{-1}(z) = 1\} \neq \emptyset$. Since $\mathbb{C}$ is connected, it follows that $\{z : \#f^{-1}(z) \geq 2\} = \emptyset$. Hence f is one-to-one, concluding the proof of (2).

The third step is to show the following topological result on graphs.

(3) Let $\mathcal{H}$ be the image of $\mathcal{G}'$ in $\mathbb{C}$ under a (graph) isomorphism f, possibly with edge crossings, which has the following properties:

(i) All edges of $\mathcal{H}$ are polygonal arcs.

(ii) For each vertex v of $\mathcal{G}'$ the images in $\mathcal{H}$ of the edges that leave v are pairwise non-crossing and leave $f(v)$ in the same clockwise order as their originals in $\mathcal{G}'$.

(iii) The image of each country cycle in $\mathcal{G}'$ is a closed Jordan curve in $\mathcal{H}$.

(iv) If C is the boundary cycle of a bounded country of $\mathcal{G}'$ and E an edge of $\mathcal{G}'$ leaving C, then the first segment of $f(E)$ is in the exterior of $f(C)$.

Then $\mathcal{H}$ is a planar representation of $\mathcal{G}'$, i.e. it has no edge crossings.

If J is a closed Jordan curve in $\mathbb{C}$, we denote by interior J the bounded component of $\mathbb{C}\backslash J$. Let C_∞ be the cycle of the exterior country of $\mathcal{G}'$.

To see (3), we first extend f to a continuous map of the *point set* of $\mathcal{G}'$, i.e. the union of all vertices and edges of $\mathcal{G}'$, onto the point set of $\mathcal{H}$ such that f is one-to-one on each edge of $\mathcal{G}'$. Next, we extend f step by step to a covering map of $\mathbb{C}$ onto $\mathbb{C}$: Noting (iii), we may extend f by the Jordan-Schönflies theorem for each country cycle $C \neq C_\infty$ of $\mathcal{G}'$ to interior C such that the extended f maps $C \cup$ interior C homeomorphically onto the compact set $f(C) \cup$ interior $f(C)$. The extended f then maps $C_\infty \cup$ interior C_∞ continuously onto the compact set $f(C_\infty) \cup f(\text{interior } C_\infty)$. Let $x \in$ interior C_∞. By distinguishing the cases where x is an interior point of a bounded country, a relatively interior point of an edge, or a vertex of $\mathcal{G}'$ and noting (iv) and (ii), we see that there are neighbourhoods U of x and W of $f(x)$ such that f maps U homeomorphically onto W. As a consequence, we see that $f(\text{interior } C_\infty)$ is an open set in $\mathbb{E}^2$. Since a boundary point of $f(\text{interior } C_\infty)$ must be a limit point of points of the form $f(x_n)$, $n = 1, 2, \dots$, where $x_n \in$ interior C_∞, it must be of the form $f(x)$ where $x \in C_\infty \cup$interior C_∞. (Consider a convergent subsequence of the sequence (x_n).) Since $x \in$ interior C_∞ is excluded, $x \in C_\infty$. Hence bd $f(\text{interior } C_\infty) \subseteq f(C_\infty)$. Thus, noting that $f(\text{interior } C_\infty)$ is open, connected and bounded, $f(\text{interior } C_\infty) =$ interior $f(C_\infty)$ follows. By a version of the Jordan-Schönflies theorem, f finally can be extended to a continuous map $f : \mathbb{C} \to \mathbb{C}$ such that f is a homeomorphism on $\mathbb{C}\backslash$interior C_∞. Let $x \in \mathbb{C}\backslash$interior C_∞. As before, we see that there are neighbourhoods U of x and W of $f(x)$ such that f maps U homeomorphically onto W. The extended f satisfies the required assumptions of (2) and thus is a homeomorphism of $\mathbb{C}$ onto $\mathbb{C}$. In particular, $\mathcal{H}$ is a planar representation of $\mathcal{G}'$, concluding the proof of (3).

In the fourth step we show that the conditions (1)(i,ii) yield a weak primal-dual circle packing of $\mathcal{G}$:

(4) Let $\rho = \big(\rho_v : v \in \mathcal{V}(\mathcal{G}')\big)$ satisfy (1)(i,ii). Then there is a weak primal-dual circle packing of $\mathcal{G}$ with radii $\rho = (\rho_v)$ and with the same local clockwise orientation as in $\mathcal{G}$ and $\mathcal{G}^*$.

We shall construct an isomorphic planar image of $\mathcal{G}'$ such that the image of an edge uv is a line segment. For each vertex $u \in \mathcal{V}(\mathcal{G}')$ we shall determine a point $\bar{u} \in \mathbb{C}$ such that the discs with centres $\bar{u}$ and radii ρ_u form the desired weak primal-dual circle packing. We start with an edge $uv \in \mathcal{E}(\mathcal{G}')$ and draw a corresponding line segment $\bar{u}\bar{v}$ in $\mathbb{C}$ of length $\big(\rho_u^2 + \rho_v^2\big)^{1/2}$. Then the position of each neighbour of $\bar{u}$ is uniquely determined. Now, given an arbitrary vertex $w \in \mathcal{V}(\mathcal{G}')$, consider a path in $\mathcal{G}'$ connecting u and w and use it to construct $\bar{w}$. For the proof that $\bar{w}$ is independent of the path chosen, it is sufficient to show that, for each simply closed path in $\mathcal{G}'$, our construction leads to a closed path in $\mathbb{C}$. This can easily be shown by induction on the number of country 4-cycles in the simply closed path. The image of $\mathcal{G}'$ in $\mathbb{C}$ thus obtained satisfies the properties specified in (3). Hence it has no crossings and thus is a planar graph. For each of its vertices $\bar{u}$ the disc D_u with centre $\bar{u}$ and radius ρ_u is contained in the convex polygon consisting of the convex *kites* which correspond

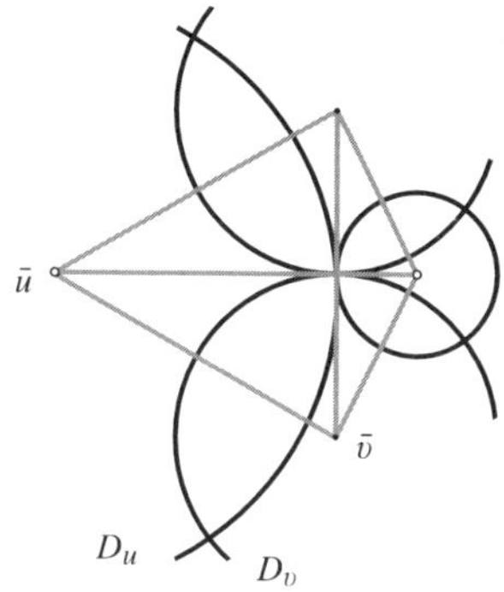

Fig. 34.4. Kite

to the country 4-cycles in $\mathcal{G}'$ with vertex u (Fig. 34.4). Since this holds for any vertex, the disc D_u touches only those discs D_v or their boundaries intersect orthogonally, for which $\bar{v}$ is a vertex of a kite with vertex $\bar{u}$. Hence the discs $D_u : u \in \mathcal{V}(\mathcal{G}')$ form a weak primal-dual circle packing of $\mathcal{G}$, concluding the proof of (4).

In the fifth step we refine (4):

(5) Let $\rho = \big(\rho_v : v \in \mathcal{V}(\mathcal{G}')\big)$ satisfy (1)(i,ii). Then there exists a primal-dual circle packing of $\mathcal{G}$.

Assume first that the outer cycle of $\mathcal{G}$ is a 3-cycle $v_1v_2v_3$. By (4) there is a weak primal-dual circle packing of $\mathcal{G}$. By applying a suitable Möbius transformation and changing notation, if necessary, we obtain a weak primal-dual circle packing of $\mathcal{G}$ and such that $\bar{v}_1, \bar{v}_2, \bar{v}_3$ are the vertices of a regular triangle with centroid 0. Since the discs $D_{v_1}, D_{v_2}, D_{v_3}$ with centres $\bar{v}_1, \bar{v}_2, \bar{v}_3$ touch pairwise, they all have the same radius. Hence there is a circular disc D_∞ with centre ∞ such that its boundary circle intersects orthogonally the boundary circles of the discs $D_{v_1}, D_{v_2}, D_{v_3}$ at the points where these discs touch pairwise. This yields the desired primal-dual circle packing of $\mathcal{G}$ (Fig. 34.5). If, second, the outer cycle of $\mathcal{G}$ has length greater than 3, then either $\mathcal{G}$ or $\mathcal{G}^*$ has a country 3-cycle by the Corollary 15.1 of the Euler polytope formula. Interchanging the roles of $\mathcal{G}$ and $\mathcal{G}^*$, if necessary, and applying a suitable Möbius transformation, if necessary, we arrive at the same situation as in the previous paragraph. Thus there exists a corresponding primal-dual circle packing. The inverse of the Möbius transformation then yields the desired primal-dual circle packing of $\mathcal{G}$.

In the sixth step we prepare the way for the seventh step:

(6) Let $S \subsetneq \mathcal{V}(\mathcal{G}^\wedge)$ with $\#S \geq 5$. Then $2\#S - \#\mathcal{E}\big(\mathcal{G}^\wedge(S)\big) \geq 5$.

Here $\mathcal{G}^\wedge(S)$ is the subgraph of $\mathcal{G}^\wedge$ with vertex set S whose edges are precisely the edges uv of $\mathcal{G}^\wedge$ with $u, v \in S$. To see (6), note first that the following holds:

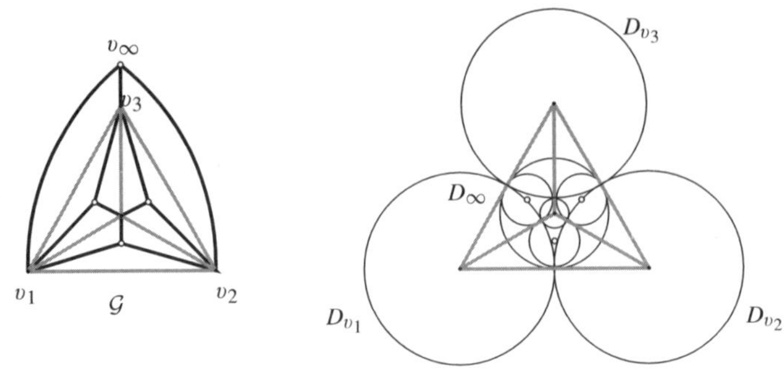

Fig. 34.5. Graph and corresponding primal-dual circle packing

(7) Every 4-cycle in $\mathcal{G}^\wedge$ is country.

If $xvyw$ is such a cycle, assume that x, y correspond to countries of $\mathcal{G}$. These countries then have the vertices v, w in common. Since $\mathcal{G}$ is 3-connected, this is possible only if v, w are connected in $\mathcal{G}$ by an edge and that this edge is the common edge of the countries. This means that $xvyw$ is a country cycle in $\mathcal{G}^\wedge$, concluding the proof of (7). Next note that by the Corollary 15.1 of the Euler polytope formula

(8) $2\#S - \#\mathcal{E}\big(\mathcal{G}^\wedge(S)\big) \geq 4$

with equality if and only if $\big(\mathcal{G}^\wedge(S)$ is 2-connected and$\big)$ all countries of $\mathcal{G}^\wedge(S)$ are quadrangles. $\mathcal{G}^\wedge$ is connected. Since $S \subsetneq \mathcal{V}(\mathcal{G}^\wedge)$, there is an edge of $\mathcal{G}^\wedge$ which is incident with a vertex of S but is not an edge of $\mathcal{G}^\wedge(S)$. This edge meets the interior of one of the countries of $\mathcal{G}^\wedge(S)$. Thus, if there is equality in (8), one of the country 4-cycles in $\mathcal{G}^\wedge(S)$ is not a country 4-cycle of $\mathcal{G}^\wedge$, in contradiction to (7). Hence, there is inequality in (8), concluding the proof of proposition (6).

In the seventh, and last, step of the proof we have to show that there is a list $\rho = \big(\rho_v : v \in \mathcal{V}(\mathcal{G}^\wedge)\big)$ of radii satisfying (1)(i,ii). More precisely, the following has to be shown:

(9) Let $v_1 v_2 \cdots v_k$ be the outer cycle of $\mathcal{G}$ and let $0 < \alpha_1, \dots, \alpha_k < \pi$ be such that $\alpha_1 + \cdots + \alpha_k = (k-2)\pi$. Then there is a list $\rho = \big(\rho_v : v \in \mathcal{V}(\mathcal{G}^\wedge)\big)$ of positive numbers such that the following statements hold:

(i) $\displaystyle\sum_{vw \in \mathcal{E}(\mathcal{G}^\wedge)} \arctan \frac{\rho_w}{\rho_v} = \pi$ for $v \in \mathcal{V}(\mathcal{G}^\wedge) \setminus \{v_1, \dots, v_k, \infty\}$.

(ii) $\displaystyle 2 \sum_{\substack{v_i w \in \mathcal{E}(\mathcal{G}^\wedge) \\ w \neq \infty}} \arctan \frac{\rho_w}{\rho_{v_i}} = \alpha_i$ for $i = 1, \dots, k$.

Given a list $\rho = \big(\rho_v : v \in \mathcal{V}(\mathcal{G}')\big)$, define

$$\vartheta_v = \vartheta_v(\rho) = \sum_{vw \in \mathcal{E}(\mathcal{G}^\wedge)} \arctan \frac{\rho_w}{\rho_v} - \pi \text{ for } v \in \mathcal{V}(\mathcal{G}^\wedge), v \neq v_1, \ldots, v_k, \infty,$$

$$\vartheta_{v_i} = \vartheta_{v_i}(\rho) = \sum_{\substack{v_i w \in \mathcal{E}(\mathcal{G}^\wedge) \\ w \neq \infty}} \arctan \frac{\rho_w}{\rho_{v_i}} - \frac{\alpha_i}{2} \text{ for } i = 1, \ldots, k.$$

The quantity

$$\mu(\rho) = \sum_{v \in \mathcal{V}(\mathcal{G}')} \vartheta_v(\rho)^2$$

is a measure for the deviation of the list ρ from a list as postulated in (9). For the proof of (9), we have to find a list ρ of positive numbers such that $\mu(\rho) = 0$.

First, the following equality will be shown:

(10) $\sum_{v \in \mathcal{V}(\mathcal{G}')} \vartheta_v(\rho) = 0$ for any list $\rho = (\rho_v)$ of positive numbers.

To see (10), note that

(11) $$\sum_{v \in \mathcal{V}(\mathcal{G}')} \vartheta_v(\rho) = \sum_{vw \in \mathcal{E}(\mathcal{G}')} \arctan \frac{\rho_w}{\rho_v} + \arctan \frac{\rho_v}{\rho_w} - \pi \left(\#\mathcal{V}(\mathcal{G}') - k\right) - \frac{1}{2} \sum_{i=1}^{k} \alpha_i .$$

Since all countries of $\mathcal{G}^\wedge$ are quadrangles, it follows from Corollary 15.1 that

(12) $2\#\mathcal{V}(\mathcal{G}') = \#\mathcal{E}(\mathcal{G}^\wedge) + 2 = \#\mathcal{E}(\mathcal{G}') + k + 2.$

Noting that $\arctan t + \arctan(1/t) = \pi/2$ for $t > 0$, Propositions (11) and (12) imply the equality (10).

Let

$$\emptyset \neq S \subsetneq \mathcal{V}(\mathcal{G}'),\ l = \#\{v_1, \ldots, v_k\} \cap S.$$

Clearly, $\#\big(S \cup \{\infty\}\big) = \#(S) + 1$ and $\#\mathcal{E}\big(\mathcal{G}^\wedge(S \cup \{\infty\})\big) = \#\mathcal{E}\big(\mathcal{G}^\wedge(S)\big) + l$. An application of (6) to $S \cup \{\infty\}$ instead of S (for $\#S \geq 4$ and $l \geq 0$) and simple arguments (for $\#S = 2, 3$ and $l = 0$) show that

(13) $2\#S - \#\mathcal{E}\big(\mathcal{G}^\wedge(S)\big) \geq l + 3$ for $\#S \geq 4,\ l \geq 0$ or $\#S = 2, 3,\ l = 0$.

Simple arguments also yield the following:

(14) $2\#S - \#\mathcal{E}\big(\mathcal{G}^\wedge(S)\big) \geq l + 2$ for $\#S = 2, 3,\ l > 0$.

Second, we show that the quantity $\mu(\cdot)$ attains its minimum. Let R be the set of all lists $\rho = \big(\rho_v : v \in \mathcal{V}(\mathcal{G}')\big)$, where $0 < \rho_v \leq 1$ and such that $\rho_v = 1$ if $\vartheta_v(\rho) > 0$. In addition, we require that $\rho_v = 1$ for at least one v. We have $R \neq \emptyset$, since the list (1) $\in R$. Unfortunately, R is not compact. Let $\big(\rho^{(n)}\big)$ be a sequence of lists in R such that $\mu\big(\rho^{(n)}\big)$ tends to the infimum of $\mu(\rho)$ for $\rho \in R$. Apply the Bolzano–Weierstrass theorem to see that, by considering a suitable subsequence and renumbering, if necessary, the sequence $\big(\rho_v^{(n)}\big)$ converges for each $v \in \mathcal{V}(\mathcal{G}')$. Let

$$S = \{v \in \mathcal{V}(\mathcal{G}') : \lim_{n\to\infty} \rho_v^{(n)} \neq 0\}.$$

We will show that

(15) $S = \mathcal{V}(\mathcal{G}')$.

For suppose not. Then, by a calculation similar to the one that led to (11), we see that

$$(16)\ \sum_{v\in S} \vartheta_v(\rho^{(n)}) = \frac{\pi}{2}\#\mathcal{E}(\mathcal{G}^\wedge(S)) - \pi(\#S - l) - \frac{1}{2}\sum_{v_i\in S}\alpha_i + \sum_{\substack{vw\in\mathcal{E}(\mathcal{G}')\\ v\in S, w\notin S}} \arctan\frac{\rho_w^{(n)}}{\rho_v^{(n)}}.$$

By the definition of S, the last sum tends to 0 as $n \to \infty$. Thus

$$(17)\ \sum_{v\in S} \vartheta_v(\rho^{(n)}) \to -\frac{\pi}{2}(2\#S - \#\mathcal{E}(\mathcal{G}^\wedge(S)) - l - 2) + \frac{1}{2}\sum_{v_i\in S}(\pi - \alpha_i) - \pi.$$

Since $\pi - \alpha_i > 0$ for $i = 1, \dots, k$, and $\sum(\pi - \alpha_i) = 2\pi$, it follows from (17) that, in case when (13) holds, $\sum\{\vartheta_v(\rho^{(n)}) : v \in S\} < 0$ for all sufficiently large n. The same is true if (14) holds with inequality. If (14) holds with equality, it is also true since then $l < k$. (To see the latter suppose that $l = k$. Since $k \geq 3$ and $\#S \geq l$ it then follows that $l = k = \#S = 3$ and thus $S = \{v_1, v_2, v_3\}$, $\#\mathcal{E}(\mathcal{G}^\wedge(S)) = 0$.) The remaining case, $\#S = 1$, trivially yields the same conclusion. Taking into account (10), we have

$$\sum_{v\notin S} \vartheta_v(\rho^{(n)}) > 0 \text{ for all sufficiently large } n.$$

By considering a suitable subsequence and renumbering, if necessary, it follows that there is a vertex $v \notin S$ with $\vartheta_v(\rho^{(n)}) > 0$ for all n. Then $\rho_v^{(n)} = 1$ for all n by the definition of R. Hence $v \in S$ by the definition of S. This contradiction concludes the proof of (15).

Let $\rho = \lim_{n\to\infty} \rho^{(n)}$. Since the $\vartheta_v(\cdot)$ are continuous, $\rho \in R$. We now show that

(18) $\mu(\rho) = 0$.

For, suppose that, on the contrary, $\mu(\rho) > 0$. Then by (10),

$$\emptyset \neq T = \{v \in \mathcal{V}(\mathcal{G}') : \vartheta_v(\rho) < 0\} \subsetneq \mathcal{V}(\mathcal{G}').$$

Let $\sigma_v = \lambda\rho_v$ for $v \in T$ and $\sigma_v = \rho_v$ for $v \notin T$ and given λ, where $0 < \lambda < 1$. Then

(19) $\vartheta_v(\rho) \leq \vartheta_v(\sigma)$ for each $v \in T$

by the definition of $\vartheta_v(\cdot)$ and the fact that arctan is non-decreasing. Since $\mathcal{G}'$ is connected, there is a $v \in T$ having one or more neighbours $w \in \mathcal{V}(\mathcal{G}')$ with $w \notin T$. Thus, by Taylor's theorem, we have the claim:

(20) There is $v \in T$ such that

$$\begin{aligned}
\vartheta_v(\sigma) &= \dots + \arctan \frac{\rho_w}{\lambda \rho_v} + \dots \\
&= \dots + \arctan \frac{\rho_w}{\rho_v} + \dots + \text{const}\left(\frac{1}{\lambda} - 1\right) + o\left(\frac{1}{\lambda} - 1\right) \\
&= \vartheta_v(\rho) + \text{const}(1-\lambda) + o(1-\lambda) \text{ as } \lambda \to 1-0,
\end{aligned}$$

where const denotes a positive constant. Similarly,

(21) $\vartheta_v(\rho) \geq \vartheta_v(\sigma)$ for each $v \notin T$

and an application of Taylor's theorem shows that the following hold:

(22) For each $v \notin T$ with $\vartheta_v(\rho) = 0$, $\vartheta_v(\sigma) = \vartheta_v(\rho) + o(1-\lambda)$ as $\lambda \to 1-0$.

For all λ sufficiently close to 1, we have $\sigma \in R$ (see the definition of R) and Propositions (19)–(22) show that

$$\begin{aligned}
\mu(\rho) - \mu(\sigma) &= \sum_{v \in \mathcal{V}(\mathcal{G}')} \vartheta_v(\rho)^2 - \vartheta_v(\sigma)^2 \\
&= \sum_{v \in T} \big(\vartheta_v(\rho) + \vartheta_v(\sigma)\big)\big(\vartheta_v(\rho) - \vartheta_v(\sigma)\big) - \sum_{\substack{v \notin T \\ \vartheta_v(\rho)=0}} \vartheta_v(\sigma)^2 \\
&\quad + \sum_{\substack{v \notin T \\ \vartheta_v(\rho)>0}} \big(\vartheta_v(\rho) + \vartheta_v(\sigma)\big)\big(\vartheta_v(\rho) - \vartheta_v(\sigma)\big) > 0.
\end{aligned}$$

This contradicts the minimality of $\mu(\rho)$, thus concluding the proof of (18) which, in turn, implies (9).

Having proved (1), (5) and (9), the proof of the theorem is complete. □

34.2 Thurston's Algorithm and the Riemann Mapping Theorem

In view of the numerous applications of the Koebe-Andreev-Thurston-Bright-well-Scheinerman theorem, the problem arises to construct the disc packings in an effective way. The first procedure to serve this purpose for graphs with triangular countries was proposed by Thurston [1000]. It is called *Thurston's algorithm*, although it is not an algorithm in the strict sense which is always finite. Rodin and Sullivan [844] and Colin de Verdière [214, 215] showed its convergence. See also Collins and Stephenson [216]. Mohar [747] considered the more general case of primal-dual circle packings and gave a polynomial time algorithm.

In the following we first describe the algorithm of Thurston, following Rodin and Sullivan. Then the relation between packing of discs and the Riemann mapping theorem will be described. No proofs are given. For the figures we are indebted to Kenneth Stephenson [968].

For more information, see the references given in the introduction of Sect. 34.

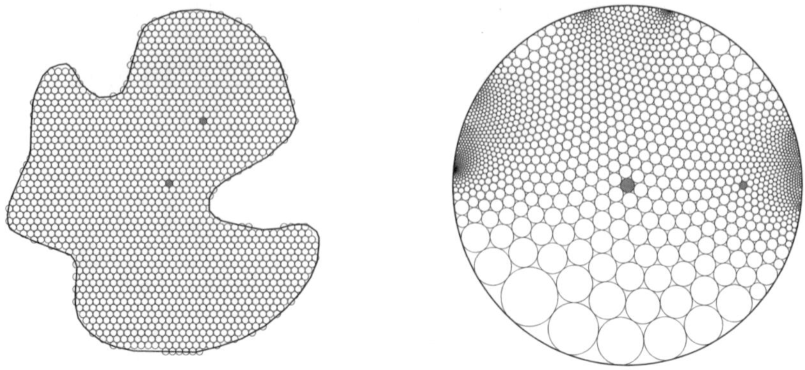

Fig. 34.6. Circle packing corresponding to a graph

Thurston's Algorithm

We need the following special case of Theorem 34.1:

(1) Let $\mathcal{G}$ be a planar, 3-connected graph all countries of which are triangular, including the exterior country. Then to $\mathcal{G}$ corresponds a packing of (circular) discs in $\mathbb{C}$. The discs are in one-to-one correspondence with the vertices of $\mathcal{G}$ and two discs touch precisely when the corresponding vertices are connected by an edge (Fig. 34.6).

Let $v_1, \dots, v_n$ be the vertices of $\mathcal{G}$ where v_1, v_2, v_3 are the vertices of the exterior country. For the construction of the corresponding packing of discs, it is sufficient to specify the radii $\varrho_1, \dots, \varrho_n$ of the discs corresponding to $v_1, \dots, v_n$ (up to Möbius transformations). The algorithm can be described as follows:

For $k = 1, 2, \dots,$ assign to $v_1, \dots, v_n$ labels $\varrho_{k1}, \dots, \varrho_{kn}$ in the following way: Let $\varrho_{11} = \varrho_{12} = \varrho_{13} = 1$ and $\varrho_{14}, \dots, \varrho_{1n} > 0$ arbitrary. Now cycle through $v_4, \dots, v_n, v_{n+1} = v_4$. Assume that at step k we have arrived at vertex v_{i-1}. If $v_i v_j v_k$ is a (triangular) country of $\mathcal{G}$, its *angle* at the vertex v_i is the angle of the Euclidean triangle with edges of lengths $\varrho_{ki} + \varrho_{kj}, \varrho_{ki} + \varrho_{kl}, \varrho_{kj} + \varrho_{kl}$ opposite to the edge of length $\varrho_{kj} + \varrho_{kl}$. The *curvature* $\kappa(v_i, \varrho_{k1}, \dots, \varrho_{kn})$ at v_i is 2π minus the sum of the angles at v_i of the triangular countries with vertex v_i. Considered as a function of $\varrho = \varrho_{ki}$, the curvature is strictly increasing. If $\varrho \to +\infty$, the curvature at v_i tends to 2π. v_i is a vertex of at least three countries. (Since the cycle $v_1 v_2 v_3$ is the exterior cycle of the graph $\mathcal{G}$, the vertex $v_i \neq v_1, v_2, v_3$ is surrounded by countries of $\mathcal{G}$. Since $\mathcal{G}$ is 3-connected, v_i cannot be a vertex of only one or two countries.) Thus, for $\varrho = 0$, the curvature at v_i is at most $-\pi$. Hence there is a unique number $\varrho > 0$ for which it vanishes. Now let $\varrho_{k+1,1} = \varrho_{k+1,2} = \varrho_{k+1,3} = 1, \varrho_{k+1j} = \varrho_{kj}$ for $j = 4, \dots, n,\ j \neq i$ and $\varrho_{k+1i} = \varrho$ and go on to vertex v_{i+1}.

It turns out that $\varrho_{k1} \to \varrho_1, \ldots, \varrho_{kn} \to \varrho_n$, with suitable numbers $\varrho_1 = \varrho_2 = \varrho_3 = 1, \varrho_4, \ldots, \varrho_m > 0$, and such that the curvatures $\kappa(v_i, \varrho_1, \ldots, \varrho_n)$ vanish for $i = 4, \ldots, n$. This, in turn, yields a packing of discs corresponding to $\mathcal{G}$ as in (1).

The algorithm of Thurston is quite effective in practice.

The Riemann Mapping Theorem

Let J be a bounded, simply connected domain in $\mathbb{C}$. By the Riemann mapping theorem there is an analytic function f which maps R in a one-to-one fashion onto int D, where D is the unit disc of $\mathbb{C}$. The problem arises to approximate f by simple functions.

For sufficiently small $\varepsilon > 0$ consider the common regular hexagonal grid in $\mathbb{C}$ of mesh-length ε. Let $\mathcal{G}$ be a part of it which almost exhausts J and is bounded by a closed Jordan polygon. $\mathcal{G}$ may be considered as a finite graph with triangular countries. Add an exterior point as a new vertex to $\mathcal{G}$ and connect it with polygonal

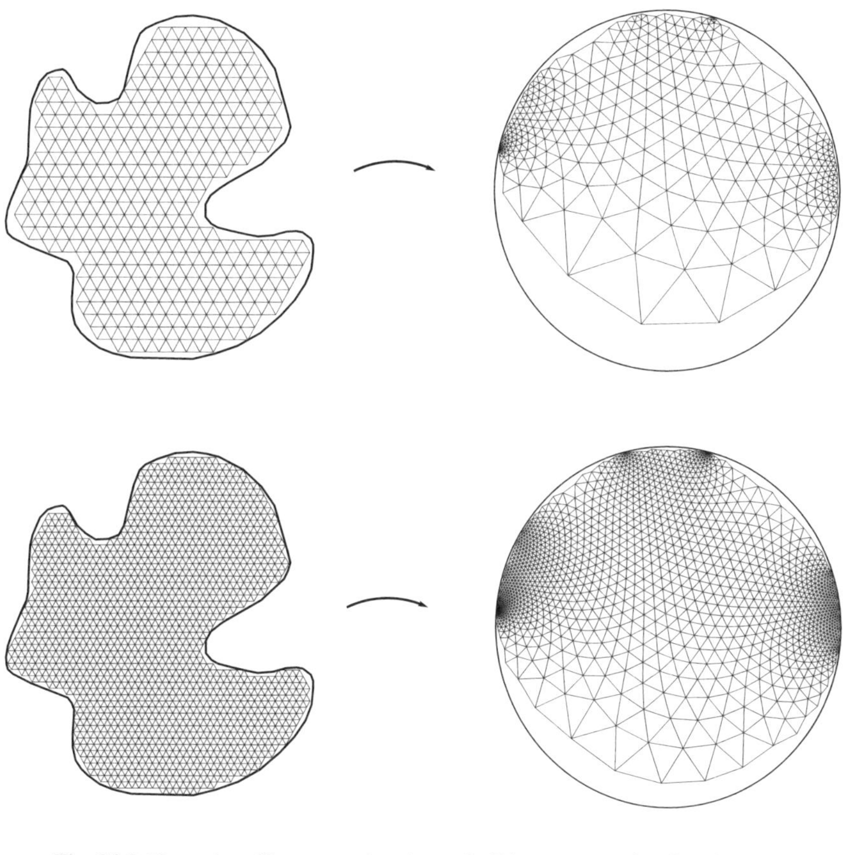

Fig. 34.7. Piecewise affine approximations of a Riemann mapping function

curves to each of the boundary vertices of $\mathcal{G}$ so as to get a new planar graph. This new graph has only triangular countries. According to (1), construct a corresponding packing of discs in $\mathbb{C}$. After a suitable Möbius transformation has been applied, we may assume that the new vertex corresponds to the exterior of D, all other discs form a packing in D and the discs which correspond to the boundary vertices of $\mathcal{G}$ touch the boundary of D (Fig. 34.7). Now define a map f_ε as follows: f_ε maps each vertex of $\mathcal{G}$ onto the centre of the corresponding disc. Then extend f_ε affinely to the triangular countries of $\mathcal{G}$. This gives a piecewise affine mapping of the union of the countries of $\mathcal{G}$, excluding the exterior country, into int D. The six circle conjecture of Fejes Tóth proved by Bárány, Füredi and Pach [70], or the hexagonal packing lemma of Rodin and Sullivan [844] proved by He [484], shows that, for small $\varepsilon > 0$, triangles deep in the interior of the graph $\mathcal{G}$ are mapped onto almost regular triangles. This, in turn, shows that the function f_ε is almost analytic deep in the interior of J.

As $\varepsilon \to +0$, the functions f_ε (possibly after suitable Möbius transformations are applied) converge to a Riemann mapping function $f : J \to \text{int}\, D$.

References

1. Achatz, H., Kleinschmidt, P., Reconstructing a simple polytope from its graph, in: Polytopes – combinatorics and computation (Oberwolfach 1997) 155–165, DMV Sem. **29**, Birkhäuser, Basel 2000
2. Agrachev, A.A., Sachkov, Y.L., Control theory from the geometric viewpoint, Springer-Verlag, Berlin 2004
3. Aharonov, D., Schiffer, M.M., Zalcman, L., Potato kugel, Israel J. Math. **40** (1981) 331–339
4. Ahmedov, N.S., The representation of square matrices in the form of a product of a diagonal, an orthogonal, a triangular and an integral unimodular matrix (in connection with the inhomogeneous Minkowski conjecture), in: Studies in number theory (LOMI) **4**, Zap. Naučn. Sem. Leningrad. Otdel. Mat. Inst. Steklov. (LOMI) **67** (1977) 86–94, 2, J. Soviet Math. **16** (1981) 820–826
5. Aichholzer, O., Alt, H., Rote, G., Matching shapes with a reference point, in: Tenth Annual ACM Symposium on Computational Geometry (Stony Brook, NY 1994), Internat. J. Comput. Geom. Appl. **7** (1997) 349–363
6. Aigner, M., Ziegler, G.M., Proofs from The Book. Including illustrations by K.H. Hofmann, 3rd ed., Springer-Verlag, Berlin 2004
7. Alesker, S., Continuous rotation invariant valuations on convex sets, Ann. of Math. (2) **149** (1999) 977–1005
8. Alesker, S., Description of translation invariant valuations on convex sets with solution of P. McMullen's conjecture, Geom. Funct. Anal. **11** (2001) 244–272
9. Alexandroff (Alexandrov), P., Hopf, H., Topologie **I**, Springer-Verlag, Berlin 1935, Reprints with corrections and addenda, Chelsea, Bronx, NY 1972, Springer-Verlag, Berlin 1974
10. Alexandrov (Aleksandrow, Alexandrow), A.D., To the theory of mixed volumes of convex bodies I. Extension of certain concepts of the theory of convex bodies, Mat. Sb. (N.S.) **2** (1937) 947–972, Selected works **I**, 31–59
11. Alexandrov, A.D., To the theory of mixed volumes of convex bodies II. New inequalities between mixed volumes and their applications, Mat. Sb. (N.S.) **2** (1937) 1205–1238, Selected works **I**, 61–97
12. Alexandrov, A.D., To the theory of mixed volumes of convex bodies III. Extension of two Minkowski theorems on convex polyhedra to all convex bodies, Mat. Sb. (N.S.) **3** (1938) 27–46, Selected works **I**, 99–117

13. Alexandrov, A.D., To the theory of mixed volumes of convex bodies IV. Mixed discriminants and mixed volumes, Mat. Sb. (N.S.) **3** (1938) 227–251, Selected works **I**, 119–148
14. Alexandrov, A.D., Almost everywhere existence of the second differential of a convex function and some properties of convex surfaces connected with it, Uchen. Zap. Leningrad. Gos. Univ. Mat. Ser. **6** (1939) 3–35
15. Alexandrov, A.D., Intrinsic geometry of convex surfaces, Gosudarstv. Izdat. Tehn.-Teor. Lit., Moscow-Leningrad 1948, Akademie-Verlag, Berlin 1955, Taylor & Francis, London 2003
16. Alexandrov, A.D., Convex polyhedra, Gosudarstv. Izdat. Tehn.-Teor. Lit., Moscow-Leningrad 1950, Akademie-Verlag, Berlin 1958, Springer-Verlag, Berlin 2005
17. Alexandrov, A.D., On filling of space by polytopes, Vestnik Leningrad. Univ. Ser. Mat. Fiz. Him. **9** (1954) 33–44, Selected works **I**, 175–185
18. Alexandrov, A.D., Selected works, Part I: Selected scientific papers, eds.: Yu.G. Reshetnyak, S.S. Kutateladze, Gordon and Breach, Amsterdam 1996
19. Alexandrov, A.D., Selected works, Part II: Intrinsic geometry of convex surfaces, ed.: S.S.Kutateladze, Taylor & Francis, New York, CRC Press, Boca Raton, Fl 2005
20. Alexandrov (Aleksandrov) A.D., Zalgaller, V.A., Intrinsic geometry of surfaces, Izdat. Akad. Nauk SSSR, Moskwa 1962, Amer. Math. Soc., Providenc, RI 1967
21. Alfsen, E.M., Compact convex sets and boundary integrals, Springer-Verlag, Berlin 1971
22. Aliev, I., Personal communication 2004
23. Aliev, I., Gruber, P.M., Lattice points in large Borel sets and successive minima, Discrete Comput. Geom. **35** (2006) 429–435
24. Almgren, F., Taylor, J.E., Flat flow is motion by crystalline curvature for curves with crystalline energies, J. Differential Geom. **42** (1995) 1–22
25. Alon, N., The number of polytopes, configurations and real matroids, Mathematika **33** (1986) 62–71
26. Ambartzumian, R.V., Factorization calculus and geometric probability, Cambridge Univ. Press, Cambridge 1990
27. Amir, D., Characterizations of inner product spaces, Birkhäuser, Basel 1986
28. Anderson, R.D., Klee, V., Convex functions and upper-semicontinuous collections, Duke Math. J. **19** (1952) 349–357
29. Andreev, E.M., On convex polyhedra in Lobačevskiĭ spaces, Mat. Sb. (N.S.) **81** (123) (1970) 445–478, Math. USSR Sb. **10** (1970) 413-440
30. Andreev, E.M., On convex polyhedra of finite volume in Lobačevskiĭ space, Mat. Sb. (N.S.) **83** (125) (1970) 256–260, Math. USSR Sb. **12** (1970) 255–259
31. Andrews, B., Entropy estimates for evolving hypersurfaces, Comm. Anal. Geom. **2** (1994) 53–64
32. Andrews, B., Gauss curvature flow: the fate of the rolling stones, Invent. Math. **138** (1999) 151–161
33. Andrews, B., Motion of hypersurfaces by Gauss curvature, Pacific J. Math. **195** (2000) 1–34
34. Andriyasyan, A.K., Il'in, I.V., Malyshev, A.V., Application of computers to estimates of Chebotarev type in the nonhomogeneous Minkowski conjecture, in: Issled. Teor. Chisel. **9** 7–25, 195, Zap. Nauchn. Sem. Leningrad. Otdel. Mat. Inst. Steklov. (LOMI) **151** (1986), J. Soviet Math. **43** (1988) 2624–2635
35. Archimedes, Sphere and cylinder. On paraboloids, hyperboloids and ellipsoids, transl. and annot. by A. Czwalina-Allenstein, reprint of 1922, 1923, Geest & Portig, Leipzig 1987, The works of Archimedes, ed. T.L.Heath, Cambridge Univ. Press, Cambridge 1897

36. Arias de Reyna, J., Ball, K., Villa, R., Concentration of the distance in finite-dimensional normed spaces, Mathematika **45** (1998) 245–252
37. Aristotle, Problemata, Book 15, Problem 6
38. Arnold, V.I., Mathematical methods of classical mechanics, Springer-Verlag, New York-Heidelberg 1978
39. Artin, E., The gamma function, Holt, Rinehart and Winston, New York 1964
40. Artstein, Z., Yet another proof of the Lyapunov convexity theorem, Proc. Amer. Math. Soc. **108** (1990) 89–91
41. Asimow, L., Roth, B., The rigidity of graphs, Trans. Amer. Math. Soc. **245** (1978) 279–289
42. Aurenhammer, F., Klein, R., Voronoi diagrams, in: Handbook of computational geometry 201–290, North-Holland, Amsterdam 2000
43. Auslander, L., MacKenzie, R.E., Introduction to differentiable manifolds, Corrected reprint, Dover, New York 1977
44. Babai, L., On Lovász' lattice reduction and the nearest lattice point problem, Combinatorica **6** (1986) 1–13
45. Baire, R., Sur les fonctions de variables réelles, Ann. Mat. Pura Appl. (3) **3** (1899) 1–123
46. Balinski, M.L., On the graph structure of convex polyhedra in n-space. Pacific, J. Math. II (1961) 431–434
47. Ball, K.M., Isometric problems in l_p and sections of convex sets, PhD thesis, Cambridge 1986
48. Ball, K.M., Some remarks on the geometry of convex sets, in: Geometric aspects of functional analysis (1986/87) 224–231, Lecture Notes Math. **1317**, Springer-Verlag, Berlin 1988
49. Ball, K.M., Volumes of sections of cubes and related problems, in: Geometric aspects of functional analysis (1987-88) 251–260, Lect. Notes Math. **1376**, Springer-Verlag, Berlin 1989
50. Ball, K.M., Volume ratios and a reverse isoperimetric inequality, J. London Math. Soc. **44** (1991) 351–359
51. Ball, K.M., Ellipsoids of maximal volume in convex bodies, Geom. Dedicata **41** (1992) 241–250
52. Ball, K.M., A lower bound for the optimal density of lattice packings, Internat. Math. Res. Notices **1992** 217–221
53. Ball, K.M., An elementary introduction to modern convex geometry, Cambridge Univ. Press, Cambridge 1997
54. Ball, K.M., Convex geometry and functional analysis, in: Handbook of the geometry of Banach spaces **1** 161–194, North-Holland, Amsterdam 2001
55. Ball, K., Personal communication 2003
56. Bambah, R.P., Packings and coverings, Indian J. Pure Appl. Math. **30** (1999) 1063–1072
57. Bambah, R.P., Dumir, V.C., Hans-Gill (Hans), R.J., Diophantine inequalities, Proc. Nat. Acad. Sci. India Sect. A **68** (1998) 101–114
58. Bambah, R.P., Dumir,V.C., Hans-Gill, R.J., Non-homogeneous problems: conjectures of Minkowski and Watson, in: R.P.Bambah, V.C.Dumir, R.J.Hans-Gill, eds., Number theory 15–41, Birkhäuser, Basel 2000
59. Bambah, R.P., Rogers, C.A., Covering the plane with convex sets, J. London Math. Soc. **27** (1952) 304–314
60. Bambah, R.P., Woods, A.C., On a theorem of Dyson, J. Number Theory **6** (1974) 422–433

61. Bambah, R.P., Woods, A.C., On the product of three inhomogeneous linear forms, in: Number theory and algebra 7–18, Academic Press, New York 1977
62. Bambah, R.P., Woods, A.C., Minkowski's conjecture for $n = 5$; a theorem of Skubenko, J. Number Theory **12** (1980) 27–48
63. Bambah, R.P., Woods, A.C., Zassenhaus, H., Three proofs of Minkowski's second inequality in the geometry of numbers, J. Austral. Math. Soc. **5** (1965) 453–462
64. Bandle, C., Isoperimetric inequalities and applications, Pitman, Boston 1980
65. Bangert, V., On the existence of closed geodesics on two-spheres, Internat. J. Math. **4** (1993) 1–10
66. Bandt, C., Baraki, G., Metrically invariant measures on locally homogeneous spaces and hyperspaces, Pacific J. Math. **121** (1986) 13–28
67. Bárány, I., Random polytopes in smooth convex bodies, Mathematika **39** (1992) 81–92
68. Bárány, I., Approximation by random polytopes is almost optimal, Rend. Circ. Mat. Palermo (2), Suppl. **50** (1997) 43–50
69. Bárány, I., Buchta, C., Random polytopes in a convex polytope, independence of shape, and concentration of vertices, Math. Ann. **297** (1993) 467–497
70. Bárány, I., Füredi, Z., Pach, J., Discrete convex functions and proof of the six circle conjecture of Fejes Tóth, Canad. J. Math. **36** (1984) 569–576
71. Barnes, E.S., The perfect and extreme senary forms, Canad. J. Math. **9** (1957) 235–242
72. Barnes, E.S., The complete enumeration of extreme senary forms, Philos. Trans. Roy. Soc. London Ser. A. **249** (1957) 461–506
73. Barnette, D.W., The minimum number of vertices of a simple polytope, Israel J. Math. **10** (1971) 121–125
74. Barnette, D.W., A proof of the lower bound conjecture for convex polytopes, Pacific J. Math. **46** (1973) 349–354
75. Bartels, S.G., The complexity of Yamnitsky and Levin's simplices algorithm, in: Polytopes—combinatorics and computation (Oberwolfach 1997) 199–225, DMV Sem. **29**, Birkhäuser, Basel 2000
76. Barthe, F., On a reverse form of the Brascamp-Lieb inequality, Invent. Math. **134** (1998) 335–361
77. Barthe, F., Autour de l'inégalité de Brunn-Minkowski, Ann. Fac. Sci. Toulouse (VI), Math. **12** (2003) 127–178
78. Barvinok, A., Computing the Ehrhart polynomial of a convex lattice polytope, Discrete Comput. Geom. **12** (1994) 35–48
79. Barvinok, A., A polynomial time algorithm for counting integral points in polyhedra when the dimension is fixed, Math. Oper. Res. **19** (1994) 769–779
80. Barvinok, A., A course in convexity, Amer. Math. Soc., Providence, RI 2002
81. Barvinok, A., Lattice points and lattice polytopes, in: Handbook of discrete and computational geometry, 2nd ed., 133–152, CRC Press, Boca Raton, FL 2004
82. Bauer, H., Measure and integration theory, de Gruyter, Berlin 2001
83. Bayer, M.M., Lee, C.W., Combinatorial aspects of convex polytopes, in: Handbook of convex geometry **A** 485–534, North-Holland, Amsterdam 1993
84. Bayer-Fluckiger, E., Nebe, G., On the Euclidean minimum of some real number fields, preprint 2004
85. Beck, M., De Loera, J.A., Develin, M., Pfeifle, J., Stanley, R.P., Coefficients and roots of Ehrhart polynomials, in: Integer points in polyhedra – geometry, number theory, algebra, optimization, Amer. Math. Soc., Providence, RLI 2005
86. Beck, M., Robins, S., Computing the continuous discretely, Springer, New York 2007
87. Beckenbach, E.F., Convex functions, Bull. Amer. Math. Soc. **54** (1948) 439–460

88. Beer, G., Topologies on closed and closed convex sets, Kluwer, Dordrecht 1993
89. Behrend, F., Über einige Affinvarianten konvexer Bereiche, Math. Ann. **113** (1937) 713–747
90. Behrend, F., Über die kleinste umbeschriebene und die grösste einbeschriebene Ellipse eines konvexen Bereichs, Math. Ann. **115** (1938) 397–411
91. Belhoste, B., Augustin–Louis Cauchy, Springer-Verlag, New York 1991
92. Bender, E.A., The number of three-dimensional convex polyhedra, Amer. Math. Monthly **94** (1987) 7–21
93. Bender, E.A., Richmond, L.B., An asymptotic expansion for the coefficients of some power series. II. Lagrange inversion, Discrete Math. **50** (1984) 135–141
94. Bender, E.A., Wormald, N.C., Almost all convex polyhedra are asymmetric, Canad. J. Math. **37** (1985) 854–871
95. Beneš, V., Rataj, J., Stochastic geometry: selected topics, Kluwer, Boston, MA 2004
96. Benson, R.V., Euclidean geometry and convexity, McGraw-Hill, New York 1966
97. Benyamini, Y., Lindenstrauss, J., Geometric nonlinear functional analysis **1**, Amer. Math, Soc., Providence, RI 2000
98. Berger, M., Seules les quadriques admettent des caustiques, Bull. Soc. Math. France **123** (1995) 107–116.
99. Berkovitz, L.D., Convexity and optimization in $\mathbb{R}^n$, Wiley-Interscience, New York 2002
100. Bernstein, D.N., The number of roots of a system of equations, Funkcional. Anal. i Priložen. **9** (1975) 1–4, Funct. Anal. Appl. **9** (1975) 183–185
101. Bernstein, D.N., The number of lattice points in integer polyhedra, Funkcional. Anal. i Priložen. **10** (1976) 72–73, Funct. Anal. Appl. **10** (1976) 223–224
102. Bertrand, D., Transcendental methods in arithmetic geometry, in: Analytic number theory (Tokyo 1988) 31–44, Lect. Notes Math. **1434**, Springer-Verlag, Berlin 1990
103. Bertraneu, A., Fichet, B., Étude de la frontière de l'ensemble des formes quadratiques positives, sur un espace vectoriel de dimension finie, J. Math. Pures Appl. (9) **61** (1982) 207–218 (1983)
104. Bertsimas, D., Weismantel, R., Optimization over integers, Dynamic ideas, Belmont MA 2005, in print
105. Betke, U., Gitterpunkte und Gitterpunktfunktionale, Habilitationsschrift, U. Siegen 1979
106. Betke, U., Das Einschließungs-Ausschließungsprinzip für Gitterpolytope, unpublished manuscript 1984
107. Betke, U., Henk, M., Densest lattice packings of 3-polytopes, Comput. Geom. **16** (2000) 157–186
108. Betke, U., Kneser, M., Zerlegungen und Bewertungen von Gitterpolytopen, J. reine angew. Math. **358** (1985) 202–208
109. Bezdek, A., Kuperberg, W., Maximum density space packing with congruent circular cylinders of infinite length, Mathematika **37** (1990) 74–80
110. Bezdek, A., Kuperberg, W., Packing Euclidean space with congruent cylinders and with congruent ellipsoids, in: Applied geometry and discrete mathematics 71–80, DIMACS Ser. Discrete Math. Theoret. Comput. Sci. **4**, Amer. Math. Soc., Providence, RI 1991
111. Bezdek, K., Improving Rogers' upper bound for the density of unit ball packings via estimating the surface area of Voronoi cells from below in Euclidean d-space for all $d \geq 8$, Discrete Comput. Geom. **28** (2002) 75–106
112. Bianchi, G., Gronchi, P., Steiner symmetrals and their distance from a ball, Israel J. Math. **135** (2003) 181–192

113. Bieberbach, L., Über eine Extremaleigenschaft des Kreises, J.-Ber. Deutsch. Math. Verein. **24** (1915) 247–250
114. Bigalke, H.-G., Charakterisierungen der Kugel im $\mathbb{E}^3$, in: Jahrbuch Überblicke Mathematik 1986, Bibl. Inst., Mannheim 1986
115. Billera, L.J., Björner, A., Face numbers of polytopes and complexes, in: Handbook of discrete and computational geometry 291–310, CRC Press, Boca Raton, FL 1997
116. Billera, L.J., Lee, C.W., A proof of the sufficiency of McMullen's conditions for f-vectors of simplicial convex polytopes, J. Combin. Theory Ser. A **31** (1981) 237–255
117. Birch, B.J., Swinnerton-Dyer, H.P.F., On the inhomogeneous minimum of the product of n linear forms, Mathematika **3** (1956) 25–39
118. Birkhoff, G., Orthogonality in linear metric spaces, Duke Math. J. **1** (1935) 169–175
119. Birkhoff, G., Tres observaciones sobre el algebra lineal, Univ. Nac. Tucumán Rev. Ser. **A 5** (1946) 147–150
120. Birkhoff, G.D., Dynamical systems. With an addendum by J. Moser, Amer. Math. Soc., Providence, R.I. 1966
121. Björner, A., The homology and shellability of matroids and geometric lattices, in: Matroid applications 226–283, Cambridge Univ. Press, Cambridge 1992
122. Björner, A., Las Vergnas, M., Sturmfels, B., White, N., Ziegler, G.M., Oriented matroids, 2nd ed., Cambridge Univ. Press, Cambridge 1999
123. Blaschke, W., Räumliche Variationsprobleme mit symmetrischer Transversalitätsbedingung, Ber. Verh. Sächs. Akad. Wiss. Leipzig, Math.-Phys. Kl. **68** (1916) 50–55, Ges. Werke **3** 167–173
124. Blaschke, W., Kreis und Kugel, Veit, Leipzig 1916, Chelsea, New York 1948, 1971, de Gruyter, Berlin 1956
125. Blaschke, W., Eine Frage über konvexe Körper, Jahresber. Deutsch. Math.-Verein. **25** (1916) 121–125, Ges. Werke **3** 302–306
126. Blaschke, W., Über affine Geometrie VII: Neue Extremeigenschaften von Ellipse und Ellipsoid, Ber. Verh. Sächs. Akad. Wiss. Leipzig, Math.-Phys. Kl. **69** (1917) 306–318, Ges. Werke **3** 246–258
127. Blaschke, W., Integralgeometrie 13: Zur Kinematik, Math. Z. **41** (1936) 465–479
128. Blaschke, W., Vorlesungen über Integralgeometrie **1, 2**, Hamburger Math. Einzelschr. **22, 20**, Leipzig, Teubner 1935, 1936, 1937, Chelsea, New York 1949, Deutsch. Verl. Wiss., Berlin 1955
129. Blaschke, W., Gesammelte Werke **2** (Kinematik und Integralgeometrie), **3** (Konvexgeometrie), **4** (Affine Differentialgeometrie), eds.: W. Burau, S.S. Chern, K. Leichtweiss, H.R. Müller, L.A. Santaló, U. Simon, K. Strubecker, Thales-Verlag, Essen 1985
130. Blichfeldt, H.F., A new principle in the geometry of numbers with some applications, Trans. Amer. Math. Soc. **15** (1914) 227–235
131. Blichfeldt, H.F., The minimum value of quadratic forms and the closest packing of spheres, Math. Ann. **101** (1929) 605–608
132. Blichfeldt, H.F., The minimum values of positive quadratic forms in six, seven and eight variables, Math. Z. **39** (1934) 1–15
133. Blind, R., Mani-Levitska, P., Puzzles and polytope isomorphisms, Aequationes Math. **34** (1987) 287–297
134. Blömer, J., Closest vectors, successive minima, and dual HKZ-bases of lattices, in: Automata, languages and programming (Geneva 2000) 248–259, Lect. Notes Comput. Sci. **1853**, Springer-Verlag, Berlin 2000
135. Böhm, J., Hertel, E., Polyedergeometrie in n-dimensionalen Räumen konstanter Krümmung, Deutsch. Verl. Wiss., Berlin 1980, Birkhäuser, Basel 1981

136. Bohr, H., Mollerup, J., Laerebog i matematisk analyse **3**, Gjellerup, København 1922, 2nd ed. 1940
137. Bokowski, J., Computational oriented matroids, Cambridge Univ. Press, Cambridge 2006
138. Bokowski, J., Hadwiger, H., Wills, J.M., Eine Ungleichung zwischen Volumen, Oberfläche und Gitterpunktanzahl konvexer Körper im n-dimensionalen euklidischen Raum, Math. Z. **127** (1972) 363–364
139. Bol, G., Beweis einer Vermutung von H. Minkowski, Abh. Math. Sem. Hansischen Univ. **15** (1943) 37–56
140. Bol, G., Über Auswahlsätze, Math.-Phys. Semesterber. **12** (1965/1966) 62–78
141. Bolker, E.D., A class of convex bodies, Trans. Amer. Math. Soc. **145** (1969) 323–345
142. Bollobás, B., Modern graph theory, Springer-Verlag, New York 1998
143. Boltyanskiĭ (Boltianski, Boltjanski), V.G., Hilbert's third problem, Winston, Washington, D.C., Wiley, New York 1978
144. Boltyanskiĭ, V.G., The century of Hilbert's third problem, Dokl. Acad. Nauk. **395** (2004) 299–302, Dokl. Math. **69** (2004) 191–194
145. Boltyanskiĭ, V.G., Martini, H., Soltan, P.S., Excursions into combinatorial geometry, Springer-Verlag, Berlin 1997
146. Bolyai, F., Tentamen, Marosvásáhely 1832/33
147. Bombieri, E., Sul teorema di Tschebotarev, Acta Arith. **8** (1962/1963) 273–281
148. Bombieri, E., Gubler, W., Heights in Diophantine geometry, Cambridge Univ. Press, Cambridge 2006
149. Bonnesen, T., Fenchel, W., Theorie der konvexen Körper, Springer-Verlag, Berlin 1934, 1974, Chelsea, New York 1948, BCS Assoc., Moscow 1987
150. Borell, C., The Brunn–Minkowski inequality in Gauss space, Invent. Math. **30** (1975) 207–216
151. Borgwardt, K.-H., The average number of pivot steps required by the simplex–method is polynomial, Z. Oper. Res. Ser. A-B **26** (1982) A157–A177
152. Borgwardt, K.-H., The simplex method. A probabilistic analysis. Algorithms and Combinatorics, Springer-Verlag, Berlin 1987
153. Böröczky, K., Jr., Approximation of general smooth convex bodies, Adv. Math. **153** (2000) 325–341
154. Böröczky, K., Jr., The error of polytopal approximation with respect to the symmetric difference metric and the L_p metric, Israel J. Math. **117** (2000) 1–28
155. Böröczky, K., Jr., Finite packing and covering, Cambridge Univ. Press, Cambridge 2004
156. Böröczky, K., Jr., The stability of the Rogers-Shephard inequality and of some related inequalities, Adv. Math. **190** (2005) 47–76
157. Böröczky, K., Jr., Oral communication 2005
158. Borwein, J.M., Lewis, A.S., Convex analysis and nonlinear optimization. Theory and examples, Springer-Verlag, New York 2000
159. Bourgain, J., On the Busemann-Petty problem for perturbations of the ball, Geom. Funct. Anal. **1** (1991) 1–13
160. Bourgain, J., Lindenstrauss, J., Milman, V., Estimates related to Steiner symmetrizations, in: Geometric aspects of functional analysis (1987-88) 264–273, Lect. Notes Math. **1376**, Springer-Verlag, Berlin 1989
161. Bourgain, J., Milman, V.D., New volume ratio properties for convex symmetric bodies in R^n, Invent. Math. **88** (1987) 319–340
162. Brascamp, H.J., Lieb, E.H., Best constants in Young's inequality, its converse, and its generalization to more than three functions, Adv. Math. **20** (1976) 151–173

163. Brascamp, H.J., Lieb, E.H., On extensions of the Brunn-Minkowski and Prékopa-Leindler theorems, including inequalities for log concave functions, and with an application to the diffusion equation, J. Funct. Anal. **22** (1976) 366–389
164. Brass, P., Moser, W.O.J., Pach, J., Research problems in discrete geometry, Springer-Verlag, New York 2005
165. Brechtken–Manderscheid, U., Heil, E., Convexity and calculus of variations, in: Handbook of convex geometry **B**, 1105–1130, North-Holland, Amsterdam 1993
166. Bricard, R., Sur une question de géométrie relative aux polyèdres, Nouv. Ann. (3) **15** (1896) 331–334
167. Bricard, R., Mémoir sur la théorie de l'octaèdre articulé, J. Math. Pures Appl. **5** (1897) 3,113–148
168. Brightwell, G.R., Scheinerman, E.R., Representations of planar graphs, SIAM J. Discrete Math. **6** (1993) 214–229
169. Brion, M., Points entiers dans les polytopes convexes, Séminaire Bourbaki **1993/94**, Astérisque **227** (1995) 145–169
170. Brion, M., Polytopes convexes entiers, Gaz. Math. **67** (1996) 21–42
171. Brøndsted, A., An introduction to convex polytopes, Springer-Verlag, New York 1983
172. Bruggesser, H., Mani, P., Shellable decompositions of cells and spheres, Math. Scand. **29** (1971) 197–205
173. Brunn, H., Über Ovale und Eiflächen, PhD Thesis, U. München 1887
174. Brunn, H., Über Kurven ohne Wendepunkte, Habilitationsschrift, U. München 1889
175. Brunn, H., Referat über eine Arbeit: Exacte Grundlagen für eine Theorie der Ovale, S.-ber. Bayer. Akad. Wiss. **1894** 93–111
176. Buchta, C., Zufällige Polyeder – eine Übersicht, in: Zahlentheoretische Analysis 1–13, Lect. Notes Math. **1114**, Springer-Verlag, Berlin 1985
177. Buffon, G.L.L. de, Essai d'arithmétique morale, in: Supplément à l'histoire naturelle **4**, Imprimerie Royale, Paris 1777
178. Burago, Yu.D., Zalgaller, V.A., Geometric inequalities, Springer-Verlag, Berlin 1988
179. Burkard, R.E., Convexity and discrete optimization, in: Handbook of convex geometry **A** 675–698, North-Holland, Amsterdam 1993
180. Busemann, H., The isoperimetric problem for Minkowski area, Amer. J. Math. **71** (1949) 743–762
181. Busemann, H., The foundations of Minkowskian geometry, Comment. Math. Helv. **24** (1950) 156–187
182. Busemann, H., Convex surfaces, Interscience, New York 1958
183. Busemann, H., Feller, W., Krümmungseigenschaften konvexer Flächen, Acta Math. **66** (1936) 1–47
184. Busemann, H., Petty, C.M., Problems on convex bodies, Math. Scand. **4** (1956) 88–94
185. Buser, P., Conway, J., Doyle, P., Semmler, K.-D., Some planar isospectral domains, Internat. Math. Res. Notices 1994, 391ff. (electronic)
186. Caffarelli, L.A., Interior $W^{2,p}$-estimates for solutions of the Monge-Ampère equation, Ann. of Math. (2) **131** (1990) 135–150
187. Caffarelli, L.A., Some regularity properties of solutions of Monge-Ampère equations, Comm. Pure Appl. Math. **44** (1991) 965–969
188. Campi, S., Colesanti, A., Gronchi, P., Blaschke-decomposable convex bodies, Israel J. Math. **105** (1998) 185–195
189. Cappell, S.E., Shaneson, J.L., Some problems in number theory I: the circle problem, manuscript 2007

190. Carathéodory, K., Über das lineare Mass von Punktmengen, eine Verallgemeinerung des Längenbegriffs, Nachr. Königl. Ges. Wiss. Göttingen, Math.-phys. Kl. **1914** 406–426
191. Carroll, F.W., A polynomial in each variable separately is a polynomial, Amer. Math. Monthly **68** (1961) 42
192. Cartier, P., Décomposition des polyèdres: le point sur le troisime problme de Hilbert, Seminar Bourbaki **1984/85**, Astérisque **133-134** (1986) 261–288
193. Cassels, J.W.S., A short proof of the Minkowski-Hlawka theorem, Proc. Cambridge Philos. Soc. **49** (1953) 165–166
194. Cassels, J.W.S., On a problem of Rankin about the Epstein zeta-function, Proc. Glasgow Math. Assoc. **4** (1959) 73–80
195. Cassels, J.W.S., An introduction to the geometry of numbers, Springer-Verlag, Berlin 1959, 2nd ed. 1971, Reprint 1997
196. Castillo, E., Ruiz Cobo, M.R., Functional equations and modelling in science and engineering, Dekker, New York 1992
197. Cauchy, A.-L., Sur les polygones et les polyèdres. Second mémoire, J. École Polytechn. **9** (1813) 87–98, Œuvres (2) **1** 26–38, Paris 1905
198. Cauchy, A.-L., Notes sur divers théorèmes relatifs à la rectification des courbes et à la quadrature des surfaces, C. R. Acad. Sci. Paris **13** (1841) 1060–1065, Œuvres (1) **6** 369–375, Paris 1888
199. Chakerian, G.D., Inequalities for the difference body of a convex body, Proc. Amer. Math. Soc. **18** (1967) 879–884
200. Chalk, J.H.H., On the positive values of linear forms, Quart. J. Math. Oxford (2) **18** (1947) 215–227
201. Chavel, I., Isoperimetric inequalities: Differential geometric and analytic perspectives, Cambridge Univ. Press, Cambridge 2001
202. Chebotarev (Tschebotaröw, Tschebotareff), N., Beweis des Minkowski'schen Satzes über lineare inhomogene Formen, Ucen. Zap. Kazansk. Gos Univ. **94** (1934) 3–16, Vierteljschr. Naturforsch. Ges. Zürich **85** (1940), Beiblatt 27–30
203. Chen, B., A simplified elementary proof of Hadwiger's volume theorem, Geom. Dedicata **105** (2004) 107–120
204. Chen, G.-L., Relations de dépendance linéaire entre des logarithmes de nombres algébriques, Ann. Fac. Sci. Toulouse Math. (6) **5** (1996) 587–597
205. Chern, S.-S., Yien, C.-T., Sulla formula principale cinematica dello spazio ad n dimensioni, Boll. U. Mat. Ital. (2) **2** (1940) 434–437
206. Chernaya (Chornaya), E.V., On the optimization of weighted cubature formulas on certain classes of continuous functions, East J. Approx. **1** (1995) 47–60
207. Chernaya, E.V., An asymptotic sharp estimate for the remainder of weighted cubature formulas that are optimal on certain classes of continuous functions, Ukrain. Mat. Zh. **47** (1995) 1405–1415, Ukrainian J. Math. **47** (1995) 1606–1618
208. Choquet, G., Unicité des représentations intégrales au moyen de points extrémaux dans les cônes convexes réticulés, C. R. Acad. Sci. Paris **243** (1956) 555–557
209. Chou (Tso), K.-S., Deforming a hypersurface by its Gauss-Kronecker curvature, Comm. Pure Appl. Math. **38** (1985), 867–882
210. Chou, K.-S., Zhu, X.-P., The curve shortening problem, Chapman & Hall/CRC Press, Boca Raton, FL 2001
211. Cohn, H., Kumar, A., The densest lattice in twenty-four dimensions, Electron. Res. Announc. Amer. Math. Soc. **10** (2004) 58–67
212. Cole, A.J., On the product of n linear forms, Quart. J. Math. Oxford (2) **3** (1952) 56–62

213. Colesanti, A., Pucci, C., Qualitative and quantitative results for sets of singular points of convex bodies, Forum Math. **9** (1997) 103–125
214. Colin de Verdière, Y., Empilements de cercles: convergence d'une méthode de point fixe, Forum Math. **1** (1989) 395–402
215. Colin de Verdière, Y., Un principe variationnel pour les empilements de cercles, Invent. Math. **104** (1991) 655–669
216. Collins, C.R., Stephenson, K., A circle packing algorithm, Comput. Geom. **25** (2003) 233–256
217. Connelly, R., A counterexample to the rigidity conjecture for polyhedra, Inst. Haut. Études Sci. Publ. Math. **47** (1978) 333–338
218. Connelly, R., Rigidity, in: Handbook of convex geometry **A** 223–271, North-Holland, Amsterdam 1993
219. Convexity and its applications, eds.: P.M. Gruber, J.M. Wills, Basel, Birkhäuser 1983
220. Conway, J.H., Sloane, N.J.A., Sphere packings, lattices and groups, Springer-Verlag, New York, 1988, 3rd ed. 1999
221. Conway, J.H., Sloane, N.J.A., Low-dimentional Lattices. III. Perfect forms, Proc. Roy. Soc. London Ser A **418** (1988) 43–80
222. Coppel, W.A., Foundations of convex geometry, Cambridge Univ. Press, Cambridge 1998
223. Coppel, W.A., Number theory: an introduction to mathematics **B**, Phalanger Press, Griffith 2002
224. Cornfeld, I.P. Fomin, S.V., Sinaĭ, Ya.G., Ergodic theory, Springer-Verlag, New York 1982
225. Corput, J.G. Van der, Konstruktion der Minimalbasis für spezielle diophantische Systeme von linear-homogenen Gleichungen und Ungleichungen, Proc. Kon. Akad. Wet. Amsterdam **34** (1931) 515–523
226. Corrádi, K., Szabó, S., A combinatorial approach for Keller's conjecture, Period. Math. Hungar. **21** (1990) 95–100
227. Courant, R., Hilbert, D., Methoden der mathematischen Physik, 2nd ed., Springer-Verlag, Berlin 1930
228. Cox, D., Little, J., O'Shea, D., Ideals, varieties, and algorithms. An introduction to computational algebraic geometry and commutative algebra, 2nd ed., Springer-Verlag, New York 1997
229. Coxeter, H.M.S., Extreme forms, Canad. J. Math. **3** (1951) 391–441
230. Coxeter, H.M.S., Regular polytopes, 2nd ed., MacMillan, New York 1963, 3rd ed., Dover, New York 1973
231. Coxeter, H.S.M., Introduction to geometry, 2nd ed., Wiley, New York 1969
232. Coxeter, H.S.M., Regular complex polytopes, 2nd ed., Cambridge Univ. Press, Cambridge 1991
233. Coxeter, H.S.M., Few, L., Rogers, C.A., Covering space with equal spheres, Mathematika **6** (1959) 147–157
234. Czerwik, S., Functional equations and inequalities in several variables, World Scientific, River Edge, NJ 2002
235. Danilov, A.N., The Cauchy-Davenport-Chowla theorem for groups, in: Semigroup varieties and semigroups of endomorphisms 50–54, Leningrad. Gos. Ped. Inst., Leningrad 1979
236. Danilova, N.I., Development of elementary methods in the theory of isoperimeters in the 18th and first third of the 19th centuries, Istor.-Mat. Issled. **29** (1985) 199–209, 348

237. Dantzig, G.B., Maximization of a linear function of variables subject to linear inequalities, in: Activity analysis of production and allocation 339–347, Wiley, New York, Chapman & Hall, London 1951
238. Dantzig, G.B., Linear programming and extensions, Princeton Univ. Press, Princeton, NJ 1963, reprint with corr. 1998
239. Dantzig, G.B., Thapa, M.N., Linear programming 1. Introduction, 2. Theory and extensions, Springer-Verlag, New York 1997, 2003
240. Danzer, L., Finite point-sets on S^2 with minimum distance as large as possible, Discrete Math. **60** (1986) 3–66
241. Danzer, L., Grünbaum, B., Klee, V., Helly's theorem and its relatives, in: Proc. Sympos. Pure Math. **VII** 101–180, Amer. Math. Soc., Providence, RI 1963
242. Danzer, L., Laugwitz, D., Lenz, H., Über das Löwnersche Ellipsoid und sein Analogon unter den einem Eikörper einbeschriebenen Ellipsoiden, Arch. Math. **8** (1957) 214–219
243. Davenport, H., A simple proof of Remak's theorem on the product of three linear forms, J. Lond. Math. Soc. **14** (1939) 47–51
244. Davenport, H., On a theorem of Tschebotareff, J. London Math. Soc. **21** (1946) 28–34, corr., ibid. **24** (1949) 316
245. Davenport, H., The higher arithmetic. An introduction to the theory of numbers, Hutchinson's Univ. Library, London, Longmans, Green and Co., New York 1952
246. Davenport, H., The collected works of Harold Davenport **1, 2**, eds.: B.J. Birch, H. Halberstam, C.A. Rogers, Academic Press, New York 1977
247. Davenport, H., Rogers, C.A., Hlawka's theorem in the geometry of numbers, Duke Math. J. **14** (1947) 367–375
248. Day, M.M., Normed linear spaces, Springer-Verlag, Berlin 1958
249. Debrunner, H.E., Helly type theorems derived from basic singular homology, Amer. Math. Monthly **77** (1970) 375–380
250. Dehn, M., Über den Rauminhalt, Math. Ann. **55** (1901) 465-478
251. Dehn, M., Die Eulersche Formel in Zusammenhang mit dem Inhalt in der nicht-Euklidischen Geometrie, Math. Ann. **61** (1905) 561–586
252. Dehn, M., Über die Starrheit konvexer Polyeder, Math. Ann. **77** (1916) 466–473
253. DeLoera, J., The many aspects of counting lattice points in polytopes, Math. Semesterber. **52** (2005) 175–195
254. DeLoera, J., Hemmecke, R., Tauzer, J., Yoshida, R., Effective lattice point counting in rational convex polytopes, J. Symb. Comput. **38** (2004) 1273–1302
255. Delone (Delaunay), B.N., Sur la sphère vide, in: Proc. Internat. Congr. Math. (Toronto 1924) **1** 695–700, Univ. Toronto Press 1928
256. Delone, B.N., Sur la partition régulière de l'espace à 4 dimensions, Izv. Akad. Nauk SSSR Otdel. Fiz-Mat. Nauk (7) **1929** 793–800
257. Delone, B.N., The geometry of positive quadratic forms, Uspekhi. Mat. Nauk **3** (1937) 16–62, **4** (1938) 102–164
258. Delone, B.N., Proof of the fundamental theorem in the theory of stereohedra, Dokl. Akad. Nauk SSSR **138** (1961) 1270–1272, Soviet Math. Dokl. **2** (1961) 812–815
259. Delone, B.N., Ryshkov, S.S., Extremal problems in the theory of positive quadratic forms, Trudy Mat. Inst. Steklov. **112** (1971) 203–223, Proc. Steklov Inst. Math. **112** (1971) 211–231
260. Delone, B.N., Sandakova, N.N., Theory of stereohedra, Trudy Mat. Inst. Steklov. **64** (1961) 28–51
261. Descartes, R., De solidorum elementis, in: Œvres **10** 265–276, Tannery, Paris 1897–1913, reprint, Paris 1964–1975; Additions, Œvres **11** 690–692

262. Descartes, R., Principia philosophiae, L. Elzevirius, Amsterdam 1644
263. Deutsch, F., Best approximation in inner product spaces, Springer-Verlag, New York 2001
264. Deutsch, J.I., Geometry of numbers proof of Götzky's four-square's theorem, J. Number Theory **96** (2002) 417–431
265. Deza, M., Grishukhin, V., Properties of parallelotopes equivalent to Voronoi's conjecture, European J. Combin. **25** (2004) 517–533
266. Diananda, P.H., Notes on two lemmas concerning the Epstein zeta-function, Proc. Glasgow Math. Assoc. **6** (1964) 202–204
267. Diestel, J., Uhl, J.J., Vector measures, Amer. Math. Soc., Providence, RI 1977
268. Dinghas, A., Über einen geometrischen Satz von Wulff für die Gleichgewichtsform von Kristallen, Z. Krist., Mineral. Petrogr. (A) **105** (1944) 304–314
269. Dinghas, A., Démonstration du théorème de Brunn-Minkowski pour des familles continues d'ensembles, C. R. Acad. Sci. Paris **239** (1954) 605–607
270. Dinghas, A., Minkowskische Summen und Integrale. Superadditive Mengenfunktionale. Isoperimetrische Ungleichungen, Mém. Sci. Math. **149**, Gauthier-Villars, Paris 1961
271. Dirichlet, G.L., Verallgemeinerung eines Satzes aus der Lehre von den Kettenbrüchen nebst einigen Anwendungen auf die Theorie der Zahlen, Ber. Verh. Königl. Preuss. Akad. Wiss. **1842** 93–95, Werke **I** 635–638, Reimer, Berlin 1889, Chelsea, New York 1964
272. Dirichlet, G.L., Über die Reduktion der positiven quadratischen Formen mit drei unbestimmten ganzen Zahlen, J. reine angew. Math. **40** (1850) 209–227, Werke **II** 27–48, Reimer, Berlin 1897, Chelsea, New York 1964
273. Discrete and computational geometry, eds.: E. Goodman, J. Pach, E. Welzl, Cambridge Univ. Press, Cambridge 2005
274. Diskant, V.I., Generalization of a theorem of Lindelöf, Mat. Fiz. Anal. Geom. **4** (1997) 59–64
275. Dobrushin (Dobrušin), R., Kotecký, R., Shlosman (Šlosman), S., Wulff construction. A global shape from local interaction, Amer. Math. Soc., Providence, RI 1992
276. Dolbilin, N.P., Rigidity of convex polyhedrons, Quantum **9** (1998) 8–13
277. Douady, R., Thèse 3ème cycle, U. Paris VII 1982
278. Du, Q., Faber, V., Gunzburger, M., Centroidal Voronoi tessellations: applications and algorithms, SIAM Rev. **41** (1999) 637–676
279. Dupont, J.L., Scissors congruences, group homology and characteristic classes, World Scientific, River Edge, NJ 2001
280. Dutour Sikiric (Dutour), M., Schürmann, A., Classification of eight dimensional perfect forms, in preparation 2005
281. Dyson, F.J., On the product of four non-homogeneous linear forms, Ann. of Math. (2) **49** (1948) 82–109
282. Ebeling, W., Lattices and codes, 2nd ed., Vieweg, Braunschweig 2002
283. Eckhoff, J., Radon's theorem revisited, in: Contributions to geometry 164–185, Birkhäuser, Basel 1979
284. Eckhoff, J., Helly, Radon, and Carathéodory type theorems, in: Handbook of convex geometry **A** 389–448, North-Holland, Amsterdam 1993
285. Edler, F., Vervollständigung der Steiner'schen elementar-geometrischen Beweise für den Satz, dass der Kreis grösseren Flächeninhalt besitzt, als jede andere ebene Figur gleich grossen Umfanges, Nachr. Königl. Ges. Wiss. Göttingen, Math.-phys. Kl. **1882** 73–80

286. Edmonds, A.L., Simplicial decompositions of convex polytopes, Pi Mu Epsilon J. **5** (1970) 124–128
287. Edmonds, J., Giles, R., A min-max relation for submodular functions on graphs, in: Studies in integer programming 185–204, North-Holland, Amsterdam 1977
288. Efimov (Efimow), N.W., Flächenverbiegung im Grossen. Mit einem Nachtrag von E. Rembs und K.P. Grotemeyer, Akademie-Verlag, Berlin 1957
289. Efimov, N.V., Qualitative problems on the theory of deformation of surfaces, in: Transl. Amer. Math. Soc. 1 **6**: Differential geometry and the calculus of variations, Amer. Math. Soc., Providence, RI 1962
290. Eggleston, H.G., Convexity, Cambridge Univ. Press, Cambridge 1958
291. Egorychev(Egoryčev), G.P., Van der Waerden conjecture and applications, in: Handbook of algebra **1** 3–26, North-Holland, Amsterdam 1996
292. Ehrhart, E., Sur un problème de géométrie diophantienne linéaire I. Polyèdres et réseaux, J. reine angew. Math. **226** (1967) 1–29
293. Ehrhart, E., Démonstration de la loi de réciprocité pour un polyèdre entier, C. R. Acad. Sci. Paris Sér. A-B **265** (1967) A5–A7
294. Ehrhart, E., Polynômes arithmétiques et méthode des polyèdres en combinatoire, Birkhäuser, Basel 1977
295. Encyclopedia of mathematics **1-10**, Kluwer, Dordrecht 1988–1994
296. Engel, P., Geometric crystallography. An axiomatic introduction to crystallography, Reidel, Dordrecht 1986
297. Engel, P., Geometric crystallography, in: Handbook of convex geometry **B** 989–1041, North-Holland, Amsterdam 1993
298. Engel, P., New investigations on parallelohedra in R^d, in: Voronoi's impact on modern science **II**, 22–60, Proc. Inst. Math. Nat. Acad. Sci. Ukraine, Kyiv 1998
299. Engel, P., On the determination of a lattice basis of facet vectors, in: IV International Conference in "Stochastic Geometry, Convex Bodies, Empirical Measures & Applications to Engineering Science" **I** (Tropea 2001), Rend. Circ. Mat. Palermo (2) Suppl. No. **70**, part I (2002) 269–278
300. Engel, P., Personal communication 2004
301. Engel, P., Michel, L., Senechal. M., Lattice geometry, in preparation
302. Engel, P., Syta, H., eds., Voronoi's impact on modern science **I, II**, Proc. Inst. Math. Nat. Acad. Sci. Ukraine, Kyiv 1998
303. Ennola, V., A lemma about the Epstein zeta-function, Proc. Glasgow Math. Assoc. **6** (1964) 198–201
304. Ennola, V., On a problem about the Epstein zeta-function, Proc. Cambridge Philos. Soc. **60** (1964) 855–875
305. Epstein, P., Zur Theorie allgemeiner Zetafunktionen **II**, Math. Ann. **63** (1906) 205–216
306. Erdahl, R.M., Zonotopes, dicings, and Voronoi's conjecture on parallelohedra, Europ. J. Combin. **20** (1999) 527–549
307. Erdös, P., Gruber, P.M., Hammer, J., Lattice points, Longman Scientific, Harlow, Essex 1989
308. Erdös, P., Rogers, C.A., The star number of coverings of space with convex bodies, Acta Arith. **9** (1964) 41–45
309. Erdös, P., Stone, A.H., On the sum of two Borel sets, Proc. Amer. Math. Soc. **25** (1970) 304–306
310. Euclid, Euclid's elements. Translated from the text of *Heiberg*. With introduction and commentary, by T.L. Heath, 2nd ed., revised with additions, **1–3**, Cambridge, Cambridge Univ. Press 1926, Green Lion Press, Santa Fe, NM 2002

311. Euler, L., Opera postuma **I** 494–496, St. Petersburg 1862
312. Euler, L., Elementa doctrinae solidorum, Novi commentarii academiae scientiarum Petropolitanae **4** (1752/3) 109–140, Opera Mathematica **26** 71–93
313. Euler, L., Demonstratio nonnullarum insignium proprietatum quibus solida hedris planis inclusa sunt praedita, Novi commentarii academiae scientiarum Petropolitanae **4** (1752/3) 140–160, Opera Mathematica **26** 94–108
314. Evans, L., Gariepy, R., Measure theory and fine properties of functions, CRC Press, Boca Raton, FL 1992
315. Ewald, G., Combinatorial convexity and algebraic geometry, Springer-Verlag, New York 1996
316. Faber, G., Beweis, dass unter allen homogenen Membranen von gleicher Fläche und gleicher Spannung die kreisförmige den tiefsten Grundton gibt, Sitzungsber. Bayer. Akad. Wiss., Math.-Phys. Kl. **1923** 169–172
317. Falconer, K.J., The geometry of fractal sets, Cambridge University Press, Cambridge 1986
318. Federico, P.J., Descartes on polyhedra. A study of the De solidorum elementis, Springer-Verlag, New York 1982
319. Fedorov, E.S., Elements of the study of figures, Zap. Mineral. Imper. S. Petersburgskogo Obšč. **21**(2) (1885) 1–279, Izdat. Akad. Nauk SSSR, Moscow 1953
320. Fejes Tóth, G., Densest packings of typical convex sets are not lattice-like, Discrete Comput. Geom. **14** (1995) 1–8
321. Fejes Tóth, G., A stability criterion to the moment theorem, Studia Sci. Math. Hungar. **34** (2001) 209–224
322. Fejes Tóth, G., Packing and covering, in: Handbook of discrete and computational geometry, 2nd ed., 19–41, CRC Press, Boca Raton, FL 2004
323. Fejes Tóth, G., Covering with fat convex discs, Discr. Comput. Geom. **34** (2005) 129–141
324. Fejes Tóth, G., Covering with convex bodies, manuscript 2006
325. Fejes Tóth, G., Kuperberg, W., Packing and covering with convex sets, in: Handbook of convex geometry **B** 799–860, North-Holland, Amsterdam 1993
326. Fejes Tóth, G., Zamfirescu, T., For most convex discs thinnest covering is not lattice-like, in: Intuitive geometry (Szeged 1991) 105–108, Colloq. Math. Soc. János Bolyai **63**, North-Holland, Amsterdam 1994
327. Fejes Tóth, L., Über einen geometrischen Satz, Math. Z. **46** (1940) 83–85
328. Fejes Tóth, L., Some packing and covering theorems, Acta Math. Acad. Sci. Hungar. **12** (1950) 62–67
329. Fejes Tóth, L., Lagerungen in der Ebene, auf der Kugel und im Raum, Springer-Verlag, Berlin 1952, 2nd ed. 1972
330. Fejes Tóth, L., Regular figures, Macmillan, New York 1964, Akadémiai Kiadó, Budapest 1965
331. Fejes Tóth, L., On the densest packing of convex discs, Mathematika **30** (1983) 1–3
332. Felsner, S., Geometric graphs and arrangements, Vieweg, Wiesbaden 2004
333. Fenchel, W., Inégalités quadratiques entre les volumes mixtes des corps convexes, C.R. Acad. Sci. Paris **203** (1936) 647–650
334. Fenchel, W., Convex cones, sets, and functions, Mimeographed lecture notes, Princeton Univ. Press, Princeton 1952
335. Fenchel, W., Jessen, B., Mengenfunktionen und konvexe Körper, Danske Vid. Selsk. Mat.-Fys. Medd. **16** (1938) 1–31
336. Finn, R., Equilibrium capillary surfaces, Springer-Verlag, New York 1986

337. Florian, A., Extremum problems for convex discs and polyhedra, in: Handbook of convex geometry **A** 177–221, North-Holland, Amsterdam 1993
338. Fonf, V.P., Lindenstrauss, J., Phelps, R.R., Infinite dimensional convexity, in: Handbook of the geometry of Banach spaces **I** 599–670, North-Holland, Amsterdam 2001
339. Fonseca, I., The Wulff theorem revisited, Proc. Roy. Soc. London Ser. A **432** (1991) 125–145
340. Fonseca, I., Müller, S., A uniqueness proof for the Wulff theorem Proc. Roy. Soc. Edinburgh Sect. A **119** (1991) 125–136
341. Fortune, S., Voronoi diagrams and Delaunay triangulations, in: Handbook of discrete and computational geometry 377–388, CRC Press, Boca Raton, FL 1977
342. Fourier, J.B.J., reported in: Analyse des travaux de l'Académie Royale des Sciences, pendant l'année 1823, Histoire de L'Académie Royale des Sciences de l'Institut de France **6** 1823 (1826) 29–41, Œuvres de Fourier **II** 321–324, Gauthier-Villars, Paris 1890
343. Fourier analysis and convexity, eds.: L. Brandolini, L. Colzani, A. Iosevich, G. Travaglini, Birkhäuser, Boston, MA 2004
344. Fricker, F., Einführung in die Gitterpunktlehre, Birkhäuser, Basel 1982
345. Fu, J.H.G., Kinematic formulas in integral geometry, Indiana Univ. Math. J. **39** (1990) 1115–1154
346. Fukuda, K., Ziegler, G., eds., Combinatorics of polytopes, European J. Combin. **21** (2000) 1–152
347. Füredi, Z., Kang, J.H., Covering the n-space by convex bodies and its chromatic number, Discrete Comput. Geom., to appear
348. Gács, P., Lovász, L., Khachiyan's algorithm for linear programming, in: Mathematical programming at Oberwolfach, Proc. Conf., Math. Forschungsinstitut Oberwolfach 1979, Math. Programming Stud. **14** (1981) 61–68
349. Gage, M.E., An isoperimetric inequality with applications to curve shortening, Duke Math. J. **50** (1983) 1225–1229
350. Gage, M.E., Curve shortening makes convex curves circular, Invent. Math. **76** (1984) 357–364
351. Gage, M.E., Hamilton, R.S., The heat equation shrinking convex plane curves, J. Differential Geom. **23** (1986) 69–96
352. Gale, D., Klee, V., Unique reducibility of subsets of commutative topological groups and semigroups, Math. Scand. **36** (1975) 174–198
353. Gale, D., Kuhn, H.W., Tucker, A.W., Linear programming and the theory of games, in: Activity analysis of production and allocation 317–329, Wiley, New York, Chapman & Hall, London 1951
354. Gal'perin, G.A., Zemlyakov, A.N., Billiard problems and related problems in mathematics and mechanics, Nauka, Moscow 1990
355. Gao, S., Absolute irreducibility of polynomials via Newton polytopes, J. Algebra **237** (2001) 501–520
356. Gao, S., Lauder, A.G.B., Decomposition of polytopes and polynomials, Discrete Comput. Geom. **26** (2001) 89–104
357. Gardner, R.J., A positive answer to the Busemann-Petty problem in three dimensions, Ann. of Math. (2) **140** (1994) 435–447
358. Gardner, R.J., Intersection bodies and the Busemann-Petty problem, Trans. Amer. Math. Soc. **342** (1994) 435-445
359. Gardner, R.J., Geometric tomography, Cambridge Univ. Press, Cambridge 1995
360. Gardner, R.J., The Brunn-Minkowski inequality, Bull. Amer. Math. Soc. **39** (2002) 355–405

361. Gardner, R.J., Koldobsky, A., Schlumprecht, T., An analytic solution to the Busemann-Petty problem on sections of convex bodies, Ann. of Math. (2) **149** (1999) 691–703
362. Gaukroger, S., Descartes' system of natural philosophy, Cambridge Univ. Press, Cambridge 2002
363. Gauss, C.F., Theorematis de resolubilitate functionum algebraicarum integrarum in factores reales demonstratio tertia, Göttingische gelehrte Anzeigen **1816** März 2, Stück **35** 337–343, Werke **3** 105–112, Königl. Ges. Wiss., Göttingen 1876
364. Gauss, C.F., Untersuchungen über die Eigenschaften der positiven ternären quadratischen Formen von Ludwig August Seeber, Göttingische gelehrte Anzeigen, 9.7.1831, Werke **2** (1836) 188–196, J. reine angew. Math. **20** (1840) 312–320
365. Gauss, C.F., Letters to Gerling, 8.4.1844 and 17.4.1844, Werke **8** 241,244
366. Gauss, C.F., Werke **2** 269–291
367. Gel'fand, I.M., Lectures on linear algebra, Interscience, Wiley, New York 1967
368. Gel'fand, I.M., Gindikin, S.G., Graev, M.I., Selected problems in integral geometry, Dobrosvet, Moscow 2000
369. Gergonne, J.D., Recherche de la surface plane de moindre contour, entre toutes celles de meme etendue, Ann. Math. **4** (1814) 338–345
370. Gericke, H., Zur Geschichte des isoperimetrischen Problems, Math. Semesterber. **29** (1982) 160–187
371. Gersho, A., Asymptotically optimal block quantization, IEEE Trans. Inform. Theory **IT-25** (1979) 373–380
372. Gerwien, P., Zerschneidung jeder beliebigen Anzahl von gleichen geradlinigen Figuren in dieselben Stücke, J. reine angew. Math. **10** (1833) 228–234
373. Giannopoulos, A.A., A note on a problem of H. Busemann and C.M. Petty concerning sections of symmetric convex bodies, Mathematika **37** (1990) 239–244
374. Giannopoulos, A.A., Milman, V.D., Euclidean structure in finite dimensional normed spaces, in: Handbook of the geometry of Banach spaces **I** 707–779, North-Holland, Amsterdam 2001
375. Giannopoulos, A.A., Milman, V.D, Asymptotic convex geometry: short overview, in: Different faces of geometry 87–162, Int. Math. Ser. (N.Y.), Kluwer/Plenum, New York 2004
376. Giannopoulos, A.A., Perissinaki, I., Tsolomitis, A., John's theorem for an arbitrary pair of convex bodies, Geom. Dedicata **84** (2001) 63–79
377. Giering, O., Über gewisse Kennzeichnungen der Kugel, Math.-Phys. Semesterber. **18** (1971) 194–204
378. Giles, F.R., Pulleyblank, W.R., Total dual integrality and integer polyhedra, Linear Algebra Appl. **25** (1979) 191–196
379. Giles, J.R., Convex analysis with application in differentiation of convex functions, Pitman, Boston 1982
380. Gluck, H., Almost all simply connected closed surfaces are rigid, in: Geometric topology (Proc. Conf. Park City, Utah 1974) 225–239, Lect. Notes Math. **438**, Springer-Verlag, Berlin 1975
381. Gluck, H., Manifolds with preassigned curvature – a survey, Bull. Amer. Math. Soc. **81** (1975) 313–329
382. Gluskin, E.D., Probability in the geometry of Banach spaces, Proc. Internat. Congr. Math. **2** (Berkeley, CA 1986) 924–938, Amer. Math. Soc., Providence, RI 1987
383. Goodey, P., Kiderlen, M., Weil, W., Section and projection means of convex bodies, Monatsh. Math. **126** (1998) 37–54
384. Goodey, P., Weil, W., Zonoids and generalisations, in: Handbook of convex geometry **B** 1297–1326, North-Holland, Amsterdam 1993

385. Goodman, J.E., Pollack, R., Upper bounds for configurations and polytopes in R^d, Discrete Comput. Geom. **1** (1986) 219–227
386. Goodman, J.E., Pollack, R., Wenger, R., Geometric transversal theory. New trends in discrete and computational geometry, 163–198, Algorithms Combin. **10**, Springer-Verlag, Berlin 1993
387. Gordan, P., Über die Auflösung linearer Gleichungen mit reellen Coefficienten, Math. Ann. **6** (1873) 23–28
388. Gordon, C., Webb, D.L., Wolpert, S., One cannot hear the shape of a drum, Bull. Amer. Math. Soc. (N.S.) **27** (1992) 134–138
389. Graf, S., Luschgy, H., Foundations of quantization of probability distributions, Lect. Notes Math. **1730**, Springer-Verlag, Berlin 2000
390. Grätzer, G., General lattice theory, 2nd ed., New appendices by the author with B.A. Davey, R. Freese, B. Ganter, M. Greferath, P. Jipsen, H.A. Priestley, H. Rose, E.T. Schmidt, S.E. Schmidt, F. Wehrung and R. Wille, Birkhäuser, Basel 1998
391. Graver, J., Servatius, B., Servatius, H., Combinatorial rigidity, Amer. Math. Soc., Providence, RI 1993
392. Graver, J. E., Counting on frameworks. Mathematics to aid the design of rigid structures, Math. Assoc. Amer., Washington, DC 2001
393. Gray, J.J., The Hilbert challenge, Oxford Univ. Press, Oxford 2000
394. Gray, R.M., Neuhoff, D.L., Quantization, IEEE Trans. Inform. Theory **IT-44** (1998) 2325–2383
395. Gritzmann, P., Hufnagel, A., On the algorithmic complexity of Minkowski's reconstruction theorem, J. London Math. Soc. (2) **59** (1999) 1081–1100
396. Gritzmann, P., Klee, V., Mathematical programming and convex geometry, in: Handbook of convex geometry **A** 627–674, North-Holland, Amsterdam 1993
397. Gritzmann, P., Wills, J.M., Lattice points, in: Handbook of convex geometry **B** 765–797, North-Holland, Amsterdam 1993
398. Groemer, H., Abschätzung für die Anzahl der konvexen Körper, die einen konvexen Körper berühren, Monatsh. Math. **65** (1960) 74–81
399. Groemer, H., Existenzsätze für Lagerungen im Euklidischen Raum, Math. Z. **81** (1963) 260–278
400. Groemer, H., Über die Zerlegung des Raumes in homothetische konvexe Körper, Monatsh. Math. **68** (1964) 21–32
401. Groemer, H., Continuity properties of Voronoi domains, Monatsh. Math. **74** (1971) 423–431
402. Groemer, H., On the extension of additive functionals on classes of convex sets, Pacific J. Math. **75** (1978) 397–410
403. Groemer, H., Stability of geometric inequalities, in: Handbook of convex geometry **A** 125–150, North-Holland, Amsterdam 1993
404. Groemer, H., Stability theorems for ellipsoids and spheres, J. London Math. Soc. (2) **49** (1994) 357–370
405. Groemer, H., Geometric applications of Fourier series and spherical harmonics, Cambridge Univ. Press, Cambridge 1996
406. Groemer, H., e-mail 2005
407. Gross, W., Die Minimaleigenschaft der Kugel, Monatsh. Math. Phys. **28** (1917) 77–97
408. Grötschel, M., Lovász, L., Schrijver, A., The ellipsoid method and its consequences in combinatorial optimization, Combinatorica **1** (1981) 169–197
409. Grötschel, M., Lovász, L., Schrijver, A., Geometric algorithms and combinatorial optimization, 2nd ed., Springer-Verlag, Berlin 1993

410. Gruber, P.M., Eine Erweiterung des Blichfeldtschen Satzes mit einer Anwendung auf inhomogene Linearformen, Monatsh. Math. **71** (1967) 143–147
411. Gruber, P.M., Über das Produkt inhomogener Linearformen, Acta Arith. **13** (1967) 9–27
412. Gruber, P.M., Zur Charakterisierung konvexer Körper. Über einen Satz von Rogers und Shephard I, II, Math. Ann. **181** (1969) 189–200, **184** (1970) 79–105
413. Gruber, P.M., Eine Bemerkung über $DOTU$-Matrizen, J. Number Theory **8** (1976) 350–351
414. Gruber, P.M., Die meisten konvexen Körper sind glatt, aber nicht zu glatt, Math. Ann. **229** (1977) 259–266
415. Gruber, P.M., Isometries of the space of convex bodies of E^d, Mathematika **25** (1978) 270–278
416. Gruber, P.M., Geometry of numbers, in: Contributions to geometry 186–225, Birkhäuser, Basel 1979
417. Gruber, P.M., Approximation of convex bodies, in: Convexity and its applications 131–162, Birkhäuser, Basel 1983
418. Gruber, P.M., In most cases approximation is irregular, Rend. Sem. Mat. Univ. Politec. Torino **41** (1983) 19–33
419. Gruber, P.M., Typical convex bodies have surprisingly few neighbors in densest lattice packings, Studia Sci. Math. Hungar. **21** (1986) 163–173
420. Gruber, P.M., Radons Beiträge zur Konvexität / Radon's contributions to convexity, in: Johann Radon, Gesammelte Abhandlungen **I** 331–342, Birkhäuser, Basel 1987, Österr. Akad. Wiss., Wien 1987
421. Gruber, P.M., Minimal ellipsoids and their duals, Rend. Circ. Mat. Palermo (2) **37** (1988) 35–64
422. Gruber, P.M., Convex billiards, Geom. Dedicata **33** (1990) 205–226
423. Gruber, P.M., A typical convex surface contains no closed geodesic!, J. reine angew. Math. **416** (1991) 195–205
424. Gruber, P.M., The endomorphisms of the lattice of convex bodies, Abh. Math. Sem. Univ. Hamburg **61** (1991) 121–130
425. Gruber, P.M., Volume approximation of convex bodies by circumscribed polytopes, in: Applied geometry and discrete mathematics 309–317, DIMACS Ser. Discrete Math. Theoret. Comput. Sci. **4**, Amer. Math. Soc., Providence, RI 1991
426. Gruber, P.M., Characterization of spheres by stereographic projection, Arch. Math. **60** (1993) 290–295
427. Gruber, P.M., Asymptotic estimates for best and stepwise approximation of convex bodies II, Forum Math. **5** (1993) 521–538
428. Gruber, P.M., The space of convex bodies, in: Handbook of convex geometry **A** 301–318, North-Holland, Amsterdam 1993
429. Gruber, P.M., Aspects of approximation of convex bodies, in: Handbook of convex geometry **A** 319–345, North-Holland, Amsterdam 1993
430. Gruber, P.M., Geometry of numbers, in: Handbook of convex geometry **B** 739–763, North-Holland, Amsterdam 1993
431. Gruber, P.M., Baire categories in convexity, in: Handbook of convex geometry **B** 1329–1346, North-Holland, Amsterdam 1993
432. Gruber, P.M., How well can space be packed with smooth bodies? Measure theoretic results, J. London Math. Soc. (2) **52** (1995) 1–14
433. Gruber, P.M., Only ellipsoids have caustics, Math. Ann. **303** (1995) 185–194
434. Gruber, P.M., Stability of Blaschke's characterization of ellipsoids and Radon norms, Discrete Comput. Geom. **17** (1997) 411–427

435. Gruber, P.M., Comparisons of best and random approximation of convex bodies by polytopes, Rend. Circ. Mat. Palermo (2), Suppl. **50** (1997) 189–216
436. Gruber, P.M., Asymptotic estimates for best and stepwise approximation of convex bodies IV, Forum Math. **10** (1998) 665–686
437. Gruber, P.M., A short analytic proof of Fejes Tóth's theorem on sums of moments, Aequationes Math. **58** (1999) 291–295
438. Gruber, P.M., In many cases optimal configurations are almost regular hexagonal, Rend. Circ. Mat. Palermo (2), Suppl. **65** (2000) 121–145
439. Gruber, P.M., Optimal configurations of finite sets in Riemannian 2-manifolds, Geom. Dedicata **84** (2001) 271–320
440. Gruber, P.M., Error of asymptotic formulae for volume approximation of convex bodies in $\mathbb{E}^3$, Trudy Mat. Inst. Steklov. **239** (2002) 106–117, Proc. Steklov Inst. Math. **239** (2002) 96–107
441. Gruber, P.M., Error of asymptotic formulae for volume approximation of convex bodies in $\mathbb{E}^d$, Monatsh. Math. **135** (2002) 279–304
442. Gruber, P.M., Optimale Quantisierung, Math. Semesterber. **49** (2002) 227–251
443. Gruber, P.M., Optimum quantization and its applications, Adv. Math. **186** (2004) 456–497
444. Gruber, P.M., Geometry of the cone of positive quadratic forms, in preparation
445. Gruber, P.M., Application of an idea of Voronoi to John type problems, in preparation
446. Gruber, P.M., Höbinger, J., Kennzeichnungen von Ellipsoiden mit Anwendungen, in: Jahrbuch Überblicke Mathematik **1976** 9–29, Bibliograph. Inst., Mannheim 1976
447. Gruber, P.M., Lekkerkerker, C.G., Geometry of numbers, 2nd ed., North-Holland, Amsterdam 1987
448. Gruber, P.M., Lettl, G., Isometries of the space of compact subsets of $\mathbb{E}^d$, Studia Sci. Math. Hungar. **14** (1979) 169–181 (1982)
449. Gruber, P.M., Lettl, G., Isometries of the space of convex bodies in Euclidean space, Bull. London Math. Soc. **12** (1980) 455–462
450. Gruber, P.M., Ramharter, G., Beiträge zum Umkehrproblem für den Minkowskischen Linearformensatz, Acta Math. Acad. Sci. Hungar. **39** (1982) 135–141
451. Gruber, P.M., Ryshkov (Ryškov), S.S., Facet-to-facet implies face-to- face, European J. Combin. **10** (1989) 83–84
452. Gruber, P.M., Schuster, F., An arithmetic proof of John's ellipsoid theorem, Arch. Math. (Basel) **85** (2005) 82–88
453. Grünbaum, B., Convex polytopes, Interscience, Wiley, London 1967, 2nd ed., prepared by V. Kaibel, V. Klee, G.M. Ziegler, Springer-Verlag, New York 2003
454. Grünbaum, B., Shephard, G.C., Tilings and patterns, Freeman, New York 1987
455. Grünbaum, B., Shephard, G.C., Some problems on polyhedra, J. Geom. **29** (1987) 182–190
456. Grünbaum, B., Shephard, G.C., Pick's theorem. Amer. Math. Monthly **100** (1993) 150–161
457. Gutkin, E., Billiard dynamics: a survey with the emphasis on open problems, Regul. Chaotic Dyn. **8** (2003) 1–13
458. Gutkin, E., Katok, A., Caustics for inner and outer billiards, Comm. Math. Phys. **173** (1995) 101–133
459. Hadamard, J., Étude sur les propriétés des fonctions entieres et en particulier d'une fonction considérée par Riemann, J. Math. Pure Appl. **58** (1893) 171–215
460. Hadamard, J., Sur certaines propriétés des trajectoires en dynamique, J. Math. Pure Appl. (5) **53** (1937) 331–387

461. Hadwiger, H., Gegenseitige Bedeckbarkeit zweier Eibereiche und Isoperimetrie, Vierteljahrschr. Naturforsch. Ges. Zürich **86** (1941) 152–156
462. Hadwiger, H., Beweis eines Funktionalsatzes für konvexe Körper, Abh. Math. Sem. Univ. Hamburg **17** (1951) 69–76
463. Hadwiger, H., Einfache Herleitung der isoperimetrischen Ungleichung für abgeschlossene Punktmengen, Math. Ann. **124** (1952) 158–160
464. Hadwiger, H., Additive Funktionale k-dimensionaler Eikörper I, II, Arch. Math. **3** (1952) 470–478, **4** (1953) 374–379
465. Hadwiger, H., Zum Problem der Zerlegungsgleichheit k-dimensionaler Polyeder, Math. Ann. **127** (1954) 170–174
466. Hadwiger, H., Altes und Neues über konvexe Körper, Birkhäuser, Basel 1955
467. Hadwiger, H., Eulers Charakteristik und kombinatorische Geometrie, J. reine angew. Math. **194** (1955) 101–110
468. Hadwiger, H., Vorlesungen über Inhalt, Oberfläche und Isoperimetrie, Springer-Verlag, Berlin 1957
469. Hadwiger, H., Glur, P., Zerlegungsgleichheit ebener Polygone, Elem. Math. **6** (1951) 97–106
470. Hadwiger, H., Wills, J.M., Neuere Studien über Gitterpolygone, J. reine angew. Math. **280** (1976) 61–69
471. Hajós, G., Über einfache und mehrfache Bedeckung des n-dimensionalen Raumes mit einem Würfelgitter, Math. Z. **47** (1942) 427–467
472. Hales, T.C., Cannon balls and honeycombs, Notices Amer. Math. Soc. **47** (2000) 440–449
473. Hales, T.C., A computer verification of the Kepler conjecture, in: Proc. Int. Congr. Math. (Beijing 2002) **III** 795–804, Higher Ed. Press, Beijing 2002
474. Hales, T.C., A proof of the Kepler conjecture, Ann. of Math. (2) **162** (2005) 1065–1185
475. Handbook of convex geometry **A, B**, eds.: P.M. Gruber, J.M. Wills, North-Holland, Amsterdam 1993
476. Handbook of discrete and computational geometry, 2nd ed., eds.: J.E. Goodman, J. O'Rourke, CRC Press, Boca Raton, FL 2004
477. Handbook of the geometry of Banach spaces **1, 2**, eds.: W.B. Johnson, J. Lindenstrauss, North-Holland, Amsterdam 2001
478. Handelman, D.E., Positive polynomials, convex integral polytopes, and a random walk problem, Lect. Notes Math. **1282**, Springer-Verlag, Berlin 1987
479. Hardy, G.H., On the expression of a number as the sum of two squares, Quart. J. Math. **46** (1915) 263–283
480. Hartogs, F., Zur Theorie der analytischen Funktionen mehrerer unabhängiger Veränderlicher, insbesondere über die Darstellung derselben durch Reihen, welche nach Potenzen einer Veränderlichen fortschreiten, Math. Ann. **62** (1906) 1–88
481. Hausdorff, F., Grundzüge der Mengenlehre, Veit & Comp., Leipzig 1914, 2nd ed., de Gruyter, Berlin 1927, Engl. transl., Chelsea, New York 1962
482. Hausdorff, F., Dimension und äusseres Mass, Math. Ann. **79** (1919) 157–179
483. Haüy, M. l'Abbé, Essai d'une théorie sur la structure de cristaux appliquée a plusier genres de substances cristallisées, Gogue et Neé de la Rochelle Libraires, Paris 1784
484. He, Z.-X., An estimate for hexagonal circle packings, J. Differential Geom. **33** (1991) 395–412
485. Heesch, H., Aufbau der Ebene aus kongruenten Bereichen, Nachr. Ges. Wiss. Göttingen (NF) **1** (1935) 115–117
486. Heesch, H., Kienzle, O., Flächenschluss. System der Formen lückenlos aneinanderschliessender Flachteile, Springer-Verlag, Berlin 1963

487. Heil, E., Über die Auswahlsätze von Blaschke und Arzelà-Ascoli, Math.-Phys. Semesterber. **14** (1967) 169–175
488. Heil, E., Martini, H., Special convex bodies, in: Handbook of convex geometry **A** 347–385, North-Holland, Amsterdam 1993
489. Helgason, S., The Radon transform, Birkhäuser, Boston 1980
490. Helly, E., Über Mengen konvexer Körper mit gemeinschaftlichen Punkten, Jahresber. Deutsch. Math. Verein. **32** (1923) 175–176
491. Henk, M., Successive minima and lattice points, Rend. Circ. Mat. Palermo (2) vol. I, Suppl. **70** (2002) 377–384
492. Henk, M., Schürmann, A., Wills, J.M., Ehrhart polynomial and successive minima, Mathematica **52** (2005) 1–16
493. Henk, M., Wills, J.M., Minkowski's successive minima, manuscript 2006
494. Henstock, R., Macbeath, A.M., On the measure of sum-sets I. The theorems of Brunn, Minkowski, and Lusternik, Proc. London Math. Soc. (3) **3** (1953) 182–194
495. Hermite, Ch., Extrait de lettre de M. Ch. Hermite à M. Jacobi sur differents objets de la théorie des nombres, Première lettre, J. reine angew. Math. **40** (1850) 261–278, Œuvres **1** 100–121
496. Hernández Cifre, M.A., Saorín, E., The roots of Steiner's polynomial for a 3-dimensional convex body. A complete characterization, in preparation
497. Hernández Cifre, M.A., Pastor, J.A., Salinas Martínez, G. Segura Gomis, S., Complete systems of inequalities for centrally symmetric convex sets in the n-dimensional space, Arch. Inequal. Appl. **1** (2003) 155–167
498. Hersch, J., Isoperimetric monotonicity: some properties and conjectures (connections between isoperimetric inequalities), SIAM Rev. **30** (1988) 551–577
499. Hewitt, E., Stromberg, K., Real and abstract analysis, Springer-Verlag, Berlin 1965, Reprint 1975
500. Hilbert, D., Über die Theorie der algebraischen Formen, Math. Ann. **36** (1890) 473–534
501. Hilbert, D., Mathematical problems. Lecture delivered before the International Congress of Mathematicians at Paris in 1900, Nachr. Königl. Ges. Wiss. Göttingen, Math.-phys. Kl. **1900** 253–297, Arch. Math. Phys. **1** (3) (1901) 44–63 and 213–237, Bull. Amer. Math. Soc. **8** (1902) 437–479, F. Browder, ed., Mathematical developments arising from Hilbert problems, Amer. Math. Soc., Providence, RI 1976
502. Hilbert, D., Gedächtnisrede auf H. Minkowski, in: H. Minkowski, Ges. Abh. **I** v–xxxi, Teubner, Leipzig 1911
503. Hill, T.P., Partitioning inequalities in probability and statistics, in: Stochastic inequalities (Seattle, WA 1991) 116–132, IMS Lecture Notes Monogr. Ser. **22**, Inst. Math. Statist., Hayward, CA 1992
504. Hingston, N., On the growth of the number of closed geodesics on the two-sphere, Internat. Math. Res. Notices **1993** 253–262
505. Hiriart-Urruty, J.-B., Lemaréchal, C., Fundamentals of convex analysis, Springer-Verlag, Berlin 2001
506. Hirschfeld, J.W.P., Projective geometries over finite fields, The Clarendon Press, Oxford University Press, New York 1979
507. Hirzebruch, F., Über vierdimensionale Riemannsche Flächen mehrdeutiger analytischer Funktionen von zwei komplexen Veränderlichen, Math. Ann. **126** (1953) 1–22, Ges. Abh. **I** 11–32, Springer-Verlag, Berlin 1987
508. History of mathematical programming. A collection of personal reminiscences, eds.: J.K. Lenstra, A.H.G. Rinnooy Kan, A. Schrijver, North-Holland, Amsterdam 1991

509. Hlawka, E., Zur Geometrie der Zahlen, Math. Z. **49** (1944) 285–312, Selecta 38–65
510. Hlawka, E., Ausfüllung und Überdeckung konvexer Körper durch konvexe Körper, Monatsh. Math. **53** (1949) 81–131, Selecta 66–117
511. Hlawka, E., Über Integrale auf konvexen Körpern. I, II, Monatsh. Math. **54** (1950) 1–36, 81–99
512. Hlawka, E., Über Gitterpunkte in Parallelepipeden, J. reine angew. Math. **187** (1950) 246–252
513. Hlawka, E., Zur Theorie der Überdeckung durch konvexe Körper, Monatsh. Math. **58** (1954) 287–291
514. Hlawka, E., Zur Geometrie der Zahlen, Jahrb. Akad. Wiss. Göttingen **1963** 23–29, Selecta 270–276
515. Hlawka, E., The theory of uniform distribution, A B Academic Publ., Berkhamsted 1984
516. Hlawka, E., Selecta, eds.: P.M. Gruber, W.M. Schmidt, Springer-Verlag, Heidelberg 1990
517. Hlawka, E., Personal communication 1996
518. Hofreiter, N., Über Extremformen, Mh. Math. Phys. **40** (1933) 129–152
519. Hölder, O., Über einen Mittelwertsatz, Nachr. Königl. Ges. Wiss. Göttingen, Math.-phys. Kl. **1889** 38–47
520. Holmes, R.B., Geometric functional analysis and its applications, Springer-Verlag, New York 1975
521. Holmes, R.D., Thompson, A.C., N-dimensional area and content in Minkowski spaces, Pacific. J. Math. **85** (1979) 77–110
522. Horváth, Á.G., Molnár, E., Densest ball packings by orbits of the 10 fixed point free Euclidean space groups, Studia Sci. Math. Hungar. **29** (1994) 9–23
523. Howard, R., The kinematic formula in Riemannian homogeneous spaces, Memoir **509**, Amer. Math. Soc. 1993
524. Hubacher, A., Instability of the boundary in the billiard ball problem, Comm. Math. Phys. **108** (1987) 483–488
525. Huber, B., Sturmfels, B., A polyhedral method for solving sparse polynomial systems, Math. Comp. **64** (1995) 1541–1555
526. Huber, B., Sturmfels, B., Bernstein's theorem in affine space, Discrete Comput. Geom. **17** (1997) 137–141
527. Hug, D., Generalized curvature measures and singularities of sets with positive reach, Forum Math. **10** (1998) 699–728
528. Hug, D., Schneider, R., Kinematic and Crofton formulae of integral geometry: recent variants and extensions, in: Homenatge al profesor Lluís Santaló i Sors 51–80, Univ. Girona 2002
529. Huisken, G., Flow by mean curvature of convex surfaces, J. Differential Geom. **20** (1984) 237–266
530. Huisken, G., Sinestrari, C., Convexity estimates for mean curvature flow and singularities of mean convex surfaces, Acta Math. **183** (1999) 45–70
531. Hunger, H., Personal communication 2003
532. Hurwitz, A., Sur quelques applications géométriques des séries de Fourier, Ann. Sci. Éc. Norm. Sup. (3) **19** (1902) 357–408, Math. Werke **1** 454–509, Birkhäuser, Basel 1932
533. Huxley, M.N., Exponential sums and lattice points. III, Proc. London Math. Soc. (3) **87** (2003) 591–609

534. Hyuga, T., An example of analogy of Nevanlinna theory and Diophantine approximation, in: Proc. Second ISAAC Congress **1** (Fukuoka 1999) 459–465, Kluwer, Dordrecht 2000
535. Illanes, A., Nadler, S.B., Jr., Hyperspaces. Fundamentals and recent advances, Dekker, New York 1999
536. Integer points in polyhedra – geometry, number theory, algebra, optimization, eds.: A. Barvinok, M. Beck, C. Haase, B. Reznick, V. Welker, Amer. Math. Soc., Providence, RI 2005
537. Istrăţescu, V.I., Inner product structures. Theory and applications, Reidel, Dordrecht 1987
538. Ivanova-Karatopraklieva, I., Sabitov, I.Kh., Deformation of surfaces I, in: Problems in geometry **23** 131–184, 187, Akad. Nauk SSSR, Vsesoyuz. Inst. Nauchn. i Tekhn. Inform., Moscow 1991, J. Math. Sci. **70** (1994) 1685–1716
539. Ivanova-Karatopraklieva, I., Sabitov, I.Kh., Bending of surfaces II, J. Math. Sci. **74** (1995) 997–1043
540. Ivić, A., Krätzel, E., Kühleitner, M., Nowak, G., Lattice points in large regions and related arithmetic functions, in: W. Schwarz, J. Steuding, eds.: Elementare und analytische Zahlentheorie. Proc. Elaz-conference 2004, Steiner, Stuttgart 2006
541. James, R.C., Inner product in normed linear spaces, Bull. Amer. Math. Soc. **53** (1947) 559–566
542. Jaquet-Chiffelle, D.-O., Énumération complète des classes de formes parfaites en dimension 7, Ann. Inst. Fourier (Grenoble) **43** (1993) 21–55
543. Jarník, V., Zwei Bemerkungen zur Geometrie der Zahlen, Vestnik Krălovské České Společnosti Nauk 12pp., 1941
544. Jensen, J.L.W.V., Sur les fonctions convexes et les inegalités entre les valeurs moyennes, Acta Math. **30** (1906) 175–193
545. Jerison, D., A Minkowski problem for electrostatic capacity, Acta Math. **176** (1996) 1–47
546. Jessen, B., A remark on the volume of polyhedra, Mat. Tidsskr. B **1941** (1941) 59–65
547. Jessen, B., Zur Algebra der Polytope, Nachr. Akad. Wiss. Göttingen Math.-Phys. Kl. II **1972** 47–53
548. Jessen, B., Thorup, A., The algebra of polytopes in affine spaces, Math. Scand. **43** (1978) 211–240
549. John, F., Extremum problems with inequalities as subsidiary conditions, in: Studies and Essays Presented to R. Courant on his 60th Birthday, January 8, 1948, 187–204, Interscience, New York 1948
550. Johnson, D.L., Symmetries, Springer-Verlag, London 2001
551. Johnson, N.W, Uniform polytopes, Cambridge Univ. Press, Cambridge, to appear
552. Johnson, W.B., Lindenstrauss, J., Basic concepts in the geometry of Banach spaces, in: Handbook of the geometry of Banach spaces **I** 1–84, North-Holland, Amsterdam 2001
553. Joswig, M., Reconstructing a non-simple polytope from its graph, in: Polytopes – combinatorics and computation 167–176, Birkhäuser, Basel 2000
554. Joswig, M., Kaibel, V., Körner, F., On the k-systems of a simple polytope, Israel J. Math. **129** (2002) 109–117
555. Jung, H.W.E., Über die kleinste Kugel, die eine räumliche Figur einschliesst, J. reine angew. Math. **123** (1901) 241–257
556. Jung, H.W.E., Darstellung der Funktionen eines algebraischen Körpers zweier unabhängigen Veränderlichen x, y in der Umgebung von $x = a$, $y = b$, J. reine angew. Math. **133** (1908) 289–314

557. Kabat'janskiĭ (Kabatyanskiĭ), G.A., Levenštein (Levenshtein), V.I., Bounds for packings on a sphere and in space, Problemy Peredachi Informatsiĭ **14** (1978) 3–25, Problems Inform. Transmission **14** (1978) 1–17
558. Kac, M., Can one hear the shape of a drum?, Amer. Math. Monthly **73** (1966) 1–23
559. Kakutani, S., Some characterizations of inner product spaces, Jap. J. Math. **16** (1939) 93–97
560. Kalai, G., A simple way to tell a simple polytope from its graph, J. Combin. Theory Ser. A **49** (1988) 381–383
561. Kalai, G., Polytope skeletons and paths, in: Handbook of discrete and computational geometry 331–344, CRCPress, Boca Raton, FL 1997
562. Kalai, G., Ziegler, G., eds., Polytopes – combinatorics and computation (Oberwolfach 1997) 167–176, DMV Sem. **29**, Birkhäuser, Basel 2000
563. Kannan, R., Algorithmic geometry of numbers, Ann. Rev. Comput. Sci. **2** (1987) 231–267
564. Kantor, J.-M., Triangulations of integral polytopes, examples and problems, in: Topology of holomorphic dynamical systems and related topics (Kyoto 1995), Sūrikaisekikenkyūsho Kōkyūroku **955** (1996) 133–144
565. Kantorovich, L.V., Mathematical methods of organizing and planning production, Publ. House Leningrad State Univ., Leningrad 1939
566. Karmarkar, N., A new polynomial-time algorithm for linear programming, Combinatorica **4** (1984) 373–395
567. Katok, A., Strelcyn, J., Ledrappier, F., Przytycki, F., Invariant manifolds, entropy and billiards; smooth maps with singularities, Lect. Notes Math. **1222** Springer-Verlag, Berlin 1986
568. Kaup, L., Kaup, B., Holomorphic functions of several variables, de Gruyter, Berlin 1983
569. Kawohl, B., Rearrangements and convexity of level sets in PDE, Lect. Notes Math. **1150**, Springer-Verlag, Berlin 1985
570. Keller, O.-H., Über die lückenlose Erfüllung des Raumes mit Würfeln, J. reine angew. Math. **163** (1930) 231–248
571. Kellerhals, R., Old and new about Hilbert's third problem, in: European women in mathematics (Loccum 1999) 179–187, Hindawi Publ. Corp., Cairo 2000
572. Kelley, J.L., General topology, Van Nostrand, New York 1955
573. Kempe, A.B., On a general method of describing plane curves of the nth degree by linkworks, Proc. London Math. Soc. **7** (1876) 213–216
574. Kendall, D.G., Simplexes and vector lattices, J. London Math. Soc. **37** (1962) 365–371
575. Kepler, J., Ad Vitellionem paralipomena quibus astronomiae pars optica, Tampach, Frankfurt 1604, Ges. Werke **2**, Beck, München 1939, Optics. Paralipomena to Witelo & optical part of astronomy, Green Lion Press, Santa Fe, NM 2000
576. Kepler, J., Strena seu de nive sexangula, Tampach, Frankfurt 1611, Ges. Werke **4**, Beck, München 1941, On the six-cornered snowflake, Clarendon Press, Oxford 1966
577. Kepler, J., Harmonice mundi, Linz 1619, Ges. Werke **6**, Beck, München 1940, Weltharmonik, Oldenbourg, München 1939
578. Kershner, R., The number of circles covering a set, Amer. J. Math. **61** (1939) 665–671
579. Kershner, R., On paving the plane, Amer. Math. Monthly **75** (1968) 839–844
580. Khachiyan, L.G., A polynomial algorithm in linear programming, Dokl. Akad. Nauk SSSR **244** (1979) 1093–1096, Soviet Math. Dokl. **20** (1979) 191–194
581. Khintchine, A.Y., Zwei Bemerkungen zu einer Arbeit des Herrn Perron, Math. Z. **22** (1925) 274–284

582. Khovanskiĭ (Hovanskiĭ), A.G., Newton polyhedra and genus of complete intersections, Funktsional. Anal. i Prilozhen. **12** (1978) 51–61, Funct. Anal. Appl. **12** (1978) 38–46
583. Khovanskiĭ, A.G., Newton polytopes (algebra and geometry), in: Theory of operators in function spaces (Kuybyshev 1988) 202–221, Saratov. Gos. Univ., Kuĭbyshev. Filial, Kuybyshev 1989
584. Kincses, J., The classification of 3- and 4-Helly-dimensional convex bodies. Geom. Dedicata **22** (1987) 283–301
585. Kirchberger, P., Über *Tschebyschef* sche Annäherungsmethoden, Math. Ann. **57** (1903) 509–540
586. Klain, D.A., A short proof of Hadwiger's characterization theorem, Mathematika **42** (1995) 329–339
587. Klain, D.A., Rota, G.-C., Introduction to geometric probability, Cambridge Univ. Press, Cambridge 1997
588. Klartag, B., Rate of convergence of Minkowski symmetrizations, Geom. Funct. Anal. **14** (2004) 1322–1338
589. Klartag B., Milman, V.D., Isomorphic Steiner symmetrization, Invent. Math. **153** (2003) 463–485
590. Klee, V., Some new results on smoothness and rotundity in normed linear spaces, Math. Ann. **139** (1959) 51–63
591. Klee, V., Review A2028, Math. Reviews **23A** (1962) 376–377
592. Klee, V., Convex polytopes and linear programming, in: Proc. IBM Sci. Comput. Sympos. Combinatorial Problems (Yorktown Heights, NY 1964) 123–158, IBM, White Plains, NY 1966
593. Klee, V., What is a convex set?, Amer. Math. Monthly **78** (1971) 616–631
594. Klee, V., Do infinite-dimensional Banach spaces admit nice tilings?, Stud. Sci. Math. Hung. **21** (1987) 415–427
595. Klee, V., Kleinschmidt, P., Convex polytopes and related complexes, in: Handbook of combinatorics 875–917, Elsevier, Amsterdam 1995
596. Klee, V., Maluta, E., Zanco, C., Tiling with smooth and rotund tiles, Fund. Math. **126** (1986) 269–290
597. Klee, V., Minty, G.J., How good is the simplex algorithm?, in: Inequalities III, Proc. Third Sympos., Univ. California, Los Angeles 1969, 159–175, Academic Press, New York 1972
598. Klein, R., e-mail 2006
599. Klíma, V., Netuka, I., Smoothness of a typical convex function, Czechoslovak Math. J. **31** (106) (1981) 569–572
600. Kneser, H., Süss, W., Die Volumina in linearen Scharen konvexer Körper, Mat. Tidsskr. B **1** (1932) 19–25
601. Kneser, M., Ein Satz über abelsche Gruppen mit Anwendungen auf die Geometrie der Zahlen, Math. Z. **61** (1955) 429–434
602. Kneser, M., Summenmengen in lokalkompakten abelschen Gruppen, Math. Z. **66** (1956) 88–110
603. Knill, O., On nonconvex caustics of convex billiards, Elem. Math. **53** (1998) 89–106
604. Köcher, M., Personal communication 1990
605. Koebe, P., Kontaktprobleme der konformen Abbildung, Ber. Verh. Sächs. Akad. Wiss. Leipzig, Math.-Phys. Kl. **88** (1936) 141–164
606. Koldobsky, A., Fourier analysis in convex geometry, American Math. Soc., Providence, RI 2005
607. Koldobsky, A., König, H., Aspects of isometric theory of Banach spaces, in: Handbook of the geometry of Banach spaces **1** 899–939, North-Holland, Amsterdam 2001

608. Koopmans, T.J., Optimum utilization of the transportation system, in: The econometric society meeting, Washington 1947, Proc. Int. Stat. Conf. **5** (1948) 136–146, Econometrica **17** suppl. (1949) 136–146, Scientific Papers 184–193, Springer-Verlag, Berlin 1970
609. Korkin, A.N., Zolotarev (Zolotareff), E.I., Sur les formes quadratiques positives quaternaires, Math. Ann. **5** (1872) 581–583
610. Korkin, A.N., Zolotarev, E.I., Sur les formes quadratiques, Math. Ann. **6** (1873) 366–389
611. Korkin, A.N., Zolotarev, E.I., Sur les formes quadratiques positives, Math. Ann. **11** (1877) 242–292
612. Kouchnirenko, A.G., Polyèdres de Newton et nombres de Milnor, Invent. Math. **32** (1976) 1–31
613. Koy, H., Schnorr, C.P., Segment LLL-reduction of lattice bases, in: Cryptography and lattices (Providence, RI 2001) 67–80, Lect. Notes Comp. Sci. **2146**, Springer-Verlag, Berlin 2001
614. Kozlov, V.V., Treshchëv, D.V., Billiards. A genetic introduction to the dynamics of systems with impacts, Amer. Math. Soc., Providence, RI 1991
615. Krahn, E., Über eine von Rayleigh formulierte Minimaleigenschaft des Kreises, Math. Ann. **94** (1925) 97–100
616. Krätzel, E., Analytische Funktionen in der Zahlentheorie, Teubner, Stuttgart 2000
617. Kreck, M., Letter to the author 2003
618. Kronecker, L., Die Periodensysteme von Functionen reeller Variablen, S.-ber. Königl. Preuss. Akad. Wiss., Berlin **1884** 1071–1080, Werke III, 31–46, Teubner, Leipzig 1895–1930, Chelsea, New York 1968
619. Kubota, T., Über die konvex-geschlossenen Mannigfaltigkeiten im n-dimensionalen Raume, Sci. Rep. Thoku Imp. Univ. (1) **14** (1925) 85–99
620. Kuczma, M., An introduction to the theory of functional equations and inequalities. Cauchy's equation and Jensen's inequality, Państwowe Wydawnictwo Naukowe (PWN), Warsaw 1985
621. Kuiper, N.H., On C^1-isometric imbeddings I, II, Nederl. Akad. Wetensch. Proc. Ser. A. **58** (1955) 545–556, 683–689, Indag. Math. **17** (1955) 545–556, 683–689
622. Kuperberg, G., A low-technology estimate in convex geometry, Internat. Math. Res. Notices **1992** 181–183
623. Kuperberg, G., The bottleneck conjecture, Geom. Topol. **3** (1999) 119–135
624. Kuperberg, G., From the Mahler conjecture to Gauss linking integrals, manuscript 2006
625. Lagarias, J.C., Point lattices, in: Handbook of combinatorics **1** 919–966, Elsevier, Amsterdam 1995
626. Lagarias, J.C., Lenstra, H.W., Jr., Schnorr, C.-P., Korkin-Zolotarev bases and successive minima of a lattice and its reciprocal lattice, Combinatorica **10** (1990) 333–348
627. Lagarias, J.C., Shor, P.W., Keller's cube-tiling conjecture is false in high dimensions, Bull. Amer. Math. Soc. (NS) **27** (1992) 279–283
628. Lagrange, J.L., Recherches d'arithmétique, Nouv. Mém. Acad. Roy. Belles Lettres Berlin **1773** 265–312, Œuvres **3** 693–758, Gauthier-Villars, Paris 1869
629. Larman, D.G., Rogers, C.A., The existence of a centrally symmetric convex body with central sections that are unexpectedly small, Mathematika **22** (1975) 164–175
630. Laugwitz, D., Differentialgeometrie in Vektorräumen, unter besonderer Berücksichtigung der unendlichdimensionalen Räume, Vieweg, Braunschweig 1965
631. Lazutkin, V.F., Existence of caustics for the billiard problem in a convex domain, Izv. Akad. Nauk SSSR Ser. Mat. **37** (1973) 186–216

632. Lazutkin, V.F., Convex billiards and eigenfunctions of the Laplace operator, Leningrad. Univ., Leningrad 1981
633. Leavitt, J., Ungar, P., Circle supports the largest sandpile, Comm. Pure Appl. Math. **15** (1962) 35–37
634. Ledoux, M., The concentration of measure phenomenon, Amer. Math. Soc., Providence, R.I. 2001
635. Lee, C.W., Regular triangulations of convex polytopes, in: Applied geometry and discrete mathematics 443–456, Amer. Math. Soc., Providence, RI 1991
636. Lee, C.W., Subdivisions and triangulations of polytopes., in: Handbook of discrete and computational geometry 271–290, CRC Press, Boca Raton, FL 1997
637. Leech, J., Some sphere packings in higher space, Canad. J. Math. **16** (1964) 657–682
638. Leech, J., Sloane, N.J.A., Sphere packings and error-correcting codes, Canad. J. Math. **23** (1971) 718–745
639. Legendre, A.-M., Éléments de géométrie, Firmin-Didot, Paris 1794
640. Leichtweiss, K., Konvexe Mengen, Springer-Verlag, Berlin 1980
641. Leichtweiss, K., Zur Affinoberfläche konvexer Körper, Manuscripta Math. **56** (1986) 429–464
642. Leichtweiss, K., Bemerkungen zur Definition einer erweiterten Affinoberfläche von E. Lutwak, Manuscripta Math. **65** (1989), 181–197
643. Leichtweiss, K., Remarks on affine evolutions, Abh. Math. Sem. Univ. Hamburg **66** (1996) 355–376
644. Leichtweiss, K., Affine geometry of convex bodies, Johann Ambrosius Barth, Heidelberg 1998
645. Leindler, L., On a certain converse of Hölder's inequality, in: Linear operators and approximation (Proc. Conf. Oberwolfach 1971), 182–184, Birkhäuser, Basel 1972
646. Lenstra, A.K., Lenstra, H.W., Jr., Lovász, L., Factoring polynomials with rational coefficients, Math. Ann. **261** (1982) 515–534
647. Lenstra, H.W., Jr., Integer programming with a fixed number of variables, Math. Oper. Res. **8** (1983) 538–548
648. Lenstra, H.W., Jr., Personal communication 1985
649. Lenstra, H.W., Jr., E-mail 2004
650. Leppmeier, M., Kugelpackungen von Kepler bis heute, Vieweg, Braunschweig 1997
651. Levenštein, V.I., The maximal density of filling an n-dimensional Euclidean space with equal balls, Mat. Zametki **18** (1975) 301–311, Math. Notes **18** (1976) 765–771
652. Levenštein, V.I., On bounds for packings in n-dimensional Euclidean space, Dokl. Akad. Nauk SSSR **245** (1979) 1299–1303, Soviet Math. Dokl. **20** (1979) 417–421
653. Lévy, P., Leçons d'analyse fonctionnelle, Gauthier-Villars, Paris 1922
654. L'Huillier, S., Polygonométrie ou de la mesure des figures rectilignes. Abrégé d'isopérimétrie élémentaire 103–124, Geneve 1789
655. Li, A.M., Simon, U., Zhao, G.S., Global affine differential geometry of hypersurfaces, de Gruyter & Co., Berlin 1993
656. Lidl, R., Niederreiter, H., Finite fields, Addison-Wesley, Reading, MA 1983
657. Liebmann, H., Eine neue Eigenschaft der Kugel, Nachr. Königl. Ges. Wiss. Göttingen, Math.-phys. Kl. **1899** 44–55
658. Lindelöf, L., Propriétés générales des polyèdres qui, sous une étendue superficielle donnée, renferment le plus grand volume, Bull. Acad. Sci. St. Petersburg **14** (1869) 258–269, Math. Ann. **2** (1870) 150–159
659. Lindenstrauss, J., A short proof of Liapounoff's convexity theorem, J. Math. Mech. **15** (1966) 971–972

660. Lindenstrauss, J., Milman, V.D., The local theory of normed spaces and its applications to convexity, in: Handbook of convex geometry **B** 1149–1220, North-Holland, Amsterdam 1993
661. Lipton, R.J., Tarjan, R.E., A separator theorem for planar graphs, SIAM J. Appl. Math. **36** (1979) 177–189
662. Little, J.J., An iterative method for reconstructing convex polyhedra from extended Gaussian images, in: Proc. Nat. Conf. on Art. Intelligence (Washington D.C. 1983) 247–250
663. Lopez, M.A., Reisner, S., A special case of Mahler's conjecture, Discrete Comput. Geom. **20** (1998) 163–177
664. Ludwig, M., Asymptotic approximation of convex curves, Arch. Math. **63** (1994) 377–384
665. Ludwig, M., A characterization of affine length and asymptotic approximation of convex discs, Abh. Math. Semin. Univ. Hamburg **69** (1999) 75–88
666. Ludwig, M., Reitzner, M., A characterization of affine surface area, Adv. Math. **147** (1999) 138–172
667. Ludwig, M., Reitzner, M., Elementary moves on triangulations, Discrete Comput. Geom. **35** (2006) 1–18
668. Lutwak, E., Intersection bodies and dual mixed volumes, Adv. Math. **71** (1988) 232–261
669. Lutwak, E., Extended affine surface area, Adv. Math. **85** (1991) 39–68
670. Lutwak, E., Yang, D., Zhang, G., A new ellipsoid associated with convex bodies, Duke Math. J. **104** (2000) 375–390
671. Lyapunov (Liapounoff, Ljapunov), A., Sur les fonctions-vecteurs completement additives, Izv. Akad. Nauk SSSR (Bull. Acad. Sci. URSS), Ser. Math. **4** (1940) 465–478
672. Lyusternik (Ljusternik, Lusternik), L.A., Die Brunn-Minkowskische Ungleichung für beliebige messbare Mengen, Dokl. Acad. Nauk. USSSR (N.S.) **1935** 55–58
673. Macbeath, A.M., On measure of sum sets II. The sum-theorem for the torus, Proc. Cambridge Philos. Soc. **49** (1953) 40–43
674. Macbeath, A.M., Factorization of matrices and Minkowski's conjecture, Proc. Glasgow Math. Assoc. **5** (1961) 86–89
675. Macdonald, I.G, The volume of a lattice polyhedron, Proc. Cambridge Philos. Soc. **59** (1963) 719–726
676. Mackey, J., A cube tiling of dimension eight with no facesharing, Discrete Comput. Geom. **28** (2002) 275–279
677. Maehara, H., Geometry of frameworks, in: Proceedings of the 10th Workshop on Topological Graph Theory (Yokohama 1998), Yokohama Math. J. **47** (1999) 41–65.
678. Magaril-Il'yaev, G.G., Tikhomirov, V.M., Convex analysis: theory and applications, Amer. Math. Soc., Providence, RI 2003
679. Mahler, K., Ein Übertragungsprinzip für konvexe Körper, Časopis Pešt. Mat. Fyz. **68** (1939) 93–102
680. Mahler, K., On lattice points in n-dimensional star bodies I. Existence theorems, Proc. Royal Soc. London **A 187** (1946) 151–187
681. Malkevitch, J., Milestones in the history of polyhedra, in: Shaping space, eds.: Senechal, M., Fleck, G., Birkhäuser, Boston 1988
682. Malyshev, A.V., Fundamental concepts and theorems of the geometry of numbers, Tr. Petrozavodsk. Gos. Univ. Ser. Mat. **4** (1997) 3–42
683. Mani-Levitska (Mani), P., Random Steiner symmetrizations, Studia Sci. Math. Hungar. **21** (1986) 373–378

684. Mani-Levitska, P., Characterizations of convex sets, in: Handbook of convex geometry **A** 19–41, North-Holland, Amsterdam 1993
685. Mani-Levitska, P., Private note 2004
686. Mankiewicz, P., Schütt, C., On the Delone triangulation numbers, J. Approx. Theory **111** (2001) 139–142
687. Mann, H., A proof of the fundamental theorem on the density of sums of sets of positive integers, Ann. of Math. (2) **43** (1942) 523–527
688. Mann, H., On the product of sets of group elements, Canad. J. Math. **4** (1952) 64–66
689. Marchaud, A., Un théorème sur les corps convexes, Ann. Sci. École Norm. Sup. (3) **76** (1959) 283–304
690. Martinet, J., Perfect and eutactic lattices: Extension of Voronoï's theory, in: Voronoï's impact on modern science **I** 186–198, Inst. Math. Proc. Inst. Math. Natl. Acad. Sci. Ukraine, Kyiv 1998
691. Martinet, J., Les réseaux parfaits des espaces euclidiens, Masson, Paris 1996, Perfect lattices in euclidean spaces, Springer-Verlag, Berlin 2003
692. Martini, H., Shadow-boundaries of convex bodies, Combinatorics (Acireale 1992), Discrete Math. **155** (1996) 161–172
693. Matérn, B., Persson, O., On the extremum properties of the equilateral triangular lattice and the regular hexagonal network, Förtech. Över. Rapp. Uppsala Stockholm **7** (1965) 15 pp.
694. Mather, J.N., Glancing billiards, Ergodic Theory Dynam. Systems **2** (1982) 397–403
695. Matoušek, J., Lectures on discrete geometry, Springer-Verlag, New York 2002
696. Mattila, P., Geometry of sets and measures in Euclidean spaces, Cambridge Univ. Press, Cambridge 1995
697. Matveev, E.M., On the successive minima of the extended logarithmic height of algebraic numbers, Mat. Sb. **190** (1999) 89–108, Sb. Math. **190** (1999) 407–425
698. Maunder, C.R.F., Algebraic topology, Cambridge Univ. Press, Cambridge 1980, Reprint, Dover, Mineola, NY 1996
699. Maurey, B., Some deviation inequalities, Geom. Funct. Anal. **1** (1991) 188–197
700. Maxwell, J. C., On reciprocal diagrams and diagrams of forces, Philos. Mag. (4) **27** (1864) 250–261
701. Mazur, S., Über konvexe Mengen in linearen normierten Räumen, Studia Math. **4** (1933) 70–84
702. McCann, R.J., Equilibrium shapes for planar crystals in an external field, Comm. Math. Phys. **195** (1998) 699–723
703. McDonald, P., Meyers, R., Isospectral polygons, planar graphs and heat content, Proc. Amer. Math. Soc. **131** (2003) 3589–3599
704. McMullen, C.T., Minkowski's conjecture, well-rounded lattices and topological dimension, Amer. J. Math. **18** (2005) 711–734
705. McMullen, P., The maximum numbers of faces of a convex polytope, Mathematika **17** (1970) 179–184
706. McMullen, P., Polytopes with centrally symmetric faces, Israel J. Math. **8** (1970) 194–196
707. McMullen, P., The numbers of faces of simplicial polytopes, Israel J. Math. **9** (1971) 559–570
708. McMullen, P., Non-linear angle-sum relations for polyhedral cones and polytopes, Math. Proc. Cambridge Philos. Soc. **78** (1975) 247–261
709. McMullen, P., Valuations and Euler-type relations on certain classes of convex polytopes, Proc. London Math. Soc. (3) **35** (1977) 113–135

710. McMullen, P., Convex bodies which tile space by translation, Mathematika **27** (1980) 113–121
711. McMullen, P., Convex bodies which tile space by translation: acknowledgement of priority, Mathematika **28** (1981) 191
712. McMullen, P., Convex bodies which tile space, in: The geometric vein. The Coxeter Festschrift 123–128, Springer-Verlag, New York 1981
713. McMullen, P., The polytope algebra, Adv. Math. **78** (1989) 76–130
714. McMullen, P., Valuations and dissections, in: Handbook of convex geometry **B** 933–988, North-Holland, Amsterdam 1993
715. McMullen, P., On simple polytopes, Invent. Math. **113** (1993) 419–444
716. McMullen, P., Schneider, R., Valuations on convex bodies, in: Convexity and its applications 170–247, Birkhäuser, Basel 1983
717. McMullen, P., Schulte, E., Abstract regular polytopes, Cambridge Univ. Press, Cambridge 2002
718. McMullen, P., Shephard, G.C., Convex polytopes and the upper bound conjecture, Cambridge Univ. Press, London-New York 1971
719. Mendez, O., Reichel, W., Electrostatic characterization of spheres, Forum Math. **12** (2000) 223–245
720. Meyer, M., Pajor, A., On the Blaschke-Santaló inequality, Arch. Math. (Basel) **55** (1990) 82–93
721. Micciancio, D., Goldwasser, S., Complexity of lattice problems. A cryptographic perspective, Kluwer, Boston, MA 2002
722. Michael, E., Topologies in spaces of subsets, Trans. Amer. Math. Soc. **71** (1951) 152–182
723. Mignotte, M., Ştefănescu, D., Polynomials. An algorithmic approach, Springer-Verlag, Singapore, Centre for Discrete Mathematics & Theoretical Computer Science, Auckland 1999
724. Miller, G.L., Teng, S.-H., Thurston, W., Vavasis, S.A., Separators for sphere-packings and nearest neighbor graphs, J. ACM **44** (1997) 1–29
725. Milman, V., Inégalité de Brunn-Minkowski inverse et applications à la théorie locale des espaces normés. C. R. Acad. Sci. Paris Sér. I Math. **302** (1986) 25–28
726. Milman, V., Phenomena that occur in high dimensions, Uspekhi Mat. Nauk **59** (2004) 157–168, Russian Math. Surveys **59** (2004) 159–169
727. Milnor, J., A Nobel Prize for John Nash, Math. Intelligencer **17** (1995) 11–17
728. Minasian, cited in: Sinai, Ya.G., Introduction to ergodic theory, Princeton Univ. Press, Princeton, N.J. 1976
729. Minc, H., Permanents, Addison-Wesley, Reading, MA 1978
730. Minc, H., Nonnegative matrices, Wiley, New York 1988
731. Minding, F., Über die Bestimmung des Grades einer durch Elimination hervorgehenden Gleichung, J. reine angew. Math. **22** (1841) 178–183, in: Topics in algebraic geometry and geometric modeling 351–362, Contemp. Math. **334**, Amer. Math. Soc., Providence, RI 2003
732. Minkowski, H., Über die positiven quadratischen Formen und über kettenbruchähnliche Algorithmen, J. reine angew. Math. **107** (1891) 278–297, Ges. Abh. **1** 243–260
733. Minkowski, H., Über Geometrie der Zahlen, Jahresber. Deutsch. Math. Verein. **1** (1892) 64–65, Ges. Abh. **1** 264–265
734. Minkowski, H., Extrait d'une lettre adressée à M Hermite, Bull. Sci. Math. **17** (2) (1893) 24–29, Ges. Abh. **1** 266–270
735. Minkowski, H., Geometrie der Zahlen, Teubner, Leipzig 1896, Chelsea, New York 1953, Johnson Reprint Corp., New York 1968

736. Minkowski, H., Allgemeine Lehrsätze über konvexe Polyeder, Nachr. Königl. Ges. Wiss. Göttingen, Math.-phys. Kl. **1897** 198–219, Ges. Abh. **2** 103–121
737. Minkowski, H., Über die Annäherung an eine reelle Grösse durch rationale Zahlen, Math. Ann. **54** (1901) 91–124, Ges. Abh. **1** 320–352
738. Minkowski, H., Über die Begriffe Länge, Oberfläche und Volumen, Jahresber. Deutsch. Math.-Ver. **9** (1901) 115–121, Ges. Abh. **2** 122–127
739. Minkowski, H., Volumen und Oberfläche, Math. Ann. **57** (1903) 447-495, Ges. Abh. **2** 230–276
740. Minkowski, H., Dichteste gitterförmige Lagerung konvexer Körper, Nachr. Königl. Ges. Wiss. Göttingen, Math.-phys. Kl. **1904** 311–355, Ges. Abh. **2** 1–42
741. Minkowski, H., Über die Körper von konstanter Breite, Mat. Sb. **25** (1904) 505–508, Ges. Abh. **2** 277–279
742. Minkowski, H., Diskontinuitätsbereich für arithmetische Äquivalenz, J. reine angew. Math. **129** (1905) 220–274, Ges. Abh. **2** 53–100
743. Minkowski, H., Diophantische Approximationen, Teubner, Leipzig 1907, 1927, Chelsea, New York 1957, Physica-Verlag, Würzburg 1961
744. Minkowski, H., Theorie der konvexen Körper, insbesondere Begründung ihres Oberflächenbegriffs, manuscript, printed in: Ges. Abh. **2** 131–229
745. Minkowski, H., Gesammelte Abhandlungen **1, 2**, ed.: D.Hilbert, Teubner, Leipzig 1911
746. Mnëv, N.E., The universality theorems on the classification problem of configuration varieties and convex polytopes varieties, Lect. Notes Math. **1346** 527–543, Springer-Verlag, Berlin 1988
747. Mohar, B., Circle packing of maps in polynomial time, Europ. J. Combinatorics **18** (1997) 785–805
748. Mohar, B., Thomassen, C., Graphs on surfaces, John Hopkins Univ. Press, Baltimore, MD 2001
749. Møller, J., Lectures on random Voronoi tessellations, Lect. Notes Stat. **87**, Springer-Verlag, New York 1994
750. Montejano, L., Convex bodies with homothetic sections, Bull. London Math. Soc. **23** (1991) 381–386
751. Montejano, L., Two applications of topology to convex geometry, Trudy Mat. Inst. Steklov. **247** (2004) 182–185, Proc. Steklov Inst. Math. **247** (2004) 1–4
752. Montgomery, H.L., Minimal theta functions, Glasgow Math. J. **30** (1988) 75–85
753. Mordell, L.J., Note on an arithmetical problem on linear forms, J. Lond. Math. Soc. **12** (1937) 34–36
754. Mordell, L.J., Diophantine equations, Academic Press, London 1969
755. Moreau, J.-J., Fonctionelles convexes, Séminaire "Équations aux dérivées partielles", Collége de France, Paris 1966
756. Morgan, F., Geometric measure theory. A beginner's guide, Academic Press, Boston, MA 1988
757. Mossino, J., Inégalités isopérimétriques et applications en physique, Hermann, Paris 1984
758. Motzkin, T.S., Comonotone curves and polyhedra, abstract 111, Bull. Amer. Math. Soc. **63** (1957) 35
759. Mukhsinov, Kh.Kh., A refinement of the estimates of the arithmetical minimum of the product of inhomogeneous linear forms for large dimensions, in: Studies in number theory **7**, Zap. Nauchn. Sem. Leningrad. Otdel. Mat. Inst. Steklov. (LOMI) **106** (1981) 82–103, 171–172, J. Soviet Math. **23** (1983) 2163–2176
760. Nachbin, L., The Haar integral, Van Nostrand, Princeton 1965, Reprint, Krieger, Huntington, NY 1976

761. Nádai, A., Theory of the plastic distortion of thick-walled cylinders, in: Reissner Anniversary Volume, Contributions to Applied Mechanics, 430–448, Edwards, Ann Arbor, Mich. 1948
762. Narzullaev, H.N., On Minkowski's problem relative to a system of inhomogeneous linear forms, Dokl. Akad. Nauk SSSR **180** (1968) 1298–1299, Soviet Math. Dokl. **9** (1968) 766–767
763. Narzullaev, H.N., Representation of a unimodular matrix in the form $DOTU$ for $n = 3$, Mat. Zametki **18** (1975) 213–221, Math. Notes **18** (1975) 713–719
764. Narzullaev, H.N., Skubenko, B.F., A refinement of the estimate of the arithmetical minimum of the product of inhomogeneous linear forms (on the inhomogeneous Minkowski conjecture), in: Studies in number theory **5**, Zap. Nauchn. Sem. Leningrad. Otdel. Mat. Inst. Steklov. (LOMI) **82** (1979) 88–94, 166, J. Soviet Math. **18** (1982) 913–918
765. Narzullaev, H.N., Ramharter, G., On the $DOTU$-factorization problem in dimension four, Monatsh. Math. **137** (2002) 213–226
766. Nash, J.F., C^1-isometric imbeddings, Ann. of Math. (2) **60** (1954) 383–396
767. Nebe, G., Gitter und Modulformen, Jahresber. Deutsch. Math. Verein. **104** (2002) 125–144
768. Neumann, J. von, Discussion of a maximum problem, unpubl. article, Inst. for Adv. Study, Princeton, N.J. 1947, Coll. Works **VI** 44–49, Pergamon Press, Oxford 1963
769. Neumann, W.D., Hilbert's 3rd problem and invariants of 3-manifolds, The Epstein birthday schrift 383–411, Geom. Topol. Publ., Coventry 1998
770. Niculescu, C., Persson, L.-E., Convex functions and their applications, Springer-Verlag, New York 2006
771. Niemeier, H.-V., Definite quadratische Formen der Diskriminante 1 und der Dimension 24, J. Number Theory **5** (1973) 141–170, Thesis, U. Göttingen 1968
772. Nishizeki, T., Chiba, N., Planar graphs: theory and algorithms, North-Holland, Amsterdam 1988
773. Nosarzewska, M., Evaluation de la différence entre l'aire d'une région plane convexe et le nombre des points aux coordonnées entiers couverts par elle, Coll. Math. **1** (1948) 305–311
774. Novikov, P.S., Sur le problème inverse du potentiel, Dokl. Acad. Sci. SSSR, N.S. **18** (1938) 165–168
775. Odlyzko, A.M., Sloane, N.J.A., New bounds on the number of unit spheres that can touch a unit sphere in n dimensions, J. Combin. Theory Ser. A **26** (1979) 210–214
776. Ohmann, D., Eine lineare Verschärfung des Brunn-Minkowskischen Satzes für abgeschlossene Mengen, Arch. Math. **6** (1954) 33–35
777. Okabe, A., Boots, B., Sugihara, K., Chiu, S.N., Spatial tessellations. Concepts and applications of Voronoi diagrams, 2nd ed., Wiley, Chichester 2000
778. Olds, C., Lax, A., Davidoff, G., The geometry of numbers, Math. Assoc. America 2000
779. Olech, C., The Lyapunov theorem: its extensions and applications, in: Methods of nonconvex analysis (Varenna 1989) 84–103, Lect. Notes Math. **1446**, Springer-Verlag, Berlin 1990
780. Osgood, W.F., Zweite Note über analytische Funktionen mehrerer Veränderlichen, Math. Ann. **52** (1900) 116–127
781. Ostrowski, A.M., On multiplication and factorization of polynomials. I. Lexicographic orderings and extreme aggregates of terms, Aequationes Math. **13** (1975) 201–228
782. Oxtoby, J.C., Measure and category. A survey of the analogies between topological and measure spaces, 2nd ed., Springer-Verlag, New York 1980

783. Pach, J., Agarwal, P.K., Combinatorial geometry, Wiley, New York 1995
784. Palamodov, V., Reconstructive integral geometry, Birkhäuser, Basel 2004
785. Papadimitrakis, M., On the Busemann-Petty problem about convex, centrally symmetric bodies in $\mathbb{R}^n$, Mathematika **39** (1992) 258–266
786. Payne, L.E., Some comments on the past fifty years of isoperimetric inequalities, in: Inequalities (Birmingham 1987) 143–161, Dekker, New York 1991
787. Peaucellier, Ch.N., Note sur une question de géométrie de compas, Nouv. Ann.(2) **12** (1873) 71–79
788. Pełczyński, A., Remarks on John's theorem on the ellipsoid of maximal volume inscribed into a convex symmetric body in $\mathbf{R^n}$, Note Mat. **10** (1990), 395–410
789. Peressini, A.L., Ordered topological vector spaces, Harper & Row, New York 1967
790. Peri, C., Intrinsic volumes and mixed volumes, Conf. Semin. Mat. Univ. Bari **280** (2001) ii+15 pp.
791. Perles, M.A., Results and problems on reconstruction of polytopes, unpublished, 1970
792. Perles, M.A., Sallee, G.T., Cell complexes, valuations, and the Euler relation, Canad. J. Math. **22** (1970) 235–241
793. Perron, O., Über diophantische Approximationen, Math. Ann. **66** (1909) 446–487
794. Perron, O., Über lückenlose Ausfüllung des n-dimensionalen Raumes durch kongruente Würfel, Math. Z. **46** (1940) 1–26
795. Petkov, V.M., Stoyanov, L.N., Geometry of reflecting rays and inverse spectral problems, Wiley, Chichester 1992
796. Petty, C.M., Projection bodies, in: Proc. Coll. Convexity (Copenhagen 1965) 234–241, Københavns Univ. Mat. Inst., Copenhagen 1967
797. Petty, C.M., Geominimal surface area, Geometriae Dedicata **3** (1974) 77–97
798. Petty, C.M., Ellipsoids, in: Convexity and its applications 264–276, Birkhäuser, Basel 1983
799. Pfender, F., Ziegler, G.M., Kissing numbers, sphere packings, and some unexpected proofs, Notices Amer. Math. Soc. **51** (2004) 873–883
800. Phelps, R.R., Convex functions, monotone operators and differentiability, Lect. Notes Math. **1363**, Springer-Verlag, Heidelberg 1993
801. Pick, G., Geometrisches zur Zahlenlehre, Naturwiss. Zeitschr. Lotus (Prag) **1899** 311–319
802. Pisier, G., The volume of convex bodies and Banach space geometry, Cambridge Univ. Press, Cambridge 1989
803. Pogorelov (Pogorelow), A.V., Unique determination of general convex surfaces, Akad. Nauk Ukrainskoĭ SSR, Kiev 1952
804. Pogorelov, A. V. Topics in the theory of surfaces in elliptic space, Gordon and Breach, New York 1961
805. Pogorelov, A.V., Extrinsic geometry of convex surfaces, Izdat. Nauka, Moscow 1969, Amer. Math. Soc., Providence, RI 1973
806. Pogorelov, A.V., The Minkowski multidimensional problem, Izdat. Nauka, Moscow 1975, Wiley, New York 1978
807. Poincaré, H., Sur la généralisation d'un théorème d'Euler relatif aux polyèdres, C. R. Acad. Sci. Paris **117** (1893) 144–145
808. Poincaré, H., Complément à l'Analysis situs, Rend. Circ. Mat. Palermo **13** (1899) 285–343
809. Pólya, G., Torsional rigidity, principal frequency, electrostatic capacity and symmetrization, Quart. J. Appl. Math. **6** (1948) 267–277
810. Pólya, G., Mathematical discovery. On understanding, learning, and teaching problem solving, Reprint, Wiley, New York 1981

811. Pólya, G., Szegö, G., Isoperimetric inequalities in mathematical physics, Princeton Univ. Press, Princeton 1951
812. Pompéiu, D., Sur la continuité des fonctions de variables complexes, Ann. Toul. (2) **7** (1905) 264–315, Thèse 59pp., Gauthier-Villars, Paris 1905
813. Pontryagin (Pontrjagin), L.S., Boltyanskiĭ (Boltjanski), V.G., Gamkrelidze, R.V., Mishchenko (Miščenko), E.F., The mathematical theory of optimal processes, Gosudarstv. Izdat. Fiz.-Mat. Lit., Moscow 1961, Wiley, New York 1962, Pergamon, Macmillan, New York 1964
814. Posicel'skiĭ, E.D., The characterization of the Steiner point, Mat. Zametki **14** (1973) 243–247
815. Preiss, D., Zajiček, L., Stronger estimates of smallness of sets of Fréchet nondifferentiability of convex functions, Rend. Circ. Mat. Palermo (2), Suppl. **3** (1984) 219–223
816. Preiss, D., Zajiček, L., Fréchet differentiation of convex functions in a Banach space with separable dual, Proc. Amer. Math. Soc. **91** (1984) 202–204
817. Prékopa, A., On logarithmic concave measures and functions, Acta Sci. Math. (Szeged) **34** (1973) 335–343
818. Protter, M.H., Can one hear the shape of a drum? revisited, SIAM Rev. **29** (1987) 185–197
819. Rademacher, H., Schoenberg, I.J., Helly's theorems on convex domains and Tchebycheff's approximation problem, Canadian J. Math. **2** (1950) 245–256
820. Rado, R., A theorem on general measure, J. London Math. Soc. **21** (1946) 291–300
821. Radon, J., Mengen konvexer Körper, die einen gemeinsamen Punkt enthalten, Math. Ann. **83** (1921) 113–115, Ges. Abh. **1** 367–369, Birkhäuser, Basel 1987, Österr. Akad. Wiss., Wien 1987
822. Ramharter, G., On Mordell's inverse problem in dimension three, J. Number Theory **58** (1996) 388–415
823. Rankin, R.A., A minimum problem for the Epstein zeta-function, Proc. Glasgow Math. Assoc. **1** (1953) 149–158.
824. Rayleigh (Strutt), J.W., Lord, The theory of sound **I, II**, 2nd ed., Macmillan, London 1894,1896, reprint 1926
825. Reeve, J.E., On the volume of lattice polyhedra, Proc. London Math. Soc. (3) **7** (1957) 378–395
826. Reeve, J.E., A further note on the volume of lattice polyhedra, J. London Math. Soc. **34** (1959) 57–62
827. Reidemeister, K., Über die singulären Randpunkte eines konvexen Körpers, Math. Ann. **83** (1921) 116–118
828. Reinhardt, K., Zur Zerlegung der euklidischen Räume in kongruente Polytope, S.-ber. Preuss. Akad. Wiss. Berlin **1928** 150–155
829. Reitzner, M., Stochastical approximation of smooth convex bodies, Mathematika, **51** (2005) 11–29
830. Remak, R., Verallgemeinerung eines Minkowskischen Satzes **I, II**, Math. Z. **17** (1923) 1–34, **18** (1924) 173–200
831. Rényi, A., Sulanke, R., Über die konvexe Hülle von n zufällig gewählten Punkten II, Z. Wahrscheinlichkeitstheorie und Verw. Gebiete **3** (1964) 138–147
832. Reynolds, O., in: British Association Report, p.897, Aberdeen 1885
833. Richter-Gebert, J., Realization spaces of polytopes, Lect. Notes Math. **1643**, Springer-Verlag, Berlin 1996
834. Richter-Gebert, J., Ziegler, G.M., Realization spaces of 4-polytopes are universal, Bull. Amer. Math. Soc. (N.S.) **32** (1995) 403–412

835. Rickert, N.W., The range of a measure, Bull. Amer. Math. Soc. **73** (1967) 560–563
836. Riener, C., Extreme Gitter, Diplomarbeit, U. Ulm 2006
837. Riener, C., On extreme forms in dimension 8, J. Théor. Nombres Bordeaux, in print
838. Rigault, G., The Bravais polar lattice and the reciprocal lattice, Atti Accad. Sci. Torino, Cl. Sci. Fis. Mat. Nat. **133** (1999) 91–98
839. Ritoré, M., Ros, A., Some updates on isoperimetric problems, Math. Intelligencer **24** (2002) 9–14
840. Roberts, A.W., Convex functions, in: Handbook of convex geometry **B** 1084–1104, North-Holland, Amsterdam 1993
841. Roberts, A.W., Varberg, D.E., Convex functions, Academic Press, New York 1973
842. Robertson, S.A., Polytopes and symmetry, Cambridge Univ. Press, Cambridge 1984
843. Rockafellar, R.T., Convex analysis, Princeton Univ. Press, Princeton 1970, Reprint 1997
844. Rodin, B., Sullivan, D., The convergence of circle packings to the Riemann mapping, J. Differential Geom. **26** (1987) 349–360
845. Rogers, C.A., Existence theorems in the geometry of numbers, Ann. of Math. **48** (1947) 994–1002
846. Rogers, C.A., The closest packing of convex two-dimensional domains, Acta Math. **86** (1951) 309–321, Corrigendum, ibid. **104** (1960) 305–306
847. Rogers, C.A., Mean values over the space of lattices, Acta Math. **94** (1955) 249–287
848. Rogers, C.A., A note on coverings, Mathematika **4** (1957) 1–6
849. Rogers, C.A., The packing of equal spheres, Proc. London Math. Soc. (3) **8** (1958) 609–620
850. Rogers, C.A., Lattice coverings of space, Mathematika **6** (1959) 33–39
851. Rogers, C.A., Packing and covering, Cambridge Univ. Press, Cambridge 1964
852. Rogers, C.A., Shephard, G.C., The difference body of a convex body, Arch. Math. **8** (1957) 220–233
853. Rojas, J.M., Toric intersection theory for affine root counting, J. Pure Appl. Algebra **136** (1999) 67–100
854. Rojas, J.M., Why polyhedra matter in non-linear equation solving, in: Topics in algebraic geometry and geometric modeling 293–320, Amer. Math. Soc., Providence, RI 2003
855. Rokhlin (Rochlin, Rohlin), V.A., Flächen- und Rauminhalt, in: Enzyklopädie der Elementarmathematik **V** 1–77, Deutsch. Verl. Wiss., Berlin 1971
856. Rose, N., Mathematical maxims and minims, Rome Press, Raleigh, NC 1988
857. Rosenthal, H., On the Choquet representation theorem, in: Functional analysis (Austin TX 1986–87) 1–32, Lect. Notes Math. **1332** Springer-Verlag, Berlin 1988
858. Rota, G.-C., Geometric probability, Math. Intelligencer **20** (1998) 11–16
859. Roth, B., Rigid and flexible frameworks, Amer. Math. Monthly **88** (1981) 6–21
860. Rudelson, M., Contact points of convex bodies, Israel J. Math. **101** (1997) 93–124
861. Rudin, W., Functional analysis, McGraw-Hill, New York 1973, 2nd ed. 1991
862. Rush, J.A., A lower bound on packing density, Invent. Math. **98** (1989) 499–509
863. Rush, J.A., Sphere packing and coding theory, in: Handbook of discrete and computational geometry, 917–932, CRC Press, Boca Raton, FL 1997
864. Ryshkov (Ryškov), S.S., A two-dimensional ζ-function with a real parameter, Dokl. Akad. Nauk SSSR **184** (1969) 288–291, Soviet Math. Dokl. **8** (1969) 73–77
865. Ryshkov, S.S., The polyhedron $\mu(m)$ and some extremal problems in the geometry of numbers, Dokl. Akad. Nauk SSSR **194** (1970) 514–517, Soviet Math. Dokl. **11** (1970) 1240–1244

866. Ryshkov, S.S., Theory of point lattices, in preparation
867. Ryshkov, S.S., Baranovskiĭ, E.P., Classical methods in the theory of lattice packings, Russian Math. Surveys **34** (1979) 1–68
868. Ryshkov, S.S., Barykinskiĭ, R.G., Kucherinenko, Y.V., Solution of the basic problems of discrete geometry in the plane, in: R.P. Bambah, V.C. Dumir, R.J. Hans-Gill, eds., Number theory 401–461, Birkhäuser, Basel 2000
869. Ryshkov, S.S., Bol'shakova, E.A., On the theory of mainstay parallelohedra, Izv. Ross. Akad. Nauk Ser. Mat. **69** (2005) 187–210, Izv. Math. **69** (2005), 1257–1277
870. Saab, E., Saab, P., Lyapunov convexity type theorems for non-atomic vector measures, Quaest. Math. **26** (2003) 371–383
871. Sabitov, I. Kh., The volume as a metric invariant of polyhedra, Discrete Comput. Geom. **20** (1998) 405–425
872. Sachs, H., Coin graphs, polyhedra, and conformal mapping, Discrete Math. **134** (1994) 133–138
873. Sah, C.–H., Hilbert's third problem, Pitman, San Francisco 1979
874. Saint-Pierre, J., Point de Steiner et sections lipschitziennes, Sém. Anal. Convexe **15** (1985) Exp. 7
875. Saint-Venant, B. de, Mémoire sur la torsion des prismes, Mém. Acad. Sci. Paris **14** (1856) 233–560
876. Sandakova, N.N., On the theory of ζ-functions of three variables, Dokl. Akad. Nauk SSSR **175** (1967) 535–538, Soviet Math. Dokl. **8** (1967) 893–896
877. Sangwine-Yager, J.R., The missing boundary of the Blaschke diagram, Amer. Math. Monthly **96** (1989) 233–237
878. Sangwine-Yager, J.R., Mixed volumes, in: Handbook of convex geometry **A** 43–71, North-Holland, Amsterdam 1993
879. Santaló, L., Integralgeometrie 5: Über das kinematische Mass im Raum, Actualités Sci. Indust. **357** (1936)
880. Santaló, L.A., Un invariante afin para los cuerpos convexos de espacio de n dimensiones, Portugal. Math. **8** (1949) 155–161
881. Santaló, L., Integral geometry and geometric probability, 2nd. ed., Cambridge Univ. Press, Cambridge 2004
882. Schäffer, J.J., Geometry of spheres in normed spaces, Marcel Dekker, New York-Basel 1976
883. Schattschneider, D., Senechal, M., Tilings, in: Handbook of discrete and computational geometry, 2nd ed., 43–61, CRC Press, Boca Raton, FL 2004
884. Schatunovsky, S.O., Über den Rauminhalt der Polyeder, Math. Ann. **57** (1903) 496–508
885. Schechtman, G., Concentration, results and applications, in: Handbook of the geometry of Banach spaces **2** 1603–1634, North-Holland, Amsterdam 2003
886. Scherrer, W., Ein Satz über Gitter und Volumen, Math. Ann. **86** (1922) 99–107
887. Schinzel, A., Polynomials with special regard to reducibility, With an appendix by U. Zannier, Cambridge Univ. Press, Cambridge 2000
888. Schläfli, L., Theorie der vielfachen Kontinuität, 1850–52, Denkschr. Schweiz. naturforsch. Ges. **38** (1901) 1–237, Ges. Math. Abh. **I** 167–387, Birkhäuser, Basel 1950
889. Schlenker, J.-M., La conjecture des soufflets (d'après I. Sabitov), Astérisque **294** (2004) 77–95
890. Schlichting, H.J., Ucke, C., Sonnentaler – Abbilder der Sonne, Physik in unserer Zeit **27** (1996) 77–78

891. Schmidt, E., Die Brunn-Minkowskische Ungleichung und ihr Spiegelbild sowie die isoperimetrische Eigenschaft der Kugel in der euklidischen und nichteuklidischen Geometrie I, II, Math. Nachr. **1** (1948) 81–157, **2** (1949) 171–244
892. Schmidt, W.M., A metrical theorem in geometry of numbers, Trans. Amer. Math. Soc. **95** (1960) 516–529
893. Schmidt, W.M., On the Minkowski-Hlawka theorem, Ill. J. Math. **7** (1963) 18–23, Corrigendum, ibid. **7** (1963) 714
894. Schmidt, W.M., Diophantine approximation and certain sequences of lattices, Acta Arith. **18** (1971) 165–178
895. Schmidt, W.M., Equations over finite fields. An elementary approach, Lect. Notes Math. **536**, Springer-Verlag, Berlin 1976
896. Schmidt, W.M., Diophantine approximation, Lect. Notes Math. **785**, Springer-Verlag, Berlin 1980
897. Schneider, R., Zu einem Problem von Shephard über die Projektionen konvexer Körper, Math. Z. **101** (1967) 71–82
898. Schneider, R., On Steiner points of convex bodies, Israel J. Math. **9** (1971) 241–249
899. Schneider, R., Krümmungsschwerpunkte konvexer Körper I, II, Abh. Math. Sem. Univ. Hamburg **37** (1972) 112–132, 204–217
900. Schneider, R., Equivariant endomorphisms of the space of convex bodies, Trans. Amer. Math. Soc. **194** (1974) 53–78
901. Schneider, R., Letter to the author, 1974
902. Schneider, R., Isometrien des Raumes der konvexen Körper, Colloq. Math. **33** (1975) 219–224, 304
903. Schneider, R., Curvature measures of convex bodies, Ann. Mat. Pura Appl. (4) **116** (1978) 101–134
904. Schneider, R., Boundary structure and curvature of convex bodies, in: Contributions to geometry (Proc. Geom. Sympos., Siegen 1978) 13–59, Birkhäuser, Basel 1979
905. Schneider, R., Zur optimalen Approximation konvexer Hyperflächen durch Polyeder, Math. Ann. **256** (1981) 289–301
906. Schneider, R., Random approximation of convex sets, J. Microscopy **151** (1988) 211–227
907. Schneider, R., Convex bodies: the Brunn-Minkowski theory, Cambridge Univ. Press, Cambridge 1993
908. Schneider, R., Convex surfaces, curvature and surface area measures, in: Handbook of convex geometry **A** 273–299, North-Holland, Amsterdam 1993
909. Schneider, R., Discrete aspects of stochastic geometry, in: Handbook of discrete and computational geometry 167–184, CRC Press, Boca Raton, FL 1997
910. Schneider, R., E-mail 2005
911. Schneider, R., Weil, W., Integralgeometrie, Teubner, Stuttgart 1992
912. Schnorr, C.-P., A hierarchy of polynomial time basis reduction algorithms, in: Theory of algorithms (Pécs 1984) 375–386, Colloq. Math. Soc. János Bolyai **44**, North-Holland, Amsterdam 1985
913. Schnorr, C.-P., Block reduced lattice bases and successive minima, Combin. Probab. Comput. **3** (1994) 507–522
914. Schramm, O., How to cage an egg, Invent. Math. **107** (1992) 543–560
915. Schrijver, A., Theory of linear and integer programming, Wiley, Chichester 1986, reprinted 1999
916. Schrijver, A., Combinatorial optimization. Polyhedra and efficiency **A,B,C**, Springer-Verlag, Berlin 2003

917. Schulte, E., Analogues of Steinitz's theorem about noninscribable polytopes, in: Intuitive geometry (Siófok 1985) 503–516, Colloq. Math. Soc. János Bolyai **48**, North-Holland, Amsterdam 1987
918. Schulte, E., Tilings, in: Handbook of convex geometry **B** 899–932, North-Holland, Amsterdam 1993
919. Schütt, C., On the affine surface area, Proc. Amer. Math. Soc. **118** (1993) 1213–1218
920. Schütt, C., Random polytopes and affine surface area, Math. Nachr. **170** (1994), 227–249
921. Schütt, C., Werner, E., The convex floating body, Math. Scand. **66** (1990) 275–290
922. Schwarz, H.A., Beweis des Satzes, dass die Kugel kleinere Oberfläche besitzt als jeder andere Körper gleichen Volumens, Nachr. Königl. Ges. Wiss. Göttingen, Math.-phys. Kl. **1884** 1–13, Ges. Abh. **2** 327–340, Berlin 1890
923. Scriba, C.J., Schreiber, P., 5000 Jahre Geometrie, Springer-Verlag, Berlin 2000
924. Seidel, J.J., Harmonics and combinatorics, in: Special functions: group theoretical aspects and applications 287–303, Reidel, Dordrecht 1984
925. Senechal, M., Quasicrystals and geometry, Cambridge Univ. Press, Cambridge 1995
926. Seneta, E., Nonnegative matrices and Markov chains, 2nd ed., Springer-Verlag, New York 1981
927. Shahgholian, H., Convexity and uniqueness in an inverse problem of potential theory, Proc. Amer. Math. Soc. **116** (1992) 1097–1100
928. Shamir, R., The efficiency of the simplex method: a survey, Management Sci. **33** (1987) 301–334
929. Shannon, C.E., A mathematical theory of communication, Bell Syst. Tech. J. **27** (1948) 379–423, 623–656
930. Shemer, I., Neighborly polytopes, Israel J. Math. **43** (1982) 291–314
931. Shephard, G.C., Shadow systems of convex bodies, Israel J. Math. **2** (1964) 229–236
932. Shlosman (Šlosman), S.B., The Wulff construction in statistical mechanics and combinatorics, Uspekhi Mat. Nauk **56** (2001) 97–128, Russian Math. Surveys **56** (2001) 709–738
933. Shor (Šor), N.Z., Use of the space expansion operation in problems of convex function minimalization, Kibernetika (Kiev) **1970** 6–12, Cybernetics **6** (1970) 7–15
934. Sidel'nikov, V.M., On the densest packing of balls on the surface of an n-dimensional Euclidean sphere and the number of binary code vectors with a given code distance, Dokl. Akad. Nauk SSSR **213** (1973) 1029–1032, Soviet Math. Dokl. **14** (1973) 1851–1855
935. Siegel, C.L., Über Gitterpunkte in konvexen Körpern und ein damit zusammenhängendes Extremalproblem, Acta Math. **65** (1935) 307–323, Ges. Abh. **I** 311–325, Springer-Verlag, Berlin 1966
936. Siegel, C.L., A mean value theorem in geometry of numbers, Ann. of Math. (2) **46** (1945) 340–347, Ges. Abh. **III** 39–46, Springer-Verlag, Berlin 1966
937. Siegel, C.L., Lectures on the geometry of numbers, ed. K. Chandrasekharan, Springer-Verlag, Berlin 1989
938. Sierpiński, W., O pewnem zagadnieniu z rachunku funckczyi asymptotycnich, Prace mat.-fiz. **17** (1906) 77–118
939. Sierpiński, W., Sur la question de la mesurabilité de la base de M. Hamel, Fund. Math. **1** (1920) 105–111
940. Simion, R., Convex polytopes and enumeration, Adv. in Appl. Math. **18** (1997) 149–180

941. Skubenko, B.F., A proof of Minkowski's conjecture on the product of n linear inhomogeneous forms in n variables for $n \leq 5$, in: Investigations in number theory **2**, Zap. Naučn. Sem. Leningrad. Otdel. Mat. Inst. Steklov. (LOMI) **33** (1973) 6–36
942. Skubenko, B.F., On Minkowski's conjecture for large n, in: Algebra, number theory and their applications, Trudy Mat. Inst. Steklov. **148** (1978) 218–224, 277, Proc. Steklov. Inst. Math. **148** (1980) 227–233
943. Skubenko, B.F., There exist square real matrices of any order $n \geq 2880$ that are not $DOTU$-matrices, in: Studies in number theory **7**, Zap. Nauchn. Sem. Leningrad. Otdel. Mat. Inst. Steklov. (LOMI) **106** (1981) 134–136, 172, J. Soviet Math. **23** (1983) 2201–2203
944. Smítal, J., On functions and functional equations, Adam Hilger, Bristol 1988
945. Sobolev, S.L., The convergence of approximate integration formulae on functions in the class ${L_2}^{(m)}$, Dokl. Akad. Nauk SSSR **162** (1965) 1259–1261, Soviet Math. Dokl. **6** (1965) 865–867
946. Sobolev, S.L., Introduction to the theory of cubature formulas, Nauka, Moscow 1974
947. Soltan, V., Choquet simplexes in finite dimension – a survey, Expo. Math. **22** (2004) 301–315
948. Sommerville, D.M.Y., The relations connecting the angle-sums and volume of a polytope in space of n dimensions, Proc. Royal Soc. London (A) **115** (1927) 103–119
949. Soundararajan, K., Omega results for the divisor and circle problems, Int. Math. Res. Not. **36** (2003) 1987–1998
950. Spielman, D.A., Teng, S.-H., Smoothed analysis of algorithms: why the simplex algorithm usually takes polynomial time, J. ACM **51** (2004) 385–463 (electronic)
951. Stanley, R.P., The number of faces of a simplicial convex polytope, Adv. in Math. **35** (1980) 236–238
952. Stanley, R.P., Enumerative combinatorics 1, Corrected reprint of the 1986 original, Cambridge Univ. Press, Cambridge 1997
953. Steele, J.M., The Cauchy-Schwarz master class. An introduction to the art of mathematical inequalities, Math. Assoc. America, Washington, DC, Cambridge Univ. Press, Cambridge 2004
954. Steffen, K., A symmetric flexible Connelly sphere with only nine vertices, Letter to IHES 1978
955. Stein, R., Additivität und Einschließungs-Ausschließungsprinzip für Funktionale von Gitterpolytopen, Ph.D. Thesis, U. Dortmund 1982
956. Stein, S.K., A symmetric star body that tiles but not as a lattice, Proc. Amer. Math. Soc. **36** (1972) 543–548
957. Stein, S.K., Szabó, S., Algebra and tiling, Math. Assoc. America 1994
958. Steiner, J., Einfache Beweise der isoperimetrischen Hauptsätze, J. reine angew. Math. **18** (1838) 289–296, Ges. Werke **II** 77–91, Reimer, Berlin 1882
959. Steiner, J., Über parallele Flächen, M.-ber. Preuss. Akad. Wiss. **1840** 114–118, Ges. Werke **II** 173–176, Reimer, Berlin 1882
960. Steiner, J., Über Maximum und Minimum bei Figuren in der Ebene, auf der Kugelfläche und im Raume überhaupt, J. Math. Pures Appl. **6** (1842) 105–170, Ges. Werke **II** 254–308, Reimer, Berlin 1882
961. Steiner, J., Gesammelte Werke **I, II**, Reimer, Berlin 1881, 1882
962. Steinitz, E., Über die *Euler*schen Polyederrelationen, Arch. Math. Phys. (3) **11** (1906) 86–88
963. Steinitz, E., Polyeder und Raumeinteilungen, in: Enz. Math. Wiss. **3** 3AB12 1–139, Teubner, Leipzig 1922

964. Steinitz, E., Über isoperimetrische Probleme bei konvexen Polyedern **I, II**, J. reine angew. Math. **158** (1927) 129–153, **159** (1928) 133–134
965. Steinitz, E., Rademacher, H., Vorlesungen über die Theorie der Polyeder unter Einschluss der Elemente der Topologie, Springer-Verlag, Berlin 1934, Reprint 1976
966. Stephenson, K., Circle packing: a mathematical tale, Notices Amer. Math. Soc. **50** (2003) 1376–1388
967. Stephenson, K., Introduction to circle packing: the theory of discrete analytic functions, Cambridge Univ. Press, Cambridge 2005
968. Stephenson, K., E-mail to the author 2005
969. Stichtenoth, H., Algebraic function fields and codes, Springer-Verlag, Berlin 1993
970. Stoer, J., Witzgall, C., Convexity and optimization in finite dimensions **I**, Springer-Verlag, Berlin 1970
971. Stoka, M.I., Geometrie integrală, Editura Acad. Rep. Soc. România, Bucharest 1967
972. Stolz, O., Grundzüge der Differential- und Integralrechnung. Erster Theil: Reelle Veränderliche und Functionen, Teubner, Leipzig 1893
973. Straszewicz, S., Über exponierte Punkte abgeschlossener Punktmengen, Fund. Math. **24** (1935) 139–143
974. Sturmfels, B., Polynomial equations and convex polytopes, Amer. Math. Monthly **105** (1998) 907–922
975. Sturmfels, B., Solving systems of polynomial equations, American Math. Soc., Providence, RI 2002
976. Su, B., Lectures on differential geometry, World Scientific, Singapore 1980
977. Süss, W. Über das Inhaltsmass bei mehrdimensionalen Polyedern, Tôhoku Math. Journ. **38** (1933) 252–261
978. Swinnerton-Dyer, H.P.F., Extremal lattices of convex bodies, Proc. Cambridge Philos. Soc. **49** (1953) 161–162
979. Sydler, J.-P., Conditions nécessaires et suffisantes pour l'équivalence des polyèdres de l'espace euclidien à trois dimensions, Comment. Math. Helv. **40** (1965) 43–80
980. Szabó, S., A symmetric star polyhedron that tiles but not as a fundamental domain, in: Intuitive geometry, Coll. Math. Soc. J. Bolyai **48** (1987) 531–544
981. Szarek, S.J., On the geometry of the Banach-Mazur compactum, Lect. Notes Math. **1470**, Springer-Verlag, Berlin 1991
982. Szekeres, G., On a problem of the lattice-plane, J. London Math. Soc. **12** (1937) 88–93
983. Szönyi, T., Some applications of algebraic curves in finite geometry and combinatorics, in: Surveys in combinatorics (London 1997) 197–236, London Math. Soc. Lect. Note Ser. **241**, Cambridge Univ. Press, Cambridge 1997
984. Tabachnikov, S.L., On the dual billiard problem, Adv. Math. **115** (1995) 221–249
985. Tabachnikov, S., Geometry and billiards, Amer. Math. Soc., Providence, RI 2005
986. Talenti, G., The standard isoperimetric theorem, in: Handbook of convex geometry **A** 73–123, North-Holland, Amsterdam 1993
987. Talenti, G., On isoperimetric theorems of mathematical physics, in: Handbook of convex geometry **B** 1131–1147, North-Holland, Amsterdam 1993
988. Tanemura, M., The forms of space division, in: Voronoï's impact on modern science **II** 153–163, Inst. Math. Natl. Acad. Sci. Ukr., Kyiv 1998
989. Tarski, A., A decision method for elementary algebra and geometry, 2nd ed., Univ. of California Press, Berkeley, CA 1951
990. Taylor, J.E., Crystalline variational problems, Bull. Amer. Math. Soc. **84** (1978) 568–588
991. Teissier, B., Du théorème de l'index de Hodge aux inégalités isopérimétriques, C. R. Acad. Sci. Paris Sér. A-B **288** (1979) 287–289

992. Teissier, B., Variétés toriques et polytopes, Sem. Bourbaki **1980/81** 71–84, Lect. Notes Math. **901**, Springer-Verlag, Berlin 1981
993. Teissier, B., Bonnesen-type inequalities in algebraic geometry I. Introduction to the problem, in: Seminar on Differential Geometry 85–105, Ann. of Math. Stud. **102**, Princeton Univ. Press, Princeton, NJ 1982
994. Thompson, A.C., Minkowski geometry, Cambridge Univ. Press, Cambridge 1996
995. Thompson, T.M., From error-correcting codes through sphere packings to simple groups, Math. Assoc. America 1983
996. Thue, A., Om nogle geometrisk-taltheoretiske Theoremer, in: Forhandlingerne ved de Skandinaviske Naturforskeres **14** 352–353, København 1892
997. Thue, A., Über die dichteste Zusammenstellung von kongruenten Kreisen in der Ebene, Skr. Vedensk-Selsk **1** 3–9, Christiania 1910
998. Thunder, J.L., An adelic Minkowski-Hlawka theorem and an application to Siegel's lemma, J. reine angew. Math. **475** (1996) 167–185
999. Thurston, P.W., The geometry and topology of 3-manifolds, Princeton Univ. Lect. Notes, Princeton, NJ 1978
1000. Thurston, P.W., The finite Riemann mapping theorem, address at the Intern. Sympos. in celebration of the proof of the Bieberbach conjecture, Purdue Univ. 1985
1001. Tomczak-Jaegermann, N., Banach-Mazur distances and finite-dimensional operator ideals, Longman Scientific, Harlow, Wiley, New York 1989
1002. Tutte, W.T., How to draw a graph, Proc. London Math. Soc. (3) **13** (1963) 743–767
1003. Tverberg, H., A generalization of Radon's theorem, J. London Math. Soc. **41** (1966) 123–128
1004. Urysohn, P.S., Mittlere Breite und Volumen der konvexen Körper im n-dimensionalen Raume, Mat. Sb. SSSR **31** (1924) 313–319
1005. Van de Vel, M.L.J., Theory of convex structures, North-Holland, Amsterdam 1993
1006. Van Tiel, J., Convex analysis, Wiley, New York 1984
1007. Van Emde Boas, P., Another NP-complete partition problem and the complexity of computing short vectors in a lattice, Report 81-04, Math. Inst. Univ. Amsterdam 1981
1008. Venkov, B.A., On a class of euclidean polytopes, Vestnik Leningrad Univ. Ser. Mat. Fiz. Him. **9** (1954) 11–31
1009. Vergil, P., Aeneid, Book 1
1010. Vietoris, L., Bereiche zweiter Ordnung, Monatsh. Math. Phys. **32** (1922) 258–280
1011. Volland, W., Ein Fortsetzungssatz für additive Eipolyederfunktionale im euklidischen Raum, Arch. Math. **8** (1957) 144–149
1012. Voronoï (Woronoi), G.F., Nouvelles applications des paramètres continus à la théorie des formes quadratiques. Première mémoire: Sur quelques propriétés des formes quadratiques positives parfaites, J. reine angew. Math. **133** (1908) 97–178, Coll. Works **II** 171–238
1013. Voronoï, G.F., Nouvelles applications des paramètres continus à la théorie des formes quadratiques. Deuxième mémoire. Recherches sur les parallélloèdres primitifs I, II, J. reine angew. Math. **134** (1908) 198–287 et **136** (1909) 67–181, Coll. Works **II** 239–368
1014. Voronoï, G.F., Collected works **I–III**, Izdat. Akad. Nauk Ukrain. SSSR, Kiev 1952
1015. Voronoï's impact on modern science **1, 2**, eds.: P.Engel, H.Syta, Inst. Math. Natl. Acad. Sci. Ukraine, Kyiv 1998
1016. Webster, R.J., Convexity, Oxford Univ. Press, Oxford 1994
1017. Weisshaupt, H., Isometries with respect to symmetric difference metrics, Studia Sci. Math. Hungar. **37** (2001) 273–318

1018. Wermer, J., Potential theory, Lect. Notes Math. **408**, Springer-Verlag, Berlin 1981
1019. Weyl, H., Über die asymptotische Verteilung der Eigenwerte, Nachr. Königl. Ges. Wiss. Göttingen **1911** 110–117
1020. Weyl, H., Über die Bestimmung einer geschlossenen konvexen Fläche durch ihr Linienelement, Vierteljahrschr. Naturforsch. Ges. Zürich **61** (1916) 40-72
1021. Weyl, H., Elementare Theorie der konvexen Polyeder, Comm. Math. Helv. **7** (1935/36) 290–306, Engl. transl. in: Contributions to the theory of games 3–18, Ann. Math. Studies **24**, Princeton Univ. Press, Princeton, NJ 1950
1022. White, B., Some recent developments in differential geometry, Math. Intelligencer **11** (1989) 41–47
1023. White, B., Evolution of curves and surfaces by mean curvature, in: Proc. Intern. Congr. Math. (Beijing 2002) **I** 525–538, Higher Ed. Press, Beijing 2002
1024. White, B., The nature of singularities in mean curvature flow of mean-convex sets, J. Amer. Math. Soc. **16** (2003) 123–138
1025. Whitney, H., Congruent graphs and the connectivity of graphs, Amer. J. Math. **54** (1932) 150–168
1026. Wills, J.M., Ein Satz über konvexe Mengen und Gitterpunkte, Monatsh. Math. **72** (1968) 451–463
1027. Wills, J.M., Lattice packings of spheres and the Wulff shape, Mathematika **43** (1996) 229–236
1028. Wong, P.-M., Diophantine approximation and the theory of holomorphic curves, in: Proc. sympos. on value distribution theory in several complex variables (Notre Dame 1990) 115–156, Univ. Notre Dame Press, Notre Dame, IN 1992
1029. Woods, A.C., On a theorem of Tschebotareff, Duke Math. J. **25** (1958) 631–637
1030. Woods, A.C., A note on dense subsets of lattices, J. London Math. Soc.**41** (1966) 631–638
1031. Wulff, G., Zur Frage der Geschwindigkeit des Wachstums und der Auflösung der Krystallflächen, Z. Kryst. **84** (1901) 449–530
1032. Yazaki, S., On an area-preserving crystalline motion, Calc. Var. Partial Differential Equations **14** (2002) 85–105
1033. Yudin, D.B., Nemirovskiĭ, A.S., Estimation of the informational complexity of mathematical programming problems, Èkonom. Mat. Metody **12** (1976) 128–142, Matekon **13** (1976/7) 3–25
1034. Zador, P.L., Topics in the asymptotic quantization of continuous random variables, Bell Lab. Tech. Memo. 1966
1035. Zador, P.L., Asymptotic quantization error of continuous signals and the quantization dimension, IEEE Trans. Inform. Theory **IT-28** (1982) 139–148
1036. Zajiček, L., On differentiation of metric projections in finite dimensional Banach spaces, Czechoslovak. Math. J. **33 (108)** (1983) 325–336
1037. Zamfirescu, T., Nonexistence of curvature in most points of most convex surfaces, Math. Ann. **252** (1980) 217–219
1038. Zamfirescu, T., The curvature of most convex surfaces vanishes almost everywhere. Math. Z. **174** (1980) 135–139
1039. Zamfirescu, T., Using Baire categories in geometry, Rend. Sem. Mat. Univ. Politec. Torino **43** (1985) 67–88
1040. Zamfirescu, T., Nearly all convex bodies are smooth and strictly convex, Monatsh. Math. **103** (1987) 57–62
1041. Zamfirescu, T., Baire categories in convexity, Atti Sem. Mat. Fis. Univ. Modena **39** (1991) 139–164

1042. Zassenhaus, H., Personal communication 1964
1043. Zhang, G., Intersection bodies and the Busemann-Petty inequalities in R^4, Ann. of Math. **140** (1994) 331–346
1044. Zhang, G., A positive solution to the Busemann-Petty problem in $\mathbf{R}^4$, Ann. of Math. **149** (1999) 535–543
1045. Ziegler, G.M., Lectures on polytopes, 2nd ed., Springer-Verlag, New York 1998
1046. Ziegler, G.M., letter 2005
1047. Ziegler, R., Morphologie von Kristallformen und symmetrischen Polyedern, Philosoph.-Anthroposoph. Verlag am Goetheanum, Dornach 1998
1048. Zong, C., Strange phenomena in convex and discrete geometry, Springer-Verlag, New York 1996
1049. Zong, C., Sphere packings, Springer-Verlag, New York 1999
1050. Zong, C., What is known about unit cubes, Bull. Amer. Math. Soc. **42** (2005) 181–211
1051. Zong, C., The cube – a window to convex and discrete geometry, Cambridge Univ. Press, Cambridge 2006
1052. Žitomirskiĭ, O.K., Verschärfung eines Satzes von Woronoi, Ž. Leningrad. Fiz.-Mat. Obšč. **2** (1929) 131–151

Index

Author Index

List of Symbols

B^d	Euclidean unit ball, 44
C^*	polar body, 171
$H_C(u)$	support hyperplane with exterior normal u, 54
$H_C(x)$	support hyperplane at x, 53
$H_C^-(x)$	support halfspace at x, 54
L^*	polar lattice, 365
$N_C(x)$	normal cone, 68
$P(Q)$	perimeter of a polygon, 318
$S(C)$	surface area, 104
S^{d-1}	Euclidean unit sphere, 54
$V(J)$	volume, Jordan measure, 120
V_i	intrinsic volume, 104
W_i	quermassintegral, 93
#	counting function, 118
$\\|\cdot\\|$	Euclidean norm, 20
aff	affine hull, 44
bd	boundary, 4
$\perp$	orthogonal complement, 45
$\mathcal{C}, \mathcal{C}(\mathbb{E}^d)$	space of convex bodies, 41
$\mathcal{C}^k$	differentiability class, 37
$\mathcal{C}_p, \mathcal{C}_p(\mathbb{E}^d)$	space of proper convex bodies, 41
$\mathcal{L}(\mathcal{C})$	space of polyconvex bodies, 117
$\mathcal{L}(\mathcal{P})$	space of polyconvex polytopes, 115
$\mathcal{P}, \mathcal{P}(\mathbb{E}^d)$	space of convex polytopes, 89
χ	Euler characteristic, 118
cl	closure, 4
conv	convex hull, 42
$\delta(C, L)$	density of a set lattice, 442
$\delta(C, T)$	density of a family of translates, 441
$\delta(T)$	density of a discrete set, 440
δ^H	Hausdorff metric, 84
δ^V	symmetric difference metric, 203
δ^{BM}	Banach–Mazur distance, 207
$\delta_L(C)$	lattice packing density, 441
$\delta_T(C)$	translative packing density, 441
diam	diameter, 49
dim	dimension, 22
$\dim_H$	Hausdorff dimension, 70
epi	epigraph, 3
ext	set of extreme points, 75
κ_k	volume of B^k, 103
λ_i	successive minimum, 376
lin	linear hull, 82
$\mathbb{E}^d$	Euclidean d-space, 2
$\mathbb{E}^d/L$	torus group, 395
$\mathbb{Z}^d$	integer lattice, 356
$\mathbb{1}_C$	characteristic function, 167
$\mathcal{D}(\delta)$	discriminant surface, 437
$\mathcal{O}(d)$	orthogonal group, 135
$\mathcal{P}$	cone of positive quadratic forms, 431
$\mathcal{P}_{\mathbb{Z}^d}$	space of lattice polytopes, 310
$\mathcal{R}(m)$	Ryshkov polyhedron, 435
$\mathcal{SL}(d)$	special linear group, 389
$\mathcal{SO}(d)$	special orthogonal group, 121
$\mu(C, L)$	covering radius, 381
$\oplus$	direct sum, 45
int	interior, 4

Grundlehren der mathematischen Wissenschaften

A Series of Comprehensive Studies in Mathematics

A Selection

246. Naimark/Stern: Theory of Group Representations
247. Suzuki: Group Theory I
248. Suzuki: Group Theory II
249. Chung: Lectures from Markov Processes to Brownian Motion
250. Arnold: Geometrical Methods in the Theory of Ordinary Differential Equations
251. Chow/Hale: Methods of Bifurcation Theory
252. Aubin: Nonlinear Analysis on Manifolds. Monge-Ampère Equations
253. Dwork: Lectures on ρ-adic Differential Equations
254. Freitag: Siegelsche Modulfunktionen
255. Lang: Complex Multiplication
256. Hörmander: The Analysis of Linear Partial Differential Operators I
257. Hörmander: The Analysis of Linear Partial Differential Operators II
258. Smoller: Shock Waves and Reaction-Diffusion Equations
259. Duren: Univalent Functions
260. Freidlin/Wentzell: Random Perturbations of Dynamical Systems
261. Bosch/Güntzer/Remmert: Non Archimedian Analysis – A System Approach to Rigid Analytic Geometry
262. Doob: Classical Potential Theory and Its Probabilistic Counterpart
263. Krasnosel'skiĭ/Zabreĭko: Geometrical Methods of Nonlinear Analysis
264. Aubin/Cellina: Differential Inclusions
265. Grauert/Remmert: Coherent Analytic Sheaves
266. de Rham: Differentiable Manifolds
267. Arbarello/Cornalba/Griffiths/Harris: Geometry of Algebraic Curves, Vol. I
268. Arbarello/Cornalba/Griffiths/Harris: Geometry of Algebraic Curves, Vol. II
269. Schapira: Microdifferential Systems in the Complex Domain
270. Scharlau: Quadratic and Hermitian Forms
271. Ellis: Entropy, Large Deviations, and Statistical Mechanics
272. Elliott: Arithmetic Functions and Integer Products
273. Nikol'skiĭ: Treatise on the shift Operator
274. Hörmander: The Analysis of Linear Partial Differential Operators III
275. Hörmander: The Analysis of Linear Partial Differential Operators IV
276. Liggett: Interacting Particle Systems
277. Fulton/Lang: Riemann-Roch Algebra
278. Barr/Wells: Toposes, Triples and Theories
279. Bishop/Bridges: Constructive Analysis
280. Neukirch: Class Field Theory
281. Chandrasekharan: Elliptic Functions
282. Lelong/Gruman: Entire Functions of Several Complex Variables
283. Kodaira: Complex Manifolds and Deformation of Complex Structures
284. Finn: Equilibrium Capillary Surfaces
285. Burago/Zalgaller: Geometric Inequalities
286. Andrianaov: Quadratic Forms and Hecke Operators
287. Maskit: Kleinian Groups
288. Jacod/Shiryaev: Limit Theorems for Stochastic Processes

289. Manin: Gauge Field Theory and Complex Geometry
290. Conway/Sloane: Sphere Packings, Lattices and Groups
291. Hahn/O'Meara: The Classical Groups and K-Theory
292. Kashiwara/Schapira: Sheaves on Manifolds
293. Revuz/Yor: Continuous Martingales and Brownian Motion
294. Knus: Quadratic and Hermitian Forms over Rings
295. Dierkes/Hildebrandt/Küster/Wohlrab: Minimal Surfaces I
296. Dierkes/Hildebrandt/Küster/Wohlrab: Minimal Surfaces II
297. Pastur/Figotin: Spectra of Random and Almost-Periodic Operators
298. Berline/Getzler/Vergne: Heat Kernels and Dirac Operators
299. Pommerenke: Boundary Behaviour of Conformal Maps
300. Orlik/Terao: Arrangements of Hyperplanes
301. Loday: Cyclic Homology
302. Lange/Birkenhake: Complex Abelian Varieties
303. DeVore/Lorentz: Constructive Approximation
304. Lorentz/v. Golitschek/Makovoz: Construcitve Approximation. Advanced Problems
305. Hiriart-Urruty/Lemaréchal: Convex Analysis and Minimization Algorithms I. Fundamentals
306. Hiriart-Urruty/Lemaréchal: Convex Analysis and Minimization Algorithms II. Advanced Theory and Bundle Methods
307. Schwarz: Quantum Field Theory and Topology
308. Schwarz: Topology for Physicists
309. Adem/Milgram: Cohomology of Finite Groups
310. Giaquinta/Hildebrandt: Calculus of Variations I: The Lagrangian Formalism
311. Giaquinta/Hildebrandt: Calculus of Variations II: The Hamiltonian Formalism
312. Chung/Zhao: From Brownian Motion to Schrödinger's Equation
313. Malliavin: Stochastic Analysis
314. Adams/Hedberg: Function spaces and Potential Theory
315. Bürgisser/Clausen/Shokrollahi: Algebraic Complexity Theory
316. Saff/Totik: Logarithmic Potentials with External Fields
317. Rockafellar/Wets: Variational Analysis
318. Kobayashi: Hyperbolic Complex Spaces
319. Bridson/Haefliger: Metric Spaces of Non-Positive Curvature
320. Kipnis/Landim: Scaling Limits of Interacting Particle Systems
321. Grimmett: Percolation
322. Neukirch: Algebraic Number Theory
323. Neukirch/Schmidt/Wingberg: Cohomology of Number Fields
324. Liggett: Stochastic Interacting Systems: Contact, Voter and Exclusion Processes
325. Dafermos: Hyperbolic Conservation Laws in Continuum Physics
326. Waldschmidt: Diophantine Approximation on Linear Algebraic Groups
327. Martinet: Perfect Lattices in Euclidean Spaces
328. Van der Put/Singer: Galois Theory of Linear Differential Equations
329. Korevaar: Tauberian Theory. A Century of Developments
330. Mordukhovich: Variational Analysis and Generalized Differentiation I: Basic Theory
331. Mordukhovich: Variational Analysis and Generalized Differentiation II: Applications
332. Kashiwara/Schapira: Categories and Sheaves. An Introduction to Ind-Objects and Derived Categories
333. Grimmett: The Random-Cluster Model
334. Sernesi: Deformations of Algebraic Schemes
335. Bushnell/Henniart: The Local Langlands Conjecture for GL(2)
336. Gruber: Convex and Discrete Geometry

Printing: Krips bv, Meppel
Binding: Stürtz, Würzburg